W9-BZE-562

CELLULAR AND MOLECULAR PATHOBIOLOGY OF CARDIOVASCULAR DISEASE

ELSEVIER
science &
technology books

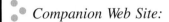

Companion Web Site:

http://store.elsevier.com/product.jsp?&isbn=9780124052062

Cellular and Molecular Pathobiology of Cardiovascular Disease
Monte S. Willis, Jonathon W. Homeister and James R. Stone

Resources:

• All figures from the book available as both Power Point slides and jpeg files

• Links to websites carefully chosen to supplement the content of the book

• Contact the editors with questions and/or suggestions

ELSEVIER

ACADEMIC
PRESS

CELLULAR AND MOLECULAR PATHOBIOLOGY OF CARDIOVASCULAR DISEASE

Edited by

MONTE S. WILLIS, MD, PhD, FCAP, FASCP, FAHA
*University of North Carolina, UNC Health Care, McAllister Heart Institute,
Department of Pathology & Laboratory Medicine, Chapel Hill, NC, USA*

JONATHON W. HOMEISTER, MD, PhD, FCAP
*University of North Carolina, UNC Health Care, McAllister Heart Institute,
Department of Pathology & Laboratory Medicine, Chapel Hill, NC, USA*

JAMES R. STONE, MD, PhD
*Harvard Medical School, Massachusetts General Hospital,
Department of Pathology, Boston, MA, USA*

AMSTERDAM • BOSTON • HEIDELBERG • LONDON
NEW YORK • OXFORD • PARIS • SAN DIEGO
SAN FRANCISCO • SINGAPORE • SYDNEY • TOKYO
Academic Press is an imprint of Elsevier

Academic Press is an imprint of Elsevier

32 Jamestown Road, London NW1 7BY, UK
225 Wyman Street, Waltham, MA 02451, USA
525 B Street, Suite 1800, San Diego, CA 92101-4495, USA

Notice
No responsibility is assumed by the publisher for any injury and/or damage to persons or property as a matter of products liability, negligence or otherwise, or from any use or operation of any methods, products, instructions or ideas contained in the material herein. Because of rapid advances in the medical sciences, in particular, independent verification of diagnoses and drug dosages should be made

Medicine is an ever-changing field. Standard safety precautions must be followed, but as new research and clinical experience broaden our knowledge, changes in treatment and drug therapy may become necessary or appropriate. Readers are advised to check the most current product information provided by the manufacturer of each drug to be administered to verify the recommended dose, the method and duration of administrations, and contraindications. It is the responsibility of the treating physician, relying on experience and knowledge of the patient, to determine dosages and the best treatment for each individual patient. Neither the publisher nor the authors assume any liability for any injury and/or damage to persons or property arising from this publication.

British Library Cataloguing-in-Publication Data
A catalogue record for this book is available from the British Library

Library of Congress Cataloging-in-Publication Data
A catalog record for this book is available from the Library of Congress

ISBN: 978-0-12-405206-2

For information on all Academic Press publications
visit our website at www.store.elsevier.com

Typeset by TNQ Books and Journals

Printed and bound in United States of America

14 15 16 17 18   10 9 8 7 6 5 4 3 2 1

 Working together
to grow libraries in
developing countries

www.elsevier.com • www.bookaid.org

Dedication

We dedicate this book to our families for their support and encouragement

– Tina, Connor, and Declan Willis
– Michelle, Hans, and Nolan Homeister
– Melinda, Jillian, Hallie and Adrian Stone

Monte S. Willis, MD, PhD
Jonathon W. Homeister, MD, PhD
James R. Stone, MD, PhD

Contents

7. Cellular and Molecular Pathobiology of the Cardiac Conduction System

THOMAS J. HUND, SAKIMA A. SMITH, MICHAEL A. MAKARA, PETER J. MOHLER

8. Molecular Pathobiology of Myocarditis

ELISA CARTURAN, CRISTINA BASSO, GAETANO THIENE

9. Calcific and Degenerative Heart Valve Disease

ELENA AIKAWA, FREDERICK J. SCHOEN

10. Vasculogenesis and Angiogenesis

JOSEPH F. ARBOLEDA-VELASQUEZ, PATRICIA A. D'AMORE

11. Diseases of Medium-Sized and Small Vessels

J. CHARLES JENNETTE, JAMES R. STONE

12. Pathophysiology of Atherosclerosis

MICHAEL A. SEIDMAN, RICHARD N. MITCHELL, JAMES R. STONE

13. Genetic Diseases of the Aorta (Including Aneurysms)

MARC K. HALUSHKA

Foreword

Cardiovascular disease continues to be a leading cause of morbidity and mortality in both women and men the world over. It is therefore a major challenge to the healthcare system that requires urgent answers to critical biomedical questions. Over the past twelve years, since Frederick J. Schoen and I coedited the third edition of *Cardiovascular Pathology* with lead editor Malcolm Silver,[1] the field of cardiovascular pathology has been transformed from one of a meticulous study of structure and anatomic pathology to a dynamic discipline in which cell and molecular pathobiology is changing our understanding of the pathogenesis of local and systemic diseases affecting the heart and blood vessels. Much has been discovered about the diagnosis, treatment, pathogenesis and prevention of cardiovascular disease. This new information, generated because of successful research investments by governments and scientific societies and agencies, is very well presented by the authors of *Cellular and Molecular Pathobiology of Cardiovascular Disease*.

The sixteen chapters provide a refreshing re-examination of cardiac and vascular disease, integrating classic anatomic pathology with state of the art basic biomedical science and advanced clinical investigation to provide an understanding of the pathogenesis of disease. This book reinforces and extends our concept put forth in 2001 in *Cardiovascular Pathology*, that in a cardiovascular textbook 'the molecular biology of disease is emphasized and newly understood pathologic mechanisms are defined'.[1] The authors of *Cellular and Molecular Pathobiology of Cardiovascular Disease* use numerous examples to show that high quality discovery science is the foundation for translational research in providing advanced diagnostic tools, new treatments and innovative prevention strategies to conquer cardiovascular disease.

It is immediately obvious from the title of the book that cells and molecules, though fascinating, in some way lead to serious cardiovascular disease when disturbed or disrupted. Cells that were known at the turn of the century are now well characterized and subsets of these cells, which have profound effects on the function of the cardiac and vascular systems, are being identified. Numerous new genes and proteins, including those developmental genes that reappear in disease, have also been identified and their function in cardiovascular health and disease has been studied. In addition, signaling pathways and metabolic pathways have been elucidated, including through the use of new, innovative,

non-invasive techniques of nuclear magnetic resonance spectroscopy and positron emission tomography. The authors show how these basic discoveries have led to the identification of new therapeutic targets that show promise in treating disease, including the discovery of the function of microRNAs in transcriptional regulation of specific targets. Additionally, there are extensive presentations on the genetics of several cardiovascular diseases, including cardiomyopathies, aortic disease, atherosclerosis and hypertension. The hundreds of gene mutations implicated in cardiomyopathies provide a fascinating story on the relation of genetic alterations, structural alterations in cardiomyocytes, functional abnormalities and ultimately clinical disease.

The advances presented on heart valve disease are a good example of the progress made in cardiovascular medicine and pathobiology since the publication of *Cardiovascular Pathology* in 2001.[1] The cell types participating in the pathogenesis of calcific aortic stenosis[2] and degenerative mitral valve disease are well described and their function in disease is considered, informed by in vitro and in vivo experimental and clinical molecular findings. Potential mechanisms of calcification are described through a comprehensive review of the literature, especially focusing on inflammatory processes in the valve. New insights on oxidative stress, bone morphogenic protein signaling and lipoproteins and Wnt/Lrp5 signaling pathways are well integrated into the complicated story of calcific aortic stenosis. The emphasis on focusing on pathogenesis through the lens of cell injury and repair has advanced the field of valve pathobiology and led to useful experimental in vitro and in vivo investigations.[3] The common pathogenic features of calcific aortic stenosis and atherosclerosis has led to attempts to identify key regulatory molecules in both these diseases, since they often appear together, including transforming growth factor-β (TGF-β).[4] Several controversies exist regarding TGF-β and other putative regulatory molecules and these are highlighted in the text.

Another field that has undergone a profound explosion of new knowledge is atherosclerosis. The role of inflammation, leukocyte biology and T-cell lymphocyte function has opened up new areas of research into the use of inflammatory biomarkers to stratify risk and for understanding the numerous pathological interactions with lipids deposited in the vessel wall. We are now able to propose several cell and molecular mechanisms to explain plaque rupture and associated acute occlusive thrombosis. Studies of the vascular and molecular biology of endothelial and smooth muscle cells have revealed how these cells become activated and contribute to the pathobiology of the plaque. We have also learned much about the interface of the endothelial surface with blood flow, especially how shear stress affects molecular function and dysfunction of both endothelial and smooth muscle cells.[5] Another new and exciting aspect of atherosclerosis is the technical innovations that have led to several forms of imaging to detect atherosclerotic plaque composition and burden, as well as novel forms of interventional treatment of coronary artery disease.

As we begin to understand the complexities of disease, it is important to convey the new pathobiology in comprehensible formats. This textbook is wonderfully illustrated, which adds much to the ability of the reader to understand the complexities of the mechanisms of cardiac and vascular disease. The cartoons and accompanying figure legends are clear and concise and highlight important concepts in each chapter. The photographs of gross and microscopic organs and tissue were chosen carefully to illustrate important anatomic pathologic features of disease. The tables are very informative and provide important integration of basic science findings and clinical disease. The references are well selected to provide evidence for the information and concepts presented in the sixteen chapters.

The strengths and weaknesses of specific animal models are considered, and data obtained in animal model systems are discussed and interpreted with respect to understanding human disease. The extensive use of animal models, especially mouse genetic models that knock out or enhance gene function, has resulted in an explosion of new discoveries that unravel some of the mysteries of organ and cell function, including those of endothelial cells, vascular smooth muscle cells, macrophages, cardiomyocytes and cardiac fibroblasts. Many specific gene products have been identified as important in the regulation of the development of, and in response to, injury pathways.

The content of the book is as valuable for the novice in the field of cardiovascular pathobiology as for the seasoned veteran, since it provides both breadth and depth in cardiovascular pathobiology. The content of the book should inspire novice undergraduate students to explore the pathobiology of cardiovascular disease by seeking hands-on opportunities to carry out mentored research projects, working alongside faculty, graduate students and postdoctoral fellows. The authors discuss many controversies in the field and direct the reader to appreciate where new breakthroughs are likely to come from. This is very useful to graduate students when choosing a topic for their research thesis.

There is therefore a seamless blend of classical cardiovascular pathology with current new cell and molecular pathobiology and a clear focus on the important problems yet to be addressed by experimental and clinical studies.

Avrum I. Gotlieb

References

1. Silver MD, Gotlieb AI, Schoen FJ. *Cardiovascular Pathology*. 3rd ed. Churchill Livingston; 2001.
2. Liu AC, Joag VR, Gotlieb AI. The emerging role of valve interstitial cell phenotypes in regulating heart valve pathobiology. *Am J Pathol* 2007;**171**:407–18.
3. Li C, Xu S, Gotlieb AI. The progression of calcific aortic valve disease through injury, cell dysfunction, and disruptive biologic and physical forces feedback loops. *Cardiovasc Pathol* 2013;**22**:1–8.
4. Xu S, Liu AC, Gotlieb AI. Common pathogenic features of atherosclerosis and calcific aortic stenosis: role of transforming growth factor-β. *Cardiovasc Pathol* 2010;**19**:236–47.
5. Gimbrone Jr MA, Garcia-Cardena G. Vascular endothelium hemodynamics and the pathobiology of atherosclerosis. *Cardiovasc Pathol* 2012;**22**:9–15.

Preface

The field of cardiovascular pathology has seen remarkable advances in the past 12 years, with valuable contributions coming from a diverse array of disciplines including molecular biology, biochemistry, metabolism, cell biology, pharmacology, and translational medicine, to name a few. Synthesizing these diverse and exciting findings with ongoing advances in the field of cardiovascular pathology is an increasing challenge. This has proven particularly true as we teach advanced cardiovascular pathology topics in medical schools to an array of medical professionals. Finding up-to-date materials to teach graduate students, post-doctoral fellows, practicing scientists in parallel with our medical colleagues at all stages of training in multiple specialties in pathology and internal medicine is challenging.

A common obstacle we face is finding a single textbook to teach from and to guide our discussions. Even attempting to use materials from multiple clinical cardiology and cardiovascular pathology texts can prove challenging. And when they did exist, they were commonly insufficiently updated or molecular-based to be adequate for students learning translational medicine. This led us to create the textbook you are currently holding, which is intended to complement outstanding anatomic cardiovascular pathology textbooks, such as *Cardiovascular Pathology* (3rd edition, edited by Malcolm D. Silver, MB, BS, MD, PhD, Avrum I. Gotlieb, MD, CM, Frederick J. Schoen, MD, PhD), that stand as a standard in the field.

This text includes a number of additional topics of emerging interest that cardiovascular pathology textbooks have not traditionally included, such as cardiac metabolism (Ch. 2), cardiac atrophy and remodeling (Ch. 3), and vasculogenesis and angiogenesis (Ch. 10). Our current understanding of the molecular basis of multiple other cardiovascular pathologies is also presented, including the pathophysiology of cardiac hypertrophy and heart failure (Ch. 4), ischemic heart disease (Ch. 5), cardiomyopathies (Ch. 6), myocarditis (Ch. 8), valvular heart disease (Ch. 9), atherosclerosis (Ch. 12), diseases of the small, medium and great vessels (Chs 11 and 13), and diseases of the conduction system (Ch. 7). Cardiac development (Ch. 1), blood pressure regulation (Ch. 14), arterial and venous thrombosis and thrombolysis (Ch. 15) are additionally discussed at the molecular level as is our increasing understanding of the diseases of the pericardium (Ch. 16).

This book covers a unique array of topics that are of interest to a broad biomedical sciences audience, including physicians, scientists, and the many specialized trainees in these fields. We are grateful for the participation of the many outstanding experts that have generously contributed to this book and hope that you find it to be a uniquely useful and valuable resource for the clinical practice and science of the pathophysiology of cardiovascular disease.

Contributors

Elena Aikawa, MD, PhD Brigham and Women's Hospital, Harvard Medical School, Boston, MA, USA

Joseph Fitzgerald Arboleda-Velasquez, MD, PhD Harvard Medical School, Boston, MA, USA

Cristina Basso, MD, PhD Department of Cardiac, Thoracic and Vascular Sciences, University of Padua Medical School, Padua, Italy

Jagdish Butany, MBBS, MS, FRCPC University of Toronto, Toronto, Ontario, Canada

John W. Calvert, PhD Emory University School of Medicine, Atlanta, GA, USA

Elisa Carturan, PhD Department of Cardiac, Thoracic and Vascular Sciences, University of Padua Medical School, Padua, Italy

Patricia Anne D'Amore, PhD Harvard Medical School, Boston, MA, USA

Laura A. Dyer, PhD McAllister Heart Institute, University of North Carolina at Chapel Hill, NC, USA

Patrick T. Ellinor, MD, PhD Massachusetts General Hospital, Boston, Massachusetts, USA; Harvard Medical School, Boston, Massachusetts, USA

Pooja Gupta, MBBS, DNB University of Toronto, Toronto, Ontario, Canada

Marc K. Halushka, MD, PhD The Johns Hopkins University School of Medicine, Baltimore, MD, USA

Pamela A. Harvey, PhD University of Colorado at Boulder, Boulder, CO, USA

Thomas J. Hund, PhD The Dorothy M. Davis Heart & Lung Research Institute, Columbus, OH, USA

Amar Ibrahim, MB, ChB, MSc University of Toronto, Toronto, Ontario, Canada

J. Charles Jennette, MD University of North Carolina at Chapel Hill, Chapel Hill, NC, USA

Leslie A. Leinwand, PhD University of Colorado at Boulder, Boulder, CO, USA

Michael A. Makara, BS The Dorothy M. Davis Heart & Lung Research Institute, Columbus, OH, USA

Richard N. Mitchell, MD, PhD Brigham & Women's Hospital/Harvard Medical School, Boston, MA, USA

Peter J. Mohler, PhD The Dorothy M. Davis Heart & Lung Research Institute, Columbus, OH, USA

Ivan Moskowitz, MD, PhD Departments of Pediatrics and Pathology, The University of Chicago, IL, USA

Lionel Opie, MD, DPhil, DSc, FRCP Professor of Medicine and the Director Emeritus of the Hatter Institute for Cardiovascular Research at the University of Cape Town, Cape Town, South Africa

Cam Patterson, MD, MBA McAllister Heart Institute, University of North Carolina at Chapel Hill, NC, USA

Avani Pendse, MBBS, PhD University of North Carolina at Chapel Hill, Chapel Hill, NC, USA

Thomas Pulinilkunnil, PhD Dalhousie University, Halifax, Nova Scotia, Canada

Mark Ranek, PhD The Johns Hopkins University, Baltimore, MD, USA

Harsimran Saini, MD Massachusetts General Hospital, Boston, Massachusetts, USA

Jonathan C. Schisler, MS, PhD University of North Carolina at Chapel Hill, Chapel Hill, NC, USA

Alvin H. Schmaier, MD Hematology and Oncology Division, Department of Medicine, Case Western Reserve University; University Hospital Case Medical Center, Cleveland, OH, USA

Frederick J. Schoen, MD, PhD Brigham and Women's Hospital, Harvard Medical School, Boston, MA, USA

Michael A. Seidman, MD, PhD Brigham & Women's Hospital/Harvard Medical School, Boston, MA, USA

Sakima A. Smith, MD The Dorothy M. Davis Heart & Lung Research Institute, Columbus, OH, USA

William E. Stansfield, MD University of North Carolina at Chapel Hill, Chapel Hill, NC, USA

Evi X. Stavrou, MD　Hematology and Oncology Division, Department of Medicine, Case Western Reserve University; Louis Stokes Veterans Administration Hospital; University Hospital Case Medical Center, Cleveland, OH, USA

James R. Stone, MD, PhD　Massachusetts General Hospital, Boston, Massachusetts, USA; Harvard Medical School, Boston, Massachusetts, USA

Sara Tabtabai, MD　Massachusetts General Hospital, Boston, Massachusetts, USA

Gaetano Thiene, MD, FRCP　Department of Cardiac, Thoracic and Vascular Sciences, University of Padua Medical School, Padua, Italy

Rhian M. Touyz, MD, PhD　University of Glasgow, Glasgow, UK

Shaobin Wang, PhD　University of North Carolina at Chapel Hill, Chapel Hill, NC, USA

Monte S. Willis, MD, PhD　University of North Carolina at Chapel Hill, Chapel Hill, NC, USA

Acknowledgments

We wish to thank our editor Mara Connor at Elsevier for her unwavering support and tenacious encouragement to take on this project. We are grateful for the tremendous generosity of the authors – for their time, expertise, and hard work in compiling and presenting their cutting edge work for the cardiovascular pathologist. We also acknowledge the support, assistance, and guidance of our developmental editors, Megan Wickline and Jeff Rossetti, throughout the publication process.

1

Molecular Basis of Cardiac Development

Laura A. Dyer, PhD[1], Ivan Moskowitz, MD, PhD[2], Cam Patterson, MD, MBA[1]

[1]McAllister Heart Institute, University of North Carolina at Chapel Hill, NC, USA; [2]Departments of Pediatrics and Pathology, The University of Chicago, IL, USA

THE HEART FIELDS AND HEART TUBE FORMATION

The heart begins simply as a bilateral field within the lateral plate mesoderm (Figure 1.1, Table 1.1). As the early embryo undergoes formation of the gut-tube, these bilateral fields migrate toward the midline, where the cranial-most aspect of the fields will fuse ventrally to form the outer curvature of the heart tube.[1] These fields will continue to migrate together, with more of the heart fields contributing to the forming tube, until the dorsal aspect of the heart fields fuse to form a closed tube.[1] The initial contributors to the heart tube are known as the first heart field.[2] The first heart field gives rise to the left ventricle, with some contributions to the atria and the right ventricle.[3] Additional heart field progenitors from the lateral plate mesoderm continue to add to the arterial and venous poles of the heart tube; these later-adding cells are known as the second heart field. The second heart field gives rise to most of the right ventricle and atria, the most distal myocardium that surrounds the aorta and pulmonary artery, and the most proximal smooth muscle that contributes to the tunica media of the great arteries.[2]

Signaling Pathways in Heart Field Specification

The Wnt Pathway

The Wnt family includes the canonical pathway, the non-canonical pathway (also known as the planar cell polarity pathway), and the Wnt/calcium pathway (Figure 1.2). Both the canonical and non-canonical pathways have well-established roles in heart field specification. Temporal waves of canonical and non-canonical Wnt signaling play distinct roles during cardiac specification and morphogenesis. As the heart field forms from the primitive streak, Wnt3a is expressed in the primitive streak and serves as a repulsive cue to the forming heart field.[4] Experiments performed in *Xenopus*, due to its ease of manipulation and genetic tractability,[5] have shown that the early repression of Wnt signaling in the *Xenopus* animal cap (i.e., in the ectodermal roof of the blastocele prior to heart field specification) via Dkk-1 and Crescent is necessary for initiation of transcription of cardiac transcription factors Nkx2.5 and Tbx5 and myocardial-specific proteins troponin I and myosin heavy chain α.[6] However, later in cardiac development, canonical Wnt signaling in embryonic mice at E8.75 promotes Nkx2.5, Islet1, and Baf60c within the entire heart.[7] Due to the genetic similarity between *Xenopus* and mouse, these differences likely reflect different temporal requirements for Wnt signaling as opposed to species-specific differences.[5] In the second heart field, non-canonical Wnts 5a and 11, which act through the non-canonical planar cell polarity pathway, are expressed slightly later in development and co-operatively repress the canonical Wnt pathway while also promoting expression of Islet1 and Hand2, whose expression serves to 'mark' the heart field;[8] as such, these genes are commonly referred to as cardiac markers. Both the repression of the canonical Wnts and the induction of the heart field markers require β-catenin in the second heart field.[8] Wnt5a and Wnt11 also promote proliferation within this progenitor region.[8] After the heart tube forms, Wnt-stabilized β-catenin is necessary in the Islet1-expressing second heart field cells to maintain their progenitor status.[9] Loss of either β-catenin or Wnt signaling in the second heart field leads to second heart field defects, including right ventricular and outflow tract defects.[9,10] Even if Wnt signaling is lost under cells expressing one of the first markers of differentiated cardiomyocytes, Mesp1, second heart field proliferation is decreased, and Islet-1

FIGURE 1.1   An overview of heart development. (A) The heart fields are specified as bilateral fields within the lateral plate mesoderm. The cranial-most aspect (asterisks) will migrate toward the midline first; this seam is depicted by the gray dashed line in B. (B) The heart tube closes ventrally, and cells continue to add from the heart fields. The dorsal aspect of the heart tube will pinch off last. After dorsal closure, additional cells can only be added via the venous (IFT, inflow tract) and arterial (OFT, outflow tract) poles. (C) As additional cells add to the heart tube, the heart tube begins to undergo looping, and the ventral midline becomes the outer curvature of the heart. During looping, the endocardial cushions begin forming in the atrioventricular canal and outflow tract. The atrioventricular canal separates the common atrium (A) from the common ventricle (V). (D) At the end of looping, the atrioventricular cushions are aligned dorsal to the outflow tract cushions, allowing connections between the left atrium and ventricle (LA and LV, respectively) and the right atrium and ventricle (RA and RV, respectively). As the outflow tract is septated, it also undergoes rotation to align the aorta with the left ventricle. Septa form between the ventricles and between the atria. (E) In the mature four-chambered heart, the aorta (Ao) serves as the outlet for the left side of the heart, and the pulmonary artery (PA) serves as the outlet for the right side of the heart.

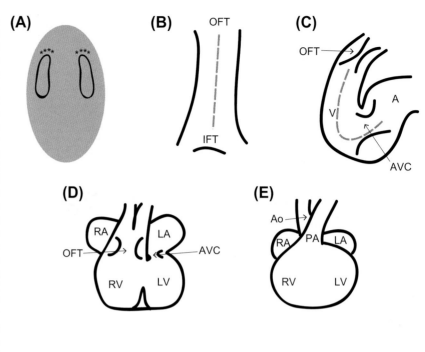

TABLE 1.1   Major Developmental Time Points in Humans and Common Experimental Models

Milestone	Human[209,210]	Mouse[211,212]	Chick[211,213]	*Xenopous*[214–216]	Zebrafish[217,218]
Heart is specified		E7.25	HH 3	Stage 15	8 somites
Heart tube forms	~CS 9/10 (2 mm)	E8	HH 9	Stage 33	24 hpf
Heart displays regular contractions	~CS 9/10	E8	HH 7*	Stage 35	24 hpf
Heart begins looping*	CS 13	E8.5	HH 9+/10-	Stage 35	30 hpf
AV cushions begin forming	CS 17	E9.5	HH 12/13-	Stage 44	48 hpf
OFT cushions begin forming	CS 15/16	E10.5	HH 12/13-	Stage 39/40	–
Outflow tract is septated	CS 17	E13.5	HH 34	–	–
Atria are septated	CS 18†	E14.5	HH 46	Stage 46	–
Ventricles are septated	CS 22	E13.5	HH 34	–	–

The major stages in cardiovascular development are presented for humans and the most commonly used animal models.
*Because the heart tube begins forming as a trough that then closes dorsally, contractions are observed prior to the pinching off of the fully formed heart tube.
†The foramen ovale is still open at this stage. This fenestration is closed at the stages listed for the other species. CS, Carnegie stage[219]; E, embryonic day; HH, Hamburger-Hamilton stage[220]; Stage, Nieuwkoop and Faber stage[221]; hpf, hours post-fertilization. See cited references for more details.

expression is down-regulated.[11] Conversely, overexpressing β-catenin under the Mesp1 promoter expands the Islet-1-positive second heart field and promotes proliferation.[11] Later, Wnt5a specifically acts upstream of the disheveled/planar cell polarity pathway to regulate the addition of the second heart field to the arterial pole.[12] In addition, Wnt signaling also promotes bone morphogenetic protein (BMP) 4 and the non-canonical

Wnt 11, which promote myocardial differentiation.[9,11] Thus, the Wnt pathway is critical for inducing heart field formation, maintaining progenitor status and promoting myocardial differentiation.

Retinoic Acid

One of the earliest required signaling pathways is the retinoic acid pathway. RALDH2, the enzyme that

Canonical   Non-canonical

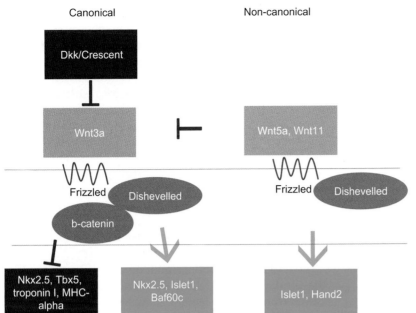

FIGURE 1.2 Wnt signaling pathway. In both the canonical and non-canonical (planar cell polarity) pathways, extracellular Wnt ligands bind to the transmembrane receptor Frizzled. In the canonical pathway, Frizzled forms a complex with Dishevelled and additional components, leading to β-catenin stabilization and translocation to the nucleus; β-catenin then promotes Wnt-induced gene expression, such as Nkx2.5 and Islet1. In contrast, in the non-canonical pathway, Frizzled and Dishevelled act on a different set of intracellular signaling molecules (e.g., Rho, Rac) to promote a different set of genes and to inhibit the canonical pathway.

synthesizes retinoic acid, is restricted within the lateral plate mesoderm to a region nearest to the heart field.[13,14] Expression of RALDH2 progresses in a cranial–caudal direction during heart field induction and heart tube formation and establishes the posterior boundary of the heart field.[15] Within the lateral plate mesoderm, retinoic acid plays an inhibitory role, where it acts both directly and indirectly to restrict cardiac transcription factors Nkx2.5 and FoxF1 to the anterior lateral plate mesoderm, Hand1 to the anterior and middle of the lateral plate mesoderm, and Sal1 to the posterior lateral plate mesoderm.[13,16,17] Retinoic acid further represses GATAs 4, 5, and 6.[18] This inhibitory role is required to limit the size of the heart field, and fate-mapping studies in the zebrafish, another experimental model that is valued for its genetic similarity to the mouse and ease of studying embryonic development,[19] have demonstrated that zebrafish embryos with decreased levels of retinoic acid exhibit an increased number of Nkx2.5-positive cells.[17] Conversely, exposing either zebrafish or *Xenopus* embryos to increasing levels of retinoic acid specifically leads to a reduced number of cardiomyocytes.[16,17] In chick embryos, which physically develop more similar to humans as compared with mice but lack the ability to manipulate the genome as in mice,[20] antagonizing retinoic acid signaling promotes the ventricular myocardial fate at the expense of the atria.[15] Together, these studies suggest that retinoic acid plays two major roles in the early heart field. First, retinoic acid generally restricts the expression of heart field markers to limit the size of the heart field. Then, it specifically promotes a 'posterior' fate within the heart field, which affects the venous precursors that will give rise to the atria.

Transcription Factors in Heart Field Specification

Several well-known cardiogenic transcription factors are expressed in the early heart fields and play a role in driving lateral plate mesoderm to a cardiac fate. In the first heart field these transcription factors include but are not limited to Nkx2-5; GATAs 4, 5, and 6; and Tbx5. The transcription factor kernel required for driving mesoderm to myocardium has been minimally defined as the transcription factors GATA4 and Tbx5 and the chromatin remodeling subunit Baf60c.[21] Furthermore, Tbx5 genetically interacts with Baf60c in cardiac morphogenesis.[22] These findings have begun to elucidate the interplay between chromatin remodeling factors and cardiogenic transcription factors during the progressive differentiation of mesoderm to cardiac progenitor to cardiomyocyte.

The homeobox transcription factor Nkx2.5 is perhaps the best-known cardiac inducer. Nkx2.5 is expressed in the first and second heart fields. During normal development, Nkx2.5 becomes turned off in the first heart field, allowing these cells to begin differentiating while maintaining the second heart field in a progenitor state.[23] During this process, Nkx2.5 represses the transcription of numerous cardiac progenitor markers, including Islet1, Mef2c, and Tgfβ2.[23] In addition to repressing cardiac progenitor markers, Nkx2.5 also represses the BMP and fibroblast growth factor (FGF) pathways, which promote differentiation into myocardium and inhibit proliferation.[23–25] This field of cells is the first to differentiate into functional myocardium and express muscle-specific markers such as the myosin light chain protein MLC3F[26] while concomitantly down-regulating cardiac progenitor transcription factors such as Islet1.[23] In the later differentiating second heart field, these pathways must be down-regulated to allow for sufficient proliferation within the progenitor pool,[23] and proliferation in this region is essential for normal development of the arterial pole.[25,27] In addition, Mef2c is required to allow the transition from specification to myocardial differentiation.[28]

Congenital Heart Disease (CHD) Resulting from Heart Field Defects: Cardia Bifida

A severe and uniformly lethal heart defect associated with the early heart fields is a failure of the bilateral heart fields to migrate and fuse at the midline, leading to two separate hearts, a condition known as cardia bifida. Though the previously mentioned signaling pathways and transcription factors are crucial for heart formation, loss of only a few specific factors can result in cardia bifida. These factors are described in detail, and Table 1.2 provides the Online Mendelian Inheritance of Man entry numbers if information on available genetic testing

TABLE 1.2 Developmental Stages of Heart Formation and their Associated Congenital Heart Defects

Developmental Process	Transcription Factors and Genes Involved	Related Human Diseases	OMIM Entry #
Heart field specification and heart tube formation	GATA4 Smad6 Nodal	Cardia bifida	600576 602931* 601265
Establishment of laterality	Pitx2 FoxC1	Axenfeld-Rieger syndrome	601542 601090
Conduction system	Pitx2c	Atrial fibrillation	601542
Valve formation	Fibrillin-1 Elastin Fibulin-4 TGFβ	Mitral valve prolapse	134797 130160 604633 190180
Outflow tract formation	Tbx1 Lhx2 Tcf21 β-catenin Neuropilin-1	DiGeorge syndrome	602054 603759* 603306* 116806 602069*
Chamber septation	GATA4 Nkx2.5 Tbx20 Tbx5	Atrial septal defects	600576 600584 606061 601620

During some steps of development, specific genes have been definitively shown to cause certain congenital heart defects in humans. These developmental steps, along with their associated genes and defects, are presented herein. More information can be obtained from the Online Mendelian Inheritance in Man database (www.omim.org), which includes clinical resources, such as whether a genetic test is available to screen for known mutations.
*No genetic test is yet available.

for these genes (as well as additional CHD-related genes presented later in the chapter) is of further interest.

Loss of GATA4 is perhaps the best-known cause of cardia bifida. The GATA4 homozygous null mouse embryo fails to form a heart tube.[29,30] Furthermore, siRNA knockdown of GATA4 in the chick heart fields results in the formation of bilateral heart tubes, and this effect is caused by a down-regulation of N-cadherin.[31] Similarly, blocking N-cadherin in chick embryos inhibits fusion of the heart fields but does not halt heart development, resulting in bilateral heart tubes.[32] However, although the N-cadherin homozygous null embryo displays heart failure, as evidenced by pericardial effusion, a single heart tube is present.[33] Thus, species-specific differences in the cadherins may underlie the contribution of N-cadherin to GATA4-mediated cardia bifida.

Loss of BMP or Nodal signaling can also result in cardia bifida, and this effect may also occur through the GATA transcription factors. In *Xenopus*, loss of BMP signaling through injection of mRNA encoding the BMP antagonist Smad6 or a mutated truncated BMP receptor leads to cardia bifida.[34] In these embryos, the heart fields are also specified but fail to fuse at the midline.[34] Consistent with BMP's previously mentioned role in promoting myocardial differentiation, these embryos also show fewer cardiomyocytes compared with untreated control embryos.[34] Similarly, the zebrafish *swirl* and *one-eyed pinhead* mutants, which lack Bmp2b and Cripto, respectively, display reduced Nkx2.5 expression, which can be rescued by the ectopic expression of GATA5.[35] Of these two zebrafish mutants, only the *one-eyed pinhead* mutant presents with cardia bifida though,[35] and genetic redundancy in the zebrafish may explain why the zebrafish *swirl* mutant has a less severe phenotype than the Smad6-injected *Xenopus* embryo. Perhaps unsurprisingly, cardia bifida is almost always incompatible with life. One case report, however, has described an infant who survived with cardia bifida in which each half heart showed characteristics of both ventricles but a single atrium.[36,37] Each half heart also had its own truncus and sinus venosus, which may have been sufficient in the intrauterine environment.[36,37]

LOOPING AND LATERALITY

The human body is asymmetric about the vertical midline, and the heart is among the visceral organs with left/right asymmetry. Left/right asymmetry is established early during embryonic development; left/right gene expression differences are observed in the lateral plate mesoderm from which cardiac progenitors emanate. The molecular mechanisms that dictate the left/right asymmetry of the heart during its morphogenesis are unknown; however, the mechanisms that underlie the initial asymmetry-breaking events in the early embryo have been described.

Left/Right Determination, Cilia, and Signaling

The cilium, a subcellular organelle that protrudes from the cellular surface, plays an important role in left/right determination. In the gastrulating embryo, rotating cilia in Hensen's node generate leftward flow that breaks left/right symmetry and establishes differential signaling and gene expression on the left and right sides of the embryo.[38] In the heart field, the cilia initially express FGF receptors. These receptors respond to FGF, which induces them to move Sonic hedgehog and retinoic acid leftward, which results in the release of calcium.[39] Despite this purported leftward movement, though, the asymmetrical expression of Sonic hedgehog and retinoic acid is controversial, with some but not all studies showing asymmetrical expression in the node.[40–42] These signaling events limit the expression of Nodal, Pitx2c, and Lefty2 to the left side of the embryo.[43] The side-specific expression of these transcriptional regulators activates a gene regulatory network on the left distinct from that on the right, causing sidedness of morphogenesis. The link from the left/right determination transcriptional kernel to the cardiac progenitor transcriptional kernel may be direct. Nkx2.5, a progenitor transcription factor, interacts with the N-terminal end of transcription factor Pitx2c, part of the left-determination transcriptional network.[44] This specific interaction is required for normal heart looping, and antagonizing Pitx2c in the left side of the heart, where it is normally expressed, leads to randomized heart looping.[44]

Sidedness of Cardiac Structures

Cardiac left/right asymmetry is first noticeable during cardiac looping, during which the ventricles undergo characteristic D-handed looping with great fidelity. This rightward looping displaces the ventral midline of the heart towards the right and establishes the outer curvature of the heart tube.[45] The formerly dorsal midline is then rotated to become the inner curvature of the heart tube. From this looped position, the outflow and inflow tracts converge, and the ventricular bend becomes displaced ventrally.[46] Finally, as the outflow tract is septated to form the aorta and pulmonary artery, the aorta is wedged between the pulmonary artery and the atrioventricular valves.[47]

Left/right determination can go awry in distinct ways with predictable results. The entire cascade can be inverted, resulting in normal but mirror image development of the embryo and heart. Consistent with this hypothesis, the *inv* mutant mice display situs inversus, and in this model, Pitx2c is selectively expressed on the right side instead of the left.[48]

The left-sided program can be disrupted, resulting in hearts with both sides developing as 'right,' or right isomerism. Work in mouse embryos suggests that Pitx2c is required for turning off the 'right-sided' developmental program on the left side. In Pitx2c homozygous null embryos, both atria display right-sided structures. Within the mature heart, left-sided structures include the left atrium with pulmonary veins, whereas right-sided structures include the right atrium with sinoatrial node. In Pitx2c mutant embryos, the sinoatrial node is duplicated on both sides, and valve and septation defects consistent with right atrial isomerism are observed.[49–51] The requirement for Pitx2c in structural morphogenesis of the heart may come after heart looping: removing Pitx2c specifically from differentiated myocardium causes right atrial isomerism, including an atrial septal defect and the duplicated sinoatrial node.[52] Thus, Pitx2c is required in the mesoderm for left/right determination and later in the myocardium to repress the 'right-sided' developmental program, including sinoatrial node formation within the left atrium.

The right-sided program can also be disrupted, resulting in hearts developing with two 'left' sides or left atrial isomerism. Left atrial isomerism is observed in embryos lacking Sonic hedgehog.[53] Sonic hedgehog null mice display isomerism of the left atrial appendages and other malformations that are frequently observed in humans with left isomerism.[53] In these embryos, Pitx2c expression is observed bilaterally, consistent with the instructive role of this transcription factor for left-sided structures.[53] Consistent with the Pitx2c expression pattern in the Sonic hedgehog null mouse, the heart fails to loop in mice lacking hedgehog signaling.[41] Finally, the left/right determination process can be randomized. Mice that lack cilia show randomized expression of left/right determination transcription factors, sidedness, cardiac looping, and sidedness of specific cardiac structures.[41,54]

CHD from Abnormal Left/Right Determination: Heterotaxy Syndrome and Abnormal Left/Right Specification

Defects in left/right determination are common in humans and are grouped into the heterotaxy syndrome. The observed cardiovascular phenotypes of humans with this syndrome are broad, as predicted from the above discussion. The molecular genetics of heterotaxy syndrome are under current investigation, and rapid progress is being made given the large number of predicted candidate genes identified from animal model studies. In particular and consistent with the importance of cilia in left/right determination, cilia gene mutations have been identified in several heterotaxy syndrome patients, causing a range of cardiac structural abnormalities from situs inversus to left or right isomerisms.[55] Furthermore, proper cardiac looping is required to align the segments of the heart for subsequent septum morphogenesis. Mis-looping would be expected to cause cardiac structural abnormalities in addition to structural isomerisms. Consistent with this hypothesis, cardiac septal defects are commonly observed in patients with heterotaxy syndrome. For example, patients with Axenfeld-Rieger syndrome display a variety of congenital heart defects, including atrial and ventricular septal defects, mitral and tricuspid defects such as prolapse and stenosis, small left ventricular outflow tract, and stenotic or bicuspid aortic and pulmonary valves.[56] An estimated 40–60% of these patients have mutations in Pitx2 or FoxC1.[56]

CHAMBER SPECIFICATION

Cardiac Chamber Versus Non-Chamber Developmental Programs

Historically, the tubular heart was thought to be composed of segments that were each progenitor structures for the components of the mature cardiac form. Thus, from posterior to anterior, the linear tube was thought to possess domains that generated the atria, the left ventricle, the right ventricle, and outflow tract myocardium. However, recent molecular developmental studies have generated overwhelming evidence for a new model: the tubular heart is the progenitor structure primarily of the atrioventricular canal and atrioventricular node myocardium in the mature heart.[57–60] The cardiac chambers, including the atria positioned dorsally and the ventricles positioned ventrally, form as rapidly proliferating structures that balloon off the primary heart tube. Thus, understanding the building blocks of the developing

heart has progressed from trying to understand segmentation of the linear heart tube (with little progress) to understanding the regulation of chamber formation off the primary heart tube. The portion of the heart tube that does not enter the 'chamber' program becomes fated to the atrioventricular canal, such as atrioventricular node myocardium, and defines the region of myocardium that encourages adjacent endocardium to enter the endocardial cushion and cardiac valve program. Thus, this 'ballooning' model defines chamber versus non-chamber fates. Because non-chamber myocardium becomes fated to the atrioventricular conduction system and determines the location of the atrioventricular valves, this model integrates chamber morphogenesis with the cardiac conduction system and cardiac valve development. BMP signaling and T-box transcription factors play fundamental roles in chamber versus non-chamber determination.

Within the heart tube, the myocardium that will be retained in the atrioventricular canal is molecularly defined by expression of BMP2, Tbx2, and Tbx3 (Figure 1.3). BMP2 is required for the development of the atrioventricular canal and to prevent its conversion to chamber myocardial program. BMP2 induces the expression of Tbx2 and Tbx3, which act as transcriptional repressors. Tbx2 and Tbx3 directly interact with and repress chamber-specific genes such as Nppa (also known as ANF) and Cx40.[58,61–64] Tbx2 and Tbx3 share redundant functions, and mice that lack at least three Tbx2/Tbx3 alleles do not form atrioventricular cushions, possibly due to reduced levels of BMP2.[61]

In contrast to the dominance of T-box transcriptional repressors Tbx2 and Tbx3 in the atrioventricular canal, the T-box transcriptional activators Tbx20 and Tbx5 dominate in cardiac chamber myocardial development. Tbx20 and Tbx5 activate the chamber myocardial

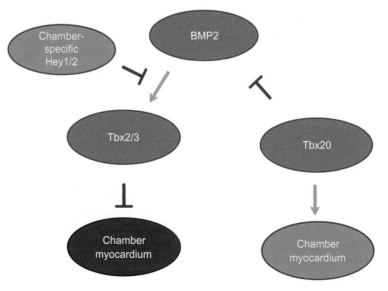

FIGURE 1.3 Chamber versus non-chamber myocardium. Myocardium that is fated to contribute to the chambers is restricted based on interactions between BMP2, a series of Tbx genes, and Hey1 and 2. In the atrioventricular canal and outflow tract, BMP2 induces Tbx2 and 3, which repress chamber myocardium-specific genes. In contrast, chamber-specific Tbx20 expression inhibits BMP signaling, thus eliminating this inhibition and promoting myocardial differentiation. In addition atrial-specific Hey1 and ventricular-specific Hey2 further inhibit BMP2 from interacting with Tbx2 and 3, providing additional feedback to demarcate the chamber myocardium from the developing valves.

program, including networks for rapid proliferation and differentiation into strongly contractile myocardium, promoting Nmyc1-induced myocardial specification and proliferation.[65–69] Tbx20 also prevents regions that are fated for cardiac chamber development from expressing the 'atrioventricular canal' program by interacting with Smad1/5 to inhibit BMP signaling and repress Tbx2 expression within chamber myocardium.[70] Additionally, atrial-specific Hey1 and ventricular-specific Hey2 also suppress Tbx2 and BMP2, thus reinforcing the boundaries between chamber myocardium and developing valves.[71,72] Importantly, if the atrioventricular region is not defined, then patterning in the atria can also be disrupted, resulting in ventricular-specific markers (such as Nppa) extending into the atria.[73]

Atrial Versus Ventricular Identity

Hey family transcription repressors act to maintain chamber identity and define atrial versus ventricular identity. Independent of the Notch pathway, atrial-restricted Hesr1 (Hey1) and ventricular-restricted Hesr2 (Hey2) both suppress Tbx2, which is expressed between these two genes to define the atrioventricular canal and thus separate the ventricles from the atria.[71] Hey2 expression in ventricular compact myocardium inhibits atrial gene expression and thus atrial identity.[74] Hey2 also promotes proliferation within the compact myocardium.[74] However, trabecular ventricular myocardium is unaffected by Hey2, based on BMP10 expression in the Hey2 homozygous null embryo. Similar to the effect of Tbx20 overexpression in differentiated cardiomyocytes, Hey2 knockout results in up-regulated Tbx5 and Cx40.[74]

VENTRICULAR SEPTATION AND MYOCARDIAL PATTERNING

To septate the ventricles, a muscular septum grows from the apex of the ventricles toward the atrioventricular canal (Figure 1.4).[75] A membranous septum forms from the cushion tissue of the atrioventricular mesenchymal septum and bridges the gap between the atrioventricular canal and the crest of the ventricular muscular septum.[76] Tbx5 and Tbx20 are crucial within the ventricles for placement of the muscular ventricular septum, independent of their role in chamber versus non-chamber determination. In Hamburger-Hamilton stage (HH) 30 chick embryos, Tbx5 is restricted to the left ventricle, and Tbx20 is restricted to the right ventricle.[68] When Tbx5 is ectopically expressed throughout both ventricles, both ventricles express left-sided markers, such as ANF.[68] The ventricular septum fails to form in these embryos.[68] Furthermore, if Tbx5 is misexpressed on the other side of the Tbx20-positive right ventricle

(i.e., within the outflow tract), a second ventricular septum-like structure appears at the juxtaposition of these two transcription factors.[68] Thus, a boundary at the edge of Tbx5 expression appears essential for patterning the placement of the ventricular septum.

Patterning across the myocardial wall is also essential for proper cardiac function. For example, the ventricles possess an outer layer of highly contractile compact myocardium, and an inner layer of poorly contractile trabeculated myocardium.[77] Neuregulin signaling is required for the proper trans-myocardial wall patterning. The neuregulin-1 homozygous null embryo forms chambers that are poorly differentiated.[78] Neuregulin-1 promotes exit from the cell cycle and cardiac differentiation and acts through the ERK intracellular pathway.[78] The neuregulin-1 knockout mouse has poor trabeculation and a thin compact myocardial layer.[78] Furthermore, most cardiac transcription factors are down-regulated, including Hand1 and Cited1, in this mouse, and the conduction system prematurely switches from base-to-apex to apex-to-base activation,[78] suggesting that the conduction system myocardium develops early and at the expense of the working ventricular myocardium. In addition to Cited2's early role in establishing left/right asymmetry, knocking out Cited2 under the Nkx2.5 promoter leads to incomplete ventricular septation and a thin compact myocardial layer.[79] Cited2 also binds to the vascular endothelial growth factor (VEGF)A promoter, thus promoting coronary vessel formation within the compact myocardium.[79] Transcription factor Tbx20 promotes

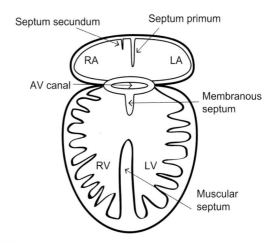

FIGURE 1.4 Chamber septation. To form the ventricular septum, the muscular component of ventricular septum elongates toward the atrioventricular canal. From the atrioventricular (AV) canal, a membranous septum forms from the cushion tissue and closes that gap between the atrioventricular canal and the muscular ventricular septum. To form the atrial septum, the septum primum grows from the cranial aspect of the common atrium toward the atrioventricular cushions. Once this septum fuses with the atrioventricular cushions, the cranial aspect detaches. This fenestration is closed by a fold in the dorsal atrial myocardium termed the septum secundum, which forms to the right of the septum primum.

the transcription of BMP10,[80] which is expressed and required for the formation of the trabeculated myocardium.[81] BMP10 in turn promotes Tbx20.[80,82] When Tbx20 is overexpressed under the early expressing Nkx2.5 promoter, Tbx5 is reduced, but ventricular chamber markers are present, and myocardial proliferation is reduced.[80] In contrast, when Tbx20 is overexpressed under the later-expressing βMHYC promoter, Tbx2 is repressed, and cardiac conduction-specific factors (e.g., Tbx5, Cx40, and Cx43) are induced; furthermore, cardiomyocyte proliferation is increased, resulting in a thickened compact myocardium.[80] These results are consistent with Tbx20 playing multiple roles during myocardial differentiation.

CONDUCTION SYSTEM DEVELOPMENT

In the primitive heart tube, all the cardiomyocytes are electrically coupled, and contractions proceed in peristaltic waves from the inflow to outflow. However, in the four-chambered heart, the conduction system paces and orients the contractions of the atria and ventricles to maximize cardiac output and efficiency. In the mature heart, an impulse initiates at the sinoatrial node at the junction of the right atrium and superior vena cava and activates electrically coupled atrial myocardium, causing the atria to contract. The impulse is rapidly transmitted to the atrioventricular node, in the atrioventricular junctional complex. The atrioventricular node is the only route of electrical continuity between the atria and ventricles and is a slowly propagating structure. The atrioventricular node slows the signal prior to its transmission to the ventricles, providing a delay to allow atrial contraction and ventricular filling. From the atrioventricular node, the impulse travels into the rapidly conducting ventricular conduction system. The impulse travels anteriorly in the atrioventricular bundle (or bundle of His), through the bundle branches coursing down each side of the interventricular septum, and ramifies into a network of Purkinje fibers at the apex of the heart. The Purkinje network activates surrounding ventricular myocardium, which triggers ventricular contraction and provides for the observed apex-to-base direction of ventricular contraction.[20]

The Atrial Nodal Conduction System

The model of chamber specification described above provides a coherent model for the localization and specification of the atrial components of the conduction system, the sinoatrial and atrioventricular nodes. Both of these conduction components are specified early and are present in the primary heart tube. Primary heart tube myocardium is poorly contractile, poorly coupled, and autonomously depolarizing. These cellular characteristics describe the precise characteristics required in the mature sinoatrial and atrioventricular nodes. Thus, primary myocardium with the desired characteristics is retained in the atrial nodes, highlighting the myocardial derivation of the atrial conduction system, and the 'chamber' myocardial program is repressed in these cells, allowing chamber positioning to retain the nodes in the appropriate locations.[83] In addition to repressing a conversion to chamber myocardium, the T-box transcription factor repressors Tbx3 and Tbx2 actively promote the cellular characteristics of the nodal conduction system.[62,84]

Early in conduction system development (HH 17 in chick), a transient sinoatrial node is observed on the left side of the sinus venosus.[85] Consistent with an incompletely differentiated phenotype, this region is negative for Nkx2.5 and differentiated marker TnI and positive for heart field marker Islet1, Tbx18, and conduction system marker hyperpolarization activated cyclic-nucleotide gated channel (HCN).[85] Within a day, the sinoatrial markers in the chick are present on the right side and remain transiently expressed on the left side of the sinus venosus. By HH 22, TnI has been up-regulated and will remain expressed in the mature sinoatrial node.[85] In addition to its earlier roles, Pitx2c expressed by TnT-positive cardiomyocytes will repress sinoatrial node formation on the left side of the sinus venosus.[52] By HH 35, the definitive sinoatrial node is present at the junction of the superior vena cava and the right atrium,[86] and it almost exclusively activates the right atrium first.[85] Only a small percentage of hearts with left-sided sinoatrial node maintain pace-making ability.[85] Gap junctions and desmosomes are rare in the sinoatrial node, leading to slow conductance.[86] Of the connexins, connexin 30.2 and connexin 45 have been identified in sinoatrial node, and the gap junctions that these connexins form exhibit slow conductance.[87]

T-box genes Tbx18 and Tbx3 are required for sinoatrial node formation. Tbx18 promotes the development of the 'head' of the sinoatrial node, whereas Tbx3 promotes the differentiation of the 'tail' that runs along the terminal crest.[88] Similar to the molecular signaling required to define the regions of the heart tube that will form the atrioventricular cushions, Tbx3 plays a role in conduction system development by repressing atrial-specific genes such as Nppa.[84] Tbx3 expression occurs as soon as atrial markers are expressed, and the Tbx3 homozygous null mouse has a small sinoatrial node as compared with wild-type mice.[84] The sinoatrial node is likely derived from the Tbx18-positive region because these two Tbx genes regulate spatially distinct portions of the sinoatrial node.[88]

Tbx2 and Tbx3 expression delineate the atrioventricular node. Whereas these genes are redundant in the atrioventricular canal for repressing the chamber myocardial program, a positive role has been elucidated for Tbx3 in conduction system fate. Tbx3 activates a genetic program for the pacemaker phenotype and is sufficient for driving atrial cells into the pacemaker program.[61,89] The atrioventricular node also shows few gap junctions and desmosomes and is insulated by connective tissue;[86] together, these properties help it slow conduction between the atria and the ventricles, allowing time for the atria to empty into the ventricles prior to the ventricular contraction. Similar to the sinoatrial node, the predominant connexins are the slow-conducting connexins 30.2 and 45, although connexin 40 is weakly expressed in the atrioventricular node.[87]

The Ventricular Conduction System

The functional requirements for the ventricular conduction system are opposite those of the atrial nodal conduction system. In the ventricles, the impulse travels through the conductive myocardium more rapidly than through normally coupled 'working' myocardium; thus, a fast conduction program must be initiated. Just as the atrial conduction system is derived from a subset of the early heart tube myocardium and utilizes the attributes of the primitive myocardium, the ventricular conduction system is similarly derived from the chamber myocardium of the heart tube. These myocardial cells are well-coupled to allow close synchronization and rapidly depolarize. Therefore, it is no surprise that the ventricular conduction system cells also originate from the original chamber myocardial cells. The ventricular conduction system is specified much later than the nodal conduction system; specification and functional emergence of the ventricular conduction system parallels the emergence of the chamber myocardial program. Classic lineage and more current fate map studies demonstrate that the ventricular conduction system derives from the ventricular chamber myocardium.[90-92]

Transcriptional regulation of ventricular conduction system development has been studied. The combinatorial action of Tbx5 and Nkx2.5 is required for emergence of a functional ventricular conduction system in mice.[93] Interestingly, expression of both of these transcriptional activators is up-regulated in the ventricular conduction system.[90,94] Furthermore, the overlap of their expression in the ventricles is specific for the ventricular conduction system.[91] Together, these observations establish that a highly expressed combination of these factors is required in the ventricular conduction system compared with the surrounding myocardium. In mice that lack a single copy of or are haploinsufficient for both Tbx5 and Nkx2.5, the ventricular conduction system fails to form. Instead, ventricular myocytes retain a normal working myocardial fate instead of being recruited to a conduction fate.[91]

Conduction is observed in a base-to-apex pattern before the His Purkinje network has fully matured,[95] highlighting the conductive properties of the early myocardium. Once the His Purkinje fibers have matured, electrical activation in the heart switches from a base-to-apex pattern (HH 25–28 in chick) to the mature to-base pattern (HH 33–36 in chick).[95] Specific channel proteins provide both rapid depolarization and large conductance connectivity between the cells of the ventricular conduction system. The large gated sodium channel Nav1.5 is unregulated in these cells, providing rapid depolarization. The large conductance channel connexin 40 is also up-regulated specifically in the ventricular conduction system. Tbx5 directly transcriptionally activates expression of both of these genes, providing a direct link from specification to functional differentiation of the ventricular conduction system.[96-98]

Among its many roles in cardiac development,[99] the vasoconstrictor peptide endothelin-1 promotes a conductive fate. Endothelin-1 induces both cardiac progenitors and cardiomyocytes to become conduction cells.[100,101] In response to endothelin-1, Nkx2.5-positive cardiac progenitors express Hcn4 and connexin 45.[101] Intriguingly, endothelin receptor type A is expressed within a subset of the heart field, specifically within the future atria and left ventricle.[102] Thus, a subset of myocardial cells that are endothelin receptor type A-positive/MESP1-negative may give rise to the conduction system myocardium.

Atrial Fibrillation

Consistent with Pitx2c's role in repressing the sinoatrial node fate within the left atrium, patients with reduced Pitx2c levels often develop atrial fibrillation.[103] In mice, if Pitx2 is knocked out in the atria, the atrial chambers are enlarged but thin, and the mice exhibit atrioventricular node block.[103] In addition, the atrioventricular node and the bundle branches have reduced insulation, which would further miscue contractions.[103]

VALVE DEVELOPMENT

The four-chambered heart contains four valves that promote unidirectional flow. The mitral and tricuspid valves separate the atria from the ventricles and derive from the atrioventricular cushions, and the semilunar valves separate the great arteries from the ventricles and derive from the outflow tract cushions. The atrioventricular cushions include the superior and inferior central cushions and the lateral cushions, and in addition to the mitral and tricuspid valves, the atrioventricular cushions also contribute to the atrioventricular septal structures (Figure 1.5). The outflow tract cushions include the proximal and distal cushions (Figure 1.5), and in addition to

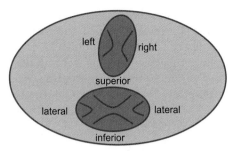

FIGURE 1.5 Atrioventricular and outflow tract cushions. The outflow tract and atrioventricular cushions are shown after the heart has looped. The major cushions in the atrioventricular canal are the superior and inferior cushions, which are formed by epithelial–mesenchymal transition (EMT). The lateral cushions develop later and are mostly populated by epicardially derived cells.[222] In the outflow tract, the proximal left and right cushions are formed through EMT; cardiac neural crest-derived cells migrate from the pharyngeal arches into the outflow tract and form two prongs in the distal left and right outflow tract cushions.

the semilunar valves, the outflow tract cushions also contribute to the aorticopulmonary septum (described below) and the outflow tract myocardium.[104]

Both sets of valves undergo similar development (Figure 1.6). Signals from the myocardium induce epicardial–mesenchymal transition of the endocardial layer and the deposition of extracellular matrix between the myocardial and endocardial layers.[105] The mesenchymal cells generated from the endocardial layer migrate into the space between the layers and are highly proliferative.[105] Finally, proliferation decreases,[104] and the extracellular matrix (ECM) is remodeled, resulting in the mature trilaminar valve structure.[105] However, while the formation of both sets of cushions shares many similarities, some key differences are present in their development.

Valve Specification and Growth

Once a subset of the myocardium has been restricted to surrounding a cardiac cushion, it begins secreting proteins between the myocardial and endocardial

layers of the early heart tube. The proteoglycans that are secreted from the myocardium and released between the cell layers are called the cardiac jelly; the resulting swellings that arise from this deposition are known as the cushions.[106,107] In the atrioventricular canal, myocardially secreted BMP2 activates the Tbx2/Msx2/Has2 pathways and induces a subset of endothelial cells to undergo epithelial–mesenchymal transition and migrate into the cardiac jelly.[73,108] Removing BMP2 from atrioventricular canal myocardium inhibits EMT and ECM deposition into the cushions.[73,108] Furthermore, Tbx2 and Tbx3 induce BMP2, suggesting that a feed-forward mechanism protects the region that will form the atrioventricular valves.[61] If Tbx2 is misexpressed in the chamber myocardium under the Myh6 promoter, where it is normally absent, ECM is deposited between the compact and trabecular myocardium, thus expanding the cardiac jelly and leading to stenotic lumens.[109] Of the numerous ECM enzymes, Tbx2 specifically induces HA synthetase via its T-box binding sites.[109] Increased Tbx2 also increases TGFβ expression and Smad2 phosphorylation,[109] suggesting that epithelial-mesenchymal transition may have also been increased. BMP signaling from the myocardium to the endocardium acts through canonical Smad transcription factors in induced endocardial epithelial-mesenchymal transition.[110]

Epithelial-mesenchymal transition establishes mesenchymal cells that populate the atrioventricular and outflow tract cushions. The transcription factor nuclear factor of activated T cells (Nfat)c1 acts downstream of VEGF via the MEK/mitogen-activated protein kinase 1 (ERK) pathway to promote proliferation of both the overlying endocardial cells and the mesenchymal cells within the cushions (Figure 1.7).[111] When VEGF is down-regulated at the beginning of the ECM remodeling period (E14.5 in mouse),[112] receptor activator of nuclear factor κB ligand (RANKL) expression increases and promotes Nfatc1 nuclear translation via the c-Jun NH2-terminal kinase (JNK) pathway to promote the expression of ECM remodeling enzyme cathepsin

FIGURE 1.6 Atrioventricular valve formation. (A) In response to signals from the myocardium (red), a subset of endothelial cells (green) undergoes epithelial–mesenchymal transition (hybrid yellow/green cells) to generate the mesenchymal cells (yellow) that populate the cushions. (B) These mesenchymal cells will undergo proliferation to expand the cushions. (C) After proliferation, these cells will remodel the valve to an elongated shape and secrete extracellular matrix. The mature mitral valve consists of three layers: the atrialis (blue), the spongiosa (yellow) composed of proteoglycans and glycosaminoglycans, and the fibrosa (red) layers. Valve interstitial cells are present in all three layers.

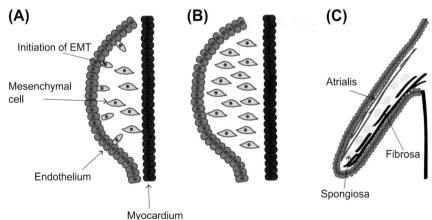

K.[111] This switch from VEGF to RANKL expression is associated with a reduction in proliferation.[111] The up-regulation of cathespin K expression is associated with increased expression of ECM proteins versican and periostin.[111]

Once the cushions are populated, they begin remodeling and elongating. The rapid proliferation in the earlier cushions is attenuated by the ErbB1 signaling pathway.[113] ErbB1 signaling also down-regulates signaling through the BMP pathway.[113] In addition to reduced proliferation, this stage is also accompanied by the deposition of ECM layers. Three distinct layers are present in the valves: an elastin-rich layer on the side of the valve exposed to flow, a proteoglycan-rich spongiosa middle layer, and a fibrillar collagen-rich layer on the non-flow-exposed side.[104] BMP2-induced Sox9 promotes the deposition of proteoglycans and cartilage differentiation, and FGF8-induced Scleraxis promotes the deposition of tenascin.[114]

Although the signaling events that define where the outflow tract cushions will form are not as well defined as for the atrioventricular cushions, there are many similarities between atrioventricular canal and outflow tract cushion epithelial-mesenchymal transition and morphogenesis. The outflow tract cushions are formed through a similar series of myocardial secretions, swelling between the myocardial and endocardial layers, and subsequent population through epithelial-mesenchymal transition.[104] Furthermore, Smad-mediated BMP signaling is required for outflow and atrioventricular canal cushion epithelial-mesenchymal transition.[110] However, the outflow tract cushions are also populated by a subset of cardiac neural crest-derived cells, which play a role in outflow tract cushion morphogenesis and valve formation.[115,116]

Mitral Valve Prolapse

To underscore the importance of the ECM in valve development, most of the genes that have been correlated with congenital valve defects include ECM proteins, such as collagens, tenascin, and elastin.[104] Among the valve defects, mitral valve prolapse is the most common.[117] In this defect, excess ECM deposition within the valve leaflets causes the leaflets to thicken or prolapse into the atrium. The valve leaflet cannot close properly, leading to regurgitation.[104] This defect is associated with mutations in fibrillin-1, elastin, and fibulin-4.[118–121] Disrupted TGFβ signaling can also lead to the development of mitral valve prolapse, as observed in patients with Loeys-Dietz syndrome.[122]

ATRIAL SEPTATION

Septation of the cardiac inflow (atrial) and outflow (arterial) poles of the heart are later stages of cardiac morphogenesis. While the poles of the heart are structurally distinct, they are both dependent on late additions of cardiac progenitors from the second heart field for septation and morphogenesis.

Atrial Septum Formation

From the second heart field, the Islet-positive/FGF10-negative posterior region gives rise to atrial myocardium.[123] Once the common atrium has formed, a septum forms to divide the left and right atria (Figure 1.4). This septum is formed from three structures, from dorsal to ventral, the septum secundum, the septum primum, and the dorsal mesenchymal protrusion.[124] The dorsal mesenchymal protrusion grows like a spine from dorsal to ventral along the posterior border of the common atrium. This structure is adjacent to the atrioventricular canal and participates in atrial and atrioventricular septation. The septum primum is first observed as a muscular crescent within the cranial aspect of the common atrium; this septum grows toward the atrioventricular canal.[124] The mesenchymal cap of the septum will fuse with the cushions and detach from the cranial aspect of the atria, leaving open a fenestration known as the ostium secundum, or foramen ovale, to maintain communication between the atria.[124] The septum secundum is a fold of the dorsal atrial myocardium, begins forming from the cranial aspect of the atria, and forms the roof of the foramen ovale. This septum will eventually fuse with the septum primum to close the foramen ovale and complete atrial septation after birth.[124]

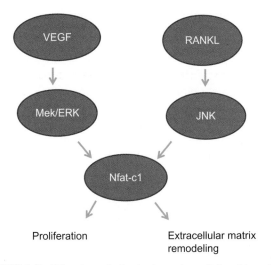

Proliferation   Extracellular matrix remodeling

FIGURE 1.7 Nfat-c1 regulation in the endocardial cushions. When VEGF is present, VEGF signals via Mek/ERK to promote Nfat-c1-induced proliferation, which is required to populate the endocardial cushions. When VEGF is down-regulated, RANKL acts via the JNK intracellular pathway, which activates Nfat-c1-induced transcription of extracellular matrix remodeling proteins. This switch allows the newly populated cushions to undergo remodeling to form the mature valve leaflets.

Atrial Septal Defects

Defects in any one of the atrial septum components can result in atrial septal defects (ASDs). ASDs therefore occur at different places in the atrial septum, with different names and pathogenesis. Defects in septum primum, the flap vale of the foramen ovale, cause ASDs of the secundum type.[124] Defects in the septum secundum also are most likely to result in ASDs of the secundum type. ASDs of the secundum type occur in the dorsal portion of the atrial septum away from the atrioventricular canal. In contrast, defects of the dorsal mesenchymal protrusion cause ASDs of the primum type.[124] This defect causes a deficiency of the atrial septum in the ventral portion of the atrial septum directly adjacent to the atrioventricular canal. ASDs of the primum type are therefore part of a class of atrioventricular septal defects and can include concomitant defects of the atrioventricular valves.

Human genetics has been successful in implicating several well-known cardiogenic transcription factors in ASD pathogenesis. Specifically, dominant mutations in GATA4, Nkx2.5, Tbx20, and Tbx5 can all cause atrial septal defects.[125–130] Loss of both copies of these essential genes results in early embryonic arrest with severe abnormalities,[98,131,132] reflecting the early requirement of these genes in the heart field described earlier. Whereas early cardiac development appears immune to haploinsufficiency or a lack of a single copy of these transcriptional regulators, atrial septation is apparently more sensitive to transcription factor dose. The mechanistic link from transcription factor haploinsufficiency to failure of atrial septum morphogenesis has recently begun to emerge.

GATA4 and Secundum and Primum ASDs

The point mutation C839T affects the DNA-binding site of GATA4; this mutation is found in familial ASDs in combination with pulmonary stenosis.[126] Mutations within the transactivation domain (TAD)-1 and the nuclear localization signal of GATA4 are also associated with familial ASD.[125,127] Missense mutations within GATA4 can affect both atrial and ventricular septation, in addition to pulmonary valve stenosis.[133] Within the 3′ untranslated region of GATA4, additional mutations are also associated with ASD and atrioventricular septal defects.[134] Furthermore, mutations in the GATA4 promoter that lead to either up- or down-regulation of GATA4 are both associated with ventricular septation defects (VSDs).[135]

Some GATA4 mutations are particularly detrimental because they disrupt interactions with other signaling pathways. Unlike the GATA4 S52F mutation, which affects the septum secundum to cause isolated ASD, the G296S and G303E mutations affect the septum primum, leading to ASD, common atrioventricular canal, and cleft mitral valve.[110] These mutations reduce the ability of GATA4 to interact with Smad4, the co-regulatory Smad that mediates BMP signaling.[110]

Although GATA4 is the best-characterized GATA transcription factor in humans with congenital heart defects, studies in mice suggest that other GATA family members also have important roles. GATA3 mutations are associated with VSDs in the mouse.[136] GATA4 and GATA6 are broadly expressed throughout the heart, and knocking out either gene singly results in early embryonic death.[29,30,137] The GATA4/6 double homozygous null mouse does not form a heart at all.[138] In contrast, GATA5 is restricted to the endocardium and endothelial cells. Mice that are doubly heterozygous for GATA4/5 or GATA5/6 show VSDs in combination with outflow tract defects.[139] However, VSDs are not seen in isolation,[139] suggesting that the GATA4 mutations observed in humans with isolated VSD may be the most crucial mutations from a human health perspective.

Nkx2.5 and ASDs

Dominant Nkx2.5 mutations have been associated with several forms of congenital heart defects, but the most frequent is ASDs. Among the identified mutations, many are within the homeobox domain or lead to truncated proteins.[127,130] Specific mutations include C568T, which is within the homeodomain, and C533T, which affects the homeodomain.[127] The G325T mutation introduces a stop codon in exon 1 and was identified in a three-generation family in which five females had ASD with additional congenital defects (e.g., VSD, atrial fibrillation, and patent foramen ovale).[140]

The T-box Transcription Factors and ASDs

Heterozygous mutations in both Tbx5 and Tbx20 are associated with ASDs and VSDs. Dominant Tbx5 mutations cause Holt-Oram syndrome, which includes skeletal and heart defects. A mutation that leads to a premature stop codon (C408A) within Tbx5 has been correlated with familial ASD.[141] Similarly, mutations within the T-box domain (e.g., A79V and Y100C) have also been associated with ASD.[142] These patients are often affected by more than ASD, with the additional complications including pulmonary stenosis and bicuspid pulmonary valve.[142] Three additional mutations within the T-box domain (P96L, L102P, and T144I) are associated with atrioventricular septation defects and complex clinical phenotypes.[142] Similarly, mutations within the T-box domain and mutations that lead to protein truncations for Tbx20 are also associated with numerous congenital heart defects, including VSDs, ASDs, and mitral valve prolapse.[128]

The requirement for Tbx5 in atrial septation occurs in progenitor cells in the second heart field, not in the heart itself.[143] Signaling pathways such as Sonic hedgehog and Wnt signaling are also required for atrial septation.[144,145] Hedgehog signaling is only present in the second heart field, and cells that receive hedgehog signaling in the second heart field migrate into the heart to form the atrial septum. These observations suggest that molecular processes in second heart field progenitor cells drive atrial septation, rather than molecular processes in the myocardium itself. Interestingly, all the genes genetically implicated in atrial septation, including GATA4 and Nkx2-5, are expressed in the second heart field.[29,146] Molecular diversity generated early among cardiac progenitors is an organizer of structural cardiac morphogenesis that occurs many days later.

ARTERIAL POLE MATURATION

After ASDs and VSDs, arterial pole defects are the next most prevalent group of congenital heart defect.[147] The arterial pole, more commonly called the outflow tract, connects the right ventricle to the pharyngeal arch arteries (in avians and mammals) or the branchial arch arteries (in fish and *Xenopus*). The arterial pole is derived from the second heart field and is added after the primitive heart tube has formed.[24,148,149] The mesodermal second heart field cells are added through a combination of active migration into the outflow tract and the movement of the outflow tract caudally across the pharyngeal arches.[150] The pharyngeal arches are a bilateral series of arches with mesenchymal cores. Initially bilateral arteries will form in arches 3, 4, and 6; these bilateral pharyngeal arch arteries will remodel, yielding the great arteries (i.e., the aorta and pulmonary artery), the carotid arteries, and the subclavian arteries.[151]

Arterial Pole Septation

In species with divided circulation, the arterial pole must be septated into the aorta and the pulmonary artery, which lead to the body and the lungs, respectively. This septation is initiated by the addition of cardiac neural crest-derived cells (Figure 1.8). These cells arise from the neuroepithelium at the border of the ectoderm, migrate through the circumpharyngeal ridge, and invade the outflow tract between the myocardial and endocardial layers.[152] As the cardiac neural crest derivatives migrate into the outflow tract, they migrate in two prongs, following a spiral pattern through the outflow tract.[153] At the distal end of these prongs is a 'shelf' that begins dividing the outflow tract into systemic and pulmonic flow. This shelf elongates into the distal outflow tract, following the spiraling prongs, and septates the distal outflow tract. The remainder of the outflow tract is septated by the 'zippering' of the proximal outflow tract cushions. In this proximal region, the outflow tract cushions are brought into close proximity, and the endocardium breaks down. The sub-endocardial cardiac neural crest-derived cells thus form the seam between the pulmonary and aortic outlets at this level of the outflow tract.[115,116]

The spiral, in combination with apoptosis predominantly at the base of the aorta,[154] helps rotate the outflow tract with respect to the ventricles. This rotation is critical for alignment of the aorta with the left ventricle and the pulmonary artery with the right ventricle.[155] Recent evidence suggests that this rotation is formed in part by asymmetrical contributions from the second heart field, with Nkx2.5-positive second heart field progenitors adding preferentially to the pulmonary side of the outflow tract to help drive this side of the outflow tract more ventrally.[156] The cardiac neural crest derivatives

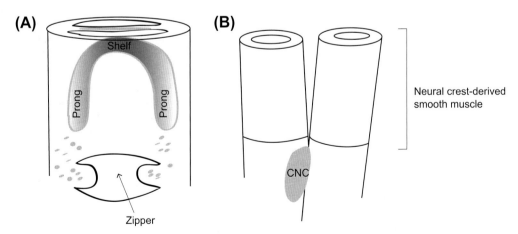

FIGURE 1.8 Cardiac neural crest and outflow tract septation. (A) As cardiac neural crest-derived cells (gray) enter the outflow tract, they form two prongs that spiral into the cushions. At the distal end of the prongs, there is a 'shelf' that septates the distal portion of the outflow tract. The proximal portion of the outflow tract is 'zippered' shut by the outflow tract cushions. (B) After septation, a plug of cardiac neural crest-derived cells (CNC) can be found between the great arteries, and cardiac neural crest-derived smooth muscle covers the great arteries distal to the semilunar valves.

populate distinct swellings in the outflow tract, termed the outflow tract cushions, that will remodel to form the semilunar valves of the aorta and pulmonary artery and the outflow tract septum.[153]

Additional Cardiac Neural Crest Contributions within the Arterial Pole

In addition to populating the pharyngeal arches and outflow tract cushions and septating the outflow tract, the cardiac neural crest derivatives also provide the smooth muscle of the great arteries and differentiate into the cardiac ganglia.[157,158] This smooth muscle abuts the smooth muscle derived from the second heart field,[24] and the seam between these two sources of smooth muscle is a common location of aortic dissection.[150]

Signaling in the Pharyngeal Arches and the Arterial Pole

Signaling pathways that are crucial for outflow tract development include Sonic hedgehog, which promotes proliferation within the second heart field progenitors,[27] and BMP, which promotes myocardial differentiation once the progenitors enter the outflow tract.[23,24] In addition, T-box and GATA transcription factors are also crucial in this region.[139,159]

Signaling between the second heart field and the migrating cardiac neural crest derivatives is essential for both populations to properly add to the outflow tract. Transcription factor Tbx3 is expressed in the pharyngeal endothelium and the cardiac neural crest derivatives.[160] However, its loss is accompanied by a failure of the second heart field to add to the outflow tract, as evidenced by abnormal heart looping beginning at E9.5 the mouse.[160] Sonic hedgehog expression is reduced in the pharyngeal endoderm of the Tbx3 homozygous null embryo at E9.5,[160] which would lead to reduced second heart field proliferation and thus fewer cells to migrate into the outflow tract.[27] In addition, the Tbx3 homozygous null also displays reduced BMP4 expression in the outflow tract at E9.5, suggesting that myocardial differentiation is impaired for any second heart field cells that successfully migrate to the outflow tract.[160] Tbx3 also has a direct effect on the pharyngeal arches, with a delay in the formation of pharyngeal arch 3 in the Tbx3 homozygous null embryo.[160] Despite its expression in the cardiac neural crest derivatives, these cells deploy normally to and septate the outflow tract of the Tbx3 homozygous null embryo.[160] However, the reduced contribution of the second heart field results in a shortened outflow tract and double outlet right ventricle, both with and without transposition of the arteries.[160]

Additional interplay is observed between Tbx1 and the TGFβ/BMP superfamily. Tbx1 can suppress BMP4 signaling via binding to Smad1 and preventing its interaction with co-activator Smad4.[161] Furthermore, if BMP4 is conditionally knocked out under the control of the Tbx1 promoter, embryos display pharyngeal arch artery remodeling defects; increased apoptosis within the outflow tract cushions; and a range of arterial pole defects, including persistent truncus arteriosus and interrupted aortic arch type B.[162] Additionally, Tbx1 genetically interacts with the repressor Smad7. Smad7 inhibits both branches of the TGFβ/BMP superfamily and is expressed in the outflow tract during cardiac neural crest migration.[163] Tbx1;Smad7 double heterozygous mice display reduced cardiac neural crest-derived smooth muscle and decreased extracellular matrix in the pharyngeal arches, and the fourth pharyngeal arch artery fails to open in most of these embryos.[163]

In addition to its role in ventricular septation, GATA3 is expressed in the outflow tract and atrioventricular cushions.[136] GATA3 homozygous null embryos show delayed arch artery formation, with the third arch partially forming by E9.5.[136] By E10.5, the arches are hypoplastic, and the cardiac neural crest derivatives do not spiral into the outflow tract, which leads to outflow tract septation defects.[136] GATA3 homozygous null embryos that survive to E15.5 display persistent truncus arteriosus (i.e., an unseptated outflow tract) or double outlet right ventricle, and both of these defects occur in combination with VSDs.[136]

The migratory cardiac neural crest cells are sensitive to Notch signaling, with either excessive or down-regulated levels of Notch activity in the migratory cardiac neural crest, reducing their ability to migrate into the pharyngeal arches and outflow tract cushions.[164] Excessive Notch signaling results in persistent truncus arteriosus, and down-regulated Notch signaling leads to double outlet right ventricle.[164] Both persistent truncus arteriosus and double outlet right ventricle result from abnormal patterning of the pharyngeal arch arteries, resulting from the absence of developing arteries that do not form cardiac neural crest-derived smooth muscle.[164]

DiGeorge Syndrome

The cardiac phenotype of DiGeorge syndrome (also known as 22q11.2 syndrome and velo-cardio-facial syndrome) is one of the most well-studied arterial pole anomalies observed in humans. DiGeorge syndrome includes craniofacial anomalies, thymus and parathyroid defects, and cardiac anomalies such as tetralogy of Fallot. This syndrome is widely attributed to a 3 Mb deletion on chromosome 22. However, this potential deletion can be observed in the absence of the syndrome,[165] and this deletion does not account for all DiGeorge patients.[166,167]

Within the 3 Mb deletion, the most promising candidate gene is Tbx1. Mice lacking either one or two copies of Tbx1 and mice that lack Tbx1 expression in the pharyngeal endoderm recapitulate the DiGeorge syndrome phenotype, including cardiac anomalies such as double outlet right ventricle, persistent truncus arteriosus, and interrupted aortic arch, and these defects are more severe in the homozygous null mice.[168–171] Furthermore, many of these same defects are also observed in the Tbx1-overexpressing mouse,[172] thus supporting the hypothesis that the dosage of Tbx1 is critical for normal cardiac development.[173] Consistent with the controversy about the relationship between the chromosomal deletion on chromosome 22 and the phenotype, however, contradictory reports also suggest that Tbx1 mutations may or may not be responsible for the phenotype. A recent analysis of 664 patients with DiGeorge syndrome including cardiac defects shows that mutations in Tbx1 were not enriched compared with patients with DiGeorge syndrome in the absence of cardiac defects.[174]

Other transcription factors interact with Tbx1 in the pharyngeal mesoderm, including the LIM domain-containing Lhx2 and Tcf21.[175] Lhx2 is expressed within the pharyngeal arch mesenchyme and is down-regulated in the Tbx1 homozygous-null embryo.[175] In turn, Tcf21 is down-regulated in the Lhx2 homozygous-null embryo, setting up a hierarchy in which Tbx1 induces Lhx2, which then induces Tcf21.[175] Tcf21 binds to the regulatory elements of Myh5 and promotes muscle differentiation in the pharyngeal arches.[175] Both the Lhx2 and the Tcf21 homozygous null embryos display the cardiovascular phenotypes associated with DiGeorge syndrome, including tetralogy of Fallot and double outlet right ventricle.[175]

Furthermore, canonical Wnt signaling via β-catenin within the pharyngeal mesenchyme represses Tbx1, and conditionally knocking out the gene encoding β-catenin from the mesenchyme recapitulates the DiGeorge phenotype.[176] This phenotype is caused in part by abnormal remodeling of the pharyngeal arch arteries and a failure of the cardiac neural crest to migrate into the arches.[176] Interestingly, FGF8, a known chemoattractant for the cardiac neural crest, is up-regulated in these conditional knockout embryos,[177] suggesting that the cardiac neural crest provides feedback that down-regulates FGF8 expression upon entering the pharyngeal arches and that this feedback is missing in the conditional knockout embryos.[176]

As an alternative model for DiGeorge syndrome, the conditional knockout of neuropilin-1 under the endothelial Tie2 promoter also leads to a DiGeorge-like phenotype, including persistent truncus arteriosus.[178] Neuropilin-1 is a coreceptor for VEGF-A and acts in conjunction with the plexins and semaphorins.[179] Additionally, VEGF

regulates Tbx1.[180] When neuropilin-1 is conditionally knocked out in endothelial cells, the pharyngeal arch arteries are abnormal, with hypoplastic and completing missing arch arteries.[178]

Together, these studies suggest the complex DiGeorge syndrome phenotype results from an equally complex signaling network. The Tbx1 pathway is crucial for the normal development of the arterial pole, and recent studies have begun addressing how other pathways, such as Sonic hedgehog and retinoic acid, exacerbate the DiGeorge phenotype.[181] However, further work is required to better understand the genotype/phenotype discrepancies observed in patients with the 22q11.2 deletion without the syndrome or with the syndrome but without defects in Tbx1.

EPICARDIAL AND CORONARY VASCULAR DEVELOPMENT

Epicardial Formation

The epicardium is primarily derived from the pro-epicardial organ, a Tbx18-positive cluster of cells caudal to the sinus venosus. In the mouse, cells from the pro-epicardial organ began covering the heart, starting at the sinus venosus and migrating ventrally and toward the outflow tract to cover the heart. Despite expressing Tbx18, this transcription factor is not required for addition of the pro-epicardium to the heart.[182] In addition, the epicardium of the distal outflow tract arises from a pro-epicardial organ-like structure in the pericardium adjacent to the outflow tract.[183,184] Maintenance of the epicardium layer is dependent on α4-integrin, and both mice lacking α4-integrin and chick embryos with virus-mediated α4-integrin knockdown lose their epicardial layer.[185,186] The epicardial cells appear to migrate across the heart but then lose their epicardial characteristics and invade the myocardium.[185,186]

As the epicardium forms, it expresses RALDH2 and secretes factors such as PDGF-A, retinoic acid, and erythropoietin toward the myocardium.[187–189] In cultured NIH3T3 cells, PDGF-A has a mitogenic effect, but the myocardium does not express receptor PDGF-R2.[188] Furthermore, retinoic acid and erythropoietin have both pro-proliferative and pro-survival effects on cardiomyocytes grown using a slice culture method.[189]

A subset of the epicardium undergoes epithelial–mesenchymal transition and invades the subepicardial space.[190] These epicardial-derived cells differentiate into interstitial and perivascular fibroblasts, coronary smooth muscle cells, and coronary endothelial cells.[191–193] Although the Snail transcription factors are crucial for epithelial–mesenchymal transition in regions such as the atrioventricular and outflow tract cushions,[194] their role in epicardial–mesenchymal transition is less clear. Snail1 is expressed in the epicardial layer and is downstream of the epicardial-specific protein WT-1.[195] Although epicardial-mesenchymal transition is disrupted in a conditional WT-1 knockdown mouse (under the GATA5 promoter),[195] epicardial-mesenchymal transition is not disrupted when Snail1 is conditionally knocked out under control of either the WT-1 promoter or the Tbx18 promoter.[196] Thus, although WT-1 promotes epicardial-mesenchymal transition, this effect does not occur through Snail1.

Tbx18 is expressed in the pro-epicardium and the epicardium but not in epicardial-derived cells after epicardial-mesenchymal transition.[182] When Tbx18 expression is ectopically maintained in these epicardial-derived cells, no changes in phenotype are observed.[182] However, when an activator form of Tbx18 is expressed constitutively, the compact myocardium thins, and the coronary plexus fails to fully penetrate the right ventricle.[182] Both phenotypes are attributed to impaired epicardial signaling, and the epicardial-derived cells prematurely differentiate into smooth muscle in these mice.[182] Thus, Tbx18 plays a role in repressing smooth muscle differentiation of epicardial-derived cells.

Coronary Vascular Development

Formation of the epicardium is closely followed by the appearance of the coronary vasculature, which follows a similar spatial pattern (Figure 1.9). Discrete patches of endothelial cells are first observed near the sinus venosus, and this plexus of endothelial cells spreads ventrally and toward the outflow tract. This spread occurs over a two- to three-day span in the chick and mouse embryo,[197,198] and arterial- and venous-specific markers are already apparent during this process.[197] Only after encompassing the heart do these endothelial cells connect to the aorta to provide blood flow to the heart. After flow initiates, the coronary vasculature begins remodeling to form larger, branched vessels that are invested with smooth muscle.[192,199,200]

The venous and arterial coronary endothelial cells appear to have different origins. The majority of the subepicardial venous coronary endothelial cells derive from a subset of pro-epicardial cells that migrate through the sinus venosus,[193,197] and the majority of the myocardially embedded arterial coronary endothelial cells arise from the endocardium.[201]

The development of the coronary vasculature is intimately connected with the other layers of the heart. Failure of the epicardium to cover the heart inhibits coronary plexus formation.[202] Furthermore, the myocardium fails to undergo trabecular compaction if the coronary plexus cannot supply the heart with nutrients, leading to thin ventricular walls. The Hey2 homozygous null mouse displays collapsed coronary veins throughout the ventricular myocardium, and smooth muscle recruitment to the larger vessels is impaired; subsequently, this mouse has thin ventricular walls.[203]

Coronary Atresia

The coronary vasculature is clearly required for continued development. However, an assortment of coronary artery anomalies are observed in humans. These coronary artery anomalies include missing coronary stems (i.e., coronary atresia) and misplaced coronary stems.[204] Although coronary atresia is linked with sudden death, particularly in young athletes,[205–207] no genetic causes have been identified as of yet.

CONCLUSIONS

Generating a complex, four-chambered heart from a mesodermal field of cells requires the coordinated actions of numerous signaling pathways and gene regulatory networks. Hundreds of genes work together to induce, specify, and differentiate the different cardiac cell types; coordinate their migration; and regulate additional behaviors such as epithelial–mesenchymal transition. A single gene may affect a number of different steps, further complicating our ability to dissect how each stage of heart development is carried out. In addition, while genetic studies in mice often rely on knocking out a gene in its entirety, humans typically present with more subtle genetic mutations, such as

FIGURE 1.9 Coronary plexus formation. (A) Starting around E11.5, a patchwork of endothelial cells (green) appears on the dorsal aspect of the ventricles, caudal to the sinus venosus. (B,C) This plexus spreads across the ventricles, completely encompassing them by E13.5. (D) After encompassing the ventricles, the plexus remodels to form the mature coronary vasculature. Based on Lavine et al.[198]

(A)
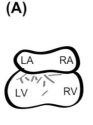
LA   RA
LV   RV
E11.5

(B)

E12.5

(C)

E13.5

(D)

Mature

the single nucleotide polymorphisms that have been associated with ASDs. From an analysis standpoint, a global knockout provides the easiest way to quickly determine whether and where a gene is important and what developmental processes go awry in its absence. However, complex genetic mouse crosses are allowing for the analysis of heart defects in response to a range of gene expression levels (e.g., Tbx1[208]), and these studies are becoming more prevalent. Even these detailed studies, though, may not fully recapitulate a defect observed in the clinical setting, where the defect may arise from a mutation within the exon that changes the resulting protein or within a promoter region, resulting in a normal protein with altered expression levels.

Over the past few decades, tremendous progress has been made identifying many genes important to cardiac development, from human genetics and animal models. We now have the challenge of understanding the precise developmental requirements of these genes and how they work cooperatively to achieve complex morphogenetic events. As our understanding of the molecular underpinnings of cardiac development improves further, the next major challenge will be to relate these genes to the etiology of congenital heart defects. Furthermore, how we translate this knowledge into better therapeutic strategies to prevent or correct congenital heart defects will pose an even greater, but highly significant, challenge.

Acknowledgments

We would first like to apologize to the many authors whose studies were excluded due to space constraints. We would like to thank Dr. Andrea Portbury and Chelsea Cyr for critical reading of the manuscript.

References

1. Abu-Issa R, Kirby ML. Patterning of the heart field in the chick. *Dev Biol* 2008;**319**(2):223–33.
2. Dyer LA, Kirby ML. The role of secondary heart field in cardiac development. *Dev Biol* 2009;**336**(2):137–44.
3. Meilhac SM, Esner M, Kelly RG, Nicolas JF, Buckingham ME. The clonal origin of myocardial cells in different regions of the embryonic mouse heart. *Dev Cell* 2004;**6**(5):685–98.
4. Yue Q, Wagstaff L, Yang X, Weijer C, Munsterberg A. Wnt3a-mediated chemorepulsion controls movement patterns of cardiac progenitors and requires RhoA function. *Development* 2008;**135**(6):1029–37.
5. Harland RM, Grainger RM. Xenopus research: metamorphosed by genetics and genomics. *Trends Genet* 2011;**27**(12):507–15.
6. Schneider VA, Mercola M. Wnt antagonism initiates cardiogenesis in Xenopus laevis. *Genes Dev* 2001;**15**(3):304–15.
7. Klaus A, Muller M, Schulz H, Saga Y, Martin JF, Birchmeier W. Wnt/beta-catenin and Bmp signals control distinct sets of transcription factors in cardiac progenitor cells. *Proc Natl Acad Sci U S A* 2012;**109**(27):10921–6.
8. Cohen ED, Miller MF, Wang Z, Moon RT, Morrisey EE. Wnt5a and Wnt11 are essential for second heart field progenitor development. *Development* 2012;**139**(11):1931–40.
9. Lin L, Cui L, Zhou W, Dufort D, Zhang X,Cai CL, et al. Beta-catenin directly regulates Islet1 expression in cardiovascular progenitors and is required for multiple aspects of cardiogenesis. *Proc Natl Acad Sci U S A* 2007;**104**(22):9313–8.
10. Cohen ED, Wang Z, Lepore JJ, Lu MM, Taketo MM, Epstein DJ, et al. Wnt/beta-catenin signaling promotes expansion of Isl-1-positive cardiac progenitor cells through regulation of FGF signaling. *J Clin Invest* 2007;**117**(7):1794–804.
11. Klaus A, Saga Y, Taketo MM, Tzahor E, Birchmeier W. Distinct roles of Wnt/beta-catenin and Bmp signaling during early cardiogenesis. *Proc Natl Acad Sci U S A* 2007;**104**(47):18531–6.
12. Sinha T, Wang B, Evans S, Wynshaw-Boris A, Wang J. Disheveled mediated planar cell polarity signaling is required in the second heart field lineage for outflow tract morphogenesis. *Dev Biol* 2012;**370**(1):135–44.
13. Deimling SJ, Drysdale TA. Retinoic acid regulates anterior-posterior patterning within the lateral plate mesoderm of Xenopus. *Mech Dev* 2009;**126**(10):913–23.
14. Koster M, Dillinger K, Knochel W. Genomic structure and embryonic expression of the Xenopus winged helix factors XFD-13/13'. *Mech Dev* 1999;**88**(1):89–93.
15. Hochgreb T, Linhares VL, Menezes DC, Sampaio AC, Yan CY, Cardoso WV, et al. A caudorostral wave of RALDH2 conveys anteroposterior information to the cardiac field. *Development* 2003;**130**(22):5363–74.
16. Collop AH, Broomfield JA, Chandraratna RA, Yong Z, Deimling SJ, Kolker SJ, et al. Retinoic acid signaling is essential for formation of the heart tube in Xenopus. *Dev Biol* 2006;**291**(1):96–109.
17. Keegan BR, Feldman JL, Begemann G, Ingham PW, Yelon D. Retinoic acid signaling restricts the cardiac progenitor pool. *Science* 2005;**307**(5707):247–9.
18. Jiang Y, Drysdale TA, Evans T. A role for GATA-4/5/6 in the regulation of Nkx2.5 expression with implications for patterning of the precardiac field. *Dev Biol* 1999;**216**(1):57–71.
19. Lieschke GJ, Currie PD. Animal models of human disease: zebrafish swim into view. *Nat Rev Genet* 2007;**8**(5):353–67.
20. Kirby ML. *Cardiac Development*. Oxford: Oxford University Press; 2007.
21. Takeuchi JK, Bruneau BG. Directed transdifferentiation of mouse mesoderm to heart tissue by defined factors. *Nature* 2009;**459**(7247):708–11.
22. Takeuchi JK, Lou X, Alexander JM, Sugizaki H, Delgado-Olguín P, Holloway AK, et al. Chromatin remodelling complex dosage modulates transcription factor function in heart development. *Nature* 2011;**2**:187.
23. Prall OW, Menon MK, Solloway MJ, Watanabe Y, Zaffran S, Bajolle F, et al. An Nkx2-5/Bmp2/Smad1 negative feedback loop controls heart progenitor specification and proliferation. *Cellule* 2007;**128**(5):947–59.
24. Waldo KL, Kumiski DH, Wallis KT, Stadt HA, Hutson MR, Platt DH, et al. Conotruncal myocardium arises from a secondary heart field. *Development* 2001;**128**(16):3179–88.
25. Dyer LA, Makadia FA, Scott A, Pegram K, Hutson MR, Kirby ML. BMP signaling modulates hedgehog-induced secondary heart field proliferation. *Dev Biol* 2010;**348**(2):167–76.
26. Kelly RG, Zammit PS, Schneider A, Alonso S, Biben C, Buckingham ME. Embryonic and fetal myogenic programs act through separate enhancers at the MLC1F/3F locus. *Dev Biol* 1997;**187**(2):183–99.
27. Dyer LA, Kirby ML. Sonic hedgehog maintains proliferation in secondary heart field progenitors and is required for normal arterial pole formation. *Dev Biol* 2009;**330**(2):305–17.
28. Hinits Y, Pan L, Walker C, Dowd J, Moens CB, Hughes SM. Zebrafish Mef2ca and Mef2cb are essential for both first and second heart field cardiomyocyte differentiation. *Dev Biol* 2012;**369**(2):199–210.

29. Kuo CT, Morrisey EE, Anandappa R, Sigrist K, Lu MM, Parmacek MS, et al. GATA4 transcription factor is required for ventral morphogenesis and heart tube formation. *Genes Dev* 1997;**11**(8):1048–60.

30. Molkentin JD, Lin Q, Duncan SA, Olson EN. Requirement of the transcription factor GATA4 for heart tube formation and ventral morphogenesis. *Genes Dev* 1997;**11**(8):1061–72.

31. Zhang H, Toyofuku T, Kamei J, Hori M. GATA-4 regulates cardiac morphogenesis through transactivation of the N-cadherin gene. *Biochem Biophys Res Commun* 2003;**312**(4):1033–8.

32. Nakagawa S, Takeichi M. N-cadherin is crucial for heart formation in the chick embryo. *Dev Growth Differ* 1997;**39**(4):451–5.

33. Radice GL, Rayburn H, Matsunami H, Knudsen KA, Takeichi M, Hynes RO. Developmental defects in mouse embryos lacking N-cadherin. *Dev Biol* 1997;**181**(1):64–78.

34. Walters MJ, Wayman GA, Christian JL. Bone morphogenetic protein function is required for terminal differentiation of the heart but not for early expression of cardiac marker genes. *Mech Dev* 2001;**100**(2):263–73.

35. Reiter JF, Verkade H, Stainier DY. Bmp2b and Oep promote early myocardial differentiation through their regulation of GATA5. *Dev Biol* 2001;**234**(2):330–8.

36. Aiello VD, de Morais CF, Ribeiro IG, Sauaia N, Ebaid M. An infant with two "half-hearts" who survived for five days: a clinical and pathological report. *Pediatr Cardiol* 1987;**8**(3):181–6.

37. Aiello VD, Xavier-Neto J. Full intrauterine development is compatible with cardia bifida in humans. *Pediatr Cardiol* 2006;**27**(3):393–4.

38. Nonaka S, Tanaka Y, Okada Y, Takeda S, Harada A, Kanai Y, et al. Randomization of left-right asymmetry due to loss of nodal cilia generating leftward flow of extraembryonic fluid in mice lacking KIF3B motor protein. *Cellule* 1998;**95**(6):829–37.

39. Tanaka Y, Okada Y, Hirokawa N. FGF-induced vesicular release of Sonic hedgehog and retinoic acid in leftward nodal flow is critical for left-right determination. *Nature* 2005;**435**(7039):172–7.

40. Levin M, Johnson RL, Stern CD, Kuehn M, Tabin C. A molecular pathway determining left-right asymmetry in chick embryogenesis. *Cellule* 1995;**82**(5):803–14.

41. Tsiairis CD, McMahon AP. An Hh-dependent pathway in lateral plate mesoderm enables the generation of left/right asymmetry. *Curr Biol* 2009;**19**(22):1912–7.

42. Tabin CJ. The key to left-right asymmetry. *Cellule* 2006;**127**(1):27–32.

43. Tsukui T, Capdevila J, Tamura K, Ruiz-Lozano P, Rodriguez-Esteban C, Yonei-Tamura S, et al. Multiple left-right asymmetry defects in Shh(-/-) mutant mice unveil a convergence of the shh and retinoic acid pathways in the control of Lefty-1. *Proc Natl Acad Sci U S A* 1999;**96**(20):11376–81.

44. Simard A, Di Giorgio L, Amen M, Westwood A, Amendt BA, Ryan AK. The Pitx2c N-terminal domain is a critical interaction domain required for asymmetric morphogenesis. *Dev Dyn* 2009;**238**(10):2459–70.

45. Manner J. Cardiac looping in the chick embryo: a morphological review with special reference to terminological and biomechanical aspects of the looping process. *Anat Rec* 2000;**259**(3):248–62.

46. Taber LA, Lin IE, Clark EB. Mechanics of cardiac looping. *Dev Dyn* 1995;**203**(1):42–50.

47. Kirby ML, Waldo KL. Neural crest and cardiovascular patterning. *Circ Res* 1995;**77**(2):211–5.

48. Ryan AK, Blumberg B, Rodriguez-Esteban C, Yonei-Tamura S, Tamura K, Tsukui T, et al. Pitx2 determines left-right asymmetry of internal organs in vertebrates. *Nature* 1998;**394**(6693):545–51.

49. Gage PJ, Suh H, Camper SA. Dosage requirement of Pitx2 for development of multiple organs. *Development* 1999;**126**(20):4643–51.

50. Liu C, Liu W, Lu MF, Brown NA, Martin JF. Regulation of left-right asymmetry by thresholds of Pitx2c activity. *Development* 2001;**128**(11):2039–48.

51. Liu C, Liu W, Palie J, Lu MF, Brown NA, Martin JF. Pitx2c patterns anterior myocardium and aortic arch vessels and is required for local cell movement into atrioventricular cushions. *Development* 2002;**129**(21):5081–91.

52. Ammirabile G, Tessari A, Pignataro V, Szumska D, Sardo FS, Benes Jr J, et al. Pitx2 confers left morphological, molecular, and functional identity to the sinus venosus myocardium. *Cardiovasc Res* 2012;**93**(2):291–301.

53. Hildreth V, Webb S, Chaudhry B, Peat JD, Phillips HM, Brown N, et al. Left cardiac isomerism in the Sonic hedgehog null mouse. *J Anat* 2009;**214**(6):894–904.

54. Okada Y, Nonaka S, Tanaka Y, Saijoh Y, Hamada H, Hirokawa N. Abnormal nodal flow precedes situs inversus in iv and inv mice. *Mol Cell* 1999;**4**(4):459–68.

55. Afzelius BA. Cilia-related diseases. *J Pathol* 2004;**204**(4):470–7.

56. Gripp KW, Hopkins E, Jenny K, Thacker D, Salvin J. Cardiac anomalies in Axenfeld-Rieger syndrome due to a novel FOXC1 mutation. *Am J Med Genet A* 2013;**161**(1):114–9.

57. Christoffels VM, Habets PE, Franco D, Campione M, de Jong F, Lamers WH, et al. Chamber formation and morphogenesis in the developing mammalian heart. *Dev Biol* 2000;**223**(2):266–78.

58. Habets PE, Moorman AF, Clout DE, van Roon MA, Lingbeek M, van Lohuizen M, et al. Cooperative action of Tbx2 and Nkx2.5 inhibits ANF expression in the atrioventricular canal: implications for cardiac chamber formation. *Genes Dev* 2002;**16**(10):1234–46.

59. Moorman AF, Christoffels VM. Development of the cardiac conduction system: a matter of chamber development. *Novartis Found Symp* 2003;**250**:25–34 discussion -43, 276-9.

60. Jensen B, Wang T, Christoffels VM, Moorman AF. Evolution and development of the building plan of the vertebrate heart. *Biochim Biophys Acta* 2012.

61. Singh R, Hoogaars WM, Barnett P, Grieskamp T, Sameer Rana M, Buermans H, et al. Tbx2 and Tbx3 induce atrioventricular myocardial development and endocardial cushion formation. *Cell Mol Life Sci* 2012;**69**(8):1377–89.

62. Aanhaanen WT, Brons JF, Dominguez JN, Rana MS, Norden J, Airik R, et al. The Tbx2+ primary myocardium of the atrioventricular canal forms the atrioventricular node and the base of the left ventricle. *Circ Res* 2009;**104**(11):1267–74.

63. Christoffels VM, Hoogaars WM, Tessari A, Clout DE, Moorman AF, Campione M. T-box transcription factor Tbx2 represses differentiation and formation of the cardiac chambers. *Dev Dyn* 2004;**229**(4):763–70.

64. Harrelson Z, Kelly RG, Goldin SN, Gibson-Brown JJ, Bollag RJ, Silver LM, et al. Tbx2 is essential for patterning the atrioventricular canal and for morphogenesis of the outflow tract during heart development. *Development* 2004;**131**(20):5041–52.

65. Lee MP, Yutzey KE. Twist1 directly regulates genes that promote cell proliferation and migration in developing heart valves. *PLoS One* 2011;**6**(12):e29758.

66. Singh MK, Christoffels VM, Dias JM, Trowe MO, Petry M, Schuster-Gossler K, et al. Tbx20 is essential for cardiac chamber differentiation and repression of Tbx2. *Development* 2005;**132**(12):2697–707.

67. Stennard FA, Costa MW, Lai D, Biben C, Furtado MB, Solloway MJ, et al. Murine T-box transcription factor Tbx20 acts as a repressor during heart development, and is essential for adult heart integrity, function and adaptation. *Development* 2005;**132**(10):2451–62.

68. Takeuchi JK, Ohgi M, Koshiba-Takeuchi K, Shiratori H, Sakaki I, Ogura K, et al. Tbx5 specifies the left/right ventricles and ventricular septum position during cardiogenesis. *Development* 2003;**130**(24):5953–64.

69. Cai CL, Zhou W, Yang L, Bu L, Qyang Y, Zhang X, et al. T-box genes coordinate regional rates of proliferation and regional specification during cardiogenesis. *Development* 2005;**132**(10):2475–87.

70. Singh R, Horsthuis T, Farin HF, Grieskamp T, Norden J, Petry M, et al. Tbx20 interacts with smads to confine tbx2 expression to the atrioventricular canal. *Circ Res* 2009;**105**(5):442–52.

71. Kokubo H, Tomita-Miyagawa S, Hamada Y, Saga Y. Hesr1 and Hesr2 regulate atrioventricular boundary formation in the developing heart through the repression of Tbx2. *Development* 2007;**134**(4):747–55.

72. Rutenberg JB, Fischer A, Jia H, Gessler M, Zhong TP, Mercola M. Developmental patterning of the cardiac atrioventricular canal by Notch and Hairy-related transcription factors. *Development* 2006;**133**(21):4381–90.

73. Rivera-Feliciano J, Tabin CJ. Bmp2 instructs cardiac progenitors to form the heart-valve-inducing field. *Dev Biol* 2006;**295**(2):580–8.

74. Koibuchi N, Chin MT. CHF1/Hey2 plays a pivotal role in left ventricular maturation through suppression of ectopic atrial gene expression. *Circ Res* 2007;**100**(6):850–5.

75. Van Mierop LH, Kutsche LM. Development of the ventricular septum of the heart. *Heart Vessels* 1985;**1**(2):114–9.

76. Conte G, Grieco M. Closure of the interventricular foramen and morphogenesis of the membranous septum and ventricular septal defects in the human heart. *Anat Anz* 1984;**155**(1-5):39–55.

77. Sedmera D, McQuinn T. Embryogenesis of the heart muscle. *Heart* 2008;**4**(3):235–45.

78. Lai D, Liu X, Forrai A, Wolstein O, Michalicek J, Ahmed I, et al. Neuregulin 1 sustains the gene regulatory network in both trabecular and nontrabecular myocardium. *Circ Res* 2010;**107**(6):715–27.

79. MacDonald ST, Bamforth SD, Bragança J, Chen CM, Broadbent C, Schneider JE, et al. A cell-autonomous role of Cited2 in controlling myocardial and coronary vascular development. *Europe* 2012.

80. Chakraborty S, Yutzey KE. Tbx20 regulation of cardiac cell proliferation and lineage specialization during embryonic and fetal development in vivo. *Dev Biol* 2012;**363**(1):234–46.

81. Chen H, Shi S, Acosta L, Li W, Lu J, Bao S, et al. BMP10 is essential for maintaining cardiac growth during murine cardiogenesis. *Development* 2004;**131**(9):2219–31.

82. Zhang W, Chen H, Wang Y, Yong W, Zhu W, Liu Y, et al. Tbx20 transcription factor is a downstream mediator for bone morphogenetic protein-10 in regulating cardiac ventricular wall development and function. *J Biol Chem* 2011;**286**(42):36820–9.

83. Bakker ML, Moorman AF, Christoffels VM. The atrioventricular node: origin, development, and genetic program. *Trends Cardiovasc Med* 2010;**20**(5):164–71.

84. Hoogaars WM, Engel A, Brons JF, Verkerk AO, de Lange FJ, Wong LY, et al. Tbx3 controls the sinoatrial node gene program and imposes pacemaker function on the atria. *Genes Dev* 2007;**21**(9):1098–112.

85. Vicente-Steijn R, Kolditz DP, Mahtab EA, Askar SF, Bax NA, Van Der Graaf LM, et al. Electrical activation of sinus venosus myocardium and expression patterns of RhoA and Isl-1 in the chick embryo. *J Cardiovasc Electrophysiol* 2010;**21**(11):1284–92.

86. Shimada T, Kawazato H, Yasuda A, Ono N, Sueda K. Cytoarchitecture and intercalated disks of the working myocardium and the conduction system in the mammalian heart. *Anat Rec A Discov Mol Cell Evol Biol* 2004;**280**(2):940–51.

87. Kreuzberg MM, Willecke K, Bukauskas FF. Connexin-mediated cardiac impulse propagation: connexin 30.2 slows atrioventricular conduction in mouse heart. *Trends Cardiovasc Med* 2006;**16**(8):266–72.

88. Wiese C, Grieskamp T, Airik R, Mommersteeg MT, Gardiwal A, Gier-de Vries C, et al. Formation of the sinus node head and differentiation of sinus node myocardium are independently regulated by Tbx18 and Tbx3. *Circ Res* 2009;**104**(3):388–97.

89. Bakker ML, Boink GJJ, Boukens BJ, Verkerk AO, van den Boogaard M, Denise den Haan A, et al. T-box transcription factor TBX3 reprogrammes mature cardiac myocytes into pacemaker-like cells. *Cardiovasc Res* 2012;**94**(3):439–49.

90. Mikawa T, Gourdie RG, Takebayashi-Suzuki K, Kanzawa N, Hyer J, Pennisi DJ, et al. Induction and patterning of the Purkinje fibre network. *Novartis Found Symp* 2003;**250**:142–53 discussion 53–6, 276–9.

91. Cheng G, Litchenberg WH, Cole GJ, Mikawa T, Thompson RP, Gourdie RG. Development of the cardiac conduction system involves recruitment within a multipotent cardiomyogenic lineage. *Development* 1999;**126**(22):5041–9.

92. Gourdie RG, Mima T, Thompson RP, Mikawa T. Terminal diversification of the myocyte lineage generates Purkinje fibers of the cardiac conduction system. *Development* 1995;**121**(5):1423–31.

93. Moskowitz IP, Kim JB, Moore ML, Wolf CM, Peterson MA, Shendure J, et al. A molecular pathway including Id2, Tbx5, and Nkx2-5 required for cardiac conduction system development. *Cellule* 2007;**129**(7) 1365-76.

94. Moskowitz IP, Pizard A, Patel VV, Bruneau BG, Kim JB, Kupershmidt S, et al. The T-Box transcription factor Tbx5 is required for the patterning and maturation of the murine cardiac conduction system. *Development* 2004;**131**(16):4107–16.

95. Chuck ET, Freeman DM, Watanabe M, Rosenbaum DS. Changing activation sequence in the embryonic chick heart. Implications for the development of the His-Purkinje system. *Circ Res* 1997;**81**(4):470–6.

96. Arnolds DE, Liu F, Fahrenbach JP, Kim GH, Schillinger KJ, Smemo S, et al. TBX5 drives Scn5a expression to regulate cardiac conduction system function. *J Clin Invest* 2012;**122**(7):2509–18.

97. van den Boogaard M, Wong LY, Tessadori F, Bakker ML, Dreizehnter LK, Wakker V, et al. Genetic variation in T-box binding element functionally affects SCN5A/SCN10A enhancer. *J Clin Invest* 2012;**122**(7):2519–32.

98. Bruneau BG, Nemer G, Schmitt JP, Charron F, Robitaille L, Caron S, et al. A murine model of Holt-Oram syndrome defines roles of the T-box transcription factor Tbx5 in cardiogenesis and disease. *Cellule* 2001;**106**(6):709–21.

99. Kurihara Y, Kurihara H, Oda H, Maemura K, Nagai R, Ishikawa T, et al. Aortic arch malformations and ventricular septal defect in mice deficient in endothelin-1. *J Clin Invest* 1995;**96**(1):293–300.

100. Gassanov N, Er F, Zagidullin N, Hoppe UC. Endothelin induces differentiation of ANP-EGFP expressing embryonic stem cells towards a pacemaker phenotype. *FASEB J* 2004;**18**(14):1710–2.

101. Zhang X, Guo JP, Chi YL, Liu YC, Zhang CS, Yang XQ, et al. Endothelin-induced differentiation of Nkx2.5(+) cardiac progenitor cells into pacemaking cells. *Mol Cell Biochem* 2012;**366**(1-2):309–18.

102. Asai R, Kurihara Y, Fujisawa K, Sato T, Kawamura Y, Kokubo H, et al. Endothelin receptor type A expression defines a distinct cardiac subdomain within the heart field and is later implicated in chamber myocardium formation. *Development* 2010;**137**(22):3823–33.

103. Chinchilla A, Daimi H, Lozano-Velasco E, Dominguez JN, Caballero R, Delpón A, et al. PITX2 insufficiency leads to atrial electrical and structural remodeling linked to arrhythmogenesis. *Circ Cardiovasc Genet* 2011;**4**(3):269–79.

104. Lincoln J, Yutzey KE. Molecular and developmental mechanisms of congenital heart valve disease. *Birth Defects Res A Clin Mol Teratol* 2011;**91**(6):526–34.

105. Combs MD, Yutzey KE. Heart valve development: regulatory networks in development and disease. *Circ Res* 2009;**105**(5):408–21.

106. Manasek FJ. Sulfated extracellular matrix production in the embryonic heart and adjacent tissues. *J Exp Zool* 1970;**174**(4):415–39.

107. Manasek FJ, Reid M, Vinson W, Seyer J, Johnson R. Glycosaminoglycan synthesis by the early embryonic chick heart. *Dev Biol* 1973;**35**(2):332–48.

108. Ma L, Lu MF, Schwartz RJ, Martin JF. Bmp2 is essential for cardiac cushion epithelial-mesenchymal transition and myocardial patterning. *Development* 2005;**132**(24):5601–11.

109. Shirai M, Imanaka-Yoshida K, Schneider MD, Schwartz RJ, Morisaki T. T-box 2, a mediator of Bmp-Smad signaling, induced hyaluronan synthase 2 and Tgfbeta2 expression and endocardial cushion formation. *Proc Natl Acad Sci U S A* 2009;**106**(44):18649–9.

110. Moskowitz IP, Wang J, Peterson MA, Pu WT, Mackinnon AC, Oxburgh L, et al. Transcription factor genes Smad4 and GATA4 cooperatively regulate cardiac valve development. [corrected]. *Proc Natl Acad Sci U S A* 2011;**108**(10):4006–11.

111. Combs MD, Yutzey KE. VEGF and RANKL regulation of NFATc1 in heart valve development. *Circ Res* 2009;**105**(6):565–74.

112. Miquerol L, Gertsenstein M, Harpal K, Rossant J, Nagy A. Multiple developmental roles of VEGF suggested by a LacZ-tagged allele. *Dev Biol* 1999;**212**(2):307–22.

113. Jackson LF, Qiu TH, Sunnarborg SW, Chang A, Zhang C, Patterson C, et al. Defective valvulogenesis in HB-EGF and TACE-null mice is associated with aberrant BMP signaling. *EMBO J* 2003;**22**(11):2704–16.

114. Lincoln J, Alfieri CM, Yutzey KE. BMP and FGF regulatory pathways control cell lineage diversification of heart valve precursor cells. *Dev Biol* 2006;**292**(2):292–302.

115. Waldo K, Miyagawa-Tomita S, Kumiski D, Kirby ML. Cardiac neural crest cells provide new insight into septation of the cardiac outflow tract: aortic sac to ventricular septal closure. *Dev Biol* 1998;**196**(2):129–44.

116. Waldo K, Zdanowicz M, Burch J, Kumiski DH, Stadt HA, Godt RE, et al. A novel role for cardiac neural crest in heart development. *J Clin Invest* 1999;**103**(11):1499–507.

117. Prunotto M, Caimmi PP, Bongiovanni M. Cellular pathology of mitral valve prolapse. *Cardiovasc Pathol* 2010;**19**(4):e113–7.

118. Ng CM, Cheng A, Myers LA, Martinez-Murillo F, Jie C, Bedja D, et al. TGF-beta-dependent pathogenesis of mitral valve prolapse in a mouse model of Marfan syndrome. *J Clin Invest* 2004;**114**(11):1586–92.

119. Hanada K, Vermeij M, Garinis GA, de Waard MC, Kunen MG, Myers L, et al. Perturbations of vascular homeostasis and aortic valve abnormalities in fibulin-4 deficient mice. *Circ Res* 2007;**100**(5):738–46.

120. Dietz HC, Mecham RP. Mouse models of genetic diseases resulting from mutations in elastic fiber proteins. *Matrix Biol* 2000;**19**(6):481–8.

121. Hinton RB, Adelman-Brown J, Witt S, Krishnamurthy VK, Osinska H, Sakthivel B, et al. Elastin haploinsufficiency results in progressive aortic valve malformation and latent valve disease in a mouse model. *Circ Res* 2010;**107**(4):549–57.

122. Neptune ER, Frischmeyer PA, Arking DE, Myers L, Bunton TE, Gayraud B, et al. Dysregulation of TGF-beta activation contributes to pathogenesis in Marfan syndrome. *Nat Genet* 2003;**33**(3):407–11.

123. Galli D, Dominguez JN, Zaffran S, Munk A, Brown NA, Buckingham ME. Atrial myocardium derives from the posterior region of the second heart field, which acquires left-right identity as Pitx2c is expressed. *Development* 2008;**135**(6):1157–67.

124. Briggs LE, Kakarla J, Wessels A. The pathogenesis of atrial and atrioventricular septal defects with special emphasis on the role of the dorsal mesenchymal protrusion. *Differentiation* 2012;**84**(1):117–30.

125. Liu XY, Wang J, Zheng JH, Bai K, Liu ZM, Wang XZ, et al. Involvement of a novel GATA4 mutation in atrial septal defects. *Int J Mol Med* 2011;**28**(1):17–23.

126. Chen Y, Mao J, Sun Y, Zhang Q, Cheng HB, Yan WH, et al. A novel mutation of GATA4 in a familial atrial septal defect. *Clin Chim Acta* 2010;**411**(21-22):1741–5.

127. Hirayama-Yamada K, Kamisago M, Akimoto K, Aotsuka H, Nakamura Y, Tomita H, et al. Phenotypes with GATA4 or NKX2.5 mutations in familial atrial septal defect. *Am J Med Genet A* 2005;**135**(1):47–52.

128. Kirk EP, Sunde M, Costa MW, Rankin SA, Wolstein O, Castro ML, et al. Mutations in cardiac T-box factor gene TBX20 are associated with diverse cardiac pathologies, including defects of septation and valvulogenesis and cardiomyopathy. *Am J Hum Genet* 2007;**81**(2):280–91.

129. Draus Jr JM, Hauck MA, Goetsch M, Austin 3rd EH, Tomita-Mitchell A, Mitchell ME. Investigation of somatic NKX2-5 mutations in congenital heart disease. *J Med Genet* 2009;**46**(2):115–22.

130. Liu XY, Wang J, Yang YQ, Zhang YY, Chen XZ, Zhang W, et al. Novel NKX2-5 mutations in patients with familial atrial septal defects. *Pediatr Cardiol* 2011;**32**(2):193–201.

131. Watt AJ, Battle MA, Li J, Duncan SA. GATA4 is essential for formation of the proepicardium and regulates cardiogenesis. *Proc Natl Acad Sci U S A* 2004;**101**(34):12573–8.

132. Lyons I, Parsons LM, Hartley L, Li R, Andrews JE, Robb L, et al. Myogenic and morphogenetic defects in the heart tubes of murine embryos lacking the homeo box gene Nkx2-5. *Genes Dev* 1995;**9**(13):1654–66.

133. Misra C, Sachan N, McNally CR, Koenig SN, Nichols HA, Guggilam A, et al. Congenital heart disease-causing GATA4 mutation displays functional deficits in vivo. *PLoS Genet* 2012;**8**(5):e1002690.

134. Reamon-Buettner SM, Cho SH, Borlak J. Mutations in the 3'-untranslated region of GATA4 as molecular hotspots for congenital heart disease (CHD). *BMC Medical genetics* 2007;**8**:38.

135. Wu G, Shan J, Pang S, Wei X, Zhang H, Yan B. Genetic analysis of the promoter region of the GATA4 gene in patients with ventricular septal defects. *Trans Res* 2012;**159**(5):376–82.

136. Raid R, Krinka D, Bakhoff L, Abdelwahid E, Jokinen E, Kärner M, et al. Lack of GATA3 results in conotruncal heart anomalies in mouse. *Mech Dev* 2009;**126**(1-2):80–9.

137. Lepore JJ, Mericko PA, Cheng L, Lu MM, Morrisey EE, Parmacek MS. GATA-6 regulates semaphorin 3C and is required in cardiac neural crest for cardiovascular morphogenesis. *J Clin Invest* 2006;**116**(4):929–39.

138. Zhao R, Watt AJ, Battle MA, Li J, Bondow BJ, Duncan SA. Loss of both GATA4 and GATA6 blocks cardiac myocyte differentiation and results in acardia in mice. *Dev Biol* 2008;**317**(2):614–9.

139. Laforest B, Nemer M. GATA5 interacts with GATA4 and GATA6 in outflow tract development. *Dev Biol* 2011;**358**(2):368–78.

140. Pabst S, Wollnik B, Rohmann E, Hintz Y, Glänzer K, Vetter H, et al. A novel stop mutation truncating critical regions of the cardiac transcription factor NKX2-5 in a large family with autosomal-dominant inherited congenital heart disease. *Clin Res Cardiol* 2008;**97**(1):39–42.

141. Gruenauer-Kloevekorn C, Froster UG. Holt-Oram syndrome: a new mutation in the TBX5 gene in two unrelated families. *Annales de genetique* 2003;**46**(1):19–23.

142. Reamon-Buettner SM, Borlak J. TBX5 mutations in non-Holt-Oram syndrome (HOS) malformed hearts. *Hum Mutat* 2004;**24**(1):104.

143. Xie L, Hoffmann AD, Burnicka-Turek O, Friedland-Little JM, Zhang K, Moskowitz IP. Tbx5-hedgehog molecular networks are essential in the second heart field for atrial septation. *Dev Cell* 2012;**23**(2):280–91.

144. Goddeeris MM, Rho S, Petiet A, Davenport CL, Johnson GA, Meyers EN, et al. Intracardiac septation requires hedgehog-dependent cellular contributions from outside the heart. *Development* 2008;**135**(10):1887–95.

145. Hoffmann AD, Peterson MA, Friedland-Little JM, Anderson SA, Moskowitz IP. Sonic hedgehog is required in pulmonary endoderm for atrial septation. *Development* 2009;**136**(10):1761–70.

146. Kasahara H, Bartunkova S, Schinke M, Tanaka M, Izumo S. Cardiac and extracardiac expression of Csx/Nkx2.5 homeodomain protein. *Circ Res* 1998;**82**(9):936–46.

147. Go AS, Mozaffarian D, Roger VL, Benjamin EJ, Berry JD, Borden WB, et al. Heart Disease and Stroke Statistics–2013 Update: A Report From the American Heart Association. *Circulation* 2013;**127**(1):e6–245.

148. Kelly RG, Brown NA, Buckingham ME. The arterial pole of the mouse heart forms from Fgf10-expressing cells in pharyngeal mesoderm. *Dev Cell* 2001;**1**(3):435–40.

149. Mjaatvedt CH, Nakaoka T, Moreno-Rodriguez RA, Norris C, Kern MJ, Eisenberg CA, et al. The outflow tract of the heart is recruited from a novel heart-forming field. *Dev Biol* 2001;**238**(1):97–109.

150. Waldo KL, Hutson MR, Ward CC, Zdanowicz M, Stadt HA, Kumiski D, et al. Secondary heart field contributes myocardium and smooth muscle to the arterial pole of the developing heart. *Dev Biol* 2005;**281**(1):78–90.

151. Hiruma T, Nakajima Y, Nakamura H. Development of pharyngeal arch arteries in early mouse embryo. *J Anat* 2002;**201**(1):15–29.

152. Kirby ML, Turnage 3rd KL, Hays BM. Characterization of conotruncal malformations following ablation of "cardiac" neural crest. *Anat Rec* 1985;**213**(1):87–93.

153. Kirby ML, Gale TF, Stewart DE. Neural crest cells contribute to normal aorticopulmonary septation. *Science* 1983;**220**(4601):1059–61.

154. Watanabe M, Jafri A, Fisher SA. Apoptosis is required for the proper formation of the ventriculo-arterial connections. *Dev Biol* 2001;**240**(1):274–88.

155. Hutson MR, Kirby ML. Neural crest and cardiovascular development: a 20-year perspective. *Birth Defects Res C Embryo Today* 2003;**69**(1):2–13.

156. Scherptong RW, Jongbloed MR, Wisse LJ, Vicente-Steijn R, Bartelings MM, Poelmann RE, et al. Morphogenesis of outflow tract rotation during cardiac development: the pulmonary push concept. *Dev Dyn* 2012;**241**(9):1413–22.

157. Bockman DE, Kirby ML. Dependence of thymus development on derivatives of the neural crest. *Science* 1984;**223**(4635):498–500.

158. Kirby ML, Stewart DE. Neural crest origin of cardiac ganglion cells in the chick embryo: identification and extirpation. *Dev Biol* 1983;**97**(2):433–43.

159. Parisot P, Mesbah K, Theveniau-Ruissy M, Kelly RG. Tbx1, subpulmonary myocardium and conotruncal congenital heart defects. *Birth Defects Res A Clin Mol Teratol* 2011;**91**(6):477–84.

160. Mesbah K, Harrelson Z, Theveniau-Ruissy M, Papaioannou VE, Kelly RG. Tbx3 is required for outflow tract development. *Circ Res* 2008;**103**(7):743–50.

161. Fulcoli FG, Huynh T, Scambler PJ, Baldini A. Tbx1 regulates the BMP-Smad1 pathway in a transcription independent manner. *PLoS One* 2009;**4**(6):e6049.

162. Nie X, Brown CB, Wang Q, Jiao K. Inactivation of Bmp4 from the Tbx1 expression domain causes abnormal pharyngeal arch artery and cardiac outflow tract remodeling. *Cells Tissues Organs* 2011;**193**(6):393–403.

163. Papangeli I, Scambler PJ. Tbx1 Genetically Interacts With the Transforming Growth Factor-beta/Bone Morphogenetic Protein Inhibitor Smad7 During Great Vessel Remodeling. *Circ Res* 2013;**112**(1):90–102.

164. Mead TJ, Yutzey KE. Notch pathway regulation of neural crest cell development in vivo. *Dev Dyn* 2012;**241**(2):376–89.

165. Marino B, Digilio MC, Toscano A, et al. Anatomic patterns of conotruncal defects associated with deletion 22q11. *Genet Med* 2001;**3**(1):45–8.

166. Markert ML, Alexieff MJ, Li J, Sarzotti M, Ozaki DA, Devlin BH, et al. Postnatal thymus transplantation with immunosuppression as treatment for DiGeorge syndrome. *Blood* 2004;**104**(8):2574–81.

167. Driscoll DA, Salvin J, Sellinger B, Budarf ML, McDonald-McGinn DM, Zackai EH, et al. Prevalence of 22q11 microdeletions in DiGeorge and velocardiofacial syndromes: implications for genetic counselling and prenatal diagnosis. *J Med Genet* 1993;**30**(10):813–7.

168. Arnold JS, Werling U, Braunstein EM, Liao J, Nowotschin S, Edelmann W, et al. Inactivation of Tbx1 in the pharyngeal endoderm results in 22q11DS malformations. *Development* 2006;**133**(5):977–87.

169. Jerome LA, Papaioannou VE. DiGeorge syndrome phenotype in mice mutant for the T-box gene, Tbx1. *Nat Genet* 2001;**27**(3):286–91.

170. Merscher S, Funke B, Epstein JA, Heyer J, Puech A, Lu MM, et al. TBX1 is responsible for cardiovascular defects in velo-cardio-facial/DiGeorge syndrome. *Cellule* 2001;**104**(4):619–29.

171. Lindsay EA, Vitelli F, Su H, Morishima M, Huynh T, Pramparo T, et al. Tbx1 haploinsufficiency in the DiGeorge syndrome region causes aortic arch defects in mice. *Nature* 2001;**410**(6824):97–101.

172. Liao J, Kochilas L, Nowotschin S, Arnold JS, Aggarwal VS, Epstein JA, et al. Full spectrum of malformations in velo-cardio-facial syndrome/DiGeorge syndrome mouse models by altering Tbx1 dosage. *Hum Mol Genet* 2004;**13**(15):1577–85.

173. Baldini A. The 22q11.2 deletion syndrome: a gene dosage perspective. *Sci World J* 2006;**6**:1881–7.

174. Guo T, McDonald-McGinn D, Blonska A, Shanske A, Bassett AS, Chow E, et al. Genotype and cardiovascular phenotype correlations with TBX1 in 1,022 velo-cardio-facial/DiGeorge/22q11.2 deletion syndrome patients. *Hum Mutat* 2011;**32**(11):1278–89.

175. Harel I, Maezawa Y, Avraham R, Rinon A, Ma HY, Cross JW, et al. Pharyngeal mesoderm regulatory network controls cardiac and head muscle morphogenesis. *Proc Natl Acad Sci U S A* 2012;**109**(46):18839–44.

176. Huh SH, Ornitz DM. Beta-catenin deficiency causes DiGeorge syndrome-like phenotypes through regulation of Tbx1. *Development* 2010;**137**(7):1137–47.

177. Sato A, Scholl AM, Kuhn EB, Stadt HA, Decker JR, Pegram K, et al. FGF8 signaling is chemotactic for cardiac neural crest cells. *Dev Biol* 2011;**354**(1):18–30.

178. Zhou J, Pashmforoush M, Sucov HM. Endothelial neuropilin disruption in mice causes DiGeorge syndrome-like malformations via mechanisms distinct to those caused by loss of Tbx1. *PLoS One* 2012;**7**(3):e32429.

179. Geretti E, Shimizu A, Klagsbrun M. Neuropilin structure governs VEGF and semaphorin binding and regulates angiogenesis. *Angiogenesis* 2008;**11**(1):31–9.

180. Stalmans I, Lambrechts D, De Smet F, Jansen S, Wang J, Maity S, et al. VEGF: a modifier of the del22q11 (DiGeorge) syndrome? *Nat Med* 2003;**9**(2):173–82.

181. Maynard TM, Gopalakrishna D, Meechan DW, Paronett EM, Newbern JM, Lamantia AS. 22q11 Gene dosage establishes an adaptive range for sonic hedgehog and retinoic acid signaling during early development. *Hum Mol Genet* 2013;**22**(2):300–12.

182. Greulich F, Farin HF, Schuster-Gossler K, Kispert A. Tbx18 function in epicardial development. *Cardiovasc Res* 2012;**96**(3):476–83.

183. Perez-Pomares JM, Phelps A, Sedmerova M, Wessels A. Epicardial-like cells on the distal arterial end of the cardiac outflow tract do not derive from the proepicardium but are derivatives of the cephalic pericardium. *Dev Dyn* 2003;**227**(1):56–68.

184. Gittenberger-de Groot AC, Winter EM, Bartelings MM, Goumans MJ, DeRuiter MC, Poelmann RE. The arterial and cardiac epicardium in development, disease and repair. *Differentiation* 2012;**84**(1):41–53.

185. Dettman RW, Pae SH, Morabito C, Bristow J. Inhibition of alpha4-integrin stimulates epicardial-mesenchymal transformation and alters migration and cell fate of epicardially derived mesenchyme. *Dev Biol* 2003;**257**(2):315–28.

186. Yang JT, Rayburn H, Hynes RO. Cell adhesion events mediated by alpha 4 integrins are essential in placental and cardiac development. *Development* 1995;**121**(2):549–60.

187. Xavier-Neto J, Shapiro MD, Houghton L, Rosenthal N. Sequential programs of retinoic acid synthesis in the myocardial and epicardial layers of the developing avian heart. *Dev Biol* 2000;**219**(1):129–41.

188. Kang J, Gu Y, Li P, Johnson BL, Sucov HM, Thomas PS. PDGF-A as an epicardial mitogen during heart development. *Dev Dyn* 2008;**237**(3):692–701.

189. Stuckmann I, Evans S, Lassar AB. Erythropoietin and retinoic acid, secreted from the epicardium, are required for cardiac myocyte proliferation. *Dev Biol* 2003;**255**(2):334–49.

190. Pérez-Pomares JM, de la Pompa JL. Signaling During Epicardium and Coronary Vessel Development. *Circ Res* 2011;**109**(12):1429–42.

191. Dettman RW, Denetclaw Jr W, Ordahl CP, Bristow J. Common epicardial origin of coronary vascular smooth muscle, perivascular fibroblasts, and intermyocardial fibroblasts in the avian heart. *Dev Biol* 1998;**193**(2):169–81.

192. Vrancken Peeters MP, Gittenberger-de Groot AC, Mentink MM, Poelmann RE. Smooth muscle cells and fibroblasts of the coronary arteries derive from epithelial-mesenchymal transformation of the epicardium. *Anat Embryol (Berl)* 1999;**199**(4):367–78.

193. Katz TC, Singh MK, Degenhardt K, Rivera-Feliciano J, Johnson RL, Epstein JA, et al. Distinct compartments of the proepicardial organ give rise to coronary vascular endothelial cells. *Dev Cell* 2012;**22**(3):639–50.

194. Romano LA, Runyan RB. Slug is a mediator of epithelial-mesenchymal cell transformation in the developing chicken heart. *Dev Biol* 1999;**212**(1):243–54.

195. Martínez-Estrada OM, Lettice LA, Essafi A, Guadix JA, Slight J, Velecela V, et al. Wt1 is required for cardiovascular progenitor cell formation through transcriptional control of Snail and E-cadherin. *Nat Genet* 2010;**42**(1):89–93.

196. Casanova JC, Travisano S, de la Pompa JL. Epithelial-to-mesenchymal transition in epicardium is independent of snail1. *Genesis* 2013;**51**(1):32–40.

197. Red-Horse K, Ueno H, Weissman IL, Krasnow MA. Coronary arteries form by developmental reprogramming of venous cells. *Nature* 2010;**464**(7288):549–53.

198. Lavine KJ, White AC, Park C, Smith CS, Choi K, Long F, et al. Fibroblast growth factor signals regulate a wave of Hedgehog activation that is essential for coronary vascular development. *Genes Dev* 2006;**20**(12):1651–66.

199. Hood LC, Rosenquist TH. Coronary artery development in the chick: origin and deployment of smooth muscle cells, and the effects of neural crest ablation. *Anat Rec* 1992;**234**(2):291–300.

200. Vrancken Peeters MP, Gittenberger-de Groot AC, Mentink MM, Hungerford JE, Little CD, Poelmann RE. The development of the coronary vessels and their differentiation into arteries and veins in the embryonic quail heart. *Dev Dyn* 1997;**208**(3):338–48.

201. Wu B, Zhang Z, Lui W, Chen X, Wang Y, Chamberlain AA, et al. Endocardial Cells Form the Coronary Arteries by Angiogenesis through Myocardial-Endocardial VEGF Signaling. *Cellule* 2012;**151**(5):1083–96.

202. Eralp I, Lie-Venema H, DeRuiter MC, Van Den Akker NM, Bogers AJ, Mentink MM, et al. Coronary artery and orifice development is associated with proper timing of epicardial outgrowth and correlated Fas-ligand-associated apoptosis patterns. *Circ Res* 2005;**96**(5):526–34.

203. Watanabe T, Koibuchi N, Chin MT. Transcription factor CHF1/Hey2 regulates coronary vascular maturation. *Mech Dev* 2010;**127**(9-12):418–27.

204. Angelini P. Coronary artery anomalies: an entity in search of an identity. *Circulation* 2007;**115**(10):1296–305.

205. Lipton MJ, Barry WH, Obrez I, Silverman JF, Wexler L. Isolated single coronary artery: diagnosis, angiographic classification, and clinical significance. *Radiology* 1979;**130**(1):39–47.

206. Turkmen N, Eren B, Fedakar R, Senel B. Sudden death due to single coronary artery. *Singapore Med J* 2007;**48**(6):573–5.

207. Debich DE, Williams KE, Anderson RH. Congenital atresia of the orifice of the left coronary artery and its main stem. *Inter J Cardiol* 1989;**22**(3):398–404.

208. Zhang Z, Baldini A. In vivo response to high-resolution variation of Tbx1 mRNA dosage. *Hum Mol Genet* 2008;**17**(1):150–7.

209. Dhanantwari P, Lee E, Krishnan A, Samtani R, Yamada S, Anderson S, et al. Human cardiac development in the first trimester: a high-resolution magnetic resonance imaging and episcopic fluorescence image capture atlas. *Circulation* 2009;**120**(4):343–51.

210. Mall FP. On the development of the human heart. *Am J Anat* 1912;**13**(3).

211. Brand T. Heart development: molecular insights into cardiac specification and early morphogenesis. *Dev Biol* 2003;**258**(1):1–19.

212. Savolainen SM, Foley JF, Elmore SA. Histology atlas of the developing mouse heart with emphasis on E11.5 to E18.5. *Toxicol Pathol* 2009;**37**(4):395–414.

213. Martinsen BJ. Reference guide to the stages of chick heart embryology. *Dev Dyn* 2005;**233**(4):1217–37.

214. Kolker SJ, Tajchman U, Weeks DL. Confocal imaging of early heart development in Xenopus laevis. *Dev Biol* 2000;**218**(1):64–73.

215. Mohun TJ, Leong LM, Weninger WJ, Sparrow DB. The morphology of heart development in Xenopus laevis. *Dev Biol* 2000;**218**(1):74–88.

216. Warkman AS, Krieg PA. Xenopus as a model system for vertebrate heart development. *Semin Cell Dev Biol* 2007;**18**(1):46–53.

217. Stainier DY, Lee RK, Fishman MC. Cardiovascular development in the zebrafish. I. Myocardial fate map and heart tube formation. *Development* 1993;**119**(1):31–40.

218. Glickman NS, Yelon D. Cardiac development in zebrafish: coordination of form and function. *Semin Cell Dev Biol* 2002;**13**(6):507–13.

219. O'Rahilly R, Muller F. *Developmental Stages in Human Embryos*. Washington, DC: Carnegie Institution of Washington; 1987.

220. Hamburger V, Hamilton H. Series of Embryonic Chicken Growth. *J Morphol* 1951;**88**:49–92.

221. Nieuwkoop PD, Faber J. *Normal Table of Xenopus laevis (Daudin)*. New York: Garland Publishing, Inc.; 1994.

222. Wessels A, van den Hoff MJ, Adamo RF, Phelps AL, Lockhart MM, Sauls K, et al. Epicardially derived fibroblasts preferentially contribute to the parietal leaflets of the atrioventricular valves in the murine heart. *Dev Biol* 2012;**366**(2):111–24.

Cardiac Metabolism in Health and Disease

Lionel H. Opie

Professor of Medicine, Hatter Institute for Cardiovascular Research in Africa, Department of Medicine,
University of Cape Town and Groote Schuur Hospital, Cape Town, South Africa

INTRODUCTION

Metabolism comes from the Greek word meaning 'change,' referring to the transformation of potential energy stored in circulating substrates such as glucose and fatty acids to adenosine triphosphate (ATP), thereby driving cardiac contraction and relaxation. Early workers delineated carbohydrate and fatty acids as two of the most important myocardial fuels. In 1914 Lovatt Evans found that about one-third of the dog heart's energy was supplied by carbohydrate oxidation.[1] The importance of fatty acids as a fuel source and their utilization in the absence of glucose for the mammalian heart was initially reported by Cruikshank and Kosterlitz in 1941.[2] The oxidative metabolism of the human heart was defined by Richard Bing in a pioneering series of papers in the 1950s.[3] In the fasted state, carbohydrates make up a relatively small part of the oxidative metabolism of the resting human heart. Focus on free fatty acids (FFA) as a major myocardial fuel started with the 1961 finding that FFA inhibited glucose oxidation in the isolated heart.[4] The close interaction of glucose and FFA in fed and fasting conditions was delineated in the classic article describing the glucose–fatty acid cycle by Randle and his associates at Cambridge.[5]

Since then, the key to substrate utilization has increasingly focused on the mitochondria.[6,7] Major reviews have defined and refined the interaction between cardiac glucose and FFA metabolism[8] and the complex but dominant role of FFAs in health and disease.[9] The practical relevance of cardiac metabolism to the cardiologist, cardiovascular surgeon, and cardiovascular pathologist has constantly grown, particularly with the introduction of new drugs that directly alter energy metabolism. Furthermore the development of new techniques such as nuclear magnetic resonance spectroscopy (NMR) and positron emission tomography (PET) has enabled the non-invasive measurement of cardiac substrate metabolism.[10] New techniques and molecular paths have led to what Taegtmeyer terms 'the new cardiac metabolism,' in that 'it finally has dawned on many that metabolism is the missing link between form and function of the heart.'[11]

ENERGY AVAILABILITY

Metabolism is critical for normal cardiac contractility, and hence for the normal cardiac output that sustains life by perfusing all the vital organs. Contractile function necessitates a high turnover of ATP in the myocardium, and hence a correspondingly high rate of mitochondrial ATP production. The two major substrates for the energy metabolism of the heart are circulating glucose and free fatty acids (Figures 2.1 and 2.2). After these substrates are taken up by muscle cells and transported to the cytosol, they are further broken down in mitochondria to an activated two-carbon substrate fragment (acetyl CoA in Figures 2.1, 2.2 and 2.3), which enters the Krebs cycle, also called the tricarboxylic acid (TCA) cycle (Figure 2.4). Next comes the complex process of oxidative phosphorylation that produces mitochondrial ATP from adenosine diphosphate (ADP) and inorganic phosphate (Figure 2.4). In the following step, ATP leaves the mitochondria in exchange for ADP entering, to become available for contractile work that in turn breaks down ATP to ADP and inorganic phosphate (Figure 2.2), which re-enter the mitochondria to be recharged back to ATP.

MAJOR SOURCES OF ENERGY

The dominance of fatty acid metabolism for the fasting state has been confirmed in the human heart[3] (Table 2.1; for review see reference 9).

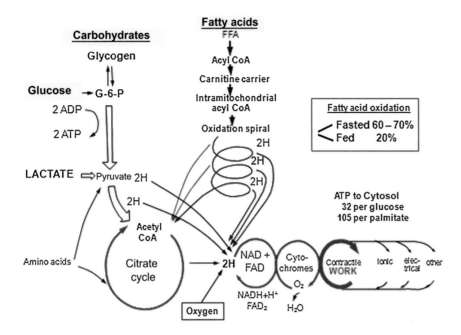

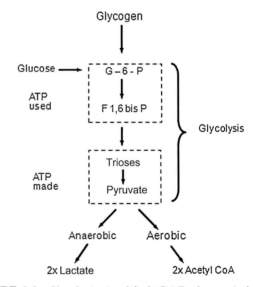

FIGURE 2.1   How carbohydrates and fatty acids yield energy for contractile work. Crucial are the processes whereby the fuels (glucose and fatty acids) are simplified for entry into the Krebs cycle that lies within the mitochondria. Their production of protons (H) within yields energy in the form of ATP (adenosine triphosphate) that is mostly used for contractile work. © LH Opie, 2004, from Opie LH. *Heart Physiology, From Cell to Circulation.* 4th ed. Philadelphia, PA: Lippincott, Williams & Wilkins; 2004:308–354.

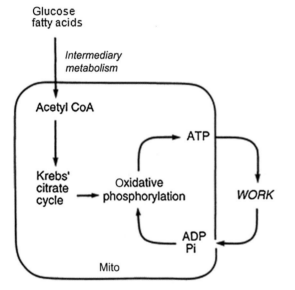

FIGURE 2.2   The ATP–ADP cycle. Energy in the form of ATP is produced by the Krebs citrate cycle (Fig. 2.3). Heart work utilizes the high energy liberated from ATP in the conversion to ADP which is then resynthesized together with inorganic phosphate (Pi) to produce ATP by oxidative phosphorylation. © LH Opie, 2004, from Opie LH. *Heart Physiology, From Cell to Circulation.* 4th ed. Philadelphia, PA: Lippincott, Williams & Wilkins; 2004:308–354.

FIGURE 2.3   Glycolysis simplified. G-6-P, glucose 6-phosphate; PFK, phosphofructokinase; F 1,6 bisP, fructose 1,6 bisphophate. © LH Opie, 2004, from Opie LH. *Heart Physiology, From Cell to Circulation.* 4th ed. Philadelphia, PA: Lippincott, Williams & Wilkins; 2004:308–354.

For its major sources of energy, the heart shifts its utilization between carbohydrate in the fed state and fatty acids in the fasted state (Table 2.1; Figure 2.1). In the carbohydrate-fed state, when circulating glucose and insulin are high, circulating fatty acid levels are suppressed. The uptake of fatty acids by the heart falls, resulting in the removal of fatty acid inhibition of glycolysis, so that glucose oxidation increases. Conversely, in the fasted state, circulating free fatty acids (FFAs) are high. The high rates of FFA uptake result in their preferential use in oxidative metabolism[4,5] (Figures 2.1, 2.5), so that fatty acids become the major source of energy.

When the heart oxidizes fatty acids, glucose oxidation is inhibited, resulting in the utilization of glucose for glycogen conversion, which is the glucose-sparing effect of fatty acid oxidation. The energy balance between fatty

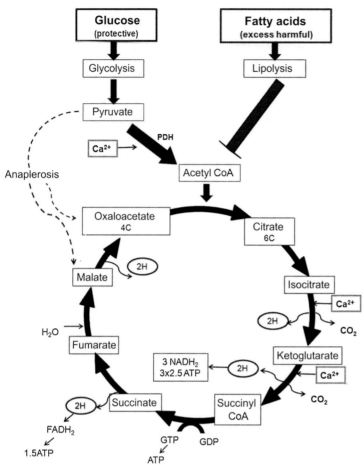

FIGURE 2.4 The citrate cycle of Krebs. In reality, the following reactions are readily reversible: citrate → isocitrate; succinate → oxaloacetate via the intervening reactions. The most important potential sites of control are citrate synthase, isocitrate dehydrogenase, α-ketoglutarate dehydrogenase, and malate dehydrogenase (by regulating the supply of oxaloacetate). Of these, isocitric dehydrogenase and α-ketoglutarate dehydrogenase are calcium-sensitive, as is pyruvate dehydrogenase (PDH). These dehydrogenases respond to increased cytosolic calcium (as in inotropic stimulation) by increased activity. © LH Opie, 1998.

TABLE 2.1 Fuels for Oxidative Metabolism of the Human Heart: Radio-Isotopic Data

Conditions	Glucose-Glycogen (OER %)	Lactate (OER %)	Total CHO (OER %)	FFA (OER %)
Glucose infusion	25–50	27	50–75	20
Exercise, moderate	14	28	42	64
Fasting overnight	3	13	16	62

CHO, carbohydrate; FFA, free fatty acids.; Oxygen extraction ratio.
For calculation details: Opie LH, *Heart Physiology, from Cell to Circulation*, Lippincott, Williams & Wilkins, 2004; p. 310.

acid β-oxidation and glucose oxidation can be pharmacologically modified at multiple levels of each metabolic pathway, as reviewed by the Lopaschuk group.[8,9]

Fatty acid oxidation is the dominant energy source for the heart when circulating FFAs are high, as after an overnight fast (Figure 2.5) or after catecholamine simulation.[10] Free fatty acid oxidation is less energy-efficient than glucose, requiring 11–12% more oxygen for a given amount of ATP produced.[12] With elevated circulating FFA levels, the actual amount of ATP produced by fatty acid oxidation may be much lower than expected.[12] Thus when fatty acid oxidation was driven by supraphysiological elevation of FFAs by lipid infusions given to dogs, the myocardial oxygen uptake increased by about 26%.[13]

Adrenergic activation increases levels of free fatty acids (FFA) by lipolysis activation in adipose tissue (see Chapters 4 and 14).[14] The FFA-driven metabolic component of the myocardial oxygen uptake, also called oxygen wastage, could experimentally be up to 90%, reduced by about half by an infusion of glucose.[14] Thus, high circulating glucose concentrations could protect the heart from FFA-induced oxygen wastage, strongly supporting the concept of adverse effects of supraphysiological FFA levels. The acute FFA-induced oxygen wastage is due to an increase in the oxygen cost for both basal metabolism and excitation–contraction coupling (EC) coupling.[15] In heart failure or myocardial infarction, high circulating levels of FFA[16] are driven by very high circulating norepinephrine levels.[17] The resulting excess FFA-induced oxygen wastage is countered by glucose and insulin[14] and by promotion of glycolysis,[18] thus closely resembling the metabolic situation after a high-carbohydrate meal (Figure 2.6).

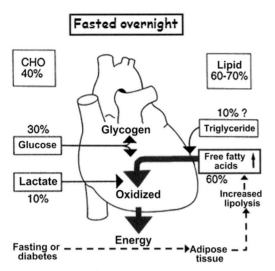

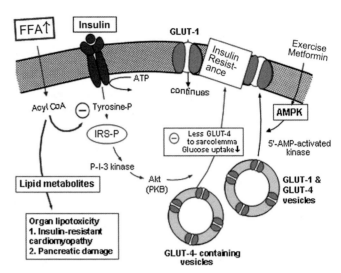

FIGURE 2.5   Substrate metabolism in the fasted state. Blood levels of free fatty acids are high in the fasted state or poorly controlled diabetes mellitus. High levels of blood free fatty acids are oxidized by the heart in preference to glucose and lactate. Use of lipid accounts for 60–70% of the oxygen uptake of the heart, whereas use of carbohydrate accounts for less than 20%. © LH Opie, 2004, from Opie LH. *Heart Physiology, From Cell to Circulation.* 4th ed. Philadelphia, PA: Lippincott, Williams & Wilkins; 2004:308–354.

FIGURE 2.7   Effects of excess FFA in muscle cells. Molecular steps that lead from increased circulating FFA to insulin resistance (top left) exist together with opposing influences exerted by exercise or the antidiabetic drug metformin (top left and top right). Excess FFA entering the cell is activated to long-chain acyl-CoA, which inhibits the insulin signaling pathway. Thus there is less translocation of glucose transporter vesicles (GLUT-4 and -1) to the cell surface. The result is that muscle glucose uptake is decreased and hyperglycemia is promoted. Exercise and metformin, by stimulating the enzyme adenosine monophosphate-activated kinase (AMPK), promote the translocation of transport vesicles to the cell surface to promote glucose entry and thereby oppose insulin resistance. Protein kinase B, also called Akt, plays a key role. IRS-P, insulin receptor substrate-1; P-I-3 kinase, phosphatidyl inositol -3-kinase; PKB, protein kinase B. © LH Opie, 2009.

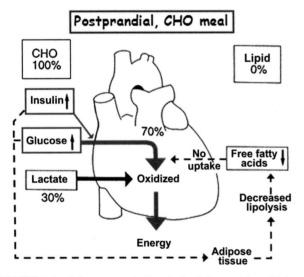

FIGURE 2.6   Substrate metabolism in the fed state. After a high carbohydrate meal or glucose feeding, blood glucose and insulin are high, and blood free fatty acids low. Glucose becomes the major fuel of the heart, and carbohydrate can account for 50–75% of the oxygen uptake. CHO, carbohydrate. © LH Opie, 2004, from Opie LH. *Heart Physiology, From Cell to Circulation.* 4th ed. Philadelphia, PA: Lippincott, Williams & Wilkins; 2004:308–354.

Glucose Transport Regulation

The uptake of glucose from the bloodstream across the sarcolemma and into the cells of the heart is controlled by the glucose transporters GLUT 1 and especially GLUT 4 (Figure 2.7). No energy is required for such glucose transport because the glucose concentration in the extracellular space is so much higher than in the cytosol. The uptake of glucose increases whenever the glucose transporter is stimulated, as during increased heart work, or by insulin as in the fed state, or during hypoxia or ischemia or by the antidiabetic drug, metformin (Figure 2.7). Conversely, the uptake of glucose is inhibited by high circulating FFA concentrations (Figure 2.7, left side).

Insulin is a circulating hormone whose levels rise in the fed state to increase glucose uptake by two major mechanisms. First, insulin decreases the release of FFA from adipose tissue, thereby removing the inhibitory effects of FFA on glucose uptake and glycolysis. Complex signaling paths convey the 'message' of insulin from the cell surface receptor to multiple internal sites of action (Figure 2.7). Insulin binds to specific insulin receptors that greatly amplify the effect of insulin and in turn phosphorylate tyrosine. Tyrosine phosphorylation increases the activity of the insulin receptor substrate-1 (IRS-P in Figure 2.7). Thereafter the enzymes phosphatidyl inositol (PI)-3-kinase and Akt (also called protein kinase B, PKB) translocate the glucose carrier GLUT-4 and to a much lesser extent GLUT-1 from internal unavailable sites to external sarcolemmal sites to increase the influx of glucose (Figure 2.7). FFAs inhibit this whole process at the level of tyrosine phosphorylation.

Beta-adrenergic stimulation acts by mobilizing FFA from adipose tissue, thereby promoting a metabolic condition similar to fasting with predominant fatty acid metabolism by the heart (Figure 2.5). Intense stimulation can cause major oxygen wastage.[14]

Pathways of Glycolysis

Glycolysis is the metabolic pathway that responds to reduced oxygen content by converting glucose to pyruvate (Figure 2.3). During normal oxidative metabolism, glycolytically produced pyruvate is then oxidized in the Krebs (tricarboxylic acid (TCA)) cycle.[12] Under anaerobic conditions, pyruvate is converted to lactate when the relatively small amounts of glycolytic ATP are of importance in preserving membrane function. Glycolysis converts glucose 6-phosphate into a compound containing two phosphate groups, fructose 1,6-diphosphate (fructose 1,6-bisphosphate) under the influence of the enzyme phosphofructokinase (PFK). Thereafter, glycolysis converts each six-carbon hexose phosphate into two three-carbon triose phosphates, using two molecules of ATP. During the next stages of glycolysis that form two molecules of pyruvate, four molecules of ATP are made independently of oxygen:

$$\text{fructose 1,6 bisphosphate} + 4\,ADP \rightarrow 2 \times \text{pyruvate} + 4\,ATP$$

Functions of Glycolysis

Normally glycolysis is a bit like water flowing in and out of a tank: what goes in at the top (glucose uptake, glycogen breakdown) passes out at the bottom (Figure 2.3). En route, the major enzymes that regulate the rate of flow to enter the Krebs (tricarboxylic acid (TCA)) cycle (Figure 2.4) are phosphofructokinase and pyruvate dehydrogenase (Table 2.2). Besides the production of protective anaerobic ATP, aerobic glycolysis provides energy for the maintenance of normal ATP-requiring membrane functions such as the sodium pump and the ATP-sensitive potassium channels. Glycolysis can also indirectly protect the mitochondria from potentially fatal calcium overload which occurs in some pathological states.

When glycolysis increases, the alternate and previously little emphasized hexose monophosphate bypass pathway is also stimulated.[18] The result is an increase in the activity of the enzyme hexokinase-II. Hexokinase-II directly inhibits the mitochondrial permeability transition pore (mPTP), thereby protecting mitochondria from calcium overload.[18] The authors of these studies found that these 'data suggest that the glycolytic enzyme HK (hexokinase-II) is an important guardian of the mitochondrion in the beating, intact heart.'[18]

TABLE 2.2 Major Factors Controlling Glycolysis and Sites of Action

Conditions	Glucose Uptake	Glycogen Content	Activity of Phospho-fructokinase	Activity of Pyruvate Dehydro-genase
Increased heart work	+	↓	+	+
Inotropic agents	+	↓	+	+
Fed state, insulin	+	↑	+	+
Starvation, high blood fatty acids or ketones	-	↑	-	-
Hypoxia, mild ischemia	+	↓	+	-
Severe ischemia	-	↓	-	-

+ stimulation; - inhibition; ↓ decreased content; ↑ increased content.
For further detail of glycolysis and its control by the above factors, see pages 314–321 in Opie LH, *Heart Physiology, from Cell to Circulation*, Lippincott, Williams & Wilkins.

Pathways of Fatty Acid Oxidation[12,19]

The chief fatty acids utilized by the human heart are long-chain fatty acids, including oleic acid, followed by palmate, of which palmitate is the better studied. In the schema shown, acyl CoA indicates both oleal CoA and palmityl CoA (Figure 2.8). The oxidation of fatty acids can be summarized as follows:

1. Extramitochondrial long-chain acyl CoA forms from long-chain FA: long-chain FA + CoA + ATP → long-chain acyl CoA + AMP + PPi;
2. Extramitochondrial long-chain acyl (oleal and palmityl) carnitine forms from extramitochondrial long-chain acyl CoA, catalyzed by the enzyme carnitine palmityl transferase 1 (CPT-1);
3. The enzyme carnitine acyl translocase transfers extramitochondrial long-chain acyl carnitine to within the space between outer and inner mitochondrial membranes;
4. Mitochondrial carnitine palmityl transferase 2 (CPT-2) located on the inner membrane allows intramitochondrial oleal and palmityl carnitine to react with CoA so as to liberate intramitochondrial oleal and palmityl CoA and carnitine; the carnitine is exported out to the mitochondrial inter-membrane space;
5. Intramitochondrial oleal and palmityl CoA enter the fatty acid β-oxidation spiral to form acetyl CoA that enters the Krebs tricarboxylic acid (TCA) cycle.

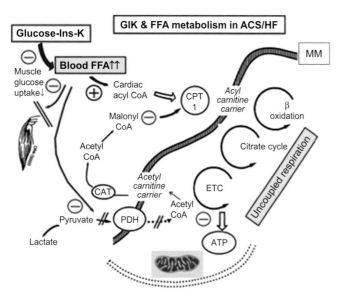

FIGURE 2.8   Long-chain free fatty acid (FFA) metabolism. Effects of excess FFA supply. Blood borne FFA is taken into the cytosol, where the acyl carnitine carrier on the mitochondrial membrane transfers activated fatty acid (acyl CoA) into the mitochondrial space for oxidation in the fatty acid oxidation cycle. ACS/HF, acute coronary syndrome, heart failure; acyl CoA, long-chain acyl CoA compounds; CPT, carnitine palmityl CoA transferase; acyl carnitine, long-chain acyl carnitine compounds; CAT, carnitine acyl translocase; FFA, free fatty acids; GIK, glucose-insulin-potassium; MM, mitochondrial PDH, pyruvate dehydrogenase. © LH Opie, 2004, from Opie LH. *Heart Physiology, From Cell to Circulation.* 4th ed. Philadelphia, PA: Lippincott, Williams & Wilkins; 2004:308–354.

6. During high rates of FFA uptake and subsequent metabolism by the above steps, more acetyl CoA may form than can enter the Krebs cycle. Such acetyl CoA can also react with intramitochondrial carnitine via the enzyme carnitine-acetyl transferase (CAT, Figure 2.6) to form acetyl carnitine. The latter is transported outwards from the mitochondria by the enzyme carnitine-acetyl translocase, and in the process cytoplasmic acetyl CoA is formed. This then undergoes transformation into malonyl CoA, which by feedback inhibition of CPT-1 helps to limit excess FFA oxidation.

7. Also during high rates of FFA uptake and metabolism, excess acyl CoA that is not converted into malonyl CoA cannot be oxidized and forms potentially adverse myocardial triglyceride and structural lipids, the latter by changes in the degree of saturation and chain length.

Fatty Acid Toxicity and Oxidation Inhibitors

In view of these adverse metabolic inhibitory effects of excess uptake and metabolism of high circulating FFA, it is noteworthy that therapeutic inhibition with practical clinical application can be achieved by infusions of glucose–insulin–potassium (GIK therapy) (Figure 2.8) and by the drugs trimetazidine and perhexiline that act by inhibition of myocardial carnitine palmityltransferase-1 (Figure 2.8).

TABLE 2.3   Comparative Energy Yields of Various Fuels Per Molecule Fully Oxidized

Molecule	ATP Yield Per Molecule		ATP Yield Per Carbon Atom		ATP Yield Per Oxygen Atom Taken up to (P/O ratio)[a]	
	'Old'	'New'	'Old'	'New'	'Old'	'New'
Glucose	38[c]	32[d]	6.3	5.2	3.17	2.58
Lactate	18	14.75	6.0	4.9	3.00	2.46
Pyruvate	15	12.25	5.0	4.1	3.00	2.50
Palmitate[b]	130	105	8.1	6.7	2.83	2.33

[a]P/O, phosphorylation/oxidation;

[b]For palmitate details see Brand, *The Biochemist,* Aug–Sept 1994, p20;

[c]36 via mitochondria, 2 via glycolysis;

[d]30 via mitochondria, 2 via glycolysis.

'old', conventional; 'new', revised. See Hinkle et al. *Biochemistry* 1991;30:3576.

Krebs Tricarboxylic Acid (TCA)/Citrate Cycle

When the Nobel prize winner Sir Hans Krebs was an unknown biochemist, he presented his concept and data of the citrate cycle to the prestigious journal *Nature* which politely rejected it. This classic then found its way to a minor journal, *Experientia.* Despite the concept's initial struggle, the Krebs cycle remains at the basis of production of essential energy in the form of ATP (Figure 2.4).

The breakdown of ATP is the only immediate source of energy for contraction, the maintenance of ion gradients, and other vital functions. The complex metabolic pathways already described transform the major fuels (glucose, free fatty acids, and lactate) to acetyl CoA, which enters the citrate cycle to produce $NADH_2$ ($NADH + H^+$). $NADH_2$ is the reduced form of the cofactor nicotinamide adenine dinucleotide. The 2H units, in turn, yield the protons pumped across the mitochondrial membrane, and the electrons that flow along the cytochrome chain, with resulting conversion of ADP into ATP by oxidative phosphorylation. Once produced in the mitochondria, ATP is transported outward to the cytosol by the ATP/ADP transport system for use in the cytoplasm, chiefly in contraction. As cytosolic ATP is used, ATP is synthesised from ADP that occurs in the mitochondria. The rates of synthesis and breakdown of ATP are, therefore, closely linked.

The rate at which the Krebs' citrate cycle operates is a major factor controlling the rate of ATP production by the heart. The standard dogma is that each turn of the Krebs cycle produces 12 molecules of ATP; in reality, allowing for technical factors often ignored it is more accurately estimated that each cycle produces closer to 10 ATP (Table 2.3). The citrate cycle accelerates with increased heart work. Conversely, decreased rates of operation of the cycle occur during states of oxygen deprivation, such as hypoxia or ischemia or during cardioplegic arrest.

When the heart must suddenly increase ATP production as during a sudden jump in work demand, it is controversial as to which one of several potential factors limits oxidative phosphorylation. Such potential factors that limit oxidative phosphorylation include the rate of ATP formation from ADP, the supply of oxygen from the coronary circulation, and the rate of the cycle (malate–aspartate) that transports cytosolic $NADH_2$ to the mitochondria, thereby helping to drive the citrate cycle.

ENERGY EXPENDITURE: WORK OF THE HEART

Why does the heart need so much energy? Because it has to work all the time, day and night. The heart also has to cope with the intermittent needs of any increases in cardiac output during exercise or catecholamine-driven emotional stress.

Myocardial Metabolic Response to Exercise

A higher rate of myocardial oxygen uptake reflects increased production of ATP. Mitochondrial metabolism is driven by a greater rate of formation of ADP and an increased mitochondrial calcium concentration. Glycogen breakdown may play a special role at the abrupt onset of increased heart work.

The following are the several factors influencing the work of the heart, each contributing to the varying energy requirements.[12,19]

1. *External work: Cardiac output.* The heart rate is the dominant factor governing the cardiac output (stroke volume × heart rate). Other major factors include preload, afterload, and contractility (Figure 2.9). Any rise in cardiac output as in exercise increases the myocardial oxygen uptake. Challenge with volume load, such as that found in dynamic exercise, results in an increase in oxygen uptake, reflecting the external work that rises in proportion to the heart rate. However, increases in pressure load as by static exercise, increase the amount of internal work.
2. *Blood pressure as an external load.* BP = cardiac output × systemic vascular resistance, or BP = CO × SVR, where SVR = peripheral arteriolar resistance (Figure 2.9). Thus as the BP rises with exercise, so does the myocardial oxygen demand.
3. *Internal work (also known as peak wall tension-internal work against an elastic element): Wall stress and oxygen uptake.* Internal work requires development of cardiac wall stress that in turn is the product of pressure and radius. Conceptually, wall stress increases with increased preload, increased afterload, and increased contractility thus embracing three of the four factors that regulate cardiac output (Figure 2.9).

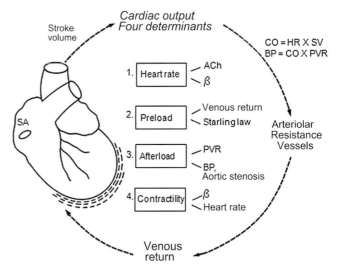

FIGURE 2.9 Major factors regulating the cardiac output (CO). The heart rate is regulated chiefly by the relative inputs of cholinergic (ACh) and β-adrenergic (β) stimulation. Preload is regulated by the venous return and also increases when the left ventricle fails to empty fully. Afterload increases when the peripheral vascular resistance (PVR) rises or when there is aortic stenosis. Contractility rises with β-stimulation or with increased fiber lengths. SV, stroke volume; HR, heart rate. © LH Opie, 2004, from Opie LH. *Heart Physiology, From Cell to Circulation.* 4th ed. Philadelphia, PA: Lippincott, Williams & Wilkins; 2004:308–354.

4. *Dynamic and static exercise both increase myocardial oxygen demand.* There are important contrasting differences between dynamic and static exercise. During dynamic exercise, the prime control mechanisms are increased cortical command and feed-forward signaling that increases sympathetic adrenergic activity and decreases vagal inhibition. A combination of increased venous return, an increased heart rate (the dominant factor), and increased contractility result in an increased cardiac output. In dynamic exercise the systolic but not diastolic pressure rises. In contrast, in static exercise both systolic and diastolic pressures rise, probably as a result of reflexes arising in the muscles and conveyed to cardiovascular centers in the medulla and hypothalamus.
5. *Athlete's heart.* In this condition, bradycardia and physiological left ventricular (LV) hypertrophy go together. Distinction from pathological LV hypertrophy is dependent upon the setting in which each occurs, and by the improved diastolic filling that can be observed in the physiological (but not pathologic) variety.
6. *Emotional stress.* Emotional stress increases myocardial oxygen demand by the combination of enhanced β- and α-adrenergic activity. In normal individuals, secretion of adrenaline dominates, with the resulting tachycardia accompanied by increased stroke volume, peripheral vasodilation, and surprisingly modest changes in the blood pressure. However, in hypertensive or borderline hypertensive individuals, emotional stress can induce substantial rises in blood pressure.

β-adrenergic signaling systems underlie the enhanced metabolic rate and positive inotropic response the heart requires to respond to increased energy demands during physiological responses (exercise) or pathological responses (e.g. fear) (Figure 2.10). These can be explained in terms of changes in the cardiac calcium cycle.[12,19] When the β-adrenergic agonist interacts with the β-receptor, a series of G-coupled protein receptor-mediated changes lead to activation of adenylate cyclase and formation of the adrenergic second messenger, cyclic AMP (cAMP). The latter acts via protein kinase A to stimulate metabolism and to phosphorylate (P) the cardiac L-type calcium (Ca^{2+}) channel protein, thus increasing the opening probability of this channel. More Ca^{2+} ions enter through the sarcolemmal (SL) channel, to release more Ca^{2+} ions from the sarcoplasmic reticulum (SR). Thus, the increased cytosolic Ca^{2+} ions also increase the rate of breakdown of adenosine triphosphate (ATP) and to adenosine diphosphate (ADP) and inorganic phosphate (P$_i$). Enhanced myosin ATPase activity explains the increased rate of contraction, with increased activation of troponin-C explaining increased peak force development. An increased rate of relaxation, the lusitropic effect, beautifully described by Arnold

Katz,[19] follows from phosphorylation of the protein phospholamban (PL), situated on the membrane of the SR, that controls the rate of calcium uptake into the SR.

PATHOLOGICAL ALTERATIONS IN MYOCARDIAL ENERGY METABOLISM

Acute Myocardial Infarction

When circulating free fatty acids (FFAs) are high enough to exceed the tight binding sites on the albumin, therefore loosely bound with an increased FFA:albumin ratio, adverse effects on the ischemic myocardium occur, including membrane damage with increased enzyme release (Figure 2.11). This scenario can be seen during the acute stress reaction with abrupt elevation of plasma catecholamines as in the first hours of acute myocardial infarction.[20]

Excess Fatty Acid Metabolites

Causes of multifactorial membrane damage include the accumulation of free fatty acids inside and outside the ischemic cells and increased amounts of potentially toxic fatty acid metabolites, such as acyl CoA and acyl carnitine. There may be a self-perpetuating

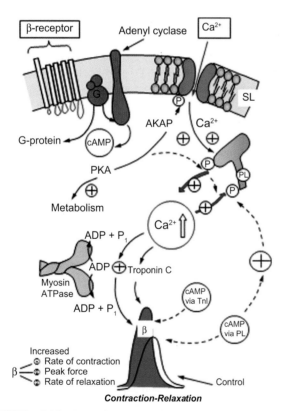

FIGURE 2.10 Beta-adrenergic signaling. β-Adrenergic signal systems involved in positive inotropic and lusitropic (enhanced relaxation) effects. © LH Opie, 2012, from Opie LH, Gersh BG (editors) *Drugs for the Heart*, 8th edition, Elsevier, Philadelphia, 2013.

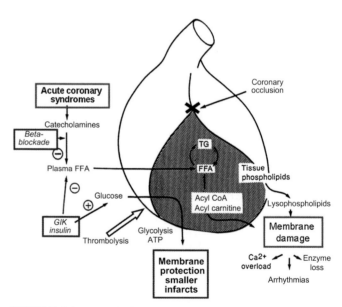

FIGURE 2.11 Metabolic cardioprotection during acute coronary syndromes. Critical is reduction of circulating free fatty acids (FFA) both by beta-blockade and by infusion of glucose-insulin-potassium (GIK). Thereby protective glycolytic metabolism by the ischemic tissue is promoted. The harm of FFA is exerted by increasing tissue fatty acid acyl CoA and carnitine, thereby promoting membrane damage and lethal arrhythmias. TG = tissue triglycerides. © LH Opie, 2004, from Opie LH. *Heart Physiology, From Cell to Circulation*. 4th ed. Philadelphia, PA: Lippincott, Williams & Wilkins; 2004:308–354.

cycle whereby part of membrane damage results from the action of phospholipases that break down membrane lipids, with formation of lysophosphoglycerides, which further promote membrane damage. The underlying metabolic paths are shown in figures 1 and 2 of Zhang et al.[21] Excess lipid metabolism predisposes to insulin resistance, which in turn is a predecessor of atherosclerosis.[21]

Mechanisms of Irreversible Ischemic Damage

The exact mechanism whereby reversible ischemia finally evolves into irreversible infarction is still controversial. Both the consequences of poor oxygen delivery and poor washout of metabolites play a role, with probably the former being the predominant factor. The irreversibility of ischemic damage is a consequence of the induction of either necrosis or apoptosis. However, both necrosis and apoptosis have different etiologies, making simplification of the concept of irreversibility in ischemic damage difficult. Briefly, four current theories have been proposed for the development of irreversible ischemic damage in the form of necrosis: (1) inhibition of glycolysis and loss of glycolytic ATP; (2) excess fatty acid metabolites; (3) sodium pump inhibition; (4) formation of excess reactive oxygen species; and (5) calcium cellular overload.

Inhibition of the Sodium Pump

Inhibition of the sodium pump may precipitate an excess of internal sodium, which in turn leads to the increased osmotic pressure that helps to 'pop' the cell membrane and to cause irreversible damage.[22] The direct proposed cause of the pump failure is inadequate synthesis of glycolytic ATP for the pump.[23]

Reactive Oxygen Species

Reactive oxygen species (ROS, also called oxygen free radicals) are a side-product of sites on mitochondrial complexes I and III of the electron transmitter chain (see later in text). In excess, ROS contribute to membrane damage by lipid peroxide formation and are part of the signaling sequence leading to apoptosis. Excess ROS are also derived from many complex sources besides damaged myocyte mitochondria, such as from uncoupled nitric oxide synthase in heart and endothelial cells, from xanthine oxidase and stimulation of membrane NADPH oxidase (by angiotensin II, endothelin, cytokines) and from neutrophils.[24] Excess ROS are formed particularly during the reperfusion phase of ischemic damage.[25]

Calcium Cellular Overload

Overload of cellular calcium as a source of ischemic heart damage irreversibility has received special prominence with respect to conditions of massive calcium overload, such as catecholamine stimulation or severe reperfusion damage or the very unusual experimental condition of the calcium paradox, whereby extracellular calcium is totally removed and then, when re-introduced, causes massive cellular damage. The basic concept of such severe degrees of calcium overload is that the mitochondria initially act as a buffer to take up calcium from the cytosol, which requires considerable energy. As a consequence, generalized cellular energy depletion is enhanced, the energy required for maintenance of ionic gradients becomes inadequate, and membrane integrity is lost. A modified form of this hypothesis is as follows. In ischemia, cytosolic calcium levels rise, possibly as a result of intracellular redistribution of calcium. Such cytosolic calcium increases can activate phospholipases, increase resting tension, and provoke fatal ventricular arrhythmias.[26]

Irreversible Ischemic Damage

Irreversibility of ischemic damage depends on no single cellular event or mechanisms but may be a complex phenomenon resulting from the simultaneous operation of many diverse mechanisms.

Molecular Cardioprotection from Reperfusion-Induced Cell Death

The extraordinary complexity of the molecular protective pathways must, like the metabolic pathways, have evolved millions of years ago when rapid cardioprotection was required after the hyperadrenergic stresses and blood losses experienced when hunting and escaping from wild animals.[27]

Today these paths can be brought into action during prompt therapy for acute coronary occlusion by rapid reperfusion, as in acute myocardial infarction (AMI). Although saving many cells otherwise threatened with ischemic cell death, rapid reperfusion kills a significant percentage of cells that could have been saved by the appropriate intervention. Indeed, the time has come to take reperfusion injury seriously.[28]

Reperfusion damage, with the sudden return of oxygen and reversal of tissue pH changes is inevitable during the optimal therapy of AMI by prompt percutaneous coronary intervention (PCI). Working on this problem, the group of Yellon and Hausenloy discovered the RISK (Reperfusion Injury Salvage Kinases) cardioprotective pathway (see Chapter 5).[28] Of major interest is that metabolic and molecular protection can go hand in hand. That concept leads to the proposal that insulin therapy, a promoter of optimal cardiac metabolic protection, can directly promote cardiac cell survival during reperfusion.[29] The site of irreversible mitochondrial damage is the mitochondrial permeability transition pore (see Figure 2.13) that promotes the entry of excess calcium into the mitochondrial space.

From the evolutionary point of view, it would be inadequate to have only one molecular cardioprotective path. Thus another path was soon discovered, namely the SAFE pathway (Survival Activating Factor Enhancement), that is the focus of our group in Cape Town.[30] This path involves cytokine protection by low levels of TNF-alpha acting on a specific series of molecular events and leading to activation of transcription factor Signal Transducer and Activator of Transcription-3 (STAT-3). These will be discussed in more detail in Chapter 5.

Metabolism in Metabolic Syndrome and Obesity

The metabolic syndrome is a combination of disorders that increase the risk of cardiovascular disease and diabetes. Five key features are characteristic of the metabolic syndrome: abdominal obesity, blood measurements of abnormalities, the fasting lipid profile, plasma glucose and triglycerides, and an elevated blood pressure (see Figure 2.12). The diagnosis of metabolic syndrome can be made when three or more of these characteristics are present.

From the metabolic point of view, obesity is a complex condition that is also associated with high blood free fatty acid (FFA) levels.[31] Even modest elevations of blood FFA levels inhibit insulin signaling[32] and are linked to other harmful signaling effects in non-diabetic persons, both obese and non-obese.[33] Such FFA elevations stimulate various proinflammatory signals such as nuclear factor kappa B (NFκB) to promote insulin resistance (see Chapter 7).[34,35] These signals in turn stimulate macrophages to provoke the chronic low-grade inflammatory response with increased plasma levels of C-reactive protein, and inflammatory cytokines such as tumor necrosis factor-alpha (TNF-alpha), interleukin-6 and -8, and the multifunctional proteins such as leptin and osteopontin.[36] Macrophages in human adipose tissue are the main but not the only source of these inflammatory mediators that evoke insulin resistance. Experimentally, a high-fat 'Western'-type diet enhances systemic cytokine production, whereas exercise diminishes it.[37] The overall sequence of events linking obesity to insulin resistance therefore has been hypothesized as follows:

Obesity → high FFA → molecular signaling (NFκB, others)
→ macrophages → inflammatory cytokines → insulin resistance

Metabolism in Diabetes

The glucose–fatty acid cycle of Randle describes an alternating interaction between glucose and FFA metabolism that may explain the metabolic abnormalities that are found in obesity and diabetes.[5] These earlier findings have now been taken further. Experimentally, high circulating FFA levels oppose the vasodilatory and protective

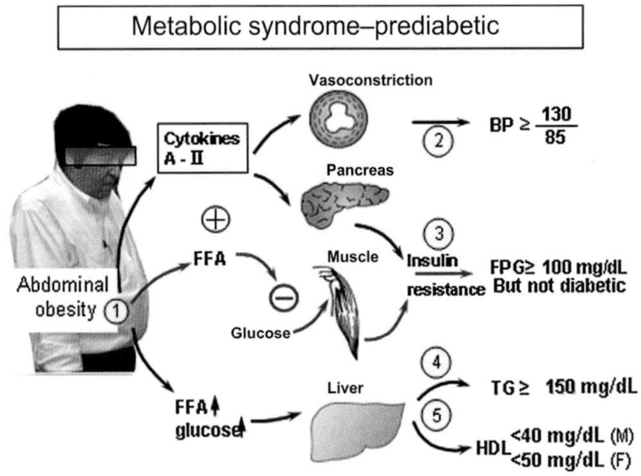

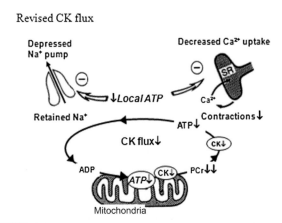

FIGURE 2.12 Metabolic syndrome. The diagnosis of metabolic syndrome, a prediabetic condition, requires a tape measure to measure abdominal obesity, blood measurements of fasting lipogram, plasma glucose and triglycerides, and the BP. Three or more of these five give the diagnosis. A-II indicates angiotensin-II; FFA, free fatty acids; BP, blood pressure; FPG, fasting plasma glucose; TG, triglycerides; and HDL, high-density lipoproteins. From: The metabolic syndrome, does it exist? © Opie LH, Kasuga M, Yellon DM, eds. *Diabetes at the Limits.* Cape Town, South Africa: University of Cape Town Press; 2006:95–110.

FIGURE 2.13 Defects of energy transfer in HF. Note depressed activity of creatine kinase (CK) and depressed flux of phosphocreatine (PCr) through CK. The decreased ATP supply impairs the uptake of Ca^{2+} ions into the sarcoplasmic reticulum (SR), thereby leading to decreased contractile force. ATP located near the cell membrane is also decreased thus depressing Na^+ pump activity which in turn leads to an increased cytosolic Na^+ and impaired Na^+/K^+ exchange. The overall result of the deficits in the energy supply is the decreased contractile force characteristic of the failing myocardium. CK, creatine kinase; PCr, phosphocreatine.

role of endothelial nitric oxide (NO) molecule.[37] Marked FFA elevation causes an increase in oxidative stress with mitochondrial uncoupling (Figure 2.13).[38–40] Experimentally abnormally high external glucose concentrations potentially promote tissue damage by stimulating the various pathological paths, activated by blocked glycolysis, that lead to increased formation of reactive oxygen species.[40] FFA also exert toxic effects on adipose tissue cells, liberating cytokines that damage blood vessels, heart and skeletal muscle, and the pancreas, thereby together with the FFA inhibition of glucose uptake by muscle, promoting formation of reactive oxygen species (ROS) [41] and thereby insulin resistance.

Lipotoxicity

Lipotoxicity is the cytosolic overload of the lipids formed from excess cytosolic acyl CoA and other adverse metabolites. Excess myocardial fatty acid metabolism, as in lipotoxicity, is thought to promote a metabolic cardiomyopathy in obesity and diabetes.[42,43] The basic mechanism is twofold. First, when circulating FFAs are excessively high as in obesity or cardiac overload,[44] such FFAs enter the cytosolic space, again in excess, so that cytosolic overload of acyl CoA results (Figure 2.8). The consequence is that the capacity of the enzyme CPT-1 (carnitine palmityl transferase-1) to transfer acyl CoA to further breakdown within the mitochondria is exceeded. [45] The final result is that there is increased excessive conversion of cytosolic acyl CoA to excess triglyceride and related lipids. Such lipid overload decreases the mechanical function of the heart.[45] The second mechanism for lipotoxicity is when excess formation of malonyl CoA production via acetyl CoA carboxylase-2 (ACC-2) inhibits the entry of long-chain fatty acids as acyl CoA into the mitochondria, as in chronic cardiac hypertrophy.[45]

In humans with type 2 diabetes mellitus the situation is made much more complex by the concurrence of an associated spectrum of metabolic abnormalities, namely insulin resistance, dyslipidemia, hypertension, visceral obesity, glucose intolerance, and endothelial dysfunction, of which insulin resistance is proposed as the key event.[46] Lipotoxicity is associated with type 2 diabetes and the metabolic syndrome.[43,46] How does this apply to the clinical entity of diabetic cardiomyopathy? Although ischemic heart disease is a major problem in diabetes, non-ischemic heart disease (diabetic cardiomyopathy) relates to the impairment of cardiac function and mortality in type 2 diabetes. The underlying etiology of diabetic cardiomyopathy can be related to cardiac lipotoxicity.[47]

Regarding the skeletal muscle in type 2 diabetics, proton nuclear magnetic resonance is used to detect intramyocellular fat in biopsies and then to measure long-chain-fatty acyl-coenzyme A (LC-FACoA).[47] Treatment of such patients to reduce insulin resistance (by the peroxisome proliferator-activated receptor-γ agonist (PPAR-γ agonist; pioglitazone) was associated with a reduction in both intramyocellular triglyceride content and skeletal muscle LC-FACoAs. The latter decrease strongly correlated with enhanced insulin sensitivity.[47]

Glucolipotoxicity

This refers to the logical concept of combined glucotoxicity and lipotoxicity. Experimentally, the direct damage of high glucose and high palmitate to cardiomyocytes is associated with insulin resistance.[47] Even modest glycemia and modest lipemia may adversely combine to increase pancreatic damage coupled with ceramide deposition and increased reactive oxygen species.[46] Glucolipotoxicity has been studied in cardiomyocytes[48] and in islet-cell-derived pancreatic cells, the latter showing that glucolipotoxicity can be alleviated by promotion of FFA oxidation or by diversion of FFA into tissue esterification.[49]

Clinically, diabetic coma is a complex and variable condition including the hyperosmolar hyperglycemic state and/or diabetic ketoacidosis (DKA), often with marked dehydration.[50] The triad of uncontrolled hyperglycemia, metabolic acidosis, and increased total body ketone concentration characterizes DKA. The hyperosmolar hyperglycemic state is characterized by severe hyperglycemia, hyperosmolality, and dehydration in the absence of significant ketoacidosis. These metabolic derangements result from the combination of absolute or relative insulin deficiency and an increase in counterregulatory hormones (glucagon, catecholamines, cortisol, and growth hormone). Common to both DKA and the hyperosmolar hyperglycemic state are high circulating FFA concentrations, with added multifactorial damage from elevations of proinflammatory cytokines (tumor necrosis factor-α, interleukin-β, -6, and -8), C-reactive protein, reactive oxygen species, lipid peroxidation, plasminogen activator inhibitor-1 and FFA in the absence of obvious infection or cardiovascular pathology.[50]

For the above reasons, glucolipotoxicity is a viable concept although it stretches far beyond abnormalities of glucose and fatty acid metabolism to include an associated spectrum of adverse mechanisms that promote disease states.

METABOLISM OF HEART FAILURE

Heart failure is an energy-poor condition with major defects in energy transfer.[51] The impaired delivery of oxygen to the tissues results in adrenergic activation with increased levels of norepinephrine,[17] increased levels of FFA[52] and lipotoxicity with myocardial insulin resistance.[53] Mechanistically, the depressed activity of creatine kinase and depressed flux of phosphocreatine through CK cause defects in the energy metabolism of the cell, including impaired delivery of local ATP to the Na$^+$ pump and the sarcoplasmic reticulum (Figure 2.13). Decreased uptake of Ca^{2+} ions into the sarcoplasmic reticulum leads to decreased contractile force.[54] The overall result of these deficits in the energy supply is the decreased contractile force characteristic of the failing myocardium.

Clinically, there is a detrimental self-perpetuating cycle:

(heart failure → adrenergic activation → altered metabolism →
increasing severity of heart failure)

that promotes the progression of heart failure.[51] The net metabolic result is depletion of myocardial ATP and phosphocreatine with decreased efficiency of mechanical work.[55,56] Deficiencies of the cardiac contents of high-energy phosphate compounds, phosphocreatine (PCr) and ATP, as shown by magnetic resonance spectroscopy, are associated with HF (for reviews see Ashrafian et al.[56] and Neubauer[57]).

Such energetic changes predict both total and cardiovascular mortality in patients with dilated cardiomyopathy.[57] Modern high-speed magnetization transfer techniques have helped to define the kinetics and thermodynamics of high-energy phosphate fluxes in the failing heart. Thus the PCr/ATP ratio falls by ≈30% in human hearts with both pressure-overload LV hypertrophy (LVH) and LVH with heart failure.[58] Especially detrimental is the impairment of energy flux through the creatine kinase system that was reduced by 30% in LVH and by 65% in LVH with HF compared with controls.[58] Of interest is the ability of allopurinol, a commonly

used medication for gout, to restore the energy flux towards normal in failing dog hearts, with an increase in contractility.[59]

Appropriate metabolic-based therapies include intense beta-blockade, which is the best tested,[60] and metabolic therapies targeting fatty acid and carbohydrate oxidation,[55,56] one of which is glucose-insulin-potassium (GIK),[51,61] with other targets at the level of malonyl CoA (Figure 2.9).[55] Of note, the metabolic abnormalities of severe heart failure may require long-term therapy by left ventricular assist devices that give partial reversal.[46]

MITOCHONDRIAL MECHANISMS IN HEART DISEASE

Ischemia Reperfusion Injury

Calcium-mediated cell death is the mechanism of lethal myocardial reperfusion injury and postulated to occur in advanced heart failure. In these conditions, opening of the mitochondrial permeability transition pore (mPTP) is regarded as crucial. Lethal programmed cell death (apoptosis) results from mitochondrial calcium overload when the mPTP opens as in reperfusion induced tissue necrosis.[57,62,63]

These basic science concepts received support from a major clinical study in which 58 patients given a bolus of intravenous cyclosporine, a blocker of the mPTP, at the time of percutaneous coronary intervention (PCI) for acute myocardial infarction, had a 20% reduction in measured infarct size.[60] Continuing research is focused on the multiple mechanisms of calcium-ion-mediated reperfusion cardiomyocyte death.[61]

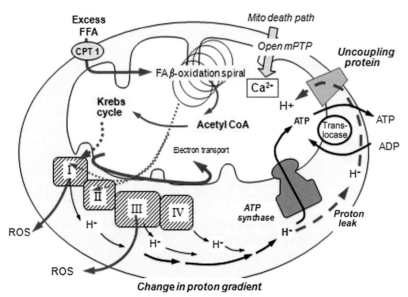

FIGURE 2.14 FFA effects on mitochondria. High circulating levels of free fatty acids (FFA), activated within the muscle cell to fatty acid CoA, enter the mitochondria by the action of CPT-1, thereby inhibiting the action of pyruvate dehydrogenase (PDH), blocking protective glycolysis that supplies cytosolic ATP used for ion pumps, and decreasing glucose uptake by muscle. Excess activated FFA increases formation of ROS. As protons are passed along the electron transport chain, ROS are formed at complexes I and III. FFA-induced proton leakage passes through uncoupling proteins which divert protons from ATP formation and hence waste oxygen. CPT 1, carnitine palmityl CoA transferase; mPTP, mitochondrial permeability transition pore; ROS, reactive oxygen species.

Heart Failure

Multiple metabolic and mitochondrial defects have been identified in heart failure. Elevated norepinephrine levels increase circulating free fatty acid levels that provoke a sequence of adverse events in mitochondria (Figure 2.14). Intramitochondrial long-chain fatty acyl-CoAs (LCFA-CoAs) increase in cardiac and peripheral muscle. When imported in excess into the mitochondrial matrix, LCFA-CoAs uncouple mitochondrial oxidative phosphorylation via UCPs. LCFA-CoAs also enter the mitochondrial fatty acid beta-oxidation (FAO) spiral to produce acetyl CoA that enters the Krebs cycle to provide acetyl CoA (Figure 2.1), which via the electron transmitter complexes I to IV, and the activity of ATP synthase, drives the synthesis of new ATP (Figure 2.14).

In heart failure, increased β-receptor stimulation and increased entry of excess FFA into the mitochondrial space, goes along with increasing the levels of mitochondrial reactive oxygen species (ROS) and activating uncoupling proteins. The specific lethal event is the opening of the mitochondrial permeability transition pore (mPTP).[62,63] These processes in turn dissipate the mitochondrial membrane potential, thereby diminishing mitochondrial ATP production and thus promoting the energy depletion characteristic of heart failure. Similar concepts to those established experimentally are also applicable to human heart failure.[64]

Looking to the future, human heart metabolism and energetics in heart failure are increasingly studied by metabolomics, an innovative technique that allows simultaneous measurements in model heart failure of all of the metabolites of glycolysis, including the pentose phosphate pathway, and of the Krebs cycle.[65] This technique promises to be the 'Face of the New Metabolism'.[11]

References

1. Evans CL. The effect of glucose on the gaseous metabolism of the isolated mammalian heart. *J Physiol* 1914;**47**:407–18.
2. Cruickshank EW, Kosterlitz HW. The utilization of fat by the aglycaemic mammalian heart. *J Physiol Lond* 1941;**99**:208.
3. Bing RJ, Siegel A, Ungar I, Gilbert M. Metabolism of the human heart. II. Studies on fat, ketone and amino acid metabolism. *Am J Med* 1954;**16**:504–15.
4. Shipp JC, Opie LH, Challoner D. Fatty acid and glucose metabolism in the perfused heart. *Nature* 1961;**189**:1018–9.
5. Randle PJ, Garland PB, Hales CN, Newsholme EA. The glucose fatty-acid cycle. Its role in insulin sensitivity and the metabolic disturbances of diabetes mellitus. *Lancet* 1963;**1**(7285):785–9.
6. Opie LH. Metabolism of the heart in health and disease. I. *Am Heart J* 1968;**76**:685–98.
7. Walters AM, Porter Jr GA, Brookes PS. Mitochondria as a drug target in ischemic heart disease and cardiomyopathy. *Circ Res* 2012;**111**:1222–36.
8. Stanley WC, Recchia FA, Lopaschuk GD. Physiol Rev. Myocardial substrate metabolism in the normal and failing heart. *Physiol Rev* 2005;**85**:1093–129.
9. Lopaschuk GD, Ussher JR, Folmes CD, Jaswal JS, Stanley WC. Myocardial fatty acid metabolism in health and disease. *Physiol Rev* 2010;**90**:207–58.
10. Rider OJ, Francis JM, Ali MK, Holloway C, Pegg T, Robson MD, et al. Effects of catecholamine stress on diastolic function and myocardial energetics in obesity. *Circulation* 2012;**125**:1511–9.
11. Taegtmeyer H. The new cardiac metabolism. *J Mol Cell Cardiol* 2013;**55**:1.
12. Opie LH. *Heart Physiology, From Cell to Circulation*. 4th ed. Philadelphia, PA: Lippincott, Williams & Wilkins; 2004. p. 308–54.
13. Mjos OD. Effect of free fatty acids on myocardial function and oxygen consumption in intact dogs. *J Clin Invest* 1971;**50**:1386–9.
14. Mjos OD. Effect of inhibition of lipolysis on myocardial oxygen consumption in the presence of isoproterenol. *J Clin Invest* 1971;**50**:1869–73.
15. Boardman NT, Larsen TS, Severson DL, Essop MF, Aasum E. Chronic and acute exposure of mouse hearts to fatty acids increases oxygen cost of excitation-contraction coupling. *Am J Physiol Heart Circ Physiol* 2011;**300**:H1631–6.
16. Oliver MF, Kurien VA, Greenwood TW. Relation between serum-free-fatty acids and arrhythmias and death after acute myocardial infarction. *Lancet* 1968;**1**(7545):710–4.
17. Cohn JN, Levine TB, Olivari MT, et al. Plasma norepinephrine as a guide to prognosis in patients with chronic congestive heart failure. *N Engl J Med* 1984;**311**:819–23.
18. Smeele KM, Southworth R, Wu R, et al. Disruption of hexokinase II-mitochondrial binding blocks ischemic preconditioning and causes rapid cardiac necrosis. *Circ Res* 2011;**108**:1165–9.
19. Katz A. *Physiology of the Heart*. 4th ed. Philadelphia: Lippincott, Williams and Wilkins. p. 365–8.
20. Vetter NJ, Strange RC, Adams W, Oliver MF. Initial metabolic and hormonal response to acute myocardial infarction. *Lancet*. 1974;**1**(7852):284–8.
21. Zhang L, Keung W, Samokhvalov V, Wang W, Lopaschuk GD. Role of fatty acid uptake and fatty acid beta-oxidation in mediating insulin resistance in heart and skeletal muscle. *Biochim Biophys Acta* 2010;**1801**:1–22.
22. Jennings RB, Reimer KA, Steenbergen C. Myocardial ischemia revisited. The osmolar load, membrane damage, and reperfusion. *J Mol Cell Cardiol* 1986;**18**:769–80.
23. Cross HR, Radda GK, Clarke K. The role of Na+/K+ ATPase activity during low flow ischemia in preventing myocardial injury: a 31P, 23Na and 87Rb NMR spectroscopic study. *Magn Reson Med* 1995;**34**:673–85.
24. Heusch G, Schultz R. A radical view on the contractile machinery in human heart failure. *J Am Coll Cardiol* 2011;**57**:310–12.
25. Vilahur G, Juan-Babot O, Peña E, Oñate B, Casaní L, Badimon L. Molecular and cellular mechanisms involved in cardiac remodeling after acute myocardial infarction. *J Mol Cell Cardiol* 2011;**50**:522–33.
26. Lubbe WF, Podzuweit T, Opie LH. Potential arrhythmogenic role of cyclic adenosine monophosphate (AMP) and cytosolic calcium overload: implications for prophylactic effects of beta-blockers in myocardial infarction and proarrhythmic effects of phosphodiesterase inhibitors. *J Am Coll Cardiol* 1992;**19**:1622–33.
27. Opie LH, Lecour S, Mardikar H, Desphande G. Cardiac survival strategies: an evolutionary hypothesis with rationale for metabolic therapy of acute heart failure. *Transact Royal Soc South Africa* 2010;**65**:185–9.
28. Yellon DM, Hausenloy DJ. Myocardial reperfusion injury. *N Engl J Med* 2007;**357**:1121–35.
29. Fuglesteg BN, Suleman N, Tiron C, et al. Signal transducer and activator of transcription 3 is involved in the cardioprotective signalling pathway activated by insulin therapy at reperfusion. *Basic Res Cardiol* 2008;**103**:444–53.

30. Lecour S. Multiple protective pathways against reperfusion injury: a SAFE path without Aktion? *J Mol Cell Cardiol* 2009;**46**:607–9.

31. Opie LH, Walfish PG. Plasma free fatty acid concentrations in obesity. *N Engl J Med* 1963;**268**:757–60.

32. Belfort R, Mandarino L, Kashyap S, et al. Dose-response effect of elevated plasma free fatty acid on insulin signaling. *Diabetes* 2005;**54**:1640–8.

33. Mathew M, Tay E, Cusi K. Elevated plasma free fatty acids increase cardiovascular risk by inducing plasma biomarkers of endothelial activation, myeloperoxidase and PAI-1 in healthy subjects. *Cardiovasc Diabetol* 2010;**9**:9. doi:10.1186/1475-2840-9-9.

34. Quilley J. Oxidative stress and inflammation in the endothelial dysfunction of obesity: a role for nuclear factor kappa B. *J Hypertens* 2010;**28**:2010–1.

35. Shoelson SE, Lee J, Goldfine AB. Inflammation and insulin resistance. *J Clin Invest* 2006;**116**:1793–801.

36. Zeyda M, Stulnig TM. Obesity, inflammation, and insulin resistance– a mini-review. *Gerontology* 2009;**55**:379–86.

37. Yi CX, Al-Massadi O, Donelan E, et al. Exercise protects against high-fat diet-induced hypothalamic inflammation. *Physiol Behav* 2012;**106**:485–90.

38. Robertson RP. Chronic oxidative stress as a central mechanism for glucose toxicity in pancreatic islet beta cells in diabetes. *J Biol Chem* 2004;**279**:42351–4.

39. Nishikawa T, Edelstein D, Du XL, et al. Normalizing mitochondrial superoxide production blocks three pathways of hyperglycaemic damage. *Nature* 2000;**404**:787–900.

40. Young ME, McNulty P, Taegtmeyer H. Adaptation and maladaptation of the heart in diabetes: Part II: potential mechanisms. *Circulation* 2002;**105**:1861–70.

41. Brownlee M. Biochemistry and molecular cell biology of diabetic complications. *Nature* 2001;**414**:813–20.

42. Bajaj M, Baig R, Suraamornkul S, et al. Effects of pioglitazone on intramyocellular fat metabolism in patients with type 2 diabetes mellitus. *J Clin Endocrinol Metab* 2010;**95**:1916–23.

43. Tuunanen H, Knuuti J. Metabolic remodelling in human heart failure. *Cardiovasc Res* 2011;**90**:251–7.

44. Kolwicz Jr SC, Olson DP, Marney LC, et al. Cardiac-specific deletion of acetyl CoA carboxylase 2 prevents metabolic remodeling during pressure-overload hypertrophy. *Circ Res* 2012;**111**:728–38.

45. He L, Kim T, Long Q, et al. Carnitine palmitoyltransferase-1b deficiency aggravates pressure overload-induced cardiac hypertrophy caused by lipotoxicity. *Circulation* 2012;**126**:1705–16.

46. Chokshi A, Drosatos K, Cheema FH, et al. Ventricular assist device implantation corrects myocardial lipotoxicity, reverses insulin resistance, and normalizes cardiac metabolism in patients with advanced heart failure. *Circulation* 2012;**125**:2844–53.

47. DeFronzo RA. Insulin resistance, lipotoxicity, type 2 diabetes and atherosclerosis: the missing links. The Claude Bernard Lecture 2009. *Diabetologia* 2010;**53**:1270–87.

48. Cao C, Chen Y, Wang W, et al. Ghrelin inhibits insulin resistance induced by glucotoxicity and lipotoxicity in cardiomyocyte. *Peptides* 2011;**32**:209–15.

49. El-Assaad W, Joly E, Barbeau A, et al. Glucolipotoxicity alters lipid partitioning and causes mitochondrial dysfunction, cholesterol, and ceramide deposition and reactive oxygen species production in INS832/13 ss-cells. *Endocrinology* 2010;**151**:3061–73.

50. Kitabchi AE, Umpierrez GE, Miles JM, et al. Hyperglycemic crises in adult patients with diabetes. *Diabetes Care* 2009;**32**:1335–43.

51. Grossman AN, Opie LH, Beshansky JR, et al. Glucose-insulin-potassium revived: current status in acute coronary syndromes and the energy-depleted heart. *Circulation* 2013;**127**:1040–8.

52. Opie LH, Knuuti J. The adrenergic-fatty acid load in heart failure. *J Am Coll Cardiol* 2009;**54**:1637–46.

53. Rame JE. Chronic heart failure: a reversible metabolic syndrome? *Circulation* 2012;**125**:2809–11.

54. Lehnart SE, Maier LS, Hasenfuss G. Abnormalities of calcium metabolism and myocardial contractility depression in the failing heart. *Heart Fail Rev* 2009;**14**:213–24.

55. Jaswal JS, Keung W, Wang W, Ussher JR, Lopaschuk GD. Targeting fatty acid and carbohydrate oxidation–a novel therapeutic intervention in the ischemic and failing heart. *Biochim Biophys Acta* 2011;**1813**:1333–50.

56. Ashrafian H, Frenneaux MP, Opie LH. Metabolic mechanisms in heart failure. *Circulation* 2007;**116**:434–48.

57. Neubauer S. The failing heart: an engine out of fuel. *N Engl J Med* 2007;**356**:1140–51.

58. Smith CS, Bottomley PA, Schulman SP, Gerstenblith G, Weiss RG. Altered creatine kinase adenosine triphosphate kinetics in failing hypertrophied human myocardium. *Circulation* 2006;**114**:1151–8.

59. Opie LH. Allopurinol for heart failure: novel mechanisms. *J Am Coll Cardiol* 2012;**59**:809–12.

60. Wallhaus TR, Taylor M, DeGrado TR, Russell DC, Stanko P, Nickles RJ, et al. Myocardial free fatty acid and glucose use after carvedilol treatment in patients with congestive heart failure. *Circulation* 2001;**103**:2441–6.

61. Howell NJ, Ashrafian H, Drury NE, et al. Glucose-insulin-potassium reduces the incidence of low cardiac output episodes after aortic valve replacement for aortic stenosis in patients with left ventricular hypertrophy: results from the Hypertrophy, Insulin, Glucose, and Electrolytes (HINGE) trial. *Circulation* 2011;**123**:170–7.

62. Hausenloy DJ, Duchen MR, Yellon DM. Inhibiting mitochondrial permeability transition pore opening at reperfusion protects against ischaemia-reperfusion injury. *Cardiovasc Res* 2003;**60**:617–25.

63. Hausenloy DJ, Ong SB, Yellon DM. The mitochondrial permeability transition pore as a target for preconditioning and postconditioning. *Basic Res Cardiol* 2009;**104**:189–202.

64. Piot C, Croisille P, Staat P, et al. Effect of cyclosporine on reperfusion injury in acute myocardial infarction. *N Engl J Med* 2008;**359**:473–81.

65. Kato T, Niizuma S, Inuzuka Y, et al. Analysis of metabolic remodeling in compensated left ventricular hypertrophy and heart failure. *Circ Heart Fail* 2010;**3**:420–30.

CHAPTER

3

Cardiac Atrophy and Remodeling

Pamela A. Harvey, PhD, Leslie A. Leinwand, PhD

University of Colorado at Boulder, Boulder, CO, USA

OVERVIEW OF ATROPHIC CARDIAC REMODELING

Cardiac remodeling encompasses the many biochemical and molecular adaptations that the heart initiates in response to altered demand (Figure 3.1). Atrophic remodeling occurs in response to ventricular or hemodynamic unloading caused by such stimuli as placement of an LVAD, surgical correction of cardiac valvular disease, bariatric surgery or pharmacological intervention in patients with hypertension. Upon unloading, the myocardium decreases in size but contractile function is generally maintained. Reduced cardiomyocyte size is responsible for the decrease in myocardial mass; cell death does not significantly contribute to the phenotype.[1] Because of the interest in the molecular and functional pathology involved in reversing cardiomyocyte size, the mechanisms responsible for atrophic remodeling have been studied in both patients and in experimental animal models of ventricular unloading. In particular, the hearts of patients receiving LVADs and experimental animals receiving a heterotopic heart transplant have been examined.

Surprisingly, many of the mechanisms and expression of molecular hallmarks that result from these stimuli are shared between hypertrophic and atrophic cardiac remodeling despite the opposing trophic adaptation. Cellular processes regulating cardiac gene expression are altered including increased signaling through pathways that promote proteolysis as well as either increased or decreased activity of pro-growth pathways. Proteolytic pathways are generally more highly activated and therefore, degradative processes are overall favored within atrophic cardiomyocytes. The mechanisms mediating cardiac atrophy are highly stimulus-specific. For example, the ubiquitin-proteasome system (UPS), which is principally responsible for protein degradation in the heart, is up-regulated while mTOR-mediated protein translation is simultaneously activated in heterotopic

heart transplantation in rats.[2] However, in rabbits that exhibit cardiac atrophic remodeling due to fasting, global protein degradation is increased while synthesis of myofibrillar proteins is inhibited.[3] Finally, fetal gene expression is up-regulated, metabolic genes mediating glucose transport are down-regulated, and expression of sarcomeric proteins is dysregulated similarly in atrophy and hypertrophy.[1]

We will present an overview of general mechanisms mediating cardiac atrophy followed by several examples of atrophic remodeling with different stimuli. We will focus first on mechanisms mediating the ability of the heart to detect alterations in hemodynamic load and the morphological and histological changes observed in atrophic remodeling. We will then discuss the biochemical and molecular pathways responsible for protein degradation and synthesis and for metabolism in the atrophic myocardium. Finally, we will present clinical and molecular data specific to LVAD- and cancer-induced atrophic remodeling.

MODELS OF ATROPHIC REMODELING

Although initially developed to examine the process of immune rejection of transplanted organs, heterotopic heart transplantation has become the main model of ventricular unloading in experimental animals. In this model, a healthy heart is transplanted into the abdominal cavity through anastomosis of the ascending aorta and pulmonary artery associated with the donor heart to the recipient abdominal aorta and inferior vena cava, respectively (Figure 3.2A). The effect of the transplantation is similar to aorto-caval fistula in that blood is diverted into the donor heart, thus reducing pressure entering the heart of the recipient.[4] This technique is commonly used to examine the temporal, molecular, and biochemical responses to ventricular unloading that parallels the cardiac

Cellular and Molecular Pathobiology of Cardiovascular Disease
http://dx.doi.org/10.1016/B978-0-12-405206-2.00003-X

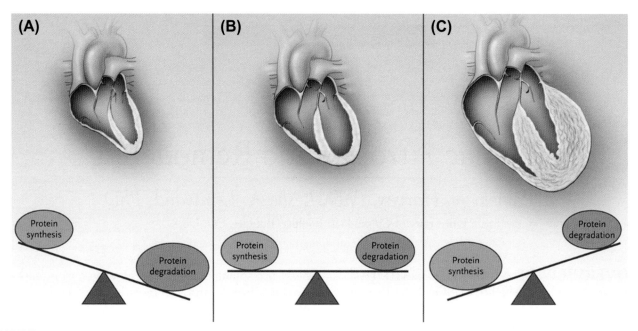

FIGURE 3.1   Cardiac remodeling involves increased and decreased growth mediated by an imbalance between protein synthesis and degradation. (A) In atrophic remodeling, protein degradation is favored to produce smaller cardiomyocytes and reduced myocardial size. (B) In the healthy heart, protein synthesis is balanced by protein degradation such that they occur equally and simultaneously. (C) In the hypertrophic heart, protein synthesis outweighs protein degradation to increase cardiomyocyte and myocardial size. Adapted from Willis et al., New Engl J Med (2013).[119]

unloading seen in patients with LVADs that mechanically unloads the heart.

Permanent implantation of portable LVADs began in the 1980s as a bridge to heart transplantation for patients diagnosed with refractory heart failure.[5] Recently, LVADs have been transplanted for longer periods with the aim of promoting reversal of cardiac pathology and removing the LVAD as an alternative to heart transplantation (see 'Atrophic remodeling as a potential therapeutic,' below for additional details).[6] LVADs are battery-operated pumps that are implanted into the patient's abdominal wall or peritoneal cavity; battery packs are worn externally (Figure 3.2B).[7] Two conduits are implanted: one into the left ventricular apex and one into the ascending aorta to unload the left ventricle by moving blood out and bypassing the systemic circulation. Generally, LVADs can respond to cardiac demand and are tolerated well; complications are mostly limited to infection, bleeding, and stroke.[6] Much of what is known about atrophic cardiac remodeling in patients was learned from studying patients receiving LVADs.

CARDIAC WORKLOAD DETERMINES CARDIAC SIZE

Although the heart has a large capacity to increase in mass, there is an upper limit mainly due to the lack of significant proliferative capacity of terminally differentiated adult cardiomyocytes.[8] Contractility may be changed, muscle mass may be altered, or actin–myosin crossbridge formation may be adjusted through differential expression of sarcomeric proteins.[9] Each of these mechanisms is in place to offset abnormal transmural pressure that causes or is a result of inefficient cardiomyocyte contractility. In atrophic remodeling, cardiomyocytes respond to reduced cardiac load by utilizing these mechanisms.

Individual myocardial cells must adapt to a new pattern of force and distortion during ventricular unloading. This response is mediated by complex interactions between the extracellular matrix (ECM) and cells in the myocardium. Components of the complex ECM that surround both myofibrils and cardiomyocytes are physically connected to the interior of the cardiomyocyte to transduce force, which also allows the translation of force into biochemical signals through the association of adaptor and signaling proteins. The signal is therefore translated through altered protein–protein interactions and is conveyed to the nucleus where gene expression is modified and functional alterations are initiated.

Numerous proteins participate in mechanosensation and mediate initiation of signaling pathways that alter cardiomyocyte form and function. At the costamere, melusin and integrin-linked kinase (ILK) interact with the cytoplasmic tail of β1-integrin, a predominant component of the cardiac ECM.[10] ILK recruits several adaptor proteins to the costamere and activates downstream signaling proteins to alter gene expression, thus

(A)

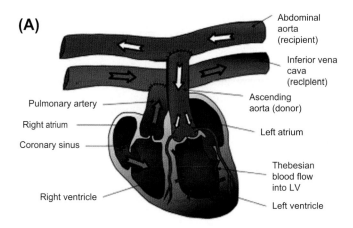

(B)

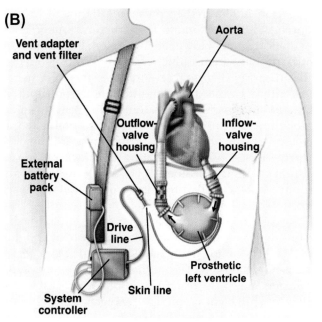

FIGURE 3.2 (A) Schematic of heterotopic heart transplant. To reduce cardiac load, a donor heart is transplanted into a recipient. Blood is shunted from the abdominal aorta or the recipient through the ascending aorta associated with the donor heart and through the donor heart. The donor heart returns blood through its associated pulmonary artery to the recipient inferior vena cava, thus reducing workload in both the donor and recipient hearts. (B) Schematic of left ventricular assist device. A prosthetic left ventricular pump is implanted that receives blood from the left ventricle and returns it via peristaltic or continuous pressure to the aorta. A battery pack attached to a harness is worn externally by the patient.

contributing to regulation of cardiomyocyte size.[11] Integrin activation also initiates focal adhesion kinase (FAK) signaling to induce transcriptional activity through mitogen-activated protein kinase (MAPK) and c-Jun N-terminal kinase (JNK) as well as Rho, which promotes actin reorganization. Activation of these signaling pathways induces alterations in gene expression that allow efficient use of metabolic substrates and maintain cardiac contractility (discussed below). As such, mechanical stress largely determines the morphological and functional plasticity of the cardiomyocyte.

Although the ability of clinicians to directly measure regional transmural pressure would assist in predicting cardiac remodeling and potentially avert functional decompensation and heart failure, accurate measurements of transmural pressure in vivo are limited by the mechanical complexity of the heart. However, mathematical modeling to quantify morphological alterations dictating intraventricular pressure and wall stress is possible using the law of Laplace. This principle demonstrates that wall thickness is indirectly proportional to ventricular wall stress and therefore makes accurate predictions about cardiomyocyte size in response to pressure changes;[1] as wall stress is reduced through a stimulus like unloading, cardiomyocyte size is reduced to normalize (increase) transmural pressure.

The law of Laplace has been exploited in the development of medical devices that are currently used to mimic increased hemodynamic load by physically changing the shape of the heart. For example, artificial alteration of transmural pressure using an Acorn support device that constricts the heart with a mesh cylinder,[12] or the Myocor Myosplint that pulls the sides of the ventricle toward one another,[13] augments signal transduction consistent with increased mechanical stress in cardiomyocytes. These studies reported promising results suggesting that physically manipulating heart size and, by extension, altering transmural pressure can limit morphological and molecular dysfunction caused by atrophic remodeling.

MORPHOLOGICAL FEATURES OF THE ATROPHIC HEART

Interestingly, in contrast to the mechanisms mediating ventricular wall thinning and chamber dilation such as cell death and cardiomyocyte elongation, atrophic remodeling is a reversible process that maintains cardiac function by reducing cardiomyocyte mass.[14] Cardiac atrophy can be identified in vivo using echocardiography to visualize loss of myocardial mass or histologically by measuring myocyte cross-sectional area. With mechanical unloading, reduced myocardial size is attributable to a reduction in cardiomyocyte mass rather than to cell death.[15,16] Reduced cardiomyocyte volume has been attributed to degradation of sarcomeric proteins in parallel, rather than preferential loss of a subset of proteins.[17] Accompanying this reduction is an alteration in the innervation of the left ventricle such that fewer than half the number of ramifying axons are observed compared to a healthy heart.[18] Myofibrillar disarray caused by the altered relationship between cardiomyocytes and the ECM is also present along with interstitial fibrosis, both considered to be deleterious to cardiac function.[19] The net effect is reduced myocardial mass and increased wall stiffness in the atrophic heart.

EXTRACELLULAR MATRIX REMODELING WITH CARDIAC ATROPHY

The ECM of cardiac tissue is comprised of type I collagen, elastin, proteoglycans, and glycoproteins and is integral to the maintenance of tensile strength and myofibrillar organization of cardiac muscle.[20] However, this collagenous matrix is not merely a passive structural component of the myocardium; it participates in mechanosensation by physically connecting the ECM to intracellular cytoskeletal elements in cardiomyocytes,[21] converting mechanical information to biochemical signals that promote functional alterations in cardiac output,[22] and organizing cardiomyocytes into functional units responsible for efficient contraction.[23] Because intimate connections between the cardiomyocyte and ECM are required to coordinate contraction of the myocardium, cardiac structural remodeling cannot be limited to the cardiomyocytes; decreased cardiomyocyte size alters the relationship between the cell and the ECM. For this type of architectural change to take place, degradation of existing matrix proteins followed by expression of new matrix proteins must occur. Expression of matrix metalloproteinases (MMPs) that degrade ECM proteins is regulated transcriptionally in response to alterations in myocardial stretch but do not act unopposed. Tissue inhibitors of metalloproteinases (TIMPs) form complexes with MMPs and inactivate the catalytic domain of MMPs. Therefore, TIMP expression in the myocardium plays an important role in modulating the activity of MMPs. Interestingly, although MMPs are up-regulated in heart failure, expression of MMPs is decreased and TIMPs are increased in rats following heterotopic heart transplant.[24]

Several soluble regulatory factors are released in the atrophic myocardium by cardiomyocytes, fibroblasts, and infiltrating immune cells, and induce increased production of the cardiac matrix with cardiac remodeling. Among these factors, the cytokine osteopontin is required for fibroblast differentiation into α-smooth muscle actin-expressing myofibroblasts that produce ECM components, which contribute to cardiac fibrosis.[25,26] Expression of osteopontin is increased in the mechanically unloaded human heart.[27] Expression of the fetal genes atrial natriuretic factor (ANF) or B-type natriuretic peptide (BNP, discussed below), which are also up-regulated in atrophic remodeling, has been shown to oppose myofibroblast activity, supporting the conclusion that initial responses to hemodynamic unloading may be cardioprotective in nature.[28]

Activated myofibroblasts enhance the synthesis and secretion of type I fibrillar collagen.[29] Matrix production is required to preserve alignment of cardiomyocytes and maintain the structural integrity of the myocardium. However, increased expression of matrix proteins in atrophic remodeling causes myofibrillar disarray (Figure 3.3), and excessive ECM negatively affects cardiac function through its contribution to tissue stiffness.[23,30,31] The transition from atrophic remodeling to heart failure is likely attributable to the development of fibrotic foci in the myocardium.[32–34] In fact, fibrotic tissue emanating from scarred myocardium entangles cardiomyocytes and can exacerbate atrophy and actively promote progression to heart failure.[35,36]

Excessive myocardial matrix production also negatively impacts cardiac function and propensity for arrhythmias due to conduction slowing or block (electrical remodeling).[37,38] The stiffness of cross-linked type I collagen may also restrict arterioles responsible for delivering sufficient amounts of oxygen to the highly metabolic cardiac tissue, leading to a metabolic crisis secondary to mild hypoxia.[39,40] Because myofibroblasts are physically connected to cardiomyocytes via gap junctions,[23] abnormal electrical coupling between myofibroblasts and cardiomyocytes can cause depolarizing

(A) 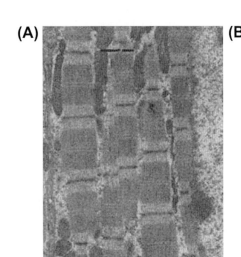   **(B)**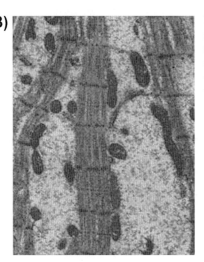

FIGURE 3.3 Electron micrographs of atrophic remodeling in feline papillary muscles after transection of the chordae tendineae. Longitudinal sections of (A) healthy papillary muscles, and (B) after one week on ventricular unloading. Adapted from Kent el al., JMCC (1985).[30]

currents in neighboring cardiomyocytes.[41] Although it may seem logical to disrupt the development of fibrosis as a therapeutic intervention to prevent increased stiffness, myofibrillar disarray, and contractile dysfunction, it is important to emphasize that the development of fibrosis is thought to be important in healing; attempts thus far to inhibit fibrosis have led to increased lethality in experimental animals.[42]

PROTEIN HOMEOSTASIS IN THE HEALTHY AND ATROPHIC HEART

In the healthy heart, protein degradation must be coordinated with protein synthesis to maintain organ size and orchestrate the process of tissue growth or regression and atrophy. Both processes are critical to maintenance of the appropriate protein turnover and stoichiometry of sarcomere components. Protein degradation is required for the maintenance of normal amounts and structures of proteins, especially when there is increased protein synthesis,[43] and protein synthesis is required to replace these degraded proteins. In atrophic cardiac remodeling, this balance is disrupted such that degradation exceeds synthesis.

Protein degradation is achieved via three proteolytic systems in the heart: the ubiquitin proteasome system (UPS), the autophagic-lysosomal pathway, and calcium-dependent cysteine proteases, the calpains. While the UPS is principally involved in the degradation of both normal and misfolded proteins in the cytoplasm, the lysosome degrades membrane-associated proteins.[44] The role of calpains in degradation of specific myofibrillar proteins is controversial in light of recent data demonstrating a requirement for prior disassembly of the sarcomere.[45] Each of these systems is involved in both hypertrophic and atrophic remodeling, however, the balance between the synthesis and degradation of proteins determines the overall cellular and tissue effects.

The UPS is the principal proteolytic system in the heart, mediating 80% of the proteolysis in the cardiomyocyte[46] and is the primary mechanism by which appropriate sarcomere stoichiometry and protein quality are maintained.[44] Ubiquitin monomers are enzymatically conjugated to misfolded or damaged proteins by ubiquitin ligases to form polymeric ubiquitin chains, marking them for degradation in the multi-subunit proteasome.[47] Ubiquitination offers the benefit of specificity thus preserving sarcomere architecture or altering the components in a productive manner.

The UPS is activated in atrophic hearts mainly through the up-regulation of expression of ubiquitin B and UbcH2, a ubiquitin-conjugating (E2) enzyme.[48] The roles of ubiquitin ligases (E3) have been demonstrated indirectly in hypertrophic remodeling,[49] however, expression of the ubiquitin ligases Atrogin-1 and muscle RING finger 1 (MuRF-1) in atrophic remodeling is decreased.[48] Although unexpected, this decrease is likely due to the increased expression of insulin-like growth factor 1 (IGF-1) that occurs in the heart with mechanical unloading;[50] IGF-1 signaling in cardiomyocytes inhibits the FoxO forkhead transcription factor family member FoxO3a, which positively regulates expression of both Atrogin-1 and MuRF-1.[51] MuRF-1 levels increase in the hearts of patients with LVAD placement who experience therapeutic cardiac atrophy, suggesting a role for MuRF-1 in atrophic remodeling. Indeed, in mice lacking expression of MuRF-1 (MuRF-1$^{-/-}$), dexamethasone, a synthetic glucocorticoid that induces cardiac atrophy, does not induce loss of cardiac mass. Unloading the heart through release of transaortic constriction that reverses cardiac hypertrophy in wild-type mice also does not result in a reduction in cardiac mass in MuRF-1$^{-/-}$ mice.[49] Interestingly, activity of the UPS also increases with cardiac hypertrophy and is required for hypertrophic remodeling, likely due to the requirement for increased surveillance of misfolded proteins in the context of increased protein production.[52] Taken together, the role of the UPS in atrophic remodeling in the heart is complex and unresolved. Increased UPS activity could also set the stage for imbalanced protein synthesis and degradation that leads to atrophy and hypertrophy, the direction of which is determined by other regulatory factors.

Several models of cardiac pathological and physiological hypertrophy implicate activity of the protein phosphatase calcineurin in the development of hypertrophic cardiac remodeling.[53,54] Persistently elevated intracellular calcium promotes interaction between calmodulin and calcineurin and activates the phosphatase.[55] Activated calcineurin dephosphorylates the transcription factor nuclear factor of activated T cells (NFAT) allowing nuclear translocation and induction of pro-growth gene expression. It is thought that calcineurin is a reciprocal regulator of the activity of pro-atrophic pathways to coordinate the cardiac response. In atrophic remodeling, FoxO1 and FoxO3a reduce calcineurin phosphorylation and expression of calcineurin/NFAT pathway targets including modulatory calcineurin-interacting protein 1.4 (MCIP1.4). Indeed, studies in transgenic mice that overexpress FoxO3a demonstrate that increased repression of calcineurin/NFAT pathway in a hypertrophic heart reduces heart mass[56] and in a healthy heart, causes atrophic remodeling.[57]

Evidence also suggests an important role for calcium-dependent calpain proteases in both UPS activity and in degradation of myofibrillar proteins during atrophic remodeling. Expression of calpains 1 and 2 is up-regulated following ventricular unloading of the rat and human heart.[58] Calpains, for example, degrade specific proteins including titin and α-actinin[59] that are responsible for sensing intercellular load; this degradation is thought to destabilize myofibrils and precede degradation mediated by the UPS.[45] However, despite increased activity of calpains during atrophic remodeling, genetic inhibition of calpains during murine ventricular unloading does not attenuate cardiac atrophy, demonstrating that the UPS is indeed the main mechanism by which proteins are degraded in atrophic cardiomyocytes.[58]

METABOLIC UNLOADING OF THE MYOCARDIUM

Recent evidence suggests that cardiac remodeling occurs primarily as a response to metabolic crisis, which is initiated first when mechanical unloading occurs.[1,60] Metabolic remodeling occurs in response to many forms of physiological and pathological stimuli to alter both the preference for substrate type and efficiency of substrate use. Both starvation and cancer induce rapid degradation of proteins to provide substrates for gluconeogenesis in the liver as well as in the heart, and both are associated with significant cardiac atrophy.

Reducing cardiomyocyte size while maintaining cardiac contractility alters the metabolic demands on the cell. As in the context of genetic or extrinsic cardiac stressors, utilization of energy substrates begins to recapitulate those of fetal development during atrophic remodeling. The adult myocardium re-expresses proteins that favor use of glucose through down-regulation of genes expressed in the adult and induction of genes expressed during fetal development.[61] Cardiomyocytes are capable of using both glucose and fatty acids, yet β-oxidation of fatty acids accounts for about 70% of the adenosine triphosphate (ATP) in the healthy adult myocardium.[62] However, per molecule of ATP, oxidation of glucose requires less oxygen, thus increasing efficiency up to 25%[63] so, ultimately, the shift to glucose may simply be more bioenergetically favorable in stressed cells. A shift in substrate preference for glucose over fatty acids is promoted by chronic cardiac stress as well as alterations in substrate availability, oxygen, contractile properties of cardiomyocytes, or hormonal signaling.[60] Therefore, in the heart undergoing atrophic remodeling, contractility improves when glucose is favored as a substrate and oxygen consumption is reduced.[63–65]

Alterations in transcription patterns of metabolic genes account for the substrate preference shift from fatty acids to glucose in the atrophic heart. Because convincing evidence for an increase in expression of glycolytic enzymes in cardiomyocytes has not been demonstrated, it has been suggested that the shift to glucose utilization is mediated post-translationally.[60] Alternatively, it has also been postulated that during atrophic remodeling, mitochondrial damage caused by increased mitochondrial uncoupling protein activity or increased production of reactive oxygen species reduces ATP production and leads to metabolic dysfunction and heart failure.[66] Although the precise temporal activation of metabolic remodeling during cardiac atrophy has remained largely elusive, it is clear that metabolic dysfunction follows alterations in expression of genes encoding transport and metabolic proteins. During cardiac atrophy, expression of peroxisome proliferator-activated receptor-α (PPARα), which promotes expression of several key enzymes involved in fatty acid esterification and oxidation as well as uptake of fatty acids,[67] is down-regulated.[68] Additionally, uncoupling protein (UCP)-2 and -3, mitochondrial transmembrane proteins that create a proton leak that uncouples ATP synthesis from the electron transport chain,[69] is also decreased in atrophic remodeling.[70] In response to mechanical unloading, the heterotopically transplanted hearts from rats also express fewer complex I, II, and IV proteins in the mitochondrial respiratory chain.[71] Indeed, PPARγ-coactivator-1α (PGC-1α) activity is decreased in both atrophic[71] and in a subset of heart failure models,[72] and this reduction leads to reduced mitochondrial gene expression and reduced oxidative capacity. This may represent the source of the metabolic crisis in atrophied hearts. Additionally, the inability to meet the immediate energetic demands of a reloaded heart may explain why explantation of LVADs is not currently a widely used therapy in lieu of heart transplantation.

Recently, a link between reduced ATP levels and protein degradation in the atrophic heart has been proposed. A reduced ATP-to-AMP ratio in cardiomyocytes activates adenosine monophosphate activated kinase (AMPK) signaling. Baskin and colleagues demonstrated in a starvation-induced model of cardiac atrophy that activated AMPK increases transcription of both MuRF1 and Atrogin-1 and inhibits the activity of mTOR.[73] This novel mechanism mediated by metabolic unloading sheds new light on the many factors regulating heart size.

SIGNALING PATHWAYS ACTIVATED DURING CARDIAC ATROPHY

Recent work suggests the pathways mediating atrophic response in the heart are unique to specific stimuli. Interestingly, examination of signaling proteins involved in cardiac hypertrophy has provided valuable information about the signaling pathways responsible for induction of cardiac atrophic remodeling. MAPKs including ERK, p38-MAPK, and JNK, for example, promote cardiac hypertrophy,[74] and these pathways form the basis of most studies in animal models of atrophic remodeling.

Mechanically unloading the heart via heterotopic transplantation in mice and rats is a common model in which to study cardiac atrophic processes. The most clinically relevant are those studies that heterotopically transplant a pressure-overloaded hypertrophic heart; this model mimics placement of an LVAD in patients with pathological cardiac hypertrophy in need of a heart transplant. In this model, there is reduced phosphorylation of several signaling proteins that mediate cardiac hypertrophy including ERK, NFκB, and Akt.[75] Heterotopic transplantation of a hypertrophic heart also induces transcriptional changes that include expression of ubiquitin B (UbB) that may be responsible for increased protein degradation by the proteasome.[48] These data suggest that reverse remodeling may be achieved by targeting the activity of these signaling proteins. However, unlike in patients receiving LVADs and despite strong evidence for an increase in UPS activity, expression of MuRF-1 and Atrogin-1 is not increased with mechanical unloading; continued examination of both hearts from patients and animal models is required to fully elucidate the signaling mechanisms responsible for atrophic remodeling with mechanical unloading.

In mice bearing colon-26 tumors, MuRF-1 and Atrogin-1 are induced, and the mechanism mediating the increase in expression involves activation of MAPKs including p38 and p44.[76] Up-regulation of these ubiquitin ligases results in degradation of sarcomeric proteins, such that all are degraded equally. Interestingly, however, others report that protein degradation in the colon-26 tumor model is not attributable to up-regulation of the UPS, but rather to increased autophagy.[17] Circulating cytokines that are increased with atrophic stimuli including some forms of cancer, as well as sepsis, starvation, and certain viral infections have been shown to directly mediate atrophic remodeling.[17,77] The precise mechanisms that mediate the effects of proinflammatory cytokines have not been fully elucidated but much has been learned from three decades' research on tumor-bearing mice that exhibit cardiac atrophy. Cytokines released from colon-26 adenocarcinomas such as IL-6, TNFα, and IL-6 bind to their respective receptors on cardiomyocytes and activate NFκB,[86] which promotes the expression of MuRF-1 and leads to cardiac atrophy that is associated with degradation of thick filament components of the cardiac sarcomere. In fact, cardiac levels of TNFα are elevated in heart failure, and TNFα is sufficient to induce much of the cardiac pathology present in this condition.[87] Certain systemic and cardiotropic viral infections that increase levels of circulating cytokines and cause myocarditis are also associated with atrophic remodeling through degradation of cardiac muscle. With exposure to interferon-γ, which is up-regulated in cardiotropic infections including Chagas' disease and human immunodeficiency virus, neonatal rat ventricular myocytes exhibit increased UPS activity and a specific degradation of myosin heavy chain.[77] Thus, although several atrophic stimuli result in increased cytokine levels both in the heart and blood, the mechanisms mediating cardiac atrophy in these settings are distinct.

MOLECULAR ALTERATIONS IN ATROPHIC REMODELING: THE FETAL GENE PROGRAM

Both positive and negative regulation of remodeling are required for appropriate adaptation to cardiac stress. Expression of genes such as ANF and BNP is a hallmark of cardiac stress induced by injury, genetic mutation of sarcomeric proteins, or metabolic alterations. Interestingly, this pattern of re-expression of genes normally expressed exclusively during development is recapitulated in both hypertrophic and atrophic remodeling.

ANF and BNP are synthesized as preprohormones and stored in granules that are released upon tissue stretch; ANF and BNP are both expressed in the atria. While ANF levels in the ventricles are constitutively low, BNP is transcriptionally regulated in the ventricles. Although BNP expression is significantly lower than ANF, expression can increase 200-fold during heart failure.[78] Both ANF and BNP are released into the circulation and affect many organ systems including the brain, kidneys, and vasculature. In addition to the commonly known functions of reducing blood pressure by inducing relaxation of smooth muscle and reducing blood volume by promoting diuresis by the kidneys, these peptides also inhibit sympathetic nervous system activity in the heart, which is activated in response to cardiac unloading.[79]

Cardiovascular effects of ANF and BNP are achieved by binding to the oligomeric natriuretic peptide receptors, which signal through cGMP. In smooth muscles of the vasculature, cGMP activates endothelial cGMP-dependent protein kinase (PKG1), which reduces intracellular calcium concentrations thereby reducing contractility. PKG1 also increases calcium sensitivity of myosin light chain phosphatase to maintain appropriate contraction of the cardiomyocytes and prevent uncontrolled relaxation.[80,81] This important adaptation is important in a milieu of reduced intracellular calcium in atrophic remodeling.

Expression of the myosin isoform normally expressed during fetal development is also up-regulated with cardiac atrophy. In the healthy adult heart, the ratio of α- and β-myosin heavy chains is relatively stable within species but changes with altered metabolic demand. During mouse fetal development, for example, β-myosin heavy chain mRNA is expressed up to 16-fold higher than α-myosin heavy chain, however, α-myosin heavy chain predominates in the adult.[82] This myosin isoform shift is associated with the significant sudden changes in oxygenation conditions and workload in the postnatal heart compared to the fetal heart. Because the ATP utilization differs between α- and β-myosin heavy chain and ATP production is reduced in the atrophic heart, it is likely that this metabolic change, which precedes other molecular alterations in atrophic remodeling, necessitates an isoform switch. Upon heterotopic heart transplantation in rats, expression of β-myosin heavy chain is up-regulated while α-myosin heavy chain is decreased, likely due to the lower requirement for ATP of β-myosin heavy chain compared to α-myosin heavy chain.[83] These data agree with expression patterns in patients; levels of β-myosin heavy chain in the left ventricular free wall were increased post-LVAD placement in 7/12 patients tested.[27] Rats receiving heterotopically transplanted hearts also exhibited decreased α-myosin heavy chain and decreased myosin ATPase activity.[84] However, regardless of species, the shift toward increased β-myosin heavy chain reduces sarcomere shortening and velocity, which is tolerated in the short term but ultimately leads to myocardial dysfunction.[85]

CONTRACTILE FUNCTION IN CARDIAC ATROPHY

Excitation–contraction coupling in cardiomyocytes requires appropriate calcium handling and sarcomeric protein expression and assembly. Data from patients and animal models of cardiac atrophic remodeling suggest that contractile function is preserved with cardiac atrophy,[15] likely due to restoration of compliance with the law of Laplace with smaller heart size. However, up-regulation of fetal natriuretic and myosin isoform genes in animal models of cardiac unloading and in patients with LVADs are hallmarks of pathological remodeling.[15]

Several mechanisms are responsible for and respond to altered intracellular concentrations of calcium induced by pathological stimuli. Cardiac contraction and relaxation is coupled to movement of calcium ions from outside the cardiomyocyte and from intracellular stores within the sarcoplasmic reticulum. Several ion channels including voltage-sensitive L-type calcium channels and sodium–calcium exchangers (NCX) on

the plasma membrane as well as ryanodine receptors (RyR2) localized to the sarcoplasmic reticulum are coordinated to increase intracellular calcium and modulate contractile function. Intracellular calcium concentrations increase in the healthy heart to permit contraction of cardiomyocytes and are followed by reduced calcium concentrations during relaxation, which is mostly mediated by activity of sarcoplasmic reticulum calcium transport ATPase (SERCA) and NCX. Contractility and calcium transients have only recently been examined as a potential mechanism for inducing heart failure in a subset of patients with cardiac atrophy. Early studies on cardiac contractility in animal models of cardiac atrophy that described preservation of function were performed at two weeks after heterotopic heart transplantation. At longer time points (4–5 weeks after transplantation), both time to relaxation and reduction in intracellular calcium were slowed,[14,15] suggesting diastolic dysfunction. Indeed, calcium handling in cardiomyocytes isolated from unloaded hearts is abnormal; intracellular calcium concentrations are 50% reduced after unloading.

Increased expression of phospholamban (PLB), which negatively regulates calcium uptake into the sarcoplasmic reticulum, could contribute to these observations.[15,53] Indeed, sarcoplasmic calcium uptake by SERCA2a, an important regulatory mechanism for the sequestration of calcium after cardiac contraction, is also reduced in atrophic remodeling, thus diminishing the intracellular stores of calcium available for release during excitation–contraction cycling of the cardiomyocyte.[1,53] The effect of this reduction is particularly apparent in the relative inability of the atrophied myocardium to respond to increased workload with increased contractility.[53] Although the density of L-type calcium channels on the plasma membrane does not appear to be altered during atrophic remodeling,[88] altered distance between L-type calcium channels and the stretch-sensitive ryanodine receptors on the sarcoplasmic reticulum may directly affect the rate of amount of calcium movement within cardiomyocytes. Additionally, the structure of t-tubules in the chronically unloaded heart is altered (Figure 3.4),

(A)   **(B)**

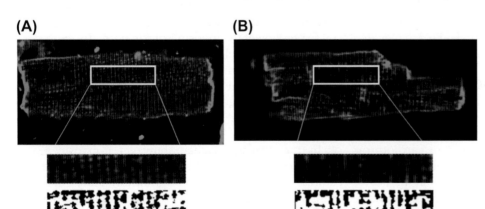

FIGURE 3.4 Structure of t-tubules in adult cardiomyocytes isolated from healthy rat (left) and from rat receiving heterotopically transplanted heart for four weeks (right). Images were obtained using confocal microscopy after staining with membrane-binding dye. Inset is 40 μm long. Binary image is shown below. Adapted from Ibrahim et al., FASEB J (2010).[88]

which uncouples L-type calcium channels from the RyRs and leads to impaired contractility.[88] Other molecular responses to unloading may indirectly impact contractile reserve such as increased ANF, which has negative ionotropic effects in cardiomyocytes.[89]

REGULATION OF ATROPHIC REMODELING BY microRNAs

MicroRNAs have emerged as important regulators of cardiac plasticity and may play a role in the shared transcriptional patterns between hypertrophy and atrophy that induce such distinct cardiac effects. These small, non-coding RNAs can be required for cardiac remodeling and recently the pattern of microRNAs expressed in hypertrophic hearts was compared with that of atrophic hearts. Surprisingly, the microRNAs that were up-regulated in rat models of hypertrophy and atrophy were highly similar.[90] These data strongly suggest that the distinct anabolic and catabolic processes associated with these pathologies are not regulated at the level of microRNAs. Interestingly, unlike other molecular markers that are shared between hypertrophic and atrophic processes in the heart, the pattern of microRNA expression was distinct from the pattern observed in fetal cardiac development.[90]

A recent report from the Olson lab indicated that miR-208a inhibition in the heart increases the expression of MED13, a component of the Mediator complex that controls hormonal regulation of transcription and increases energy expenditure.[91] miR-208a is expressed under the control of the promoter for α-myosin heavy chain, which is reduced in atrophic hearts. This important finding demonstrates that the heart not only responds to systemic alterations in inflammatory cytokines as in cancer (discussed below) to induce protein degradation but also to determine the degree to which the organ and other tissues in the body respond to insulin. This regulatory mechanism could explain why atrophic hearts have reduced insulin responsiveness, thus reducing growth signaling in the context of increased protein degradation by the UPS.

ATROPHIC REMODELING DUE TO CARDIAC PATHOLOGY

Experiments in animals with heterotopically transplanted hearts suggest that cardiac hypertrophic remodeling or heart failure may be reversed through atrophic remodeling. However, most data using this model were obtained by transplanting healthy hearts. These experiments have provided valuable data on the molecular and biochemical mechanisms of cardiac atrophy following ventricular unloading, however, there are specific cases of atrophic remodeling that involve prior or concomitant cardiac pathology that reveal mechanisms distinct from those associated with mechanically unloading a healthy heart. We therefore discuss the effects of hemodynamically unloading a failing heart with LVAD placement and of the inflammatory milieu of cancer.

Atrophic Remodeling with Left Ventricular Assist Devices

A strategy for reducing mechanical load in the failing heart is implantation of pulsatile or, more recently, continuous-flow LVADs that actively pump blood through the left ventricle. Unloading the failing heart via LVAD placement is mainly a short-term solution for patients waiting to receive heart transplants. However, longer-term use is implemented in end-stage heart failure, during shortages of donor hearts, as a bridge to transplant or as a destination therapy for those not eligible for transplant.[92] In patients with heart failure who receive LVADs, decreased ventricular dilation,[93] reduced fetal gene expression,[94] and increased left ventricular function likely due to normalized calcium handling rather than to reduced cardiomyocyte size are observed.[95,96] In fact, up to 25% of patients with LVAD placement experience functional recovery that is sufficient to explant the LVAD. In light of the less than 2% of patients in the United States in need of and eligible for heart transplantation,[97] destination therapy in which an LVAD permanently placed may address the growing need for treatments for patients in end-stage heart failure. Indeed, destination therapy using LVADs improves two-year survival to nearly 80% with continuous-flow devices and is therefore on the rise compared to LVAD placement as a bridge-to-transplant.[98]

However, LVAD placement for extended periods of time and with higher flow rates is associated with significant reductions in cardiomyocyte volume, indicating that increased time may cause atrophy.[99] Heterotopically transplanted hearts in mice mimic mechanical unloading produced by an LVAD. This mechanical unloading increases activation of the UPS via the activity of Foxo3a and the ubiquitin ligases Atrogin-1 and MuRF-1. Heterotopic transplantation is also associated with increased growth signals through IGF-1 and FGF-2 with concomitant activation of extracellular signal-regulated kinase 1 (ERK1), signal transducer and activator of transcription 3 (STAT3), and p70S6K signaling.[50] It is thought that growth signaling increases to balance the increase in protein degradation and to restrict atrophy.

Recent studies have demonstrated reversal of heart failure in a subset of patients after LVAD explantation and combinatorial therapy consisting of an angiotensin

II receptor antagonist, an angiotensin-converting enzyme inhibitor, an aldosterone antagonist, and a steroidal mineralocorticoid.[100] However, although initial reports suggested that atrophic remodeling in patients with hypertrophy leads to improved function, it appears that failed reversal of calcium handling abnormalities thwarts full recovery. Complete reversal of heart failure remains elusive likely due to unbalanced atrophic and hypertrophic mechanisms in the heart with unloading. Indeed, only 5–10% of patients with LVADs recover sufficient ventricular function to allow explantation of the device, although the mechanisms preventing full recovery are not yet known.[101,102] It appears likely that 'bridge-to-recovery' weaning of LVADs is highly time-dependent, as suggested in animal studies with heterotopic transplantation.[53]

As emphasized in previous sections, anabolic and catabolic mechanisms are coordinately activated to achieve an optimal heart size to meet functional demands. Disruption of this balance in a pathological milieu leads to reversible and irreversible cardiac dysfunction. A complicating factor in the use of LVADs is the up-regulation of cardiomyocyte atrophy with unloading, despite a reversal of pathological hypertrophic remodeling. Additionally, increased fibrosis and persistent myofibrillar disarray have been observed during LVAD treatment, although these data are contentious as both reduced and increased fibrosis after LVAD placement have been reported.[93,103,104] It is agreed, however, that the ratio of MMPs/TIMPs is altered in favor of normal activity, and there is in fact no change in fibrosis after placement of an LVAD.[105] Additionally, despite the apparent dysregulation of proteolysis and protein synthesis, a mouse model of cardiac unloading which uses heterotopic heart transplantation demonstrated that the increased presence of polyubiquitinated proteins may be balanced by the activation of mTOR.[2]

Loss of Cardiac Mass Induced by Cancer Cachexia

Many chronic illnesses are associated with increased systemic inflammation, increased circulating neurohormonal factors, and loss of body and cardiac mass. A significant proportion of patients with cancer, for example, experience body wasting that is mainly characterized by loss of skeletal muscle (cachexia) as well as cardiac insufficiency that is a result of cardiac atrophy. The mechanisms mediating cancer cachexia appear to stem from the complex metabolic crisis that occurs in the tissues of patients with cancer. Although the cause of cachexia has been studied extensively, the processes leading to cardiac atrophy induced by cancer are less well understood.

As in other forms of atrophic remodeling, cardiac fibrosis increases and gene expression changes are observed including reduced PPAR-α, which reduces lipid metabolism in favor of glucose, and increased levels of fetal genes like BNP.[76] In mice bearing tumors caused by injection of colon-26 cells, cachexia develops quickly and cardiac dysfunction develops within two weeks of tumor inoculation. The number of studies reporting cardiac dysfunction with cancer is limited due to the strain- and tumor-specific development of cachexia. However, these studies report variable functional alterations including significantly reduced left ventricular ejection fraction and fractional shortening[76] as well as reduced aortic velocity.[17] Additionally, cardiac atrophy has been demonstrated by determining cell volume, performing echocardiography, and recording heart weight.[17]

The 20–30% reduction in cardiac mass[17,106] and impaired functional output is directly attributable to an imbalance between protein synthesis and degradation; an equal reduction in expression of sarcomeric proteins including myosin heavy chain and troponin I is observed.[17,76] Interestingly, the involvement of the main proteolytic mechanism in the heart, the UPS, is disputable. One study measured significant increases in MuRF-1 and Atrogin-1 levels in the hearts of tumor-bearing mice but did not directly measure UPS activity in the same experimental model of cardiac atrophy induced by inoculation with colon-26 tumor-forming cells.[76] However, despite little evidence for autophagy in the atrophy caused by LVAD use, microtubule-associated protein 1A/1B-light chain 3-II (LC3-II) levels in the heart are increased in tumor-bearing mice; indeed, tumor-induced atrophic remodeling is mediated by autophagy.[17] Interestingly, cardiac atrophy induced by cancer is also unique in that it is more prevalent in male than in female mice; the protective effect in females is mediated by estrogen receptor signaling.[17]

Many of the molecular hallmarks observed in animal models of atrophic remodeling induced by ventricular unloading are also up-regulated in the hearts of animals with tumors, and these appear to be mediated by factors secreted by the tumor itself. For example, pro-inflammatory circulating cytokines, for example, including interleukin-6 and TNF-α are increased systemically in cancer and activate downstream signaling pathways mediated by MAPK.[76] Indeed, activation of p42/44 MAPK was observed in the hearts of tumor-bearing mice, thus linking the tumor to the atrophic phenotype in the heart. However, this pathway may mediate increased cardiac fibrosis rather than cardiomyocyte atrophy due to its requirement in fibroblast proliferation;[107] this mechanism requires further investigation.

ATROPHIC REMODELING AS A POTENTIAL THERAPEUTIC

Although cardiac atrophy induced by alterations in neurohormonal, nutritional, or mechanical stimuli is generally thought to be maladaptive, decades of research on mechanically unloaded hearts suggest that atrophic remodeling may represent a therapeutic intervention when pathological cardiac hypertrophy is present. Promoters of cardiac atrophy (like caloric restriction representing 60% of the normal dietary intake) reduce cardiomyocyte hypertrophy and other markers of pathological hypertrophy including apoptosis, ANF expression, and fibrosis.[108] Molecular responses to mechanical unloading that induce atrophic remodeling including metabolic, protein homeostatic, and mechanical load responses may also be exploited as therapeutic targets. Additionally, expression of key regulators of cardiac atrophy may be activated to promote atrophy in the context of cardiac hypertrophy.

Therapeutic strategies that directly or indirectly reduce cardiac load are currently approved and in use to treat patients with pathological cardiac hypertrophy and end-stage heart failure. For example, antihypertensive drugs such as ACE inhibitors, calcium channel antagonists, and sympatholytic agents each reduce cardiac load and reduce cardiac mass.[109] However, not all pressure-reducing agents promote regression of hypertrophy. Despite antihypertensive effects, use of diuretics does not reduce cardiac hypertrophy.[110] LVAD implantation holds promise as a therapeutic for patients in end-stage heart failure either as a bridge-to-transplant or destination therapy. Due to the limited number of donor hearts, destination therapy is of interest to the majority of patients who are ineligible for transplant or for whom there are no hearts available. The HeartMate XVE LVAD was approved for destination therapy by the FDA in 2002. A large subsequent clinical study comparing it to the second-generation HeartMate II performed between 2005 and 2007 demonstrated further improvement with the HeartMate II; 2-year survival increased by 34%, and post-placement complications were significantly reduced.[111,112] Additionally, the Fourth Annual Report of the Interagency Registry for Mechanically Assisted Circulatory Support (INTERMACS) that considered more than 6000 patients receiving LVADs reported two-year survival in patients with LVAD placement intended as destination therapy is nearly 80%.[113] However, LVADs have limited longevity and require replacement at 2–4 years post-placement.[114] Complications including infection, stroke and gastrointestinal bleeding that result from reduced arterial pressure also continue to affect patients with these devices.[115] Predictive models that estimate the success of LVAD candidates are critical to the improved utility of these devices.[116]

Atrophic remodeling of the heart involves several common mechanisms that could serve as therapeutic targets of virally delivered peptides or small molecules in the context of pathological hypertrophy. Atrogin-1 and MuRF-1, for example, are required for atrophic remodeling, and forced expression could induce atrophy of the hypertrophic heart. Activating FOXO3a, which regulates expression of these ubiquitin ligases could increase selective proteolytic degradation of sarcomeric proteins that contribute to cardiac hypertrophy. Additionally, activity of signaling proteins like NK-κB, which also specifically leads to degradation of components of the sarcomere, could be induced to promote atrophy in the hypertrophic heart.[117] Molecularly recapitulating the pro-atrophic transcriptional profile including UbB, UbcH2, Atrogin-1, and MuRF-1 may represent a strategy for reducing pathological hypertrophy.[118] However, much work remains in understanding the precise temporal patterns that distinguish them from those observed with hypertrophy.

SUMMARY

Atrophic cardiac remodeling occurs in response to many stimuli, including those that occur with disease or with clinical mechanical intervention. Protein degradation in the atrophic heart is increased such that mechanical strain is reduced and cardiomyocyte volume decreases to maintain compliance with the law of Laplace. Cellular atrophic remodeling is mainly mediated by the UPS, however, in some cases of cardiac or systemic pathology such as cancer, autophagy is implicated in increased myofibrillar protein degradation. The myocardium therefore experiences an overall reduction in size along with an increase in fibrosis yet initially maintains cardiac function. Mechanical unloading also reduces metabolic demand and alters expression of metabolic genes such that the heart favors the use of glucose over lipids as an energy substrate. Evidence suggests that atrophic remodeling could represent a potentially important therapeutic for the treatment of some forms of cardiomyopathy or heart failure. However, hallmarks of pathological cardiac remodeling including re-expression of fetal genes are present in the atrophic heart and adaptive cardiac function is not maintained over the longer term. Further investigation is required to characterize the temporal and molecular aspects of pathological remodeling reversal using mechanical unloading to optimally promote permanent recovery of cardiac function with disease.

References

1. Baskin KK, Taegtmeyer H. Taking pressure off the heart: the ins and outs of atrophic remodelling. *Cardiovasc Res* 2011;**90**(2):243–50.

2. Razeghi P, Sharma S, Ying J, et al. Atrophic remodeling of the heart in vivo simultaneously activates pathways of protein synthesis and degradation. *Circulation* 2003;**108**(20):2536–41.

3. Samarel AM, Parmacek MS, Magid NM, Decker RS, Lesch M. Protein synthesis and degradation during starvation-induced cardiac atrophy in rabbits. *Circ Res* 1987;**60**(6):933–41.

4. Liu F, Kang SM. Heterotopic heart transplantation in mice. *J Visualized Exp* 2007(6):238.

5. McCarthy PM, Smedira NO, Vargo RL, et al. One hundred patients with the HeartMate left ventricular assist device: evolving concepts and technology. *J Thorac Cardiovasc Surg* 1998;**115**(4):904–12.

6. Pagani FD, Miller LW, Russell SD, et al. Extended mechanical circulatory support with a continuous-flow rotary left ventricular assist device. *J Am Coll Cardiol* 2009;**54**(4):312–21.

7. Rose EA, Gelijns AC, Moskowitz AJ, et al. Long-term use of a left ventricular assist device for end-stage heart failure. *N Engl J Med* 2001;**345**(20):1435–43.

8. Ahuja P, Sdek P, MacLellan WR. Cardiac myocyte cell cycle control in development, disease, and regeneration. *Physiol Rev* 2007;**87**(2):521–44.

9. Lorell BH, Carabello BA. Left ventricular hypertrophy: pathogenesis, detection, and prognosis. *Circulation* 2000;**102**(4):470–9.

10. Brancaccio M, Fratta L, Notte A, et al. Melusin, a muscle-specific integrin beta1-interacting protein, is required to prevent cardiac failure in response to chronic pressure overload. *Nat Med* 2003;**9**(1):68–75.

11. Hannigan GE, Coles JG, Dedhar S. Integrin-linked kinase at the heart of cardiac contractility, repair, and disease. *Circ Res* 2007;**100**(10):1408–14.

12. Mann DL, Acker MA, Jessup M, et al. Rationale, design, and methods for a pivotal randomized clinical trial for the assessment of a cardiac support device in patients with New York health association class III–IV heart failure. *J Card Fail* 2004;**10**(3):185–92.

13. McCarthy PM, Takagaki M, Ochiai Y, et al. Device-based change in left ventricular shape: a new concept for the treatment of dilated cardiomyopathy. *J Thorac Cardiovasc Surg* 2001;**122**(3):482–90.

14. Kolar F, MacNaughton C, Papousek F, Korecky B. Systolic mechanical performance of heterotopically transplanted hearts in rats treated with cyclosporin. *Cardiovasc Res* 1993;**27**(7):1244–7.

15. Welsh DC, Dipla K, McNulty PH, et al. Preserved contractile function despite atrophic remodeling in unloaded rat hearts. *Am J Physiol Heart Circ Physiol* 2001;**281**(3):H1131–6.

16. Schena S, Kurimoto Y, Fukada J, et al. Effects of ventricular unloading on apoptosis and atrophy of cardiac myocytes. *J Surg Res* 2004;**120**(1):119–26.

17. Cosper PF, Leinwand LA. Cancer causes cardiac atrophy and autophagy in a sexually dimorphic manner. *Cancer Res* 2011;**71**(5):1710–20.

18. Muhlfeld C, Das SK, Heinzel FR, et al. Cancer induces cardiomyocyte remodeling and hypoinnervation in the left ventricle of the mouse heart. *PloS One* 2011;**6**(5):e20424.

19. Brinks H, Tevaearai H, Muhlfeld C, et al. Contractile function is preserved in unloaded hearts despite atrophic remodeling. *J Thorac Cardiovasc Surg* 2009;**137**(3):742–6.

20. Tulloch NL, Muskheli V, Razumova MV, et al. Growth of engineered human myocardium with mechanical loading and vascular coculture. *Circ Res* 2011;**109**(1):47–59.

21. Hoshijima M. Mechanical stress-strain sensors embedded in cardiac cytoskeleton: Z disk, titin, and associated structures. *Am J Physiol Heart Circ Physiol* 2006;**290**(4):H1313–25.

22. Calderone A, Bel-Hadj S, Drapeau J, et al. Scar myofibroblasts of the infarcted rat heart express natriuretic peptides. *J Cell Physiol* 2006;**207**(1):165–73.

23. Weber KT, Sun Y, Bhattacharya SK, Ahokas RA, Gerling IC. Myofibroblast-mediated mechanisms of pathological remodelling of the heart. *Nature Rev Cardiol* 2013;**10**(1):15–26.

24. Wang WJ, Meng ZL, Mo YC, et al. Unloading the infarcted heart affect MMPs-TIMPs axis in a rat cardiac heterotopic transplantation model. *Mol Biol Rep* 2012;**39**(1):277–83.

25. Lenga Y, Koh A, Perera AS, McCulloch CA, Sodek J, Zohar R. Osteopontin expression is required for myofibroblast differentiation. *Circ Res* 2008;**102**(3):319–27.

26. Jinnin M, Ihn H, Mimura Y, Asano Y, Yamane K, Tamaki K. Regulation of fibrogenic/fibrolytic genes by platelet-derived growth factor C, a novel growth factor, in human dermal fibroblasts. *J Cell Physiol* 2005;**202**(2):510–7.

27. Rodrigue-Way A, Burkhoff D, Geesaman BJ, et al. Sarcomeric genes involved in reverse remodeling of the heart during left ventricular assist device support. *J Heart Lung Transplant* 2005;**24**(1):73–80.

28. Watson CJ, Phelan D, Xu M, et al. Mechanical stretch up-regulates the B-type natriuretic peptide system in human cardiac fibroblasts: a possible defense against transforming growth factor-beta mediated fibrosis. *Fibrogenesis Tissue Repair* 2012;**5**(1):9.

29. Weber KT, Janicki JS, Shroff SG, Pick R, Chen RM, Bashey RI. Collagen remodeling of the pressure-overloaded, hypertrophied nonhuman primate myocardium. *Circ Res* 1988;**62**(4):757–65.

30. Kent RL, Uboh CE, Thompson EW, et al. Biochemical and structural correlates in unloaded and reloaded cat myocardium. *J Mol Cell Cardiol* 1985;**17**(2):153–65.

31. O'Hanlon R, Grasso A, Roughton M, et al. Prognostic significance of myocardial fibrosis in hypertrophic cardiomyopathy. *J Am Coll Cardiol* 2010;**56**(11):867–74.

32. Marijianowski MM, Teeling P, Mann J, Becker AE. Dilated cardiomyopathy is associated with an increase in the type I/type III collagen ratio: a quantitative assessment. *J Am Coll Cardiol* 1995;**25**(6):1263–72.

33. Lopez B, Gonzalez A, Querejeta R, Larman M, Diez J. Alterations in the pattern of collagen deposition may contribute to the deterioration of systolic function in hypertensive patients with heart failure. *J Am Coll Cardiol* 2006;**48**(1):89–96.

34. Hein S, Arnon E, Kostin S, et al. Progression from compensated hypertrophy to failure in the pressure-overloaded human heart: structural deterioration and compensatory mechanisms. *Circulation* 2003;**107**(7):984–91.

35. Fidzianska A, Bilinska ZT, Walczak E, Witkowski A, Chojnowska L. Autophagy in transition from hypertrophic cardiomyopathy to heart failure. *J Electron Microsc* 2010;**59**(2):181–3.

36. Jalil JE, Janicki JS, Pick R, Abrahams C, Weber KT. Fibrosis-induced reduction of endomyocardium in the rat after isoproterenol treatment. *Circ Res* 1989;**65**(2):258–64.

37. de Bakker JM, van Capelle FJ, Janse MJ, et al. Slow conduction in the infarcted human heart. 'Zigzag' course of activation. *Circulation* 1993;**88**(3):915–26.

38. Spach MS, Boineau JP. Microfibrosis produces electrical load variations due to loss of side-to-side cell connections: a major mechanism of structural heart disease arrhythmias. *Pacing Clin Electrophysiol* 1997;**20**(2 Pt 2):397–413.

39. Maron BJ, Wolfson JK, Epstein SE, Roberts WC. Intramural ("small vessel") coronary artery disease in hypertrophic cardiomyopathy. *J Am Coll Cardiol* 1986;**8**(3):545–57.

40. Olivotto I, Girolami F, Sciagra R, et al. Microvascular function is selectively impaired in patients with hypertrophic cardiomyopathy and sarcomere myofilament gene mutations. *J Am Coll Cardiol* 2011;**58**(8):839–48.

41. Janse MJ, van Capelle FJ, Morsink H, et al. Flow of "injury" current and patterns of excitation during early ventricular arrhythmias in acute regional myocardial ischemia in isolated porcine and canine hearts. Evidence for two different arrhythmogenic mechanisms. *Circ Res* 1980;**47**(2):151–65.

42. Fan D, Takawale A, Lee J, Kassiri Z. Cardiac fibroblasts, fibrosis and extracellular matrix remodeling in heart disease. *Fibrogenesis Tissue Repair* 2012;**5**(1):15.

43. Wang X, Robbins J. Heart failure and protein quality control. *Circ Res* 2006;**99**(12):1315–28.

44. Goldberg AL. Protein degradation and protection against misfolded or damaged proteins. *Nature* 2003;**426**(6968):895–9.

45. Portbury AL, Willis MS, Patterson C. Tearin' up my heart: proteolysis in the cardiac sarcomere. *J Biol Chem* 2011;**286**(12):9929–34.

46. Powell SR. The cardiac 26S proteasome: regulating the regulator. *Circ Res* 2006;**99**(4):342–5.

47. Lecker SH, Goldberg AL, Mitch WE. Protein degradation by the ubiquitin-proteasome pathway in normal and disease states. *J Am Soc Nephrol* 2006;**17**(7):1807–19.

48. Razeghi P, Baskin KK, Sharma S, et al. Atrophy, hypertrophy, and hypoxemia induce transcriptional regulators of the ubiquitin proteasome system in the rat heart. *Biochem Biophys Res Commun* 2006;**342**(2):361–4.

49. Willis MS, Rojas M, Li L, et al. Muscle ring finger 1 mediates cardiac atrophy in vivo. *Am J Physiol Heart Circ Physiol* 2009;**296**(4):H997–1006.

50. Sharma S, Ying J, Razeghi P, Stepkowski S, Taegtmeyer H. Atrophic remodeling of the transplanted rat heart. *Cardiology* 2006;**105**(2):128–36.

51. Skurk C, Izumiya Y, Maatz H, et al. The FOXO3a transcription factor regulates cardiac myocyte size downstream of AKT signaling. *J Biol Chem* 2005;**280**(21):20814–23.

52. Depre C, Wang Q, Yan L, et al. Activation of the cardiac proteasome during pressure overload promotes ventricular hypertrophy. *Circulation* 2006;**114**(17):1821–8.

53. Ito K, Nakayama M, Hasan F, Yan X, Schneider MD, Lorell BH. Contractile reserve and calcium regulation are depressed in myocytes from chronically unloaded hearts. *Circulation* 2003;**107**(8):1176–82.

54. Nagata K, Somura F, Obata K, et al. AT1 receptor blockade reduces cardiac calcineurin activity in hypertensive rats. *Hypertension* 2002;**40**(2):168–74.

55. Olson EN, Williams RS. Calcineurin signaling and muscle remodeling. *Cellule* 2000;**101**(7):689–92.

56. Ni YG, Berenji K, Wang N, et al. Foxo transcription factors blunt cardiac hypertrophy by inhibiting calcineurin signaling. *Circulation* 2006;**114**(11):1159–68.

57. Schips TG, Wietelmann A, Hohn K, et al. FoxO3 induces reversible cardiac atrophy and autophagy in a transgenic mouse model. *Cardiovasc Res* 2011;**91**(4):587–97.

58. Razeghi P, Volpini KC, Wang ME, Youker KA, Stepkowski S, Taegtmeyer H. Mechanical unloading of the heart activates the calpain system. *J Mol Cell Cardiol* 2007;**42**(2):449–52.

59. Chen B, Zhong L, Roush SF, et al. Disruption of a GATA4/Ankrd1 signaling axis in cardiomyocytes leads to sarcomere disarray: implications for anthracycline cardiomyopathy. *PloS One* 2012;**7**(4):e35743.

60. van Bilsen M, Smeets PJ, Gilde AJ, van der Vusse GJ. Metabolic remodelling of the failing heart: the cardiac burn-out syndrome? *Cardiovasc Res* 2004;**61**(2):218–26.

61. Razeghi P, Young ME, Abbasi S, Taegtmeyer H. Hypoxia in vivo decreases peroxisome proliferator-activated receptor alpha-regulated gene expression in rat heart. *Biochem Biophys Res Commun* 2001;**287**(1):5–10.

62. Stanley WC, Lopaschuk GD, McCormack JG. Regulation of energy substrate metabolism in the diabetic heart. *Cardiovasc Res* 1997;**34**(1):25–33.

63. Kjekshus JK, Mjos OD. Effect of free fatty acids on myocardial function and metabolism in the ischemic dog heart. *J Clin Investigation* 1972;**51**(7):1767–76.

64. Burkhoff D, Weiss RG, Schulman SP, Kalil-Filho R, Wannenburg T, Gerstenblith G. Influence of metabolic substrate on rat heart function and metabolism at different coronary flows. *Am J Physiol* 1991;**261**(3 Pt 2):H741–50.

65. Hutter JF, Schweickhardt C, Piper HM, Spieckermann PG. Inhibition of fatty acid oxidation and decrease of oxygen consumption of working rat heart by 4-bromocrotonic acid. *J Mol Cell Cardiol* 1984;**16**(1):105–8.

66. Porter RK. Mitochondrial proton leak: a role for uncoupling proteins 2 and 3? *Biochim Biophys Acta* 2001;**1504**(1):120–7.

67. Stanley WC, Recchia FA, Lopaschuk GD. Myocardial substrate metabolism in the normal and failing heart. *Physiol Rev* 2005;**85**(3):1093–129.

68. Taegtmeyer H, Razeghi P, Young ME. Mitochondrial proteins in hypertrophy and atrophy: a transcript analysis in rat heart. *Clinical and Experimental Pharmacology & Physiology* 2002;**29**(4):346–50.

69. Allard MF, Schonekess BO, Henning SL, English DR, Lopaschuk GD. Contribution of oxidative metabolism and glycolysis to ATP production in hypertrophied hearts. *Am J Physiol* 1994;**267**(2 Pt 2):H742–50.

70. Young ME, Laws FA, Goodwin GW, Taegtmeyer H. Reactivation of peroxisome proliferator-activated receptor alpha is associated with contractile dysfunction in hypertrophied rat heart. *J Biol Chem* 2001;**276**(48):44390–5.

71. Bugger H, Leippert S, Blum D, et al. Subtractive hybridization for differential gene expression in mechanically unloaded rat heart. *Am J Physiol Heart Circ Physiol* 2006;**291**(6):H2714–22.

72. Garnier A, Fortin D, Delomenie C, Momken I, Veksler V, Ventura-Clapier R. Depressed mitochondrial transcription factors and oxidative capacity in rat failing cardiac and skeletal muscles. *J Physiol* 2003;**551**(Pt 2):491–501.

73. Baskin KK, Taegtmeyer H. AMP-activated protein kinase regulates E3 ligases in rodent heart. *Circ Res* 2011;**109**(10):1153–61.

74. Sugden PH. Signalling pathways in cardiac myocyte hypertrophy. *Annu Mediaev* 2001;**33**(9):611–22.

75. Xu R, Lin F, Zhang S, Chen X, Hu S, Zheng Z. Signal pathways involved in reverse remodeling of the hypertrophic rat heart after pressure unloading. *Int J Cardiol* 2010;**143**(3):414–23.

76. Tian M, Asp ML, Nishijima Y, Belury MA. Evidence for cardiac atrophic remodeling in cancer-induced cachexia in mice. *Int J Oncol* 2011;**39**(5):1321–6.

77. Cosper PF, Harvey PA, Leinwand LA. Interferon-gamma causes cardiac myocyte atrophy via selective degradation of myosin heavy chain in a model of chronic myocarditis. *Am J Pathol* 2012;**181**(6):2038–46.

78. Mukoyama M, Nakao K, Hosoda K, et al. Brain natriuretic peptide as a novel cardiac hormone in humans. Evidence for an exquisite dual natriuretic peptide system, atrial natriuretic peptide and brain natriuretic peptide. *J Clin Investigation* 1991;**87**(4):1402–12.

79. Potter LR, Abbey-Hosch S, Dickey DM. Natriuretic peptides, their receptors, and cyclic guanosine monophosphate-dependent signaling functions. *Endocr Rev* 2006;**27**(1):47–72.

80. Cornwell TL, Pryzwansky KB, Wyatt TA, Lincoln TM. Regulation of sarcoplasmic reticulum protein phosphorylation by localized cyclic GMP-dependent protein kinase in vascular smooth muscle cells. *Mol Pharmacol* 1991;**40**(6):923–31.

81. Nakamura M, Ichikawa K, Ito M, et al. Effects of the phosphorylation of myosin phosphatase by cyclic GMP-dependent protein kinase. *Cell Signal* 1999;**11**(9):671–6.

82. Ng WA, Grupp IL, Subramaniam A, Robbins J. Cardiac myosin heavy chain mRNA expression and myocardial function in the mouse heart. *Circ Res* 1991;**68**(6):1742–50.

83. Depre C, Shipley GL, Chen W, et al. Unloaded heart in vivo replicates fetal gene expression of cardiac hypertrophy. *Nat Med* 1998;**4**(11):1269–75.

84. Geenen DL, Malhotra A, Buttrick PM, Scheuer J. Increased heart rate prevents the isomyosin shift after cardiac transplantation in the rat. *Circ Res* 1992;**70**(3):554–8.

85. Herron TJ, McDonald KS. Small amounts of alpha-myosin heavy chain isoform expression significantly increase power output of rat cardiac myocyte fragments. *Circ Res* 2002;**90**(11):1150–2.

86. Yasumoto K, Mukaida N, Harada A, et al. Molecular analysis of the cytokine network involved in cachexia in colon 26 adenocarcinoma-bearing mice. *Cancer Res* 1995;**55**(4):921–7.

87. Pajak B, Orzechowska S, Pijet B, et al. Crossroads of cytokine signaling–the chase to stop muscle cachexia. *J Physiol Pharmacol* 2008;**59**(Suppl. 9):251–64.

88. Ibrahim M, Al Masri A, Navaratnarajah M, et al. Prolonged mechanical unloading affects cardiomyocyte excitation-contraction coupling, transverse-tubule structure, and the cell surface. *FASEB J* 2010;**24**(9):3321–9.

89. Tajima M, Bartunek J, Weinberg EO, Ito N, Lorell BH. Atrial natriuretic peptide has different effects on contractility and intracellular pH in normal and hypertrophied myocytes from pressure-overloaded hearts. *Circulation* 1998;**98**(24):2760–4.

90. El-Armouche A, Schwoerer AP, Neuber C, et al. Common microRNA signatures in cardiac hypertrophic and atrophic remodeling induced by changes in hemodynamic load. *PloS One* 2010;**5**(12):e14263.

91. Grueter CE, van Rooij E, Johnson BA, et al. A cardiac microRNA governs systemic energy homeostasis by regulation of MED13. *Cellule* 2012;**149**(3):671–83.

92. Garbade J, Bittner HB, Barten MJ, Mohr FW. Current trends in implantable left ventricular assist devices. *Cardiol Res Pract* 2011;**2011**:290561.

93. McCarthy PM, Nakatani S, Vargo R, et al. Structural and left ventricular histologic changes after implantable LVAD insertion. *Ann Thorac Surg* 1995;**59**(3):609–13.

94. Bruggink AH, de Jonge N, van Oosterhout MF, et al. Brain natriuretic peptide is produced both by cardiomyocytes and cells infiltrating the heart in patients with severe heart failure supported by a left ventricular assist device. *J Heart Lung Transplant* 2006;**25**(2):174–80.

95. Zafeiridis A, Jeevanandam V, Houser SR, Margulies KB. Regression of cellular hypertrophy after left ventricular assist device support. *Circulation* 1998;**98**(7):656–62.

96. Terracciano CM, Hardy J, Birks EJ, Khaghani A, Banner NR, Yacoub MH. Clinical recovery from end-stage heart failure using left-ventricular assist device and pharmacological therapy correlates with increased sarcoplasmic reticulum calcium content but not with regression of cellular hypertrophy. *Circulation* 2004;**109**(19):2263–5.

97. Lietz K. Destination therapy: patient selection and current outcomes. *J Cardiovasc Surg* 2010;**25**(4):462–71.

98. Kirklin JK, Naftel DC, Kormos RL, et al. Third INTERMACS Annual Report: the evolution of destination therapy in the United States. *J Heart Lung Transplant* 2011;**30**(2):115–23.

99. Kinoshita M, Takano H, Taenaka Y, et al. Cardiac disuse atrophy during LVAD pumping. *ASAIO Trans/Am Soc Artif Intern Organs* 1988;**34**(3):208–12.

100. Birks EJ, Tansley PD, Hardy J, et al. Left ventricular assist device and drug therapy for the reversal of heart failure. *N Engl J Med* 2006;**355**(18):1873–84.

101. Mancini DM, Beniaminovitz A, Levin H, et al. Low incidence of myocardial recovery after left ventricular assist device implantation in patients with chronic heart failure. *Circulation* 1998;**98**(22):2383–9.

102. Maybaum S, Mancini D, Xydas S, et al. Cardiac improvement during mechanical circulatory support: a prospective multicenter study of the LVAD Working Group. *Circulation* 2007;**115**(19):2497–505.

103. de Jonge N, van Wichen DF, Schipper ME, et al. Left ventricular assist device in end-stage heart failure: persistence of structural myocyte damage after unloading. An immunohistochemical analysis of the contractile myofilaments. *J Am Coll Cardiol* 2002;**39**(6):963–9.

104. Muller J, Wallukat G, Weng YG, et al. Weaning from mechanical cardiac support in patients with idiopathic dilated cardiomyopathy. *Circulation* 1997;**96**(2):542–9.

105. Li YY, Feng Y, McTiernan CF, et al. Downregulation of matrix metalloproteinases and reduction in collagen damage in the failing human heart after support with left ventricular assist devices. *Circulation* 2001;**104**(10):1147–52.

106. Zhou X, Wang JL, Lu J, et al. Reversal of cancer cachexia and muscle wasting by ActRIIB antagonism leads to prolonged survival. *Cellule* 2010;**142**(4):531–43.

107. Pages G, Lenormand P, L'Allemain G, Chambard JC, Meloche S, Pouyssegur J. Mitogen-activated protein kinases p42mapk and p44mapk are required for fibroblast proliferation. *Proc Natl Acad Sci U S A* 1993;**90**(18):8319–23.

108. Finckenberg P, Eriksson O, Baumann M, et al. Caloric restriction ameliorates angiotensin II-induced mitochondrial remodeling and cardiac hypertrophy. *Hypertension* 2012;**59**(1):76–84.

109. Motz W, Strauer BE. Therapeutic effect on left ventricular hypertrophy by different antihypertensive drugs. *Clin Investigator* 1992;**70**(Suppl. 1):S87–92.

110. Wollam GL, Hall WD, Porter VD, et al. Time course of regression of left ventricular hypertrophy in treated hypertensive patients. *Am J Med* 1983;**75**(3A):100–10.

111. Fang JC. Rise of the machines–left ventricular assist devices as permanent therapy for advanced heart failure. *N Engl J Med* 2009;**361**(23):2282–5.

112. Slaughter MS, Rogers JG, Milano CA, et al. Advanced heart failure treated with continuous-flow left ventricular assist device. *N Engl J Med* 2009;**361**(23):2241–51.

113. Kirklin JK, Naftel DC, Kormos RL, et al. Fifth INTERMACS annual report: risk factor analysis from more than 6,000 mechanical circulatory support patients. *J Heart Lung Transplant* 2013;**32**(2):141–56.

114. Lietz K, Long JW, Kfoury AG, et al. Outcomes of left ventricular assist device implantation as destination therapy in the post-REMATCH era: implications for patient selection. *Circulation* 2007;**116**(5):497–505.

115. Wang Q, Chen B, Liu P, et al. Calmodulin binds to extracellular sites on the plasma membrane of plant cells and elicits a rise in intracellular calcium concentration. *J Biol Chem* 2009;**284**(18):12000–7.

116. Slaughter MS, Meyer AL, Birks EJ. Destination therapy with left ventricular assist devices: patient selection and outcomes. *Curr Opin Cardiol* 2011;**26**(3):232–6.

117. Aulino P, Berardi E, Cardillo VM, et al. Molecular, cellular and physiological characterization of the cancer cachexia-inducing C26 colon carcinoma in mouse. *BMC Cancer* 2010;**10**:363.

118. Razeghi P, Taegtmeyer H. Hypertrophy and atrophy of the heart: the other side of remodeling. *Ann N Y Acad Sci* 2006;**1080**:110–9.

119. Willis MS, Patterson C. Proteotoxicity and cardiac dysfunction–Alzheimer's disease of the heart? *N Engl J Med* 2013;**368**(5):455–64.

The Pathophysiology of Cardiac Hypertrophy and Heart Failure

William E. Stansfield, MD[1], Mark Ranek, PhD[2],
Avani Pendse, MBBS, PhD[1], Jonathan C. Schisler, MS, PhD[1],
Shaobin Wang, PhD[1], Thomas Pulinilkunnil, PhD[3], Monte S. Willis, MD, PhD[1]

[1]University of North Carolina at Chapel Hill, Chapel Hill, NC, USA, [2]The Johns Hopkins University, Baltimore, MD, USA, [3]Dalhousie University, Halifax, Nova Scotia, Canada

INTRODUCTION

Heart disease is a common global cause of morbidity and mortality. In the US alone, an estimated 83 million individuals carry the diagnosis and 1 in every 3 deaths are believed due to heart disease. More than 75% of patients have hypertension-related heart disease with associated cardiac enlargement.[1–3] At the cellular level, chronic hypertension results in physiologic and pathological changes that culminate in adaptive changes in left ventricular mass (LVM) or cardiac growth. Despite the overwhelming presence of non-cardiac cell types in the heart – fibroblasts, circulating blood cells, endothelial cells, smooth muscle cells and adipocytes – cardiomyocyte size accounts for at least two-thirds of cardiac mass and is the determining factor regulating LVM.

Increases in left ventricular mass, also known as cardiac hypertrophy or left ventricular hypertrophy (LVH), occur as an adaptive response to stress. This includes physiologic stresses such as exercise or pregnancy, or pathological stimuli such as pressure- or volume-overload. In early LVH, left ventricular function is conserved. With progression, LVH results in ventricular dysfunction (i.e. heart failure). Disproportionate enlargement of the left ventricle relative to its functional efficiency renders the myocardium more sensitive to ischemia and arrhythmia. Morphologically, hypertrophy is categorized as either eccentric or concentric. Eccentric hypertrophy is more commonly associated with endurance exercise training, pregnancy, and volume overload. Concentric hypertrophy is most often the result of chronic pressure overload, but is possible to a minor degree with weight training.

Left ventricular mass index (LVMI), measured using echocardiography, has long been established as one of the most robust, independent predictors of cardiovascular morbidity and mortality.[4] In the Framingham heart study, each 50 g/m^2 increase in LVMI caused a 1.5-fold increment in adjusted relative risk of cardiovascular disease, heart failure, and death.[5] Electrocardiographic detection of LVH, although less sensitive, is an equally powerful predictor.[4] Moreover, subtle increases in LVMI that do not meet the threshold criteria for LVH are still associated with increased cardiovascular disease.[6–8] LVH prevalence rates range from 58–77% in patients with hypertension,[9] obesity, and diabetes mellitus.[10,11]

Over the last several decades, molecular-level research has elaborated numerous mechanisms of LVH progression.[12–15] In spite of current pharmacological therapy, however, few patients achieve regression of LVH. Most patients' disease continues to progress while on therapy, and they suffer an ever-increasing risk of cardiovascular events. Only by continuing to explore these mechanisms and advancing translational research will we devise new, more effective therapies.

ETIOLOGY OF HEART FAILURE

Heart failure is the clinical syndrome that describes the physiologic effects of acutely or chronically decreased cardiac function. It can result from a wide range of pathologic processes. More specifically, failure is the inability to maintain a sufficient cardiac output to fulfill the metabolic requirements of organs or accommodate systemic venous return. Occasionally, a failing heart can maintain necessary cardiac output, but only at an abnormally elevated filling pressure or volume.

Approximately 5.7 million people in the United States have heart failure. In 2008, it was the underlying cause of death in more than 56 000 patients.[3] The clinical presentation of a patient with heart failure includes signs and symptoms of volume overload such as dyspnea, lower extremity edema, and ascites. Additional stigmata of insufficient tissue perfusion include fatigue, exercise intolerance, and renal hypo-perfusion. Depending on the underlying disease process leading to heart failure, both volume overload and peripheral malperfusion may be present. Heart failure usually presents with an acute onset of symptoms in a person with known chronic heart disease. Alternately, symptoms may appear abruptly with rapid progression to pulmonary edema and resting dyspnea. Diagnoses such as acute myocardial infarction, valvular heart disease, and even acute myocarditis must be identified as early as possible in order for patients to receive effective treatments.

Ischemic heart disease, resulting from either acute myocardial infarction or chronic ischemia, accounts for most cases of heart failure with systolic dysfunction.[16] Non-ischemic causes of systolic heart failure can be classified as arising from chronic pressure overload, chronic volume overload, and dilated cardiomyopathy. Conditions resulting in pressure overload include predominantly hypertension, followed by aortic stenosis (or less commonly pulmonary stenosis), coarctation, or hypertrophic cardiomyopathy. Importantly, systolic failure is a relatively late occurrence in pressure overload – patients typically present first with diastolic dysfunction. Causes of volume overload include aortic or mitral regurgitation, intracardiac shunts (such as an atrial or ventricular septal defect), and extracardiac shunts (such as a high-flow arterio-venous dialysis fistula). Dilated cardiomyopathy is most often idiopathic, but has been linked with a host of other conditions including myocarditis, ischemic heart disease, peripartum cardiomyopathy, connective tissue disease, and HIV, among others.[17]

Diastolic dysfunction results most commonly from pressure overload conditions that lead to a pathologically hypertrophied and stiff ventricle that is unable to relax. Restrictive cardiomyopathy is an alternate pathogenesis that gives rise to the same functional outcome. Diagnoses include endo-myocardial fibrosis, endocardial fibroelastosis, cardiac amyloidosis, hemochromatosis, and radiation injury.

Less common causes of heart failure include severe chronic anemia, metabolic disorders (such as beri-beri), endocrine derangements (such as thyrotoxicosis), arrhythmias, and pulmonary heart disease. Rare instances of heart failure are reported as side effects of treatments for unrelated conditions. For instance, the cardiotoxic effects of Doxorubicin/Adriamycin can culminate in heart failure, especially if potentiated by concurrent cardiotoxic drugs, mediastinal radiotherapy, or chronic hypertension.[18]

Left Ventricular Failure

Under physiologic conditions, the stroke volume is regulated by preload, which is the measure of myocardial fiber stretch at the end of diastole. Afterload is the resistance that needs to be overcome by the ventricle to eject blood. Contractility is the inotropic state of the heart independent of the preload and the afterload. Disease processes that result in heart failure are known to modulate one or more of these factors that affect cardiac output.[19] Left- or right-sided heart failure may result, depending on the site of major damage. Left ventricular dysfunction can be characterized as systolic dysfunction – reduced ejection fraction due to compromised ventricular contraction, or diastolic dysfunction – reduced ventricular filling due to inadequate relaxation. Ejection fraction (EF), is defined as the fraction (%) of the end diastolic volume that is pumped by the ventricle during systole. Systolic dysfunction is typically classified by an EF <40%. In contrast, diastolic dysfunction typically maintains an EF >40%. Although 70% of cases of left ventricular heart failure are considered to be a result of systolic dysfunction, recent data suggest that a significant percentage of cardiac dysfunction occurs in the presence of preserved left ventricular systolic function.[20] Ischemic heart disease (including myocardial infarction), and chronic uncontrolled hypertension with associated pressure overload, are leading causes of systolic left heart dysfunction, and culminate in heart failure. The impaired contractility of the left ventricle in systolic dysfunction leads to a decrease in stroke volume (SV) and cardiac output (CO), with resultant global hypoperfusion. Decrease in SV is also associated with an increase in end-systolic and end-diastolic ventricular volumes and an increase in left ventricular end-diastolic pressure (LV-EDP). These changes in left ventricular indices cause an increase in left atrial pressure and subsequent backpressure in the pulmonary capillary circulation. Dyspnea is the clinical manifestation of impaired alveolar gas exchange secondary to pulmonary venous congestion.

Diastolic dysfunction is seen most commonly in the setting of hypertension, and also complicates heart disease in diabetes mellitus, obesity, and cardiomyopathy. Contrary to systolic failure, the contractility of the heart and ejection fraction is maintained close to physiologic levels in diastolic dysfunction. Diastolic dysfunction characteristically results from abnormal stiffness of the ventricular wall and the inability of the left ventricle to relax adequately during diastole, such as is seen in cardiac pathologies with extensive fibrosis. Under conditions of increased metabolic demand, such as exercise, the heart is unable to increase cardiac output. The inability of the ventricle to adequately expand during diastole results in an increase in ventricular filling pressure with subsequent elevation of pulmonary venous pressure. Rapid changes may cause acute-onset pulmonary edema, with dyspnea and impaired exercise tolerance. Contemporary literature indicates increased awareness of heart failure with preserved systolic function (including diastolic dysfunction), especially in elderly and female populations.[20]

Right Ventricular Failure

Most often, right-sided heart failure occurs at a late stage in patients with left-sided heart failure, when the elevated pressure in the pulmonary circuit affects the right ventricle and atrium. Pure right-sided heart failure is a rare event that occurs secondary to pulmonary diseases and is termed as cor pulmonale. Pulmonary diseases that result in cor pulmonale are associated with vasoconstriction, pulmonary hypertension, and increased afterload of the right ventricle. These include interstitial lung disease, primary pulmonary hypertension, and pulmonary thromboembolic disease. Conditions such as chronic sleep apnea and altitude sickness cause pulmonary vasoconstriction through hypoxia. In right-sided heart failure, hypertrophy of the right ventricle helps to overcome the elevated pulmonary vasculature resistance, and reduce congestion of systemic and portal venous circulations (which are proximal to the right heart). In a pure right-sided heart failure, there is minimal pulmonary congestion – instead the systemic venous and portal venous systems become congested. The clinical presentation of a right-sided failure is thus characterized by edema, ascites, pleural effusions, hepatosplenomegaly, renal hypoperfusion, and azotemia. Because left ventricular failure is the most common cause of right-sided heart failure, a clinical syndrome of biventricular failure is common.

Neurohormonal Adaptation

Many neurohormonal compensatory mechanisms, including the sympathetic nervous system (Fig. 4.1) and the renin–angiotensin–aldosterone axis, increase the mean arterial pressure and total peripheral resistance by vasoconstriction (Fig. 4.2). In addition, by augmenting sodium and water retention, these processes contribute to increasing cardiac output via the Frank-Starling mechanism. Although initially beneficial, chronically elevated activity of these systems eventually adds to pressure and volume overload, with resultant cardiac decompensation.[21]

PHYSIOLOGIC HYPERTROPHY

Hypertrophy is derived from the Greek *hyper*, meaning over, and *trophy*, meaning growth. It is widely believed to be an adaptive response to increased workload. By undergoing hypertrophy, ventricular wall stress remains constant at higher intraventricular pressures (LaPlace's law). Since cardiomyocytes make up 80–85% of the ventricular volume and are largely thought to be terminally differentiated, the bulk of cardiac hypertrophy results from cardiomyocyte growth (i.e. increase in size). From both clinical and mechanistic standpoints, two fundamental types of cardiac hypertrophy occur: physiologic and pathologic.

Physiologic hypertrophy occurs in very limited circumstances. The most dramatic example is postnatal, or maturational, where the heart grows more than twofold in size. Although some cardiomyocytes become binucleate, most growth results from an increase in cardiomyocyte length and diameter.[23] Ventricular hypertrophy observed in pregnant women and professional endurance athletes results in more limited growth. Typical changes are only about a 10–20% increase in size compared to age-matched, sedentary, non-pregnant controls.[24,25]

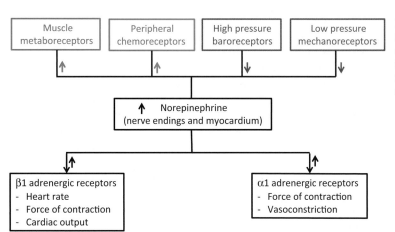

FIGURE 4.1 Increases in heart rate, cardiac output and force of contraction occur early during the course of heart failure and aid in maintaining tissue perfusion close to physiologic levels. Green, normal excitatory stimulus; Red, normal inhibitory stimulus; up-arrow, increase in heart failure; down-arrow, decrease in heart failure.

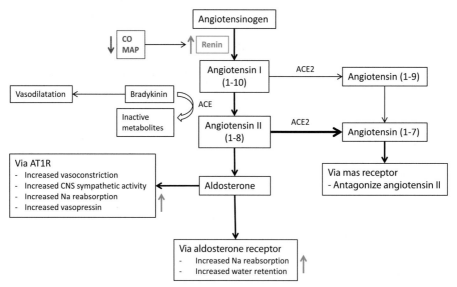

FIGURE 4.2   Activation of the renin–angiotensin–aldosterone system in heart failure. The physiologic response to decreased cardiac output and mean arterial pressure is mediated by a number of neuro-endocrine intermediates, including angiotensinogen, angiotensin, aldosterone, and renin. In the context of heart failure, activation of the renin–angiotensin–aldosterone axis results in exacerbation of heart failure. Low cardiac output results in decreased renal perfusion, triggering activation of the system. Angiotensin causes vasoconstriction, increasing peripheral vascular resistance and mean arterial pressure, while aldosterone results in volume retention. In a pressure- and volume-overloaded heart, these mechanisms exacerbate pressure and volume overload, resulting in further declines in cardiac output and further decreases in renal perfusion, reactivating the system, and taking the patient downward. ACE, angiotensin-converting enzyme, AT1R, angiotensin II receptor type 1, CO, cardiac output, MAP, mean arterial pressure, green arrow, increase in heart failure, red arrow, decrease in heart function.[22]

Ventricular Function

The most important characteristic of physiologic hypertrophy, compared with pathologic hypertrophy, is that ventricular function remains normal or even improved, rather than impaired. Both systolic and diastolic functions are normal or enhanced in both athletes and pregnancy when measured by echocardiogram. In further contrast to pathologic hypertrophy, both states are fully reversible. Post-partum women undergo complete mass regression within 8 weeks, and athletes regress even faster, losing most additional mass within a few weeks of deconditioning.

Angiogenesis, Fibrosis, Energy Substrates, and Gene Activation

Critical cellular and molecular events further separate physiologic and pathologic hypertrophy. Angiogenesis is significantly increased in the myocardium during exercise training, as measured by coronary blood flow capacity, coronary artery diameter, and capillary density. Pathologic models are associated with increased fibroblast activity and fibrosis, while physiologic hypertrophy is associated with unchanged levels of fibroblast activity and collagen deposition. In mitochondria, fatty acid oxidation (FAO) accounts for 80–85% of the energy production in the adult cardiomyocyte. In pathologic hypertrophy, there is increased

utilization of less efficient glycolytic pathways. In physiologic hypertrophy, the ratio of FAO to glycolysis is preserved. At the gene expression level, pathologic hypertrophy models classically demonstrate induction of a fetal gene expression program including atrial natriuretic factor (ANF), brain natriuretic peptide (BNP), skeletal muscle α-1-actin, (SMα1actin) and β-myosin heavy chain (β-MHC) – all of which are absent in exercise models of hypertrophy.

Thyroid Hormone

The role of thyroid hormone tri-iodothyronine (T3) on physiologic growth is best understood in the context of postnatal cardiac growth. Within a few weeks after birth, T3 levels spike 2000-fold, and then fall back down by the third week.[26] Rodent studies demonstrate that T3 regulates the perinatal change in transcription from β-MHC to α-MHC.[27] T3 additionally increases the expression of SERCA (sarcoplasmic/endoplasmic reticulum calcium ATPase-2, critical for maintaining Ca^{++} concentrations in the sarcoplasmic reticulum), the β1 adrenergic receptor, cardiac troponin I (cTNI), atrial natriuretic factor (ANF), sodium/calcium exchanger (NCX), thyroid receptor alpha (TRα1) and adenylyl cyclase subtypes.[28] Given this array of proteins whose function enhances cardiac performance, it is logical that T3 stimulation of cardiomyocytes can result in enhanced cardiac performance.

Insulin

Insulin acts by binding to the tyrosine kinase insulin receptor (IR), which ultimately activates the phosphatidylinositol 3'-kinase–protein kinase B (PI3K-AKT) signaling pathway. Cardiac specific IR knockout mice show smaller hearts with smaller individual cardiomyocyte volumes, indicating that physiologic hypertrophy is inhibited.[29] When challenged with aortic constriction, however, these mice are more prone to the development of pathologic hypertrophy.[30] In short, the insulin-signaling pathway is essential for normal cardiac growth, and its absence may promote or enable pathologic hypertrophy.

Insulin-like Growth Factor 1

Insulin-like growth factor 1 (IGF1) has roles in both systemic and organ-specific regulatory mechanisms. IGF1 binds to the insulin receptor (IR) and the IGF1 receptor (IGF1R). IGF1R is a transmembrane tyrosine kinase receptor that activates PI3K-AKT-phosphoinositide-dependent protein kinase 1 (PDK1) and subsequently glycogen synthase kinase 3β (GSK3b). IGF1 and IGF1R knockout mice have severe growth retardation and die at birth.[31] IGF1 transgenic mice, in which IGF1 is linked to the α-MHC or SM-α-1-actin promoters, show early development of physiologic hypertrophy, but over time the phenotype becomes pathologic, with development of fibrosis and decreased function.[32] Transgenic overexpression of IGF1R using the α-MHC promoter results in development of physiologic hypertrophy without subsequent development of pathology.[33] Conversely, IGF1R conditional deletion does not affect cardiac growth, but does make mice resistant to exercise-induced hypertrophy.[34]

Mechanotransduction

Mechanotransduction is a well-known phenomenon in the cardiomyocyte in which physical contacts are converted into intracellular signals by transmembrane proteins. One stretch receptor expressed by all cells (including myocytes) is the transient receptor potential channel (TRPC). Two subtypes of this receptor – TRPC1 and TRPC6 – are each activated by stretching and are overexpressed in hypertrophy. When knocked out, mice are more resistant to pathologic hypertrophic stimuli.[35] Integrins are another class of transmembrane protein that transmit stretch-related changes in the extracellular matrix through an intracytoplasmic tail. This signals intracellular focal adhesion complexes that include focal adhesion kinase (FAK) and integrin-linked kinase (ILK). These kinases then phosphorylate and activate RHO GTPases, PI3K, and protein kinase C (PKC).

Cardiac-specific ablation of the intracytoplasmic integrin signaling tail exacerbates pressure-overload-induced hypertrophy.[36] Within the cardiomyocyte, numerous proteins at the Z-line are involved in stretch sensing including: muscle LIM protein,[37] myopalladin, palladin, ankyrin, and cardiac ankyrin repeat domain protein (CARP). Of these, muscle LIM protein and CARP have known associations with hypertrophy. CARP overexpression transgene is resistant to the development of isoproterenol and pressure-overload-induced hypertrophy.[38] Titin is a protein that spans the length of the sarcomere, from Z-line to Z-line, with over 20 known ligands, many of which are believed to be stretch receptors, yet its precise relationship with hypertrophy remains to be explored.

Intracellular Pathways

PI3K

PI3K is one of the common effectors of insulin, insulin-like growth factor, and integrin signaling pathways (Fig. 4.3). Overexpression of the catalytic subunit of PI3K, p110α, in mouse hearts promotes physiologic hypertrophy.[39] Conversely, overexpression of a dominant negative form of p110α results in atrophy. Phosphatase and tensin homolog (PTEN) is a lipid phosphatase that acts to inhibit phosphatidylinositol 3,4,5 triphosphate (PIP3). Cardiac-specific PTEN deletion has also been shown to promote cardiac growth.[40]

AKT

AKT, also known as protein kinase B, is activated by 3-phosphoinositide-dependent protein kinase-1 (PDK1), another kinase recruited to the cell membrane by PIP3 synthesis. PDK1 inactivation reduces cardiomyocyte volume and heart mass. Similarly, Akt null mice are resistant to physiologic hypertrophy in response to swimming. Constitutively active Akt1 mutant mice initially develop physiologic LVH, although pathologic conversion occurs over time. Similar effects are observed in a membrane-localized mutant of Akt1. By comparison, nuclear-targeted Akt1 results in hyperplasia without hypertrophy. One of the mechanisms of AKT is to promote protein translation by inhibiting glycogen synthase kinase 3-β (GSK3β), itself a negative regulator of protein translation. Mouse overexpression models of GSK3β fail to hypertrophy in the post-natal period and die shortly thereafter from heart failure.[41] Lastly, AKT shifts the balance of protein turnover to anabolism by phosphorylating and inactivating the pro-catabolic transcription factor forkhead box protein 03 (FOX03). This prevents transcription of the pro-catabolites ubiquitin ligase atrogin-1 and

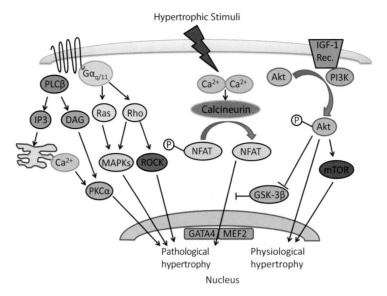

FIGURE 4.3 Intracellular signaling pathways. Intracellular signaling pathways involved in pathological and physiologic hypertrophy. Activation of a Gα$_{q/11}$ G-protein coupled receptor (Gα$_{q/11}$) leads to activation of the small GTP-binding proteins, Ras and Rho, which promote pathological hypertrophy through activation of the mitogen-activated protein kinase (MAPK) signaling cascade. Rho also activates Rho kinase (ROCK), another activator of pathologic hypertrophy. Activation of a Gα$_{q/11}$ coupled receptor additionally activates phospholipase-Cβ (PLCβ), resulting in inosital-1,4,5-trisphosphate (IP3) and diacylglycerol (DAG) production. IP3 binds to an IP3 receptor on the sarcoplasmic reticulum stimulating calcium release. Calcium and DAG activate protein kinase Cα (PKCα), which promotes pathological hypertrophy. Many forms of hypertrophic stimuli increase the amount of intracellular calcium, leading to the activation of the protein phosphatase, calcineurin. Activated calcineurin dephosphorylates the nuclear factor of activated T-cells (NFAT), allowing NFAT to enter the nucleus, interact with GATA4 and myocyte enhancer factor-2 (MEF2) leading to increased protein synthesis and pathological hypertrophy. Glycogen synthase kinase-3β (GSK-3β) can phosphorylate and thereby inhibit NFAT nuclear translocation. Stimulation of the insulin-like growth factor 1 receptor activates phosphatidylinositide 3-kinase (PI3K), which phosphorylates and activates Akt to promote physiologic hypertrophy. Akt further activates the mammalian target of rapamycin (mTOR) and inhibits GSK-3β.

muscle-specific RING finger protein-1 (MURF1).[42,43] Altogether, AKT appears to promote hypertrophic growth of the heart; the timing, duration, and precise nature of the action determine if this is ultimately beneficial or pathologic.

mTOR

The mammalian target of rapamycin (mTOR) regulates adaptive growth of the heart at the level of mRNA translation. mTOR and regulatory associated protein of mTOR (RAPTOR) combine with other proteins to make up mTOR complex-1 and -2 (mTORC1 and mTORC2). mTORC1 is activated via an AKT-led pathway, as well as by certain amino acids, and is inhibited by 5′ adenosine monophosphate-activated protein kinase (AMPK).[44] Activated mTORC1 initiates translation activity by directly regulating S6K ribosomal proteins, and by liberating eukaryotic translation initiation factor 4E from its binding protein.[44] Experimentally, treatment with rapamycin is effective in reversing hypertrophy produced through Akt overexpression.[45] However, blocking the mTOR pathway by overexpression of a dominant negative mTOR is insufficient to inhibit the hypertrophic response in exercised mice.[46] In summary, the mTOR pathway is one of several redundant pathways that contribute to the development of physiologic LVH.

C/EBPβ

CCAAT/enhancer binding protein-β (C/EBPβ) is a transcription factor that is commonly associated with regulation of cellular proliferation, but has recently been tied to the regulation of physiologic hypertrophy.[47] C/EBPβ is down-regulated during exercise-induced physiologic hypertrophy, but remains constant during pressure-overload-induced pathologic hypertrophy. siRNA silencing of C/EBPβ in rat neonatal cardiomyocytes induces both cardiomyocyte proliferation and hypertrophy. In adult mice, C/EBPβ heterozygotes are resistant to the pathologic effects of pressure overload. Relative to wild-type mice, C/EBPβ heterozygotes have a comparable increase in cardiomyocyte size, but with improved fractional shortening and decreased pulmonary weight (signifying less heart failure).[47] C/EBPβ inhibition, as a means of inducing physiologic hypertrophy, represents a potential therapeutic modality for patients with heart failure.

ERK1/2

Extracellular signal related kinases 1/2 (ERK1/2) are kinases activated by extracellular signals that translocate to the nucleus, phosphorylate targets, and initiate transcription. Also called mitogen activated kinase 3/1 (MAPK3/1), these kinases are stimulated by growth factors and stretching. Overexpression of an active mutant

ERK1 induces physiologic hypertrophy that is protective from ischemia reperfusion injury.[48,49] Conversely, inhibition of ERK1/2 leads to increased dilated cardiomyopathy in the face of pressure overload.[50] ERK1/2 therefore represents an important aspect of physiologic hypertrophic signaling, both in the response to exercise and in the balance of response to pathologic stresses.

AMPK

5' Adenosine monophosphate-activated protein kinase (AMPK) is a metabolic switch that balances energy supply with metabolic demand. During exercise, activated AMPK increases the available energy supply by stimulating catabolic pathways including fatty acid oxidation, glucose uptake and glycolysis, and shuttering anabolic pathways like fatty acid synthesis and protein transcription. Although similarly named to cyclic AMP-activated protein kinase (protein kinase A), the actions of AMPK are very different and should not be confused. Long-term inhibition of AMPK leads to pathologic hypertrophy and heart failure.[51] Treatment with a constitutively active mutant or rapamycin restores normal ventricular shape and function.[52] AMPK is thus a vital control in maintaining the heart's ability to respond to different stresses, both physiologic and pathologic.

PATHOLOGIC HYPERTROPHY

Risk Factors

The most common risk factor for LVH is advanced age, making pathologic LVH one of the most common conditions of elderly North Americans. Intuitively, longer exposure to a physiologic stress will increase the likelihood of symptoms resulting from that stress.[53] Coincident with age are the additional risk factors of increased blood pressure and increased body weight.[54] Both have an increased incidence in the elderly, and are themselves independent risk factors for LVH. Additional independent risk factors for LVH include hypercholesterolemia,[55] prior myocardial infarction,[56] and diabetes.[57] Other forms of pressure overload, such as aortic stenosis, are similarly powerful predictors. Valvular insufficiency is more commonly associated with myocardial dilation, but may involve hypertrophy as well.[58] The African-American race is also linked to hypertrophy,[59] as are dietary preferences such as high sodium intake (independent of blood pressure), and social stressors such as 'job strain.'[60]

Clinical Sequelae of Pathologic Hypertrophy

Ventricular Arrhythmias

Left ventricular hypertrophy is strongly associated with both atrial and ventricular arrhythmias. When detected by electrocardiogram, there is a significant increase in arrhythmias leading to sudden cardiac death (SCD).[61] LVH is one of the biggest risk factors for ventricular tachycardia; there is a 40-fold increase in ventricular tachyarrhythmia in patients with electrocardiographic LVH.[62] In patients with both ventricular tachycardia and LVH, the risk of SCD is increased 10-fold.[63] More recent evidence suggests that with regression of LVH, the risk of SCD is reduced.[64] The incidence and prevalence of atrial fibrillation are also increased with LVH. In one study, for each standard deviation increase in LV mass, there was a 20% increase in the incidence of atrial fibrillation.[65]

Coronary Flow Reserve

Coronary flow reserve (CFR) is a descriptor of myocardial blood supply, specifically the ability of the coronaries to increase blood flow under stress. Patients with LVH have decreased CFR, especially in the context of pressure overload.[66] Essentially, the muscular growth of the heart outstrips the vascular supply. This leads to myocardial ischemia, even in the context of normal epicardial coronary anatomy. When atherosclerotic coronary disease is combined with LVH, there is a significantly increased risk of mortality.[67] Decreased coronary flow reserve may also explain the increased prevalence of ventricular arrhythmia and sudden death in the LVH patient population.

Ventricular Function

Although ejection fraction (EF) is the most widely used descriptor of cardiac function, EF may overstate cardiac function in the setting of LVH. With LVH, patients may even present with clinical heart failure with a normal EF.[68] Mechanistically, increased ventricular wall thickness and increased interstitial fibrosis create a stiffer ventricle with impaired diastolic relaxation. Less filling of the ventricle during diastole results in a smaller stroke volume, and therefore less cardiac output is produced at a given heart rate, even though the ejection fraction may be within normal range.[69,70] Progressive ventricular wall thickness and fibrosis may go on to impair contraction as well as relaxation, resulting in both a small stroke volume and a depressed ejection fraction. This strong association between LVH and heart failure and mortality has been borne out in numerous studies from the last several decades.[5,71–73]

Animal Models of Pathologic LVH

Experimental models used to induce cardiac hypertrophy in animals have remained remarkably constant over the last 50 years.[74] The underlying constants are pressure overload and increased myocardial work, buttressed by different types of neurologic, hormonal, and biochemical stresses. Aortic constriction is by far the most widely used method for the initiation of pathologic LVH. The robust nature of the response enables the technique to be applied to any area of the thoracic aorta, including the ascending, transverse, or descending thoracic aorta. Over time, transverse aortic constriction (TAC) has become the dominant model in the mouse because it is the most technically feasible. Pulmonary artery constriction yields a similarly consistent response with right ventricular hypertrophy. Renal artery constriction activates the renin–angiotensin–aldosterone axis, causes hypertension, and yields rapid progression of LVH. Hyperthyroidism, like the preceding three mechanisms, also produces rapid results, with measurable hypertrophy developing in mere days. A more gradual hypertrophy may be induced by treatment with sympathomimetic agents, repeat bleeding to produce anemia, certain nutritional deficiencies, and low environmental oxygen similar to extreme elevation.[75] In rats, the spontaneously hypertensive rat is a well-established model of cardiovascular disease including hypertrophy.[76] The model is limited by the absence of many knockout strains, as well as the longer lifespan, increased gestation time, increased size, and housing costs.

Measurement of LVH

Clinical Measurement of LVH

Initial diagnosis of LVH, apart from autopsy, was performed using electrocardiographic criteria. With the development of ultrasound, echocardiography has become the new standard, although electrocardiographic criteria have been refined using echocardiographic results. Both have a high degree of specificity, and both have well-established prognostic significance. The primary limitation is reproducibility due to operator skill. Newer imaging modalities include 3D echo and magnetic resonance imaging (MRI). Both have the advantage of high sensitivity, high specificity, and high reproducibility. However, both have highly limited availability, and MRI in particular is quite expensive.[77]

Measurement of LVH in Experimental Models

In living animals, echocardiography remains the common standard, regardless of animal size. Echocardiograms are routinely performed in mice, and have even been reproducibly performed in fruit flies.[78] Invasive measurements have also stood the test of time, including gross measurements such as heart weight and the ratio of heart weight to body weight. Microscopic measurements include the cardiomyocyte cross-sectional area as well as quantitative measurements of fibrosis. Biochemical measures predominate in the evaluation of whole animals and at the cellular level. Synonymous with pressure overload hypertrophy is the term fetal gene expression program. Protein studies in the rat first showed that pressure overload induces a switch from the adult α-myosin heavy chain (MHC) to the fetal β-MHC.[79] Later publications confirmed that this is part of a broader transition to a mitogenic growth program that mimics that of the fetal heart.[80] The genes typically characterized as part of the hypertrophic growth phase include β-MHC, atrial natriuretic peptide (ANP), brain natriuretic peptide (BNP), skeletal α-actin (ACTA1), and smooth muscle α-actin (ACTA2).

MOLECULAR MECHANISMS OF PATHOLOGIC LVH

Cell Signaling Processes

Calcineurin/NFAT

Induction of the calcineurin/NFAT pathway is a common hallmark of pathological hypertrophy (Fig. 4.3). Calcineurin (also called protein phosphatase 2B) is a calcium/calmodulin-activated serine/threonine phosphatase that, once stimulated, de-phosphorylates and thereby activates the NFAT transcription factor.[81] De-phosphorylated NFAT translocates to the nucleus and associates with the GATA4 and myocyte enhancer factor-2 (MEF2) transcription factors.[81,82] Ultimately, activated NFAT leads to the transcription of hypertrophy-associated genes (fetal gene program), including α-actin, endothelin-1, atrial natriuretic factor (ANF), and β-myosin heavy chain.[83] Calcineurin can be regulated by the ubiquitin ligase, atrogin-1, which can ubiquitinate it, leading to its degradation by the proteasome. Calcineurin-induced NFAT translocation is antagonized by the protein kinases PI3K, AKT, and GSK-3β, thereby attenuating pathological hypertrophy.[81,84]

Small GTP-binding Proteins – Ras and Rho

Small guanosine triphosphate (GTP)-binding proteins are involved in a variety of cellular processes including cell differentiation, migration, and division (Fig. 4.3). They can be divided into five main subfamilies: Ras, Rho, Rab, Arf, and Ran. Ras and Rho have been implicated in the progression of pathological hypertrophy.[85] Also called GTPases, these small GTP-binding proteins act as molecular switches by cycling between an active GTP-bound state and an inactive GDP-bound state.[86,87] They are activated by stimulation of cardiomyocyte receptors coupled to $G\alpha_{q/11}$ (Ras and Rho) and $G\alpha_{12/13}$ (Rho only) G proteins with angiotensin II, endothelin-1, or phenylephrine.[88,89] Once activated, Ras and Rho modulate the activity of the mitogen-activated protein kinases (MAPKs) and ERK along with Rho-activating Rho kinase (ROCK). These events activate the hypertrophic gene program, increase protein synthesis, and increase cardiomyocyte size. ROCK also increases the amount of actin production and actin organization, hallmarks of cardiac hypertrophy.[87,90–92] Importantly, inhibition of Ras and Rho is protective against cardiac hypertrophy.[93–97] Ras and Rho activity is inhibited by cGMP-dependent protein kinase (PKG), guanine nucleotide-dissociation inhibitors (GDI), and guanine nucleotide exchange factors (GEFs).[86,87,98,99]

PKC

Neurohormonal signals such as angiotensin II, endothelin-1, and catecholamines bind to α-adrenergic receptors on the surface of cardiomyocytes, which are coupled to heterotrimeric G proteins named $G\alpha_{q/11}$ (Fig. 4.3). These $G\alpha_{q/11}$-coupled receptors are associated with phospholipase Cβ (PLCβ). Activation of PLCβ results in the production of diacylglycerol (DAG) and inositol-1,4,5-trisphosphate (IP3).[100,101] IP3 binds to an IP3 receptor on the sarcoplasmic reticulum, stimulating the release of calcium.[81] DAG and calcium then bind to and activate protein kinase Cα (PKCα), a key mediator of cardiac hypertrophy and contractility.[100,101] Once activated, PKCα phosphorylates many intracellular proteins including myofilament proteins (sensitizing them to calcium) and promotes calcium release. Additionally, PKCα phosphorylates and inhibits the antihypertrophic histone deacetylases (HDAC) 4, 5, 7, and 9. This enhances protein synthesis and cardiomyocyte growth.[81] PKCα inhibition in mice is protective against cardiac hypertrophy and reduces cardiac remodeling.[101–105] Furthermore, the cardiomyocyte expression of a dominant negative PKCα inhibits progression to heart failure.[106]

Transcriptional Regulation of Hypertrophy

SRF

Serum response factor (SRF) is a MAD-box containing transcription factor known to regulate many muscle-specific genes.[107] SRF has been implicated in regulating the hypertrophic genes cardiac α-actin, α-MHC, and β-MHC, and other transcription factors like GATA4.[107,108] Cardiac overexpression of SRF results in hypertrophy.[108] Interestingly, mice with cardiac-specific deletion of SRF show dilated cardiomyopathy and reduced cardiac contractility, along with defects in their cardiac structural proteins (regulated by SRF) and early-onset heart failure.[109] SRF is regulated by the ubiquitin ligase, MuRF1. Genetic deletion of MuRF1 yields mice with exaggerated hypertrophy and enhanced expression of the SRF-dependent hypertrophy genes.[110] These findings suggest that SRF plays a pivotal role during the development of hypertrophy and that SRF is needed at baseline to maintain adequate cardiac function.

MEF2

Myocyte enhancer factor-2 (MEF2) is another MAD-box containing transcription factor that promotes myocyte differentiation and hypertrophic gene expression (Fig. 4.3).[111] The transcriptional activity of MEF2 is enhanced by phosphorylation from p38 MAPK and BMK-1, and dephosphorylation by calcineurin. Class II histone deacetylases (HDACs) inhibit the actions of MEF2. This inhibition can be relieved by calcium-calmodulin kinase and protein kinase D-mediated phosphorylation of the class II HDAC.[81,111] Overexpression of MEF2 in the heart results in increased hypertrophy following stress stimuli.[112] MEF2 is a primary target of the calcineurin/NFAT pathway that induces expression of hypertrophic genes.[82,113] Inhibition of MEF2 effectively blunts calcineurin-induced cardiac hypertrophy.[113] Along with MEF2, GATA4 is the other primary target of the calcineurin/NFAT pathway.[81,82]

GATA4

GATA4 belongs to the GATA family of transcription factors characterized by their ability to bind to the DNA base pairs G, A, T, A and by the presence of a zinc finger. This allows them to bind DNA to regulate cardiac development, differentiation, proliferation, and survival.[114,115] GATA4 mediates the induction of a set of genes including: α-MHC, myosin light chain 1/3 (MLC1/3), cardiac troponin C, cardiac troponin I, atrial natriuretic peptide (ANP), brain natriuretic peptide (BNP), cardiac-restricted ankyrin repeat protein (CARP), cardiac sodium–calcium exchanger (NCX1), cardiac m2 muscarinic acetylcholine receptor, A1 adenosine receptor, and carnitine palmitoyl transferase I β, many during cardiac hypertrophy (Fig. 4.3).[107,116] Overexpression of GATA4 alone is sufficient to cause hypertrophy.[117] GATA4 phosphorylation on serine105 by ERK1/2 induces GATA4 DNA binding and subsequent hypertrophic gene expression.[118] During hypertrophy, NFAT associates with GATA4 to induce expression of hypertrophic genes.[81] Genetic inhibition of GATA4 attenuates hypertrophy induced by pathological and physiologic stimuli.[119,120] GSK-3β can phosphorylate GATA4, which induces its nuclear export, eventual ubiquitination, and subsequent degradation by the 26S proteasome.[121] Collectively, these studies indicate the pro-hypertrophic effects of GATA4 activation, and suggest that GATA4 may represent a worthy target for therapeutic pharmacological intervention.

Cell Surface/Membrane Level Control

G-protein-coupled Receptors

G-protein-coupled receptors (GPCRs) compose the largest receptor family in mammals and are responsible for the regulation of most physiologic functions (Fig. 4.3). Due to their vast expression and physiologic impact, GPCRs are preferentially targeted with therapeutic regimens.[122,123] GPCRs are coupled to four different families of heterotrimeric G proteins, G_s, $G_{i/o}$, G_q/G_{11}, and G_{12}/G_{13}. Once the receptor is stimulated by a ligand, the G-protein becomes active and regulates the activity of an effector such as second-messenger-producing enzymes or ion channels.[122]

β-Adrenergic – Gs

β1 adrenergic receptors (β1-AR) are coupled to the G_s G-protein/adenyl cyclase signal transduction pathway, a central pathway regulating cardiac function (Fig. 4.4).[123] This pathway activates cAMP-dependent protein kinase (PKA) to mediate β1-AR's actions, mainly increased heart rate and force of contraction.[122] During cardiac stress there is dysregulation and uncoupling of the β1-AR pathway.[122] Overexpression of β1-AR results in cardiac hypertrophy, increased apoptosis, and eventually heart failure, which are attenuated by β1-AR inhibition.[123] Notably, β1-AR blockade is one of the most prevalent and effective treatments of human heart failure patients.[122,123]

α-Adrenergic – Gq/G11

The α1 adrenergic receptor (α1-AR) is coupled to a G_q/G_{11} G-protein/phospholipase Cβ1 pathway. Stimulation of this pathway generates IP3, subsequent release of intracellular calcium, and DAG, followed by PKC activation (Fig. 4.4).[122,124] There is increased stimulation of α1-AR during cardiac stress, resulting in increased cardiac force of contraction, vasoconstriction, and protein synthesis. Cardiac hypertrophy is a common feature of α1-AR stimulation, whereas α1-AR inhibition effectively attenuates the onset of hypertrophy.[124] Overexpression of α1-AR in mice yields hypertrophy, increased fibrosis, and early death. α1-AR blockade has had mixed results in clinical practice, demonstrating the complexity of this system.[124]

Renin–angiotensin System

The renin–angiotensin system (RAS) and its primary effector, angiotensin II (AngII), are principal mediators of cardiac hypertrophy. Once produced, AngII binds AngII type1 (AT1) or AngII type2 (AT2) receptors on cardiomyocytes. The AT1 receptor is coupled to a G_q/G_{11} G-protein and activates ERK1/2, p38 MAPK, and protein synthesis (Fig. 4.4).[125,126] The RAS, through aldosterone, promotes water retention. This increases blood pressure and cardiac stress. Additionally, AngII can be cleaved by angiotensin-converting enzyme 2 (ACE2) to yield Ang1-7, which is thought to be cardioprotective by counterbalancing the harmful effects of AngII.[125] Inhibition of RAS in mice exposed to pressure overload reduces hypertrophy, decreases fibrosis, and increases lifespan.[125,126] Inhibition of RAS, either with an ACE inhibitor, an AT1 receptor blocker, or an aldosterone inhibitor, generally has beneficial effects, although combination therapy is limited by systemic hypotension and diminished renal perfusion.[125]

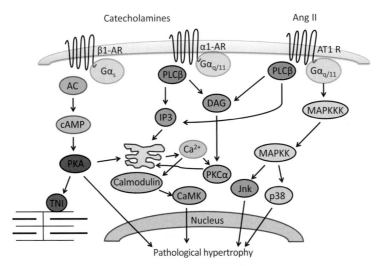

FIGURE 4.4　Cardiomyocyte G-protein-coupled receptor regulation of hypertrophy. The catecholamines epinephrine and norepinephrine can stimulate the cardiomyocyte β1 and α1 adrenergic receptors (β1- and α1-AR), which are coupled to a Gα$_s$ and Gα$_{q/11}$ G-protein, respectively. Activation of the β1-AR stimulates adenylate cyclase (AC) to produce cyclic adenosine monophosphate (cAMP), which activates protein kinase A (PKA). PKA promotes increased chronotropy, inotropy, and hypertrophy. Stimulation of the α1-AR activates phospholipase-Cβ (PLCβ) to produce inositol-1,4,5-trisphosphate (IP3) and diacylglycerol (DAG). IP3 stimulates calcium release from sarcoplasmic reticulum by binding to an IP3 receptor. Intracellular calcium and DAG activate protein kinase Cα (PKCα). Calcium can similarly activate calmodulin leading to the stimulation of calcium–calmodulin protein kinase II (CaMK). CaMK and PKCα promote pathological hypertrophy. The angiotensin 1 receptor (AT1R) is coupled to a Gα$_{q/11}$ G-protein, with similar effects on IP3 and DAG. Additionally, stimulation of the AT1R activates the MAPK signaling cascade to promote hypertrophy: mitogen-activated protein kinase kinase kinase (MAPKKK) phosphorylates MAPKK, which phosphorylates and activates p38 MAPK and c-Jun N terminal kinase (JNK).

MAPK

The mitogen-activated protein kinase (MAPK) pathway involves a sequence of phosphorylation events from protein kinases. The pathway is activated by G-protein-coupled receptors, receptor tyrosine kinases, transforming growth factor-β, cardiotrophin-1 (gp130 receptor), and stretch (Fig. 4.4). The MAPK pathway has three main divisions: p38 kinases, c-Jun N-terminal kinases (JNK), and extracellular regulated protein kinase 1/2 (ERK1/2). These pathways are stimulated by upstream MAPK kinases (MEKs). MEKs 4/7, MEKs 3/6, and MEKs 1/2 are respectively responsible for activating p38 kinases, JNKs, and ERK1/2.[127,128] Most MAPK pathways are activated during pathological cardiac hypertrophy and end-stage human heart failure.[128–131] Constitutive activation of ERK1/2 results in the formation of concentric hypertrophy, however the heart does not go into failure and cardiomyocytes are protected from cell death.[48,49] The pro-hypertrophic actions of ERK1/2 are in part mediated by enhanced transcriptional activity of NFAT.[132] ERK1/2 inhibition results in cardiac dilation with increased myocyte length.[50,133,134] Overt stimulation of the JNK and p38 MAPK pathways leads to cardiac dilation, while genetic inhibition of JNK and p38 MAPK demonstrates exaggerated hypertrophy in response to pressure overload in a calcineurin/NFAT-dependent mechanism.[135–140] Apparently, crosstalk occurs between all three divisions of the MAPK pathway and the hypertrophic calcineurin/NFAT pathway.

Gp130/STAT3

Glycoprotein 130 (Gp130) is a common receptor subunit of the interleukin-6 family of cytokines. The Gp130 receptor can be stimulated by leukemia inhibitor factor or cardiotrophin-1, leading to the activation of the janus kinase/signal transducer and activator of transcription (JAK/STAT) pathway, ERK, and PI3K/Akt pathways. Gp130 signaling through JAK/STAT and PI3K promotes cell survival and is cardioprotective. ERK-mediated Gp130 signaling stabilizes sarcomeric cytoskeletal organization.[141,142] There is accumulating evidence that Gp130 activation, through JAK/STAT, also stimulates the production of anti-apoptotic and anti-oxidant proteins. Gp130 knockout mice show exaggerated hypertrophy in response to pressure overload. Furthermore, human cardiac hypertrophy and heart failure patients have reduced circulating interleukin-6, therefore reduced Gp130 stimulation. Activation of the Gp130/JAK/STAT pathway has been proposed as a treatment option.[141,142]

Na/H Exchanger

The sodium/hydrogen exchanger-1 (NHE1) is a ubiquitously expressed transporter implicated in the pathogenesis of cardiac hypertrophy.[143] NHE1 expression is increased in both in vitro and in vivo models of hypertrophy and heart failure. NHE-1 inhibition using selective small molecule inhibitors has been shown to decrease NFAT signaling in fibroblasts, decrease hypertrophic gene expression in neonatal cardiomyocytes, and improve contractility in murine models.[143–145] Conversely, overexpression of high-activity NHE1 in mice leads to exaggerated hypertrophy, contractile dysfunction, and heart failure.[144] These mice are also characterized by activation of the pro-hypertrophic CaMKII and calcineurin proteins. Calcineurin binds NHE1, and in the presence of calcium, becomes activated causing translocation of NFAT to activate gene transcription.[143]

Novel/new Signaling Pathways Regulating LVH

Anchoring Proteins

A-kinase anchoring proteins (AKAPs) are scaffold proteins that bind to the regulatory domain of PKA. AKAPs localize PKA to specific subcellular domains depending on which PKA regulatory subunit is bound, regulatory subunit 1 or regulatory subunit 2. Localizing PKA ensures that PKA is coupled to upstream regulator adenylyl cyclase, and downstream phosphodiesterase termination enzymes. AKAPs are known to play roles in heart development, contractility, cytoskeletal organization, β1-AR signaling, cardiac hypertrophy, and heart failure among others.[146,147] AKAPs are believed to function as a signaling hub where multiple signals converge, primarily including mAKAP, AKAP-Lbc, and AKAP79/150. In response to hypertrophic signaling, mAKAP (AKAP6) forms a perinuclear macromolecular complex consisting of ERK5, calcineurin, and PLCε, that regulates gene transcription. MEF2 and NFAT are activated by ERK5 and calcineurin, respectively. PLCε integrates many upstream signaling cascades including those activated by endothelin-1, norepinephrine, insulin-like growth factor-1, and isoproterenol. Disruption of mAKAP from the nuclear envelope results in reduced hypertrophy.[146] AKAP-Lbc, commonly stimulated by α1-AR activation, is not only a scaffold for PKA, PKC, and PKD but is also a guanine nucleotide exchange factor for RhoA. Suppression of AKAP-Lbc is correlated with reduced hypertrophic signaling. AKAP-Lbc has also been proposed to be a point of cross-talk between PKD and RhoA signaling pathways.[146] The role of AKAP79/150 remains incompletely understood, although it does appear to have a significant role in hypertrophy. AKAP79/150 has a calcineurin-binding domain that inhibits calcineurin activity. Cardiomyocyte overexpression of AKAP79/150 has reduced hypertrophy in response to pressure overload.[146,147]

Calcium-regulated Pathways

CALCINEURIN

Calcineurin activates NFAT by dephosphorylation, which in turn stimulates the transcription of pro-hypertrophic genes, including MEF2 and GATA4, thereby promoting pathological hypertrophy. Cardiac-specific activation of the calcineurin/NFAT pathway induces cardiac hypertrophy in transgenic mice, while inhibition of calcineurin/NFAT effectively attenuates cardiac hypertrophy.[81,82] Preclinical evidence suggests that calcineurin inhibition may be a valuable therapeutic modality to reduce progression of cardiac hypertrophy. Results are promising but novel calcineurin inhibitors with improved safety profiles are required.[148]

CALPAINS

Calpains are a family of ubiquitously expressed calcium-activated cysteine proteases. Without calcium, calpains may be activated through ERK-mediated phosphorylation (Fig. 4.5). Calpains are unique as they can regulate protein degradation in the cytosol or specific intracellular compartments, or be secreted to regulate the extracellular milieu.[149,150] Calpains activate pathological hypertrophy through the degradation of IκBα and the activation of calcineurin. This allows NF-κB and NFAT, respectively, to enter the nucleus and induce transcription of hypertrophic genes.[150] Additionally, calpains inhibit physiologic hypertrophy by suppressing Akt.

Calpain inhibitors attenuate the apoptotic pathway and AngII-induced hypertension.[149] The net result of calpain activation is pathological hypertrophy.[150]

PKG

Inhibition of phosphodiesterase 5 (PDE5) activates cGMP-dependent protein kinase (PKG) and is strongly cardioprotective (Fig. 4.6). Accumulating evidence is so impressive that PDE5 inhibition, and by extension PKG activation, is now being assessed as a novel treatment option for cardiac hypertrophy and heart failure patients in clinical trials (e.g. RELAX trial). PDE5 expression increases during cardiac pathology, which suppresses PKG leading to further cardiac dysfunction.[151] Hypertrophic signaling is repressed in vitro and in vivo by various PKG activators, natriuretic peptides, NO, and PDE5 inhibitors. PKG activation is known to inhibit the calcineurin/NFAT pathway, the RhoA pathway, and the transient receptor potential canonical 6 channel. It further activates regulation of G protein signaling proteins 2/4 among other protective effects.[152] PDE5 inhibition effectively prevents the onset of cardiac hypertrophy and even reverses pre-existing hypertrophic remodeling in mice.[153] One clinical trial reported that PDE5 inhibition, via sildenafil, reversed both systolic and diastolic dysfunction, along with decreased left ventricular volume and mass in heart failure patients.[154] The PDE5/PKG axis is a very promising target for the treatment of hypertrophy and heart failure patients.

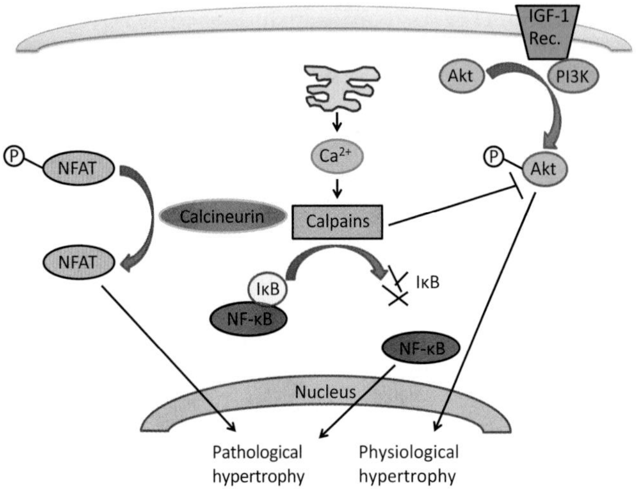

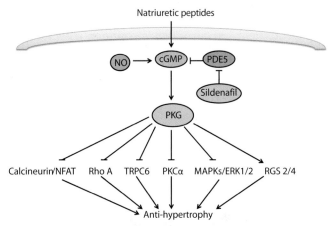

FIGURE 4.5 Calpains modulate hypertrophic signaling pathways. Calpains are activated by intracellular calcium, including calcium released from the sarcoplasmic reticulum. Activated calpains are known to interact with and activate calcineurin, which dephosphorylates NFAT, and lead to pathological hypertrophy. Nuclear factor kappa B (NF-κB) is known to induce pathological hypertrophy but must first be released from its inhibitor, inhibitor of NF-κB (IκB). IκB is degraded by calpains, thus allowing NF-κB to enter the nucleus and promote pathological hypertrophy. Physiologic hypertrophy is repressed by calpains through inhibition of Akt.

FIGURE 4.6 The anti-hypertrophic mechanisms of PKG. Natriuretic peptide binding to the natriuretic peptide receptor and nitric oxide (NO) binding to nitric oxide synthase (NOS) stimulate particulate or soluble guanylate cyclase respectively to produce cyclic guanosine monophosphate (cGMP). Phosphodiesterase-5 (PDE5) is responsible for the breakdown of cGMP. PDE5 inhibitors, such as sildenafil, inhibit cGMP breakdown thereby increasing intracellular cGMP. Once produced, cGMP can bind to and activate cGMP-dependent protein kinase (PKG). Activated PKG has many anti-hypertrophic effects including the activation of G-protein signaling 2/4 (RGS 2/4), and the inhibition of calcineurin/NFAT, Rho A, transient receptor potential cation channel 6 (TRPC6), PKCα, and MAPKs (ERK1/2).

Endothelial Cell Regulation of LVH

Endothelial cells are known to play a critical role in modulating cardiovascular function by producing and secreting factors that act on cardiomyocytes (Fig. 4.7). Endothelial NOS (eNOS) knockout mice show enhanced hypertrophy and fibrosis and reduced capillary density in response to pressure overload.[155] Overexpression of related transcription enhancer factor-1 (RTEF1) in endothelial cells causes exaggerated hypertrophy in response to pressure overload. Presumably RTEF1 mediates increased transcription of vascular endothelial growth factor-B and ERK1/2 activation in adjacent cardiomyocytes.[156] Additionally, overexpression of endothelin-1 (ET1) in endothelial cells is sufficient to cause pathological cardiac hypertrophy.[157] Collectively these findings suggest that endothelial cells have both a pro-hypertrophic (RTEF1 and ET1) and anti-hypertrophic effect (eNOS/NO) on cardiomyocytes.

HDAC

Histone deacetylases (HDACs) are a pivotal element in the cardiac hypertrophic response. They function by removing acetyl groups from chromatin, a process called chromatin remodeling, which enables increased gene expression.[158] HDACs are a large family divided into three main classes: class I (HDACs 1, 2, 3, and 8), class II (HDACs 4, 5, 6, 7, and 9), and class III (sirtuins). Class II HDACs are anti-hypertrophic, while class I HDACs

are pro-hypertrophic.[159,160] Pharmacological inhibition of class I HDACs effectively inhibits cardiac hypertrophy following 2 weeks of pressure overload.[161,162] PKC and CaMK phosphorylate class II HDACs, inducing their nuclear export and de-repressing targeted genes. Among other targets, class II HDACs bind to and repress the transcription factor MEF2 (Fig. 4.8).[163] HDAC5 and HDAC9 constitutive nuclear mutants demonstrate typical class II HDAC anti-hypertrophic effects by inhibiting hypertrophic growth. Conversely, Hdac5 and Hdac9 gene-deleted mice show spontaneous hypertrophy with age and enhanced hypertrophy with pathological stimuli.[159,164] Sirtuins, mainly Sirt3, are elevated during hypertrophy and are thought to act as a negative regulator of hypertrophy. Sirt3-expressing mice are protected from hypertrophic stimuli, whereas Sirt3-deficient mice develop cardiac hypertrophy. Sirt3 activates FOXO3a and the antioxidant genes encoding manganese superoxide dismutase and catalase. This leads to suppressed hypertrophy and decreased cellular levels of reactive oxygen species (ROS).[165]

CHAMP

Cardiac helicase activated by MEF2C protein (CHAMP) is an inhibitor of cardiomyocyte proliferation and hypertrophy. CHAMP overexpression significantly reduces cardiomyocyte hypertrophy during phenylephrine stimulation, blocks activation of the hypertrophic

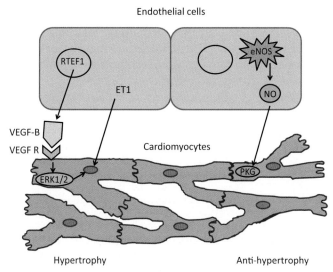

FIGURE 4.7 Endothelial cell regulation of cardiomyocyte hypertrophy. Expression of related transcription enhancer factor-1 (RTEF1) in endothelial cells is associated with increased production of vascular endothelial growth factor-B, which can bind to a vascular endothelial growth factor receptor on the cardiomyocytes leading to pathological hypertrophy via ERK1/2 activation. Endothelial cell overexpression of endothlin-1 (ET1) is sufficient to cause pathological hypertrophy in adjacent cardiomyocytes. Alternatively, nitric oxide synthase (NOS) in endothelial cells can produce nitric oxide (NO), which can diffuse into neighboring cardiomyocytes to stimulate the anti-hypertrophic PKG.

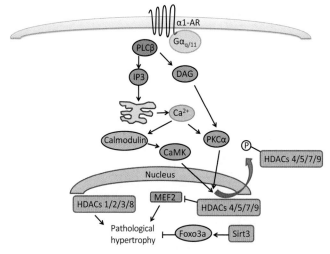

FIGURE 4.8 HDAC modulation of cardiac hypertrophy. Histone deacetylases (HDACs) participate in both forms of hypertrophy. Class I HDACs (HDACs 1, 2, 3, and 8) are thought to be pro-hypertrophic. Class II HDACs (HDACs 4, 5, 7, and 9) exert anti-hypertrophic effects by suppressing the pro-hypertrophic MEF2. Increased intracellular calcium activates PKCα and the calmodulin/calcium–calmodulin protein kinase II pathways. This phosphorylates class II HDACs, causing their nuclear export, and thus removing MEF2 inhibition. Sirtuin3 (Sirt3) activates FOXO3A (among other targets) to produce anti-hypertrophic effects.

gene program, and decreases apoptosis. Additionally, CHAMP overexpression inhibits cell proliferation in non-cardiomyocytes, suggesting that CHAMP acts via a common mechanism for inhibition of both cardiomyocyte hypertrophy and cell cycle proliferation.[166] CHAMP over-expression is associated with up-regulation of p21[CIP1], a cyclin-dependent protein kinase inhibitor that may act to mediate the anti-hypertrophic effect of CHAMP. Together, both CHAMP and p21[CIP1] represent possible therapeutic targets.[166]

Mechanotransduction

Mechanotransduction is the process of converting mechanical stimuli into cellular signals, enabling cells to regulate a wide range of physiologic responses for balancing cardiac functions and structures (Fig. 4.9).[167,168] Mechanotransduction is a highly conserved process found in many cell types, including endothelial cells, fibroblasts, and cardiomyocytes.[169] In myocardial tissue, stretch-induced signaling may give rise to either physiologic or pathologic hypertrophy, or may be transient with no lasting effect. Several common pathways mediate both physiologic and pathologic hypertrophy, where the intensity, frequency, and duration of action determine the end result. Pathways that are distinct to pathologic hypertrophy include FAK, β3 integrin, thrombospondin, and BMP an activin-membrane bound inhibitor.

FOCAL ADHESION KINASE AND RHO KINASE

Focal adhesion kinase (FAK) is a 125-kDa non-receptor protein tyrosine kinase that was originally identified in association with fibroblast adhesions.[170] Subsequent investigation has shown FAK to possess a wide range of biological functions including control of cell motility,

proliferation, and migration.[171,172] FAK is now believed to play an important role in cardiac hypertrophy. For example, FAK is activated in both cellular models (using pulsatile stretch) and animal models of pressure overload.[173,174] Furthermore, cardiac-specific FAK knockout in mice attenuates pressure-overload-induced hypertrophy.[175] Together these findings confirm FAK as a critical component of the response to chronic mechanical stress.

The Rho kinase (ROCK) is an effector of the small GTPase *Rho* and belongs to the AGC family of kinases (PKA, PKG, PKC). Studies using ROCK inhibitors indicate a pivotal role for ROCK signaling in many cardiovascular conditions including cardiac hypertrophy.[176] ROCK is rapidly activated by pressure overload in rat myocardium.[177] Inhibition of ROCK abolishes ventricular hypertrophy and ameliorates cardiac function in Dahl salt-sensitive hypertensive rats.[178] Consistently, long-term inhibition of ROCK suppresses left ventricular hypertrophy and improves cardiac function after myocardial infarction in mice.[179] RhoA and ROCK activation are believed to act as upstream regulators of stretch-induced FAK activation.[180]

β3 INTEGRIN

Integrins are a class of non-covalently associated heterodimeric transmembrane receptors composed of α and β subunits.[181] Integrins govern cellular survival and proliferation by their physical association with several growth factor receptors and non-receptor tyrosine kinases like FAK and Src. These non-receptor tyrosine kinases modulate downstream signals including the MAPK and PI3K pathways[182] – key players in physiologic and pathological hypertrophy. The β1 integrin subgroup is highly expressed in cardiomyocytes and it functions in controlling

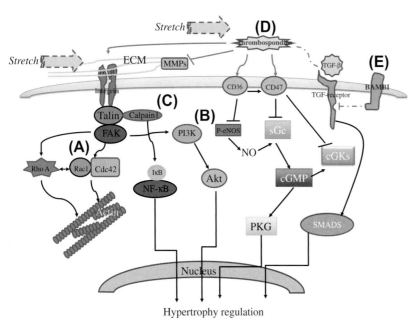

FIGURE 4.9 Mechanotransduction signaling pathways in cardiac remodeling. Mechanical stretch directly activates mechanosensors, like integrins, which up-regulate the expression of signal molecules such as thrombospondins. Thrombospondins activate signaling transduction cascades involving PI3K, Rho GTPase, NO, Smads, and transcription factors that regulate cardiac remolding. (A) Activation of FAK and Rho GTPase cascade mediates cardiac remodeling. (B) Integrin and FAK activate PI3K and AKT and lead to cardiac hypertrophy. (C) Integrin-mediated calpain and NF-κB cascades control cardiomyocyte survival. (D) Thrombospondins regulate integrin, MMPs and growth factor activation and subsequent cascades to control cardiac hypertrophy and remodeling. (E) BAMBI negatively regulates TGF-β receptor activation to modulate cardiac hypertrophy. ECM, extracellular matrix; MMPs, matrix metalloproteinases; FAK, focal adhesion kinase; NF-κB, nuclear factor kappa B; PKG, protein kinase G; eNOS, endothelial nitric oxide synthase; NO, nitric oxide; BAMBI, BMP and activin membrane-bound inhibitor.

hypertrophic signaling.[181,183,184] The β3 subgroup of integrins plays an additional role in myocardial hypertrophy. β3 integrin signaling mediates protein ubiquitination during pressure overload, resulting in improved cardiomyocyte survival and preserved ventricular function.[185] The β3 integrins also up-regulate NF-κB and inhibit μ-calpain, enabling compensatory hypertrophy.[186]

THROMBOSPONDINS

Thrombospondins (TSPs) are a small family of secreted, calcium-binding glycoproteins that play an essential role in regulating cell–cell and cell–matrix interactions.[187] TSPs are divided into trimeric subgroup A (TSP-1 and TSP-2) and pentameric subgroup B (TSP-3–5). The actions of TSPs depend on their binding partners in a given environment.[188] Expression of TSPs is increased in the hearts of hypertensive and mechanical pressure overload animal models.[189,190] This indicates an important role for TSPs in hypertrophic cardiac remodeling. Multiple proposed mechanisms are reported. For example, TSPs regulate matrix metalloproteinase (MMP) activity in extracellular matrix remodeling during cardiac hypertrophy.[191,192] TSPs may also protect the heart from adverse remodeling by interacting with the TGF-β signaling pathway during left ventricle hypertrophy.[193]

BAMBI

BMP and activin membrane-bound inhibitor (BAMBI) is a transmembrane protein that is highly similar to transforming growth factor-β (TGF-β) receptor. BAMBI is important for mouse embryonic development and post-natal survival.[194] Like other components of the TGF-β signaling pathway, BAMBI plays a significant role in cardiac remodeling. For example, BAMBI expression is increased in severe aortic stenosis patients. BAMBI knockout mice show exacerbated hypertrophy after transverse aortic constriction.[195] Clearly important to hypertrophy, the mechanism of action of BAMBI remains to be determined.

Protein Synthesis and Protein Degradation

UBIQUITIN PROTEASOME SYSTEM

Degradation of approximately 80% of intracellular proteins, including sarcomeric proteins, is mediated by the ubiquitin proteasome system (UPS).[196] This gives the UPS an intricate role in many cellular processes and in maintaining homeostasis (Fig. 4.10). UPS-mediated proteolysis can be broken down into two general steps: targeting of the substrate protein through the attachment of a polyubiquitin chain (ubiquitination) and its subsequent degradation by the 26S proteasome.[196] Protein ubiquitination occurs through a series of enzymatic reactions involving a ubiquitin-activating enzyme (E1), ubiquitin-conjugating enzyme (E2), and a ubiquitin ligase (E3).[197] The 26S proteasome is a highly regulated, dynamic complex consisting of a 19S cap and a 20S core proteasome. The 19S cap functions to recognize, bind, deubiquitinate, and translocate the protein into the 20S core for degradation.[198] Alternative lids for the 20S proteasome have been implicated to play a major role in protein degradation, especially during disease, such as proteasome activator 28.[199]

The UPS has been increasingly implicated in the development of pathological left ventricular hypertrophy.[200,201] The exact role of the dynamic UPS in cardiac hypertrophy is complex and is not yet fully elucidated. Detailed in the following section is a breakdown of what is currently known about the roles of the proteasome and protein ubiquitination during pathological hypertrophy.

During pressure-overload-induced cardiac hypertrophy, the UPS is highly active. Proteasome activities increase, as does gene expression of proteasome

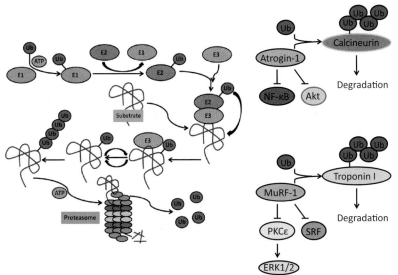

FIGURE 4.10 The ubiquitin proteasome system mitigates hypertrophy through targeted degradation. Protein degradation by the ubiquitin proteasome system (UPS) involves a series of ATP-dependent enzymatic reactions involving an ubiquitin-activating enzyme (E1), ubiquitin-conjugating enzyme (E2), and ubiquitin ligase (E3). This process attaches an ubiquitin moiety (Ub) to a substrate protein, thereby targeting the protein for degradation by the 26S proteasome. The proteasome removes and recycles ubiquitin. Atrogin-1 is an E3 that is capable of ubiquitinating calcineurin for proteasome-mediated degradation, which inhibits pathological hypertrophy. Atrogin-1 can also suppress NF-κB and Akt to inhibit physiologic hypertrophy. Muscle ring finger-1 (MuRF1) is known to target a key sarcomeric protein, troponin I for degradation. Additionally, MuRF1 can associate with and inhibit PKCε and serum response factor (SRF) to suppress hypertrophic growth.

subunits.[202] Most studies demonstrate that proteasome inhibition effectively inhibits hypertrophy and cardiac remodeling,[203–205] although a few conflicting reports describe reduced proteasome activity after pressure overload.[206] Proteasome inhibition increases NFAT nuclear translocation, enabling increased NFAT activity.[207] Furthermore, during proteasome functional insufficiency there is an increase in both calcineurin protein level and NFAT activity.[208] Interestingly, protein kinase A (PKA), a downstream protein kinase of the hypertrophic β1-adrenergic receptor, is known for its ability to enhance proteasome activities through direct phosphorylation.[209,210] The progression from hypertrophy to end-stage heart failure in humans is characterized by reduced proteasome activity and increased amounts of ubiquitinated protein, suggesting dysfunction or inadequacy of the proteasome.[211–213] Proteasome function is restored in patients undergoing pressure unloading by a left ventricular assist device (LVAD).[214] Collectively these studies strongly implicate a role for the proteasome in the development and progression of cardiac hypertrophy. Not to be overlooked, much of the specificity of the UPS during pathological hypertrophy is due to the presence/activity/substrate selection of the ubiquitin ligases.

Atrogin-1 (also known as MAFbx1) is a cardiac- and skeletal-muscle-specific ubiquitin ligase. This F-box protein has anti-hypertrophic effects, and is able to inhibit both pathological and physiologic cardiac hypertrophy.[84] FOXO proteins, members of the Forkhead family of transcription factors, regulate the expression of atrogin-1. Interestingly, atrogin-1 controls its own expression by catalyzing the addition of a non-canonical K63-linked ubiquitin chain on FOXO proteins, mainly FOXO1 and FOXO3a. This creates a feed-forward mechanism by which activation of FOXO proteins increases atrogin-1 expression, which further enhances FOXO activity.[84,215] Atrogin-1 forms a complex with Skp1, Cul1, F-box (e.g. Roc1), and the SCF ubiquitin ligase complex. The SCF complex ubiquitinates the pro-hypertrophic protein calcineurin, targeting it for degradation by the proteasome. This action inhibits pathologic hypertrophy.[129,216] A recent confounding study found that knockdown of atrogin-1 is protective during hypertrophy due to the stabilization of IκB, which then inhibits NF-κB.[217] Atrogin-1 inhibits physiologic hypertrophy by ubiquitinating FOXO1 and FOXO3a, enhancing their activity, and enabling them to inhibit Akt-induced cardiac hypertrophy.[218] Atrogin-1 ubiquitination of FOXO proteins creates a positive feedback loop in which FOXO proteins increase the expression of atrogin-1 as well as another anti-hypertrophic protein, muscle ring finger-1 (MuRF1).[84]

The muscle ring finger proteins (MuRFs) are a family of ubiquitin ligases composed of MuRF1, MuRF2, and MuRF3. While all MuRFs are involved in contractile regulation and the myogenic response to stress, only MuRF1 appears to have the ability to inhibit cardiac hypertrophy.[110,219,220] Analysis of cardiac samples taken from human hearts (at time of transplant) that have been pressure off-loaded with LVAD therapy reveals that MuRF1 protein levels are increased during regression of cardiac hypertrophy. These results are confirmed by studies in mice in which hypertrophy is induced by pressure overload and then released to induce atrophy.[221] MuRF1 is known to ubiquitinate the cardiac sarcomeric protein troponin I for degradation.[222] To suppress cardiac hypertrophy, MuRF1 also interacts with the receptor for activated protein kinase C to inhibit PKCε signaling, which suppresses focal adhesion kinase and ERK1/2.[223] MuRF1 associates with the transcription factor, SRF, inhibiting hypertrophic growth. Mice lacking MuRF1 display an enhanced expression of the SRF-dependent genes BNP, smooth muscle actin, and β-MHC.[110] Additionally, MuRF1 knockout mice are unable to undergo cardiac atrophy.[219,224] Collectively these studies demonstrate that MuRF1 is a critical regulator of atrophy and hypertrophy regression.

mTOR

Activated Akt phosphorylates and activates the mammalian target of rapamycin (mTOR) (Fig. 4.11).[92,121] Once activated, mTOR stimulates protein synthesis through p70/85 S6 kinase-1 (S6K1) and p54/56 (S6K2), which enhances protein translation and cardiomyocyte growth. Additionally, mTOR stimulates the dissociation of 4E-binding protein-1 from eukaryotic translation initiation factor 4E (eIF4E), allowing eIF4E to bind to eIF4G, which results in increased translation.[92,121] Inhibition of mTOR by rapamycin has cardioprotective effects by attenuating cardiac hypertrophy, regressing existing hypertrophy, reducing fibrosis, and improving cardiac contractile function.[225–228] mTOR inhibition can also stimulate autophagy, which is associated with cardioprotection.[229–231]

PROTEIN TURNOVER

The balance between protein synthesis and protein degradation determines if the heart will atrophy, hypertrophy, or neither. During acute cardiac hypertrophy, an increase in protein synthesis and a parallel suppression of protein degradation has been reported.[202,231,232] One hypothesis is that increased protein synthesis is responsible for acute hypertrophy – within one month – while reduced protein degradation is responsible for chronic hypertrophy.[233] Prolonged fasting leads to cardiac atrophy, which is characterized by a 40–50% reduction in protein synthesis. Protein degradation rates overall remain stable, tipping the balance towards atrophy.[234,235]

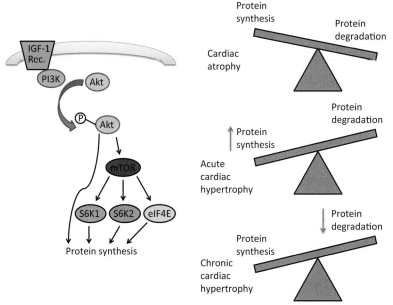

FIGURE 4.11 Cardiac atrophy vs. hypertrophy: an unbalancing act. The mammalian target of rapamycin (mTOR), activated by Akt, can increase protein synthesis by targeting p70/85 S6 kinase-1 (S6K1) and p54/56 (S6K2). Protein translation is enhanced by mTOR by stimulating the dissociation of 4E-binding protein-1 from eIF4E, allowing eIF4E to bind to eIF4G. The heart is constantly being remodeled. However, a healthy heart maintains balance of protein synthesis and degradation. During cardiac atrophy an increase in protein degradation predominates over a relatively unchanged protein synthesis, indicating the primary role of the ubiquitin proteasome system (UPS) in mediating protein (including sarcomere) degradation. Acute cardiac hypertrophy is characterized by increased protein synthesis, while chronic hypertrophy shows impaired protein degradation. Both conditions tip the balance in favor of protein synthesis.

MicroRNA

MicroRNAs were first discovered in 1993[236] and have since been identified as important post-transcriptional regulators in a wide range of biologic processes. They are short, single-stranded, non-coding RNAs that bind to complementary sequences of mRNA and either inhibit translation or initiate degradation. Although tissue-based gene array studies demonstrated the presence of microRNA in cardiac tissue as early as 2002,[237] it was not until 2007 that the microRNA miR-1 was shown to play an essential role in the regulation of cardiac hypertrophy.[238]

Since then, several microRNAs have been identified to have key regulatory roles within canonical hypertrophic pathways, including thyroid hormone, IGF1, TGF-β, and calcineurin.[239] Thyroid hormone induces physiologic hypertrophy by up-regulating expression of α-MHC and miR-208a, and reduces expression of β-MHC and miR-208b, and Myh7b and miR-499.[240] MiR-499 and miR-208b both activate slow myofiber gene programs and facilitate pathologic hypertrophy.[241] MiR-1 disrupts the IGF-1/PI3K/Akt hypertrophic pathway by directly targeting IGF-1 and IGF1R. However, IGF-1 stimulation increases Foxo3a expression. This, in turn, represses miR-1 resulting in the activation of the IGF-1 pathway.[242] TGF-β signaling is known to have contradictory roles in the development of hypertrophy. New evidence shows that TGF-β/Smad pathways prevent hypertrophy, in part by down-regulation of miRNAs that induce hypertrophy.[243] Specifically, TGF-β signaling inhibits miR-27b, a miRNA known to induce hypertrophy by blocking transcription of PPAR-gamma.[244]

Calcineurin signaling and downstream NFAT activation have been shown to induce hypertrophy. Recent reports show miR-23a is up-regulated by NFATc3 and targets MURF-1, a known anti-hypertrophic protein.[245] Calcineurin inhibition with cyclosporine A prevents the down-regulation of miR-133, which acts to decrease NFAT expression, thus limiting hypertrophic gene activation. MiR-199b is induced by calcineurin signaling and acts as part of a positive feedback loop within the calcineurin-NFAT pathway.[246] MiR-199b attacks the dual-specificity tyrosine kinase (Y) phosphorylation-regulated kinase 1a (Dyrk1a), which blocks NFATc. In this way, miR-199b enables NFATc activity, resulting in increased hypertrophy.[247]

Fibroblasts are increasingly recognized for their role in the progression of hypertrophy and associated fibrosis. MicroRNAs elaborated by fibroblasts have been similarly implicated in these processes. Within the TGF-β signaling pathway, miR-21 increases fibroblast survival by repressing the sprout homolog 1 (Spry1), enhancing the MAPK/extracellular signal-related kinase (ERK) pathway, and ultimately promoting pathologic hypertrophy.[248] MiR-29 is linked to the TGF-β signaling pathway, but targets RNA encoding collagens, fibrillins, and elastin. Overexpression of miR-29 limits cardiac fibrosis, while TGF-β signaling down-regulates miR-29, enabling fibrosis.[249] MiR-133 and miR-30 both inhibit cardiac fibrosis by targeting connective tissue growth factor, another downstream product of TGF-β signaling in both cardiac fibroblasts and cardiomyocytes (Table 4.1).[250]

Mitochondria in Left Ventricular Hypertrophy

As a continuously active muscle, myocardium is one of the highest consumers of energy in the body. The adult heart is adapted to generate 80–90% of its ATP from oxidative metabolism, making mitochondrial function a

TABLE 4.1   MicroRNA Implicated in Left Ventricular Hypertrophy

miRNA	Pro-Hypertrophic	Anti-Hypertrophic
miR-1		X
miR-21	X	
miR-23a	X	
miR-27b	X	
miR-29		X
miR-30		X
miR-133		X
miR-199b	X	
miR-208a	X	
miR-208b	X	
miR-499	X	

critical component of cardiomyocyte physiology. During times of stress, demands are even greater due to the need to respond to reactive oxygen species (ROS) and manage intracellular calcium levels. Therefore, the heart becomes extremely dependent on the function of mitochondria. Not surprisingly, recent studies link critical changes in mitochondrial function with major cardiovascular stresses including ventricular hypertrophy.[251]

Hypertrophy is a fundamental part of the response to increased demand for myocardial work. During physiologic growth and hypertrophy, the increase in required ATP is supplied by mitochondrial biogenesis (i.e., mitochondrial growth and division). The mitochondrial genome encodes only 37 proteins. The rest are encoded by the host nucleus and transported to mitochondria as needed.[252]

MITOCHONDRIAL BIOGENESIS

Peroxisome proliferator-activated receptor gamma coactivator-1α (PGC-1α) is currently the most studied regulator of mitochondrial biogenesis. ROS and other stimuli induce up-regulation of PGC-1α. PGC-1α then binds and activates nuclear respiratory factor 1&2 (NRF-1/2).[253] NRF1 and 2 are transcription factors that activate nuclear genes coding for mitochondrial import machinery and components of the respiratory chain, as well as transcription factor A, mitochondrial (TFAM). TFAM is exported from the nucleus to the mitochondrion where it activates mtDNA transcription.[254]

PHYSIOLOGIC HYPERTROPHY

PGC-1α is up-regulated during exercise and physiologic hypertrophy. PGC-1α promotes increased fatty acid oxidative (FAO) capacity and increased mitochondrial proliferation. It is believed that exercise training in heart failure may improve myocardial energetics through increases in PGC-1α.

PATHOLOGIC HYPERTROPHY

Although pathologic hypertrophy is associated with increased numbers of mitochondria, there is a shift away from FAO to increased glucose utilization.[255,256] This is a markedly less efficient form of ATP generation, and is associated with decreased levels of PPARα and PGC-1α.[257] Although increases in PPARα have been associated with worsening function, PGC-1α augmentation enables reversal in some animal models of pathologic hypertrophy.[258]

RECIPROCITY/RECIPROCAL EFFECTS

Changes in mitochondrial number and function closely follow systemic changes that result in ventricular hypertrophy. Conversely, changes in mitochondrial function have been shown to independently induce LVH. For example, knockout of the GLUT-4 glucose transporter in mice, and knockout of the adenine nucleotide translocator (ANT-1) (which exports ATP from mitochondria) both result in mitochondrial proliferation and hypertrophy.[259,260] These findings suggest that decreases in mitochondrial function in response to systemic stimuli may form a positive feedback loop for LVH.

ROS

Reactive oxygen species (ROS) are normal bi-products of mitochondrial metabolism and energy production. Numerous hypertrophic stimuli including endothelin 1, angiotensin II, TNF-α, and mechanical stretch induce increased production of ROS.[251] Cells are protected from oxidative stress by several antioxidant mechanisms both inside and outside the mitochondria. These include: manganese superoxide dismutase (Mn-SOD) in the mitochondria, copper/zinc-SOD in the cytosol, and extracellular-SOD in the interstitium. Mn-SOD, when temporally knocked out of adult mouse hearts, induces dilated cardiomyopathy.[261] PGC-1α is believed to up-regulate transcription of several antioxidants including Mn-SOD, Cu/Zn-SOD, and peroxisomal catalase.[251] Lastly, anti-oxidant therapy attenuates ATII-induced LVH.[262]

CALCIUM

Mitochondrial permeability transition pore (MPTP) is a transmembrane protein residing in the mitochondrial inner membrane. Normally closed, this large protein pore opens when stimulated by mitochondrial matrix Ca^{2+} accumulation, adenine nucleotide depletion, increased phosphate concentration or oxidative stress. Opening of the pore is linked to apoptosis.[263] During progression of pathologic hypertrophy, increased cytoplasmic calcium transients drive calcium into mitochondria. Progressive Ca^{2+} loading of the mitochondria, together with decreased FAO and increased ROS generation ultimately cause the MPTP to open

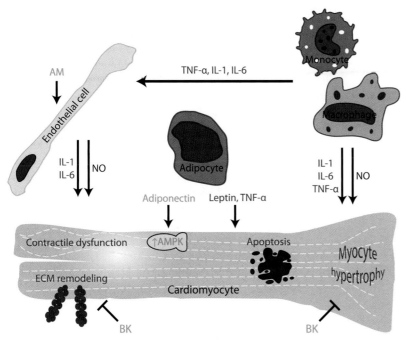

FIGURE 4.12 Effects of inflammatory mediators on cardiomyocyte function. Pro-inflammatory mediators are associated with cardiomyocyte dysfunction (red). TNF-α, IL-1, and IL-6 are produced by multiple cell types including endothelial cells, monocytes, macrophages, and adipocytes. The primary signaling mediator utilized by inflammatory molecules is nitric oxide (NO) produced by inducible nitric oxide synthase (iNOS). High levels or chronic exposure of pro-inflammatory molecules have detrimental effects on cardiomyocytes (both through NO-dependent and NO-independent pathways) including contractile dysfunction, remodeling of the extracellular matrix (ECM), apoptosis, and hypertrophy. Pro-inflammatory molecules such as bradykinin (BK) and adrenomedullin (AM), as well as anti-inflammatory molecules such as adiponectin, have apparent cardioprotective effects (green). These include modulation of vascular tone and endothelial cell function (AM), inhibition of ECM remodeling and cardiomyocyte hypertrophy (BK), and activation of critical signaling enzymes such as AMP-activated kinase (AMPK).[297]

and release apoptotic factors into the cytoplasm.[264] In compensated LVH, the MPTP is closed but is more susceptible to opening in response to stress. Because opening of the MPTP is closely associated with progression to heart failure, this event is believed to be one of the cardinal events in the transition from compensated to decompensated LVH.[264]

The Inflammatory Response and Cardiomyocte Function

Inflammation has been known for decades to contribute to the pathogenesis of heart failure[265] but has only more recently been shown to contribute to LVH.[266] Importantly, inflammation associated with cardiac diseases often involves other organs, and is associated with disturbances of the neurohormonal axis, the parasympathetic and sympathetic nervous systems, and the angiotensin–aldosterone system (Fig. 4.12).[267–269] As in other organs, the inflammatory response within the myocardium initially functions as an adaptive response that leads to the activation of several cellular protective mechanisms such as free radical scavengers, multiple heat-shock proteins, and activation of anti-apoptotic protein gp130.[270] Chronic inflammatory conditions, such as metabolic syndrome and hypertension eventually lead to maladaptive responses in the cardiomyocyte including cellular hypertrophy, contractile dysfunction, programmed cell death, and myocardial extracellular matrix remodeling.[270,271] In summary, the inflammatory system is responsible for both harmful and beneficial effects on myocardial physiology. Only by better understanding this complex system will we be able to devise meaningful therapeutic interventions.

NITRIC OXIDE

Nitric oxide (NO) is a signaling molecule that modulates many functions of the cardiomyocyte, from the generation of ATP to contraction of the sarcomere.[272] The localization and synthesis of NO depends on the specific inflammatory signaling pathway involved. NO plays an important role in maintaining cardiomyocyte homeostasis, but excessive NO production induced by inflammatory stressors leads to a detrimental increase in reactive oxygen species (ROS) and significantly perturbs oxidative signaling within the heart. Oxidative/redox signaling is integral to myocardial excitation–contraction coupling, and cardiomyocyte differentiation and proliferation.[273] ROS are normally counter-balanced by antioxidants, but pro-inflammatory molecules lead to increased production of ROS that overwhelms the antioxidants and contributes to the pathophysiology of LVH and heart failure.[272,274]

TNF-α AND NF-κB/JNK ACTIVATION

Mechanical stretch induces secretion of TNF-α by both cardiomyocytes and myocardial fibroblasts.[275,276] Additional TNF-α is produced by circulating monocytes and macrophages (Fig. 4.12). Prolonged increases in TNF-α in rodent models lead to both cardiac hypertrophy[277] and activation of programmed cell death of cardiomyocytes.[278] TNF-α stimulates both the anti-apoptotic NF-κB and pro-apoptotic JNK pathways. The balance between the two opposing pathways determines the extent of cell death during maladaptive remodeling.[278] In addition, TNF-α works transiently though calcium-dependent increases in NO production by nitric oxide synthase (NOS). Elevated TNF-α activates the inducible form of NOS (iNOS) and subsequent calcium desensitization of the cardiomyocyte leads to contractile dysfunction.[279]

IL-1 AND IL-6

Several interleukin molecules, both pro- and anti-inflammatory, are elevated in heart failure patients.[280,281] IL-1 and IL-6 are produced by macrophages and monocytes. In mouse models, IL-1 and IL-6 stimulate cardiomyocyte hypertrophy and myocardial fibrosis.[280] In the cardiomyocyte, IL-1 signaling results in decreased expression of SERCA. This results in impaired cytosolic calcium removal, diminished sarcoplasmic reticulum calcium release, and therefore impaired contractile function. In the fibroblast, IL-1β induces increased gene expression of pro-matrix metalloproteinase(MMP)-2 and proMMP-3, as well as increased activation of MMPs. By increasing MMP production and activation, IL-1 causes increased turnover and remodeling of the extracellular matrix (ECM).[270] IL-6 similarly causes decreased calcium transients and impaired myocardial contractility, but through a different mechanism. IL-6 signals through the JAK/STAT3 pathway to activate calcium-dependent iNOS as well as up-regulation of calcium-independent iNOS (Fig. 4.12).[271,282]

BRADYKININ

Bradykinin (BK) is a circulating peptide derived from high-molecular-weight kininogen. It acts primarily on endothelial cells in the peripheral and coronary vasculature. There are two subtypes of BK receptor, bradykinin receptor 1 (B1R) and bradykinin receptor 2 (B2R). Both receptors are in the G-protein-coupled receptor (GPCR) family, but mediate different actions. B1R is an inducible receptor that is expressed as a result of tissue injury. B1R signaling induces activation of calcium-dependent iNOS. B1R signaling promotes local tissue inflammation by recruiting neutrophils, dilating capillaries, and constricting venous outflow. Endothelial cells throughout the vascular system constitutively express B2R. B2R signaling causes increased intracellular calcium, causing activation of eNOS.[283] Vasodilation occurs by release of NO to surrounding vascular smooth muscle cells (VSMC), where it induces formation of cGMP, a potent VSMC relaxant.[284] B2R signaling is also believed to result in negative transcriptional regulation of proliferation and growth in VSMC and cardiomyocytes. This action counters both hypertrophy and arterial wall thickening.[285] Importantly, BK is metabolized by the angiotensin-converting enzyme. Treatment with ACE inhibitors acts to potentiate serum levels of BK, resulting in a systemic vasodilatory effect (in addition to the direct effect of blocking production of the vasoconstrictor ATII).

ADRENOMEDULLIN

Adrenomedullin (AM) is similar to BK in that it functions as a vasodilatory peptide. It is produced and secreted predominantly by vascular endothelial cells in response to oxidative stress, hypoxia, and low shear stress. It acts through the PI3K/AKT pathway to phosphorylate eNOS and increase production of NO.[286] In the heart, AM vasodilates the coronary bed, and AM is present in the intima and media of atherosclerotic lesions.[287] Systemic infusion of AM increases cardiac output in patients with heart failure, possibly through improved coronary blood flow and decreased systemic vascular resistance. Indeed, experiments with isolated cardiomyocytes fail to show any positive inotropic effect of AM.[288,289] Notably, serum AM is elevated in patients with hypertension and heart failure.[290] Therapeutic use of AM will require a better understanding of the complex role of adrenomedullin in cardiac physiology.

ADIPOKINES

Adipokines are a family of hormones and cytokines with both pro- and anti-inflammatory effects that are secreted by adipose tissue. Family members include the aforementioned TNF-α and IL-6 (pro-inflammatory), leptin, and adiponectin, among others.[291] An excess of pro-inflammatory adipokines is hypothesized to explain the metabolic syndrome that accompanies obesity. In addition, several cell types within the myocardium express adipokines resulting in both autocrine and paracrine influences.[292] Pro-inflammatory molecules, such as TNF-α and IL-6, inhibit adiponectin production. Decreased circulating adiponectin is associated with LVH,[291,293] suggesting that adiponectin may be cardioprotective. The cardioprotective effects may be due to activation of AMPK and Sirt1 resulting in enhanced activity of PGC-1α.[291,294–296] There is a large collection of clinical studies that demonstrate associations between adiponectin, LVH, and heart failure. However, direct clinical evidence to support a protective role for adiponectin awaits further study.[296]

References

1. Go AS, Mozaffarian D, Roger VL, Benjamin EJ, Berry JD, Borden WB, et al. Heart disease and stroke statistics–2013 update: A report from the American Heart Association. *Circulation* 2013;**127**:e6–245.

2. Roger VL, Go AS, Lloyd-Jones DM, Benjamin EJ, Berry JD, Borden WB, et al. Executive summary: Heart disease and stroke statistics–2012 update: A report from the American Heart Association. *Circulation* 2012;**125**:188–97.

3. Roger VL, Go AS, Lloyd-Jones DM, Benjamin EJ, Berry JD, Borden WB, et al. Heart disease and stroke statistics–2012 update: A report from the American Heart Association. *Circulation* 2012;**125**:e2–20.

4. Casale PN, Devereux RB, Milner M, Zullo G, Harshfield GA, Pickering TG, et al. Value of echocardiographic measurement of left ventricular mass in predicting cardiovascular morbid events in hypertensive men. *Ann Intern Med* 1986;**105**:173–8.

5. Levy D, Garrison RJ, Savage DD, Kannel WB, Castelli WP. Prognostic implications of echocardiographically determined left ventricular mass in the Framingham Heart Study. *N Engl J Med* 1990;**322**:1561–6.

6. Bombelli M, Facchetti R, Carugo S, Madotto F, Arenare F, Quarti-Trevano F, et al. Left ventricular hypertrophy increases cardiovascular risk independently of in-office and out-of-office blood pressure values. *J Hypertens* 2009;**27**:2458–64.

7. Tsioufis C, Vezali E, Tsiachris D, Dimitriadis K, Taxiarchou E, Chatzis D, et al. Left ventricular hypertrophy versus chronic kidney disease as predictors of cardiovascular events in hypertension: A Greek 6-year-follow-up study. *J Hypertens* 2009;**27**:744–52.

8. Cipriano C, Gosse P, Bemurat L, Mas D, Lemetayer P, N'Tela G, et al. Prognostic value of left ventricular mass and its evolution during treatment in the Bordeaux cohort of hypertensive patients. *Am J Hypertens* 2001;**14**:524–9.

9. Wachtell K, Bella JN, Liebson PR, Gerdts E, Dahlof B, Aalto T, et al. Impact of different partition values on prevalences of left ventricular hypertrophy and concentric geometry in a large hypertensive population: The life study. *Hypertension* 2000;**35**:6–12.

10. Eguchi K, Kario K, Hoshide S, Ishikawa J, Morinari M, Shimada K. Type 2 diabetes is associated with left ventricular concentric remodeling in hypertensive patients. *Am J Hypertens* 2005;**18**:23–9.

11. Nardi E, Palermo A, Mule G, Cusimano P, Cottone S, Cerasola G. Impact of type 2 diabetes on left ventricular geometry and diastolic function in hypertensive patients with chronic kidney disease. *J Hum Hypertens* 2011;**25**:144–51.

12. Barry SP, Davidson SM, Townsend PA. Molecular regulation of cardiac hypertrophy. *Int J Biochem Cell Biol* 2008;**40**:2023–39.

13. Lorell BH, Carabello BA. Left ventricular hypertrophy: Pathogenesis, detection, and prognosis. *Circulation* 2000;**102**:470–9.

14. Oka T, Komuro I. Molecular mechanisms underlying the transition of cardiac hypertrophy to heart failure. *Circ J* 2008;**72**(Suppl. A):A13–6.

15. Yamamoto S, Kita S, Iyoda T, Yamada T, Iwamoto T. New molecular mechanisms for cardiovascular disease: Cardiac hypertrophy and cell-volume regulation. *J Pharmacol Sci* 2011;**116**:343–9.

16. Jugdutt BI. Ischemia/infarction. *Heart Fail Clin* 2012;**8**:43–51.

17. Felker GM, Thompson RE, Hare JM, Hruban RH, Clemetson DE, Howard DL, et al. Underlying causes and long-term survival in patients with initially unexplained cardiomyopathy. *N Engl J Med* 2000;**342**:1077–84.

18. Minow RA, Benjamin RS, Lee ET, Gottlieb JA. Adriamycin cardiomyopathy – risk factors. *Cancer* 1977;**39**:1397–402.

19. Kemp CD, Conte JV. The pathophysiology of heart failure. *Cardiovasc Pathol* 2012;**21**:365–71.

20. Masoudi FA, Havranek EP, Smith G, Fish RH, Steiner JF, Ordin DL, et al. Gender, age, and heart failure with preserved left ventricular systolic function. *J Am Coll Cardiol* 2003;**41**:217–23.

21. Packer M. Neurohormonal interactions and adaptations in congestive heart failure. *Circulation* 1988;**77**:721–30.

22. Regoli D, Plante GE, Gobeil Jr F. Impact of kinins in the treatment of cardiovascular diseases. *Pharmacol Ther* 2012;**135**:94–111.

23. Hew KW, Keller KA. Postnatal anatomical and functional development of the heart: A species comparison. *Birth Defects Res Part B Dev Reprod Toxicol* 2003;**68**:309–20.

24. Schannwell CM, Zimmermann T, Schneppenheim M, Plehn G, Marx R, Strauer BE. Left ventricular hypertrophy and diastolic dysfunction in healthy pregnant women. *Cardiology* 2002;**97**:73–8.

25. Sugishita Y, Koseki S, Matsuda M, Yamaguchi T, Ito I. Myocardial mechanics of athletic hearts in comparison with diseased hearts. *Am Heart J* 1983;**105**:273–80.

26. Hadj-Sahraoui N, Seugnet I, Ghorbel MT, Demeneix B. Hypothyroidism prolongs mitotic activity in the post-natal mouse brain. *Neurosci Lett* 2000;**280**:79–82.

27. Morkin E. Regulation of myosin heavy chain genes in the heart. *Circulation* 1993;**87**:1451–60.

28. Arsanjani R, McCarren M, Bahl JJ, Goldman S. Translational potential of thyroid hormone and its analogs. *J Mol Cell Cardiol* 2011;**51**:506–11.

29. Belke DD, Betuing S, Tuttle MJ, Graveleau C, Young ME, Pham M, et al. Insulin signaling coordinately regulates cardiac size, metabolism, and contractile protein isoform expression. *J Clin Invest* 2002;**109**:629–39.

30. Hu P, Zhang D, Swenson L, Chakrabarti G, Abel ED, Litwin SE. Minimally invasive aortic banding in mice: Effects of altered cardiomyocyte insulin signaling during pressure overload. *Am J Physiol Heart Circ Physiol* 2003;**285**:H1261–9.

31. Liu JP, Baker J, Perkins AS, Robertson EJ, Efstratiadis A. Mice carrying null mutations of the genes encoding insulin-like growth factor 1 (igf-1) and type 1 igf receptor (igf1r). *Cell* 1993;**75**:59–72.

32. Delaughter MC, Taffet GE, Fiorotto ML, Entman ML, Schwartz RJ. Local insulin-like growth factor 1 expression induces physiologic, then pathologic, cardiac hypertrophy in transgenic mice. *FASEB J* 1999;**13**:1923–9.

33. McMullen JR, Shioi T, Huang WY, Zhang L, Tarnavski O, Bisping E, et al. The insulin-like growth factor 1 receptor induces physiological heart growth via the phosphoinositide 3-kinase(p110alpha) pathway. *J Biol Chem* 2004;**279**:4782–93.

34. Kim J, Wende AR, Sena S, Theobald HA, Soto J, Sloan C, et al. Insulin-like growth factor 1 receptor signaling is required for exercise-induced cardiac hypertrophy. *Mol Endocrinol* 2008;**22**:2531–43.

35. Seth M, Zhang ZS, Mao L, Graham V, Burch J, Stiber J, et al. Trpc1 channels are critical for hypertrophic signaling in the heart. *Circ Res* 2009;**105**:1023–30.

36. Shai SY. Cardiac myocyte-specific excision of the beta1 integrin gene results in myocardial fibrosis and cardiac failure. *Circ Res* 2002;**90**:458–64.

37. Boateng SY, Senyo SE, Qi L, Goldspink PH, Russell B. Myocyte remodeling in response to hypertrophic stimuli requires nucleocytoplasmic shuttling of muscle lim protein. *J Mol Cell Cardiol* 2009;**47**:426–35.

38. Song Y, Xu J, Li Y, Jia C, Ma X, Zhang L, et al. Cardiac ankyrin repeat protein attenuates cardiac hypertrophy by inhibition of erk1/2 and tgf-beta signaling pathways. *PLoS One* 2012;**7**:e50436.

39. Shioi T, Kang PM, Douglas PS, Hampe J, Yballe CM, Lawitts J, et al. The conserved phosphoinositide 3-kinase pathway determines heart size in mice. *The EMBO Journal* 2000;**19**:2537–48.

40. Crackower MA, Oudit GY, Kozieradzki I, Sarao R, Sun H, Sasaki T, et al. Regulation of myocardial contractility and cell size by distinct pi3k-pten signaling pathways. *Cell* 2002;**110**:737–49.

41. Michael A, Haq S, Chen X, Hsich E, Cui L, Walters B, et al. Glycogen synthase kinase-3beta regulates growth, calcium homeostasis, and diastolic function in the heart. *J Biol Chem* 2004;**279**:21383–93.

42. Maillet M, van Berlo JH, Molkentin JD. Molecular basis of physiological heart growth: Fundamental concepts and new players. *Nat Rev Mol Cell Biol* 2012;**14**:38–48.

43. Skurk C, Izumiya Y, Maatz H, Razeghi P, Shiojima I, Sandri M, et al. The foxo3a transcription factor regulates cardiac myocyte size downstream of akt signaling. *J Biol Chem* 2005;**280**:20814–23.

44. Shiojima I, Walsh K. Regulation of cardiac growth and coronary angiogenesis by the akt/pkb signaling pathway. *Genes Dev* 2006;**20**:3347–65.

45. Shioi T, McMullen JR, Kang PM, Douglas PS, Obata T, Franke TF, et al. Akt/protein kinase b promotes organ growth in transgenic mice. *Mol Cell Biol* 2002;**22**:2799–809.

46. Shen WH, Chen Z, Shi S, Chen H, Zhu W, Penner A, et al. Cardiac restricted overexpression of kinase-dead mammalian target of rapamycin (mtor) mutant impairs the mtor-mediated signaling and cardiac function. *J Miol Chem* 2008;**283**:13842–9.

47. Bostrom P, Mann N, Wu J, Quintero PA, Plovie ER, Panakova D, et al. C/ebpbeta controls exercise-induced cardiac growth and protects against pathological cardiac remodeling. *Cell* 2010;**143**: 1072–83.

48. Bueno OF, De Windt LJ, Tymitz KM, Witt SA, Kimball TR, Klevitsky R, et al. The mek1-erk1/2 signaling pathway promotes compensated cardiac hypertrophy in transgenic mice. *The EMBO J* 2000;**19**:6341–50.

49. Lips DJ, Bueno OF, Wilkins BJ, Purcell NH, Kaiser RA, Lorenz JN, et al. Mek1-erk2 signaling pathway protects myocardium from ischemic injury in vivo. *Circulation* 2004;**109**:1938–41.

50. Purcell NH, Wilkins BJ, York A, Saba-El-Leil MK, Meloche S, Robbins J, et al. Genetic inhibition of cardiac erk1/2 promotes stress-induced apoptosis and heart failure but has no effect on hypertrophy in vivo. *Proc Natl Acad Sci U S A* 2007;**104**:14074–9.

51. Zhang P, Hu X, Xu X, Fassett J, Zhu G, Viollet B, et al. Amp activated protein kinase-alpha2 deficiency exacerbates pressure-overload-induced left ventricular hypertrophy and dysfunction in mice. *Hypertension* 2008;**52**:918–24.

52. Ikeda Y, Sato K, Pimentel DR, Sam F, Shaw RJ, Dyck JR, et al. Cardiac-specific deletion of lkb1 leads to hypertrophy and dysfunction. *J Biol Chem* 2009;**284**:35839–49.

53. Lavie CJ, Milani RV, Messerli FH. Prevention and reduction of left ventricular hypertrophy in the elderly. *Clin Geriatr Med* 1996;**12**:57–68.

54. Messerli FH. Cardiovascular effects of obesity and hypertension. *Lancet* 1982;**1**:1165–8.

55. de Simone G, Palmieri V, Bella JN, Celentano A, Hong Y, Oberman A, et al. Association of left ventricular hypertrophy with metabolic risk factors: The hypergen study. *J Hypertens* 2002;**20**:323–31.

56. Jilaihawi H, Greaves S, Rouleau JL, Pfeffer MA, Solomon SD. Left ventricular hypertrophy and the risk of subsequent left ventricular remodeling following myocardial infarction. *Am J Cardiol* 2003;**91**:723–6.

57. Lee M, Gardin JM, Lynch JC, Smith VE, Tracy RP, Savage PJ, et al. Diabetes mellitus and echocardiographic left ventricular function in free-living elderly men and women: The Cardiovascular Health Study. *Am Heart J* 1997;**133**:36–43.

58. Carabello BA. The relationship of left ventricular geometry and hypertrophy to left ventricular function in valvular heart disease. *J Heart Valve Dis* 1995;**4**(Suppl. 2):S132–8 discussion S138–139.

59. Drazner MH, Dries DL, Peshock RM, Cooper RS, Klassen C, Kazi F, et al. Left ventricular hypertrophy is more prevalent in blacks than whites in the general population: The Dallas Heart Study. *Hypertension* 2005;**46**:124–9.

60. Artham SM, Lavie CJ, Milani RV, Patel DA, Verma A, Ventura HO. Clinical impact of left ventricular hypertrophy and implications for regression. *Prog Cardiovasc Dis* 2009;**52**:153–67.

61. Kannel WB, Schatzkin A. Sudden death: Lessons from subsets in population studies. *J Am Coll Cardiol* 1985;**5**:141B–9B.

62. Messerli FH, Ventura HO, Elizardi DJ, Dunn FG, Frohlich ED. Hypertension and sudden death. Increased ventricular ectopic activity in left ventricular hypertrophy. *Am J Med* 1984;**77**:18–22.

63. Aronow WS, Epstein S, Koenigsberg M, Schwartz KS. Usefulness of echocardiographic left ventricular hypertrophy, ventricular tachycardia and complex ventricular arrhythmias in predicting ventricular fibrillation or sudden cardiac death in elderly patients. *Am J Cardiol* 1988;**62**:1124–5.

64. Wachtell K, Okin PM, Olsen MH, Dahlof B, Devereux RB, Ibsen H, et al. Regression of electrocardiographic left ventricular hypertrophy during antihypertensive therapy and reduction in sudden cardiac death: The Life Study. *Circulation* 2007;**116**:700–5.

65. Verdecchia P, Reboldi G, Gattobigio R, Bentivoglio M, Borgioni C, Angeli F, et al. Atrial fibrillation in hypertension: Predictors and outcome. *Hypertension* 2003;**41**:218–23.

66. Marcus ML, Koyanagi S, Harrison DG, Doty DB, Hiratzka LF, Eastham CL. Abnormalities in the coronary circulation that occur as a consequence of cardiac hypertrophy. *Am J Med* 1983;**75**:62–6.

67. Cooper RS, Simmons BE, Castaner A, Santhanam V, Ghali J, Mar M. Left ventricular hypertrophy is associated with worse survival independent of ventricular function and number of coronary arteries severely narrowed. *Am J Cardiol* 1990;**65**:441–5.

68. Devereux RB, Roman MJ, Liu JE, Welty TK, Lee ET, Rodeheffer R, et al. Congestive heart failure despite normal left ventricular systolic function in a population-based sample: The Strong Heart Study. *Am J Cardiol* 2000;**86**:1090–6.

69. Diez J, Querejeta R, Lopez B, Gonzalez A, Larman M, Martinez Ubago JL. Losartan-dependent regression of myocardial fibrosis is associated with reduction of left ventricular chamber stiffness in hypertensive patients. *Circulation* 2002;**105**:2512–7.

70. Villari B, Campbell SE, Hess OM, Mall G, Vassalli G, Weber KT, et al. Influence of collagen network on left ventricular systolic and diastolic function in aortic valve disease. *J Am Coll Cardiol* 1993;**22**:1477–84.

71. Kannel WB, Gordon T, Castelli WP, Margolis JR. Electrocardiographic left ventricular hypertrophy and risk of coronary heart disease. The framingham study. *Ann Intern Med* 1970;**72**:813–22.

72. Kannel WB. Prevalence and natural history of electrocardiographic left ventricular hypertrophy. *Am J Med* 1983;**75**:4–11.

73. Okin PM, Devereux RB, Nieminen MS, Jern S, Oikarinen L, Viitasalo M, et al. Electrocardiographic strain pattern and prediction of new-onset congestive heart failure in hypertensive patients: The losartan intervention for endpoint reduction in hypertension (life) study. *Circulation* 2006;**113**:67–73.

74. Norman TD. The pathogenesis of cardiac hypertrophy. *Prog Cardiovasc Dis* 1962;**4**:439–63.

75. Fanburg BL. Experimental cardiac hypertrophy. *N Engl J Med* 1970;**282**:723–32.

76. Okamoto K, Aoki K, Nosaka S, Fukushima M. Cardiovascular diseases in the spontaneously hypertensive rat. *Jpn Circ J* 1964;**28**:943–52.

77. Agabiti-Rosei E, Muiesan ML, Salvetti M. New approaches to the assessment of left ventricular hypertrophy. *Ther Adv Cardiovasc Dis* 2007;**1**:119–28.

78. Wolf MJ, Amrein H, Izatt JA, Choma MA, Reedy MC, Rockman HA. Drosophila as a model for the identification of genes causing adult human heart disease. *Proc Natl Acad Sci U S A* 2006;**103**:1394–9.

79. Izumo S, Lompre AM, Matsuoka R, Koren G, Schwartz K, Nadal-Ginard B, et al. Myosin heavy chain messenger rna and protein isoform transitions during cardiac hypertrophy. Interaction between hemodynamic and thyroid hormone-induced signals. *J Clin Invest* 1987;**79**:970–7.

80. Izumo S, Nadal-Ginard B, Mahdavi V. Protooncogene induction and reprogramming of cardiac gene expression produced by pressure overload. *Proc Natl Acad Sci U S A* 1988;**85**:339–43.

81. Heineke J, Molkentin JD. Regulation of cardiac hypertrophy by intracellular signalling pathways. *Nat Rev Mol Cell Biol* 2006;**7**:589–600.

82. Suzuki E, Nishimatsu H, Satonaka H, Walsh K, Goto A, Omata M, et al. Angiotensin ii induces myocyte enhancer factor 2- and calcineurin/nuclear factor of activated t cell-dependent transcriptional activation in vascular myocytes. *Circ Res* 2002;**90**:1004–11.

83. Chin ER, Olson EN, Richardson JA, Yang Q, Humphries C, Shelton JM, et al. A calcineurin-dependent transcriptional pathway controls skeletal muscle fiber type. *Genes Dev* 1998;**12**:2499–509.

84. Schisler JC, Willis MS, Patterson C. You spin me round: Mafbx/atrogin-1 feeds forward on foxo transcription factors (like a record). *Cell Cycle* 2008;**7**:440–3.

85. Clerk A, Sugden PH. Small guanine nucleotide-binding proteins and myocardial hypertrophy. *Circ Res* 2000;**86**:1019–23.

86. Bos JL, Rehmann H, Wittinghofer A. Gefs and gaps: Critical elements in the control of small G proteins. *Cell* 2007;**129**:865–77.

87. Lezoualc'h F, Metrich M, Hmitou I, Duquesnes N, Morel E. Small gtp-binding proteins and their regulators in cardiac hypertrophy. *J Mol Cell Cardiol* 2008;**44**:623–32.

88. Maruyama Y, Nishida M, Sugimoto Y, Tanabe S, Turner JH, Kozasa T, et al. Galpha(12/13) mediates alpha(1)-adrenergic receptor-induced cardiac hypertrophy. *Circ Res* 2002;**91**:961–9.

89. Appert-Collin A, Cotecchia S, Nenniger-Tosato M, Pedrazzini T, Diviani D. The a-kinase anchoring protein (akap)-lbc-signaling complex mediates alpha1 adrenergic receptor-induced cardiomyocyte hypertrophy. *Proc Natl Acad Sci U S A* 2007;**104**: 10140–5.

90. Muslin AJ. Role of raf proteins in cardiac hypertrophy and cardiomyocyte survival. *Trends Cardiovasc Med* 2005;**15**:225–9.

91. Clerk A, Cullingford TE, Fuller SJ, Giraldo A, Markou T, Pikkarainen S, et al. Signaling pathways mediating cardiac myocyte gene expression in physiological and stress responses. *J Cell Physiol* 2007;**212**:311–22.

92. Proud CG. Ras, pi3-kinase and mtor signaling in cardiac hypertrophy. *Cardiovasc Res* 2004;**63**:403–13.

93. Sala V, Gallo S, Leo C, Gatti S, Gelb BD, Crepaldi T. Signaling to cardiac hypertrophy: Insights from human and mouse rasopathies. *Mol Med* 2012;**18**:938–47.

94. Gelb BD, Tartaglia M. Ras signaling pathway mutations and hypertrophic cardiomyopathy: Getting into and out of the thick of it. *J Clin Invest* 2011;**121**:844–7.

95. Thorburn J, Thorburn A. The tyrosine kinase inhibitor, genistein, prevents alpha-adrenergic-induced cardiac muscle cell hypertrophy by inhibiting activation of the ras-map kinase signaling pathway. *Biochem Biophys Res Commun* 1994;**202**:1586–91.

96. Hattori T, Shimokawa H, Higashi M, Hiroki J, Mukai Y, Kaibuchi K, et al. Long-term treatment with a specific rho-kinase inhibitor suppresses cardiac allograft vasculopathy in mice. *Circ Res* 2004;**94**:46–52.

97. Higashi M, Shimokawa H, Hattori T, Hiroki J, Mukai Y, Morikawa K, et al. Long-term inhibition of rho-kinase suppresses angiotensin ii-induced cardiovascular hypertrophy in rats in vivo: Effect on endothelial nad(p)h oxidase system. *Circ Res* 2003;**93**: 767–75.

98. Sauzeau V, Le Jeune H, Cario-Toumaniantz C, Smolenski A, Lohmann SM, Bertoglio J, et al. Cyclic gmp-dependent protein kinase signaling pathway inhibits rhoa-induced ca2+ sensitization of contraction in vascular smooth muscle. *J Biol Chem* 2000;**275**:21722–9.

99. Loirand G, Guilluy C, Pacaud P. Regulation of rho proteins by phosphorylation in the cardiovascular system. *Trends Cardiovasc Med* 2006;**16**:199–204.

100. Rockman HA, Koch WJ, Lefkowitz RJ. Seven-transmembrane-spanning receptors and heart function. *Nature* 2002;**415**:206–12.

101. Braz JC, Gregory K, Pathak A, Zhao W, Sahin B, Klevitsky R, et al. Pkc-alpha regulates cardiac contractility and propensity toward heart failure. *Nat Med* 2004;**10**:248–54.

102. Hambleton M, York A, Sargent MA, Kaiser RA, Lorenz JN, Robbins J, et al. Inducible and myocyte-specific inhibition of pkcalpha enhances cardiac contractility and protects against infarction-induced heart failure. *Am J Physiol Heart Circ Physiol* 2007;**293**:H3768–71.

103. Ladage D, Tilemann L, Ishikawa K, Correll RN, Kawase Y, Houser SR, et al. Inhibition of pkcalpha/beta with ruboxistaurin antagonizes heart failure in pigs after myocardial infarction injury. *Circ Res* 2011;**109**:1396–400.

104. Boyle AJ, Kelly DJ, Zhang Y, Cox AJ, Gow RM, Way K, et al. Inhibition of protein kinase c reduces left ventricular fibrosis and dysfunction following myocardial infarction. *J Mol Cell Cardiol* 2005;**39**:213–21.

105. Liu Q, Chen X, Macdonnell SM, Kranias EG, Lorenz JN, Leitges M, et al. Protein kinase c{alpha}, but not pkc{beta} or pkc{gamma}, regulates contractility and heart failure susceptibility: Implications for ruboxistaurin as a novel therapeutic approach. *Circ Res* 2009;**105**:194–200.

106. Hambleton M, Hahn H, Pleger ST, Kuhn MC, Klevitsky R, Carr AN, et al. Pharmacological- and gene therapy-based inhibition of protein kinase calpha/beta enhances cardiac contractility and attenuates heart failure. *Circulation* 2006;**114**:574–82.

107. Akazawa H, Komuro I. Roles of cardiac transcription factors in cardiac hypertrophy. *Circ Res* 2003;**92**:1079–88.

108. Zhang X, Azhar G, Chai J, Sheridan P, Nagano K, Brown T, et al. Cardiomyopathy in transgenic mice with cardiac-specific overexpression of serum response factor. *Am J Physiol Heart Circ Physiol* 2001;**280**:H1782–92.

109. Parlakian A, Charvet C, Escoubet B, Mericskay M, Molkentin JD, Gary-Bobo G, et al. Temporally controlled onset of dilated cardiomyopathy through disruption of the srf gene in adult heart. *Circulation* 2005;**112**:2930–9.

110. Willis MS, Ike C, Li L, Wang DZ, Glass DJ, Patterson C. Muscle ring finger 1, but not muscle ring finger 2, regulates cardiac hypertrophy in vivo. *Circ Res* 2007;**100**:456–9.

111. Czubryt MP, Olson EN. Balancing contractility and energy production: The role of myocyte enhancer factor 2 (mef2) in cardiac hypertrophy. *Recent Prog Horm Res* 2004;**59**:105–24.

112. Xu J, Gong NL, Bodi I, Aronow BJ, Backx PH, Molkentin JD. Myocyte enhancer factors 2a and 2c induce dilated cardiomyopathy in transgenic mice. *J Biol Chem* 2006;**281**:9152–62.

113. van Oort RJ, van Rooij E, Bourajjaj M, Schimmel J, Jansen MA, van der Nagel R, et al. Mef2 activates a genetic program promoting chamber dilation and contractile dysfunction in calcineurin-induced heart failure. *Circulation* 2006;**114**:298–308.

114. Molkentin JD. The zinc finger-containing transcription factors gata-4, -5, and -6. Ubiquitously expressed regulators of tissue-specific gene expression. *J Biol Chem* 2000;**275**:38949–52.

115. Peterkin T, Gibson A, Loose M, Patient R. The roles of gata-4, -5 and -6 in vertebrate heart development. *Semin Cell Dev Biol* 2005;**16**:83–94.

116. Liang Q, Molkentin JD. Divergent signaling pathways converge on gata4 to regulate cardiac hypertrophic gene expression. *J Mol Cell Cardiol* 2002;**34**:611–6.

117. Liang Q, De Windt LJ, Witt SA, Kimball TR, Markham BE, Molkentin JD. The transcription factors gata4 and gata6 regulate cardiomyocyte hypertrophy in vitro and in vivo. *J Biol Chem* 2001;**276**:30245–53.

118. Suzuki YJ, Evans T. Regulation of cardiac myocyte apoptosis by the gata-4 transcription factor. *Life Sci* 2004;**74**:1829–38.

119. Xia Y, McMillin JB, Lewis A, Moore M, Zhu WG, Williams RS, et al. Electrical stimulation of neonatal cardiac myocytes activates the nfat3 and gata4 pathways and up-regulates the adenylosuccinate synthetase 1 gene. *J Biol Chem* 2000;**275**:1855–63.

120. Oka T, Maillet M, Watt AJ, Schwartz RJ, Aronow BJ, Duncan SA, et al. Cardiac-specific deletion of gata4 reveals its requirement for hypertrophy, compensation, and myocyte viability. *Circ Res* 2006;**98**:837–45.

121. Cantley LC. The phosphoinositide 3-kinase pathway. *Science* 2002;**296**:1655–7.

122. Johnson JA, Liggett SB. Cardiovascular pharmacogenomics of adrenergic receptor signaling: Clinical implications and future directions. *Clin Pharmacol Ther* 2011;**89**:366–78.

123. Wachter SB, Gilbert EM. Beta-adrenergic receptors, from their discovery and characterization through their manipulation to beneficial clinical application. *Cardiology* 2012;**122**:104–12.

124. Jensen BC, O'Connell TD, Simpson PC. Alpha-1-adrenergic receptors: Targets for agonist drugs to treat heart failure. *J Mol Cell Cardiol* 2011;**51**:518–28.

125. Dell'Italia LJ. Translational success stories: Angiotensin receptor 1 antagonists in heart failure. *Circ Res* 2011;**109**:437–52.

126. Lijnen P, Petrov V. Renin-angiotensin system, hypertrophy and gene expression in cardiac myocytes. *J Mol Cell Cardiol* 1999;**31**:949–70.

127. Rose BA, Force T, Wang Y. Mitogen-activated protein kinase signaling in the heart: Angels versus demons in a heart-breaking tale. *Physiol Rev* 2010;**90**:1507–46.

128. Ravingerova T, Barancik M, Strniskova M. Mitogen-activated protein kinases: A new therapeutic target in cardiac pathology. *Mol Cell Biochem* 2003;**247**:127–38.

129. Portbury AL, Ronnebaum SM, Zungu M, Patterson C, Willis MS. Back to your heart: Ubiquitin proteasome system-regulated signal transduction. *J Mol Cell Cardiol* 2012;**52**:526–37.

130. Lazou A, Sugden PH, Clerk A. Activation of mitogen-activated protein kinases (p38-mapks, sapks/jnks and erks) by the G-protein-coupled receptor agonist phenylephrine in the perfused rat heart. *Biochem J* 1998;**332**(Pt 2):459–65.

131. Haq S, Choukroun G, Lim H, Tymitz KM, del Monte F, Gwathmey J, et al. Differential activation of signal transduction pathways in human hearts with hypertrophy versus advanced heart failure. *Circulation* 2001;**103**:670–7.

132. Sanna B, Bueno OF, Dai YS, Wilkins BJ, Molkentin JD. Direct and indirect interactions between calcineurin-nfat and mek1-extracellular signal-regulated kinase 1/2 signaling pathways regulate cardiac gene expression and cellular growth. *Mol Cell Biol* 2005;**25**:865–78.

133. Bueno OF, De Windt LJ, Lim HW, Tymitz KM, Witt SA, Kimball TR, et al. The dual-specificity phosphatase mkp-1 limits the cardiac hypertrophic response in vitro and in vivo. *Circ Res* 2001;**88**:88–96.

134. Kehat I, Davis J, Tiburcy M, Accornero F, Saba-El-Leil MK, Maillet M, et al. Extracellular signal-regulated kinases 1 and 2 regulate the balance between eccentric and concentric cardiac growth. *Circ Res* 2011;**108**:176–83.

135. Liao P, Georgakopoulos D, Kovacs A, Zheng M, Lerner D, Pu H, et al. The in vivo role of p38 map kinases in cardiac remodeling and restrictive cardiomyopathy. *Proc Natl Acad Sci U S A* 2001;**98**:12283–8.

136. Wang Y, Su B, Sah VP, Brown JH, Han J, Chien KR. Cardiac hypertrophy induced by mitogen-activated protein kinase kinase 7, a specific activator for c-jun nh2-terminal kinase in ventricular muscle cells. *J Biol Chem* 1998;**273**:5423–6.

137. Nicol RL, Frey N, Pearson G, Cobb M, Richardson J, Olson EN. Activated mek5 induces serial assembly of sarcomeres and eccentric cardiac hypertrophy. *EMBO J* 2001;**20**:2757–67.

138. Marber MS, Rose B, Wang Y. The p38 mitogen-activated protein kinase pathway – a potential target for intervention in infarction, hypertrophy, and heart failure. *J Mol Cell Cardiol* 2011;**51**:485–90.

139. Nishida K, Yamaguchi O, Hirotani S, Hikoso S, Higuchi Y, Watanabe T, et al. P38alpha mitogen-activated protein kinase plays a critical role in cardiomyocyte survival but not in cardiac hypertrophic growth in response to pressure overload. *Mol Cell Biol* 2004;**24**:10611–20.

140. Braz JC, Bueno OF, Liang Q, Wilkins BJ, Dai YS, Parsons S, et al. Targeted inhibition of p38 mapk promotes hypertrophic cardiomyopathy through upregulation of calcineurin-nfat signaling. *J Clin Invest* 2003;**111**:1475–86.

141. Fujio Y, Maeda M, Mohri T, Obana M, Iwakura T, Hayama A, et al. Glycoprotein 130 cytokine signal as a therapeutic target against cardiovascular diseases. *J Pharmacol Sci* 2011;**117**:213–22.

142. Fischer P, Hilfiker-Kleiner D. Survival pathways in hypertrophy and heart failure: The gp130-stat3 axis. *Basic Res Cardiol* 2007;**102**:279–97.

143. Hisamitsu T, Nakamura TY, Wakabayashi S. Na(+)/h(+) exchanger 1 directly binds to calcineurin a and activates downstream nfat signaling, leading to cardiomyocyte hypertrophy. *Mol Cell Biol* 2012;**32**:3265–80.

144. Nakamura TY, Iwata Y, Arai Y, Komamura K, Wakabayashi S. Activation of na+/h+ exchanger 1 is sufficient to generate ca2+ signals that induce cardiac hypertrophy and heart failure. *Circ Res* 2008;**103**:891–9.

145. Shibata M, Takeshita D, Obata K, Mitsuyama S, Ito H, Zhang GX, et al. Nhe-1 participates in isoproterenol-induced downregulation of serca2a and development of cardiac remodeling in rat hearts. *Am J Physiol Heart Circ Physiol* 2011;**301**:H2154–60.

146. Perino A, Ghigo A, Scott JD, Hirsch E. Anchoring proteins as regulators of signaling pathways. *Circ Res* 2012;**111**:482–92.

147. Diviani D, Maric D, Perez Lopez I, Cavin S, Del Vescovo CD. A-kinase anchoring proteins: Molecular regulators of the cardiac stress response. *Biochim Biophys Acta* 2013;**1833**:901–8.

148. Finckenberg P, Mervaala E. Novel regulators and drug targets of cardiac hypertrophy. *J Hypertens* 2010;**28**(Suppl. 1):S33–8.

149. Muller AL, Dhalla NS. Role of various proteases in cardiac remodeling and progression of heart failure. *Heart Fail Rev* 2012;**17**:395–409.

150. Letavernier E, Zafrani L, Perez J, Letavernier B, Haymann JP, Baud L. The role of calpains in myocardial remodelling and heart failure. *Cardiovasc Res* 2012;**96**:38–45.

151. Lee DI, Kass DA. Phosphodiesterases and cyclic gmp regulation in heart muscle. *Physiology (Bethesda)* 2012;**27**:248–58.

152. van Berlo JH, Maillet M, Molkentin JD. Signaling effectors underlying pathologic growth and remodeling of the heart. *J Clin Invest* 2013;**123**:37–45.

153. Blanton RM, Takimoto E, Lane AM, Aronovitz M, Piotrowski R, Karas RH, et al. Protein kinase g ialpha inhibits pressure overload-induced cardiac remodeling and is required for the cardioprotective effect of sildenafil in vivo. *J Am Heart Assoc* 2012;**1**:e003731.

154. Guazzi M, Vicenzi M, Arena R, Guazzi MD. Pde5 inhibition with sildenafil improves left ventricular diastolic function, cardiac geometry, and clinical status in patients with stable systolic heart failure: Results of a 1-year, prospective, randomized, placebo-controlled study. *Circ Heart Failure* 2011;**4**:8–17.

155. Kazakov A, Muller P, Jagoda P, Semenov A, Bohm M, Laufs U. Endothelial nitric oxide synthase of the bone marrow regulates myocardial hypertrophy, fibrosis, and angiogenesis. *Cardiovasc Res* 2012;**93**:397–405.

156. Xu M, Jin Y, Song Q, Wu J, Philbrick MJ, Cully BL, et al. The endothelium-dependent effect of rtef-1 in pressure overload cardiac hypertrophy: Role of vegf-b. *Cardiovasc Res* 2011;**90**:325–34.

157. Leung JW, Wong WT, Koon HW, Mo FM, Tam S, Huang Y, et al. Transgenic mice overexpressing et-1 in the endothelial cells develop systemic hypertension with altered vascular reactivity. *PLoS One* 2011;**6**:e26994.

158. Haberland M, Montgomery RL, Olson EN. The many roles of histone deacetylases in development and physiology: Implications for disease and therapy. *Nat Rev Genet* 2009;**10**:32–42.

159. Zhang CL, McKinsey TA, Chang S, Antos CL, Hill JA, Olson EN. Class ii histone deacetylases act as signal-responsive repressors of cardiac hypertrophy. *Cell* 2002;**110**:479–88.

160. Trivedi CM, Luo Y, Yin Z, Zhang M, Zhu W, Wang T, et al. Hdac2 regulates the cardiac hypertrophic response by modulating gsk3 beta activity. *Nat Med* 2007;**13**:324–31.

161. Kee HJ, Sohn IS, Nam KI, Park JE, Qian YR, Yin Z, et al. Inhibition of histone deacetylation blocks cardiac hypertrophy induced by angiotensin ii infusion and aortic banding. *Circulation* 2006;**113**:51–9.

162. Kong Y, Tannous P, Lu G, Berenji K, Rothermel BA, Olson EN, et al. Suppression of class i and ii histone deacetylases blunts pressure-overload cardiac hypertrophy. *Circulation* 2006;**113**:2579–88.

163. McKinsey TA, Zhang CL, Olson EN. Mef2: A calcium-dependent regulator of cell division, differentiation and death. *Trends Biochem Sci* 2002;**27**:40–7.

164. Chang S, McKinsey TA, Zhang CL, Richardson JA, Hill JA, Olson EN. Histone deacetylases 5 and 9 govern responsiveness of the heart to a subset of stress signals and play redundant roles in heart development. *Mol Cell Biol* 2004;**24**:8467–76.

165. Sundaresan NR, Gupta M, Kim G, Rajamohan SB, Isbatan A, Gupta MP. Sirt3 blocks the cardiac hypertrophic response by augmenting foxo3a-dependent antioxidant defense mechanisms in mice. *J Clin Invest* 2009;**119**:2758–71.

166. Liu ZP, Olson EN. Suppression of proliferation and cardiomyo cyte hypertrophy by champ, a cardiac-specific rna helicase. *Proc Natl Acad Sci U S A* 2002;**99**:2043–8.

167. Hidalgo C, Donoso P. Cell signaling. Getting to the heart of mechanotransduction. *Science* 2011;**333**:1388–90.

168. McCain ML, Parker KK. Mechanotransduction: The role of mechanical stress, myocyte shape, and cytoskeletal architecture on cardiac function. *Pflugers Arch* 2011;**462**:89–104.

169. Knoll R, Hoshijima M, Chien K. Cardiac mechanotransduction and implications for heart disease. *J Mol Med (Berl)* 2003;**81**:750–6.

170. Schaller MD, Borgman CA, Cobb BS, Vines RR, Reynolds AB, Parsons JT. Pp125fak a structurally distinctive protein-tyrosine kinase associated with focal adhesions. *Proc Natl Acad Sci U S A* 1992;**89**:5192–6.

171. Cox BD, Natarajan M, Stettner MR, Gladson CL. New concepts regarding focal adhesion kinase promotion of cell migration and proliferation. *J Cell Biochem* 2006;**99**:35–52.

172. Mitra SK, Hanson DA, Schlaepfer DD. Focal adhesion kinase: In command and control of cell motility. *Nat Rev Mol Cell Biol* 2005;**6**:56–68.

173. Seko Y, Takahashi N, Tobe K, Kadowaki T, Yazaki Y. Pulsatile stretch activates mitogen-activated protein kinase (mapk) family members and focal adhesion kinase (p125(fak)) in cultured rat cardiac myocytes. *Biochem Biophys Res Commun* 1999;**259**: 8–14.

174. Franchini KG, Torsoni AS, Soares PH, Saad MJ. Early activation of the multicomponent signaling complex associated with focal adhesion kinase induced by pressure overload in the rat heart. *Circ Res* 2000;**87**:558–65.

175. DiMichele LA, Doherty JT, Rojas M, Beggs HE, Reichardt LF, Mack CP, et al. Myocyte-restricted focal adhesion kinase deletion attenuates pressure overload-induced hypertrophy. *Circ Res* 2006;**99**:636–45.

176. Loirand G, Guerin P, Pacaud P. Rho kinases in cardiovascular physiology and pathophysiology. *Circ Res* 2006;**98**:322–34.

177. Torsoni AS, Fonseca PM, Crosara-Alberto DP, Franchini KG. Early activation of p160rock by pressure overload in rat heart. *Am J Physiol Cell Physiol* 2003;**284**:C1411–9.

178. Mita S, Kobayashi N, Yoshida K, Nakano S, Matsuoka H. Cardioprotective mechanisms of rho-kinase inhibition associated with enos and oxidative stress-lox-1 pathway in dahl salt-sensitive hypertensive rats. *J Hypertens* 2005;**23**:87–96.

179. Hattori T, Shimokawa H, Higashi M, Hiroki J, Mukai Y, Tsutsui H, et al. Long-term inhibition of rho-kinase suppresses left ventricular remodeling after myocardial infarction in mice. *Circulation* 2004;**109**:2234–9.

180. Torsoni AS, Marin TM, Velloso LA, Franchini KG. Rhoa/rock signaling is critical to fak activation by cyclic stretch in cardiac myocytes. *Am J Physiol Heart Circ Physiol* 2005;**289**:H1488–96.

181. Ross RS, Borg TK. Integrins and the myocardium. *Circ Res* 2001;**88**:1112–9.

182. Stupack DG, Cheresh DA. Get a ligand, get a life: Integrins, signaling and cell survival. *J Cell Sci* 2002;**115**:3729–38.

183. Brancaccio M, Hirsch E, Notte A, Selvetella G, Lembo G, Tarone G. Integrin signalling: The tug-of-war in heart hypertrophy. *Cardiovasc Res* 2006;**70**:422–33.

184. Dabiri BE, Lee H, Parker KK. A potential role for integrin signaling in mechanoelectrical feedback. *Prog Biophys Mol Biol* 2012;**110**:196–203.

185. Johnston RK, Balasubramanian S, Kasiganesan H, Baicu CF, Zile MR, Kuppuswamy D. Beta3 integrin-mediated ubiquitination activates survival signaling during myocardial hypertrophy. *FASEB J* 2009;**23**:2759–71.

186. Suryakumar G, Kasiganesan H, Balasubramanian S, Kuppuswamy D. Lack of beta3 integrin signaling contributes to calpain-mediated myocardial cell loss in pressure-overloaded myocardium. *J Cardiovasc Pharmacol* 2010;**55**:567–73.

187. Bornstein P. Thrombospondins as matricellular modulators of cell function. *J Clin Invest* 2001;**107**:929–34.

188. Mustonen E, Ruskoaho H, Rysa J. Thrombospondins, potential drug targets for cardiovascular diseases. *Basic Clin Pharmacol Toxicol* 2013;**112**:4–12.

189. Mustonen E, Aro J, Puhakka J, Ilves M, Soini Y, Leskinen H, et al. Thrombospondin-4 expression is rapidly upregulated by cardiac overload. *Biochem Biophys Res Commun* 2008;**373**:186–91.

190. Schroen B, Heymans S, Sharma U, Blankesteijn WM, Pokharel S, Cleutjens JP, et al. Thrombospondin-2 is essential for myocardial matrix integrity: Increased expression identifies failure-prone cardiac hypertrophy. *Circ Res* 2004;**95**:515–22.

191. Donnini S, Morbidelli L, Taraboletti G, Ziche M. Erk1-2 and p38 mapk regulate mmp/timp balance and function in response to thrombospondin-1 fragments in the microvascular endothelium. *Life Sci* 2004;**74**:2975–85.

192. Yang Z, Strickland DK, Bornstein P. Extracellular matrix metalloproteinase 2 levels are regulated by the low density lipoprotein-related scavenger receptor and thrombospondin 2. *J Biol Chem* 2001;**276**:8403–8.

193. Frangogiannis NG, Ren G, Dewald O, Zymek P, Haudek S, Koerting A, et al. Critical role of endogenous thrombospondin-1 in preventing expansion of healing myocardial infarcts. *Circulation* 2005;**111**:2935–42.

194. Chen J, Bush JO, Ovitt CE, Lan Y, Jiang R. The tgf-beta pseudoreceptor gene bambi is dispensable for mouse embryonic development and postnatal survival. *Genesis* 2007;**45**:482–6.

195. Villar AV, García R, Llano M, Cobo M, Merino D, Lantero A, et al. BAMBI (BMP and activin membrane-bound inhibitor) protects the murine heart from pressure-overload biomechanical stress by restraining TGF-β signaling. *Biochim Biophys Acta* 2012;**1832**:323–35.

196. Wang X, Su H, Ranek MJ. Protein quality control and degradation in cardiomyocytes. *J Mol Cell Cardiol* 2008;**45**:11–27.

197. Willis MS, Patterson C. Into the heart: The emerging role of the ubiquitin-proteasome system. *J Mol Cell Cardiol* 2006;**41**:567–79.

198. Wang X, Robbins J. Heart failure and protein quality control. *Circ Res* 2006;**99**:1315–28.

199. Li J, Horak KM, Su H, Sanbe A, Robbins J, Wang X. Enhancement of proteasomal function protects against cardiac proteinopathy and ischemia/reperfusion injury in mice. *J Clin Invest* 2011;**121**:3689–700.

200. Willis MS, Townley-Tilson WH, Kang EY, Homeister JW, Patterson C. Sent to destroy: The ubiquitin proteasome system regulates cell signaling and protein quality control in cardiovascular development and disease. *Circ Res* 2010;**106**:463–78.

201. Powell SR. The ubiquitin-proteasome system in cardiac physiology and pathology. *Am J Physiol Heart Circ Physiol* 2006;**291**:H1–9.

202. Depre C, Wang Q, Yan L, Hedhli N, Peter P, Chen L, et al. Activation of the cardiac proteasome during pressure overload promotes ventricular hypertrophy. *Circulation* 2006;**114**:1821–8.

203. Stansfield WE, Tang RH, Moss NC, Baldwin AS, Willis MS, Selzman CH. Proteasome inhibition promotes regression of left ventricular hypertrophy. *Am J Physiol Heart Circ Physiol* 2008;**294**:H645–50.

204. Meiners S, Dreger H, Fechner M, Bieler S, Rother W, Gunther C, et al. Suppression of cardiomyocyte hypertrophy by inhibition of the ubiquitin-proteasome system. *Hypertension* 2008;**51**:302–8.

205. Hedhli N, Lizano P, Hong C, Fritzky LF, Dhar SK, Liu H, et al. Proteasome inhibition decreases cardiac remodeling after initiation of pressure overload. *Am J Physiol Heart Circ Physiol* 2008;**295**:H1385–93.

206. Tsukamoto O, Minamino T, Okada K, Shintani Y, Takashima S, Kato H, et al. Depression of proteasome activities during the progression of cardiac dysfunction in pressure-overloaded heart of mice. *Biochem Biophys Res Commun* 2006;**340**:1125–33.

207. Fan Y, Xie P, Zhang T, Zhang H, Gu D, She M, et al. Regulation of the stability and transcriptional activity of nfatc4 by ubiquitination. *FEBS Lett* 2008;**582**:4008–14.
208. Tang M, Li J, Huang W, Su H, Liang Q, Tian Z, et al. Proteasome functional insufficiency activates the calcineurin-nfat pathway in cardiomyocytes and promotes maladaptive remodelling of stressed mouse hearts. *Cardiovasc Res* 2010;**88**:424–33.
209. Zong C, Gomes AV, Drews O, Li X, Young GW, Berhane B, et al. Regulation of murine cardiac 20s proteasomes: Role of associating partners. *Circ Res* 2006;**99**:372–80.
210. Drews O, Tsukamoto O, Liem D, Streicher J, Wang Y, Ping P. Differential regulation of proteasome function in isoproterenol-induced cardiac hypertrophy. *Circ Res* 2010;**107**:1094–101.
211. Hein S, Arnon E, Kostin S, Schonburg M, Elsasser A, Polyakova V, et al. Progression from compensated hypertrophy to failure in the pressure-overloaded human heart: Structural deterioration and compensatory mechanisms. *Circulation* 2003;**107**:984–91.
212. Predmore JM, Wang P, Davis F, Bartolone S, Westfall MV, Dyke DB, et al. Ubiquitin proteasome dysfunction in human hypertrophic and dilated cardiomyopathies. *Circulation* 2010;**121**:997–1004.
213. Weekes J, Morrison K, Mullen A, Wait R, Barton P, Dunn MJ. Hyperubiquitination of proteins in dilated cardiomyopathy. *Proteomics* 2003;**3**:208–16.
214. Wohlschlaeger J, Sixt SU, Stoeppler T, Schmitz KJ, Levkau B, Tsagakis K, et al. Ventricular unloading is associated with increased 20s proteasome protein expression in the myocardium. *J Heart Lung Transplant* 2010;**29**:125–32.
215. Sandri M, Sandri C, Gilbert A, Skurk C, Calabria E, Picard A, et al. Foxo transcription factors induce the atrophy-related ubiquitin ligase atrogin-1 and cause skeletal muscle atrophy. *Cell* 2004;**117**:399–412.
216. Li HH, Kedar V, Zhang C, McDonough H, Arya R, Wang DZ, et al. Atrogin-1/muscle atrophy f-box inhibits calcineurin-dependent cardiac hypertrophy by participating in an scf ubiquitin ligase complex. *J Clin Invest* 2004;**114**:1058–71.
217. Usui S, Maejima Y, Pain J, Hong C, Cho J, Park JY, et al. Endogenous muscle atrophy f-box mediates pressure overload-induced cardiac hypertrophy through regulation of nuclear factor-kappab. *Circ Res* 2011;**109**:161–71.
218. Li HH, Willis MS, Lockyer P, Miller N, McDonough H, Glass DJ, et al. Atrogin-1 inhibits akt-dependent cardiac hypertrophy in mice via ubiquitin-dependent coactivation of forkhead proteins. *J Clin Invest* 2007;**117**:3211–23.
219. Fielitz J, Kim MS, Shelton JM, Latif S, Spencer JA, Glass DJ, et al. Myosin accumulation and striated muscle myopathy result from the loss of muscle ring finger 1 and 3. *J Clin Invest* 2007;**117**:2486–95.
220. Witt CC, Witt SH, Lerche S, Labeit D, Back W, Labeit S. Cooperative control of striated muscle mass and metabolism by murf1 and murf2. *The EMBO J* 2008;**27**:350–60.
221. Willis MS, Rojas M, Li L, Selzman CH, Tang RH, Stansfield WE, et al. Muscle ring finger 1 mediates cardiac atrophy in vivo. *Am J Physiol Heart Circ Physiol* 2009;**296**:H997–1006.
222. Frey N, Olson EN. Cardiac hypertrophy: The good, the bad, and the ugly. *Annu Rev Physiol* 2003;**65**:45–79.
223. Arya R, Kedar V, Hwang JR, McDonough H, Li HH, Taylor J, et al. Muscle ring finger protein-1 inhibits pkc{epsilon} activation and prevents cardiomyocyte hypertrophy. *J Cell Biol* 2004;**167**:1147–59.
224. Bodine SC, Latres E, Baumhueter S, Lai VK, Nunez L, Clarke BA, et al. Identification of ubiquitin ligases required for skeletal muscle atrophy. *Science* 2001;**294**:1704–8.
225. Shiojima I, Sato K, Izumiya Y, Schiekofer S, Ito M, Liao R, et al. Disruption of coordinated cardiac hypertrophy and angiogenesis contributes to the transition to heart failure. *J Clin Invest* 2005;**115**:2108–18.
226. McMullen JR, Sherwood MC, Tarnavski O, Zhang L, Dorfman AL, Shioi T, et al. Inhibition of mtor signaling with rapamycin regresses established cardiac hypertrophy induced by pressure overload. *Circulation* 2004;**109**:3050–5.
227. Shioi T, McMullen JR, Tarnavski O, Converso K, Sherwood MC, Manning WJ, et al. Rapamycin attenuates load-induced cardiac hypertrophy in mice. *Circulation* 2003;**107**:1664–70.
228. Gao XM, Wong G, Wang B, Kiriazis H, Moore XL, Su YD, et al. Inhibition of mtor reduces chronic pressure-overload cardiac hypertrophy and fibrosis. *J Hypertens* 2006;**24**:1663–70.
229. Su H, Wang X. Autophagy and p62 in cardiac protein quality control. *Autophagy* 2011;**7**:1382–3.
230. Zheng Q, Su H, Ranek MJ, Wang X. Autophagy and p62 in cardiac proteinopathy. *Circ Res* 2011;**109**:296–308.
231. Su H, Wang X. P62 stages an interplay between the ubiquitin-proteasome system and autophagy in the heart of defense against proteotoxic stress. *Trends Cardiovasc Med* 2011;**21**:224–8.
232. McDermott PJ, Baicu CF, Wahl SR, Van Laer AO, Zile MR. In vivo measurements of the contributions of protein synthesis and protein degradation in regulating cardiac pressure overload hypertrophy in the mouse. *Mol Cell Biochem* 2012;**367**:205–13.
233. Magid NM, Wallerson DC, Borer JS. Myofibrillar protein turnover in cardiac hypertrophy due to aortic regurgitation. *Cardiology* 1993;**82**:20–9.
234. Samarel AM, Parmacek MS, Magid NM, Decker RS, Lesch M. Protein synthesis and degradation during starvation-induced cardiac atrophy in rabbits. *Circ Res* 1987;**60**:933–41.
235. Preedy VR, Smith DM, Kearney NF, Sugden PH. Rates of protein turnover in vivo and in vitro in ventricular muscle of hearts from fed and starved rats. *Biochem J* 1984;**222**:395–400.
236. Lee RC, Feinbaum RL, Ambros V. The c. Elegans heterochronic gene lin-4 encodes small rnas with antisense complementarity to lin-14. *Cell* 1993;**75**:843–54.
237. Lagos-Quintana M, Rauhut R, Yalcin A, Meyer J, Lendeckel W, Tuschl T. Identification of tissue-specific micrornas from mouse. *Curr Biol* 2002;**12**:735–9.
238. Sayed D, Hong C, Chen IY, Lypowy J, Abdellatif M. Micrornas play an essential role in the development of cardiac hypertrophy. *Circ Res* 2007;**100**:416–24.
239. Wang J, Yang X. The function of mirna in cardiac hypertrophy. *Cell Mol Life Sci* 2012;**69**:3561–70.
240. Callis TE, Pandya K, Seok HY, Tang RH, Tatsuguchi M, Huang ZP, et al. Microrna-208a is a regulator of cardiac hypertrophy and conduction in mice. *J Clin Invest* 2009;**119**:2772–86.
241. van Rooij E, Sutherland LB, Qi X, Richardson JA, Hill J, Olson EN. Control of stress-dependent cardiac growth and gene expression by a microrna. *Science* 2007;**316**:575–9.
242. Elia L, Contu R, Quintavalle M, Varrone F, Chimenti C, Russo MA, et al. Reciprocal regulation of microrna-1 and insulin-like growth factor-1 signal transduction cascade in cardiac and skeletal muscle in physiological and pathological conditions. *Circulation* 2009;**120**:2377–85.
243. Xu J, Kimball TR, Lorenz JN, Brown DA, Bauskin AR, Klevitsky R, et al. Gdf15/mic-1 functions as a protective and antihypertrophic factor released from the myocardium in association with smad protein activation. *Circ Res* 2006;**98**:342–50.
244. Wang J, Song Y, Zhang Y, Xiao H, Sun Q, Hou N, et al. Cardiomyocyte overexpression of mir-27b induces cardiac hypertrophy and dysfunction in mice. *Cell Research* 2012;**22**:516–27.
245. Lin Z, Murtaza I, Wang K, Jiao J, Gao J, Li PF. Mir-23a functions downstream of nfatc3 to regulate cardiac hypertrophy. *Proc Natl Acad Sci U S A* 2009;**106**:12103–8.
246. Dong DL, Chen C, Huo R, Wang N, Li Z, Tu YJ, et al. Reciprocal repression between microrna-133 and calcineurin regulates cardiac hypertrophy: A novel mechanism for progressive cardiac hypertrophy. *Hypertension* 2010;**55**:946–52.
247. da Costa Martins PA, Salic K, Gladka MM, Armand AS, Leptidis S, el Azzouzi H, et al. Microrna-199b targets the nuclear kinase dyrk1a in an auto-amplification loop promoting calcineurin/nfat signalling. *Nat Cell Biol* 2010;**12**:1220–7.

248. Thum T, Gross C, Fiedler J, Fischer T, Kissler S, Bussen M, et al. Microrna-21 contributes to myocardial disease by stimulating map kinase signalling in fibroblasts. *Nature* 2008;**456**:980–4.

249. van Rooij E, Sutherland LB, Thatcher JE, DiMaio JM, Naseem RH, Marshall WS, et al. Dysregulation of micrornas after myocardial infarction reveals a role of mir-29 in cardiac fibrosis. *Proc Natl Acad Sci U S A* 2008;**105**:13027–32.

250. Duisters RF, Tijsen AJ, Schroen B, Leenders JJ, Lentink V, van der Made I, et al. Mir-133 and mir-30 regulate connective tissue growth factor: Implications for a role of micrornas in myocardial matrix remodeling. *Circ Res* 2009;**104**:170–8.

251. Zhou LY, Liu JP, Wang K, Gao J, Ding SL, Jiao JQ, et al. Mitochondrial function in cardiac hypertrophy. *Int J Cardiol* 2013;**167**:1118–25.

252. DiMauro S, Schon EA. Mitochondrial respiratory-chain diseases. *N Engl J Med* 2003;**348**:2656–68.

253. Irrcher I, Ljubicic V, Hood DA. Interactions between ros and amp kinase activity in the regulation of pgc-1alpha transcription in skeletal muscle cells. *Am J Physiol Cell Physiol* 2009;**296**:C116–23.

254. Wu Z, Puigserver P, Andersson U, Zhang C, Adelmant G, Mootha V, et al. Mechanisms controlling mitochondrial biogenesis and respiration through the thermogenic coactivator pgc-1. *Cell* 1999;**98**:115–24.

255. Asayama K, Dobashi K, Hayashibe H, Megata Y, Kato K. Lipid peroxidation and free radical scavengers in thyroid dysfunction in the rat: A possible mechanism of injury to heart and skeletal muscle in hyperthyroidism. *Endocrinology* 1987;**121**:2112–8.

256. Bishop SP, Altschuld RA. Increased glycolytic metabolism in cardiac hypertrophy and congestive failure. *Am J Physiol* 1970;**218**:153–9.

257. Lehman JJ, Kelly DP. Gene regulatory mechanisms governing energy metabolism during cardiac hypertrophic growth. *Heart Fail Rev* 2002;**7**:175–85.

258. Sano M, Wang SC, Shirai M, Scaglia F, Xie M, Sakai S, et al. Activation of cardiac cdk9 represses pgc-1 and confers a predisposition to heart failure. *EMBO J* 2004;**23**:3559–69.

259. Domenighetti AA, Danes VR, Curl CL, Favaloro JM, Proietto J, Delbridge LM. Targeted glut-4 deficiency in the heart induces cardiomyocyte hypertrophy and impaired contractility linked with ca(2+) and proton flux dysregulation. *J Mol Cell Cardiol* 2010;**48**:663–72.

260. Graham BH, Waymire KG, Cottrell B, Trounce IA, MacGregor GR, Wallace DC. A mouse model for mitochondrial myopathy and cardiomyopathy resulting from a deficiency in the heart/muscle isoform of the adenine nucleotide translocator. *Nat Genet* 1997;**16**:226–34.

261. Nojiri H, Shimizu T, Funakoshi M, Yamaguchi O, Zhou H, Kawakami S, et al. Oxidative stress causes heart failure with impaired mitochondrial respiration. *J Biol Chem* 2006;**281**:33789–801.

262. Bugger H, Schwarzer M, Chen D, Schrepper A, Amorim PA, Schoepe M, et al. Proteomic remodelling of mitochondrial oxidative pathways in pressure overload-induced heart failure. *Cardiovasc Res* 2010;**85**:376–84.

263. Bopassa JC, Michel P, Gateau-Roesch O, Ovize M, Ferrera R. Low-pressure reperfusion alters mitochondrial permeability transition. *Am J Physiol Heart Circ Physiol* 2005;**288**:H2750–5.

264. Matas J, Young NT, Bourcier-Lucas C, Ascah A, Marcil M, Deschepper CF, et al. Increased expression and intramitochondrial translocation of cyclophilin-d associates with increased vulnerability of the permeability transition pore to stress-induced opening during compensated ventricular hypertrophy. *J Mol Cell Cardiol* 2009;**46**:420–30.

265. Levine B, Kalman J, Mayer L, Fillit HM, Packer M. Elevated circulating levels of tumor necrosis factor in severe chronic heart failure. *N Engl J Med* 1990;**323**:236–41.

266. Simko F. Statins: A perspective for left ventricular hypertrophy treatment. *Eur J Clin Invest* 2007;**37**:681–91.

267. Marchant DJ, Boyd JH, Lin DC, Granville DJ, Garmaroudi FS, McManus BM. Inflammation in myocardial diseases. *Circ Res* 2012;**110**:126–44.

268. Taqueti VR, Mitchell RN, Lichtman AH. Protecting the pump: Controlling myocardial inflammatory responses. *Annu Rev Physiol* 2006;**68**:67–95.

269. Lange LG, Schreiner GF. Immune mechanisms of cardiac disease. *N Engl J Med* 1994;**330**:1129–35.

270. Hedayat M, Mahmoudi MJ, Rose NR, Rezaei N. Proinflammatory cytokines in heart failure: Double-edged swords. *Heart Fail Rev* 2010;**15**:543–62.

271. Mann DL. Inflammatory mediators and the failing heart: Past, present, and the foreseeable future. *Circ Res* 2002;**91**:988–98.

272. Taylor AL. Nitric oxide modulation as a therapeutic strategy in heart failure. *Heart Fail Clin* 2012;**8**:255–72.

273. Burgoyne JR, Mongue-Din H, Eaton P, Shah AM. Redox signaling in cardiac physiology and pathology. *Circ Res* 2012;**111**:1091–106.

274. Seddon M, Looi YH, Shah AM. Oxidative stress and redox signalling in cardiac hypertrophy and heart failure. *Heart* 2007;**93**:903–7.

275. Wang BW, Hung HF, Chang H, Kuan P, Shyu KG. Mechanical stretch enhances the expression of resistin gene in cultured cardiomyocytes via tumor necrosis factor-alpha. *Am J Physiol Heart Circ Physiol* 2007;**293**:H2305–12.

276. Yokoyama T, Sekiguchi K, Tanaka T, Tomaru K, Arai M, Suzuki T, et al. Angiotensin ii and mechanical stretch induce production of tumor necrosis factor in cardiac fibroblasts. *Am J Physiol Cell Physiol* 1999;**276**:H1968–76.

277. Dibbs ZI, Diwan A, Nemoto S, DeFreitas G, Abdellatif M, Carabello BA, et al. Targeted overexpression of transmembrane tumor necrosis factor provokes a concentric cardiac hypertrophic phenotype. *Circulation* 2003;**108**:1002–8.

278. Haudek SB, Taffet GE, Schneider MD, Mann DL. Tnf provokes cardiomyocyte apoptosis and cardiac remodeling through activation of multiple cell death pathways. *J Clin Invest* 2007;**117**:2692–701.

279. Goldhaber JI, Kim KH, Natterson PD, Lawrence T, Yang P, Weiss JN. Effects of tnf-alpha on [ca2+]i and contractility in isolated adult rabbit ventricular myocytes. *Am J Physiol Cell Physiol* 1996;**271**:H1449–55.

280. Gullestad L, Ueland T, Vinge LE, Finsen A, Yndestad A, Aukrust P. Inflammatory cytokines in heart failure: Mediators and markers. *Cardiology* 2012;**122**:23–35.

281. Oikonomou E, Tousoulis D, Siasos G, Zaromitidou M, Papavassiliou AG, Stefanadis C. The role of inflammation in heart failure: New therapeutic approaches. *Hellenic J Cardiol* 2011;**52**:30–40.

282. Yu X, Kennedy RH, Liu SJ. Jak2/stat3, not erk1/2, mediates interleukin-6-induced activation of inducible nitric-oxide synthase and decrease in contractility of adult ventricular myocytes. *J Biol Chem* 2003;**278**:16304–9.

283. Kuhr F, Lowry J, Zhang Y, Brovkovych V, Skidgel RA. Differential regulation of inducible and endothelial nitric oxide synthase by kinin b1 and b2 receptors. *Neuropeptides* 2010;**44**:145–54.

284. Domenico R. Pharmacology of nitric oxide: Molecular mechanisms and therapeutic strategies. *Curr Pharm Des* 2004;**10**:1667–76.

285. Chao J, Chao L. Kallikrein-kinin in stroke, cardiovascular and renal disease. *Exp Physiol* 2005;**90**:291–8.

286. Kato J, Tsuruda T, Kita T, Kitamura K, Eto T. Adrenomedullin: A protective factor for blood vessels. *Arterioscler Thromb Vasc Biol* 2005;**25**:2480–7.

287. Marutsuka K, Hatakeyama K, Sato Y, Yamashita A, Sumiyoshi A, Asada Y. Immunohistological localization and possible functions of adrenomedullin. *Hypertens Res* 2003;**26**(Suppl.):S33–40.

288. Saetrum Opgaard O, Hasbak P, de Vries R, Saxena PR, Edvinsson L. Positive inotropy mediated via cgrp receptors in isolated human myocardial trabeculae. *Eur J Pharmacol* 2000;**397**:373–82.

289. Ikenouchi H, Kangawa K, Matsuo H, Hirata Y. Negative inotropic effect of adrenomedullin in isolated adult rabbit cardiac ventricular myocytes. *Circulation* 1997;**95**:2318–24.
290. Nagaya N, Satoh T, Nishikimi T, Uematsu M, Furuichi S, Sakamaki F, et al. Hemodynamic, renal, and hormonal effects of adrenomedullin infusion in patients with congestive heart failure. *Circulation* 2000;**101**:498–503.
291. Ouchi N, Parker JL, Lugus JJ, Walsh K. Adipokines in inflammation and metabolic disease. *Nat Rev Immunol* 2011;**11**:85–97.
292. Schram K, Sweeney G. Implications of myocardial matrix remodeling by adipokines in obesity-related heart failure. *Trends Cardiovasc Med* 2008;**18**:199–205.
293. Hong SJ, Park CG, Seo HS, Oh DJ, Ro YM. Associations among plasma adiponectin, hypertension, left ventricular diastolic function and left ventricular mass index. *Blood Pressure* 2004;**13**:236–42.
294. Shibata R, Ouchi N, Ito M, Kihara S, Shiojima I, Pimentel DR, et al. Adiponectin-mediated modulation of hypertrophic signals in the heart. *Nat Med* 2004;**10**:1384–9.
295. Liao Y, Takashima S, Maeda N, Ouchi N, Komamura K, Shimomura I, et al. Exacerbation of heart failure in adiponectin-deficient mice due to impaired regulation of ampk and glucose metabolism. *Cardiovasc Res* 2005;**67**:705–13.
296. Park M, Sweeney G. Direct effects of adipokines on the heart: Focus on adiponectin. *Heart Fail Rev* 2013;**18**:631–44.
297. Kelly RA, Smith TW. Cytokines and cardiac contractile function. *Circulation* 1997;**95**:778–81.

Ischemic Heart Disease and its Consequences

John W. Calvert, PhD

Emory University School of Medicine, Atlanta, GA, USA

INTRODUCTION

Worldwide, cardiovascular disease, including cardiac disease, vascular diseases of the brain and kidney, and peripheral artery disease, is on the rise and continues to be the leading cause of morbidity and mortality in both men and women. Despite numerous advances in healthcare practices, it is estimated that 83.6 million American adults (about one in three) have one or more types of cardiovascular disease and that one of every three deaths in the United States is attributed to cardiovascular disease.[1] Economically, it is estimated that over $500 billion is spent a year to cover the associated healthcare-related expenses making cardiovascular disease both a social and economic burden.

The effects of cardiovascular disease are commonly attributable to the damaging effects of acute myocardial ischemia–reperfusion injury, which typically arises in patients presenting with either a ST-segment elevation myocardial infarction (STEMI) or a non-ST-segment elevation myocardial infarction (NSTEMI). Both types are often induced by the disruption of an atherosclerotic plaque with superimposed thrombus, which results in either a subtotal occlusion (NSTEACS) or total occlusion (STEMI) of the culprit coronary artery.[2] Currently, the most effective therapeutic intervention for diminishing myocardial ischemic injury and limiting the degree of myocardial infarction (i.e. cell death) is timely and effective reperfusion of the blocked artery using either thrombolytic therapy or primary percutaneous coronary intervention. However, both experimental and clinical investigations have demonstrated that reperfusion salvages ischemic myocardium while further inducing cardiomyocyte death, a phenomenon known as myocardial reperfusion injury. This suggests that a cell's fate is determined by events that occur during both the ischemic and reperfusion periods. As such, myocardial ischemia–reperfusion injury is a complex pathophysiological event characterized by a cascade of responses that ultimately leads to left ventricular remodeling.[3]

Although myocardial reperfusion outcomes improve with more timely and effective reperfusion (using advances in percutaneous coronary intervention technology and pharmacological agents to maintain coronary patency), effective therapy for preventing myocardial reperfusion injury does not exist. Moreover, the major determinant of the long-term prognosis of a patient who experiences myocardial ischemia is the amount of myocardium that is destroyed by ischemic injury (i.e., the size of infarction). Thus, it is believed that a significant reduction in myocardial infarct size will decrease subsequent morbidity and mortality. As such, interventions designed to effectively reduce myocardial infarction following an ischemic event are still sorely needed. This chapter summarizes our evolving understanding of the pathophysiology of ischemia–reperfusion injury and reviews the current state of experimental therapeutic interventions.

PATHOPHYSIOLOGY OF ISCHEMIA– REPERFUSION INJURY

Risk Factors

Ischemic heart disease develops as a consequence of multiple etiological risk factors and coexists with other disease states.[4] An individual's age, gender, and genetics (determined most frequently by family history) are important risk factors associated with the development of cardiovascular disease. However, it is increasingly apparent that the modern lifestyle plays a significant role in a person's susceptibility to ischemic heart disease. Broadly speaking, these detrimental lifestyles that are associated with increased risk in ischemic heart disease are sedentary, associated with smoking and diets comprising saturated fats and sugar and devoid of fruits and vegetables, among other critical nutrients.[5] Hypertension, hyperlipidemia, insulin resistance, obesity, and diabetes are also major risk factors for the development of cardiovascular disease. These systemic diseases are more common with age and act as a modifying condition that exerts multiple biochemical effects on the heart that can potentially enhance the development and/or severity of ischemia–reperfusion injury.[4] For instance, patients with Type 2 diabetes mellitus (T2DM) have up to a fourfold increased risk of developing coronary heart disease compared to non-diabetic patients. Moreover, patients with T2DM have a higher risk of mortality following myocardial ischemia compared with non-diabetics due in part to an increased size of myocardial infarction.[6] The good news is that for the most part, with the exception of age, gender, and genetics, the other major risk factors for developing cardiovascular disease can be targeted with preventive measures. Indeed, advances in medicine over the last 50 years have led to the use of numerous pharmacological agents to effectively treat the risk factors of cardiovascular disease. For example, statins are used to lower circulating cholesterol levels, metformin and other blood glucose-lowering drugs are used in the management of diabetes, and a wide variety of anticoagulants and antiplatelet drugs are used to reduce the occurrence of coronary thrombosis. However, the complete treatment of cardiovascular disease and its consequences have been difficult to extrapolate from the experimental laboratory to the clinical setting and it appears that while these drugs are effective in reducing many of the symptoms that patients present with, they typically do not treat the underlying cause of symptoms. As such, patients continue to develop myocardial ischemia.

Myocardial Ischemia and Reperfusion

The reduction of coronary blood supply to the myocardium results in the development of myocardial ischemia. The reduction can either be in terms of absolute flow rate (low-flow or no-flow ischemia) or relative to increased tissue demand (demand ischemia).[4] A pivotal feature of ischemia is that the deprivation of oxygen and nutrient supply results in a series of abrupt biochemical and metabolic changes within the myocardium.[7] First, the absence of oxygen ceases oxidative phosphorylation, leading to mitochondrial membrane depolarization, ATP depletion, and inhibition of myocardial contractile function. Second, this process is exacerbated by the breakdown of any available ATP, resulting in ATP hydrolysis and a surge in mitochondrial inorganic phosphate. Third, in the absence of oxygen, cellular metabolism switches to anaerobic glycolysis, resulting in the accumulation of lactate, which

reduces intracellular pH. Finally, these changes contribute to intracellular calcium overload.

In experimental models and in human patients, cardiac ischemia is usually followed by reperfusion – the restoration of oxygen and metabolic substrates with washout of ischemic metabolites.[4] As noted above, although reperfusion salvages the ischemic myocardium it also induces further cardiomyocyte death. This phenomenon, known as reperfusion injury, was first described by Jennings and colleagues[8] in 1960 and may in part explain why despite optimal myocardial reperfusion, the rate of patient death after an acute myocardial infarction approaches 10% and why the incidence of cardiac failure after an acute myocardial infarction is almost 25%.[9] Additionally, it has been estimated that reperfusion injury accounts for up to 50% of the final size of a myocardial infarction (Figure 5.1).[7] However, the concept that reperfusion injury is an independent mediator of cardiomyocyte death, distinct from ischemic injury, has been hotly debated since its description. Some argue that the occurrence of injury after reperfusion only exacerbates the injury that was sustained during the ischemic period,[9] while others have argued that one must consider ischemia–reperfusion injury as a composite entity with distinct components of injury associated specifically with ischemia and with reperfusion since reperfusion can never occur independently of ischemia.[4] Regardless of these arguments, the simple truth is that there is evidence to suggest that cells alive before reperfusion die during the reperfusion period. Importantly, therapeutic interventions applied solely at the onset of myocardial reperfusion reduce infarct size by 40–50%,[7] suggesting that reperfusion injury can be targeted for therapeutic interventions (Figure 5.1).

Consequences of Ischemia–Reperfusion Injury

The development of myocardial stunning, reperfusion arrhythmias, endothelial dysfunction and irreversible cell death leading to infarction are all relevant as clinical consequences of myocardial ischemia–reperfusion injury and also represent important experimental correlates and endpoints (Figure 5.2). Here we define and discuss the mechanisms leading to myocardial stunning, reperfusion arrhythmias, endothelial dysfunction, and irreversible cell death endpoints in the context of cardiac ischemia–reperfusion injury.

Myocardial Stunning

Myocardial stunning is defined as 'the mechanical dysfunction that persists after reperfusion despite the absence of irreversible damage'.[10] It was first described by Heyndrickx and colleagues in 1975,[11] when the authors reported that coronary occlusions in the dog lasting for periods not long enough to induce cell death

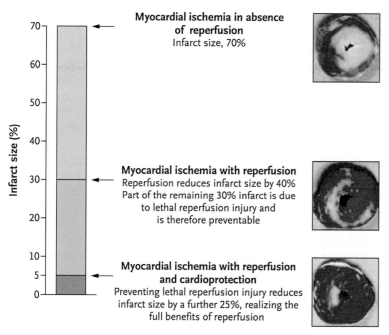

Infarct size (%)

70 —

60 —

50 —

40 —

30 —

20 —

10 —

5 —

0 —

Myocardial ischemia in absence of reperfusion
Infarct size, 70%

Myocardial ischemia with reperfusion
Reperfusion reduces infarct size by 40%
Part of the remaining 30% infarct is due
to lethal reperfusion injury and
is therefore preventable

Myocardial ischemia with reperfusion and cardioprotection
Preventing lethal reperfusion injury reduces
infarct size by a further 25%, realizing the
full benefits of reperfusion

FIGURE 5.1 Contribution of reperfusion injury to size of myocardial infarction. This hypothetical scheme illustrates ischemia–reperfusion injury as an independent mediator of cardiomyocyte death distinct from ischemia injury. This hypothesis explains why the full benefits of reperfusion are not realized in myocardial infarction. This hypothesis is supported by the findings that therapeutic interventions applied solely at the onset of myocardial reperfusion reduced infarct size by 40–50%. The areas in white depict infarcted myocardium and the areas in red depict viable at-risk myocardium. Infarct size is expressed as the percentage of the myocardium at risk. From Yellon and Hausenloy *N Engl J Med* 2007;357:1121–35.

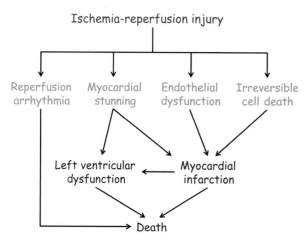

FIGURE 5.2 Consequences of myocardial ischemia–reperfusion injury. The development of myocardial stunning, reperfusion arrhythmias, endothelial dysfunction and irreversible cell death leading to myocardial infarction, left ventricular dysfunction, and death are all consequences of myocardial ischemia–reperfusion injury.

produced prolonged periods of contractile dysfunction. Subsequent studies demonstrated that reversible contractile dysfunction occurs in experimental models of demand ischemia in the hypertrophic heart, in isolated hearts subjected to global ischemia, and after cardiac surgery. Although the traditional definition of cardiac stunning indicated that the ischemic insult was non-lethal (i.e. not of sufficient duration to induce cell death), cardiac stunning also occurs after prolonged coronary artery occlusion resulting from myocardial infarction, as evidenced by myocardium salvaged by reperfusion, which can remain stunned with delayed contractile

recovery.[12] Importantly, experimental work has demonstrated the occurrence of stunning in all animal species investigated, suggesting that, despite the differences in contractile physiology, this is a universal phenomenon.

Reperfusion Arrhythmias

Reperfusion arrhythmias result from complex cellular and humoral reactions that accompany the opening of a blocked coronary artery.[13] Early insights into the conditions that produce reperfusion arrhythmias came from experiments that subjected the hearts of large animals to varying periods of coronary artery occlusion followed by reperfusion. In these experiments, it was noted that the incidence of reperfusion-induced ventricular fibrillation increased when the ischemic period extended from five minutes to 20 or 30 minutes. In contrast, the incidence of reperfusion-induced ventricular fibrillation declined when reperfusion was delayed beyond 30 to 60 minutes. These studies also found that reperfusion-induced ventricular fibrillation was more frequent when severe arrhythmias developed during the ischemic period. In humans, the most common reperfusion arrhythmia is an accelerated idioventricular rhythm. However, ventricular tachycardia and ventricular fibrillation remain the most important causes of sudden death following spontaneous restoration of antegrade flow.[14]

Endothelial Dysfunction

Endothelial dysfunction is a systemic pathological state of the endothelium characterized by an impairment in endothelium-dependent vasodilation and an exacerbation in the response to endothelium-dependent vasoconstrictors.[14] Endothelial dysfunction occurs during the early reperfusion period and persists. Experimental studies have identified that vascular injury begins with an endothelial triggering phase followed by a neutrophil amplification phase.[15] For instance, within five minutes of reperfusion the endothelium becomes dysfunctional as the production of nitric oxide (NO) decreases. By 20 minutes of reperfusion, leukocytes start to adhere to the endothelium and neutrophils begin to migrate across the endothelium into the damaged tissue. Once in the tissue, activated neutrophils release cytotoxic and chemotactic substances, such as cytokines, proteases, leukotrienes, and oxygen-derived free radicals. In the experimental setting, endothelial dysfunction after ischemia–reperfusion injury can be identified after 4–12 weeks. A combination of endothelial dysfunction, microvascular obstruction (downstream microembolism of platelets, de novo thrombosis, and/or neutrophil capillary plugging), edema, and oxidative stress is responsible for the pathogenesis of microvascular dysfunction following ischemia–reperfusion injury.[14] Sometimes, severe microvascular dysfunction limits adequate perfusion after reperfusion. This is known as the 'no-reflow' phenomenon and is characterized by the absence of tissue perfusion despite both epicardial coronary artery patency and flow.[16] Although the underlying mechanisms of this phenomenon have not been fully elucidated, the result is microvascular damage produced by microvascular vasoconstriction and obstruction associated with reperfusion injury. No-reflow after coronary revascularization therapy is associated with incomplete ST-segment recovery and increases the incidence of acute myocardial infarction, myocardial rupture, and death. Therefore, the recognition of no-reflow affords an opportunity for therapeutic intervention designed to augment tissue perfusion and maintain the viability of myocardium at risk for infarction.

Irreversible Cell Death

While arrhythmias, contractile impairment, and impaired coronary blood flow can be either reversible or irreversible, myocardial cell death or infarction is irreversible.[17] As such, myocardial infarction is the most robust endpoint of all studies designed to investigate myocardial ischemia–reperfusion injury. Studies demonstrate that the myocardium undergoes rapid ultrastructural changes after the onset of ischemia. These alterations tend to be reversible if reperfusion is established promptly. However, if blood flow is not rapidly restored (20–30 minutes of ischemia), these reversible alterations undergo a transition to a state of irreversible tissue injury characterized by cell death. Originally, cardiac myocyte cell death following ischemia–reperfusion injury was hypothesized to occur through the process of necrosis. However, studies over the last 20 years have challenged this dogma and established that apoptotic cell death also contributes in a meaningful way to the development of myocardial infarction after ischemia–reperfusion injury.[18]

Apoptosis is characterized by cytoplasmic shrinkage, plasma membrane blebbing, nuclear condensation, and fragmentation of both the cytoplasm and nucleus into membrane-enclosed apoptotic bodies. These bodies are subjected to phagocytosis by macrophages or even neighboring cells, thereby avoiding an inflammatory response. Apoptosis is an actively regulated form of cell death that is mediated by two pathways: extrinsic and intrinsic pathways (Figures 5.3 and 5.4).[19] The extrinsic pathway utilizes cell surface receptors to signal through death effector domains to cleave and activate downstream procaspases. The activated caspases then amplify apoptotic signaling by cutting and stimulating a variety of additional pro-apoptotic mediators. In contrast, the intrinsic pathway involves the mitochondria and endoplasmic reticulum. A key feature of the intrinsic pathway involves the permeabilization of the outer mitochondrial membrane, which permits the release of several apoptogens (e.g. bcl-1, bax) into the cytosol. In the cytosol, these apoptogens trigger downstream targets to facilitate the activation of caspases. Therefore, both pathways lead to the activation of caspases.[19] Interestingly, most of the proteins that are released from the mitochondria in response to apoptotic signaling have important physiological functions in healthy cells, but exhibit pathological properties when discharged into the cytosol. Both the extrinsic and intrinsic apoptosis pathways have been shown to play an important role in the development of myocardial infarction following ischemia–reperfusion injury. For example, mice deficient in the death receptor, Fas, display distinct reductions in infarct size after myocardial ischemia–reperfusion when compared to wild-type controls.[20] Additionally, mice with a cardiac-specific overexpression of the anti-apoptogene, Bcl-2, also display distinct reductions in infarct size, as well as cardiac myocyte apoptosis and left ventricular dysfunction after myocardial ischemia–reperfusion when compared to wild-type mice.[21] Similarly, mice deficient in the pro-apoptogene, Bax, display smaller infarcts following myocardial ischemia–reperfusion and treatment with various caspase inhibitors leads to a reduction in infarct size.[19]

In contrast to apoptosis, necrosis has been the conventional example of a type of unregulated cell death. Although a significant proportion of necrotic deaths are passive, evidence has emerged to indicate that much like apoptosis, necrosis can also be regulated.[22] The exact parts of the necrotic cell death pathway that are unregulated as opposed to those parts that are regulated have not been established. However, regulated necrosis has clearly been shown to be an important component of myocardial cell death after ischemia–reperfusion injury.[23] Currently, efforts are underway to determine the molecular components of the necrosis-signaling cascade and to determine if it can be targeted with

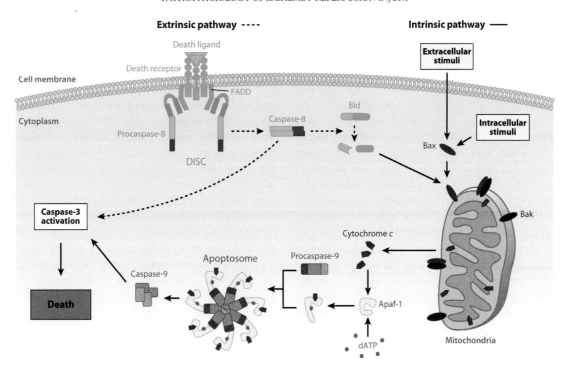

FIGURE 5.3   Schematic representation of the extrinsic and intrinsic cell death pathways that regulate cellular apoptosis. Apoptosis is mediated by the extrinsic pathway involving cell surface death receptors and by the intrinsic pathway that utilizes the mitochondria and endoplasmic reticulum. Both the extrinsic and intrinsic apoptosis pathways have been shown to play an important role in the development of myocardial infarction following ischemia-reperfusion injury. FADD, Fas-associated via death domain; Apaf1, apoptotic protease activating factor-1; Bax, Bcl-2-associated X protein. From Whelan et al. *Annu Rev Physiol* 2010;72:19–44.

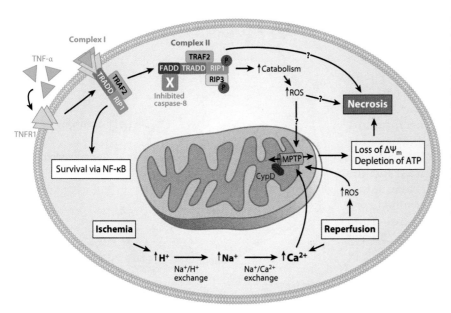

FIGURE 5.4   Schematic representation of the pathways that regulate cellular necrosis. Currently, necrosis signaling is limited to two pathways. The first involves death receptors, as exemplified by TNFR1 (tumor necrosis factor-α receptor 1) and the second involves the mitochondrial permeability transition pore (MPTP) in the inner mitochondrial membrane and its regulation by cyclophilin D (CypD). TRADD, TNF receptor superfamily 1A-associated via death domain; RIP, receptor interacting protein; TRAF2, TNF receptor-associated factor 2; FADD, Fas-associated via death domain; ROS, reactive oxygen species. From Whelan et al. *Annu Rev Physiol* 2010;72:19–44.

pharmacological intervention. A key feature of necrosis is the opening of the mitochondrial permeability transition pore (MPTP), a non-selective channel found in the inner mitochondrial membrane. While the components of the MPTP still remain unknown, cyclophilin D clearly plays a regulatory role in its opening.[24] The MPTP opens in response to elevated matrix calcium (Ca^{2+}) concentration, oxidative stress, elevated phosphate concentration, and adenine nucleotide depletion. As a result, there is a loss of the electrical potential difference ($\Delta\psi_m$) across the inner mitochondrial membrane, which ultimately leads to ATP depletion. Additionally, opening of the MPTP causes mitochondrial swelling due to an influx of water down its osmotic gradient into the mitochondrial matrix. Furthermore, swelling of the inner mitochondrial membrane can cause the rupture of the outer mitochondrial membrane resulting in the release of mitochondrial apoptogens into the cytosol.[19] Whether or not the release of these factors adds to necrotic cell death is currently not known. However, the rupturing of the outer mitochondrial membrane under these conditions is different from the permeabilization that occurs during apoptosis. Opening of the MPTP is the most apparent link between myocardial ischemia with or without reperfusion and necrosis. Studies have demonstrated that events during ischemia and events that occur after reperfusion contribute to the opening of the MPTP.

Cyclophilin D regulates the opening of the MPTP without being a component of it. Studies have shown that hearts from cyclophilin-D-deficient mice display smaller infarct sizes after myocardial ischemia–reperfusion when compared to wild-type control mice.[18] Necrotic cell death can also be activated through serine/threonine protein kinases in the receptor interacting protein (RIP) family. Inhibition of RIP1 with the small molecule inhibitor, necrostatin-1, has been reported to decrease infarct size after myocardial ischemia–reperfusion injury.[25] Interestingly, necrostatin-1 did not reduce infarct size further in mice deficient in cyclophilin D, suggesting that there is a link between the RIP kinase-dependent necrotic pathway and the opening of the MPTP. Although the molecular nature of such a connection remains to be determined, one possibility is the generation of oxidative stress from the activation of metabolic pathways by RIP during necrosis.[19]

Development of Myocardial Infarction

Several factors influence the extent of myocardial infarction after ischemia–reperfusion injury: (1) the location of the blockage; (2) the duration of the ischemic event; and (3) the extent of the collateral blood flow that is available.[4,17] The location of the blockage will determine the extent of myocardium risk for developing myocardial infarction (see Figure 5.5). If the coronary artery is occluded at Point A, then the area-at-risk is relatively small. However, if the coronary artery is occluded at Point B, then the area-at-risk is much larger. The second major determining factor of the size of infarction is the duration of ischemia to which the area-at-risk is subjected. As one might expect, longer periods of ischemia produce larger infarct areas, as illustrated in Figure 5.6, experimentally in mice subjected to either 45 minutes (right) or 60 minutes (left) of left coronary artery occlusion followed by 24 hours of reperfusion. Here the dark blue area in the heart slices represents the non-ischemic zone; the non-blue area, consisting of both the red and white tissue represents the area-at-risk; the white area represents the infarcted tissue within the area-at-risk; the red area represents the viable myocardium in the area-at-risk. The heart subjected to 60 minutes demonstrates a larger infarcted area, illustrating the relationship between ischemic time and infarct size. Early in this process, myocardial infarction begins in the inner myocardial layer (endocardium) of the core of the area-at-risk and extends out in a wavefront laterally and transmurally towards the epicardium. While this concept holds true for large mammals, it is somewhat different in mice, since the left ventricular free wall of a mouse is so thin that the inner layers receive oxygen by diffusion from the lumen.[17] Therefore, an infarct in a mouse is mostly midmyocardial or subepicardial. The third major determining factor of the size of infarction is the extent

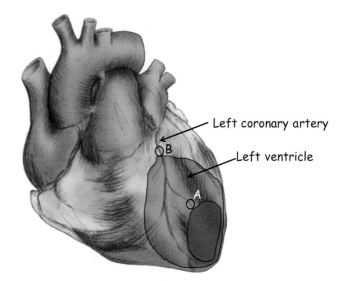

FIGURE 5.5 Location of coronary artery blockage is a factor that contributes to the size of infarction. Several major factors influence the extent of myocardial infarction after ischemia–reperfusion injury, including the location of the blockage, which determines how much of the myocardium is at risk for developing myocardial infarction. If the coronary artery is occluded at Point A, then the area-at-risk is rather small (dark area). However, if the coronary artery is occluded at Point B, then the area-at-risk is much larger (light area).

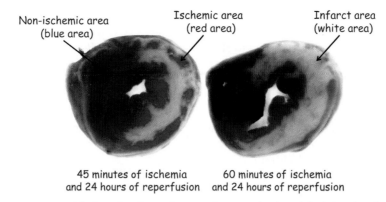

Non-ischemic area
(blue area)

Ischemic area
(red area)

Infarct area
(white area)

45 minutes of ischemia
and 24 hours of reperfusion

60 minutes of ischemia
and 24 hours of reperfusion

FIGURE 5.6   **Duration of coronary artery blockage is a factor that contributes to the size of the infarction.** Another major determining factor of the size of infarction is the duration of ischemia to which the area-at-risk is subjected. Longer periods of ischemia produce larger areas of infarct, as illustrated in these heart sections taken from mice subjected to left coronary artery occlusion for either 45 minutes (left) or 60 minutes (right), followed by 24 hours of reperfusion. The infarct size was determined by the Evan's Blue dye – triphenyl tetrazolium chloride (TTC) method. The dark blue area in the heart slices represents the non-ischemic zone. The non-blue area, consisting of both the red and white tissue represents the area-at-risk. The white area represents the infarcted tissue within the area-at-risk and the red area represents the viable myocardium in the area-at-risk. As one can see, the heart subjected to 60 minutes of ischemia displays a larger infarcted area.

of collateral blood flow or residual flow through the infarct-related artery. Additionally, it has been suggested that systemic hemodynamic influences during ischemia (i.e. heart rate) also contribute to the rate of development of infarction.[17]

The interaction of blockage location, the duration of the blockage, and the extent of the collateral flow results in the spatial and progressive expansion of infarction. Interestingly, the contribution of each factor appears to be species-dependent. For instance, in a small rodent heart with low collateral blood flow, the final infarct size is reached within 30–60 minutes, whereas in large animals with complete coronary occlusion and little collateral flow, infarction begins after 30–40 minutes and develops over several hours. This is in part due to the increased heart rates found in small rodents.[17] Given that the onset of coronary occlusion is rarely observable in humans, few and somewhat unreliable data exist for infarct development in patients. However, based on the available data, it appears that infarct development in humans seems to be between that found in dogs and that found in pigs, which are the preclinical models the FDA requires for future testing in humans.[17]

THERAPEUTIC STRATEGIES TO COMBAT MYOCARDIAL ISCHEMIA–REPERFUSION INJURY

The term cardioprotection has become synonymous with strategies aimed at attenuating myocardial ischemia–reperfusion injury. Studies throughout the years have demonstrated that cardioprotection is realized through the recruitment of endogenous mechanisms that are activated by physical interventions or chemical substances.[17] The remainder of this chapter will focus on

the signaling cascades that have been associated with cardioprotection, as well as provide some insights into the current state of therapeutic interventions.

Brief History of Cardioprotection

The study of cardioprotection began in the 1970s when Braunwald and colleagues first established that it was possible to limit infarct size and achieve cardioprotection. Remarkably, the effort to identify therapeutic interventions to induce cardioprotection has continued relentlessly for almost 40 years. Along the way, several hundred interventions, both pharmacological and non-pharmacological, have been reported to limit myocardial infarct size experimentally. Unfortunately, the vast majority of the results from these experimental studies proved to be non-reproducible and none, with the exception of reperfusion therapy, have made a successful transition to the clinic. Therefore, four decades after its inception, cardioprotection in the setting of myocardial ischemia–reperfusion injury remains an unfulfilled promise.

The development and testing of therapeutic interventions in the late 1970s and early 1980s were troubled by several confounding factors.[4,26] For instance, during those early years, the cellular mechanisms that contributed to cell death during ischemia and reperfusion were poorly understood. Thus, it was unclear which interventions should be tested or if myocardial salvage was even possible.[26] As a result, early strategies aimed at reducing infarct size centered on agents that either decreased myocardial oxygen demand or increased myocardial oxygen and metabolic substrates delivery to the injury. Consequently, agents such as calcium channel blockers, β-adrenoceptor antagonists, and glyceryl trinitrate were studied without any reproducible indication that infarct size decreased.[4,27] Additionally, it was not fully recognized until the late 1970s and early 1980s that reperfusion of the blocked artery was the crucial intervention to stop the progressive wave front of the developing infarction. As such, most of the studies of that time investigated potential cardioprotective agents in animal models of infarction with permanent coronary occlusion.[4]

The discovery and characterization of preconditioning (PC) in the late 1980s has proven to be the most important advancement in the pursuit to identify rational strategies to limit infarct size following ischemia–reperfusion injury. Ischemic PC (IPC) refers to the observation that one or several short intermittent periods of ischemia protect tissue against the injury caused by a subsequent, prolonged period of ischemia, usually lasting at least 30 minutes.[28] IPC was first described in the heart in 1986 when Murry and colleagues[29] demonstrated that short, intermittent periods of ischemia paradoxically limited infarct size when the hearts of dogs were subjected to subsequent prolonged ischemic insults. Since this first groundbreaking report, IPC has been observed in all species from mice to man and has been observed in other organ systems besides the heart. More importantly, IPC has proven to be the only intervention besides reperfusion that consistently reduces infarct size. Importantly, exposing the heart to various drugs prior to myocardial ischemia–reperfusion injury can mimic the protective effects of IPC, a phenomenon known as pharmacological PC (PPC). Some of the drugs with PPC effects include K^+ channel openers, volatile anesthetics, opioids, bradykinin, and nitric oxide (NO) donors.[28] Despite the clear cardioprotective effects of PC reported in experimental studies, there are two drawbacks to PC strategies that lessen their clinical utility. First, it is unknown when patients will experience a myocardial infarction. Therefore, individuals would have to be on medication chronically (to induce PC) to protect them from subsequent myocardial infarction that may or may not be imminent. Second, while many factors appear to be potent in reducing myocardial infarct size when administered acutely, repetitive administration of the PC stimuli results in a loss of efficacy. Therefore, the clinical utility of PC as infarct-sparing strategies continues to be in question.

However, despite questions regarding the clinical applicability of PC strategies, scientists have used this approach as an effective tool to identify cellular targets that are involved in the pathophysiology of myocardial ischemia–reperfusion injury. Research efforts have identified multiple molecular mechanisms pertinent to cell death and cytoprotection that form the basis of current experimental approaches being investigated. For example, studies have implicated the following signaling molecules, among others, as potential cardioprotective targets:[28] protein kinase C (PKC),[30] heat shock proteins (HSPs),[31] tyrosine kinases,[30] mitogen activated protein kinases (MAPKs),[30] protein kinase A (PKA),[32] nuclear factor κB (NF-κB),[33] adenosine,[34] and NO.[35] With the identification of these cellular targets and signaling cascades, scientists have for the past several years started to focus on the ability of therapeutic strategies given just prior to reperfusion to activate these same targets and confer additional protection to reperfusion alone.

Cardioprotective Signaling

Cellular signaling often involves complex systems, whereby interactions between membrane-bound proteins and signaling molecules lead to the activation of intracellular molecules. In this regard three hierarchical levels of signaling transduction exist: (1) triggers; (2) an intracellular mediator cascade; and (3) effectors (see Figure 5.7). Cardioprotective signaling has been shown to have many of these same aspects. For instance, the trigger that induces the cardioprotective signaling would be the drug or substance administered before, during or after the onset of ischemia and/or reperfusion that acts upon a sarcolemmal membrane receptor. Stimulation of the receptor in turn initiates a downstream signaling cascade involving more often than not protein kinases, which ultimately act on subcellular elements that act as the effectors to stabilize the cardiomyocyte and protect it from cell death.[17] An example of this signaling cascade is depicted in Figure 5.7: adenosine (triggers) binds to the A_{2A} receptor to stimulate Akt-endothelial nitric oxide synthase (eNOS)-NO signaling (downstream signaling cascade), which then acts on the mitochondria (effector). Discussion of all of the known cardioprotective signaling cascades and potential cellular targets would be exhaustive and beyond the scope of this chapter. Therefore, a few examples will be discussed below.

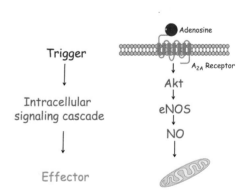

FIGURE 5.7 Schematic representation of a cardioprotective signaling cascade. Three hierarchical levels of signaling transduction exist: (1) triggers; (2) an intracellular mediator cascade; and (3) effectors. Cardioprotective signaling involves all of these levels. For example, adenosine (triggers) binds to the A_{2A} receptor to stimulate Akt-endothelial nitric oxide synthase (eNOS)-NO signaling (downstream signaling cascade), which then acts on the mitochondria.

THERAPEUTIC STRATEGIES TO COMBAT MYOCARDIAL ISCHEMIA–REPERFUSION INJURY

Mitochondria

The mitochondria are by far the most common effector for the different signaling cascades identified. Mitochondria are essential for cell survival, both because of their role as metabolic energy producers and as regulators of programmed cell death. Roughly 98% of inhaled oxygen is consumed by mitochondria, and without mitochondria, we would have no need for the oxygen transfer machinery of the lungs, red cells, hemoglobin, or even the circulatory system that delivers oxygen to the tissues. Likewise, the organization of food intake, digestion, and processing is designed primarily to supply substrates destined for mitochondrial oxidation.[36] Mitochondria also provide the energy required for numerous cellular processes such as muscle contraction, accumulation of secreted material into vesicles, vesicle fusion and cycling necessary for the secretion of hormones and neurotransmitters, and maintenance of ionic gradients in excitable cells.[36] So, it follows predictably that impaired mitochondrial function will lead to disease, ranging from subtle alterations in cell function to cell death and from minor to major disability, or even death.

The healthy heart relies on a sophisticated regulation of substrate metabolism to meet its continual energy demand. Under normal conditions, the heart derives roughly 70% of its energy from the oxidation of fatty acids, with the remainder primarily from pyruvate oxidation, derived from glucose.[37] In order for ATP to be produced from fatty acids or glucose, an intact respiratory chain for oxidative phosphorylation in the mitochondria is essential. Components of the respiratory complex include: complex I, NADH ubiquinone oxidoreductase; complex II, succinate ubiquinone oxidoreductase;

complex III ubiquinone–cytochrome c oxidase; complex IV, cytochrome c oxidase; complex V, ATP synthase; ubiquinone; cytochrome c (Figure 5.8). Electron-producing substrates enter the respiratory chain at complexes I and II. From there, ubiquinone shuttle the electrons to complex III where the electrons are passed to complex IV by cytochrome C. The passing of electrons through the respiratory complexes provides energy to pump protons from the matrix across the inner membrane into the inner membrane space. Complex V allows the protons to flow back into the matrix, using the energy released to synthesize ATP. The incomplete processing of oxygen and/or release of free electrons from the mitochondria can result in the production of oxygen radicals, such as superoxide (O_2^-), hydrogen peroxide (H_2O_2), and hydroxyl radical (OH-).[38] An increase in superoxide formation can have both direct and indirect consequences on cellular metabolism. Electron leakage of the respiratory chain occurs normally (0.1–2% reducing equivalents handled by mitochondria). Damage due to electron leak is kept in check by an efficient array of redundant antioxidant defense systems (catalase, glutathione reductase). However, during pathological conditions, the oxidant versus antioxidant balance is altered in favor of the latter, either primarily or secondarily. Oxidative damage will occur when the production of reactive oxygen species exceeds the antioxidant defense systems. Being a major source of reactive oxygen species, mitochondria are at risk for a direct attack by reactive oxygen species and very susceptible to oxidative damage. The first radical produced in the mitochondria is O_2^-; since O_2^- is a highly reactive species that does not diffuse relatively very far, it has been postulated that its main target in the mitochondria is the mitochondrial DNA.[39] Damage to mitochondrial

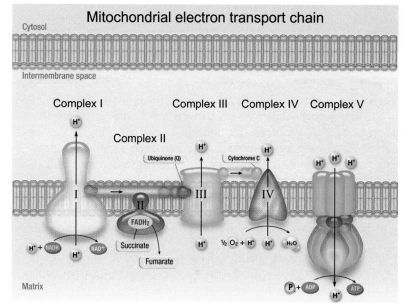

FIGURE 5.8 Schematic representation of the mitochondrial electron transport chain. The mitochondrial electron transport chain consists of five complexes: complex I, NADH ubiquinone oxidoreductase; complex II, succinate ubiquinone oxidoreductase; complex III, ubiquinone-cytochrome c oxidase; complex IV, cytochrome c oxidase; complex V, ATP synthase. Electron-producing substrates enter the respiratory chain at complexes I and II. From there, ubiquinone shuttles the electrons to complex III where the electrons are passed to complex IV by cytochrome C. The passing of electrons through the respiratory complexes provides energy to pump protons from the matrix across the inner membrane into the inner membrane space. Complex V allows the protons to flow back into the matrix, using the energy released to synthesize ATP.

DNA could result in dysfunctional respiratory chain components, thus resulting in more ROS production.[38] H_2O_2 is the next player in mitochondrial-derived ROS. Due to its small size and relative benign reactivity, H_2O_2 can freely diffuse across several cell radii and mediate toxic events relatively distant from its production site. H_2O_2 is not a major source of damage, but it can react with free transition metals by Fenton chemistry and form the potent hydroxyl radical. Superoxide can interact with nitric oxide (NO) to form peroxynitrite (ONOO-), a highly damaging agent with many targets and detrimental effects. Targets in the mitochondria include complexes I, II, IV, and VI, aconitase, creatine kinase, superoxide dismutase, mitochondrial membranes, and mitochondrial DNA.[38] Damage to these molecules can cause mitochondrial swelling, depolarization, calcium release, and permeability transition resulting in cell death.

Under normal conditions, the mitochondrial network of the myocyte must have properties of both constancy and flexibility, first providing a steady supply of ATP to fuel contraction, and second, to adapt the rate of energy production to meet the changing metabolic demand as workload varies.[40] The global response of the cardiac myocyte depends on the coordinated action of thousands of mitochondria arranged in a lattice-like network of non-linearly coupled elements. When the normal function of the heart cell becomes severely compromised under pathophysiological conditions, the inherent vulnerability of the mitochondrial network is revealed. At this point of mitochondrial criticality, bioenergetics becomes unstable and can undergo a rapid transition to mitochondrial membrane potential ($\Delta\Psi_m$) depolarization that involves a coordinated response of almost the entire population of mitochondria in the network.[40] Additionally, MPTP occupy a fundamental role in determining cellular survival in the setting of myocardial ischemia–reperfusion injury because MPTP opening also causes $\Delta\Psi_m$ depolarization.[40] Early reperfusion following ischemia represents a period when $\Delta\Psi_m$ is most likely to become unstable due to the production of high levels of ROS and ensuing oxidative stress. As a result, the loss of $\Delta\Psi_m$ during this time causes a rapid impairment of mitochondrial function, which ultimately leads to apoptotic cell death through the release of pro-apoptotic proteins or can initiate necrotic cell death. Thus, maintaining $\Delta\Psi_m$ is of paramount importance during the period of early reperfusion, as it is a major determinant of cell fate following ischemia.[40] Given that mitochondria lie at the core of the existence of cellular life, it is of no surprise that they are the most common effector for numerous cardioprotective-signaling cascades.

Reperfusion Injury Salvage Kinase Pathway

The Reperfusion Injury Salvage Kinase (RISK) pathway is the term given to a signaling cascade involving prosurvival kinases, which confer cardioprotection

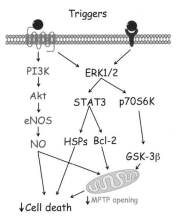

RISK pathway

FIGURE 5.9   Schematic representation of the Reperfusion Injury Salvage Kinase pathway. The Reperfusion Injury Salvage Kinase (RISK) pathway is the term given to a signaling cascade involving a number of pro-survival kinases, which confer cardioprotection when specifically activated at the onset of reperfusion following myocardial ischemia. Activation of this pathway is triggered by insulin, urocortin, atorvastatin, bradykinin, opioid receptor agonists, and Glucagon-Like Peptide-1. The common effector of the different arms of the RISK pathway is the mitochondria, where the signaling cascades converge to inhibit the opening of the MPTP and decrease cell death. STAT3, signal transducer and activator of transcription 3; PI3K, phosphatidylinositol-3 kinase; Erk1/2, extracellular regulated kinase 1/2; p70S6K, p70 ribosomal S6 kinase; GSK-3β, glycogen synthase kinase 3β; eNOS, endothelial nitric oxide synthase; NO, nitric oxide; HSPs, heat shock proteins; MPTP, mitochondrial permeability transition pore.

when specifically activated at the onset of reperfusion following myocardial ischemia (Figure 5.9). The concept for the RISK pathway is based on the evidence that apoptosis contributes to myocyte cell death following ischemia–reperfusion injury and that activation of certain kinases exerts anti-apoptotic effects.[41] Therefore, it has been postulated that targeting these kinases at the time of reperfusion with pharmacological agents would protect the myocardium.[42] As shown in Figure 5.9, the original members reported to be a part of the RISK pathway were the phosphatidylinositol-3 kinase (PI3K), Akt, and extracellular regulated kinase 1/2 (Erk1/2). Additional studies have expanded this list to include other kinases such as protein kinase C (PKC; primarily the PKC-ε isoform), protein kinase G (PKG), p70 ribosomal S6 kinase (p70S6K), and glycogen synthase kinase 3β (GSK-3β).[41,42] It has now been demonstrated that insulin urocortin, atorvastatin, bradykinin, opioid receptor agonists, and Glucagon-Like Peptide-1 (GLP-1) reduce myocardial infarct size when administered at the time of myocardial reperfusion through the activation of the RISK pathway.[41]

While the downstream effectors of the RISK pathway have not been fully elucidated, the Erk1/2 component of the RISK pathway has been shown to signal through signal transducer and activator of transcription 3 (STAT-3), p90RSK, Bcl-2, Bcl-xL, and HSPs.[42–44]

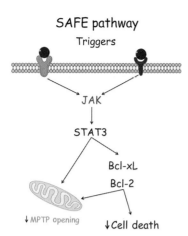

FIGURE 5.10 Schematic representation of the Survivor Activating Factor Enhancement pathway. The Survivor Activating Factor Enhancement (SAFE) pathway involves the activation of Janus kinase (JAK) and STAT-3. SAFE has been reported to be activated by tumor necrosis factor α, opioids, insulin, and sphingosine-1. Much like the RISK pathway, the signaling cascade in the SAFE pathway converges on the mitochondria.[125]

Additionally, as shown by the convergence of the cascades in Figure 5.9, a common target of the signaling activated by the RISK pathway is the mitochondria. Specifically, activation of the RISK pathway has been shown to inhibit the opening of the MPTP.[45] The mechanism through which the RISK pathway inhibits the opening of the MPTP is unclear, although there are several hypotheses that have been proposed. First, GSK-3β signaling could play a role, since it has been shown to inhibit MPTP opening in the context of cardioprotection.[46] Second, endothelial nitric oxide synthase, another identified downstream target of the RISK pathway, has the potential to inhibit the opening of the MPTP through the generation of NO, which either directly inhibits opening or acts via a PKG-PKC-ε-mKATP channel signaling pathway. Third, the inhibition of Bax translocation to the mitochondria and/or the activation of mitochondrial hexokinase II may act together to inhibit MPTP opening. Finally, PI3K activation reduces the uptake of calcium by the sarcoplasmic reticulum, which may in turn act to inhibit MPTP opening and subsequent cell death.

Survivor Activating Factor Enhancement Pathway

Tumor necrosis factor alpha (TNFα) is a cytokine traditionally thought to contribute to the pathophysiology of reperfusion injury.[47] However, emerging evidence suggests that TNFα contributes to the myocardial protective adaptation in response to injury stimuli.[48] In fact, recent studies suggest that TNFα initiates the activation of a novel protective signaling cascade termed the Survivor Activating Factor Enhancement (SAFE) pathway (Figure 5.10), which involves the activation of Janus kinase (JAK) and STAT-3.[49] JAKs are a family of tyrosine kinases that are associated with membrane receptors and

play a major role in conducting signals from the cytosol to the nucleus.[50] STATs are phosphorylated by activated JAKs, which enables them to form homodimers or heterodimers that translocate to the nucleus, resulting in gene transcription. As such, the activation of the JAK/STAT pathway plays a crucial role in the expression of stress-responsive genes. The STAT pathway has recently been shown to be an integral part of the response of the myocardium to various cardiac insults, including myocardial infarction.[51] In particular, the overexpression of STAT-3 results in cardioprotection, whereas cardiac-specific deficiency of STAT-3 exacerbates cardiac injury. Several downstream targets of STAT-3 have been identified including proteins that are involved in cell survival and proliferation (Bcl-2, Bcl-xL, Mcl-1, and p21), some growth factors (vascular endothelial growth factor) and other transcription factors.[49] Also, STAT-3 mediates cardioprotection via the phosphorylation and inactivation of the pro-apoptotic factor Bad.[52] Importantly, triggers other than TNFα activate the SAFE pathway via STAT-3, including opioids, insulin, and sphingosine-1.

Activation of the SAFE pathway occurs alone or in parallel to the RISK pathway during the early phase of reperfusion following myocardial ischemia. Although there is evidence to suggest that both pathways protect independently from each other,[52] there is also evidence for cross-talk under certain conditions. For example, both pathways seem to have a STAT-3 signaling component. However, there are still a number of questions regarding the exclusivity and convergence of these pathways that remain to be addressed.[49] For instance, do these two pathways converge on the same targets such as the mitochondrial permeability transition pore? Does the activation of the two pathways provide an additive effect to maximize the protection? Can the SAFE pathway protect beyond the RISK pathway or has the threshold of protection already been achieved by the activation of either pathway acting alone?

AMP-Activated Protein Kinase

Multiple individual proteins that do not necessarily belong to a particular signaling cascade have been identified as potential cardioprotective targets, including AMP-activated protein kinase (AMPK) (see Figure 5.11). AMPK is a serine–threonine kinase that exists as a heterotrimeric complex containing an α catalytic subunit, as well as β and γ regulatory subunits. AMPK has emerged as an important regulator of diverse cellular pathways in the setting of energetic stress. Metabolic activators of AMPK include hypoxia, ischemia, oxidative and hyperosmotic stresses, as well as exercise and glucose deprivation.[53] Its activation is mediated by increases in AMP-to-ATP ratios through mechanisms involving allosteric regulation of AMPK subunits, making it a better substrate for upstream AMPK kinases (AMPKK) and a worse substrate for competing protein

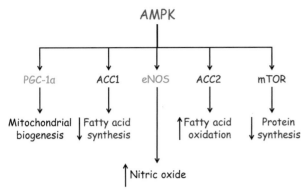

FIGURE 5.11 Schematic representation of the downstream targets of AMP-activated protein kinase. AMP-activated protein kinase (AMPK) is a serine-threonine kinase that exists as a heterotrimeric complex containing an α catalytic subunit as well as β and γ regulatory subunits. AMPK has emerged as an important regulator of diverse cellular pathways in the setting of energetic stress. Activators of AMPK include hypoxia, ischemia, oxidative and hyperosmotic stresses, as well as exercise, glucose deprivation and metformin. Activation of AMPK leads to an increase in mitochondrial biogenesis, fatty acid oxidation, and generation of nitric oxide, as well as a decrease in fatty acid synthesis and protein synthesis; PGC-1α, peroxisome proliferator-activated receptor-γ coactivator; ACC1, acetyl-CoA carboxylase 1; ACC2, Acetyl-CoA carboxylase 2; eNOS, endothelial nitric oxide synthase; mTOR, mammalian target of rapamycin.

phosphatases.[54] In addition, multiple conditions/treatments activate AMPK in a manner that is not dependent on changes in the AMP to ATP or the phosphocreatine-to-creatine ratios. In other words, certain conditions/treatments lead to the activation of AMPK through direct interaction or via upstream signals. For example, leptin, adiponectin, hyperosmotic stress, metformin, AICAR, and long chain fatty acids have been reported to activate AMPK. AMPK activation stimulates fatty acid oxidation, promotes glucose transport, accelerates glycolysis, and inhibits triglyceride and protein synthesis.[53] By increasing ATP synthesis and decreasing ATP utilization, AMPK functions to maintain normal energy stores during ischemia. Chronic activation of AMPK also phosphorylates transcription factors altering gene expression and modulates muscle mitochondrial biogenesis. Additionally, it has been shown that the anti-hyperglycemic actions of metformin are modulated by the activity of AMPK in hepatocytes and skeletal myocytes.[55] Long-term activation of AMPK by AICAR has also been shown to improve glucose tolerance, improve lipid profile, and reduce systolic blood pressure in an insulin-resistant animal model.[56]

Cardiac AMPK activity increases in response to a wide array of stimuli, including ischemia.[57] Studies in isolated perfused heart preparations experimentally have demonstrated that AMPK is activated by both severe no-flow ischemia and partial ischemia.[58,59] In vivo, the phosphorylation and activity of AMPK are increased

within minutes of the onset of myocardial ischemia and remain elevated for at least 48 hours following reperfusion.[60] For the most part, the evidence indicates that activation of AMPK serves a cardioprotective role. Studies using two separate lines of genetically engineered mice that express inactive forms of the α2-subunit of AMPK (K45R and D157A) have reported that these mice display impaired glucose uptake and ATP homeostasis during ischemia and have increased apoptosis and necrosis following ischemia–reperfusion injury.[58,61] Additionally, hearts from global AMPK-α2-deficient mice display reduced glycolysis, greater ATP depletion, more rapid development of contracture, and impaired contractile function after ischemia. Importantly, pharmacological activation of AMPK has also been shown to provide cardioprotection after myocardial ischemia–reperfusion injury. For instance, the administration of metformin activated AMPK and reduced myocardial infarct size in non-diabetic and diabetic.[61] Additionally, the administration of the AMPK activator, A769662, also provides cardioprotection against myocardial ischemia.[62]

The exact mechanisms by which AMPK activation facilitates cardioprotection remain to be elucidated, however there is considerable evidence that it possesses pleiotropic effects.[57] AMPK activation has also been shown to increase the phosphorylation of eNOS at serine residue 1177, which in turn increases its activity and increases the bioavailability of NO.[61] The activation of eNOS and NO appear to be critical for the AMPK-mediated cardioprotective effects of metformin, but less important for the AMPK-mediated cardioprotective effects of A769662. Finally, activation of AMPK has also been linked to inhibition of MPTP opening.[62]

Despite all of the evidence suggesting that activation of AMPK provides cardioprotection in the setting of ischemia–reperfusion injury, there are still those that argue that AMPK activation during early reperfusion may be detrimental to the myocardium, because of enhanced high fatty acid oxidation (FAO) rates and subsequent inhibition of glucose oxidation.[63] For instance, a major metabolic consequence resulting from the activation of AMPK during and following ischemia is the stimulation of fatty acid oxidation.[59] While overall oxidative rates during ischemia are limited by oxygen supply to the muscle, during and following ischemia, AMPK activation can lower cardiac malonyl CoA levels, secondary to phosphorylation and inhibition of acetyl-CoA carboxylase (ACC).[64] Unfortunately, in the presence of an ischemic-induced limitation of overall oxidative metabolism, the AMPK-dependent acceleration of fatty acid oxidation occurs at the expense of glucose oxidation, and has the potential to decrease cardiac function and cardiac efficiency during the critical period of reperfusion.[65] While these potential detrimental effects must be taken into consideration, the evidence stemming from the studies in which AMPK-inactivated mice were used suggests a cardioprotective role. Furthermore, the administration of metformin and subsequent activation of AMPK reduced infarct size in db/db diabetic mice, which have high circulating fatty acid levels. Therefore, if metformin and AMPK stimulate excessive FAO at the expense of glucose oxidation during early reperfusion, the beneficial effects of AMPK signaling seem to outweigh the potential detrimental effects associated with FAO. Taken together, it appears that the activation of AMPK during ischemia–reperfusion injury is an endogenous protective mechanism that can be augmented by pharmacological intervention during reperfusion to provide substantial infarct-lowering effects.

Therapeutic Interventions that Activate Multiple Signaling Cascades

Up to this point, the discussion on cardioprotection has focused on specific signaling cascades and specific targets that have been associated with cardioprotection. However, the events that occur during ischemia–reperfusion injury are complex. So, targeting one specific pathway or signaling molecule may not be the most effective strategy. Strategies that have the ability to activate multiple cellular targets would certainly be advantageous. For example, the activation of the RISK and SAFE pathways, as well as AMPK may prove to reduce infarct size better than activation of a single pathway alone. The discussion will now turn to several strategies that have the potential to activate multiple pathways at the same time.

Gasotransmitters

As mentioned above, cellular signaling often involves complex systems, whereby interactions between membrane-bound proteins and signaling molecules lead to the activation of intracellular molecules. These intracellular molecules act as secondary messengers, which then relay a signal to a specific destination. A set of endogenous gaseous molecules called gasotransmitters possesses similar signaling capabilities as other signaling molecules but does not require the regular string of regulatory mechanisms to transmit a signal.[66] Gasotransmitters are small, labile, endogenously produced gaseous molecules that play important roles in cellular signaling.[67] There are three members of the gasotransmitter family; nitric oxide (NO), carbon monoxide (CO) and hydrogen sulfide (H_2S). Historically, these gases were considered to be highly toxic and hazardous to the environment. However, under normal physiological conditions in mammals, these molecules are enzymatically regulated and endogenously produced. Because of this discovery, the biological and physiological role of these gases has been re-evaluated. As such, an extensive

amount of work has been conducted over the last several decades and has led to the discovery that each gasotransmitter possesses multiple beneficial physiological actions.

Three isoforms of nitric oxide synthase (NOS) have been characterized, purified, and cloned: (1) the endothelial isoform (eNOS); (2) the neuronal isoform (nNOS); and the (3) inducible isoform (iNOS). All three isoforms are able to catalyze the production of NO from an assortment of precursors and co-factors, including L-arginine, NADPH, and oxygen. NOS isoforms are found in a variety of cell types and tissues, including neuronal cells and the vascular endothelium. The localization of eNOS in the vascular endothelium is of particular importance for cardiovascular physiology, as eNOS maintains basal vascular tone through its release of low levels of NO.[68] NO has been extensively studied in the setting of myocardial ischemia–reperfusion injury. The role of endogenously derived NO has been studied using pharmacological inhibitors against NOS and by genetically targeting each NOS. The role of exogenously derived NO has been studied through the administration of NO in the form of authentic NO gas, NO donors, and more recently nitrite and nitrate. Perhaps the most clear-cut evidence for a protective role of endogenously derived NO in the setting of myocardial injury comes from studies aimed at investigating eNOS. Studies of mice deficient in eNOS (eNOS$^{-/-}$) have overwhelmingly shown that eNOS is cardioprotective; eNOS$^{-/-}$ mice experience exacerbated infarct sizes and increased myocardial dysfunction in response to myocardial ischemia.[69] In contrast, the overexpression of eNOS reduces myocardial infarction size and improves myocardial function in the same experimental models of ischemic injury.[70] Early studies reported that a deficiency of nNOS or iNOS did not affect infarct size in response to acute myocardial ischemia.[70–73] However, more recent evidence suggests that nNOS plays a crucial role in preventing adverse left ventricular remodeling and ventricular arrhythmias and maintaining myocardial β-adrenergic reserve after myocardial infarction.[74,75] Likewise, new evidence has emerged to suggest that gene transfer of iNOS affords cardioprotection against myocardial ischemia–reperfusion injury.[76] Taken together, these studies clearly demonstrate that endogenously produced NO has the ability to protect the heart from ischemia–reperfusion injury.[77] In terms of its cytoprotective mechanisms, NO possesses a number of physiological properties that makes it a potent cardioprotective-signaling molecule.[78] First, NO is a potent vasodilator in the ischemic myocardium, which allows for essential perfusion of injured tissue. Second, NO reversibly inhibits mitochondrial respiration during early reperfusion, counterintuitively leading to a decrease in mitochondrial-driven injury by extending the zone of adequate tissue cellular oxygenation away from vessels.[79] Third, NO is a potent inhibitor of neutrophil adherence to vascular endothelium. Neutrophil adherence is an important event initiating further leukocyte activation and superoxide radical generation, which in turn leads to injury to the endothelium and perivascular myocardium. Fourth, NO also prevents platelet aggregation, which together with the antineutrophil actions of NO attenuates capillary plugging.[16] The cytoprotective signaling pathways activated by NO in various experimental systems including isolated cardiac cells and intact hearts have been evaluated. NO has been shown to activate cellular targets linked to cardioprotection, including components of the RISK and SAFE pathway, including PKC, Erk1/2, JAKs, and STAT1/3.[80] Additionally, NO has been shown to inhibit apoptosis either directly or indirectly by inhibiting caspase-3-like activation via a cGMP-dependent mechanism and by direct inhibition of caspase-3-like activity through protein S-nitrosylation.[81]

Following the discovery of NO as a gasotransmitter, carbon monoxide (CO) was found to have similar roles.[82–84] CO is formed endogenously by the enzyme heme oxygenase (HO) through the degradation of heme. There are three known isoforms of HO: HO-1, inducible under cellular stress; HO-2, homeostatic form; and HO-3, recently found in rat brain but no gene for HO-3 has yet been found.[85] HO-1 is responsible for degrading heme into biliverdin and CO. Biliverdin is quickly reduced in the cell to bilirubin, which is a very important antioxidant. Enhanced expression of HO-1 and its degradation products have been shown to augment multiple intracellular cytoprotective pathways. In particular, HO-1 protein expression is significantly up-regulated in myocardial infarction, and hypoxia-induced up-regulation of HO-1 in the heart has been shown to significantly increase CO production.[86] Predictably, studies investigating myocardial damage in HO-1 knockout mice following myocardial ischemia have reported[87] exacerbated myocardial injury, increased ROS production, and decreased endogenous CO production. However, at low levels exogenous CO has been shown to stimulate cardioprotection in HO-1 knockout mice, and rat hearts during ischemia–reperfusion injury.[88] The role of endogenous CO in cardioprotection has also been demonstrated using carbon-monoxide-releasing molecules (CO-RMs) to elicit pharmacological activities in myocardial cells against ischemia–reperfusion injury.[89] Taken together, these studies suggest that HO-1-induced CO production and direct administration of CO provide potential therapeutic alternatives for the pharmacological regulation of myocardial ischemia–reperfusion injury.

H_2S was the third gasotransmitter identified after NO and CO. Enzymatically produced in all mammalian species at low micromolar levels, H_2S results from the action of three cysteine enzymes: cystathionine-γ-lyase (CSE), cystathionine-β-synthase (CBS), and 3-mercaptopyruvate sulfutransferase (3-MST).[67] An increasing number of studies also provide evidence that both exogenous and endogenous H_2S exert cytoprotective effects, especially against myocardial ischemia–reperfusion injury. Targeted deletion and genetic manipulation of CGL leads to modification of H_2S expression in the aorta, heart, and serum.[66] Studies using murine models of ischemia–reperfusion injury have shown that treatment with H_2S prior to myocardial ischemia significantly reduces infarct size, and H_2S administered at the time of reperfusion has been shown to reduce infarct size and exerts dose-dependent cardioprotection.[90,91] However, when the production of H_2S is reduced by pharmacological inhibition prior to myocardial ischemia, mice experience exacerbated myocardial injury.[91] Further evidence that H_2S confers cardioprotection has been shown by genetically altering CGL expression. Mice deficient in CGL (CGL$^{-/-}$) have been reported to experience decreased myocardial function, reduced serum H_2S levels, pronounced hypertension, diminished endothelium-dependent vasodilation, and significantly larger areas of myocardial infarction compared to wild-type control animals.[44,92] Recent studies have also demonstrated that H_2S protects against acute myocardial ischemia–reperfusion injury via the induction of components of the RISK and SAFE pathway. Specifically, H_2S provides antiapoptotic effects via PI3K/Akt, PKC and ERK 1/2 signaling. It also provides antioxidant actions via the activation and translocation of Nrf-2 to the nucleus including an increase in antioxidant response element-related antioxidants.[93] Moreover, previous experimental studies suggest that H_2S augments angiogenesis under ischemic conditions both in vitro and in vivo. Finally, H_2S exhibits potent anti-inflammatory actions and modulates mitochondrial respiration in part by reversible inhibition of cytochrome c oxidase.[67]

MicroRNA

MicroRNAs (miRNA) are short, single-stranded RNAs that anneal with complementary sequences in messenger RNAs (mRNA). The manipulation of miRNA expression or function can have a profound impact on cellular phenotypes given that individual miRNAs can engage multiple mRNA targets. As such, miRNAs are able to suppress the protein expression of multiple components of complex intracellular networks. Depending on the abundance of a miRNA and its targets, as well as the physiological state of a cell, a miRNA can act as a fine-tuner of gene expression or an on/off switch. Under conditions of pathophysiological stress and in response to diseases, the functions of miRNAs are heightened, making them attractive candidates for therapeutic manipulation.[94,95] In response to myocardial ischemia, studies have shown that some miRNAs protect the heart against ischemic injury, whereas others contribute to ischemic injury (Figure 5.12). For instance, miRNA-24 and miR-499 protect the heart against myocardial ischemia–reperfusion injury by controlling calcium overload, mitochondrial fission and cell death.[96,97] In contrast, therapeutic targeting of miR-15 and miR-92a in mice reduced infarct size and cardiac remodeling in response to myocardial ischemia–reperfusion injury.[98,99]

Stem Cells

The discussion up to this point has focused on therapeutic strategies that target events during the later parts of ischemia and early reperfusion. As noted above, despite our best efforts, nothing, with the exception of timely reperfusion, has been shown to have a clinical benefit. Without going into too much detail here, one of the arguments that has been made is that we may already be at the therapeutic ceiling with current reperfusion therapy. Simply opening up the blocked artery may be the most effective way to decrease infarct size. This has led both scientists and physicians to investigate therapeutic strategies that can be initiated days, weeks, or even months after reperfusion to decrease the extent of the development of heart failure. Heart failure is basically the inability of the heart to meet hemodynamic demands and represents the end stage of various forms of cardiovascular disease, including myocardial infarction or ischemia associated with coronary artery disease. Interestingly, the prevalence of heart failure has increased dramatically as modern therapies have reduced the in-hospital mortality of acute myocardial infarction.[100] In the United States,

it has become the most common discharge diagnosis in patients aged 65 years or older and the primary cause of readmission within 60 days of discharge. It is also estimated that 5.7 million people in the United States have heart failure resulting in about $37.2 billion spent a year to cover associated healthcare-related costs. Current treatments for heart failure are woefully inadequate, and the availability of hearts for transplantation is severely limited.[100] Therefore, adjunct pharmacotherapies designed to coincide with the standard means of care are needed to decrease the extent of injury leading to the development of heart failure. One such strategy that has been advocated for almost a decade now is stem cell therapy.

Every single cell in the body is derived from a stem cell, which is defined as a primitive, undifferentiated, undefined pluripotent multilineage cell that retains the ability to renew itself through mitotic cell division and can divide and create a cell more differentiated than itself.[101] Stem cells were originally believed to be present only in embryos and fetuses; however they can actually be found in most organs throughout the adult body. Adult stem cells are defined as undifferentiated progenitor cells from an individual after embryonic development. While they are similar to those found in embryos, adult stem cells seem to need assistance to differentiate. Stem cells are generally classified according to the following criteria: origin, type of organ or tissue from which the cells are derived, surface markers, and final differentiation fate.[101] With respect to our discussion of ischemic heart disease therapy, these stem cells can be classified broadly as pluripotent, cardiac-derived, and bone-marrow-derived stem cells, which are discussed next.

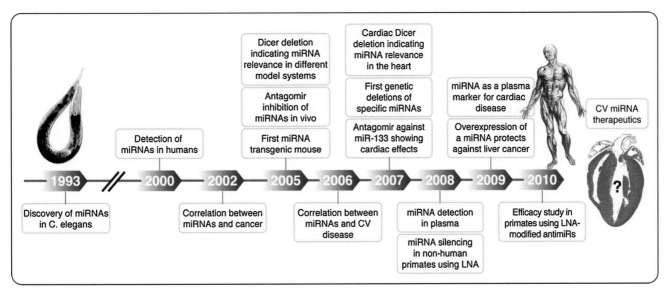

FIGURE 5.12 Role of microRNAs in myocardial ischemia–reperfusion injury. MicroRNAs (miRNA) are short, single-stranded RNAs that anneal with complementary sequences in messenger RNAs. The manipulation of miRNA expression or function can have a profound impact on cellular phenotypes given that individual miRNAs can engage multiple mRNA targets. In response to myocardial ischemia, studies have shown that some miRNAs protect the heart against ischemic injury, whereas others contribute to ischemic injury. The schematic here depicts a time line indicating seminal discoveries in miRNA biology with a special focus on the cardiovascular field. From van Rooij Circ Res 2011;108:219–34.

Pluripotent stem cells, such as embryonic stem cells (ESCs) can differentiate into all adult cell types.[102] They are derived from the inner mass of the blastocyst stage late in the first week of fertilization. They differentiate into multicellular embryoid bodies containing differentiating cells from all three germ layers, ectoderm, mesoderm, and endoderm. Experimental studies have shown that ESCs have the ability to differentiate into beating cardiac myocytes and couple themselves electromechanically to the host cardiac cells.[103] Importantly, those cardiac myocytes that are derived from ESCs closely resemble embryonic-derived cardiac myocytes and express the complete repertoire of cardiac-restricted transcription factors including GATA4, Nkx2.5, MEF2C, and Irx4. When transplanted into the infarcted myocardium of a rodent, ESCs-derived cardiomyocytes have been shown to engraft and to improve cardiac function.[104] There are, however, several risks associated with the use of ESCs that limit their use, including teratoma formation at the implantation site and worries of graft rejection that might necessitate immunosuppression. Additionally, the source of ESCs has raised substantial ethical apprehensions and led to passionate deliberations among scientists and the general public.[101] The recent discovery that it is possible to generate ESC-like cells, called inducible pluripotent stem (iPS) cells, by reprogramming adult somatic cells with genes regulating ESC pluripotency, may resolve the ethical and immunogenic issues associated with the use of ESCs.[105] iPS cells are thought to be therapeutically equivalent to ESCs in many respects and are genetically identical to the donor cells.[105] Strategically, the expansion of iPS in stem cell media can produce an abundant number of cells that can then be used for studies on cardiac differentiation. While the use of iPS cells clearly has its advantages, the full extent of their potential and possible toxicity is still being assessed.

Several groups of investigators have reported that the postnatal heart includes niches of cardiac stem cells (CSCs) and/or cardiac progenitors that possess the capability to replicate and differentiate into cardiac myocytes.[106] There are approximately four populations of CSC: side population (SP) cells, cells expressing the stem cell factor c-Kit, cells expressing the stem cell antigen 1 (Sca-1+), cardiosphere-derived cells, and cells expressing the protein Islet-1.[101] Cardiac stem cells appear to be located in specialized niches throughout the myocardium, which support the growth and maintenance of the stem cell pool. After an ischemic insult, it appears that CSCs increase in number and migrate to the areas that are injured. As such, it is now thought that CSCs are a source of new cells in normal organ homeostasis as well as under conditions of stress.[107] Given that these cells reside in the heart, it is thought that they may provide a mechanically and electrophysiologically compatible source of cells for transplantation. Strategies designed

with this in mind have demonstrated that CSCs isolated and cloned from rodent hearts are effective in the treatment of myocardial ischemia.[108] This suggests that the heart itself is a viable source of stem cells for myocardial repair. Despite the evidence that CSCs are effective in treating the heart, there are some limitations to their use. Specifically, there is evidence that the cardiac stem cell pool diminishes with aging, which could possibly contribute to the lack of efficacy of regeneration in elderly individuals.

The bone marrow contains different cell populations that have the potential to migrate and transdifferentiate into cells of diverse phenotypes including: hematopoietic stem cells (HSCs), endothelial progenitor cells (EPCs), and mesenchymal stem cells (MSCs).[101,102] HSCs represent the prototypic adult stem cell population and have been shown to differentiate into cardiomyocytes in culture, making them of particular interest in the treatment of cardiac disease because they represent a well-characterized and ample source of progenitor cells.[109] Experimental studies have shown that HSCs provide a significant improvement in cardiac function that in some instances is associated with regeneration of contracting cardiomyocytes.[110] EPCs can be isolated directly from the bone marrow or from the peripheral circulation and represent a subset of HSCs that are able to develop an endothelial phenotype. Experimental studies have shown that EPCs contribute to the formation of new vessels after ischemic injury by secreting pro-angiogenic growth factors and stimulating reendothelialization. In addition, the injection of EPCs into the infarcted myocardium not only improves left ventricular function, but it also inhibits fibrosis. MSCs are found in the bone marrow and in adipose tissue and are less immunogenic than other cells due to the lack of MHC-II and B-7 co-stimulatory molecule expression.[101] Experimental studies have shown that MSCs have the potential for site-specific differentiation into heart muscle cells, vascular-like structures, as well as gap junction protein, suggesting that MSCs act by regenerating functionally effective, integrated cardiomyocytes and possibly new blood vessels. The injection of allogeneic MSCs with the use of a catheter-based approach in pigs resulted in the regeneration of myocardium, a reduction in infarct size, and improvement in regional and global cardiac contractile function.[111] Given that MSC clones can be expanded in vitro from the bone marrow or adipose tissue and that they have a low immunogenicity, they represent an abundant and easily accessible source of stem cells that have the potential to regenerate the heart after ischemia–reperfusion injury.

In light of the positive finding regarding the ability of stem cells to differentiate into cardiomyocytes, endothelial cells, and smooth muscle cells in the injured heart and provide some evidence for protection in experimental studies, there have been multiple clinical trials initiated to test the effectiveness of this strategy in humans. The results of four trials using CSCs and BMCs are discussed here. In the Stem Cell Infusion in Patients with Ischemic Cardiomyopathy (SCIPIO) trial, patients undergoing coronary artery bypass grafting were consecutively enrolled in the treatment and control groups.[112] Those in the treatment group received about one million autologous CSCs by intracoronary infusion. At the one-year follow-up, patients in the treatment group showed an overall improvement in left ventricular systolic function. Interestingly, some patients even showed a reduction in infarct size. In the prospective, randomized CArdiosphere-Derived aUtologous stem CElls to reverse ventricUlar dySfunction (CADUCEUS) trial, patients with a left ventricular ejection fraction of 25–45% were infused with cardiosphere-derived autologous cells grown from endomyocardial biopsy specimens into the infarct-related artery 1.5–3 months after suffering from a myocardial infarction.[113] By six months, no patients had died, developed cardiac tumors, or suffered from a major adverse coronary event in either group. Compared with controls at six months, MRI analysis of patients treated with CDCs showed reductions in scar mass, increases in viable heart mass and regional contractility, and regional systolic wall thickening. However, changes in end-diastolic volume, end-systolic volume, and left ventricular ejection fraction did not differ between groups by six months. In the Transplantation of Progenitor Cells and Regeneration Enhancement in Acute Myocardial Infarction (TOPCARE-AMI) trial, patients with an acute myocardial infarction received an intracoronary infusion of ex vivo expanded bone marrow derived mononuclear cells or culture-enriched EPCs derived from peripheral blood.[114] With this strategy, patients who received treatment displayed sustained improvements in left ventricular function and also experienced a reduction in end-systolic volume and prevention of left ventricular remodeling that has persisted for up to five years.[102,114] In the BOne marrOw transfer to enhance ST-elevation infarct regeneration (BOOST) trial, patients received an intracoronary transfer of autologous BMCs after successful percutaneous coronary intervention.[115] At the six-month follow-up, patients who received the BMCs infusion displayed improvements in left ventricular ejection fraction. However, it was recently reported that at the five-year follow-up, these improvements were not sustained.

The results of these trials are very encouraging and provide the rationale for larger randomized trials to test the effectiveness of stem cells in the treatment of myocardial ischemia-reperfusion injury and chronic ischemic cardiomyopathy. Although the results of the trial were slightly inconsistent with the sustained effects on left ventricular ejection fraction, they demonstrate that these strategies are safe and do not cause any injurious effects. The next phase of work will need to focus on determining the best cell types to use and the mechanisms by which these cells interact with host cells and elicit their therapeutic effects.[102]

CLINICAL TRIAL FAILURE

Despite the best efforts of scientists and clinicians to identify therapeutic interventions to induce cardioprotection over the last 40 years, no therapy has made a successful transition to the clinic with the exception of reperfusion. This has led to a great deal of discussion on the reasons behind these failures and led to the convening of a Working Group by the National Heart, Lung, and Blood Institute of the National Institutes of Health with the sole purpose of recommending new approaches.[116] James Downey and Michael Cohen from the University of South Alabama recently published a review article, in which they examined why drugs have not been translated into clinical practice and provided in their opinion the five most compelling reasons, summarized below.[27]

It is Quite Possible that We have been Testing the Wrong Drugs all Along

The early literature is littered with reputed cardioprotective drugs that made their way to clinical testing before their efficacy in experimental studies was actually proven. As it turns out, many of these drugs did not consistently reduce infarct size in animals and not surprisingly turned out to be failures in the clinic. For example, the infusion of low dose adenosine was reported to reduce infarct size by acting as an anti-inflammatory agent without having any hemodynamic effect.[117] Subsequent studies failed to reproduce these findings.[118] Nevertheless, clinical investigators opted to ignore these negative effects and organized the AMISTAD I trial, which failed to report a positive effect.[119] As mentioned earlier in this chapter, the cellular mechanisms of ischemia-reperfusion injury were not completely known when these early studies were conducted and it was not fully appreciated that reperfusion was necessary to reduce infarct size. Certainly, these two factors can be attributed to some of the early negative findings of the clinical trials. But it seems that the lack of reproducibility of animal studies is the biggest reason clinical trials have failed as the drugs that were chosen for clinical trials either did not work at all or were not potent enough for their beneficial effect to be resolved in the clinical trial setting.[27]

It is Possible that Reperfusion of the Blocked Artery is so Effective at Limiting Infarct Size that there is Relatively Little Room for Improvement

A recent review of studies in which anatomical infarct size and ischemic zones were measured using imaging methods in patients with acute myocardial infarction undergoing reperfusion indicated that on average 50% of the ischemic tissue survived in the patients following reperfusion.[120] In contrast, about 23% of the ischemic tissue survives in the absence of reperfusion.[121] Based on these findings and the notion that a critical threshold of infarction of 20% of the left ventricle is necessary for the appearance of adverse symptoms, it appears that reperfusion therapy keeps infarct size at or below the 20% threshold in 75% of patients. In this scenario the administration of a cardioprotective agent would possibly decrease infarct size, but would probably not provide a benefit to symptoms (i.e. left ventricular ejection fraction) given the already small pre-intervention infarct size. This leaves only about 25% of patients who would benefit from a cardioprotective drug during reperfusion. This reasoning could explain why most clinical trials have been failures, since any improvements found in the 25% of patients who would benefit from the treatment could be diluted by the inability of the treatment to improve symptoms in the remaining 75% of patients who are non-responders.[27]

Clinical Trials May have been Designed Inadequately

This goes back to the previous point suggesting that only 25% of patients who experience myocardial ischemia-reperfusion may benefit from therapeutic intervention at the time of reperfusion. Clinical trials may benefit from selecting the most at risk patients and targeting them for treatment (i.e. patients with large areas at risk). For instance, in a recent small-scale clinical trial, cyclosporine was given to patients just prior to reperfusion.[122] When the authors plotted their data as a relationship between area-at-risk and serum creatine kinase and a relationship between area-at-risk and serum troponin-I, they noted that treated and control patients with small risk zones were intermingled, but in patients with large risk zones the benefit of treatment was clearly seen (Figure 5.13). This suggests that many past clinical trials may have degraded the sensitivity of their findings by not incorporating a predictor of infarct size. Additionally, dose and scheduling of a cardioprotective agent has not always been rigorously studied in pre-clinical trials. Therefore, the clinical trial that is based on the animal studies could be using the wrong dose and delivery method. For example, AMP579 is a mixed adenosine agonist that produced

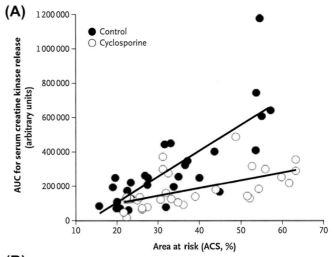

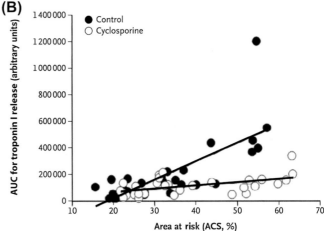

FIGURE 5.13   Infarct size as a function of area-at-risk. Linear relationship between area-at-risk measured by left ventriculography and infarct size assessed by enzyme release in patients with acute myocardial infarction. The filled closed circles represent the control patients and the open circles represent patients treated with cyclosporine at the time of reperfusion. From Piot et al. *N Engl J Med* 2008;359:473–81.

cardioprotective effects when given to animals just prior to reperfusion. In patients, delivery of AMP579 was started when the blocked artery was opened to achieve a therapeutic dose at about 30 minutes of reperfusion.[123] This study turned out to be negative and it was subsequently shown in animal studies years later that AMP579 must be present during the first few minutes of reperfusion to have a benefit.[124]

Experimental Studies Aimed at Testing the Effectiveness of Cardioprotective Agents have Used Various Animal Models, Including Mice, Rats, Rabbits, Dogs, Pigs, and Non-Human Primates

These studies have uniformly used protocol of ischemia ranging from 30 to 60 minutes. However, no patient's heart is reperfused after just 30 minutes of ischemia. The current optimal door to balloon time is 90 minutes, but in practice often exceeds 120 minutes. Additionally, patients often do not come to the hospital after the onset of symptoms for a heart attack, suggesting that ischemic times in the patient population are more in the order of three hours.[27] Additionally, virtually all studies on cardioprotection have been conducted in young healthy animals. Yet, most patients who experience myocardial ischemia are not young and have one or more risk factors such as diabetes.[27] Therefore, most of the cardioprotective agents that have been tested in the clinic have not undergone an extensive evaluation in the pre-clinical setting. Agents have not been tested in multiple species at clinically relevant periods of ischemia and in animals that have co-morbidities. Coupled with the lack of consistency among the multiple studies for select agents that have been conducted it is no wonder that most clinical trials have failed.

These points offer some possible explanations as to why cardioprotective drugs have not found their way into the clinic. While there are certainly other reasons and explanations, this list provides a helpful starting point for discussion. Importantly, it suggests that the design of future clinical trials should take into consideration the rigorousness of pre-clinical studies as it pertains to dosage and time of delivery, as well as if the agent in question was examined in more than one species and in the setting of co-morbidities. Additionally, it also suggests that clinical trials should at least include an estimate of the ischemic zone size so that the impact of the cardioprotective agent in high-risk patients can be appreciated.[27]

SUMMARY AND CONCLUDING REMARKS

The field of cardioprotection in the context of ischemic heart disease has come a long way since its inception almost 40 years ago. Extensive work by scientists and clinicians has established that reperfusion of a blocked artery is a necessity for reducing the size of myocardial infarction. We have also learned a great deal about the cellular mechanism that occurs during ischemia and reperfusion, which has led to the identification of signaling cascades (i.e. RISK and SAFE Pathways) and cellular targets (i.e. Akt, Erk, STAT-3, AMPK, eNOS, etc.) that have cardioprotective actions. However, despite this great progress, there is still a great deal of work to be done given that an effective therapeutic strategy to accompany reperfusion has yet to make its way to the clinic. It is time to re-think and re-evaluate the strategy of moving potential cardioprotective agents from the lab to the clinic. First, we must determine which animal(s) are the most appropriate to use to test potential agents. It is possible that agents could first be tested in mice before

moving to larger animals such as pigs. Alternatively, agents may need to be tested in mice, then pigs, then in non-human primates before being tested in humans. We must decide upon a standard period of ischemia to use in our experimental studies. We must also make sure that potential therapeutic agents are tested in animal models that have co-morbidities, such as diabetes, obesity, and high cholesterol. Hopefully the adaptation of such methods will not only combat the lack of reproducibility that continues to plague the field of cardioprotection, but will lead to the identification of an intervention that will fulfill the promise of reperfusion therapy in the setting of myocardial ischemia-reperfusion injury.

References

1. Go AS, Mozaffarian D, Roger VL, et al. Heart disease and stroke statistics–2013 update: a report from the American Heart Association. *Circulation* 2013;**127**:e6–245.
2. Ortega-Gil J, Perez-Cardona JM. Unstable angina and non ST elevation acute coronary syndromes. *P R Health Sci J* 2008;**27**:395–401.
3. Sutton MG, Sharpe N. Left ventricular remodeling after myocardial infarction: pathophysiology and therapy. *Circulation* 2000;**101**:2981–8.
4. Ferdinandy P, Schulz R, Baxter GF. Interaction of cardiovascular risk factors with myocardial ischemia/reperfusion injury, preconditioning, and postconditioning. *Pharmacol Rev* 2007;**59**:418–58.
5. Ignarro LJ, Balestrieri ML, Napoli C. Nutrition, physical activity, and cardiovascular disease: an update. *Cardiovasc Res* 2007;**73**:326–40.
6. Rennert G, Saltz-Rennert H, Wanderman K, Weitzman S. Size of acute myocardial infarcts in patients with diabetes mellitus. *Am J Cardiol* 1985;**55**:1629–30.
7. Hausenloy DJ, Yellon DM. Myocardial ischemia-reperfusion injury: a neglected therapeutic target. *J Clin Invest* 2013;**123**:92–100.
8. Jennings RB, Sommers HM, Smyth GA, Flack HA, Linn H. Myocardial necrosis induced by temporary occlusion of a coronary artery in the dog. *Arch Pathol* 1960;**70**:68–78.
9. Yellon DM, Hausenloy DJ. Myocardial reperfusion injury. *N Engl J Med* 2007;**357**:1121–35.
10. Bolli R. Mechanism of myocardial "stunning". *Circulation* 1990;**82**:723–38.
11. Heyndrickx GR, Millard RW, McRitchie RJ, Maroko PR, Vatner SF. Regional myocardial functional and electrophysiological alterations after brief coronary artery occlusion in conscious dogs. *J Clin Invest* 1975;**56**:978–85.
12. Kloner RA, Arimie RB, Kay GL, et al. Evidence for stunned myocardium in humans: a 2001 update. *Coron Artery Dis* 2001;**12**:349–56.
13. Jurkovicova O, Cagan S. [Reperfusion arrhythmias]. *Bratisl Lek Listy* 1998;**99**:162–71.
14. Moens AL, Claeys MJ, Timmermans JP, Vrints CJ. Myocardial ischemia/reperfusion-injury, a clinical view on a complex pathophysiological process. *Int J Cardiol* 2005;**100**:179–90.
15. Tsao PS, Aoki N, Lefer DJ, Johnson 3rd G, Lefer AM. Time course of endothelial dysfunction and myocardial injury during myocardial ischemia and reperfusion in the cat. *Circulation* 1990;**82**:1402–12.
16. Calvert JW, Lefer DJ. Statin therapy and myocardial no-reflow. *Br J Pharmacol* 2006;**149**:229–31.
17. Heusch G. Cardioprotection: chances and challenges of its translation to the clinic. *Lancet* 2013;**381**:166–75.

18. Baines CP, Kaiser RA, Purcell NH, et al. Loss of cyclophilin D reveals a critical role for mitochondrial permeability transition in cell death. *Nature* 2005;**434**:658–62.

19. Whelan RS, Kaplinskiy V, Kitsis RN. Cell death in the pathogenesis of heart disease: mechanisms and significance. *Ann Rev Physiol* 2010;**72**:19–44.

20. Jeremias I, Kupatt C, Martin-Villalba A, et al. Involvement of CD95/Apo1/Fas in cell death after myocardial ischemia. *Circulation* 2000;**102**:915–20.

21. Kurrelmeyer KM, Michael LH, Baumgarten G, et al. Endogenous tumor necrosis factor protects the adult cardiac myocyte against ischemic-induced apoptosis in a murine model of acute myocardial infarction. *Proc Natl Acad Sci U S A* 2000;**97**:5456–61.

22. Kung G, Konstantinidis K, Kitsis RN. Programmed necrosis, not apoptosis, in the heart. *Circ Res* 2011;**108**:1017–36.

23. Lim SY, Davidson SM, Mocanu MM, Yellon DM, Smith CC. The cardioprotective effect of necrostatin requires the cyclophilin-D component of the mitochondrial permeability transition pore. *Cardiovasc Drugs Ther* 2007;**21**:467–9.

24. Halestrap AP. What is the mitochondrial permeability transition pore? *J Mol Cell Cardiol* 2009;**46**:821–31.

25. Smith CC, Davidson SM, Lim SY, Simpkin JC, Hothersall JS, Yellon DM. Necrostatin: a potentially novel cardioprotective agent? *Cardiovasc Drugs Ther* 2007;**21**:227–33.

26. Cohen MV, Downey JM. Is it time to translate ischemic preconditioning's mechanism of cardioprotection into clinical practice? *J Cardiovasc Pharmacol Ther* 2011;**16**:273–80.

27. Downey JM, Cohen MV. Why do we still not have cardioprotective drugs? *Circ J* 2009;**73**:1171–7.

28. Hanley PJ, Daut J. K(ATP) channels and preconditioning: a reexamination of the role of mitochondrial K(ATP) channels and an overview of alternative mechanisms. *J Mol Cell Cardiol* 2005;**39**:17–50.

29. Murry CE, Jennings RB, Reimer KA. Preconditioning with ischemia: a delay of lethal cell injury in ischemic myocardium. *Circulation* 1986;**74**:1124–36.

30. Carini R, Grazia De Cesaris M, Splendore R, et al. Signal pathway responsible for hepatocyte preconditioning by nitric oxide. *Free Radic Biol Med* 2003;**34**:1047–55.

31. Doi Y, Hamazaki K, Yabuki M, Tanaka N, Utsumi K. Effect of HSP70 induced by warm ischemia to the liver on liver function after partial hepatectomy. *Hepatogastroenterology* 2001;**48**:533–40.

32. Kulhanek-Heinze S, Gerbes AL, Gerwig T, Vollmar AM, Kiemer AK. Protein kinase A dependent signalling mediates anti-apoptotic effects of the atrial natriuretic peptide in ischemic livers. *J Hepatol* 2004;**41**:414–20.

33. Ricciardi R, Schaffer BK, Kim RD, et al. Protective effects of ischemic preconditioning on the cold-preserved liver are tyrosine kinase dependent. *Transplantation* 2001;**72**:406–12.

34. Nilsson B, Friman S, Wallin M, Gustafsson B, Delbro D. The liver protective effect of ischemic preconditioning may be mediated by adenosine. *Transpl Int* 2000;**13**(Suppl. 1):S558–61.

35. Peralta C, Serafin A, Fernandez-Zabalegui L, Wu ZY, Rosello-Catafau J. Liver ischemic preconditioning: a new strategy for the prevention of ischemia-reperfusion injury. *Transplant Proc* 2003;**35**:1800–2.

36. Duchen MR. Roles of mitochondria in health and disease. *Diabetes* 2004;**53**(Suppl. 1):S96–102.

37. Schroeder MA, Cochlin LE, Heather LC, Clarke K, Radda GK, Tyler DJ. In vivo assessment of pyruvate dehydrogenase flux in the heart using hyperpolarized carbon-13 magnetic resonance. *Proc Natl Acad Sci U S A* 2008;**105**:12051–6.

38. Kirkinezos IG, Moraes CT. Reactive oxygen species and mitochondrial diseases. *Semin Cell Dev Biol* 2001;**12**:449–57.

39. Madsen-Bouterse SA, Zhong Q, Mohammad G, Ho YS, Kowluru RA. Oxidative damage of mitochondrial DNA in diabetes and its protection by manganese superoxide dismutase. *Free Radic Res* 2010;**44**:313–21.

40. Aon MA, Cortassa S, Akar FG, O'Rourke B. Mitochondrial criticality: a new concept at the turning point of life or death. *Biochim Biophys Acta* 2006;**1762**:232–40.

41. Hausenloy DJ, Yellon DM. Reperfusion injury salvage kinase signalling: taking a RISK for cardioprotection. *Heart Fail Rev* 2007;**12**:217–34.

42. Yellon DM, Baxter GF. Reperfusion injury revisited: is there a role for growth factor signaling in limiting lethal reperfusion injury? *Trends Cardiovasc Med* 1999;**9**:245–9.

43. Boengler K, Schulz R, Heusch G. Loss of cardioprotection with ageing. *Cardiovasc Res* 2009;**83**:247–61.

44. Calvert JW, Coetzee WA, Lefer DJ. Novel insights into hydrogen sulfide–mediated cytoprotection. *Antioxid Redox Signal* 2010;**12**:1203–17.

45. Churchill EN, Mochly-Rosen D. The roles of PKCdelta and epsilon isoenzymes in the regulation of myocardial ischaemia/reperfusion injury. *Biochem Soc Trans* 2007;**35**:1040–2.

46. Juhaszova M, Zorov DB, Kim SH, et al. Glycogen synthase kinase-3beta mediates convergence of protection signaling to inhibit the mitochondrial permeability transition pore. *J Clin Invest* 2004;**113**:1535–49.

47. Mann DL. Stress-activated cytokines and the heart: from adaptation to maladaptation. *Annu Rev Physiol* 2003;**65**:81–101.

48. Skyschally A, Gres P, Hoffmann S, et al. Bidirectional role of tumor necrosis factor-alpha in coronary microembolization: progressive contractile dysfunction versus delayed protection against infarction. *Circ Res* 2007;**100**:140–6.

49. Lecour S. Activation of the protective Survivor Activating Factor Enhancement (SAFE) pathway against reperfusion injury: Does it go beyond the RISK pathway? *J Mol Cell Cardiol* 2009;**47**:32–40.

50. Imada K, Leonard WJ. The Jak-STAT pathway. *Mol Immunol* 2000;**37**:1–11.

51. Barry SP, Townsend PA, Latchman DS, Stephanou A. Role of the JAK-STAT pathway in myocardial injury. *Trends Mol Med* 2007;**13**:82–9.

52. Lecour S, Suleman N, Deuchar GA, et al. Pharmacological preconditioning with tumor necrosis factor-alpha activates signal transducer and activator of transcription-3 at reperfusion without involving classic prosurvival kinases (Akt and extracellular signal-regulated kinase). *Circulation* 2005;**112**:3911–8.

53. Hardie DG. Minireview: the AMP-activated protein kinase cascade: the key sensor of cellular energy status. *Endocrinology* 2003;**144**:5179–83.

54. Hawley SA, Boudeau J, Reid JL, et al. Complexes between the LKB1 tumor suppressor, STRAD alpha/beta and MO25 alpha/beta are upstream kinases in the AMP-activated protein kinase cascade. *J Biol* 2003;**2**:28.

55. Zhou G, Myers R, Li Y, et al. Role of AMP-activated protein kinase in mechanism of metformin action. *J Clin Invest* 2001;**108**:1167–74.

56. Buhl ES, Jessen N, Pold R, et al. Long-term AICAR administration reduces metabolic disturbances and lowers blood pressure in rats displaying features of the insulin resistance syndrome. *Diabetes* 2002;**51**:2199–206.

57. Zaha VG, Young LH. AMP-activated protein kinase regulation and biological actions in the heart. *Circ Res* 2012;**111**:800–14.

58. Russell 3rd RR, Li J, Coven DL, et al. AMP-activated protein kinase mediates ischemic glucose uptake and prevents postischemic cardiac dysfunction, apoptosis, and injury. *J Clin Invest* 2004;**114**:495–503.

59. Kudo N, Gillespie JG, Kung L, et al. Characterization of 5'AMP-activated protein kinase activity in the heart and its role in inhibiting acetyl-CoA carboxylase during reperfusion following ischemia. *Biochimica et biophysica acta* 1996;**1301**:67–75.

60. Nguyen TT, Stevens MV, Kohr M, Steenbergen C, Sack MN, Murphy E. Cysteine 203 of cyclophilin D is critical for cyclophilin D activation of the mitochondrial permeability transition pore. *J Biol Chem* 2011;**286**:40184–92.

61. Calvert JW, Gundewar S, Jha S, et al. Acute metformin therapy confers cardioprotection against myocardial infarction via AMPK-eNOS-mediated signaling. *Diabetes* 2008;**57**:696–705.

62. Graef IA, Chen F, Chen L, Kuo A, Crabtree GR. Signals transduced by Ca(2+)/calcineurin and NFATc3/c4 pattern the developing vasculature. *Cell* 2001;**105**:863–75.

63. Melling CW, Thorp DB, Milne KJ, Krause MP, Noble EG. Exercise-mediated regulation of Hsp70 expression following aerobic exercise training. *Am J Physiol Heart Circ Physiol* 2007;**293**:H3692–8.

64. Kudo N, Barr AJ, Barr RL, Desai S, Lopaschuk GD. High rates of fatty acid oxidation during reperfusion of ischemic hearts are associated with a decrease in malonyl-CoA levels due to an increase in 5'-AMP-activated protein kinase inhibition of acetyl-CoA carboxylase. *J Biol Chem* 1995;**270**:17513–20.

65. Dyck JR, Lopaschuk GD. AMPK alterations in cardiac physiology and pathology: enemy or ally? *J Physiol* 2006;**574**:95–112.

66. Mustafa AK, Gadalla MM, Snyder SH. Signaling by gasotransmitters. *Sci Signal* 2009:2 re2.

67. Nicholson CK, Calvert JW. Hydrogen sulfide and ischemia-reperfusion injury. *Pharmacol Res* 2010;**62**:289–97.

68. Loscalzo J, Welch G. Nitric oxide and its role in the cardiovascular system. *Prog Cardiovasc Dis* 1995;**38**:87–104.

69. Sharp BR, Jones SP, Rimmer DM, Lefer DJ. Differential response to myocardial reperfusion injury in eNOS-deficient mice. *Am J Physiol Heart Circ Physiol* 2002;**282**:H2422–6.

70. Jones SP, Lefer DJ. Myocardial Reperfusion Injury: Insights Gained from Gene-Targeted Mice. *News Physiol Sci* 2000;**15**:303–8.

71. Sumeray MS, Rees DD, Yellon DM. Infarct size and nitric oxide synthase in murine myocardium. *J Mol Cell Cardiol* 2000;**32**:35–42.

72. Jones SP, Greer JJ, Kakkar AK, et al. Endothelial nitric oxide synthase overexpression attenuates myocardial reperfusion injury. *Am J Physiol Heart Circ Physiol* 2004;**286**:H276–82.

73. Xi L. Nitric oxide-dependent mechanism of anti-ischemic myocardial protection induced by monophosphoryl lipid A. *Zhongguo Yao Li Xue Bao* 1999;**20**:865–71.

74. Burger DE, Xiang FL, Hammoud L, Jones DL, Feng Q. Erythropoietin protects the heart from ventricular arrhythmia during ischemia and reperfusion via neuronal nitric-oxide synthase. *J Pharmacol Exp Ther* 2009;**329**:900–7.

75. Dawson D, Lygate CA, Zhang MH, Hulbert K, Neubauer S, Casadei B. nNOS gene deletion exacerbates pathological left ventricular remodeling and functional deterioration after myocardial infarction. *Circulation* 2005;**112**:3729–37.

76. Li Q, Guo Y, Tan W, et al. Cardioprotection afforded by inducible nitric oxide synthase gene therapy is mediated by cyclooxygenase-2 via a nuclear factor-kappaB dependent pathway. *Circulation* 2007;**116**:1577–84.

77. Calvert JW, Condit ME, Aragon JP, et al. Exercise protects against myocardial ischemia-reperfusion injury via stimulation of beta(3)-adrenergic receptors and increased nitric oxide signaling: role of nitrite and nitrosothiols. *Circ Res* 2011;**108**:1448–58.

78. Lefer DJ. Myocardial protective actions of nitric oxide donors after myocardial ischemia and reperfusion. *New Horiz* 1995;**3**:105–12.

79. Thomas DD, Liu X, Kantrow SP, Lancaster Jr JR. The biological lifetime of nitric oxide: implications for the perivascular dynamics of NO and O2. *Proc Natl Acad Sci U S A* 2001;**98**:355–60.

80. Bolli R, Dawn B, Xuan YT. Role of the JAK-STAT pathway in protection against myocardial ischemia/reperfusion injury. *Trends Cardiovasc Med* 2003;**13**:72–9.

81. Kim YM, Talanian RV, Billiar TR. Nitric oxide inhibits apoptosis by preventing increases in caspase-3-like activity via two distinct mechanisms. *J Biol Chem* 1997;**272**:31138–48.

82. Thorup C, Jones CL, Gross SS, Moore LC, Goligorsky MS. Carbon monoxide induces vasodilation and nitric oxide release but suppresses endothelial NOS. *Am J Physiol* 1999;**277**:F882–9.

83. Furchgott RF, Jothianandan D. Endothelium-dependent and -independent vasodilation involving cyclic GMP: relaxation induced by nitric oxide, carbon monoxide and light. *Blood Vessels* 1991;**28**:52–61.

84. Hobbs AJ. Soluble guanylate cyclase: the forgotten sibling. *Trends Pharmacol Sci* 1997;**18**:484–91.

85. Kikuchi G, Yoshida T, Noguchi M. Heme oxygenase and heme degradation. *Biochem Biophys Res Commun* 2005;**338**:558–67.

86. Grilli A, De Lutiis MA, Patruno A, et al. Inducible nitric oxide synthase and heme oxygenase-1 in rat heart: direct effect of chronic exposure to hypoxia. *Ann Clin Lab Sci* 2003;**33**:208–15.

87. Liu X, Chapman G, Peyton K, Schafer A, Durante W. Carbon monoxide inhibits apoptosis in vascular smooth muscle cells. *Cardiovasc Res* 2002;**55**:396–405.

88. Mei DS, Du YA, Wang Y. [Cardioprotection and mechanisms of exogenous carbon monoxide releaser CORM-2 against ischemia/reperfusion injury in isolated rat hearts]. *Zhejiang Da Xue Xue Bao Yi Xue Ban* 2007;**36**:291–7.

89. Clark JE, Kottam A, Motterlini R, Marber MS. Measuring left ventricular function in the normal, infarcted and CORM-3-preconditioned mouse heart using complex admittance-derived pressure volume loops. *J Pharmacol Toxicol Methods* 2009;**59**:94–9.

90. Calvert JW, Jha S, Gundewar S, et al. Hydrogen sulfide mediates cardioprotection through Nrf2 signaling. *Circ Res* 2009;**105**:365–74.

91. Elrod JW, Calvert JW, Morrison J, et al. Hydrogen sulfide attenuates myocardial ischemia-reperfusion injury by preservation of mitochondrial function. *Proc Natl Acad Sci U S A* 2007;**104**:15560–5.

92. Yang G, Wu L, Jiang B, et al. H2S as a physiologic vasorelaxant: hypertension in mice with deletion of cystathionine gamma-lyase. *Science* 2008;**322**:587–90.

93. Calvert JW, Jha S, Gundewar S, et al. Hydrogen sulfide mediates cardioprotection through Nrf2 signaling. *Circ Res* 2009;**105**:365–74.

94. van Rooij E, Olson EN. MicroRNA therapeutics for cardiovascular disease: opportunities and obstacles. *Nat Rev Drug Discov* 2012;**11**:860–72.

95. van Rooij E. The art of microRNA research. *Circ Res* 2011;**108**:219–34.

96. Wang JX, Jiao JQ, Li Q, et al. miR-499 regulates mitochondrial dynamics by targeting calcineurin and dynamin-related protein-1. *Nature Med* 2011;**17**:71–8.

97. Aurora AB, Mahmoud AI, Luo X, et al. MicroRNA-214 protects the mouse heart from ischemic injury by controlling Ca(2)(+) overload and cell death. *J Clin Invest* 2012;**122**:1222–32.

98. Bonauer A, Carmona G, Iwasaki M, et al. MicroRNA-92a controls angiogenesis and functional recovery of ischemic tissues in mice. *Science* 2009;**324**:1710–3.

99. Hullinger TG, Montgomery RL, Seto AG, et al. Inhibition of miR-15 protects against cardiac ischemic injury. *Circ Res* 2012;**110**:71–81.

100. Foo RS, Mani K, Kitsis RN. Death begets failure in the heart. *J Clin Invest* 2005;**115**:565–71.

101. Shah VK, Shalia KK. Stem Cell Therapy in Acute Myocardial Infarction: A Pot of Gold or Pandora's Box. *Stem Cells Int* 2011;**2011**:536758.

102. Karantalis V, Balkan W, Schulman IH, Hatzistergos KE, Hare JM. Cell-based therapy for prevention and reversal of myocardial remodeling. *Am J Physiol Heart Circ Physiol* 2012;**303**:H256–70.

103. Kehat I, Kenyagin-Karsenti D, Snir M, et al. Human embryonic stem cells can differentiate into myocytes with structural and functional properties of cardiomyocytes. *J Clin Invest* 2001;**108**:407–14.

104. Min JY, Yang Y, Converso KL, et al. Transplantation of embryonic stem cells improves cardiac function in postinfarcted rats. *J Appli Physiol* 2002;**92**:288–96.

105. Wernig M, Meissner A, Foreman R, et al. In vitro reprogramming of fibroblasts into a pluripotent ES-cell-like state. *Nature* 2007;**448**:318–24.

106. Limana F, Zacheo A, Mocini D, et al. Identification of myocardial and vascular precursor cells in human and mouse epicardium. *Circ Res* 2007;**101**:1255–65.

107. Torella D, Ellison GM, Mendez-Ferrer S, Ibanez B, Nadal-Ginard B. Resident human cardiac stem cells: role in cardiac cellular homeostasis and potential for myocardial regeneration. *Nat Clin Pract Cardiovasc Med* 2006;**1**(Suppl 3):S8–13.

108. Parra VM, Macho P, Domenech RJ. Late cardiac preconditioning by exercise in dogs is mediated by mitochondrial potassium channels. *J Cardiovasc Pharmacol* 2010;**56**:268–74.

109. Orlic D, Kajstura J, Chimenti S, et al. Bone marrow cells regenerate infarcted myocardium. *Nature* 2001;**410**:701–5.

110. Murry CE, Soonpaa MH, Reinecke H, et al. Haematopoietic stem cells do not transdifferentiate into cardiac myocytes in myocardial infarcts. *Nature* 2004;**428**:664–8.

111. Amado LC, Saliaris AP, Schuleri KH, et al. Cardiac repair with intramyocardial injection of allogeneic mesenchymal stem cells after myocardial infarction. *Proc Natl Acad Sci U S A* 2005;**102**:11474–9.

112. Bolli R, Chugh AR, D'Amario D, et al. Cardiac stem cells in patients with ischaemic cardiomyopathy (SCIPIO): initial results of a randomised phase 1 trial. *Lancet* 2011;**378**:1847–57.

113. Makkar RR, Smith RR, Cheng K, et al. Intracoronary cardiosphere-derived cells for heart regeneration after myocardial infarction (CADUCEUS): a prospective, randomised phase 1 trial. *Lancet* 2012;**379**:895–904.

114. Assmus B, Schachinger V, Teupe C, et al. Transplantation of Progenitor Cells and Regeneration Enhancement in Acute Myocardial Infarction (TOPCARE-AMI). *Circulation* 2002;**106**:3009–17.

115. Wollert KC, Meyer GP, Lotz J, et al. Intracoronary autologous bone-marrow cell transfer after myocardial infarction: the BOOST randomised controlled clinical trial. *Lancet* 2004;**364**:141–8.

116. Bolli R, Becker L, Gross G, Mentzer Jr R, Balshaw D, Lathrop DA. Myocardial protection at a crossroads: the need for translation into clinical therapy. *Circ Res* 2004;**95**:125–34.

117. Olafsson B, Forman MB, Puett DW, et al. Reduction of reperfusion injury in the canine preparation by intracoronary adenosine: importance of the endothelium and the no-reflow phenomenon. *Circulation* 1987;**76**:1135–45.

118. Vander Heide RS, Reimer KA. Effect of adenosine therapy at reperfusion on myocardial infarct size in dogs. *Cardiovas Res* 1996;**31**:711–8.

119. Mahaffey KW, Puma JA, Barbagelata NA, et al. Adenosine as an adjunct to thrombolytic therapy for acute myocardial infarction: results of a multicenter, randomized, placebo-controlled trial: the Acute Myocardial Infarction STudy of ADenosine (AMISTAD) trial. *J Am Coll Cardiol* 1999;**34**:1711–20.

120. Miura T, Miki T. Limitation of myocardial infarct size in the clinical setting: current status and challenges in translating animal experiments into clinical therapy. *Basic Res Cardiol* 2008;**103**:501–13.

121. Parodi G, Ndrepepa G, Kastrati A, et al. Ability of mechanical reperfusion to salvage myocardium in patients with acute myocardial infarction presenting beyond 12 hours after onset of symptoms. *Am Heart J* 2006;**152**:1133–9.

122. Piot C, Croisille P, Staat P, et al. Effect of cyclosporine on reperfusion injury in acute myocardial infarction. *N Engl J Med* 2008;**359**:473–81.

123. Kopecky SL, Aviles RJ, Bell MR, et al. A randomized, double-blinded, placebo-controlled, dose-ranging study measuring the effect of an adenosine agonist on infarct size reduction in patients undergoing primary percutaneous transluminal coronary angioplasty: the ADMIRE (AmP579 Delivery for Myocardial Infarction REduction) study. *Am Heart J* 2003;**146**:146–52.

124. Xu Z, Downey JM, Cohen MV. Timing and duration of administration are crucial for antiinfarct effect of AMP 579 infused at reperfusion in rabbit heart. *Heart Dis* 2003;**5**:368–71.

125. Penna C, Settanni F, Tullio F, et al. GH-releasing hormone induces cardioprotection in isolated male rat heart via activation of RISK and SAFE pathways. *Endocrinology* 2013;**154**:1624–35.

6

Pathophysiology of Cardiomyopathies

Harsimran Saini, MD[1], Sara Tabtabai, MD[1], James R. Stone, MD, PhD[1, 2], Patrick T. Ellinor, MD, PhD[1, 2]

[1]Massachusetts General Hospital, Boston, Massachusetts, USA; [2]Harvard Medical School, Boston, Massachusetts, USA

INTRODUCTION

Cardiomyocytes, the contractile cells of the heart, exist in a complex environment comprised of endothelial cells, vascular smooth muscle, fibroblasts, and immune cells.[1] The components that form the extracellular matrix provide the structural framework and include protein-rich gap junctions that coordinate the contraction of individual myocytes to the extracellular matrix. Intracellularly, the myocytes are comprised of repeating units of sarcomeres that form the contractile units.[1] Given this complex milieu, even subtle alterations in the function of any number of these proteins can lead to cardiac disease. Over time, microscopic changes can have a more pronounced effect and ultimately lead to the disruption of normal cardiac function. Although the heart is an adaptive organ that is able to withstand minor perturbations in homeostasis, sustained pathological insults can result in severe functional abnormalities.

The cardiomyopathies are a group of disorders that can be acquired or have a congenital basis.[2] The different forms of cardiomyopathy exhibit overlapping manifestations that may be a reflection of the limited number of adaptive changes the heart can withstand, including cardiac hypertrophy, necrosis and deposition with fibrosis, and metabolic derangements. These changes are mediated by molecular processes that serve to maintain cardiac integrity. However, the persistent activation of these molecular pathways chronically can lead to cardiac dysfunction and/or increased susceptibility to sudden cardiac death.

The definition of cardiomyopathy has undergone significant revision as our knowledge of the etiologies has expanded. Cardiomyopathy was originally defined as an idiopathic disorder, resulting from cardiovascular diseases ranging from hypertension and ischemic heart disease to valvular disease. Conceptually, the cardiomyopathies have evolved to refer to a range of diseases defined by a set of morphological and functional characteristics as well as clinical presentation.[3] The clinical manifestations of cardiomyopathy vary and can include reduced cardiac output (from either systolic or diastolic dysfunction) severe enough to result in heart failure. The five major forms of cardiomyopathy currently recognized include the following:

- Dilated cardiomyopathy (DCM)
- Hypertrophic cardiomyopathy (HCM)
- Restrictive cardiomyopathy (RCM)
- Arrhythmogenic right ventricular cardiomyopathy/dysplasia (ARVC/D)
- Left ventricular non-compaction (LVNC).

Several of these recognized disease categories present clinically in a similar manner. Over 900 mutations that result in the different types of cardiomyopathies have been reported to date,[4] many with overlapping manifestations. Below we review the evolving definitions of the types of cardiomyopathies recognized today, then discuss the molecular mechanisms involved in the pathogenesis of each type.

DILATED CARDIOMYOPATHY

Dilated cardiomyopathy (DCM), also known as congestive systolic cardiomyopathy, is recognized by impaired systolic function and global dilatation of either one or both ventricular chambers (Fig. 6.1). DCM is the most prevalent form of cardiomyopathy with an incidence of one in approximately 2500 individuals. It can also be the result of a multitude of cardiovascular diseases.[5] DCM can present with overt signs of congestive heart failure (HF), atrial and/or ventricular arrhythmias, or sudden cardiac death, which can occur at any point in the disease progression. The diagnosis of DCM is based on an assessment of cardiovascular function and structure using both invasive and non-invasive studies. Clinically, the main criteria for diagnosis are reduced ejection

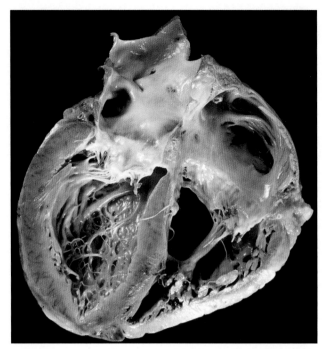

FIGURE 6.1   Gross appearance of a dilated heart. DCM is characterized by impaired left ventricular function. Examination of the heart with DCM is representative of all disease states that cause DCM. The characteristic decrease in the ventricular wall thickness masks the overall increase in left ventricular mass, due to myocyte hypertrophy and extracellular fibrosis.

fraction <45% and ventricular dilatation with left ventricular diastolic dimensions >2.7 cm/m^2 (body surface area). The histopathological changes that may be observed in DCM include myocyte hypertrophy with loss of myofibrils, fibrosis, and alterations in the nuclear and mitochondrial architecture, features that are common and non-specific.

Multiple mechanisms have been hypothesized to underlie the onset and development of DCM. For example, it has been hypothesized that DCM results from myocyte damage from chronic viral myocarditis. This cell damage in turn may induce an aberrant activation of the immune system against myocardial 'self proteins' and subsequent autoimmune damage. There are multiple conditions and causes associated with secondary forms of DCM (Box 6.1); however, more than 50% of cases are not associated with other disorders and are classified as idiopathic. This has led to a second hypothesis that DCM has a genetic cause that is directly involved in its pathogenesis. Increasing evidence for this hypothesis comes from molecular studies of patients and families with primary or idiopathic DCM, which has demonstrated an expanding number of mutations purported to cause DCM. In this chapter, we focus on the molecular and biochemical mechanisms of representative disease processes identified in DCM.

ISCHEMIC CARDIOMYOPATHY

Coronary artery disease is the leading cause of death with overall mortality of approximately 245 per 100 000 individuals in 2008.[6] Although mortality from acute myocardial infarction declined by 30% from 1998 to 2008, in part due to early reperfusion strategies and newer medications, the prevalence of heart failure continues to rise with an estimated 5.7 million reported cases per year.[6,7] Much of our current understanding about the pathophysiological processes that lead to heart failure comes from animal models of ischemia

---

BOX 6.1

SECONDARY CAUSES OF DILATED CARDIOMYOPATHY (DCM)

- Idiopathic cardiomyopathy
- Ischemic cardiomyopathy
- Stress-induced cardiomyopathy (Takotsubo's cardiomyopathy)
- Hypertensive cardiomyopathy
- Infectious cardiomyopathies
  - Viral myocarditis
  - HIV
  - Chagas disease
  - Lyme disease
- Toxic cardiomyopathies
  - Alcohol
  - Cocaine-induced cardiomyopathy

- Medication-induced cardiomyopathy (anthracycline)
  - Trace mineral deficiencies
- Autoimmune cardiomyopathies
  - Systemic lupus erythematosus (SLE)-associated cardiomyopathy
  - Sarcoidosis
- Peripartum cardiomyopathy
- Uremic cardiomyopathy in end-stage renal disease
- Endocrine dysfunction
  - Diabetic cardiomyopathy
  - Thyroid dysfunction
- Obstructive sleep apnea

that highlight the critical nature of post-ischemic remodeling. Post-ischemic cardiac remodeling can progressively lead to ventricular dysfunction, and result in the generation of clinical symptoms of heart failure that include resting and exertional dyspnea, pulmonary vascular congestion, and in extreme cases malignant, life-threatening arrhythmias. In this section, we will discuss the mechanisms that are known to be involved at the onset of myocardial dysfunction after ischemic injury.

The initial insult in cardiac ischemic injury is myocardial hypoxia, which results in irreversible cardiomyocyte death in as little as 20 minutes.[8] This is mediated by aberrant activation of neurohumeral mechanisms, inflammation, and necrotic/apoptotic pathways that induce myocardial death.[9] Apoptotic pathways involving caspase-3-mediated mechanisms predominate in the first 24 hours post-injury. This mechanism involves an imbalance between the levels of pro-apoptotic proteins Bax and Bad, and anti-apoptotic protein Bcl-2.[10] This altered Bax/Bcl-2 rheostat leads to degradation of the mitochondrial membrane, release of cytochrome c, with subsequent activation of the caspase-3-mediated nuclear degradation and programmed cell death.[11] The persistence of the Bax/Bcl-2 imbalance can last up to 12 weeks after a myocardial infarction.[10] Moreover, additional pathways are activated that involve the mitochondrial permeability transition pore (MPTP), and lead to reactive oxygen species (ROS)-mediated degradation of the mitochondria and release of pro-apoptotic proteins.[12] The mechanisms of ROS-mediated myocardial death are multifaceted, as influx of intracellular calcium leads to cleavage and activation of phospholipases and calpain that in turn potentiate mitochondrial damage with the release of pro-apoptotic molecules and ultimately cell death.[13] Moreover, high concentrations of angiotensin II (Ang II) and overstimulation of the sympathetic mechanisms that are initially triggered as protective measures in acute ischemic injury play a role in accelerating the rate of myocardial apoptosis. Indeed, blocking of renin–angiotensin–aldosterone (RAAS) slows down myocardial death, further supporting its role in acute ischemic injury.[14] The remodeling process continues with mobilization of inflammatory cells that infiltrate the affected myocardium and promote collagen degradation and synthesis.[7] The next phase of cardiac remodeling involves the release of pro-inflammatory and chemo-attractant molecules, such as TNF-α, interleukin (IL)-6, IL-1β by the myocytes and inflammatory cells. These chemokines attract macrophages, lymphocytes, and neutrophils to the necrosed myocardium.[15] In addition to phagocytosing cellular debris, these leukocytes initiate the process of scar formation. Neutrophils release matrix metalloproteinases (MMPs),

enzymes that degrade collagen, while macrophages consume necrotic cell debris and release cytokines that stimulate fibroblast proliferation and collagen synthesis.[16] The process of collagen degradation is regulated by several factors in the acute period. Selective expression of MMPs, tissue inhibitor of matrix metalloproteinases (TIMP), transforming growth factor-β (TGF-β) and connective tissue growth factor (CTGF) all contribute to fine-tune this process by selectively regulating expression of genes responsible for cardiac remodeling.[15] The newly remodeled myocardial tissue has a predominance of fibrosis with stretching and thinning of infracted areas with significant alterations in the key cytoplasmic and membrane proteins that are vital to normal function. Furthermore, up-regulation of compensatory mechanisms causes the viable myocytes to hypertrophy, a process that reduces diastolic wall stress in the acute phases of injury. In the long-term, however, such a response can prove deleterious as altered cellular processes lead to aberrant myocardial contraction. Calcium channel expression in the sarcoplasmic reticulae in the post-infarction myocardium is affected with subsequent changes to the electrophysiological properties of the viable myocardium, giving rise to the increased susceptibility to arrhythmias in some patients.[17]

In summary, after an infarction, the heart undergoes complex remodeling processes that occur at both cellular and sub-cellular levels, resulting in the development of ischemic cardiomyopathy. A further discussion of ischemic heart disease, as well as its present and future therapies, can be found in Chapter 5.

IDIOPATHIC DILATED CARDIOMYOPATHY

In London's Hammersmith Hospital, Goodwin first documented three different categories of heart failure in the 1950s: (1) heart failure associated with generalized disease; (2) heart failure without a specific pathology; and (3) heart failure associated with an unknown etiology.[5] While this classification has been revised multiple times, cardiomyopathy without systemic causes (Goodwin's category 2), also known as idiopathic DCM (IDCM), is increasingly common. IDCM is a chronic disease with a prevalence of around 20 per 100 000 individuals in the United States. The prevalence of IDCM increases with age, is more common among males, and varies with ethnicity.[18] The diagnosis of IDCM is one of exclusion, as it relies on the absence of other cardiovascular diseases. However, the development of IDCM is multifactorial, and the presence of other cardiac abnormalities can lead to early onset of disease with poor prognosis. Approximately

25–50% of patients with DCM have a family history and although multiple modes of inheritance have been reported, autosomal dominant pattern is the most common.[19] To date, over 50 single genes with dominant mutations have been identified in familial forms of DCM. With a few exceptions, assessing the clinical effect of any given mutation is challenging, as most mutations are specific to each family. Because mutations in many individual genes lead to IDCM, genetic testing commonly employs multi-gene panels. Even in the rare case where the same mutation has been observed in unrelated families, the penetrance and clinical presentation of idiopathic DCM can be highly variable. Given the heterogeneity associated with IDCM, this section will provide a discussion of our current understanding of the molecular mechanisms that are known to cause familial and idiopathic DCM.

The earliest insights into the pathophysiology of DCM came from observations that muscular dystrophy patients were commonly affected with DCM. Specific mutations in the dystrophin gene (DMD) were identified as the underlying cause in patients with (idiopathic) X-linked DCM. Patients with X-linked DCM can have only subtle peripheral muscle weakness but have a high incidence of cardiac involvement, and the severity of muscle involvement appears to be dictated by mutations in different domains of the DMD gene.[20]

The DMD gene encodes a protein dystrophin that plays a key role in anchoring muscle cells and is part of a larger structure that interacts with the sarcomere. The sarcomere is the basic contractile unit for both striated and cardiac muscle and is made up of a complex mesh of thick filaments, thin filaments, and a giant protein titin. The Z-disk forms an anchor site for thick and thin filaments and functions to align adjacent myofibrils; it therefore acts as the primary conduit for coordination of contraction and is highly involved with mechano-transduction and signaling. The thick filament core is formed by β-myosin heavy chain (β-MHC) with myosin-binding protein C. The thin filaments emanate from the Z-disk with filamentous actin forming the core with tropomyosin and the troponin subunits providing calcium regulation of the actino-myosin interface. A third major filament, titin, functions as a stretch sensor that transmits signals from its anchor at the Z-disk to its carboxyterminal kinase zone that helps to maintain the tensile integrity of the sarcomere. An intermediate filament protein desmin tethers the Z-disks to the plasma membrane and maintains structural integrity of myofibrils in both skeletal and cardiac tissues. In addition, a complex of laminin and dystroglycan proteins links myocytes to the extracellular matrix (ECM). Identified mutations in familial forms of DCM to date implicate each of the aforementioned molecules with an estimated 35–40% of IDCMs resulting from sarcomere gene mutations.[21]

Mutations in β-myosin heavy chain (MYH7) and troponin T are more common in familial IDCM, than mutations in α-tropomyosin and troponin-C.[22] Recently, mutations in the titin (TTN) gene were identified to be a leading cause of DCM. Mutations that led to a truncated titin protein were found to occur in over 25% of familial and 18% of sporadic cases of IDCM.[21,23–27] Still some IDCM patients can be identified that do not have any detectable mutations in the sarcomeric apparatus, suggesting that other unidentified mechanisms also underlie this cardiomyopathy.

An interesting subset of DCM that includes mutations in the nuclear membrane protein lamin is classified as laminopathies. X-linked Emery-Dreifuss muscular dystrophy (EDMD) is the most common laminopathy that occurs due to mutations in the lamin (LMNA) gene on chromosome 1q21.2–21.3. More than 200 LMNA mutations have been reported in the LMNA gene, which encodes nuclear lamin A and C.[19] Clinically, patients with EDMD have progressive skeletal muscle weakness and contracture formation. Interestingly, cardiac phenotypes predominate early, including early-onset DCM and severe conduction diseases that can manifest as sino-atrial nodal dysfunction, atrio-ventricular blocks, and bundle branch blocks.[28] In young LMNA carriers, either atrial or ventricular arrhythmias can be the initial presentation as chamber dilatation is usually a late and independent phenotype. In a study of patients with IDCM and prominent conduction block, 33% had LMNA mutations.[29] A recent report found that frame-shifting LMNA mutations have the highest risk for ventricular arrhythmias.[30] Laminopathies follow a very aggressive natural course, often causing premature death, thus underscoring the importance of identifying the etiology earlier in these disease processes.

Mutations in multiple genes have been implicated in causing idiopathic DCM (Table 6.1). The pathways uncovered from molecular studies highlight the importance of sarcomere components and the signaling mechanisms that regulate myocardial contraction. As with other genetic disorders, the study of IDCM is currently undergoing a revolution, with implementation of newer, massively parallel next-generation sequencing. Whole-exome and whole-genome sequencing approaches that are becoming prominent in the clinical domain will continue to enhance our understanding of this heterogeneous and complex disease. Ultimately, further studies on the genetic basis of IDCM and the longitudinal course of mutation carriers can be expected to provide useful information about prognosis and treatment of this morbid condition.

TABLE 6.1   Gene Mutations Implicated in the Cardiomyopathies

Gene	Protein	Reference
DILATED CARDIOMYOPATHY		
TTN	Titin	[119]
DES	Desmin	[120]
ACTC	Cardiac actin	[121]
LMNA	Lamin A/C	[28]
DMD	Dystrophin	[20]
MYH7	β-Myosin heavy chain	[23]
TPM1	α-Tropomyosin 1	[22]
TNNI3	Troponin I	[24]
TNNT2	Troponin T2	[27]
TNNC	Troponin C	[26]
SYNE 1/2	Nesprin 1/2	[122]
HYPERTROPHIC CARDIOMYOPATHY		
MYH7	β-Myosin heavy chain	[123]
MYH6	α-Myosin heavy chain	[124]
TPMI	α-Tropomyosin	[26]
TNNT	Troponin T	[125]
TNNI	Troponin I	[125]
TTN	Titin	[126]
TTNC	Troponin C	[26]
VCL	Vinculin	[127]
MBPC	Myosin-binding protein C	[124]
RESTRICTIVE CARDIOMYOPATHY		
TNNI	Troponin I	[128]
DES	Desmin	[129]
ARRHYTHMOGENIC RIGHT VENTRICULAR CARDIOMYOPATHY/DYSPLASIA		
PKP2	Plakophilin	[130]
JUP	Plakoglobin	[96]
DSC2	Desmocollin-2	[131]
DSG2	Desmoglein-2	[103]
DSP	Desmoplakin	[103]
TGFβ	Tissue growth factor β	[104]
RYR2	Ryanodine receptor 2	[105]
LEFT VENTRICULAR NONCOMPACTION		
DTNA	α-Dystrobrevin	[117]
TAZ/G4.5	Taffazin	[132]
ZASP/LDBP	LIM domain binding protein	[116]
LMNA	Lamin A/C	[118]

CHAGAS DISEASE-RELATED CARDIOMYOPATHY

Recent WHO reports estimate that 10 million people are affected by Trypanosoma cruzi (T. cruzi), making this protozoal parasite-causing Chagas disease a significant public health issue in its endemic regions of Central and South America.[31] Even Western Europe and the United States are experiencing a surge in Chagas disease cases due to evolving immigration patterns.[32] Most cases of acute T. cruzi infection result in a benign asymptomatic myocarditis, which is rarely life-threatening. Although most people infected remain free of any clinical symptoms, approximately 20% of disease carriers develop chronic Chagas-disease-related cardiomyopathy 10–20 years after the initial infection.[33] The clinical progression and prognosis are significantly worse in patients with Chagas disease than those with DCM caused by other etiologies (Fig. 6.2).

Chagas disease is often slow to progress. Cardiac involvement can be seen over years to decades causing a Chagas-disease-related cardiomyopathy. The presence of minor electrocardiographic changes such as right bundle-branch and/or hemi-blocks as well as premature ventricular ectopy, usually precede gross anatomic changes, and correlate with increased risk for sudden cardiac death from arrhythmias. Morphological changes such as ventricular thinning can lead to apical aneurysms; wall motion abnormalities can lead to thromboembolic events and can present during the intermediate phases of the disease. End-stage disease is characterized by severe global dysfunction, the onset of DCM, and refractory heart failure.[34]

The mechanisms involved in the pathogenesis of Chagas cardiomyopathy remain poorly understood. Six mechanisms have been hypothesized to explain the disease process: (1) micro-vascular spasm;[35] (2) ischemia; (3) immune-mediated infiltration; (4) parasite-mediated myocardial injury; (5) heightened immune response to T. cruzi parasites or parasitic antigens that persist in the heart;[36,37] and (6) T.-cruzi-induced autoimmunity.[38] Histological hallmarks of Chagas disease are myocarditis with T-cell and macrophage predominance, followed by hypertrophy and fibrosis, providing support for the role of chronic inflammation and remodeling. However, the exact antigens responsible for eliciting such a destructive immune response remain elusive. Data in support of the autoimmune theory come from studies showing complete absence of the T. cruzi parasite but low concentrations of the parasite antigens and DNA, as well as presence of parasite-specific CD8+ T-cells in the endomyocardium of affected individuals.[39] The current hypothesis proposes that the protective processes that begin with macrophage- and dendritic-cell-mediated endocytosis of the parasite in the initial phases of

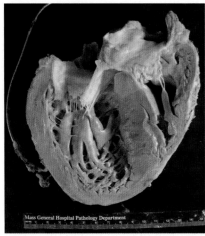
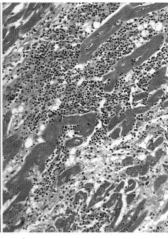

FIGURE 6.2 Chronic changes in Chagas cardiomyopathy. Gross heart specimen (left) with left ventricular dilatation in a patient with chronic Chagas disease, with histological evidence of extensive mononuclear infiltration of the myocardium (right) as visualized using H&E stain.

infection ultimately result in the production of destructive auto-antibodies. Although supportive data in human subjects are lacking, animal model studies showing that T.-cruzi-infected mice display auto-antibodies specific for the myocite proteins, including desmin, myosin, actin, β-adrenergic and M2-muscarinic cholinergic receptors further add to the validity of this theory.[40] The cross-reactivity and auto-immune response is complemented with an up-regulation of the inflammatory cytokines tumor necrosis factor-α (TNF-α), interferon-γ (IFN-γ), chemokine ligand 3 (CCL3) and Toll-like receptors (TLR).[41] In the acute phase of the disease, the T and B cells that are generated to recognize parasite antigens cross-react with host antigens with similar epitopes, thus causing an auto-immune response and a resultant polyclonal activation and auto-antibody production. It is this balance of pro- and anti-inflammatory host responses that is hypothesized to give rise to the chronic inflammatory process and development of Chagas-related cardiomyopathy. Much however remains to be elucidated as it is unclear how the parasite modulates the immune response and manages to survive unnoticed in the body for such prolonged periods.

ALCOHOLIC CARDIOMYOPATHY

The prevalence of cardiomyopathy among alcoholics varies between 23–40% with a slight male predominance.[42] Furthermore, the incidence of DCM in alcoholics is greater compared to the general population. Alcoholic cardiomyopathy accounts for a large proportion of DCM, accounting for 21–32% of all DCM cases at some referral centers. Alcoholic cardiomyopathy is characterized by dilation of all four cardiac chambers, decreased cardiac output, and normal or depressed left ventricular wall thickness in the absence of other causes of DCM. The clinical presentation is indistinguishable from other forms of dilated cardiomyopathies, and must be determined based on the absence of other likely causes of cardiac failure and a history of excessive alcohol intake. Importantly, the majority of alcoholics have also been noted to have preclinical structural heart disease thought to be due to excessive alcohol intake. Given the evidence for a cardioprotective effect of moderate alcohol consumption, there has been considerable debate about the dose and chronicity of alcohol use necessary to result in cardiotoxicity.

Numerous studies report that mild–moderate alcohol consumption is associated with a reduced risk of stroke and coronary artery disease.[43-45] Investigators from the Framingham Heart Study found a marked reduced risk for developing cardiomyopathy in men who consumed 8–14 units of alcohol per week.[44] One study of nearly half a million individuals found that moderate alcohol consumption reduced cardiovascular deaths, especially in middle-aged individuals, although all-cause mortality due to heavy alcohol use was increased.[46] A study of more than 10 000 women revealed a reduced risk of coronary artery disease and stroke with moderate alcohol use.[47] A large meta-analysis confirmed that consumption of 2–4 drinks/day was significantly associated with a decreased incidence of cardiovascular and all-cause mortality.[48]

Moderate alcohol is proposed to have these beneficial effects by increasing high-density lipoprotein cholesterol, reducing plasma viscosity by reducing platelet-mediated coagulation, increasing fibrin degradation, and improving endothelial function.[49] However, the potential cardiovascular benefits of alcohol intake decline as the amount consumed increases. Interestingly, the cardiovascular effect of alcohol is not specific to beverage or quantity, but can vary based on genetic and environmental factors.[50] For most alcoholic patients, consumption of more than six drinks per day for at least five years is common before the detection of cardiovascular effects.[51,52]

Animal model studies have elucidated multiple potential mechanisms for alcoholic cardiomyopathy, including damage to proteins in the contractile apparatus, organelle dysfunction, disturbances in calcium handling, and activation of neurohormonal systems.[50] Biochemically, alcohol has been shown to have detrimental effects on protein metabolism, resulting in a decrease in protein synthesis and ultimately a decreased ventricular mass. Specifically, production of myosin heavy chain and desmin are reduced. This appears to result from inhibition of mRNA translation and a decrease in translational efficiency. Additionally, the ethanol molecule is able to diffuse across cellular membranes and cause direct toxic damage. Studies of animal models have reported that long-term alcohol consumption leads to myocyte apoptosis and necrosis, ultimately causing ventricular dysfunction.[50] Alcohol exposure causes disruption of myocardial sarcoplasmic reticulum, degradation of the contractile elements, disruption of the intercalated disc, and deposition of fat-filled inclusion bodies in the cytoplasm.[51] In vitro studies of myocytes treated with high concentrations of ethanol show over-expression of Bax, a pro-apoptotic protein, and augmented caspase-3 activity.[53] Endomyocardial biopsies of alcoholic patients have a disruption in the mitochondrial cristae, suggesting a role of impaired calcium homeostasis as well as oxidative processes. Furthermore, prolonged and heavy alcohol consumption leads to alterations in the neurohormonal pathways that result in up-regulation of the renin–angiotensin system.[54] The renin–angiotensin activation in turn initiates a cascade of events that results in the development of heart failure symptoms.[55] Clinical manifestations are those of heart failure with presence of a pulmonary vascular congestion, jugulovenous distention and a third heart sound. Over time, cardiomegaly ensues, and the disease course can be complicated by the onset of ventricular arrhythmias and sudden cardiac death in alcoholic cardiomyopathy patients.

To date, no formal guidelines exist regarding optimal treatment strategies; however, abstinence from alcohol has been shown to have a high tendency toward improvement in left ventricular function. The prognosis in patients who continue to drink remains poor, with mortality reaching up to 50% at 4 years. There is also evidence supporting the role of conventional medical therapies when combined with alcohol cessation providing the best improvement in echocardiographic parameters.[56] Heart transplantation is often not an option given the concern for relapse, but remains a consideration.

In conclusion, alcohol if used in moderation may have cardioprotective effects, while its excessive use is associated with a subtype of dilated cardiomyopathy that carries a poor prognosis. The diagnosis of alcoholic cardiomyopathy poses a challenge to clinicians as it relies heavily on determining the history of alcohol usage. Echocardiographic findings that suggest systolic and diastolic impairment usually precede the clinical manifestation of alcoholic cardiomyopathy and are present in the majority of heavy alcohol drinkers. While abstinence from alcohol appears to slow down the course of the disease process, treating these patients with medications such as angiotensin-converting enzyme inhibitors, beta blockers, and aldosterone antagonists appears to reduce symptoms and improve survival.

PERIPARTUM CARDIOMYOPATHY

Peripartum cardiomyopathy (PPCM) is a rare but serious disease that occurs in 1:3000 to 1:15 000 women during late pregnancy in the United States with a high incidence in the African-American population. Globally, the incidence appears to be much higher in developing countries, with WHO estimates reporting the incidence to be as high as one in 299 live births in Haiti.[57]

PPCM presents with clinical signs of systolic heart failure in the last month of pregnancy or within five months following delivery. PPCM can be challenging to recognize and diagnose, as the clinical symptoms associated with early congestive heart failure and routine late pregnancy are often indistinguishable due to the physiologic fluid shifts that can occur concomitantly during pregnancy. Therefore, diagnosis of PPCM relies on a high index of suspicion, clear identification of the timing of symptom onset, and documented new left ventricular dysfunction by echocardiography. Once left ventricular dysfunction is identified, other causes of heart failure must be excluded prior to making a diagnosis of PPCM. Epidemiologic studies have identified risk factors for the development of PPCM that include African-American ancestry, multiparity, maternal age greater than 30 years of age, pregnancy-induced hypertension, eclampsia, the use of tocolytic therapy, and twin gestation.[58] The pathogenesis of PPMC is not very well understood but likely involves a combination of genetic and environmental factors. Small-scale investigations have resulted in multiple hypotheses implicating infectious, nutritional, and autoimmune causes, while the slightly higher incidence of PPCM in families with idiopathic DCM suggests a genetic predisposition for the disease. The data supporting a role for infectious etiologies comes from endomyocardial biopsies of patients that have high rates of myocarditis. Furthermore, animal studies with high serum titers of Coxsackie and Echoviruses in pregnant mice with PPCM provide additional support. It is likely that pregnancy-induced immune suppression increases susceptibility to viral-mediated infections. The increase in physiological cardiac output is thought to worsen the effect of this viral burden on cardiac function.

Nutritional mechanisms leading to PPCM are supported by the increased incidence of PPCM in Nigeria, where post-partum traditions involve ingesting large amounts of *kanwa*, a form of dried lake salt. This increased salt load leads to volume expansion and hemodynamic stress, which might help explain the high incidence of PPCM in this region. Other theories implicate trace mineral deficiencies such as selenium, as low levels have been demonstrated in patients with PPCM. Although our understanding of these pathways is still limited, they all appear to involve alterations in the immune responses that may be altered during the puerperium. Support for the autoimmune hypothesis in PPCM includes high concentrations of pro-inflammatory cytokines, mainly TNF-α and IL-6,[59] and suppression of the normal pathways that rescue the cells from oxidative stress.[60]

Perhaps the most compelling potential mechanism for PPCM comes from elegant recent work by the Arany laboratory. They found that mice lacking a regulator of angiogenesis, PGC-1α, unexpectedly have PPCM and that treatment with pro-angiogenic therapies can correct PPCM.[61]

Clinically, the prognosis for PPCM is dependent upon the recovery of systolic function. Mortality rates range from 7 to 50%, with the vast majority of patients recovering within 3–6 months of disease onset. Persistence of left ventricular dysfunction for more than six months renders a poor prognosis. In general, survivors are shown to have higher left ventricular function at the time of diagnosis compared to non-survivors, making left ventricular ejection fraction the strongest predictor of disease outcome. Even after complete recovery, patients are at higher risk for recurrence in subsequent pregnancies, depending on the degree of recovery. High mortality rates are seen in women with persistence of left ventricular dysfunction, increased occurrence of heart failure symptoms, and higher levels of premature births and therapeutic abortions in future pregnancies, when compared to those that had recovered left ventricular function. Outcomes are improved by aggressive medical therapy, with cardiac transplantation as a consideration for refractory cases.

HYPERTROPHIC CARDIOMYOPATHY

Hypertrophic cardiomyopathy (HCM) is defined as an increase in cardiac mass usually due to left ventricular hypertrophy and a corresponding increase in left ventricular mass (Fig. 6.3). Since long-standing hypertension and infiltrative diseases can also lead to cardiac hypertrophy, the diagnosis of primary HCM is only considered when such causes have been ruled out. HCM is the most common genetically determined cardiac disease with a prevalence of approximately 1 in 500 individuals. The clinical presentation of HCM is variable and includes chest pain, congestive heart failure symptoms, syncope or pre-syncope, palpitations, and sudden cardiac death. Indeed, HCM is

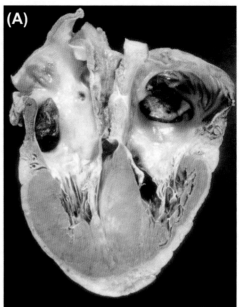

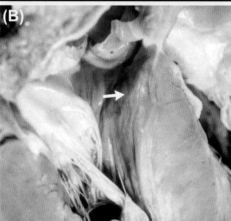

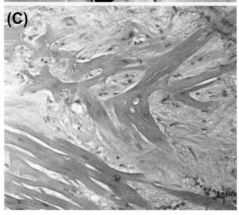

FIGURE 6.3 Gross and histologic appearance of hypertrophic cardiomyopathy. (A) Gross section reveals hypertrophy of the left ventricle, with increased septal thickening. (B) Arrow highlights fibrosis of left ventricular outflow tract. (C) H&E histology of interventricular septum with interstitial fibrosis and enlarged myocytes showing disarray. (Reproduced with permission from Binder WD, Fifer MA, King ME, Stone JR. A 48-year-old man with sudden loss of consciousness while jogging. *N Engl J Med* 2005;353; 824–32).

the most common cause of sudden cardiac death in young patients and competitive athletes. Patients with HCM have enlarged left ventricles with walls that are hypertrophied in either a symmetric or asymmetric fashion. The classic form of HCM involves hypertrophy of the interventricular septum with systolic anterior motion of the mitral valve leading to outflow tract obstruction.[62–64] The following section will review both the macroscopic and microscopic processes that are involved in the development of HCM.

Microscopically, HCM is characterized by enlargement or hypertrophy of the cardiac myocytes. The classical histopathologic findings of cardiac biopsy are myofiber disarray, myocyte hypertrophy, and interstitial fibrosis. The development of fibrosis is attributed to premature apoptotic death of myocytes, which are then replaced by expansion of the interstitial matrix. Increased expression of pro-fibrotic molecules are also thought to be involved in the development of HCM. Deposition of collagen results in fibrosis, which typically appears in a patchy pattern and can be identified using cardiac MRI, an imaging modality that can aid with the diagnosis.[65]

On the macroscopic level, while hypertrophy of any part of the left ventricle is possible, symmetric hypertrophy accounts for about 40% of cases, and is more common in the elderly population. Symmetric hypertrophy is characterized by concentric thickening of the left ventricular wall with a small cavity. The asymmetric variant is more common in younger patients. With both forms there is right ventricular involvement about 15% of the time. Another morphologic variant is localized apical hypertrophy. This variant appears to have geographic variation, as it was first described in Japan in the 1970s, though it is recognized increasingly in Western regions. Apical hypertrophy portends a better prognosis than other variants. Ischemia or infarction in the apical region resulting in aneurysm formation is a common complication. All morphologic variants of HCM can progress to a dilated or 'burned-out' stage, where left ventricular dilatation predominates and is seen in about 10% of patients. Dilatation is secondary to fibrous replacement of the myocardium and is attributed to occlusion of the intramural vasculature causing ischemic injury.

The histopathologic findings of HCM have largely been attributed to mutations in genes encoding the sarcomere proteins, and are found in 70% of affected patients with a family history of HCM, and in up to 40% of cases thought to be sporadic. To date, over 1000 specific mutations are known to account for HCM with myosin heavy chain (MYH7) and myosin binding protein C (MYBPC3) accounting for about 80% of cases. Additional mutations in other sarcomere-related proteins, including cardiac troponin T and I, tropomyosin, myosin light chain, titin and actinin-2, are commonly inherited in an autosomal dominant fashion. There are also reports of multiple mutations in the same patient

and these findings confer a more aggressive clinical course for the disease.[66]

Sarcomere mutations lead to HCM by several proposed mechanisms, including altered reaction to biomechanical stress and energy handling. Mutations in the sarcomere architecture can lead to an altered sensing of biomechanical stress and ultimately an exaggerated response to mechanical load, resulting in excessive hypertrophy. Energy efficiency by the sarcomere is compromised with HCM mutations, causing the sarcomere to be unable to effectively perform critical homeostatic functions including calcium reuptake. This can be particularly important in situations of high-energy demand, such as exercise, and is thought to account for diastolic dysfunction and ventricular arrhythmias induced by exercise seen in the HCM.[67,68] The phenotypic expression of specific mutations is quite variable, with a wide range in the age of onset, degree and pattern of left ventricular changes, and ultimately different clinical outcomes even in individuals with identical mutations.

RESTRICTIVE CARDIOMYOPATHY

Impairment of ventricular contraction gives rise to congestive or systolic dysfunction, while abnormalities in ventricular relaxation lead to diastolic dysfunction and restrictive physiology. Restrictive cardiomyopathy (RCM) is a rare form of cardiomyopathy, without well-accepted diagnostic criteria (Box 6.2). It is characterized by a stiffened ventricle, which causes an abnormal rise in chamber pressure in response to relatively small increases in volume. Systolic function is almost always preserved except in late phases of the disease; however, diastolic volume of the ventricle can be either normal or reduced, and wall thickness may or may not be affected. The definition of RCM is more descriptive than categorical, since the restrictive physiology can be caused by multiple different cardiovascular and systemic diseases. In general, RCM can be defined as restrictive ventricular physiology with normal or reduced diastolic volumes and normal ventricular wall thickness in the absence of ischemic heart disease, hypertension, valvular disease, or congenital heart disease.

The physiology of restrictive cardiomyopathy is classically a hemodynamic state characterized by rapid ventricular filling in early diastole, with little or no filling in late diastole due to impaired ventricular relaxation and expansion. This ventricular filling pattern is responsible for the classic 'dip and plateau' pattern noted on invasive hemodynamic monitoring (square-root sign). The diastolic phase is a two-step process with an energy-dependent active phase that involves isovolumetric relaxation of the ventricle, coupled with a passive phase governed by the intrinsic compliance of the muscle

---

BOX 6.2
_____

SECONDARY CAUSES OF RESTRICTIVE CARDIOMYOPATHY

Myocardial

- Infiltrative
  - Amyloidosis
  - Sarcoidosis
  - Hemosiderosis
- Storage disorders
  - Gaucher's disease
  - Hurler's disease
  - Glycogen storage disease
  - Fabry's disease
  - Hemochromatosis

Endomyocardial

- Endomyocardial fibroelastosis
- Hypereosinophilic syndrome
- Carcinoid heart disease
- Metastatic cancer
- Radiation
- Anthracycline-induced cardiotoxicity
- Drug induced fibrous endocarditis

---

fibers. The intraventricular pressures, which remain low during the diastolic phases to allow for proper filling, rise precipitously if ventricular compliance is abnormal. This phenomenon results in the representative anatomic features of restrictive disease, i.e., the presence of a normal left ventricle, a markedly dilated left atrium, and normal systolic function.[69]

Restrictive physiology can occur due to two independent mechanisms that need to be differentiated: either the myocardium becomes non-compliant (stiff), or the pericardium limits the expansion of ventricles during the filling process, also referred to as constrictive disease. Restriction due to pericardial involvement often follows inflammation, fibrosis and thickening and can be sequelae of multiple etiologies including chronic inflammatory state, irradiation, hemorrhage, acute infections and metabolic abnormalities such as uremia. Patients suffering from constrictive disease can recover following pericardectomy or surgical removal of the fibrotic pericardium. Therefore, differentiation between constrictive pericarditis and restrictive cardiomyopathy is essential. Without clinical and echocardiographic signs of constrictive pericarditis, a physiology consistent with restrictive disease is suggestive of RCM. Secondary causes of restrictive disease can be classified into myocardial or endomyocardial pathologies. Myocardial causes of restrictive cardiomyopathy are associated with systemic diseases, and fall into two major categories: infiltrative disorders and storage disorders. Additionally, a pure form without any secondary cause, also referred to as idiopathic restrictive cardiomyopathy, has also been described. This idiopathic hereditary restrictive disease has recently been shown to involve mutations in sarcomeric proteins in some cases.

A familial inheritance of a disease resembling hypertrophic and restrictive clinical features was first described in 1992. Linkage analysis in an extended family with 12 individuals who had died of sudden cardiac death, led to identification of a missense mutation in a conserved region of troponin I gene (TNNI3) that co-segregated in the surviving members with cardiac manifestations.[70] Subsequent studies of other patients with idiopathic restrictive disease further identified similar de novo mutations within the same conserved region of the gene. The proposed mechanism suggests that the mutated form of cardiac troponin I, which forms the cardiac troponin complex and regulates the calcium-mediated interaction of actin and myosin, alters the phosphorylation of the thin filament, and leads to diminished calcium desensitizing effects. Microscopically, these interactions translate to a lack of inhibition of actin and myosin release and a lack of muscle relaxation, as that observed in the case of restrictive cardiomyopathy.

Systemic disease states are more commonly associated with restrictive cardiac filling pattern;[71] cardiac amyloidosis the most common cause and the most well understood and is discussed in further detail below.

Cardiac Amyloidosis

Amyloidosis is the most common cause of restrictive cardiomyopathy and refers to a family of disorders characterized by protein misfolding. Amyloidosis involves deposition of misfolded extracellular proteins as soluble or insoluble β-sheet fibrils with resultant impairment of tissue function.[72] Although myocardial amyloid deposition is associated with restrictive filling patterns, dysfunction is not just limited to abnormal diastolic filling, but can also result in systolic dysfunction.

Interstitial infiltration of atrial and ventricular tissue as observed in cardiac amyloidosis results in a firm and rubbery consistency to the myocardium.[73] To date more than 25 different precursor proteins with a propensity to form amyloid fibrils have been identified. Specific precursor proteins define the amyloid type and can help to predict the clinical course.[74] The diagnosis of amyloidosis is a histological one and involves Congo red staining to identify the salmon pink amyloid deposits under light microscopy, with a characteristic apple green birefringence under polarized light. Methods to identify the exact precursor proteins include immuno-fluorescence microscopy, immunogold electron micros-copy, and mass spectrometry.[75]

Amyloidosis occurs in both acquired and genetic forms that are further classified based on the fibril precursor protein and the nature of the associated clinical features. Of the acquired forms, primary or AL amyloidosis is a systemic illness frequently associated with plasma cell dyscrasias such as monoclonal gammopathies and multiple myeloma. It is characterized by monoclonal proliferation of plasma cells with production and deposition of kappa and lambda light chains into the myocardium.[76] With an aging population, the incidence of AL amyloidosis is on the rise with more than 2500 newly diagnosed cases each year. Cardiac manifestations are observed in >90% of patients, with prognosis of typically less than one year in patients with untreated AL amyloidosis.

A secondary form, also known as reactive or AA amyloidosis causes less severe structural and functional cardiac impairment. This form of acquired amyloidosis is observed in the setting of chronic inflammatory disease states such as rheumatoid arthritis, ankylosing spondy-litis, and familial Mediterranean fever.[77] In this disorder, amyloid fibrils are composed of an acute-phase reactant, serum amyloid A protein. It is thought that chronic inflammatory state increases one's risk for developing AA amyloidosis. Patients with end-stage renal disease requiring hemodialysis also have an increased prevalence of amyloidosis; in this subtype, β-2 microglobulin is deposited in the periarticular spaces and is the cause of serious arthritic pathologies in hemodialysis patients.[77]

Another type of amyloidosis called transthyretin (TTR) arises from misfolding and deposition of a mutated or variant TTR, and is further classified into familial and sporadic variants. The familial disease variant of TTR also known as familial amyloid cardiomyopathy or familial amyloidotic polyneuropathy, typically has an autosomal dominant pattern of inheritance with a variable penetrance and significant cardiac manifesta-tions. Unlike AA amyloidosis, transthyretin amyloid cardiomyopathy is an indolent disease with slow pro-gression with prognosis ranging from years to decades in untreated patients. The non-familial or sporadic disease variant occurs due to abnormal tissue deposition of the wild-type transthyretin protein and is known as senile systemic amyloidosis (SSA).[78] TTR is a transport protein previously characterized as pre-albumin, and is primarily synthesized by the liver. It is the product of a single gene comprised of 7 kilobase region on chro-mosome 18 (18q21.1) spanning 4 exons and 5 introns, with over 100 single nucleotide polymorphisms encod-ing variant TTR and 80 pathogenic mutations that have been associated with onset of cardiac amyloidosis.[73] In its native state, TTR circulates as a homo-tetramer with symmetrical funnel-shaped thyroxine binding sites at its dimer–dimer interface. However, due to processes that are still poorly understood, the tetrameric assembly is disrupted and the monomers misassemble into amyloid fibrils that are deposited in target organs. In vitro studies show that tissues exposed to elevated levels of amyloid proteins have increased levels of ROS, which alter the cardiomyocyte ion fluxes, contractility and programmed cell death through the p38 mitogen-activated protein kinase pathway.[79] These molecular aberrations lead to myocardial necrosis, dysfunction and ultimately to a restrictive physiology. Untreated, TTR disease is associ-ated with significantly longer median survival than AL amyloidosis, but can progress to overt heart failure and death from systolic heart failure or dysrhythmia.[73]

In summary, cardiac amyloidosis is an under-appre-ciated contributor of heart failure and is increasingly becoming more common with the advancing age of our population. Understanding the molecular basis of this disease will be essential for identifying new avenues to combat the morbidity and mortality associated with amyloid cardiomyopathy.

ANDERSON-FABRY'S CARDIOMYOPATHY

Fabry's disease is only one of nearly 40 lysosomal stor-age disorders that can result from the intracellular accu-mulation of glycoproteins (Fig. 6.4). Fabry's disease is characterized histologically by deposition of glycosphin-golipids (globotriaosylceramide) within the lysosomes.[80] This abnormal accumulation results from a non-func-tional protein alpha-galactosidase A.[81] Fabry's disease has an estimated incidence of approximately one in 55 000 male births; however, recent studies identifying milder variants of this disease have suggested an incidence as high as one in 3000 male births.[82] Fabry's disease is clas-sically an X-linked disorder, yet heterozygous females and hemizygous males can also present with clinical disease.[80] Cardiac manifestations are more common in carriers or heterozygous females via the X-inactivation mechanism.[83] Clinical manifestations of Fabry's disease are highly dependent on the organs that are affected. The disease affects predominantly the kidneys, periph-eral nerves, skin, and heart. The initial presentation of

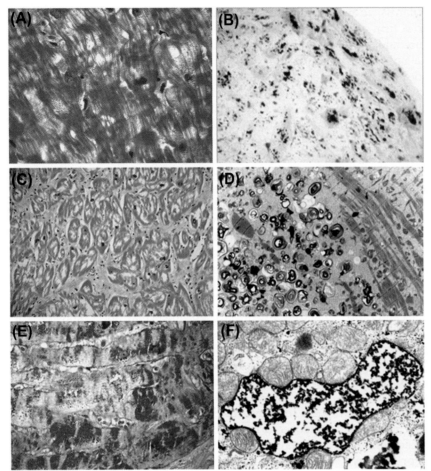

FIGURE 6.4 Cardiomyopathy secondary to storage disease. (A, B) Intracytoplasmic hemosiderin granules in the myocardium of a patient with hemochromatosis as visualized by Perle blue staining. (C, D) Hypertrophied myocytes, containing vacuoles rich with globotriaosylceramide are characteristic of cardiomyopathy associated with Anderson-Fabry's disease. (E, F) Pompe's disease is characterized by the presence of myocytes with glycogen deposits as visualized by PAS staining and electron microscopy. (Reproduced with permission from Leone O, Veinot JP, Angelini A, et al. 2011 consensus statement on endomyocardial biopsy from the Association for European Cardiovascular Pathology and the Society for Cardiovascular Pathology. *Cardiovasc Pathol* 2012;21:245–74).

Fabry's is typically in early childhood, when children present with painful neuropathy of extremities and gastrointestinal symptoms. Skin manifestations include angiokeratomas of the trunk and renal failure in early childhood. Cardiac manifestations can be variable and range from palpitations, acute heart failure symptoms and valvular disease, to severe conduction abnormalities and arrhythmias.[84]

Classically, cardiac manifestations of Fabry's disease become apparent in the fourth or fifth decades, and are observed in over 60% of affected patients. Cardiac dysfunction follows a similar pathophysiological mechanism, with the accumulation of glycosphingolipid-laden vacuoles within the myocytes, followed by myocyte hypertrophy and fibrosis.[85] Although the cellular processes leading to cardiomyocyte infiltration have been well-studied, the mechanisms that govern hypertrophy and fibrosis are not fully understood. Interestingly, glycosphingolipid-filled lysosomes only make up about 1–3% of the mass in hypertrophic hearts, suggesting that hypertrophy and lysosomal infiltration are two different processes.[80] In vitro studies indicate that presence of lysosomal galactoaosyltranferase increases intracellular oxidative stress and mitochondrial metabolism,

ultimately leading to microvascular lesions and further myocardial necrosis.[85]

Although cardiac hypertrophy is commonly observed, infiltration with lipid-rich vacuoles alters both systolic and diastolic dynamics. Global left ventricular hypertrophy or asymmetrical septal hypertrophy similar to obstructive HCM can be seen, but systolic function is preserved.[86] Furthermore, only mild to moderate impairment in diastolic function is present and restrictive filling may or may not be present. Therefore, Fabry's cardiomyopathy can also be classified as 'pseudo'-hypertrophic cardiomyopathy in that myocyte hypertrophy and fibrosis play a more prominent role than restrictive physiology in contributing to heart failure.[80]

CARDIAC HEMOSIDEROSIS

Cardiac hemosiderosis, also known as iron overload syndrome, is a syndrome of increased iron deposition into the myocardial tissue and is a common manifestation of hemochromatosis. Iron overload can result from two different mechanisms, one involving increased catabolism of erythrocytes (e.g. transfusion-mediated iron overload, hematologic malignancies), or a second in which plasma iron concentration exceeds the iron-binding capacity of transferrin while erythropoiesis remains normal. The second etiology is an autosomally recessive inherited disorder called hereditary hemochromatosis and will be the focus of discussion in this chapter.

Hereditary hemochromatosis has a high prevalence of 0.2–0.5% and results from homozygous mutations in the hemochromatosis gene *HFE*, although other genes have also been implicated.[87] The *HFE* gene encodes a 343-amino-acid human leukocyte antigen type molecule that regulates iron uptake in the duodenum and upper intestine. Missense mutations in *HFE* gene result in a non-functional protein product that causes inappropriate iron absorption by the duodenal enterocytes. Multiple variants of hereditary hemochromatosis have been identified with genetic loci mapping to chromosomes 6, 1, 7, and 2, with chromosome 6 being the most common locus involved.[87] A juvenile form, HFE2, is a rare but highly aggressive disease with early presentation and multi-organ involvement and dysfunction.[88]

Clinically, echocardiographic findings for cardiac hemosiderosis are non-specific and require an appropriate clinical context. Left ventricular size is usually normal in the initial stages with thickened ventricular walls and regional or global hypokinesis. While global systolic function may remain unchanged, impairment of diastolic indices with a restrictive pattern due to elevated filling pressures usually predominates. The late stages of cardiac involvement include the development of dilated cardiomyopathy as well as conduction abnormalities ranging from ventricular and supraventricular arrhythmias to heart block. Removal of excess iron via phlebotomy and iron chelation therapy remains the mainstay of treatment and has been reported in some cases to lead to improvement in the cardiac manifestations of the disease.

ARRHYTHMOGENIC RIGHT VENTRICULAR CARDIOMYOPATHY/ DYSPLASIA

In 1952 Henry Uhl described a 'parchment right ventricle' in an eight-month-old child as a paper-thin right ventricular surface without visible myocardium that had become attached to the epicardium.[89] In subsequent years, a milder form of the disease with less extensive right ventricular pathology became known as arrhythmogenic right ventricular cardiomyopathy/dysplasia (ARVC/D). The estimated prevalence of ARVC/D worldwide ranges from one in 2000 to one in 5000 individuals.[90] ARVD/C has a male predominance and usually presents between the ages of 30 and 50 years.[91]

Anatomically, the right ventricle is characterized microscopically by an atrophic or scarred appearance with fibrous or fibro-fatty infiltration that may involve inflow tract, outflow tract, and/or apex.[92] Clinically, these anatomic changes result in right ventricular electrical instability and ventricular arrhythmias causing sudden death. In subsets of individuals, rapid progression can result in right ventricular enlargement and ultimately heart failure.[90] The most common presenting symptoms consist of palpitations, lightheadedness, syncope, or cardiac arrest due to ventricular arrhythmias. Ventricular tachycardias in ARVC/D will typically have a left bundle morphology indicating a right ventricular origin.[91]

The diagnosis of ARVC/D is often challenging since the clinical features can be non-specific and the diagnosis is based on an extensive set of guideline criteria (Fig. 6.5). The natural course of the disease is protracted with an early 'concealed' phase, where subtle electrocardiographic findings are present without overt clinical symptoms. Palpitations and ventricular arrhythmias develop next and progress to detectable structural abnormalities in the third phase. The final disease stage is characterized by biventricular impairment and may be hard to differentiate from DCM.[93] Echocardiographic findings include an enlarged right ventricle with depressed function and regional wall motion abnormalities. The definitive diagnosis is made using comprehensive criteria that include structural abnormalities, histopathological findings, electrocardiographic abnormalities, family history, and genetic mutations.[94,95]

Initially hypothesized as a developmental defect of the right ventricular myocardium, ARVD/C is now known to be due to abnormalities in the desmosomal complex in most individuals. The desmosomes consist of a complex of cell surface proteins that both connect to the intracellular cytoskeletal network and link adjacent myocytes. Thus, desmosomes serve a critical role in maintaining the structural integrity of the heart. In recent years, mutations in multiple desmosome-related proteins, including plakoglobin, desmoplakin, plakophilin-2, and desmoglein-2, have been found to underlie ARVC/D.[96–98]

Family studies initially found that ARVC/D follows an autosomal dominant pattern with incomplete penetrance and variable expression.[92] The first genetic locus for ARVC/D was identified on chromosome 14q23–q24 in a large Italian family, and subsequently multiple other loci have been identified on chromosomes 1, 2, 3, 6, 10, 12, and 14.[90] An autosomal recessive form of ARVC/D associated with extracardiac manifestations that include palmoplantar keratosis and woolly hair was identified in individuals living on the Greek island of Naxos.[99] Pathogenic mutations were identified in the *JUP* gene that is located on chromosome 17q21 and encodes the desmosomal protein plakoglobin.[100] A similar pattern of inheritance and cutaneous manifestations was identified in separate families in India and Ecuador. However, individuals with this variant

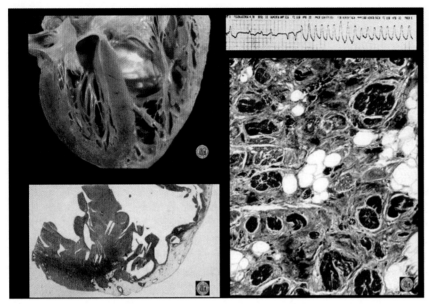

FIGURE 6.5 Gross, histological and electrographic patterns of arrhythmogenic cardiomyopathy (ARVC). The characteristic paper-thin right ventricular wall with no visible myocardium (A, B). Histologically, infiltration with fatty deposition is pathognomonic, giving rise to non-contractile right ventricle and predisposes to ventricular tachyarrhythmias. (Reproduced with permission from Thiene G. Arrhythmogenic cardiomyopathy: from autopsy to genes and transgenic mice. *Cardiovasc Pathol* 2012;21:229–39).

known as Caravajal syndrome showed predominantly left-sided cardiomyopathy.[101,102] Genome mapping of patients afflicted with Caravajal syndrome identified desmoplakin gene (*DSP*) mutations as the primary cause. Since these initial studies, numerous other disease-causing genes have been identified in other desmosomal proteins including plakophilin-2 and desmoglein-2.[96–98,103]

Non-desmosomal forms of ARVC/D have also been reported due to mutations in the genes encoding transforming growth factor beta-3 (TGFβ3)[104] and the ryanodine receptor 2 (RYR2).[105] Mechanisms that have been postulated for ARVC/D include the loss of myocyte adhesion resulting in altering the rheostat between pro- and anti-apoptotic signaling and cell death, causing fibrosis.[106] Another theory implicates plakoglobin, a member of the β-catenin family, in promoting adipogenesis and fibrosis by inhibiting the Wnt signaling pathway.[107]

LEFT VENTRICULAR NON-COMPACTION CARDIOMYOPATHY

The first case of left ventricular non-compaction (LVNC) was described in 1997, and has since become more widely recognized as a clinical entity. LVNC is a rare form of cardiomyopathy with a prevalence of less than 0.14% in adults referred for echocardiography (Fig. 6.6).[108] LVNC can be associated with heart failure in the pediatric population and with conduction abnormalities such as Wolff-Parkinson-White syndrome and facial dysmorphisms.[109] Although still not recognized by the European Society of Cardiology, the World Health Organization has considered LVNC a primary cardiomyopathy, with a distinct phenotype from congenital abnormalities that occur in the presence or absence of neuromuscular disorders.[110]

While asymptomatic LVNC has a protracted and indolent course over several years with an approximately 2% incidence of heart failure, symptomatic disease renders a rapid clinical decline and high mortality with over 60% of patients developing heart failure.[109] The first reported case of LVNC was associated with congenital heart disease with outflow tract obstruction, cyanotic congenital malformations, and anomalous coronary anatomy.[111] However, subsequent cases suggested a wide range of clinical manifestations, including arrhythmias and thromboembolism.[112] Implantable defibrillator therapy may be considered in symptomatic patients.

The diagnosis of LVNC is often made by two-dimensional echocardiography, although magnetic resonance imaging is becoming a more common technique. Several definitions are used to describe the morphology of LVNC and these criteria rely heavily on the ratios of compacted and non-compacted myocardium at end-diastole.[113] The current echocardiographic criteria for diagnosing LVNC are as follows:[113,114]

- Absence of valvular abnormalities
- Thickened ventricular myocardium with a visibly distinct thin epicardial and a thick endocardial layer with trabeculations and recesses. End systolic ratio of non-compact inner layer to compact outer layer >2:1.
- Localization of trabeculations at the apical/lateral wall of the left ventricle with hypokinesis of non-compacted segments
- Color flow Doppler indicating flow between the trabeculations.

LVNC has an autosomal dominant pattern of inheritance in over half of the cases. An X-linked mode of inheritance is also recognized but is rare. Barth syndrome is an X-linked infantile disorder of lipid metabolism

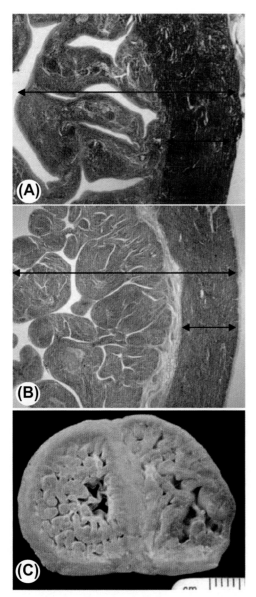

FIGURE 6.6 Characteristic pattern of LV non-compaction cardiomyopathy. Microscopically, on hematoxylin-&-eosin-stained sections, the endocardium is relatively normal, but has alternating trabeculations with deep irregular invaginations forming staghorn-like recesses that line the endocardial spaces (A). A fibrous band separating the spongy and compact portion of the myocardium is visible on trichrome stain (B). Replacement of papillary muscles with interlacing muscle bundles that form anastomosing trabeculae give a spongy characteristic appearance to the heart grossly (C). (Reproduced with permission from Burke A, Mont E, Kutys R, Virmani R. Left ventricular noncompaction: a pathologic study of 14 cases. *Human Pathol* 2005;36:403–11).

that is characterized by dilated cardiomyopathy, skeletal muscle weakness, growth retardation, and aciduria. Atypical forms of Barth syndrome with LVNC have also been reported. The syndrome is due to mutations in the taffazin gene (TAZ/G4.5) on chromosome X (locus Xq28).[115] Mutations in the taffazin gene result in decreased levels of the integral mitochondrial membrane protein cardiolipin.

There has been increasing evidence for overlap in the genes related to LVNC and other cardiomyopathies. For example, a missense mutation in the cardiac troponin T gene (TNNT2) has been described in a family with LVNC;[116] however, transgenic mice with mutated TNNT2 allele have molecular and histologic signs of heart failure, but LVNC was not observed. Similarly, mutations in the α-dystrobrevin gene (DTNA), lamin A/C (LMNA),[117] and ZASP,[118] have also been described in patients with LVNC.

In summary, LVNC is a unique form of cardiomyopathy due to mutations in multiple cytoskeletal, sarcomeric, and mitochondrial proteins; however, the exact mechanisms that lead to pathological alterations underlying LVNC still remain to be elucidated.

ROLE OF GENETIC TESTING AND FUTURE DIRECTIONS

Current practice guidelines recommend that when an inherited cardiomyopathy is suspected, an initial step should include a detailed family history that encompasses at least three generations. All first-degree relatives of patients with a cardiomyopathy should be screened with a combination of a detailed history and physical examination, electrocardiogram, and echocardiogram. Affected patients should be counseled on the potential for disease transmission and genetic testing should be considered for the index case in a family. If a genetic mutation is identified in the proband, targeted testing or cascade screening for that specific mutation can then be performed on other family members to identify carriers. The identification of carriers in a family can enable more careful clinical monitoring and potentially the initiation of medical therapies in the early stages of disease.

Given the genetic heterogeneity of the cardiomyopathies, clinical genetic testing has traditionally been costly and often incomplete. Conventional sequencing technology resulted in the testing of a limited panel of genes with a variable yield. With the advent of next-generation sequencing technologies it is now possible to test a more comprehensive set of genes at a reduced cost. Increasingly the field is moving from specific gene panels to more extensive approaches that include whole-exome and whole-genome sequencing. While the comprehensive nature of exome or genome sequencing is appealing, such an approach is limited by the enormous number of variants that can be detected and the subsequent difficulty in identifying a single pathogenic variant.

Within the last 20 years, cardiovascular genetics has revolutionized our understanding of the etiology for a wide range of cardiomyopathies. Despite this rapid progress, many challenges remain. Ongoing work will help to clarify the full complement of genes and biological pathways that can lead ultimately to a cardiomyopathy. We continue to need a greater understanding of the relation between genotype, mechanism, and outcomes. And perhaps most importantly, we will need continued work to translate these genetic findings into novel therapeutic approaches that in turn lead to improved patient care.

References

1. Harvey PA, Leinwand LA. The cell biology of disease: cellular mechanisms of cardiomyopathy. *J Cell Biol* 2011;**194**(3):355–65.
2. McKenna WJ, Beiras AC, Lado MP. The cardiomyopathies. *Br Heart J* 1994;**72**(6 Suppl): S1.
3. Maron BJ, Towbin JA, Thiene G, Antzelevitch C, Corrado D, Arnett D, et al. Contemporary definitions and classification of the cardiomyopathies: an American Heart Association Scientific Statement from the Council on Clinical Cardiology, Heart Failure and Transplantation Committee; Quality of Care and Outcomes Research and Functional Genomics and Translational Biology Interdisciplinary Working Groups; and Council on Epidemiology and Prevention. *Circulation* 2006;**113**(14):1807–16.
4. Wang L, Seidman JG, Seidman CE. Narrative review: harnessing molecular genetics for the diagnosis and management of hypertrophic cardiomyopathy. *Ann Intern Med* 2010;**152**(8):513–20, W181.
5. Dellefave L, McNally EM. The genetics of dilated cardiomyopathy. *Curr Opin Cardiol* 2010;**25**(3):198–204.
6. Roger VL, Go AS, Lloyd-Jones DM, Benjamin EJ, Berry JD, Borden WB, et al. Heart disease and stroke statistics–2012 update: a report from the American Heart Association. *Circulation* 2012;**125**(1): e2–20.
7. Mill JG, Stefanon I, dos Santos L, Baldo MP. Remodeling in the ischemic heart: the stepwise progression for heart failure. *Braz J Med Biol Res* 2011;**44**(9):890–8.
8. Kalogeris T, Baines CP, Krenz M, Korthuis RJ. Cell biology of ischemia/reperfusion injury. *Int Rev Cell Mol Biol* 2012;**298**:229–317.
9. dos Santos L, Santos AA, Goncalves GA, Krieger JE, Tucci PJ. Bone marrow cell therapy prevents infarct expansion and improves border zone remodeling after coronary occlusion in rats. *Int J Cardiol* 2010;**145**(1):34–9.
10. Palojoki E, Saraste A, Eriksson A, Pulkki K, Kallajoki M, Voipio-Pulkki LM, et al. Cardiomyocyte apoptosis and ventricular remodeling after myocardial infarction in rats. *Am J Physiol Heart Circ Physiol* 2001;**280**(6):H2726–31.
11. Gottlieb RA. Mitochondrial signaling in apoptosis: mitochondrial daggers to the breaking heart. *Basic Res Cardiol* 2003;**98**(4):242–9.
12. Nakagawa T, Shimizu S, Watanabe T, Yamaguchi O, Otsu K, Yamagata H, et al. Cyclophilin D-dependent mitochondrial permeability transition regulates some necrotic but not apoptotic cell death. *Nature* 2005;**434**(7033):652–8.
13. Susin SA, Daugas E, Ravagnan L, Samejima K, Zamzami N, Loeffler M, et al. Two distinct pathways leading to nuclear apoptosis. *J Exp Med* 2000;**192**(4):571–80.
14. Resende MM, Mill JG. Alternate angiotensin II-forming pathways and their importance in physiological or physiopathological conditions. *Arq Bras Cardiol* 2002;**78**(4):425–38.
15. Dean RG, Balding LC, Candido R, Burns WC, Cao Z, Twigg SM, et al. Connective tissue growth factor and cardiac fibrosis after myocardial infarction. *J Histochem Cytochem* 2005;**53**(10): 1245–56.
16. Spruill LS, Lowry AS, Stroud RE, Squires CE, Mains IM, Flack EC, et al. Membrane-type-1 matrix metalloproteinase transcription and translation in myocardial fibroblasts from patients with normal left ventricular function and from patients with cardiomyopathy. *Am J Physiol Cell Physiol* 2007;**293**(4):C1362–73.
17. Prunier F, Kawase Y, Gianni D, Scapin C, Danik SB, Ellinor PT, et al. Prevention of ventricular arrhythmias with sarcoplasmic reticulum Ca2+ ATPase pump overexpression in a porcine model of ischemia reperfusion. *Circulation* 2008;**118**(6):614–24.
18. Petretta M, Pirozzi F, Sasso L, Paglia A, Bonaduce D. Review and metaanalysis of the frequency of familial dilated cardiomyopathy. *Am J Cardiol* 2011;**108**(8):1171–6.
19. McNally EM, Golbus JR, Puckelwartz MJ. Genetic mutations and mechanisms in dilated cardiomyopathy. *J Clin Invest* 2013;**123**(1):19–26.
20. Kaspar RW, Allen HD, Ray WC, Alvarez CE, Kissel JT, Pestronk A, et al. Analysis of dystrophin deletion mutations predicts age of cardiomyopathy onset in Becker muscular dystrophy. *Circ Cardiovasc genet* 2009;**2**(6):544–51.
21. Herman DS, Lam L, Taylor MR, Wang L, Teekakirikul P, Christodoulou D, et al. Truncations of titin causing dilated cardiomyopathy. *N Engl J Med* 2012;**366**(7):619–28.
22. Olson TM, Kishimoto NY, Whitby FG, Michels VV. Mutations that alter the surface charge of alpha-tropomyosin are associated with dilated cardiomyopathy. *J Mol Cell Cardiol* 2001;**33**(4):723–32.
23. Villard E, Duboscq-Bidot L, Charron P, Benaiche A, Conraads V, Sylvius N, et al. Mutation screening in dilated cardiomyopathy: prominent role of the beta myosin heavy chain gene. *Eur Heart J* 2005;**26**(8):794–803.
24. Carballo S, Robinson P, Otway R, Fatkin D, Jongbloed JD, de Jonge N, et al. Identification and functional characterization of cardiac troponin I as a novel disease gene in autosomal dominant dilated cardiomyopathy. *Circ Res* 2009;**105**(4):375–82.
25. Hanson EL, Jakobs PM, Keegan H, Coates K, Bousman S, Dienel NH, et al. Cardiac troponin T lysine 210 deletion in a family with dilated cardiomyopathy. *J Card Fail* 2002;**8**(1):28–32.
26. Mogensen J, Murphy RT, Shaw T, Bahl A, Redwood C, Watkins H, et al. Severe disease expression of cardiac troponin C and T mutations in patients with idiopathic dilated cardiomyopathy. *J Am Coll Cardiol* 2004;**44**(10):2033–40.
27. Hershberger RE, Parks SB, Kushner JD, Li D, Ludwigsen S, Jakobs P, et al. Coding sequence mutations identified in MYH7, TNNT2, SCN5A, CSRP3, LBD3, and TCAP from 313 patients with familial or idiopathic dilated cardiomyopathy. *Clin Transl Sci* 2008;**1**(1):21–6.
28. Wolf CM, Wang L, Alcalai R, Pizard A, Burgon PG, Ahmad F, et al. Lamin A/C haploinsufficiency causes dilated cardiomyopathy and apoptosis-triggered cardiac conduction system disease. *J Mol Cell Cardiol* 2008;**44**(2):293–303.
29. Arbustini E, Pilotto A, Repetto A, Grasso M, Negri A, Diegoli M, et al. Autosomal dominant dilated cardiomyopathy with atrioventricular block: a lamin A/C defect-related disease. *J Am Coll Cardiol* 2002;**39**(6):981–90.
30. van Rijsingen IA, Arbustini E, Elliott PM, Mogensen J, Hermans-van Ast JF, van der Kooi AJ, et al. Risk factors for malignant ventricular arrhythmias in lamin a/c mutation carriers a European cohort study. *J Am Coll Cardiol* 2012;**59**(5):493–500.
31. Schofield CJ, Jannin J, Salvatella R. The future of Chagas disease control. *Trends Parasitol* 2006;**22**(12):583–8.
32. Bilate AM, Cunha-Neto E. Chagas disease cardiomyopathy: current concepts of an old disease. *Rev Inst Med Trop Sao Paulo* 2008;**50**(2):67–74.
33. Parada H, Carrasco HA, Anez N, Fuenmayor C, Inglessis I. Cardiac involvement is a constant finding in acute Chagas' disease: a clinical, parasitological and histopathological study. *Int J Cardiol* 1997;**60**(1):49–54.
34. Henao-Martinez AF, Schwartz DA, Yang IV. Chagasic cardiomyopathy, from acute to chronic: is this mediated by host susceptibility factors? *Trans R Soc Trop Med Hyg* 2012;**106**(9):521–7.
35. Machado FS, Jelicks LA, Kirchhoff LV, Shirani J, Nagajyothi F, Mukherjee S, et al. Chagas heart disease: report on recent developments. *Cardiol Rev* 2012;**20**(2):53–65.

36. Bonney KM, Engman DM. Chagas heart disease pathogenesis: one mechanism or many? *Curr Mol Med* 2008;**8**(6):510–8.

37. Cunha-Neto E, Bilate AM, Hyland KV, Fonseca SG, Kalil J, Engman DM. Induction of cardiac autoimmunity in Chagas heart disease: a case for molecular mimicry. *Autoimmunity* 2006;**39**(1):41–54.

38. Kierszenbaum F. Where do we stand on the autoimmunity hypothesis of Chagas disease? *Trends Parasitol* 2005;**21**(11):513–6.

39. Fonseca SG, Reis MM, Coelho V, Nogueira LG, Monteiro SM, Mairena EC, et al. Locally produced survival cytokines IL-15 and IL-7 may be associated to the predominance of CD8+ T cells at heart lesions of human chronic Chagas disease cardiomyopathy. *Scand J Immunol* 2007;**66**(2-3):362–71.

40. Cunha-Neto E, Teixeira PC, Nogueira LG, Mady C, Lanni B, Stolf N, et al. [New concepts on the pathogenesis of chronic Chagas cardiomyopathy: myocardial gene and protein expression profiles]. *Rev Soc Bras Med Trop* 2006;**39**(Suppl. 3):59–62.

41. Talvani A, Rocha MO, Ribeiro AL, Correa-Oliveira R, Teixeira MM. Chemokine receptor expression on the surface of peripheral blood mononuclear cells in Chagas disease. *J Infect Dis* 2004;**189**(2):214–20.

42. Liao Y, McGee DL, Cao G, Cooper RS. Alcohol intake and mortality: findings from the National Health Interview Surveys (1988 and 1990). *Am J Epidemiol* 2000;**151**(7):651–9.

43. Nova E, Baccan GC, Veses A, Zapatera B, Marcos A. Potential health benefits of moderate alcohol consumption: current perspectives in research. *Proc Nutr Soc* 2012;**71**(2):307–15.

44. Walsh CR, Larson MG, Evans JC, Djousse L, Ellison RC, Vasan RS, et al. Alcohol consumption and risk for congestive heart failure in the Framingham Heart Study. *Ann Intern Med* 2002;**136**(3):181–91.

45. Jones A, McMillan MR, Jones RW, Kowalik GT, Steeden JA, Pruessner JC, et al. *Habitual alcohol consumption is associated with lower cardiovascular stress responses - a novel explanation for the known cardiovascular benefits of alcohol? Stress*; 2013.

46. Thun MJ, Peto R, Lopez AD, Monaco JH, Henley SJ, Heath Jr CW, et al. Alcohol consumption and mortality among middle-aged and elderly U.S. adults. *N Engl J Med* 1997;**337**(24):1705–14.

47. Bos S, Grobbee DE, Boer JM, Verschuren WM, Beulens JW. Alcohol consumption and risk of cardiovascular disease among hypertensive women. *Eur J Cardiovasc Prev Rehabil* 2010;**17**(1):119–26.

48. Costanzo S, Di Castelnuovo A, Donati MB, Iacoviello L, de Gaetano G. Alcohol consumption and mortality in patients with cardiovascular disease: a meta-analysis. *J Am Coll Cardiol* 2010;**55**(13):1339–47.

49. Kloner RA, Rezkalla SH. To drink or not to drink? That is the question. *Circulation* 2007;**116**(11):1306–17.

50. George A, Figueredo VM. Alcoholic cardiomyopathy: a review. *J Card Fail* 2011;**17**(10):844–9.

51. Fauchier L, Babuty D, Poret P, Casset-Senon D, Autret ML, Cosnay P, et al. Comparison of long-term outcome of alcoholic and idiopathic dilated cardiomyopathy. *Eur Heart J* 2000;**21**(4):306–14.

52. Iacovoni A, De Maria R, Gavazzi A. Alcoholic cardiomyopathy. *J Cardiovasc Med (Hagerstown)* 2010;**11**(12):884–92.

53. Chen DB, Wang L, Wang PH. Insulin-like growth factor I retards apoptotic signaling induced by ethanol in cardiomyocytes. *Life Sci* 2000;**67**(14):1683–93.

54. Figueredo VM, Chang KC, Baker AJ, Camacho SA. Chronic alcohol-induced changes in cardiac contractility are not due to changes in the cytosolic Ca2+ transient. *Am J Physiol* 1998;**275**(1 Pt 2):H122–30.

55. Piano MR. Alcohol and heart failure. *J Card Fail* 2002;**8**(4):239–46.

56. Nicolas JM, Fernandez-Sola J, Estruch R, Pare JC, Sacanella E, Urbano-Marquez A, et al. The effect of controlled drinking in alcoholic cardiomyopathy. *Ann Intern Med* 2002;**136**(3):192–200.

57. Bhattacharyya A, Basra SS, Sen P, Kar B. Peripartum cardiomyopathy: a review. *Tex Heart Inst J* 2012;**39**(1):8–16.

58. Elkayam U, Akhter MW, Singh H, Khan S, Bitar F, Hameed A, et al. Pregnancy-associated cardiomyopathy: clinical characteristics and a comparison between early and late presentation. *Circulation* 2005;**111**(16):2050–5.

59. Sliwa K, Fett J, Elkayam U. Peripartum cardiomyopathy. *Lancet* 2006;**368**(9536):687–93.

60. Hilfiker-Kleiner D, Kaminski K, Podewski E, Bonda T, Schaefer A, Sliwa K, et al. A cathepsin D-cleaved 16 kDa form of prolactin mediates postpartum cardiomyopathy. *Cell Calcium* 2007;**128**(3):589–600.

61. Patten IS, Rana S, Shahul S, Rowe GC, Jang C, Liu L, et al. Cardiac angiogenic imbalance leads to peripartum cardiomyopathy. *Nature* 2012;**485**(7398):333–8.

62. Ashrafian H, McKenna WJ, Watkins H. Disease pathways and novel therapeutic targets in hypertrophic cardiomyopathy. *Circ Res* 2011;**109**(1):86–96.

63. Ho CY. Genetics and clinical destiny: improving care in hypertrophic cardiomyopathy. *Circulation* 2010;**122**(23):2430–40; discussion 40.

64. Ho CY. Hypertrophic cardiomyopathy in 2012. *Circulation* 2012;**125**(11):1432–8.

65. Semsarian C, Group CCGDCW. Guidelines for the diagnosis and management of hypertrophic cardiomyopathy. *Heart Lung Circ* 2011;**20**(11):688–90.

66. Hughes SE. The pathology of hypertrophic cardiomyopathy. *Histopathology* 2004;**44**(5):412–27.

67. Marian AJ. Hypertrophic cardiomyopathy: from genetics to treatment. *Eur J Clin Invest* 2010;**40**(4):360–9.

68. Frey N, Luedde M, Katus HA. Mechanisms of disease: hypertrophic cardiomyopathy. *Nat Rev Cardiol* 2012;**9**(2):91–100.

69. Kushwaha SS, Fallon JT, Fuster V. Restrictive cardiomyopathy. *N Engl J Med* 1997;**336**(4):267–76.

70. Mogensen J, Kubo T, Duque M, Uribe W, Shaw A, Murphy R, et al. Idiopathic restrictive cardiomyopathy is part of the clinical expression of cardiac troponin I mutations. *J Clin Invest* 2003;**111**(2):209–16.

71. Nihoyannopoulos P, Dawson D. Restrictive cardiomyopathies. *Eur J Echocardiogr* 2009;**10**(8):iii23–33.

72. Merlini G, Seldin DC, Gertz MA. Amyloidosis: pathogenesis and new therapeutic options. *J Clin Oncol* 2011;**29**(14):1924–33.

73. Ruberg FL, Berk JL. Transthyretin (TTR) cardiac amyloidosis. *Circulation* 2012;**126**(10):1286–300.

74. Sipe JD, Benson MD, Buxbaum JN, Ikeda S, Merlini G, Saraiva MJ, et al. Amyloid fibril protein nomenclature: 2010 recommendations from the nomenclature committee of the International Society of Amyloidosis. *Amyloid* 2010;**17**(3-4):101–4.

75. Vrana JA, Gamez JD, Madden BJ, Theis JD, Bergen 3rd HR, Dogan A. Classification of amyloidosis by laser microdissection and mass spectrometry-based proteomic analysis in clinical biopsy specimens. *Blood* 2009;**114**(24):4957–9.

76. Meier-Ewert HK, Sanchorawala V, Berk JL, Ruberg FL. Cardiac amyloidosis: evolving approach to diagnosis and management. *Curr Treat Options Cardiovasc Med* 2011;**13**(6):528–42.

77. Falk RH, Dubrey SW. Amyloid heart disease. *Prog Cardiovasc Dis* 2010;**52**(4):347–61.

78. Dubrey SW, Hawkins PN, Falk RH. Amyloid diseases of the heart: assessment, diagnosis, and referral. *Heart* 2011;**97**(1):75–84.

79. Brenner DA, Jain M, Pimentel DR, Wang B, Connors LH, Skinner M, et al. Human amyloidogenic light chains directly impair cardiomyocyte function through an increase in cellular oxidant stress. *Circulation* 2004;**94**(8):1008–10.

80. Morrissey RP, Philip KJ, Schwarz ER. Cardiac abnormalities in Anderson-Fabry disease and Fabry's cardiomyopathy. *Cardiovasc J Africa* 2011;**22**(1):38–44.

81. Zarate YA, Hopkin RJ. Fabry's disease. *Lancet* 2008;**372**(9647):1427–35.

82. Spada M, Pagliardini S, Yasuda M, Tukel T, Thiagarajan G, Sakuraba H, et al. High incidence of later-onset fabry disease revealed by newborn screening. *Am J Hum Genet* 2006;**79**(1):31–40.

83. Wang RY, Lelis A, Mirocha J, Wilcox WR. Heterozygous Fabry women are not just carriers, but have a significant burden of disease and impaired quality of life. *Genet Med* 2007;**9**(1):34–45.

84. Schiffmann R, Warnock DG, Banikazemi M, Bultas J, Linthorst GE, Packman S, et al. Fabry disease: progression of nephropathy, and prevalence of cardiac and cerebrovascular events before enzyme replacement therapy. *Nephrol Dial Transplant* 2009;**24**(7):2102–11.

85. Linhart A, Lubanda JC, Palecek T, Bultas J, Karetova D, Ledvinova J, et al. Cardiac manifestations in Fabry disease. *J Inherit Metab Dis* 2001;**24**(Suppl. 2):75–83; discussion 65.

86. Sadick N, Thomas L. Cardiovascular manifestations in Fabry disease: a clinical and echocardiographic study. *Heart Lung Circ* 2007;**16**(3):200–6.

87. Adams PC, Barton JC. Haemochromatosis. *Lancet* 2007;**370**(9602): 1855–60.

88. Roetto A, Daraio F, Alberti F, Porporato P, Cali A, De Gobbi M, et al. Hemochromatosis due to mutations in transferrin receptor 2. *Blood Cells Mol Dis* 2002;**29**(3):465–70.

89. Uhl HS. A previously undescribed congenital malformation of the heart: almost total absence of the myocardium of the right ventricle. *Bull Johns Hopkins Hosp* 1952;**91**(3):197–209.

90. Corrado D, Thiene G. Arrhythmogenic right ventricular cardiomyopathy/dysplasia: clinical impact of molecular genetic studies. *Circulation* 2006;**113**(13):1634–7.

91. Marcus FI, Abidov A. Arrhythmogenic right ventricular cardiomyopathy 2012: diagnostic challenges and treatment. *J Cardiovasc Electrophysiol* 2012;**23**(10):1149–53.

92. Fei L, Keeling PJ, Gill JS, Bashir Y, Statters DJ, Poloniecki J, et al. Heart rate variability and its relation to ventricular arrhythmias in congestive heart failure. *Br Heart J* 1994;**71**(4):322–8.

93. Sen-Chowdhry S, Morgan RD, Chambers JC, McKenna WJ. Arrhythmogenic cardiomyopathy: etiology, diagnosis, and treatment. *Annu Rev Med* 2010;**61**:233–53.

94. Marcus FI, McKenna WJ, Sherrill D, Basso C, Bauce B, Bluemke DA, et al. Diagnosis of arrhythmogenic right ventricular cardiomyopathy/dysplasia: proposed modification of the Task Force Criteria. *Eur Heart J* 2010;**31**(7):806–14.

95. Murray B. Arrhythmogenic right ventricular dysplasia/cardiomyopathy (ARVD/C): a review of molecular and clinical literature. *J Genet Couns* 2012;**21**(4):494–504.

96. Protonotarios N, Tsatsopoulou A, Anastasakis A, Sevdalis E, McKoy G, Stratos K, et al. Genotype-phenotype assessment in autosomal recessive arrhythmogenic right ventricular cardiomyopathy (Naxos disease) caused by a deletion in plakoglobin. *J Am Coll Cardiol* 2001;**38**(5):1477–84.

97. Coonar AS, Protonotarios N, Tsatsopoulou A, Needham EW, Houlston RS, Cliff S, et al. Gene for arrhythmogenic right ventricular cardiomyopathy with diffuse nonepidermolytic palmoplantar keratoderma and woolly hair (Naxos disease) maps to 17q21. *Circulation* 1998;**97**(20):2049–58.

98. Gerull B, Heuser A, Wichter T, Paul M, Basson CT, McDermott DA, et al. Mutations in the desmosomal protein plakophilin-2 are common in arrhythmogenic right ventricular cardiomyopathy. *Nat Genet* 2004;**36**(11):1162–4.

99. Protonotarios N, Tsatsopoulou A, Fontaine G. Naxos disease: keratoderma, scalp modifications, and cardiomyopathy. *J Am Acad Dermatol* 2001;**44**(2):309–11.

100. McKoy G, Protonotarios N, Crosby A, Tsatsopoulou A, Anastasakis A, Coonar A, et al. Identification of a deletion in plakoglobin in arrhythmogenic right ventricular cardiomyopathy with palmoplantar keratoderma and woolly hair (Naxos disease). *Lancet* 2000;**355**(9221):2119–24.

101. Carvajal-Huerta L. Epidermolytic palmoplantar keratoderma with woolly hair and dilated cardiomyopathy. *J Am Acad Dermatol* 1998;**39**(3):418–21.

102. Rao BH, Reddy IS, Chandra KS. Familial occurrence of a rare combination of dilated cardiomyopathy with palmoplantar keratoderma and curly hair. *Indian Heart J* 1996;**48**(2):161–2.

103. Syrris P, Ward D, Asimaki A, Evans A, Sen-Chowdhry S, Hughes SE, et al. Desmoglein-2 mutations in arrhythmogenic right ventricular cardiomyopathy: a genotype-phenotype characterization of familial disease. *Eur Heart J* 2007;**28**(5):581–8.

104. Beffagna G, Occhi G, Nava A, Vitiello L, Ditadi A, Basso C, et al. Regulatory mutations in transforming growth factor-beta3 gene cause arrhythmogenic right ventricular cardiomyopathy type 1. *Cardiovasc Res* 2005;**65**(2):366–73.

105. Tiso N, Stephan DA, Nava A, Bagattin A, Devaney JM, Stanchi F, et al. Identification of mutations in the cardiac ryanodine receptor gene in families affected with arrhythmogenic right ventricular cardiomyopathy type 2 (ARVD2). *Hum Mol Genet* 2001;**10**(3):189–94.

106. Mallat Z, Tedgui A, Fontaliran F, Frank R, Durigon M, Fontaine G. Evidence of apoptosis in arrhythmogenic right ventricular dysplasia. *N Engl J Med* 1996;**335**(16):1190–6.

107. Simcha I, Shtutman M, Salomon D, Zhurinsky J, Sadot E, Geiger B, et al. Differential nuclear translocation and transactivation potential of beta-catenin and plakoglobin. *J Cell Biol* 1998;**141**(6):1433–48.

108. Aras D, Tufekcioglu O, Ergun K, Ozeke O, Yildiz A, Topaloglu S, et al. Clinical features of isolated ventricular noncompaction in adults long-term clinical course, echocardiographic properties, and predictors of left ventricular failure. *J Cardiac Fail* 2006;**12**(9):726–33.

109. Malla R, Sharma R, Rauniyar B, Kc MB, Maskey A, Joshi D, et al. Left ventricular noncompaction. JNMA. *J Nepal Med Assoc* 2009;**48**(174):180–4.

110. Paterick TE, Tajik AJ. Left ventricular noncompaction: a diagnostically challenging cardiomyopathy. *Circ J* 2012;**76**(7):1556–62.

111. Chin TK, Perloff JK, Williams RG, Jue K, Mohrmann R. Isolated noncompaction of left ventricular myocardium. A study of eight cases. *Circulation* 1990;**82**(2):507–13.

112. Lofiego C, Biagini E, Pasquale F, Ferlito M, Rocchi G, Perugini E, et al. Wide spectrum of presentation and variable outcomes of isolated left ventricular non-compaction. *Heart* 2007;**93**(1):65–71.

113. Frischknecht BS, Attenhofer Jost CH, Oechslin EN, Seifert B, Hoigne P, Roos M, et al. Validation of noncompaction criteria in dilated cardiomyopathy, and valvular and hypertensive heart disease. *J Am Soc Echocardiogr* 2005;**18**(8):865–72.

114. Jenni R, Oechslin E, Schneider J, Attenhofer Jost C, Kaufmann PA. Echocardiographic and pathoanatomical characteristics of isolated left ventricular non-compaction: a step towards classification as a distinct cardiomyopathy. *Heart* 2001;**86**(6):666–71.

115. Ichida F, Tsubata S, Bowles KR, Haneda N, Uese K, Miyawaki T, et al. Novel gene mutations in patients with left ventricular noncompaction or Barth syndrome. *Circulation* 2001;**103**(9):1256–63.

116. Klaassen S, Probst S, Oechslin E, Gerull B, Krings G, Schuler P, et al. Mutations in sarcomere protein genes in left ventricular noncompaction. *Circulation* 2008;**117**(22):2893–901.

117. Xing Y, Ichida F, Matsuoka T, Isobe T, Ikemoto Y, Higaki T, et al. Genetic analysis in patients with left ventricular noncompaction and evidence for genetic heterogeneity. *Mol Genet Metab* 2006;**88**(1):71–7.

118. Hermida-Prieto M, Monserrat L, Castro-Beiras A, Laredo R, Soler R, Peteiro J, et al. Familial dilated cardiomyopathy and isolated left ventricular noncompaction associated with lamin A/C gene mutations. *Am J Cardiol* 2004;**94**(1):50–4.

119. Itoh-Satoh M, Hayashi T, Nishi H, Koga Y, Arimura T, Koyanagi T, et al. Titin mutations as the molecular basis for dilated cardiomyopathy. *Biochem Biophys Res Commun* 2002;**291**(2):385–93.

120. Park KY, Dalakas MC, Goebel HH, Ferrans VJ, Semino-Mora C, Litvak S, et al. Desmin splice variants causing cardiac and skeletal myopathy. *J Med Genet* 2000;**37**(11):851–7.

121. Kamisago M, Sharma SD, DePalma SR, Solomon S, Sharma P, McDonough B, et al. Mutations in sarcomere protein genes as a cause of dilated cardiomyopathy. *N Engl J Med* 2000;**343**(23):1688–96.

122. Puckelwartz MJ, Kessler E, Zhang Y, Hodzic D, Randles KN, Morris G, et al. Disruption of nesprin-1 produces an Emery Dreifuss muscular dystrophy-like phenotype in mice. *Hum Mol Genet* 2009;**18**(4):607–20.

123. Geisterfer-Lowrance AA, Kass S, Tanigawa G, Vosberg HP, McKenna W, Seidman CE, et al. A molecular basis for familial hypertrophic cardiomyopathy: a beta cardiac myosin heavy chain gene missense mutation. *Cell* 1990;**62**(5): 999–1006.

124. Niimura H, Patton KK, McKenna WJ, Soults J, Maron BJ, Seidman JG, et al. Sarcomere protein gene mutations in hypertrophic cardiomyopathy of the elderly. *Circulation* 2002;**105**(4): 446–51.

125. Mogensen J, Murphy RT, Kubo T, Bahl A, Moon JC, Klausen IC, et al. Frequency and clinical expression of cardiac troponin I mutations in 748 consecutive families with hypertrophic cardiomyopathy. *J Am Coll Cardiol* 2004;**44**(12):2315–25.

126. Matsumoto Y, Hayashi T, Inagaki N, Takahashi M, Hiroi S, Nakamura T, et al. Functional analysis of titin/connectin N2-B mutations found in cardiomyopathy. *J Muscle Res Cell Motil* 2005;**26**(6-8):367–74.

127. Vasile VC, Ommen SR, Edwards WD, Ackerman MJ. A missense mutation in a ubiquitously expressed protein, vinculin, confers susceptibility to hypertrophic cardiomyopathy. *Biochem Biophys Res Commun* 2006;**345**(3):998–1003.

128. Huang XP, Du JF. Troponin I, cardiac diastolic dysfunction and restrictive cardiomyopathy. *Acta Pharmacol Sin* 2004;**25**(12):1569–75.

129. Dalakas MC, Park KY, Semino-Mora C, Lee HS, Sivakumar K, Goldfarb LG. Desmin myopathy, a skeletal myopathy with cardiomyopathy caused by mutations in the desmin gene. *N Engl J Med* 2000;**342**(11):770–80.

130. Awad MM, Dalal D, Tichnell C, James C, Tucker A, Abraham T, et al. Recessive arrhythmogenic right ventricular dysplasia due to novel cryptic splice mutation in PKP2. *Hum Mutat* 2006;**27**(11):1157.

131. Heuser A, Plovie ER, Ellinor PT, Grossmann KS, Shin JT, Wichter T, et al. Mutant desmocollin-2 causes arrhythmogenic right ventricular cardiomyopathy. *Am J Hum Genet* 2006;**79**(6):1081–8.

132. Chen R, Tsuji T, Ichida F, Bowles KR, Yu X, Watanabe S, et al. Mutation analysis of the G4.5 gene in patients with isolated left ventricular noncompaction. *Mol Genet Metab* 2002;**77**(4):319–25.

Cellular and Molecular Pathobiology of the Cardiac Conduction System

Thomas J. Hund, PhD[1,4], Sakima A. Smith, MD[1,3],
Michael A. Makara, BS[1,2], Peter J. Mohler, PhD[1,2,3]

[1]The Dorothy M. Davis Heart & Lung Research Institute, [2]Department of Physiology and Cell Biology,
[3]Department of Internal Medicine, The Ohio State University Wexner Medical Center; [4]Department of Biomedical
Engineering, The Ohio State University College of Engineering, Columbus, OH, USA

OVERVIEW OF THE CARDIAC CONDUCTION SYSTEM

The cardiac conduction system is a network of specialized cells responsible for the initiation and co-ordination of the heartbeat. Relative to the myocytes responsible for regulating cardiac contraction in a normal heart (~1×10^9), the cells that make up the cardiac conduction system are relatively few in number, but are essential for cardiac electrical signaling and normal physiology. The three main components of the system are the sinoatrial node (SAN), the atrioventricular node (AVN) and the His–Purkinje system (HPS). The SAN, located in the right atrium, is the primary pacemaker of the heart and thus is responsible for the normal initiation of the cardiac action potential (sinus rhythm; Figure 7.1). Cells in the conduction system (SAN, AVN, and Purkinje cells) are unique in their ability to generate an electrical impulse or action potential without an external stimulus. This property of automaticity requires a distinct ion channel profile and a finely controlled intracellular electrical coupling. With respect to SAN function, pacemaking requires a synchronized effort because of the principle of entrainment (whereby faster discharging cells are slowed by cells firing more slowly) in a highly heterogeneous complex.[1] The innervation of the sinus node consists of post-ganglionic adrenergic and cholinergic terminals.[2] Most of the efferent vagal fibers converge at the superior vena cava–aortic root fat pad in the right atria, which is also the site of the highest concentrations of norepinephrine.[3] Subsequently the SAN is triggered to discharge after catecholamines bind to sympathetic nerve terminals,

causing a positive heart-rate-dependent response via β-adrenergic receptors and the cyclic adenosine monophosphate signaling pathway. A negative chronotropic response is caused by vagal stimulation via acetylcholine binding to muscarinic receptors.[4] Internodal tracts then lead from the SAN to the AVN to continue conduction.

Following initiation in the sinoatrial node, the cardiac action potential propagates to the atrioventricular node (AVN; Figure 7.1), located at the apex of a triangle formed by the tricuspid annulus and the tendon of Todaro.[5,6] The atria and the ventricles are separated by a ring of fibrous tissue, and the only conduction pathway between the two sets of chambers is the AVN, which is located at the base of the right atrium. Conduction of the action potential through the atrioventricular node is relatively slow, consistent with its role as a functional delay between atrial and ventricular systole to allow the atria to pump blood into the ventricles before they in turn contract. Due to automaticity of AVN cells, the AVN may serve as a secondary pacemaker in case of SAN failure (e.g. due to aging or disease). In the event of atrial tachyarrhythmia (e.g. atrial fibrillation or atrial flutter), the AVN plays an important role in limiting the number of action potentials conducted to the ventricles. The main function of the AVN is transmission of the atrial impulse to the His–Purkinje system (HPS) and the ventricles to stimulate chamber contractions.

The HPS in the ventricles consists of a common bundle (the bundle of His), the left and right bundle branches (which arise from the bundle of His), and a network of terminal Purkinje fibers (which arise from the bundle branches; Figure 7.1). The function of the HPS is

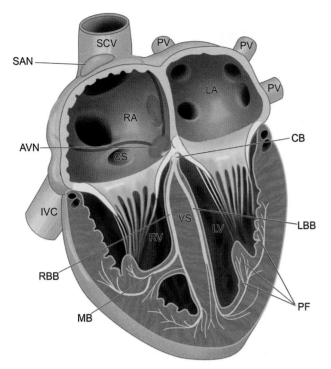

FIGURE 7.1 Organization of the human cardiac conduction system. The sinoatrial node (SAN), generally positioned near the confluence of the superior caval vein (SCV) and the right atrium (RA), is the primary cardiac pacemaker. An electrical impulse generated by the SAN is conducted across the atria to the atrioventricular node (AVN), to the bundle of His (CB), and the left and right bundle branches (LBB, RBB). Finally, cardiac Purkinje fibers (PF) transmit the cardiac action potential throughout the myocardium. CS, coronary sinus; IVC, inferior vena cava; MB, moderator band; LV, RV, left and right ventricle; LA, RA, left and right atrium; PV, pulmonary vein; VS, ventricular septum. Reprinted from MRM Jongbloed, RV Steijn, ND Hahurij, TP Kelder, MJ Schalij, AC Gittenberger-de Groot and NA Blom. Normal and abnormal development of the cardiac conduction system; implications for conduction and rhythm disorders in the child and adult. *Differentiation* 2012;84:131–148, with permission from Elsevier.

to conduct the action potential rapidly (at velocities up to four meters/second; compared to 0.3–1 meter/second in ventricle[7]) to the ventricles to ensure that the ventricular muscle contracts simultaneously.[8] The bundle of His or the penetrating portion of the AVN sends out extensions to the actual bundle branches.[9] These fibers connect with the terminal Purkinje fibers on the endocardial surface of the ventricles. Here they form multicellular bundles in longitudinal strands that transmit the atrial action potential to the ventricle to stimulate myocyte contraction.[10]

THE SINOATRIAL NODE

Sinoatrial Node Anatomy

Originally described by the pioneering work of Keith, Mackenzie, Leipzig, Aschoff, Tawara, and Flack over a century ago,[11,12] the human sinoatrial node (Figure 7.2) is

called a 'spindle'- or 'cresent'-shaped collection of excitable cells located in a subepicardial location juxtaposed with the crista terminalis (terminal crest). The cells are surrounded by a dense and fibrous tissue environment usually 10–20 mm in length and 2–3 mm in width. The position of the node within the right atria is most often located 1 mm from the surface of the epicardium, laterally in the sulcus terminalis of the right atria (near the junction of the right atrium and the superior vena cava; Figure 7.2).[13] The precise position of the SAN may vary by individual, but is usually localized in an environment that includes a nodal artery.[13]

Sinoatrial Node Cell Morphology

The pacemaker is a collection of weakly coupled, heterogeneous cells, including pacemaker cells as well as non-pacemaker cells such as atrial myocytes, adipocytes and fibroblasts. Within the node, pacemaker cells vary by size and electrophysiological properties with smaller cells showing slower intrinsic firing concentrated in the SAN center and larger, faster-firing cells in the periphery. Pacemaker cells have been successfully isolated from a wide range of species and can be separated into three main categories based on gross morphology.[14] Spindle-shaped cells are relatively small with total membrane capacitance less than 30 pF and sparse apparent myofilaments (Figure 7.3). Elongated cells are larger (membrane capacitance between 35 pF and 50 pF) especially in the longitudinal axis and have a higher myofilament density. Finally, spider cells are distinguished by their branched appearance, although the functional consequences are unclear. It is commonly believed that spindle-shaped cells are concentrated in the central SAN and serve as the leading pacemakers, while elongated cells are more likely found in the periphery. As discussed in the next section, these morphologically distinct cells possess unique electrophysiological properties that are critical for synchronized pacemaking.

Sinoatrial Node Action Potential and Electrical Components

The SAN action potential is distinct from that measured in atrial or ventricular cells (Figure 7.4).[14] Most importantly, unlike ventricular or atrial cells, the SAN action potential shows a phase 4 spontaneous depolarization phase with a maximum diastolic value around −60 mV (compared to rest potential close to −90 mV for ventricular myocytes) that eventually reaches the threshold of action potential generation without an external depolarizing stimulus. Control of this spontaneous depolarization phase is critical for pacemaking and involves the coordinated effort of multiple ion channels/transporters/exchangers.

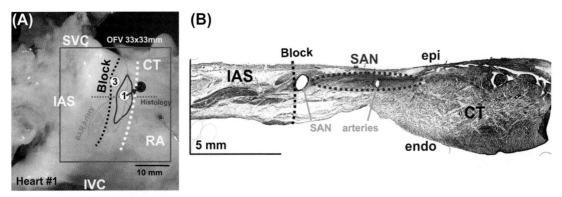

FIGURE 7.2 Organization of human sinoatrial node. Image of human right atrium depicting sinoatrial node arteries (dashed blue lines), sulcus terminalis (dashed white line), crista terminalis (CT), and the approximate border of the central sinoatrial node region (pink line). The black dashed line depicts location of the intra-atrial septum (IAS). **(B)** Transmural histological section from central sinoatrial node. Epicardium (epi), endocardium (endo). Reprinted from VV Fedorov, AV Glukhov, R Chang, G Kostecki, H Aferol, WJ Hucker, JP Wuskell, LM Loew, RB Schuessler, N Moazami, and IR Efimov. Optical mapping of the isolated coronary-perfused human sinus node. *Journal of the American College of Cardiology* 2010;56(17):1386–1394, with permission from Elsevier.

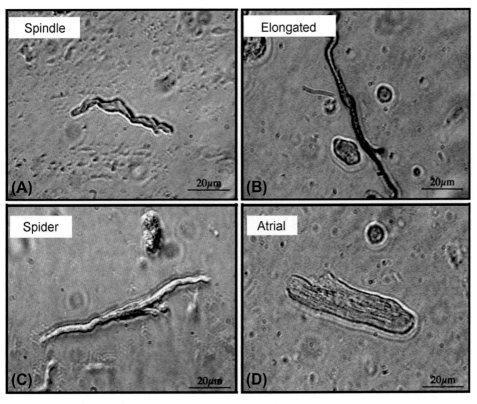

FIGURE 7.3 Sinoatrial node cell structures. The morphology of isolated sinoatrial node cells varies widely from spindle **(A)** and elongated structures **(B)** to spider-type cell **(C)**. For reference, an atrial cell is shown in **(D)**. Reprinted from ME Mangoni and J Nargeot. Properties of the hyperpolarization-activated current (I(f)) in isolated mouse sino-atrial cells. *Cardiovascular Research* 2001;52(1):51–64, with permission from Oxford University Press.

Pacemaking in automatic cells from the SAN depends on a unique electrophysiological profile that allows for generation of spontaneous action potentials. One key channel highly expressed in pacemaker cells is the f- or 'funny' channel (primarily HCN4 in the SAN), a hyperpolarization-activated channel that is permeable to both Na$^+$ and K$^+$.[14] F-channel current (I_f) is unique in that it was the first voltage- and time-dependent current that was discovered to be activated upon hyperpolarization of the membrane (rather than depolarization like most other channels). The mixed conductance of Na$^+$ and K$^+$ provides a reversal potential for the channel around −10 mV so that during diastole, the current is depolarizing. I_f is believed to be a key determinant of intrinsic excitability of SAN pacemaker cells. Importantly, these channels are heavily regulated by β-adrenergic and muscarinic receptor activation mediated in large part by direct activation by cAMP.

Expression of Na$^+$ channels is highly heterogeneous in the pacemaker with variability across species. In general,

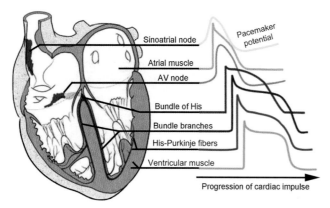

FIGURE 7.4   Action potential morphologies of the cardiac conduction system. Left, image of the rabbit heart with predicted locations of the sinoatrial node, atrioventricular node, and His-Purkinje network. Right, action potential morphologies of different cardiac cell types. Note that different cardiac regions display differences in diastolic membrane potential (more positive in SAN vs. ventricle), action potential morphology, and duration. Reprinted from O Monfredi, H. Dobrzynski, T Mondal, MR Boyett, and GM Morris. The anatomy and physiology of the sinoatrial node – A contemporary review. *Pacing and Clinical Electrophysiology*, 2010;33(11):1392–1406, with permission from John Wiley and Sons.

I_{Na} is higher in mouse than in larger animals (e.g. rabbit). Both tetrodotoxin (TTX)-sensitive neuronal-type channels and TTX-resistant cardiac isoforms have been found in adult mouse SAN cells. In rabbit, I_{Na} is found in larger cells from the SAN periphery but not in smaller, central SAN. Furthermore, I_{Na} measured in larger rabbit cells shows increased TTX sensitivity compared to that measured in ventricular myocytes, indicating a greater contribution from neuronal isoforms (e.g. $Na_v1.1$). These findings are significant for the design of Na+ channel-based therapies for cardiac conduction disease.

L-type voltage-gated Ca^{2+} channels are essential for generating the SAN action potential upstroke, which as a consequence is much slower than the upstroke recorded in ventricular or atrial myocytes. In SAN cells, Ca^{2+} current is contributed to by a variety of channels including multiple isoforms of L- and T-type channels. L-type Ca^{2+} current in SAN cells is likely the result of both $Ca_v1.2$ and $Ca_v1.3$ channels, unlike ventricular I_{Ca}, which is predominantly $Ca_v1.2$. $Ca_v1.3$ channels activate earlier (at more negative potentials) and inactivate more slowly compared to $Ca_v1.2$.[14]

An important electrophysiological feature of SAN cells is their low expression of inward rectifier K+ channels that regulate I_{K1}, a current responsible for maintaining the stable resting potential in atrial and ventricular myocytes (Figure 7.4).[14] The lack of this major repolarizing current facilitates the spontaneous diastolic depolarization phase in these cells and increases the importance of other K+ channel currents (e.g. I_{Kr}, I_{Ks}) in controlling the maximum diastolic potential. In mouse and rabbit SAN, I_{Kr} block

using E-4031 depolarizes the maximum diastolic potential, decreases the action potential amplitude, and delays repolarization leading to a decrease in SAN cell spontaneous firing rate, although to varying degrees depending on species. In contrast, larger animals such as the pig seem to depend more heavily on I_{Ks} for control of maximum diastolic potential. The transient outward K+ current (I_{to}) is also expressed in SAN and displays a heterogeneous distribution with increasing density as cells become larger (i.e. from SAN center to periphery). I_{to} likely regulates SAN repolarization but not maximum diastolic potential or firing rate.[14] In addition to I_{Kr}, I_{Ks}, and I_{to}, multiple additional K+ channel currents play important regulatory roles in the SAN. Notably, SAN cells express acetylcholine-activated K+ channels (Kir3.1/Kir3.4) that give rise to an inward rectifying current (I_{KACh}), that is activated by muscarinic and adenosine receptor agonists. Also found in SAN cells are K_{ATP} channels (I_{KATP}) that activate under conditions of low ATP (e.g. myocardial ischemia) to help decrease pacemaker activity and slow heart rate.[14]

Normal homeostasis of intracellular ions (Na+, Ca^{2+}, K+, Cl−) is tightly controlled in SAN cells through similar mechanisms as found in atrial and ventricular cells. In particular, just as in other cell types, the sarcolemmal Na+/Ca^{2+} exchanger (NCX) exploits the Na+ gradient to extrude Ca^{2+} from inside the cell (forward mode) and help maintain normal homeostasis of intracellular Ca^{2+}. Importantly, in the forward mode, NCX generates a depolarizing current, providing a potential link between intracellular Ca^{2+}/Na+ levels and membrane excitability. In particular, I_{NCX} has been proposed to be an important determinant of the diastolic depolarization rate. In contrast, the Na+/K+ ATPase extrudes three Na+ from the cell for two K+ generating a net repolarizing current that is thought to help set the maximum diastolic potential.[14] Finally, while diastolic depolarization rate (DDR) is heavily influenced by activity of I_f, recent studies demonstrate that intracellular Ca^{2+} cycling is yet an additional important pathway for controlling DDR and thus heart rate.[8] Specifically, local Ca^{2+} release from sarcoplasmic reticulum (SR) ryanodine receptor Ca^{2+} channels has been measured during diastole and is believed to facilitate spontaneous depolarization through forward-mode Na/Ca^{2+} exchange, which generates a depolarizing current that can affect the rate of spontaneous diastolic depolarization.

Catecholamine-dependent increases in heart rate (positive chronotropy) depend on activation of beta-adrenergic receptors and subsequent activation of a myriad of sarcolemmal and SR ion channels, exchangers and transporters. Increases in cAMP associated with beta-adrenergic receptor activation directly modulate the activities of several channels, including I_f, while downstream activation of protein kinase A leads to enhanced phosphorylation and activity of L-type Ca^{2+} channels, SR uptake, and release.[8,13]

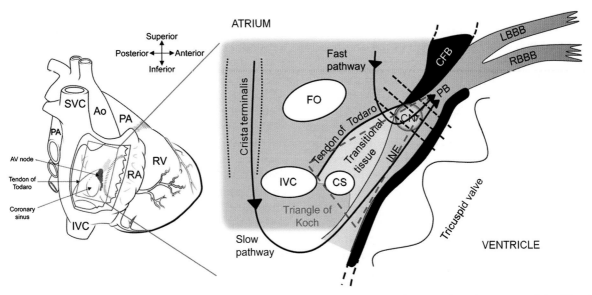

FIGURE 7.5 Structural view of the atrioventricular node. Organization of the human heart depicting the right atrium and the AV node (red). The image on the right represents a magnified image of the region of interest on the left. Ao, aorta; CN, compact node; CS, coronary sinus; FO, fossa ovalis; INE, inferior nodal extension; IVC, inferior vena cava; LBBB, left bundle branch block; PA, pulmonary artery; PB, penetrating bundle; RA, right atrium; RBBB, right bundle branch block; RV, right ventricle; SVC, superior vena cava; CFB, central fibrous body. Reprinted from IP Temple, S Inada, H Dobrinski, and MR Boyett. Connexins and the atrioventricular node. *Heart Rhythm* 2013;10:297–301, with permission from Elsevier.

Heart Rate Regulation by CaMKII

In addition to protein kinase A (PKA),[15] activity of Ca^{2+}/calmodulin-dependent kinase II (CaMKII) has been shown to regulate SAN cell excitability and pacemaking.[16–20] In particular, CaMKII appears to be critical for mediating acute 'fight-or-flight' changes in SAN cell firing, pacemaking, and heart rate in response to beta-adrenergic stimulation. Acute CaMKII activation increases SAN excitability in large part by increasing SR Ca^{2+} uptake and release, which in turn promotes forward-mode (depolarizing current) Na^+/Ca^{2+} exchanger activity that increases the rate of diastolic depolarization. In contrast, chronic CaMKII activity (for example under heart failure conditions) has a deleterious effect on SAN pacemaking and heart rate likely by increasing SAN cell death and disrupting the source–sink relationship between SAN and surrounding atrial myocardium. As CaMKII has multiple functions in the myocyte, it will be critical in the future to understand the relationship between the kinase and specific intracellular targets.

Sinoatrial Node Conduction

The degree of electrical coupling within the node and to the surrounding atrial myocardium is a critical determinant of pacemaker function. In general, SAN cells are poorly coupled to maintain the critical source–sink balance necessary for a relatively small mass of automatic cells (SAN) to excite the encompassing atrial mass. Similar to other electrophysiological properties, coupling of SAN cells is heterogeneous throughout the pacemaker, and is determined largely by variable expression of connexin isoforms important for forming gap junctions between adjacent cells. The high conductance Cx43 channel is found only in peripheral pacemaker cells, while central SAN cells express more of the low-conductance Cx45 channel, resulting in a functional transition in coupling and conduction velocity moving from the SAN center out to the surrounding atrial myocardium.

THE ATRIOVENTRICULAR NODE

Rapid impulse propagation from the atria to the ventricles is critical to produce the highly synchronized sequence of atrial and ventricular systole necessary for proper cardiac function. Action potentials generated in the sinoatrial (SA) node are quickly conducted through the atria, causing the depolarization and contraction of atrial myocytes leading to atrial systole (Figure 7.4). This wave of depolarization must next reach the ventricles to produce ventricular systole. However, electrical impulses cannot spread directly from the atria to the ventricles due to the presence of fibrous tissue separating the two chambers. Instead, action potentials must propagate through a single conduction pathway located at the base of the right atria. This bridge electrically uniting the atria and the ventricles is known as the atrioventricular (AV) node (Figure 7.5). The AV node was first characterized by Tawara in 1906.[21] In his seminal work, Tawara details how he traced the Purkinje fibers in human hearts back to a compact 'knoten' (node) of cells located in the anterior interatrial septum.[21] Tawara also identified that this structure was conserved and nearly identical across a wide variety of species.[22] In his autobiography, Sir Arthur Keith appropriately noted that 'With the discovery of the conduction system of Tawara, heart research entered a new epoch.'[12]

The AV node has three primary functions. First and most importantly, the AV node acts to slow the conduction of the impulse passing from the atria to the ventricles.[3] This slowing of conduction, giving rise to the P–R interval seen in the electrocardiogram (ECG), allows the atria to enter the systole ahead of the ventricles.[8,22] This conduction-slowing is crucial to maintain the proper synchronization of atrial and ventricular contraction. Second, the AV node possesses an intrinsic pacemaking ability.[23] In response to sinoatrial pacemaker failure or ablation, the AV node assumes the role of primary pacemaker.[24] However, intrinsic pacemaker function of the AV node, known as AV junctional rhythm, occurs at a slower rate than the SAN.[25] Lastly, the AV node also serves to protect the ventricles from atrial tachyarrhythmias and other irregularities in atrial rhythm. The molecular substrates responsible for producing these three functions are not clearly defined, however. Although over a century has passed since Tawara first characterized what he termed the 'atrioventricular connecting system', functional insight into the electrophysiology of the AV node remains elusive. This section will provide an overview of the gross and cellular anatomy of the AV node as well as present the current understanding of AV nodal electrophysiology and conduction properties.

AV Node Structure and Function

The mammalian AV node is located in the right anterior interatrial septum in an area known as Koch's triangle. This area is demarcated by the septal leaflet of the tricuspid valve, the tendon of Todaro, and the coronary sinus ostium (Figure 7.5).[26] A functional layer of myocardium superficially surrounds the AV node. Deep to the myocardium lies a layer of transitional cells, which are smaller than atrial cells. The AV node, along with the bundle of His, comprises what is known as the atrioventricular junction. The AV node can be delineated from the bundle of His (penetrating bundle) by the central fibrous body as suggested by Tawara. The AV node is composed of the lower nodal bundle and the compact node, each with a distinct population of cells.[22] Cells of the compact node are small, spindle-like cells with no discernible arrangement or orientation. In contrast, cells of the lower nodal bundle are larger and are arranged in parallel. Connected to the lower nodal bundle are the right and left posterior projections.[23] The right posterior projection (also termed the inferior nodal extension) projects from

the lower nodal bundle toward the coronary sinus and tricuspid annulus.[23] Unique to the human AV node, the left nodal extension projects toward the mitral annulus and is shorter in length than the right nodal extension.[23] There is also an extensive network of nervous tissue within the AV node due to the presence of both sympathetic and parasympathetic neuronal inputs.[8]

The heterogeneous population of cells that make up the AV node may alternatively be defined by the shape of the action potential (Figure 7.4). These definitions arose due to microelectrode studies from the rabbit AV node. The AV node has been shown to contain three functionally different cell types: atrionodal (AN), nodal (N), and nodo-His (NH) cells. Nodal cells, found primarily in the compact node, have a low resting membrane potential, a low-amplitude action potential, and pacemaker function. It is the nodal cells that are responsible for the slowing of the action potential propagation necessary for synchronization of atrial and ventricular contraction. AN cells, also called transitional cells, have an action potential morphology intermediate between atrial and nodal cells. This layer comprises a transitional zone between the fast-conducting atrial myocytes and the nodal cells. Similarly, NH cells have an action potential shape intermediate between nodal cells and cells of the bundle of His. The NH cells are primarily located in the lower nodal bundle, a finding determined by the combination of histology and microelectrode study. A prevailing problem in the study of the AV node is correlating the morphology of the various cell types with the differences in the action potential waveform. The varying morphology of the action potential propagating through the AV node is very likely to be dependent on the complement of ion channels expressed in each cell.

Atrioventricular Node Action Potential and Electrical Components

Similar to SAN cells, AV node cells demonstrate automaticity, albeit at a slower intrinsic rate than that found in the leading pacemaker, the SAN (Figure 7.4).[6,23] Many of the same channels important for automaticity in the SAN are also expressed in the AV node although at different levels. Similar to the SAN, the AV node is a highly heterogeneous structure with morphologically distinct regions characterized by unique electrophysiological profiles. Cells in the compact node (N cells) demonstrate relatively low-amplitude action potentials with slower upstrokes compared to cells from AN (constituents of AV node transitional zone) or NH (constituents of penetrating bundle) regions. In general, AN and NH cells have action potentials that are between those of a nodal cell and an atrial or His cell, respectively. N cells are largely devoid of I_{Na}, which contributes to the smaller and slower action potentials measured in this region. At the same time, abundant I_f has been found in N cells and likely determines their intrinsic pacemaking activity. Both T-type and L-type Ca^{2+} channels are expressed in AV node, although less is known about the specific molecular profile and regional distribution. Like SAN, AV node cells lack I_{K1} to anchor the resting membrane potential. Instead, delayed rectifier K^+ current, I_{Kr} in particular, controls AV node cell maximum diastolic potential with vagal regulation mediated by I_{KAch}. Transient outward K^+ current is also expressed in the AV node for control of repolarization phase with much higher expression in the transitional AN and NH cells that also express I_{Na}.

Atrioventricular Node Conduction

The AV node is a critical gateway as electrical activation passes from the atria to the ventricles and integrates fast- and slow-conduction pathways from the atria. Importantly, the AV node serves as a functional frequency-dependent delay that allows for optimal filling of ventricles before their contraction and provides limited protection from inappropriate ventricular activation, for example in the case of atrial tachyarrhythmias. Thus, conduction through the AV node occurs through elaborate but well-defined pathways that integrate dual inputs from atrial conduction pathways (so-called dual-pathway electrophysiology). The slow pathway is formed by the right inferior nodal extension and N cells that express Cx45 in addition to Cx40, which is highly expressed throughout the AV node. The fast pathway occurs through the anterior transitional zone that also expresses Cx43 to support more rapid conduction than that found in the deeper layers. While this dual pathway provides a critical substrate for AV nodal re-entry, its role in normal AV node electrophysiology is less clear. However, it is likely that the fast pathway carries the impulse during normal sinus rhythm.

BUNDLE OF HIS AND BUNDLE BRANCHES

The cardiac bundle of His and right and left bundle branches are an essential ventricular component of the AV conduction axis (Figures 7.1, 7.5).[26] These cells, surrounded by a defined fibrous matrix serve to transmit AV activation to the ventricular myocardium. The organizational architecture of the system has evolved to precisely synchronize ventricular contraction from apex to base to ensure efficient and optimized ejection. As illustrated in Figure 7.4, action potentials within this region of the heart display rapid upstroke velocities, and prolonged phases of repolarization. As noted in the next section, conduction velocity in this network is rapid, due to the

high expression of voltage-gated Na^+ channels as well as connexin proteins. As reviewed by others, this region of the heart has been linked with human pathophysiology,[26] most notably conduction block and bundle branch re-entrant ventricular tachycardias.[27]

CARDIAC PURKINJE FIBERS

As noted above, and reviewed by others,[14,28,29] the Purkinje fiber/cell network is critical for cardiac action potential propagation across the entire ventricular myocardium. The network of fibers is vast, with groups of free-running cell fibers ('false tendons')[28] and a sub-endocardial network (Figure 7.6). Purkinje fiber/cell action potential conduction is remarkably rapid, due to the enrichment of connexin proteins and voltage-gated Na^+ channels (resulting in action potential upstroke velocities of ~1000 V/s[30,31]) as well as low intercellular resistance (as low as 100 ohms/cm).[31,32] The Purkinje fiber network has intrinsic pacemaking ability (usually only relevant in pathophysiological conditions), albeit at much lower rates than the SAN or AVN network (partially regulated by high expression of I_{K1}). However, this pacing may be sufficient (25–40 beats/min) for survival in many species.[14] For example, Purkinje fiber pacing may be critical in case of complete AV block in response to either pathological or genetic conditions.[14] Importantly, the cardiac Purkinje network is a source of potentially fatal ventricular tachyarrhythmias in response to either acquired cardiovascular disease or genetic defects. As reviewed by Boyden and colleagues, re-entrant bundle branch circuits have been associated with multiple forms of heart disease including left ventricular dysfunction, valvular disease, dilated cardiomyopathy, and other forms of heart disease.[28]

Structurally, electrically, and functionally, cardiac Purkinje cells are distinct from sinus node, atrial, and ventricular cardiomyocytes (Figure 7.7). Purkinje cells lack transverse-tubules and obvious structures of the cardiac dyad by electron microscopy (Figure 7.7). Due to enhanced expression of glycogen and reduced concentrations of myofibrils, Purkinje cells show light histological signatures (Figure 7.7).[28] Upon dissociation, Purkinje cells are more often larger and more rod-shaped than their SAN and AVN counterparts.[14] Electrically, the action potential duration of the Purkinje cell is longer than the ventricular cell, and Purkinje cells generally display a more negative diastolic membrane potential compared with other conduction system cell types (Figure 7.4). Related to Purkinje-system-based arrhythmias described above, early after-depolarizations may be observed in either Purkinje cells or fibers to trigger ventricular arrhythmias.[14]

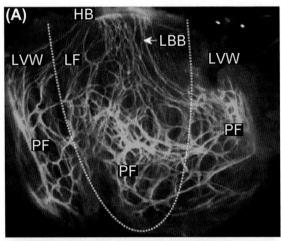

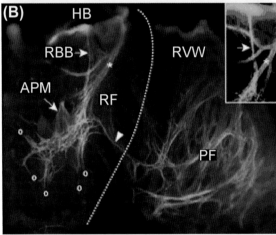

FIGURE 7.6 Organizational complexity of the cardiac His-Purkinje fiber system. Images of left and right ventricle **(A,B)** of mouse heart with transgenic expression enhanced green fluorescent protein under the control of the connexin40 gene promoter to illuminate the conduction system. The dashed line in **(A)** denotes the limits between the left flank (LF) of the IVS and the left ventricular wall (LVW). In **(B)**, the dashed white line denotes the border between the right ventricular wall (RVW) and the right flank (RF) of the IVS. RVW, right ventricular wall; PF, Purkinje fibers; LBB, RBB, left and right bundle branch; HB, His bundle; APM, anterior papillary muscle; circles denote connecting fibers and asterisk denotes the septal artery. Reprinted from L Miquerol, S Meysen, M Mangoni, P Bois, HVM van Rijen, P Abran, H Jongsma, J Nargeot, and D Gros. Architectural and functional asymmetry of the His-Purkinje system of the murine heart. *Cardiovascular Research*, 2004;63:77-86, with permission from Oxford University Press.

ROLE OF AUTONOMIC NERVOUS SYSTEM

Sympathetic autonomic influences on cardiovascular function are mediated by: cardiac sympathetic nerves, vascular sympathetic nerves, the adrenal medulla that releases epinephrine and norepinephrine into the circulation, and sympathetic stimulation of renal juxtaglomerular cells that activate the renin–angiotensin–aldosterone system (RAAS). Most sympathetic innervation in the human cardiovascular system is noradrenergic. Norepinephrine is the primary neurotransmitter, while epinephrine and other co-transmitters (e.g. aldosterone, dopamine) perform secondary functions.[33] The cardiac sympathetic nerve fibers represent the predominant autonomic component in the ventricles. They have a sub-epicardial location, and impulses are conducted along the major coronary arteries. The parasympathetic fibers run in parallel with the vagus nerve sub-endocardially after it crosses the atrioventricular groove and are mainly present in the atrial myocardium.[34] The ventricular sympathetic innervation runs as an electrical impulse from base to apex.[35] The cardiac neuronal system is made up of spatially distributed cell bundles comprising afferent, efferent, and interconnecting neurons working in unison.[36] The neurons are in constant communication with each other and each neuronal cell bundle is involved in reflexes that control spatially organized cardiac regions. The sympathetic outflow to the heart and peripheral circulation is regulated by cardiovascular reflexes. Afferent fiber signals are propagated to the central nervous system by autonomic nerves in contrast to efferent impulses, which are conducted from the central nervous system toward different organs either in autonomic or somatic nerves.[2]

The primary recognized roles of the sympathetic nervous system in cardiovascular control are the maintenance of blood pressure and the regulation of blood flow for seconds to minutes via the arterial baroreflex (although there may be some long-term influences as well).[37] The arterial baroreflex senses changes in blood pressure via baroreceptors, which are sensory afferent nerve endings located in the carotid sinus and the aortic arch. The baroreceptors respond to stretching of the vessel wall. The arterial wall stretching (the result of a short-term increase in blood pressure) leads to an increase in afferent input into central autonomic nuclei. Therefore, the increase in afferent input results in a reflex decrease in sympathetic neural outflow, which subsequently leads to a decreased vasoconstrictor tone and myocardial contractility. Collectively, these changes lead to decreased stroke volume and heart rate. All of these influences then result in a correction of the original 'error signal' of increased blood pressure. These sympathetic influences work in conjunction with parasympathetic influences on the SA node to decrease heart rate. During a short-term decrease in blood pressure, the opposite occurs, and the autonomic nervous system acts to increase vasoconstriction, increase stroke volume, and increase heart rate.[38]

The sympathetic nervous system may also have chronic effects on the cardiovascular system. There may be central neural mechanisms that set the long-term level of blood pressure by providing input to sympathetic premotor neurons in the rostral ventrolateral medulla via mechanisms independent of baroreceptor afferent

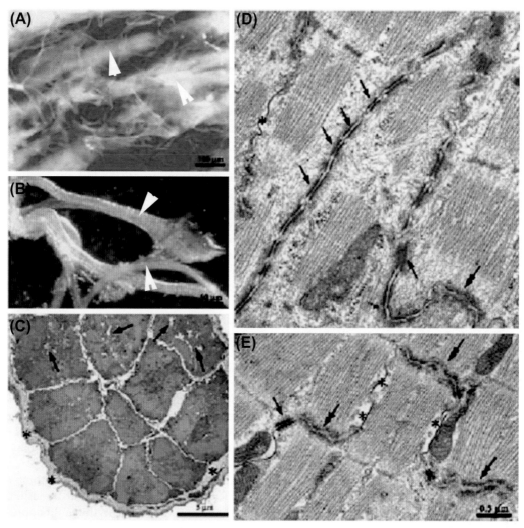

FIGURE 7.7   Specialized morphology of cardiac conduction system Purkinje fibers. **(A,B)** Rat Purkinje fibers (arrowheads) with surrounding layer of collagen. **(C)** Cross-section of Purkinje fiber. Asterisks show collagen where arrows denote glycogen. **(D,E)** Comparison of Purkinje fiber **(D)** and ventricular myocyte **(E)** intercalated disc junction. Arrows denote intercalated disc region. Note the elongated border between adjacent Purkinje cells. Reprinted from W Dun and PA Boyden. The Purkinje cell. *Heart Rhythm* 2008;45:617–624, with permission from Elsevier.

input.[39] These central nervous system areas appear to receive input from volume-regulatory hormones such as angiotensin II.[40] Additionally, recent studies of interindividual variability in the action of sympathetic neural mechanisms in regulating blood pressure provide some initial evidence that the sympathetic nervous system plays an important role in long-term blood pressure control in humans. Collectively, these findings may have implications for the treatment of chronic cardiovascular disease.

HUMAN CONDUCTION SYSTEM DISEASE

The pathologies encompassing human conduction system disease are broad, including ischemia, infarction, aging-related degeneration, drug toxicity, and

complications due to procedures and have been reviewed in detail.[41–44] 'Sick sinus syndrome' is a broadly defined umbrella encompassing multiple etiologies including severe sinus bradycardia, sinus pause and exit block, atrial tachyarrhythmias, brady/tachyarrhythmias, and aberrant rate regulation following increased catecholaminergic tone.[45] The incidence of sinus node disease is related to age (1:600 cardiac patients >65 years of age),[43] but may be present in newborns and children affected with congenital forms of the disease (see below).[13] Failure of the conduction system is a significant cause of morbidity and mortality in the United States, and even a decade ago was linked to over two billion dollars in expenditures on permanent, surgically implanted pacemakers annually.[46] Pacemakers are a successful but highly imperfect therapy that are subject to mechanical failure, infection and require repetitive surgical replacement.[47,48] Furthermore, SAN failure is a frequent finding in patients with atrial fibrillation[49] and is associated with increased mortality in heart failure.[50,51]

While conduction disease is most common in adults with acquired heart disease, following surgical correction for congenital heart disease, or during anti-arrhythmic therapy, it may also be present in patients without identifiable cardiac abnormalities or associated conditions ('idiopathic' disease).[52–55] These findings and twin studies support the role of inherited variables in human conduction disease. Variants in multiple human genes, mostly encoding ion channels, predispose a significant fraction of the population to atrial and ventricular arrhythmias.[56] The identification and functional characterization of these gene variants has enabled early diagnosis and treatment of potentially fatal cardiovascular disease. However, despite the clear importance of conduction disease, critical knowledge gaps remain for understanding the biology and pathology of automaticity and conduction. Fortunately, the past decade has seen the emergence of a number of new molecular pathways directly linked with human conduction disease affecting both automaticity and block of impulse propagation.

As noted above, the hyperpolarization-activated cyclic nucleotide-gated potassium channel 4 (HCN4) regulates sinus node cell diastolic depolarization. Over the past few years, multiple human HCN4 gene variants have been linked with familial sinus node disease. While these variants are localized across the six transmembrane-spanning ion channels, mechanistically these mutations have been linked with aberrant channel function. For example, work from Schulze-Bahr and colleagues demonstrated that the HCN4 1631ΔC variant found in individuals with sick sinus syndrome, lacks sensitivity to cAMP.[57] Additional variants, including HCN4 G480R and S672R mutations display decreased currents during the diastolic depolarization phase of the action potential due to activation at hyperpolarized membrane voltages (Table 7.1).[58,59]

SCN5A encodes $Na_v1.5$, the primary voltage-gated Na^+ channel of the vertebrate ventricle. This channel is widely expressed throughout the heart, and has been linked with familial forms of cardiac electrical disease including type 3 long QT syndrome and Brugada syndrome.[56] While the role of this channel in the central sinus node is controversial (likely more important in peripheral node and conduction system), clear links between $Na_v1.5$ dysfunction and conduction disease are present as demonstrated by SCN5A gene mutations in multiple forms of sick sinus syndrome and conduction disease. In fact, over a dozen different human SCN5A variants are now linked with human conduction disease.[60,61] Mechanistically, independent work from the Benson and Wilde laboratories have demonstrated that

aberrant I_{Na} properties of multiple SCN5A gene variants are linked with familial sinus node disease.[62,63] Furthermore, work from Schott, Le Marec, Escande and colleagues have linked SCN5A mutations with a large kindred harboring Lev-Lenègre, a disease characterized by degeneration of the His–Purkinje system, as well as familial conduction block associated with syncope.[64–66] Notably, more recent work from Bezzina and colleagues has linked the $Na_v1.5$ accessory subunit SCN1B gene variants with three families displaying conduction disease.[67] Mechanistically, these SCN1B variants phenocopy SCN5A conduction disease variants resulting in reduced I_{Na}.[67]

While not an ion channel, the cytoskeletal adapter protein ankyrin-B, encoded by ANK2, is critical for normal cardiac automaticity and conduction (Table 7.1). This role is illustrated by multiple large families with ankyrin-B loss of function and phenotypes including severe sinus node disease (bradycardia and heart rate variability requiring pacemakers), atrial fibrillation, and conduction disease.[68–70] These individuals may display ventricular arrhythmia and syncope following stress similar to patients with catecholaminergic polymorphic ventricular tachycardia (Figure 7.8).[71] Mechanistically, ankyrin-B in the sinus node has been linked to membrane targeting of the Na/Ca exchanger, the Na/K ATPase, and the L-type calcium channel, $Ca_v1.3$. Notably, animal models lacking $Ca_v1.3$ display severe sinus and AV node dysfunction.[72–75]

Andersen-Tawil syndrome is a complex syndrome characterized by developmental abnormalities, muscle paralysis, and cardiac arrhythmia. In approximately 60% of cases of Andersen-Tawil, individuals harbor a mutation in the KCNJ2 gene that encodes Kir2.1 (I_{K1}). In the cardiovascular system, individuals harboring KCNJ2 mutations may display conduction phenotypes including AV block and distal conduction defects.[76,77] While the role of Kir2.1 is still not completely understood in the cells of the conduction system, work from cell and animal models demonstrates a requirement of I_{K1} for normal action potential morphology, heart rate, and conduction.[78,79]

Critical for ventricular sarcoplasmic reticulum calcium regulation, both the ryanodine receptor (RYR2) and calsequestrin (CASQ2) have now also been linked to defects in automaticity in individuals with catecholaminergic polymorphic ventricular tachycardia (CPVT; Table 7.1). Specifically, while the phenotypes are complex and not limited to the conduction system, individuals with genetic linkages with mutant RYR2 and CASQ2 alleles display sinus bradycardia.[80–82] Furthermore connexin40 (GJA5), an important cardiac gap junction protein, has been associated with individuals with atrial standstill and atrial fibrillation[83,84] although the precise mechanisms of these pathologies are still not completely resolved.

TABLE 7.1 Genes Implicated in Human Conduction Disease

Conduction Disease	Gene	Protein	References	Other Disease Conditions
Sinus bradycardia	HCN4	HCN4	57–59	
Rate variability, sinus bradycardia, altered conduction	ANK2	Ankyrin-B	68–70	Atrial fibrillation, ventricular arrhythmia
Sinus node disease, atrial standstill, conduction block, AV and bundle branch block	SCN5A	Nav1.5	60–66	Ventricular arrhythmia (LQT3, Brugada syndrome)
Conduction defects	SCN1B	Scn1b	67	Brugada syndrome
AV and bundle branch block	KCNJ2	Kir2.1	76,77	Andersen-Tawil syndrome
Atrial standstill	GJA5	Cx40	83,84	Atrial fibrillation
Sinus node disease, AV block, atrial standstill	RYR2	RyR2	80,81	Catecholaminergic polymorphic ventricular tachycardia
Sinus bradycardia	CASQ2	Calsequestrin 2	82	Catecholaminergic polymorphic ventricular tachycardia
AV block	LMNA	Lamin A/C	94	Muscular dystrophy
AV block, Wolff-Parkinson-White (WPW)	PRKAG2	AMPK subunit	95,96	Glycogen storage disease
Wolff-Parkinson-White (WPW)	BMP2	Bone morphogenic protein 2	97	
AV block, conduction disease	DMPK	Myotonin-protein kinase	98	Myotonic dystrophy
Sinus bradycardia, AV and bundle branch block	TBX5	Tbx5	99	Holf-Oram syndrome
AV and bundle branch block	NKX1-5	Nkx2-5	66	

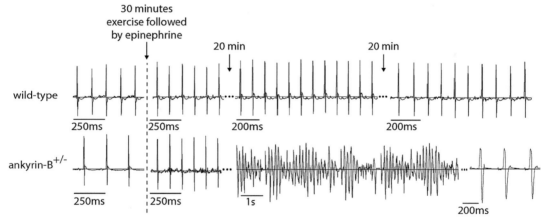

FIGURE 7.8 Ankyrin-B dysfunction results in human conduction disease and arrhythmia. Humans and mice harboring ankyrin-B (ANK2) loss of function mutations display bradycardia and heart rate variability at rest. Moreover, humans and mice with ankyrin-B loss of function mutations display increased susceptibility to severe ventricular arrhythmia and susceptibility to death following elevated catecholamine levels. Top panel depicts wild-type mouse, lower panel depicts ankyrin-B heterozygous mouse. Reprinted from S DeGrande, D Nixon, O Koval, JW Curran, P Wright, Q Wang, F Kashef, D Chiang, N Li, XHT Wehrens, ME Anderson, TJ Hund, and PJ Mohler. CaMKII inhibition rescues proarrhythmic phenotypes in the model of human ankyrin-B syndrome. *Heart Rhythm* 2012;9:2034–2041, with permission from Elsevier.

Finally, gene variants associated with human accessory conduction disease (i.e. Wolff-Parkinson-White), cardiac conduction developmental disorders, and cardiac conduction phenotypes associated with neuromuscular disorders have been identified over the past decade (Table 7.1) and have been well reviewed by Park and Fishman.[42]

Notably, these pathways are not limited to ion channels and transporters, but instead involve multiple surprising molecular pathways including transcription factors, proteins of the nuclear lamina, and kinase regulatory proteins. These findings clearly demonstrate the complexity of the pathways underlying vertebrate conduction.

MicroRNAs AND CARDIAC CONDUCTION

Cardiac conduction is tightly regulated to preserve cardiac function in health and disease. Over the past decade, the discovery of microRNAs (miRs) has revealed a new mechanism for transcriptional regulation of the cardiac conduction system. MiRs are ~20 basepair single-strand non-coding RNAs that modify the expression of downstream target RNAs and ultimately protein expression. While originally discovered in Drosophila, miRs are now known to regulate transcription across metazoans and play key roles in the cardiovascular system. For example, miRs are tightly coupled to cardiac development[85] as demonstrated by Srivastava and colleagues. Additionally, miRs are key new biomarkers for a host of cardiovascular conditions including heart failure and hypertrophy.[86–90] Directly related to this chapter, both miR-1 and miR-133 have been directly linked with defects in cardiac conduction. For example, overexpression of miR-1 in the rat heart results in suppressed cardiac conduction and arrhythmias that the authors hypothesize is linked with reduced expression of connexin 43 and the K+ channel Kcnd2.[91] Interestingly, miR-1-2 also targets Irx5 to repress Kcnd2 and is associated with arrhythmia.[92] Finally, miR-133 has been linked with the repression of the ether-a-go-go (ERG) K+ channel in the heart. Mice that overexpress miR-133 display repolarization defects and prolongation of the QT interval, both associated with arrhythmia.[93] While future experiments will be critical for directly linking miR function with specific targets, these data support the new and growing list of regulatory elements that regulate cardiac conduction in the vertebrate.

References

1. Kwong KF, Schuessler RB, Green KG, Laing JG, Beyer EC, Boineau JP, et al. Differential expression of gap junction proteins in the canine sinus node. *Circ Res* 1998;**82**(5):604–12.

2. Triposkiadis F, Karayannis G, Giamouzis G, Skoularigis J, Louridas G, Butler J. The sympathetic nervous system in heart failure physiology, pathophysiology, and clinical implications. *J Am Coll Cardiol* 2009;**54**(19):1747–62.

3. Tsuboi H, Ohno O, Ogawa K, Ito T, Hashimoto H, Okumura K, et al. Acetylcholine and norepinephrine concentrations in the heart of spontaneously hypertensive rats: a parasympathetic role in hypertension. *J Hypertens* 1987;**5**(3):323–30.

4. Steinberg SF. The molecular basis for distinct beta-adrenergic receptor subtype actions in cardiomyocytes. *Circ Res* 1999;**85**(11):1101–11.

5. Wu J, Olgin J, Miller JM, Zipes DP. Mechanisms underlying the reentrant circuit of atrioventricular nodal reentrant tachycardia in isolated canine atrioventricular nodal preparation using optical mapping. *Circ Res* 2001;**88**(11):1189–95.

6. Efimov IR, Nikolski VP, Rothenberg F, Greener ID, Li J, Dobrzynski H, et al. Structure-function relationship in the AV junction. *Anat Rec A Discov Mol Cell Evol Biol* 2004;**280**(2):952–65.

7. Bers DM. *Excitation-Contraction Coupling and Cardiac Contractile Force*. 2nd ed. Dordrecht: Kluwer Academic Publishers; 2001.

8. Boyett MR. 'And the beat goes on.' The cardiac conduction system: the wiring system of the heart. *Exp Physiol* 2009;**94**(10):1035–49.

9. Wu J, Zipes DP. Mechanisms underlying atrioventricular nodal conduction and the reentrant circuit of atrioventricular nodal reentrant tachycardia using optical mapping. *J Cardiovasc Electrophysiol* 2002;**13**(8):831–4.

10. Tamaddon HS, Vaidya D, Simon AM, Paul DL, Jalife J, Morley GE. High-resolution optical mapping of the right bundle branch in connexin40 knockout mice reveals slow conduction in the specialized conduction system. *Circ Res* 2000;**87**(10):929–36.

11. Keith A, Flack MW. The form and nature of the muscular connections between the primary divisions of the vertebrate heart. *J Anat Physiol* 1907;**41**:172–89.

12. Keith A. *An Autobiography: Sir Arthur Keith*. London: Watts and Co; 1950.

13. Dobrzynski H, Boyett MR, Anderson RH. New insights into pacemaker activity: promoting understanding of sick sinus syndrome. *Circulation* 2007;**115**(14):1921–32.

14. Mangoni ME, Nargeot J. Genesis and regulation of the heart automaticity. *Physiol Rev* 2008;**88**(3):919–82.

15. Vinogradova TM, Lyashkov AE, Zhu W, Ruknudin AM, Sirenko S, Yang D, et al. High basal protein kinase A-dependent phosphorylation drives rhythmic internal Ca^{2+} store oscillations and spontaneous beating of cardiac pacemaker cells. *Circ Res* 2006;**98**(4):505–14.

16. Swaminathan PD, Purohit A, Soni S, Voigt N, Singh MV, Glukhov AV, et al. Oxidized CaMKII causes cardiac sinus node dysfunction in mice. *J Clin Invest* 2011;**121**(8):3277–88.

17. Swaminathan PD, Purohit A, Hund TJ, Anderson ME. Calmodulin-dependent protein kinase II: linking heart failure and arrhythmias. *Circ Res* 2012;**110**(12):1661–77.

18. Luo M, Guan X, Luczak ED, Lang D, Kutschke W, Gao Z, et al. Diabetes increases mortality after myocardial infarction by oxidizing CaMKII. *J Clin Invest* 2013;**123**(3):1262–74.

19. Vinogradova TM, Zhou YY, Bogdanov KY, Yang D, Kuschel M, Cheng H, et al. Sinoatrial node pacemaker activity requires Ca(2+)/calmodulin-dependent protein kinase II activation. *Circ Res* 2000;**87**(9):760–7.

20. Wu Y, Gao Z, Chen B, Koval OM, Singh MV, Guan X, et al. Calmodulin kinase II is required for fight or flight sinoatrial node physiology. *Proc Natl Acad Sci U S A* 2009;**106**(14):5972–7.

21. Akiyama T. Sunao Tawara: discoverer of the atrioventricular conduction system of the heart. *Cardiol J* 2010;**17**(4):428–34.

22. Meijler FL, Janse MJ. Morphology and electrophysiology of the mammalian atrioventricular node. *Physiol Rev* 1988;**68**(2):608–47.

23. Kurian T, Ambrosi C, Hucker W, Fedorov VV, Efimov IR. Anatomy and electrophysiology of the human AV node. *Pacing Clin Electrophysiol* 2010;**33**(6):754–62.

24. Lee PC, Chen SA, Hwang B. Atrioventricular node anatomy and physiology: implications for ablation of atrioventricular nodal reentrant tachycardia. *Curr Opin Cardiol* 2009;**24**(2):105–12.

25. Hucker WJ, Nikolski VP, Efimov IR. Autonomic control and innervation of the atrioventricular junctional pacemaker. *Heart Rhythm* 2007;**4**(10):1326–35.

26. Anderson RH, Yanni J, Boyett MR, Chandler NJ, Dobrzynski H. The anatomy of the cardiac conduction system. *Clinical anatomy*, New York, NY 2009;**22**(1):99–113.

27. Schmidt B, Chun KR, Kuck KH, Ouyang F. Ventricular tachycardias originating in the his-purkinje system. Bundle branch reentrant ventricular tachycardias and fascicular ventricular tachycardias. *Herz* 2009;**34**(7):554–60.

28. Boyden PA, Hirose M, Dun W. Cardiac Purkinje cells. *Heart Rhythm* 2010;**7**(1):127–35.

29. Dun W, Boyden PA. The Purkinje cell; 2008 style. *J Mol Cell Cardiol* 2008;**45**(5):617–24.

30. Coraboeuf E, Deroubaix E, Coulombe A. Effect of tetrodotoxin on action potentials of the conducting system in the dog heart. *Am J Physiol* 1979;**236**(4):H561–7.

31. Callewaert G, Carmeliet E, Vereecke J. Single cardiac Purkinje cells: general electrophysiology and voltage-clamp analysis of the pacemaker current. *J Physiol* 1984;**349**:643–61.

32. Mobley BA, Page E. The surface area of sheep cardiac Purkinje fibres. *J Physiol* 1972;**220**(3):547–63.

33. Kellogg Jr DL, Pergola PE, Piest KL, Kosiba WA, Crandall CG, Grossmann M, et al. Cutaneous active vasodilation in humans is mediated by cholinergic nerve cotransmission. *Circ Res* 1995;**77**(6):1222–8.

34. Zipes DP. Heart-brain interactions in cardiac arrhythmias: role of the autonomic nervous system. *Cleve Clin J Med* 2008;**75**(Suppl. 2):S94–6.

35. Pierpont GL, DeMaster EG, Reynolds S, Pederson J, Cohn JN. Ventricular myocardial catecholamines in primates. *J Lab Clin Med* 1985;**106**(2):205–10.

36. Armour JA. Cardiac neuronal hierarchy in health and disease. *Am J Physiol Regul Integr Comp Physiol* 2004;**287**(2):R262–71.

37. Osborn JW. Hypothesis: set-points and long-term control of arterial pressure. A theoretical argument for a long-term arterial pressure control system in the brain rather than the kidney. *Clin Exp Pharmacol Physiol* 2005;**32**(5-6):384–93.

38. Joyner MJ, Charkoudian N, Wallin BG. A sympathetic view of the sympathetic nervous system and human blood pressure regulation. *Exp Physiol* 2008;**93**(6):715–24.

39. Osborn JW, Jacob F, Guzman P. A neural set point for the long-term control of arterial pressure: beyond the arterial baroreceptor reflex. *Am J Physiol Regul Integr Comp Physiol* 2005;**288**(4):R846–55.

40. Osborn JW, Fink GD, Sved AF, Toney GM, Raizada MK. Circulating angiotensin II and dietary salt: converging signals for neurogenic hypertension. *Curr Hypertens Rep* 2007;**9**(3):228–35.

41. Bharati S. Pathology of the Conduction System. In: Silver G, Schoen, editors. *Cardiovascular Pathology*. Philadelphia: Churchill Livingstone; 2001. p. 607–27.

42. Park DS, Fishman GI. The cardiac conduction system. *Circulation* 2011;**123**(8):904–15.

43. Benditt D, Sakaguchi S, Goldstein M, Lurie K, Gornick C,S. A. Sinus node dysfunction: pathophysiology, clinical features, evaluation, and treatment. In: Zipes DP, Jalife J, editors. *Cardiac Electrophysiology: From Cell to Bedside*. Philadelphia: WB Saunders Company; 1995. p. 1215–47.

44. Adan V, Crown LA. Diagnosis and treatment of sick sinus syndrome. *Am Fam Physician* 2003;**67**(8):1725–32.

45. Mangrum JM, DiMarco JP. The evaluation and management of bradycardia. *N Engl J Med* 2000;**342**(10):703–9.

46. Lamas GA, Lee K, Sweeney M, Leon A, Yee R, Ellenbogen K, et al. The mode selection trial (MOST) in sinus node dysfunction: design, rationale, and baseline characteristics of the first 1000 patients. *Am Heart J* 2000;**140**(4):541–51.

47. Ellenbogen KA, Hellkamp AS, Wilkoff BL, Camunas JL, Love JC, Hadjis TA, et al. Complications arising after implantation of DDD pacemakers: the MOST experience. *Am J Cardiol* 2003;**92**(6):740–1.

48. Hauser RG, Hayes DL, Kallinen LM, Cannom DS, Epstein AE, Almquist AK, et al. Clinical experience with pacemaker pulse generators and transvenous leads: an 8-year prospective multicenter study. *Heart Rhythm* 2007;**4**(2):154–60.

49. Hocini M, Sanders P, Deisenhofer I, Jais P, Hsu LF, Scavee C, et al. Reverse remodeling of sinus node function after catheter ablation of atrial fibrillation in patients with prolonged sinus pauses. *Circulation* 2003;**108**(10):1172–5.

50. Luu M, Stevenson WG, Stevenson LW, Baron K, Walden J. Diverse mechanisms of unexpected cardiac arrest in advanced heart failure. *Circulation* 1989;**80**(6):1675–80.

51. Saxon LA, Stevenson WG, Middlekauff HR, Stevenson LW. Increased risk of progressive hemodynamic deterioration in advanced heart failure patients requiring permanent pacemakers. *Am Heart J* 1993;**125**(5 Pt 1):1306–10.

52. Sarachek NS, Leonard JL. Familial heart block and sinus bradycardia. Classification and natural history. *Am J Cardiol* 1972;**29**(4):451–8.

53. Spellberg RD. Familial sinus node disease. *Chest* 1971;**60**(3):246–51.

54. Lehmann H, Klein UE. Familial sinus node dysfunction with autosomal dominant inheritance. *Br Heart J* 1978;**40**(11):1314–6.

55. Bertram H, Paul T, Beyer F, Kallfelz HC. Familial idiopathic atrial fibrillation with bradyarrhythmia. *Eur J Pediatr* 1996;**155**(1):7–10.

56. Lehnart SE, Ackerman MJ, Benson Jr DW, Brugada R, Clancy CE, Donahue JK, et al. Inherited arrhythmias: a National Heart, Lung, and Blood Institute and Office of Rare Diseases workshop consensus report about the diagnosis, phenotyping, molecular mechanisms, and therapeutic approaches for primary cardiomyopathies of gene mutations affecting ion channel function. *Circulation* 2007;**116**(20):2325–45.

57. Schulze-Bahr E, Neu A, Friederich P, Kaupp UB, Breithardt G, Pongs O, et al. Pacemaker channel dysfunction in a patient with sinus node disease. *J Clin Invest* 2003;**111**(10):1537–45.

58. Milanesi R, Baruscotti M, Gnecchi-Ruscone T, DiFrancesco D. Familial sinus bradycardia associated with a mutation in the cardiac pacemaker channel. *N Engl J Med* 2006;**354**(2):151–7.

59. Nof E, Luria D, Brass D, Marek D, Lahat H, Reznik-Wolf H, et al. Point mutation in the HCN4 cardiac ion channel pore affecting synthesis, trafficking, and functional expression is associated with familial asymptomatic sinus bradycardia. *Circulation* 2007;**116**(5):463–70.

60. Lei M, Zhang H, Grace AA, Huang CL. SCN5A and sinoatrial node pacemaker function. *Cardiovasc Res* 2007;**74**(3):356–65.

61. Ruan Y, Liu N, Priori SG. Sodium channel mutations and arrhythmias. *Nat Rev Cardiol* 2009;**6**(5):337–48.

62. Benson DW, Wang DW, Dyment M, Knilans TK, Fish FA, Strieper MJ, et al. Congenital sick sinus syndrome caused by recessive mutations in the cardiac sodium channel gene (SCN5A). *J Clin Invest* 2003;**112**(7):1019–28.

63. Smits JP, Koopmann TT, Wilders R, Veldkamp MW, Opthof T, Bhuiyan ZA, et al. A mutation in the human cardiac sodium channel (E161K) contributes to sick sinus syndrome, conduction disease and Brugada syndrome in two families. *J Mol Cell Cardiol* 2005;**38**(6):969–81.

64. Kyndt F, Probst V, Potet F, Demolombe S, Chevallier JC, Baro I, et al. Novel SCN5A mutation leading either to isolated cardiac conduction defect or Brugada syndrome in a large French family. *Circulation* 2001;**104**(25):3081–6.

65. Probst V, Kyndt F, Potet F, Trochu JN, Mialet G, Demolombe S, et al. Haploinsufficiency in combination with aging causes SCN5A-linked hereditary Lenegre disease. *J Am Coll Cardiol* 2003;**41**(4):643–52.

66. Schott JJ, Alshinawi C, Kyndt F, Probst V, Hoorntje TM, Hulsbeek M, et al. Cardiac conduction defects associate with mutations in SCN5A. *Nat Genet* 1999;**23**(1):20–1.

67. Watanabe H, Koopmann TT, Le Scouarnec S, Yang T, Ingram CR, Schott JJ, et al. Sodium channel beta 1 subunit mutations associated with Brugada syndrome and cardiac conduction disease in humans. *J Clin Invest* 2008;**118**(6):2260–8.

68. Cunha SR, Hund TJ, Hashemi S, Voigt N, Li N, Wright P, et al. Defects in ankyrin-based membrane protein targeting pathways underlie atrial fibrillation. *Circulation* 2011;**124**(11):1212–22.

69. Le Scouarnec S, Bhasin N, Vieyres C, Hund TJ, Cunha SR, Koval O, et al. Dysfunction in ankyrin-B-dependent ion channel and transporter targeting causes human sinus node disease. *Proc Natl Acad Sci U S A* 2008;**105**(40):15617–22.

70. Mohler PJ, Schott JJ, Gramolini AO, Dilly KW, Guatimosim S, duBell WH, et al. Ankyrin-B mutation causes type 4 long-QT cardiac arrhythmia and sudden cardiac death. *Nature* 2003;**421**(6923):634–9.

71. Mohler PJ, Le Scouarnec S, Denjoy I, Lowe JS, Guicheney P, Caron L, et al. Defining the cellular phenotype of "ankyrin-B syndrome" variants: human ANK2 variants associated with clinical phenotypes display a spectrum of activities in cardiomyocytes. *Circulation* 2007;**115**(4):432–41.

72. Mangoni ME, Couette B, Bourinet E, Platzer J, Reimer D, Striessnig J, et al. Functional role of L-type Cav1.3 Ca^{2+} channels in cardiac pacemaker activity. *Proc Natl Acad Sci U S A* 2003;**100**(9):5543–8.

73. Mangoni ME, Traboulsie A, Leoni AL, Couette B, Marger L, Le Quang K, et al. Bradycardia and slowing of the atrioventricular conduction in mice lacking CaV3.1/alpha1G T-type calcium channels. *Circ Res* 2006;**98**(11):1422–30.

74. Zhang Z, He Y, Tuteja D, Xu D, Timofeyev V, Zhang Q, et al. Functional roles of Cav1.3(alpha1D) calcium channels in atria: insights gained from gene-targeted null mutant mice. *Circulation* 2005;**112**(13):1936–44.

75. Zhang Z, Xu Y, Song H, Rodriguez J, Tuteja D, Namkung Y, et al. Functional Roles of Ca(v)1.3 (alpha(1D)) calcium channel in sinoatrial nodes: insight gained using gene-targeted null mutant mice. *Circ Res* 2002;**90**(9):981–7.

76. Andelfinger G, Tapper AR, Welch RC, Vanoye CG, George Jr AL, Benson DW. KCNJ2 mutation results in Andersen syndrome with sex-specific cardiac and skeletal muscle phenotypes. *Am J Hum Genet* 2002;**71**(3):663–8.

77. Zhang L, Benson DW, Tristani-Firouzi M, Ptacek LJ, Tawil R, Schwartz PJ, et al. Electrocardiographic features in Andersen-Tawil syndrome patients with KCNJ2 mutations: characteristic T-U-wave patterns predict the KCNJ2 genotype. *Circulation* 2005;**111**(21):2720–6.

78. Zaritsky JJ, Redell JB, Tempel BL, Schwarz TL. The consequences of disrupting cardiac inwardly rectifying K(+) current (I(K1)) as revealed by the targeted deletion of the murine Kir2.1 and Kir2.2 genes. *J Physiol* 2001;**533**(Pt 3):697–710.

79. McLerie M, Lopatin AN. Dominant-negative suppression of I(K1) in the mouse heart leads to altered cardiac excitability. *J Mol Cell Cardiol* 2003;**35**(4):367–78.

80. Bhuiyan ZA, van den Berg MP, van Tintelen JP, Bink-Boelkens MT, Wiesfeld AC, Alders M, et al. Expanding spectrum of human RYR2-related disease: new electrocardiographic, structural, and genetic features. *Circulation* 2007;**116**(14):1569–76.

81. Postma AV, Denjoy I, Kamblock J, Alders M, Lupoglazoff JM, Vaksmann G, et al. Catecholaminergic polymorphic ventricular tachycardia: RYR2 mutations, bradycardia, and follow up of the patients. *J Med Genet* 2005;**42**(11):863–70.

82. Postma AV, Denjoy I, Hoorntje TM, Lupoglazoff JM, Da Costa A, Sebillon P, et al. Absence of calsequestrin 2 causes severe forms of catecholaminergic polymorphic ventricular tachycardia. *Circ Res* 2002;**91**(8):e21–6.

83. Groenewegen WA, Firouzi M, Bezzina CR, Vliex S, van Langen IM, Sandkuijl L, et al. A cardiac sodium channel mutation cosegregates with a rare connexin40 genotype in familial atrial standstill. *Circ Res* 2003;**92**(1):14–22.

84. Firouzi M, Ramanna H, Kok B, Jongsma HJ, Koeleman BP, Doevendans PA, et al. Association of human connexin40 gene polymorphisms with atrial vulnerability as a risk factor for idiopathic atrial fibrillation. *Circ Res* 2004;**95**(4):e29–33.

85. Zhao Y, Samal E, Srivastava D. Serum response factor regulates a muscle-specific microRNA that targets Hand2 during cardiogenesis. *Nature* 2005;**436**(7048):214–20.

86. van Rooij E, Sutherland LB, Liu N, Williams AH, McAnally J, Gerard RD, et al. A signature pattern of stress-responsive microRNAs that can evoke cardiac hypertrophy and heart failure. *Proc Natl Acad Sci U S A* 2006;**103**(48):18255–60.

87. Thum T, Galuppo P, Wolf C, Fiedler J, Kneitz S, van Laake LW, et al. MicroRNAs in the human heart: a clue to fetal gene reprogramming in heart failure. *Circulation* 2007;**116**(3):258–67.

88. Ikeda S, Kong SW, Lu J, Bisping E, Zhang H, Allen PD, et al. Altered microRNA expression in human heart disease. *Physiological genomics* 2007;**31**(3):367–73.

89. Naga Prasad SV, Duan ZH, Gupta MK, Surampudi VS, Volinia S, Calin GA, et al. Unique microRNA profile in end-stage heart failure indicates alterations in specific cardiovascular signaling networks. *J Biol Chem* 2009;**284**(40):27487–99.

90. Matkovich SJ, Van Booven DJ, Youker KA, Torre-Amione G, Diwan A, Eschenbacher WH, et al. Reciprocal regulation of myocardial microRNAs and messenger RNA in human cardiomyopathy and reversal of the microRNA signature by biomechanical support. *Circulation* 2009;**119**(9):1263–71.

91. Yang B, Lin H, Xiao J, Lu Y, Luo X, Li B, et al. The muscle-specific microRNA miR-1 regulates cardiac arrhythmogenic potential by targeting GJA1 and KCNJ2. *Nat Med* 2007;**13**(4):486–91.

92. Zhao Y, Ransom JF, Li A, Vedantham V, von Drehle M, Muth AN, et al. Dysregulation of cardiogenesis, cardiac conduction, and cell cycle in mice lacking miRNA-1-2. *Cellule* 2007;**129**(2):303–17.

93. Xiao J, Luo X, Lin H, Zhang Y, Lu Y, Wang N, et al. MicroRNA miR-133 represses HERG K+ channel expression contributing to QT prolongation in diabetic hearts. *J Biol Chem* 2007;**282**(17): 12363–7.

94. Bonne G, Di Barletta MR, Varnous S, Becane HM, Hammouda EH, Merlini L, et al. Mutations in the gene encoding lamin A/C cause autosomal dominant Emery-Dreifuss muscular dystrophy. *Nat Genet* 1999;**21**(3):285–8.

95. Gollob MH, Green MS, Tang AS, Gollob T, Karibe A, Ali Hassan AS, et al. Identification of a gene responsible for familial Wolff-Parkinson-White syndrome. *N Engl J Med* 2001;**344**(24):1823–31.

96. Arad M, Benson DW, Perez-Atayde AR, McKenna WJ, Sparks EA, Kanter RJ, et al. Constitutively active AMP kinase mutations cause glycogen storage disease mimicking hypertrophic cardiomyopathy. *J Clin Invest* 2002;**109**(3):357–62.

97. Lalani SR, Thakuria JV, Cox GF, Wang X, Bi W, Bray MS, et al. 20p12.3 microdeletion predisposes to Wolff-Parkinson-White syndrome with variable neurocognitive deficits. *J Med Genet* 2009;**46**(3):168–75.

98. Mahadevan M, Tsilfidis C, Sabourin L, Shutler G, Amemiya C, Jansen G, et al. Myotonic dystrophy mutation: an unstable CTG repeat in the 3′ untranslated region of the gene. *Science* 1992;**255**(5049):1253–5.

99. Basson CT, Bachinsky DR, Lin RC, Levi T, Elkins JA, Soults J, et al. Mutations in human TBX5 [corrected] cause limb and cardiac malformation in Holt-Oram syndrome. *Nat Genet* 1997;**15**(1): 30–5.

Molecular Pathobiology of Myocarditis

Elisa Carturan, PhD, Cristina Basso, MD, PhD,
Gaetano Thiene, MD, FRCP London

Department of Cardiac, Thoracic and Vascular Sciences, University of Padua Medical School, Padua, Italy

INTRODUCTION

In the 1995 World Health Organization (WHO) *Definition of Cardiomyopathies,*[1] myocarditis (also called inflammatory cardiomyopathy), is defined as an 'inflammatory disease of the myocardium associated with cardiac dysfunction' and is listed among 'specific cardiomyopathies.' Myocarditis is diagnosed in vivo on endomyocardial biopsy (EMB) by established histological, immunological, and immunohistochemical criteria, and molecular testing on EMB specimens is recommended to identify viral etiology. Infectious, autoimmune, and idiopathic forms of inflammatory cardiomyopathy are recognized as distinct subtypes of myocarditis that may lead to dilated cardiomyopathy (DCM).[1]

In the 2006 American Heart Association consensus document on cardiomyopathies,[2] myocarditis is listed among primary cardiomyopathies under the acquired subgroup, with infectious, autoimmune, and toxic causes; moreover, it is also identifiable under the mixed subgroup of cardiomyopathies, in the form of inflammatory DCM (i.e. predominantly non-genetic, with infectious, autoimmune, and toxic causes). The actual incidence of myocarditis is difficult to determine since EMB, the diagnostic gold standard, is performed in a minority of cases. Studies of sudden cardiac death in young people report a highly variable autopsy prevalence of myocarditis, ranging from 2 to 42% of cases.[3,4] Similarly, biopsy-proven myocarditis is reported in 9–16% of adult patients with unexplained non-ischaemic DCM[5,6] and in 46% of children with a known cause of DCM.[7]

The clinical presentation of myocarditis is highly variable, ranging from subclinical to severe, and includes unexplained congestive heart failure (i.e., exertional dyspnea, fatigue) or cardiogenic shock, chest pain with myocardial enzyme release mimicking myocardial infarction, arrhythmias, syncope, or even sudden death. A viral prodrome including fever, rash, myalgias, arthralgias, fatigue, and respiratory or gastrointestinal symptoms often precedes the onset of myocarditis by several days to a few weeks. The disease may affect individuals of all ages, although it is most frequent in the young. The wide spectrum of clinical scenarios implies that the diagnosis of myocarditis requires a high level of suspicion and the use of appropriate investigations. In all cases of suspected myocarditis, it is mandatory to exclude coronary artery disease and other cardiovascular or extracardiac diseases that could explain the clinical presentation. Rarely patients with other cardiovascular disorders (such as coronary artery disease, cardiomyopathy, and hypertensive heart disease) present with a clinical deterioration caused by myocarditis that is wrongly ascribed to the pre-existing disease. If this is strongly suspected by the clinician, further investigation including EMB may be appropriate.

Heterogeneity of clinical presentation, including subclinical or asymptomatic forms, explains why the incidence and prevalence of myocarditis is still unknown and probably underestimated. The clinical outcome is highly variable and may be related to the different causes and/or genetic susceptibility, thus clarifying why there are patients who resolve completely, those who have deterioration, and those who progress to DCM.

Most affected patients who present with acute onset of left ventricular dysfunction have a relatively mild myocarditis that resolves with few short-term sequelae. Children often have a more fulminant presentation. Certain clinical 'red flags' often identify those at higher risk.[8] For instance, rash, fever, blood eosinophilia, or history of recent medications could suggest a hypersensitivity myocarditis. Giant-cell myocarditis (GCM) should be considered in patients with acute left ventricular dysfunction associated with thymoma, autoimmune disorder, or ventricular tachycardia. An unusual cause of myocarditis, such as cardiac sarcoidosis, should be suspected in patients who present with chronic heart failure, DCM, and new ventricular arrhythmias, second-degree or third-degree heart block, or those who do not respond to standard care.

Diagnosis of myocarditis requires a multiparametric approach including laboratory tests, non-invasive and invasive tools. Biomarkers of cardiac injury are elevated in a minority of patients with acute myocarditis but are useful to confirm the diagnosis.[8] The electrocardiogram may show sinus tachycardia with non-specific ST segment and T-wave abnormalities and, occasionally, the changes mimic an acute myocardial infarction.

Echocardiography is needed to exclude other causes of heart failure, but there are no specific features of acute myocarditis. Segmental or global wall-motion abnormalities in myocarditis can simulate myocardial infarction. A small left ventricular cavity size with increased wall thickness and diastolic disfunction is often observed in fulminant myocarditis.

Cardiac magnetic resonance (CMR) is being used with increasing frequency as a diagnostic test in suspected acute myocarditis and may serve to localize sites for EMB. Based on pre-clinical and clinical studies, an 'International Consensus Group on CMR Diagnosis of Myocarditis' published detailed recommendations for appropriate CMR techniques in the non-invasive diagnosis of myocarditis (Lake Louise criteria).[9] CMR findings are consistent with myocardial inflammation, if at least two of the following criteria are present: (a) regional or global myocardial signal intensity increase in T2-weighted edema images; (b) increased global myocardial early gadolinium enhancement ratio between myocardium and skeletal muscle in gadolinium-enhanced T1-weighted images; and (c) there is at least one focal lesion with non-ischemic regional distribution in inversion recovery-prepared gadolinium-enhanced T1-weighted images (late gadolinium enhancement).[9]

ETIOLOGY

Myocarditis has multiple causes, both infectious and non-infectious (Table 8.1).

Infectious – Virus

Viral infection is the most common cause of myocarditis in Western Europe and North America. Using molecular techniques, viral genomic material can be identified in a varying subset of patients with acute and chronic myocarditis (Table 8.2).[10,11] The prevalence of enteroviral RNA in EMB specimens from patients with myocarditis, assessed by reverse transcription polymerase chain reaction (RT-PCR) or nested RT-PCR, ranged from 1%[12] to 50%.[13] In another study, enteroviral RNA was detected in 14% of EMB specimens from patients with

myocarditis.[14] Kühl et al. showed that in 56/172 patients (32.6%) with histology-proven myocarditis, enteroviral RNA was detectable; in 28/56 (50%) of those patients, viral RNA was eliminated spontaneously with a significant improvement in ejection fraction, whereas in the remaining 28 patients, persistence of viral genomes on subsequent biopsies was correlated with a significant decrease in ejection fraction.[15]

Several respiratory tract viruses may cause myocarditis at variable frequencies. In particular, adenoviruses (type 2 and 5) have been shown to be an important cause of myocarditis and DCM both in childhood[16,17] and adulthood.[14] Rhinovirus-associated myocarditis has also been reported rarely.[18] Many cases of acute myocarditis have been described in association with pandemic H1N1 influenza virus infections, especially in young patients. Genomes of influenza A/H1N1 virus have been detected by RT-PCR analysis in blood as well as in myocardial tissue in patients with lethal influenza virus infection.[19]

Cytomegalovirus (CMV) is a recognized cause of acute infectious myocarditis. CMV-specific genome has been detected in the myocytes of EMBs from 3%[20] to 38%[21] of patients with myocarditis. Human herpes virus (HHV) 6-induced myocarditis has been reported in a small number of patients, occasionally with a fatal outcome.[22] In large studies of patients with inflammatory heart diseases, analyses for HHV6 and Epstein-Barr virus (EBV) have been included. Prevalence for HHV6 genomes detected in patients with myocarditis are up to 20%[15] and for EBV genomes are up to 5%.[20]

Parvovirus B19 (PV B19), the causative agent of erythema infectiosum, also called fifth disease, has been reported to be a rare but severe cause of myocarditis in infants and children.[23] PV B19 genomes have been detected in up to 56% of patients with myocarditis.[24] As indicated above for enterovirus, the persistence of PV B19 in patients with left ventricle dysfunction was also found to be associated with progressive impairment of left ventricle ejection fraction, whereas spontaneous viral elimination was associated with a significant improvement in left ventricle function. However, in contrast to enterovirus, spontaneous virus elimination of PV B19 was observed in only 22% of patients.[15] In situ hybridization revealed that endothelial cells are the principal cell target for PV B19, even though myocyte tropism has also been demonstrated.[25] Human immunodeficiency virus and hepatitis C virus have also been seldom associated with myocarditis.[26]

While adenovirus and enterovirus have long been considered the most important viruses leading to myocarditis, recent multicenter analyses showed a wide discrepancy in results. In a US multicenter analysis of histologically identified myocarditis, adenovirus, enterovirus and CMV were the most commonly identified viruses by PCR from EMB samples.[14] In studies

TABLE 8.1 Myocarditis: Causes

Category	Subcategory	Causes
Infective	Viruses	DNA: Adenovirus, Chikungunya virus, Flavivirus, Hepatitis B virus, Herpes viruses (Human Herpes virus type 1, Cytomegalovirus, Varicella virus, Epstein-Barr virus, Human herpesvirus-6), Parvovirus B-19, Poliovirus, Rabies virus, Rubella virus, Variola virus. RNA: Arborvirus, Hepatitis C virus, Orthomyxovirus (Influenza A and B viruses), Paramyxovirus (rubeola virus, mumps virus, respiratory syncytial virus), Picornavirus (entero, echo, rhino), retrovirus (HIV)
	Bacteria	Burkholderia pseudomallei, Brucella, Chlamydia, Clostridium, Corynebacterium diphtheriae, Francisella tularensis, Haemophilus influenzae, Gonococcus, Legionella pneumophila, Mycobacterium (avium intracellulare, leprae, tuberculosis), Mycoplasma, Neisseria meningitidis, Salmonella, Staphylococcus, Streptococcus A, Streptococcus pneumoniae, Vibrio cholera
	Spirochaete	Borrelia burgdorferi, Borrelia recurrentis, Leptospira, Treponema pallidum
	Rickettsiae	Coxiella burnetii, Rickettsia prowazekii, Rickettsia rickettsii
	Fungi	Actinomyces, Aspergillus, Blastomyces, Candida, Coccidioides, Cryptococcus Histoplasma, Mucor species, Nocardia, Sporothrix schenckii, Strongyloides stercoralis
	Protozoa	Balantidium, Entamoeba histolytica, Leishmania, Plasmodium falciparum, Sarcocystis, Trypanosoma cruzi, Trypanosoma brucei, Toxoplasma gondii
	Helminths	Ascaris, Echinococcus granulosus, Heterophyes, Paragonimus westermani, Schistosoma, Strongyloides stercoralis, Taenia solium, Toxocara canis, Trichinella spiralis, Wuchereria bancrofti
Immune mediated	Allergens	Serum sickness, Tetanus toxoid, Vaccines. Drugs: amitriptyline, cefaclor, colchicine, furosemide, isoniazid, lidocaine, methyldopa, penicillin, phenylbutazone, phenytoin, sulfonamides, tetracycline, thiazide diuretics
	Alloantigens	Heart transplant rejection
	Autoantigens	Infection-negative lymphocytic, infection-negative giant cell. Associated with autoimmune or immune-oriented disorders: Churg-Strauss syndrome, inflammatory bowel disease, insulin-dependent diabetes mellitus, Kawasaki's disease, myasthenia gravis, polymyositis, rheumatoid arthritis, thyrotoxicosis, rheumatic heart disease (rheumatic fever), sarcoidosis, scleroderma, systemic lupus erythematosus, Wegener's granulomatosis
Toxic	Drugs	Aminophylline, amphetamines, anthracyclines, catecholamines, chloramphenicol, cocaine, cyclophosphamide, doxorubicin, ethanol, 5-flurouracil, imatimib mesylate, interleukin-2, methysergide, phenytoin, trastuzumab, zidovudin
	Heavy Metals	Copper, iron, lead (rare, more commonly cause intramyocyte accumulation)
	Miscellaneous	Arsenic, bee and wasp stings, carbon monoxide, inhalants, phosphorus, scorpion sting, snake and spider bites, sodium azide
	Hormones	Phaeochromocytoma
	Vitamins	Beri-beri
	Physical agents	Electric shock, radiation

TABLE 8.2 Cardiotropic Viruses, Target Myocardial Cells Receptors and Co-receptors, and Prevalence of Viral Infections Detected by Molecular Biology-Based Techniques in Active Myocarditis

Virus (Viral Genome)	Cell Target in the Heart	Receptors - Co-receptors	Prevalence (%)
Coxsackievirus (RNA)	Cardiomyocytes, B-cells, CD4+ T-cells, macrophages, fibroblasts	CAR - DAF	3–50
Adenovirus (DNA)	Cardiomyocytes, B-cells, CD4+ T-cells, macrophages, fibroblasts	CAR - $\alpha_V\beta3$ and $\alpha_V\beta5$ integrins	1–23
Parvovirus B19 (DNA)	Endothelial cells, cardiomyocytes (?)	Erythrocyte P antigen - $\alpha5\beta1$ integrin	1–56
Human herpesvirus-6 (DNA)	T-cells, endothelial cells	CD46	<18
Epstein-Barr virus (DNA)	B cells, T-cells, macrophages	MHCII- CD21	<1
Cytomegalovirus (DNA)	Cardiomyocytes, macrophages, fibroblasts, endothelial cells	HSPG, EGFR - $\alpha_V\beta3$ integrin	6–15
Influenza virus (RNA)	Cardiomyocytes, macrophages, lymphocytes	Sialic acid	<1
Hepatitis C virus (RNA)	Cardiomyocytes	CD81 - SR-B1	<1
HIV (RNA)	CD4+ T-cells, macrophages	CD4 - CCR5, CXCR4	<1

CAR, Coxsackievirus and adenovirus receptor; CCR, β-chemokine receptor; CXCR, α-chemokine receptor; DAF, decay-accelerating factor; EGFR, epidermal growth factor receptor; HSPG, Heparin sulfate proteoglycan; MHC II, major histocompatibility complex; SR-B1, Scavenger receptor class B member 1. *(Modified from Reference 10).*

performed in Germany, a high prevalence of PV B19 has been found in patients with a history of clinically suspected myocarditis or idiopathic DCM.[15] In Italy, in a series of 120 adult EMBs with a diagnosis of myocarditis collected from 1992 to 2005, enterovirus was the most frequent virus followed by adenovirus and EBV.[20] This wide discrepancy in viral prevalences could have several explanations, such as differences in detection procedures, changing prevalence of viruses in local populations, or the differences in the stage of the disease. This emphasizes the need for standardization of protocols to achieve comparable results among molecular laboratories.

Infectious – Non-Virus

Although numerous bacterial infections can cause myocarditis, in industrialized countries bacterial-induced myocarditis is far less common than viral-induced myocarditis. Toxin-producing bacteria, including Clostridium and diphtheria, can cause severe myocardial damage. Bacteremia from any source may result in myocarditis, with the most common pathogens being Meningococcus, Streptococcus, Mycobacterium, and Listeria.[27,28] The spirochete Borrelia burgdorferi causes Lyme disease, which can result in both acute and chronic myocarditis.[29] Infection with the protozoa Trypanosoma cruzi (Chagas disease), common in Central and South America, can present as acute myocarditis or chronic DCM.[30]

Toxoplasma gondii poses significant problems among recipients of cardiac transplants. A study reported that 57% of transplanted patients lacking antibodies to that agent developed Toxoplasma myocarditis.[31] However, toxoplasmosis may become reactivated in antibody-positive transplant recipients, and myocarditis has been reported in 4–53% of transplant cases.[31] This large variation in rate is probably due to differences in antibody testing methods. After the introduction of pyrimethamine prophylaxis, this complication has decreased substantially.

Fungal myocarditis frequently occurs in the setting of disseminated disease. The major fungal pathogen responsible for myocardial infection is Aspergillus fumigatus. The incidence of invasive fungal disease has dramatically increased over the past few decades corresponding to the increasing number of immunocompromised patients.[32]

Immune-Mediated

Allergens

Drug-induced hypersensitivity reactions can cause an eosinophilic myocarditis that often responds to withdrawal of the offending medication, though adjuvant corticosteroid therapy is often necessary.[33] Drugs associated with hypersensitivity myocarditis include clozapine, sulfonamide antibiotics, methyldopa, and some anti-seizure drugs. The US civilian vaccination program reported myocarditis after smallpox vaccination.[34] Fortunately, myocarditis after other vaccines is rare. Eosinophilic myocarditis has been diagnosed in explanted hearts from patients treated with dobutamine or other inotrope drugs. Whether it is caused by dobutamine itself or the preservative sodium bisulfate is still unclear.[35] De novo development of eosinophilic myocarditis with left ventricular assist device support as a bridge to transplantation was described by Pereira et al.,[36] although the patient was also receiving a leukotriene-receptor antagonist for asthma. In parasitic infections, eosinophilic myocarditis is not infrequent. It is often associated with protozoal infections such as Trypanosoma cruzi, Toxoplasma gondii, Trichinella spiralis, Entamoeba fragilis, Echinococcus. It is thought that the heart is infiltrated with eosinophils because of the significant and persistent blood eosinophilia caused by the parasites.

Alloantigens

Acute rejection is caused by antigen-specific Th1 and cytotoxic T-lymphocytes. They recognize transplanted tissue because of expression of alloantigens.

Autoantigens

Idiopathic GCM is a rare, autoimmune form of myocarditis histologically defined by the presence of multinucleated giant cells, lymphocytic inflammatory infiltrate, and myocyte necrosis.[37,38] This disease usually occurs in young adults and typically causes fulminant heart failure, arrhythmias, or rarely heart block, necessitating aggressive immunosuppression, ventricular assist device insertion, or cardiac transplantation. Fatal outcome occurs frequently unless cardiac transplantation is performed. In some patients, GCM is associated with other autoimmune disorders, and drug hypersensitivity reactions. GCM is rarer in children and, in this setting, is often associated with immune-mediated disease in other organs.[37,38] A variant of GCM has been reported that primarily involves the atria, displays distinctive clinical features, and follows a more benign course than typical ventricular GCM.[39,40]

Cardiac sarcoidosis is another unusual form of idiopathic myocarditis that is distinct from GCM in that it is characterized histologically by granulomas with epithelioid macrophages and giant cells but without caseous necrosis. It has a lower fatality rate than GCM.[37,41] Clinical manifestations of cardiac sarcoidosis are dependent on the location, extent, and activity of disease, and range from clinically silent disease to sudden cardiac death. Conduction abnormalities, ventricular arrhythmias, and heart failure may occur. Ventricular tachycardia is the second most common manifestation of cardiac sarcoidosis and has been reported in nearly one-fourth of patients even mimicking arrhythmogenic cardiomyopathy.[41,42]

Eosinophilic myocarditis has been reported as a consequence of systemic diseases, such as Churg-Strauss syndrome, celiac disease, cancer, and hypereosinophilic syndrome.[43–45] Churg-Strauss syndrome is a multisystem disease characterized by rhinitis, asthma, peripheral blood eosinophilia, and necrotizing vasculitis (notably glomerulonephritis).[46] Cardiac involvement has been found in 64% of autopsy cases.[47] Cardiac pathologic findings include myocardial inflammatory infiltration with eosinophils and myocytolysis, intracavitary endocardial thrombi, coronary vasculitis, acute fibrinous pericarditis, and pericardial fibrosis.[48] Eosinophilic myocarditis can be seen in patients with eosinophilic neoplastic disorders as well as T-cell lymphoma and carcinomas (notably lung and biliary tract cancers).[49] In hypereosinophilic syndrome there is marked unexplained blood and tissue eosinophilia in combination with a variety of clinical manifestations. Tissue infiltration with eosinophils is found in many organs, i.e. skin, brain, liver, spleen, heart, lungs, etc. In the heart, a spectrum of disease is observed, from an acute inflammatory to a fibrotic stage.[50]

Toxic

Drugs can cause myocardial inflammation by either a direct toxic effect on the heart or by inducing hypersensitivity reactions. Among the various drugs, anthracyclines and cocaine in particular have often been implicated in acute myocarditis due to myocardial toxicity[51,52] (Table 8.1).

DIAGNOSIS

Histopathology

In order to develop uniform and reproducible criteria for the pathologic diagnosis of myocarditis, the so-called Dallas criteria were put forward in 1987, based upon histological features on EMB.[53] Accordingly, myocarditis is defined as an 'inflammatory infiltrate of the myocardium with necrosis and/or degeneration of adjacent myocytes, not typical of ischemic damage associated with coronary artery disease.' Two distinct diagnoses are then used for the first and subsequent biopsies (Table 8.3). Several limitations of the Dallas criteria have been raised: (a) inflammatory cell characterization by immunohistochemistry

TABLE 8.3 Dallas Criteria: Morphological Diagnosis of Myocarditis[53]

First biopsy	Myocarditis with/without fibrosis
	Borderline myocarditis (repeat biopsy may be indicated)
	No myocarditis
Subsequent biopsy	Ongoing (persistent) myocarditis with/without fibrosis
	Resolving (healing) myocarditis with/without fibrosis
	Resolved (healed) myocarditis with/without fibrosis

and the degree of fibrosis are not part of the criteria; (b) the type and extent of myocyte damage is not taken into account; (c) the phenotypic characteristics of the inflammatory cell infiltrate (i.e., lymphocytic, eosinophilic, polymorphous, giant cell, and granulomatous myocarditis) are only listed secondarily as a descriptive modifier; (d) a diagnosis of healing or healed myocarditis was not recommended for the first EMB, but only when unequivocal myocarditis has been previously diagnosed; (e) the recommended term 'borderline myocarditis.' for cases with limited amounts of inflammation including chronic forms of myocarditis, is not useful for treating and managing patients; (f) finally, any reference to etiological agents, such as viruses, was lacking because molecular techniques were not available at that time.[4,54]

EMB represents a fundamental step in the diagnosis of myocarditis. Although the disease is a diffuse process, the intensity and distribution of the inflammatory infiltrate are highly variable, including solitary small foci and multifocal aggregates, causing a low diagnostic yield on EMB due to sampling error.[55–58] The number, size, and processing of EMB samples have been demonstrated to influence the diagnostic sensitivity of EMB in the setting of myocarditis.[59,60] The Stanford-Caves and Cordis bioptomes have been shown to yield greater sensitivity compared with other smaller bioptomes.[57,58] It has also been shown that serial sectioning of the specimens and examination of multiple histologic levels increases the sensitivity of EMB in the evaluation of myocarditis.[61,62] Although biventricular EMB, rather than simply right ventricular EMB, has been demonstrated to improve the sensitivity in the detection of myocarditis[63]; this approach is more aggressive and may entail greater risk of complications.

Histopathological assessment not only remains the gold standard for the diagnosis of myocarditis but it is essential to reach a classification of myocarditis based upon histological criteria, i.e. lymphocytic, eosinophilic, polymorphous, granulomatous, and giant cell, where the histology often reflects a different etiopathogenesis of the myocardial inflammatory process (Figure 8.1). To overcome the limits of the Dallas criteria, a classification of myocarditis based upon semi-quantitative histological criteria has been proposed (Table 8.4), which includes assessment of the inflammatory cell type, grading the extent of myocyte damage and inflammation, and staging of the disease by semi-quantitative assessment of fibrosis (Table 8.4).[64] Grading and staging have been traditionally applied to tumor pathology for many years, as well as to non-neoplastic conditions such as chronic hepatitis. Grading can be used in inflammatory cardiomyopathy to describe the intensity of necro-inflammatory activity; staging, on the other hand, reflects the degree of fibrosis and architectural changes, which are the consequence of tissue injury and repair (Figure 8.2).

Immunohistochemistry

An actual increase in the sensitivity of EMB can be achieved by using immunohistochemistry together with histology. It is well recognized that many interstitial cells may be difficult to characterize on routine hematoxylin-and-eosin-stained sections, and normal components such as mast cells, fibroblast nuclei cut in cross-section, pericytes, histiocytes, and endothelial cells may resemble lymphocytes.[65–67] Moreover, a small number of inflammatory cells, including lymphocytes, may be found in the normal myocardium.[56–58] A large panel of monoclonal and polyclonal antibodies can be used, in addition to routine histology, to identify and characterize the inflammatory cell population as well as the activated immunological processes.[68–72] The main antibodies used for immunophenotype cell characterization are CD45 (leukocyte antigen common), CD43 (T lymphocytes), CD3 (T-cell marker), CD4 (helper T-cell marker), CD8 (cytotoxic T-cell marker), CD45RO (memory subset of CD8+ T-cell marker, expressed on activated T-cells), CD20 (B-lymphocytes), and CD68 (macrophages). A value of >14 leukocytes/mm^2 with the presence of T lymphocytes >7 cells/mm^2 has been considered a realistic cut-off to reach a diagnosis of myocarditis on EMB[72] (Figure 8.3). Additional stains relevant to immune activation include HLA-ABC and HLA-DR.[73]

The diagnosis of myocarditis at postmortem examination deserves a separate discussion. Myocarditis has been traditionally considered an important cause not only of heart-failure-related death but also of sudden death at any age, particularly in young individuals.[3,74] However, a patchy inflammatory infiltrate, in the form of a starry-sky-like feature (<14 leukocytes/mm^2) not associated with myocyte necrosis, is a frequent observation at autopsy in patients dying from other causes. Thus, the prevalence of myocarditis may well have been exaggerated in some studies because of over-interpretation of histological data and lack of standardized morphologic criteria. A review of major autopsy series of sudden cardiac death in the young demonstrated that myocarditis accounts for 4% to 42% of fatal events.[3] For these reasons, the Association for European Cardiovascular Pathology (AECVP) put forward guidelines which represent the minimum standard that is required in the routine autopsy practice for the adequate assessment of sudden cardiac death, including not only a protocol for heart examination and histological sampling, but also for toxicology and molecular investigation.[75] Accordingly, scattered small inflammatory foci with or without small spots of fibrosis are non-specific, and should be distinguished from borderline/focal myocarditis. In the AECVP document, it has been emphasized that, in the absence of myocyte necrosis or injury, small foci of inflammatory cells (even after immunohistochemistry) are not sufficient evidence of myocarditis, nor are scattered small spots of fibrosis.

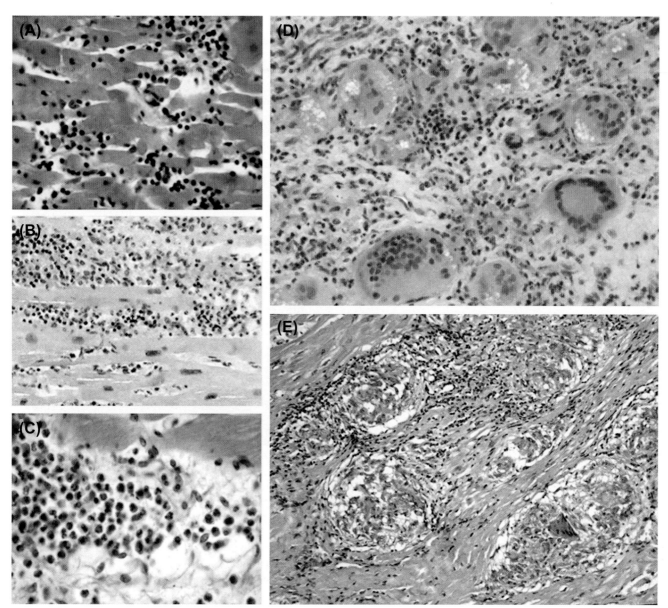

FIGURE 8.1 Histological subtypes of myocarditis: **(A)** lymphocytic; **(B)** polymorphous; **(C)** eosinophilic; **(D)** giant cell; **(E)** sarcoid (hematoxylin eosin stain).

Molecular Biology

Lymphocytic myocarditis is the most common form of myocarditis in Western countries, and most cases are documented or presumed to be viral in origin (Table 8.1). Non-viral infective agents are often morphologically distinctive and identifiable by routine histology and special stains.[4,11,76] Classical morphological analysis (histology and immunohistochemistry) has great limits in the detection of viral agents, with the rare exception of some forms of cytomegalovirus myocarditis. Moreover, the cytopathic effects, such as cytoplasmic degenerative changes or myocyte nuclear hyperplasia, are often not specific.

Many infective agents, particularly viruses, cannot be cultured as they are very difficult to isolate. Virus isolation procedures are poorly standardized, and results vary considerably between laboratories. The type and quality of the specimen, the timing of specimen collection, and the choice of cell types can highly influence the sensitivity.[77] Traditional serological studies and peripheral viral cultures have been used in the past to identify the most frequent pathogens associated with viral myocarditis, but these methods lack sensitivity and specificity. Today the serological analyses for viral infection in suspected myocarditis are still widely used; however the 'futility' of this test has been recently demonstrated emphasizing that EMB remains the gold standard in the etiological diagnosis of viral myocarditis.[78]

Before the molecular biology era, the infectious nature of myocarditis was only suspected on the basis

TABLE 8.4   Inflammatory Cardiomyopathy Classification: Grading and Staging[64]

Grading (Burden of Myocyte Damage and Inflammation)		
(A) Myocyte damage	Absent	0
	Focal	1
	Plurifocal	2
(B) Interstitial inflammation	≤7 T-cells/mm^2	0
	>7 to ≤14 T-cells/mm^2	1
	>14 T-cells/mm^2	2
(C) Endocardial involvement (inflammation, thrombosis)	Absent	0
	Present	1
	Maximum score	5
Staging (fibrosis)		
(A) Interstitial/replacement fibrosis	Absent	0
	>10 to ≤20 %	1
	>20 to ≤40 %	2
	>40 %	3
(B) Subendocardial fibrosis	Absent	0
	Present	1
(C) Endocardial fibroelastosis	Absent	0
	Present	1
	Maximum score	5

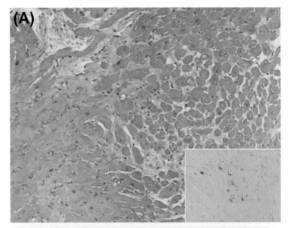

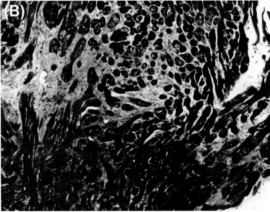

FIGURE 8.2   Chronic myocarditis: note the spotty replacement-type fibrosis associated with few focal inflammatory infiltrates ((A), hematoxylin eosin stain; (B), trichrome stain. In (A), positive T lymphocytes (CD3)). Fibrosis stains blue on trichrome stain.

of indirect criteria.[79,80] Historically, the first molecular technique utilized for detection of virus in heart tissue was the dot blot hybridization. With this technique, heart tissue extracts are applied directly on a membrane, followed by detection using specific radiolabel nucleotide probes.[81] Subsequently, southern and northern blots which involve the electrophoretic separation of specific nucleic acids, and in situ hybridization, which localizes a specific DNA or RNA sequence in a tissue section, have replaced the previous technique reducing non-specific inter-reactions.[82]

The development of molecular biological amplification methods, such as polymerase chain reaction (PCR), allows for the detection of a low copy of viral genomes even from an extremely small amount of tissue such as an EMB specimen. PCR is an enzymatic amplification technique whereby very few copies of RNA or DNA sequences can be amplified more than a million-fold. This allows transformation of a target sequence (i.e. the pathogen in question) from very low numbers to literally millions of copies. Several studies of patients with myocarditis have demonstrated the usefulness of PCR analysis for etiologic diagnosis. The decision to develop and apply PCR for routine diagnosis of infective myocarditis

has been made in relation to the speed, safety, sensitivity, and specificity of this test.[16,17] The main viruses to be considered when routinely performing molecular pathology studies in the myocardium of patients with a suspicion of myocarditis are adenovirus, CMV, EBV, enterovirus, hepatitis C virus, HHV6, herpes simplex virus 1 and 2, influenza viruses A and B, and PVB19. Other infectious agents may be investigated according to the clinical indication[4,11,76] (Table 8.5; Figure 8.4). However, it has been emphasized that the presence of viral genomes does not automatically imply a direct role of viruses in the pathogenesis of myocarditis, since an infective agent detected by PCR/RT-PCR/nested-PCR may be just an innocent bystander. Therefore, it is recommended to use molecular techniques as diagnostic tools ancillary to other mandatory investigations, either clinical or morphological, and apply it with skilled expertise. On the other hand, while interpreting the viral genome data, a limitation is still represented by the number of EMB specimens needed to obtain an acceptable sensitivity for the detection of viruses: only a positive PCR result is diagnostic, whereas a negative PCR result does

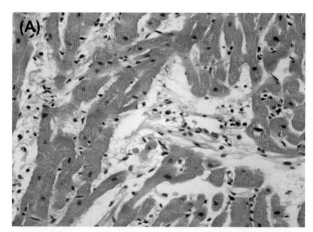

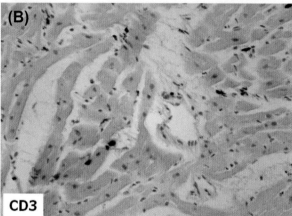

FIGURE 8.3 Histological and immunohistochemical diagnosis of lymphocytic myocarditis on endomyocardial biopsy. **(A)** Active lymphocytic myocarditis with interstitial edema and inflammatory infiltrates (hematoxylin eosin stain). **(B)** Immunophenotypic characterization of inflammatory cells as active T lymphocytes (CD3 antibody).

not exclude viral disease, particularly when a very small sample of tissue is analyzed, which is not representative of all the myocardium.

A peculiar population is represented by pediatric patients with a clinical suspicion of myocarditis, since EMB may be technically difficult to carry out at this age. However, it has been demonstrated that tracheal aspirates may be a useful surrogate for identification of causative viral agents by PCR analysis in young patients with suspected myocarditis and presumed pneumonitis. Molecular studies have demonstrated high concordance of viral genomes detected in EMB specimens and those obtained in tracheal aspirate.[83,84]

Additional Molecular Biology Tools

Other molecular strategies have been developed and applied to improve the identification of infective agents.

• **Blood sample.** Positive PCR results obtained on EMB should always be accompanied by a parallel

investigation on blood samples collected at the time of the EMB. The absence of a viral genome in the blood sample rules out the possibility of passive blood contamination of the myocardium, while viral blood positivity requires additional investigation by using quantitative PCR analysis.[11]

• **Viral gene sequencing.** Among the molecular biology techniques used to differentiate viral genomes, gene sequencing allows not only the precise characterization of the infective agent but can also help in assessing the molecular basis of cardiotropism as well as cardiovirulence.[85] Genomic sites determining the virulence phenotypes are now considered the principal factors responsible for the severity of the disease. Most of the efforts by the experts in the field using strains engineered in the laboratory have principally addressed the coxsackie virus B (CVB) genome. Dunn et al.[85] have identified cardiovirulence determinants in the CVB 3 5′ non-translated region (5′NTR), using two phenotypically and genotypically distinct clinical CVB 3 strains for the construction of six intratypic chimeric viruses. Identification of cardiovirulent strains could provide not only important prognostic indications on the clinical progression of the disease but also useful therapeutic suggestions on responsiveness to specific medical treatment. Moreover, in the future the knowledge of genomic sites which determine the cardiovirulence could lead to the development of vaccines for patients with a major risk of lethal complications.[86]

• **Viral infective status.** A further step in the molecular pathology investigation can be represented by the identification of the 'infective status' of viruses. The detection of viral replication has been demonstrated as useful not only for those DNA viruses known to be in the latent form but also for RNA viruses. The detection of the mRNA transcripts (early and late gene) of DNA virus using the RT-PCR may be considered diagnostic markers of the disease with expression of early genes preceding the antigenemia detection in most cases.[87] To characterize the activity of RNA viruses, an assay based on RT-PCR was developed to detect plus-strand-specific and minus-strand-specific viral RNA. The method relies on the first step of enteroviral replication, i.e., transcription of minus-strand RNA from the plus-strand enteroviral genomic template. Selective RT-PCR detection of minus-strand viral RNA is, therefore, an indicator of active viral replication. Using this method, Paushinger et al.[88] studied the EMB specimens from 45 patients with clinically suspected myocarditis. Eighteen of these patients (40%) were positive for plus-strand enteroviral RNA in the myocardium, and 56% of these 18 patients also had minus-strand enteroviral RNA, indicating that a significant fraction of patients with clinically suspected myocarditis had active enteroviral replication in the myocardium.[88] Similar data are

TABLE 8.5   Suggested PCR Primers for Viral PCR

Virus	Sequence (5′→3′)	T Anealing	Gene Target	Size bp
AV	GCCGCAGTGGTCTTACATGCACATC CAGCACGCCGCGGATGTCAAAGT	65°C	Exon protein	308
CMV (DNA)	CACCTGTCACCGCTGCTATATTGC CACCACGCAGCGGCCCTTGATGTTT	52°C	Phosphorylated matrix protein (pp65 e pp71)	399
CMV (RNA)	GTGACCTTGACGGTGGCTTT CGTCATACCCCCCGGAGTAA	57°C	Early gene	275
EBV	TTCGGGTTGGAACCTCCTTG GTCATCATCATCCGGGTCTC	64°C	Nuclear antigen 1 (EBNA 1)	268
EV/RV	AAGCACTTCTGTTTCC CATTCAGGGGCCGGAGGA	50°C	5′ untranslated region (5′-UTR)	297
HCV	GGAACTACTGTCTTCACGCAGA TGCTCATGGTGCACGGTCTA	54°C	5′ untranslated region (5′-UTR)	255
	GTGCAGCCTCCAGGACCC GGCACTCGCAAGCACCCTAT	56°C		210
HHV6	GTGAAAACTACGATTCAGGC TTTCGGAACATTGTTGAGC	55°C	major DNA binding protein (U41) gene	264
HSV	CATCACCGACCCGGAGAGGGA GGGCCAGGCGCTTGTTGGTA	60°C	DNA polymerase	92
INF A	AAGGGCTTTCACCGAAGAGG CCCATTCTCATTACTGCTTC	50°C	Non structural protein 1 and 2	190
INF B	ATGGCCATCGGATCCTCAAC TGTCAGCTATTATGGAGCTG	57°C	Non structural protein 1 and 2	241
PVB19	GGTAAGAAAAATACACTGT TTGCCCGCCTAAAATGGCTTT	57°C	Non structural protein 1	218

AV, adenovirus; bp, base pair; CMV, cytomegalovirus; EBV, Epstein-Barr virus; EV/RV, enterovirus/rinovirus; HCV, hepatitis C virus; HHV6, human herpes; 6 HSV, herpes simplex virus; INF A, Influenza A virus; INF B, Influenza B virus; PV B19, parvovirus B19.

FIGURE 8.4   Histological and molecular diagnosis of lymphocytic viral myocarditis on endomyocardial biopsy. **(A)** Active myocarditis with lymphocytic infiltrates and foci of myocyte injury (arrow) (hematoxylin eosin stain). **(B)** Gel electrophoresis, positive PCR for enterovirus. Line 1, DNA marker (factor VIII); line 2, positive control for enterovirus (180 bp); line 3, EMB positive for enterovirus; line 4, negative control.[4]

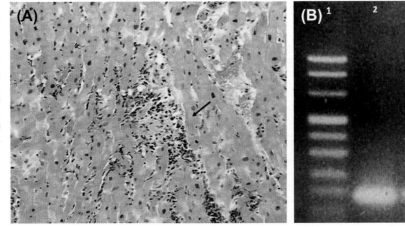

reported in HCV myocarditis where both positive and negative strands of HCV RNA were detected in cardiac tissue from patients with acute and chronic disease.[89]

- **Viral load.** More recently, the need for viral genome load quantification has been advanced. Real-time PCR amplifies a specific target sequence in a very small sample, then monitors the amplification progress using fluorescent technology determining the presence and quantity of pathogen-specific nucleotide. In particular, in recent years, several studies revealed a high prevalence, up to 60%, of PVB 19. However, PVB 19 was also detected in healthy transplant donors, in autopsy samples without myocarditis, and in patients undergoing EMB for other reasons.[90–94] Recently, Bock et al.[95] measured viral loads in EMB specimens obtained from 498 patients with myocarditis or chronic DCM who were positive for PVB 19 on immunohistologic analysis. PVB-19-associated acute myocarditis was characterized by a high mean viral load of 316000 genome equivalents (ge) per microgram of isolated nucleic acids, whereas the mean viral load of 709 ge per microgram in EMB specimens obtained from patients with chronic myocarditis was consistent with a persistent PVB 19 infection ($P < 0.001$). In contrast, viral loads in EMB specimens obtained from patients with chronic DCM were 392 ge per microgram, and viral loads in normal control hearts were 84 ge per microgram ($P < 0.001$). Of note, PVB 19 replication in myocardial endothelial cells was substantiated by the detection of PVB19 RNA replicative intermediates in EMB specimens only from myocarditis, whereas viral RNA was not detected in assays in chronic DCM without inflammation or in control hearts. On the basis of these data, the authors suggested that a viral load of more than 500 ge per microgram in EMB specimens is a clinically relevant threshold for the maintenance of myocardial inflammation. PCR analysis, including detection of viral replicative status and viral load, can be applied on follow-up EMB of patients with viral myocarditis and may represent the best way to verify the efficacy of specific antiviral therapy and to evaluate viral clearance.

- **Viral localization.** A role for laser microdissection has also been suggested in specific types of myocarditis. Identification of viral myocarditis requires unequivocal demonstration of the viral genome or virus gene products within the cardiomyocytes and their absence in unaffected hearts. In particular, EBV has been detected occasionally in myocardial tissue of patients with myocarditis, but no conclusive data concerning its causal role are available since this virus may persist for life in circulating B cells after a primary infection, and contamination of blood might be the source of the identified viral DNA. To this aim, Chimenti et al.[96] studied a series

of patients with inflammatory cardiomyopathy by laser capture microdissection. Selection of cells for laser was guided by immuhistochemical detection of lymphocytes and of myocytes. The different cells were dissected separately from tissue sections, followed by extraction of nucleotides and assessment of EBV genome by PCR. EBV was detected in microdissected myocytes of all patients and was absent in infiltrating lymphocytes, suggesting a cytopathic role for EBV in these patients.

Anti-Heart Autoantibodies Testing

Myocarditis is also associated with anti-heart auto-antibodies (AHA) in the serum in some patients. AHA to various cardiac and muscle-specific autoantigens (Table 8.6) are found in myocarditis patients suggesting immune-mediated myocarditis.[97] Antibodies of IgG class, which are shown to be cardiac- and disease-specific for myocarditis/DCM, can be used as auto-immune biomarkers for identifying at-risk relatives and those patients in whom, in the absence of active infection of the myocardium, immunosuppression

TABLE 8.6 Reported Frequencies of Antiheart Autoantibodies in Human Myocarditis

Antibody	Technique	Percentage Antibody-Positive
MUSCLE-SPECIFIC		
ASA	s-I IFL	47
AMLA	AMC	41
AFA	s-I IFL	28
IFA	s-I IFL	32
Heart-reactive	s-I IFL	59
Organ-specific cardiac	s-I IFL +abs	41
ANTIMITOCHONDRIAL		
M7	ELISA	13
ANT	SPRIA	91
BCKD-E2	ELISA	10
Antilaminin	ELISA	73
Anti-β1 receptor	Bioassay	96
Anti-β MYHC	ELISA	37
Anti-α MYHC	ELISA	17

+abs, +absorption; AFA, antifibrillary antibody; AMLA, antimyolemmal antibody; ANT, adenine nucleotide translocator; ASA, antisarcolemmal antibody; BCKD-E2, branched chain α-ketoacid dehydrogenase dihydrolipoyl transacylase; ELISA, enzyme-linked immunosorbent assay; IFA, antiinterfibrillary; MYHC, myosin heavy chain; s-I IFL, standard indirect immunofluorescence; M7, heart mitochondria; SPRIA, micro solid-phase radioimmunoassay.
(Modified from 97).

and/or immunomodulation may be beneficial.[97] Some AHA have been described to predict poor outcome in patients with myocarditis or DCM. At present, no commercially available cardiac autoantibody tests have been validated against the results obtained in research laboratories. For the detection of AHA, sera from 2 ml of blood are tested by standard indirect immunofluorescence at one-tenth dilution on 4-mm-thick unfixed fresh frozen cryostat sections of blood group O normal human atrium and skeletal muscle. AHA titers are measured by doubling dilutions of sera in phosphate-buffered saline solution. AHA patterns are classified based on the staining pattern. 'Organ-specific' antibodies produce a diffuse cytoplasmic staining of myocytes. 'Cross-reactive 1' antibodies give a fine striational immunofluorescence on cardiac tissue, but stain skeletal muscle fibers only weakly, and 'cross-reactive 2' antibodies stain both heart and skeletal muscle sections with a striational pattern.[98,99]

In our experience at Padua University, we found AHA in a high proportion (56%) of myocarditis patients[20]; the low frequency in controls and the association of AHA with family history for cardiomyopathy are in keeping with what is observed in other autoimmune diseases. AHA are more often detected in the acute phase of myocardial inflammation than in the chronic stage. AHA of the IgG class were detected in 73 (56%) of the study patients: 54 (41%) of the organ-specific type and 19 (15%) of the partially organ-specific type. The frequency of organ-specific AHA was higher in myocarditis (41%) than in non-inflammatory heart disease (1%), ischemic heart disease (1%), or normal blood donors (2.5%) ($P < 0.0001$). Similarly, the frequency of the partially organ-specific AHA was higher in myocarditis (15%) than in non-inflammatory heart disease (4%), ischemic heart disease (1%), or normal subjects (3%) ($P < 0.0001$). Of the 98 patients for whom both PCR and AHA were available, myocarditis was classified as autoimmune (positive AHA and virus-negative PCR) in 48% of patients, viral (virus-positive PCR and negative AHA) in 9%, viral and immune (virus-positive PCR and positive AHA) in 12%, and idiopathic and/or cell-mediated (virus-negative PCR and negative AHA) in 31%.[20]

A prospective randomized trial, recruiting AHA-positive and PCR-negative patients for immunosuppression, is warranted. The so-called 'idiopathic myocarditis' could reflect viral myocarditis, owing to yet-unknown pathogens, or, most likely, a cell-mediated autoimmune form that might also benefit from immunosuppression. Finally, for the 12% of patients with positive PCR and AHA, these patients might be candidates for antiviral therapy and, after virus clearance, immunosuppression or combined antiviral and immunosuppressive therapy.

PATHOGENIC MECHANISMS

Myocarditis is most commonly initiated by the introduction of a virus of potentially pathogenic strain (e.g., enteroviruses such as coxsackievirus), or reactivation of a dormant pathogen (e.g., PVB 19). The virus can proliferate in the permissive tissues of the susceptible host and ultimately reaches the myocardium or blood vessels through hematogenous or lymphangitic spread, or both. The pathophysiological progression of virus-induced myocarditis is mainly derived from experiments with CVB in animal models of different genetic backgrounds (Figure 8.5).[100] It is important to note that there are considerable differences in the outcome and course of myocarditis depending on both host factors and viral genetics factors.

Host Factors

Factors that determine susceptibility to viral myocarditis are not fully known, although a variety of factors such as malnutrition, pregnancy, sex hormones, and age have been implicated. Genetic host factors (major histocompatibility haplotype, HLA-DQ locus, CD45 polymorphisms) may be important determinants of early viral infection.[101] Other host factors including mercury exposure, selenium deficiency, vitamin E deficiency, increase the propensity to viral infection and the progression towards chronic myocarditis.[102,103]

Viral Genetics

Viral factors, including genome phenotype, have been shown to affect cardiovirulence as well.

Different serotypes within the CVB group, as well as variants of a single serotype, vary in virulence and the pattern of disease that they produce. Variants of each serotype have enabled mapping of mutations associated with defined phenotypic characteristics in a number of instances. For example, in a laboratory-adapted strain of CVB3, a myocarditis phenotype has been associated with a C-to-U mutation at nucleotide 234 in the 5' NTR.[104] In addition, a study of clinical isolates has identified a number of mutations in the 5' NTR that appear to be associated with cardiovirulence; these mutations are localized in stem-loop II of the predicted secondary structure of this region.[85] Intuitively, one would expect structural proteins to be major determinants of tissue tropism, through their binding affinity for the cell receptor. It would seem likely that structural gene mutations are responsible for capsid stability and tissue tropism. In addition, nucleotide substitutions in the 5' NTR as well as perhaps in non-structural genes may play key roles in the ability of a virus to propagate in different cell types.[105]

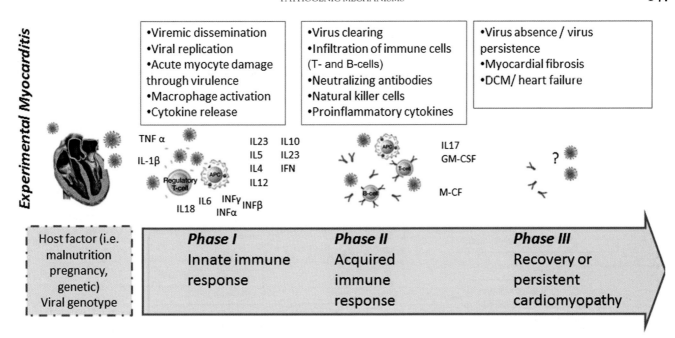

FIGURE 8.5 Pathophysiological phases of experimental viral myocarditis. DCM, dilated cardiomyopathy; GM-CSF, Granulocyte-macrophage colony-stimulating factor; IFN, interferon; IL, interleukin; M-CF, macrophage colony-stimulating factor; TNF-α, tumor necrosis factor-alpha.

Phases of Viral Myocarditis: From Virus Infection to Autoimmunity

Phase I: Viral Entry into Cardiomyocytes – Innate Immune Response

The virus enters the host through the gastrointestinal or respiratory tract (Figure 8.6). The infective agent is hidden in the immune cells of lymphoid organs, thus temporally escaping immune clearance and is then transported to other target organs such as the heart and pancreas, as a 'Trojan horse.' It is important to note that the different infection sites in cardiac tissues, in addition to different virus variants and viral load, may explain the heterogeneity of viral heart disease with respect to the expression of its phenotype, clinical presentation, and prognosis.

Coxsackievirus infects primarily cardiomyocytes through internalization of the coxsackie–adenoviral receptor (CAR) and its co-receptor, decay-accelerating factor (DAF).[106] CAR is preferentially localized at the intercellular tight junction in the adult cardiac myocyte. The virus initially links to DAF and this facilitates virus–CAR interaction. Specifically, the virus binds to DAF triggering actin rearrangement through AbI kinase activation, thus allowing viral particle movement into the tight junction and facilitating the interaction of the virus with CAR. The internalized virus, in turn, produces protease 2A, which cleaves dystrophin, a protein that functionally connects the cell surface sarcoglycan complex with extracellular lamin and intracellular actin.[107] Subsequent activation of the immune system may be accomplished by direct activation of the co-receptor associated signaling pathway, such as the

tyrosine kinase associated with DAF. On the other hand, some viruses often detected in the human heart cannot infect myocytes, due to the absence of the corresponding viral receptors; these viruses are usually localized to the endothelial cells or inflammatory cells (Table 8.2). As a consequence, these viruses do not damage the heart via cytolysis of cardiomyocytes but most likely by expression of cardiotoxic chemokines and cytokines.[10]

Equally, both types of virus trigger the innate immune response through pro-inflammatory cytokines, suppressors of cytokine signaling proteins and Toll-like receptors (TLRs). In concert, these cytokine signals activate local macrophages and up-regulate endothelial adhesion molecules as well as chemokines and chemokine receptors to collectively trigger the recruitment of innate immune cells. The innate immune response provides a fast, non-specific response to tissue damage and infection. The duration and degree of the innate immune response to viral infection has a crucial role in the development of myocarditis, and the inflammatory mediators are major players. This immune response enhances cardiac cell injury but also may result in efficient clearance of the virus, which corresponds clinically to a non-symptomatic or subclinical myocarditis.[108]

The direct proteolytic damage by the virus and recognition of viral pathogen motifs through cardiac TLRs contribute to the increasing expression of proinflammatory cytokines, including interleukin-1β (IL-1β), IL-6, IL-18, tumor necrosis factor-α (TNF-α), and interferons (IFNs) type I and type II. These cytokines are produced by resident cardiac cells, i.e. myocytes, fibroblasts, endothelial cells, and dendritic cells.

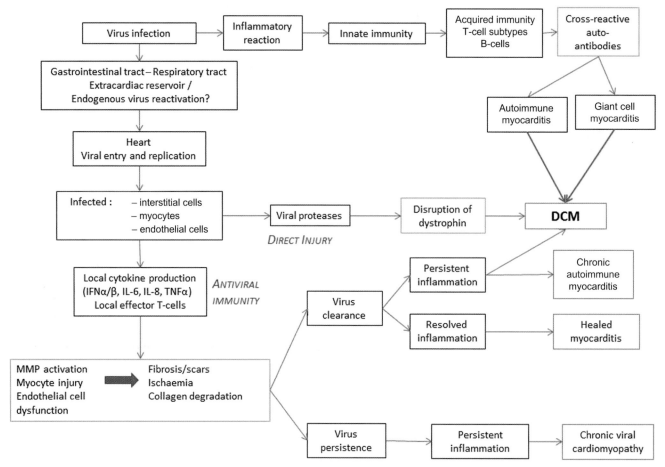

FIGURE 8.6   Myocardial tissue injury during different phases of infectious myocarditis, from direct virus-associated and/or immune-mediated tissue injury to recovery or persistent cardiomyopathy. DCM, dilated cardiomyopathy; IFN, interferon; IL, interleukin; LV, left ventricular; MMP, matrix metalloproteinase; TNF-α, tumor necrosis factor-alpha.

Phase II: Acquired Immune Response

The second phase is characterized by a shift towards a specific immune response. Cell-mediated immunity plays a critical role in the development of both viral and autoimmune myocarditis. The viral particles are captured by antigen-presenting cells, degraded within the Golgi apparatus, and presented on the cell surface with the major histocompatibility complex (MHC). The acquired or adaptive immune system responds to specifically recognized antigenic epitopes through clonal expansion of antigen-specific T and B lymphocytes, causing cellular and humoral antibody-mediated immune responses, respectively.[109]

Effector T-cells further differentiate into CD4+ and CD8+ subsets (Figure 8.7). CD4+ T helper cells produce cytokines that can be directly toxic to the target cells or can stimulate other T-cell effector functions and B-cell antibody production, as well as mobilize powerful inflammatory mechanisms. To date, four main T-cell populations have been reported to develop from naïve CD4+ T lymphocytes. T-cells differentiate through distinct pathways after activation and acquire

specialized properties and effector functions. Traditionally, T helper cells are thought to differentiate into T helper (Th) 1 and Th2 cell subsets. Th1 cells enhance the eradication of intracellular pathogens (by producing IFN-γ), and the Th2 cell lineage is important for enhancing the elimination of extracellular organisms (by producing IL-4, IL-5, and IL-13). Severity of autoimmune myocarditis is determined by the profile of Th1 and Th2 cytokines. The cellular immune response mediated by Th1 cells plays a critical role in the adaptive immune response in the viral myocarditis. Th1 cells produce mainly IFN-γ, but also IL-2 and TNF-α. Th2 cells secrete IL-4, -5, -6, -10, and -13 and promote non-inflammatory immediate immune responses; they have been shown to be essential in B-cell production of IgG, IgA, and IgE. The humoral immunity mediated by Th2 cells has been associated with the development of myocarditis and DCM with the discovery of the AHA, including the autoantibodies against the myosin heavy chain, the adenine nucleotide translocator, the β1-adrenergic receptor, the M2-cholinergic receptor, and others.

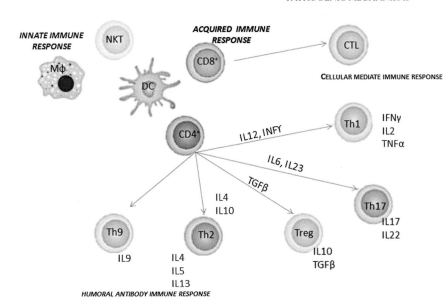

FIGURE 8.7 Differentiation of effector T-cells in infective myocarditis. DC, dendritic cell; CTL, cytotoxic T lymphocytes; IL, interleukin; INF, interferon; MΦ, macrophage; NKT, natural killer T-cell; TGF, transforming growth factor; TNF, tumor necrosis factor.

A key role in inflammatory tissue injury was recently advanced for a subset of IL-17-producing cells, distinct from the Th1 or Th2 cells. Th17 cells are the first cell subset that is produced during infection that induce the onset and development of acute viral myocarditis. Fibroblasts, epithelial cells, and keratinocytes express the IL-17 receptor, and the interaction with IL17 leads to the production of IL 6, chemokines (i.e. CXCL8 and CXCL2), and granulocyte macrophage colony stimulating factors (GM-CSF).[110]

Furthermore, regulatory T (Treg) cells are known as the regulators of effector responses, and these cells can down-modulate autoimmune responses and protect against inflammatory tissue injury. Treg cells and related cytokines (transforming growth factor (TGF)-β, IL10) are important in the maintenance of immune homeostasis in many infectious diseases. The protective role that Treg cells play in inflammation of the heart during viral myocarditis suggests that there is a balance to be struck between clearance of infection and immune-associated damage to the myocardium. Although the majority of Treg cells appear within the CD4+ T-cell set, suppressor activity was also reported among CD8+ T-cells.[108]

A new subset of CD4+ Th cells that predominantly secrete cytokine IL-9 has been identified, termed Th9 cells. Driven by the combined effects of TGF-β and IL-4, Th9 cells produce large quantities of IL-9 which has long been thought to be a Th2 cytokine. It has been hypothesized that the same precursor can be induced to differentiate into either the Th1 or Th2 population, similar to the differentiation of Th17 and Treg cells.[109] Th17 and Treg cells and the developmental programs of T-cells are reciprocally interconnected.

Cytotoxic CD8+ T-cells are very effective in directly binding and eliminating infected cells by recognizing MHC class I antigens on infected cardiomyocytes, assisted by the effects of TNF-α and IFN-γ which promote MHC class I presentation and facilitate cell–cell contact between T-cells and myocytes. Cytotoxic CD8+ T-cells are able to destroy the infected cardiac cells through secretion of toxic cytokines or perforin, which perforates the plasma membrane of the infected cells.[109]

Phase III: Recovery or Persistent Cardiomyopathy

The last phase is characterized by cardiac repair and remodeling, probably in the absence of viral replication or viral genome within the myocardium. The resolution of the immune response and replacement of dead tissue by a fibrotic scar is promoted by anti-inflammatory cytokines (TGF-β and IL-10) secreted by Treg cells or alternatively activated macrophages. In most patients with viral myocarditis, the pathogen is cleared, and the immune system is down-regulated with no further adverse effects. However, in a minority of patients, the virus is not cleared, resulting in persistent myocyte damage due to the immune response with cardiac autoantibodies. The contractile dysfunction in the long term depends on the extent of myocardial damage, cardiac dilatation and compensatory hypertrophy during this phase.[111]

Key Inflammatory Immune Effectors and Regulators

Cytokines

Cytokines have essential roles in immune cell development, immune regulation, and immune effector functions. A combination of cytokines exerts variable effects at different times during the evolution of disease. Cytokines are important in controlling T-cells responsive to self-antigens and are critical in shifting the immune response toward different Th patterns. The Th1 response shifts the cytokine profile toward delayed hypersensitivity, macrophage activation, and a proinflammatory T-cell response associated with IFN-γ and IL12; whereas the Th2 response is associated with B-cell activation and humoral immunity, and IL-4, -5, -10, and IgE production; Th17 cells lead to production of IL 6 and Tregs produce TGFβ and IL10. Th1 T-cells secrete IL-2 and IFN-γ that suppresses Th2 responses, while Th2 T-cells secrete IL-4 and IL-10 that inhibit Th1 responses. The expression of the mRNA of Th1-related cytokines, such as IFN-α and TNF-α, occurs throughout the early phase of infection, and the mRNA of Th2-related cytokines, such as IL-4 and IL-10, is expressed when massive cell infiltration appears in the myocardium. In addition, the timing of cytokine production may enhance or abrogate autoimmune disease. Large amounts of IL-12 or TNF-α produced early in the disease may lead to progression of autoimmunity, whereas their production later in disease could institute terminal differentiation and death of T-cells and abrogation of disease.[112]

IL-1 has multiple effects on cardiac function. In synergy with TNF-α, IL-1 exacerbates cardiac myocyte and intact heart contraction producing delayed and prolonged myocardial contractility. It also acts by interfering with calcium homeostasis. Along with TNF-α, IL-1 promotes collagen synthesis and induction of matrix metalloproteinase (MMP)2 and MMP9.

TNF-α acts through two receptors: TNF-R1 and TNF-R2. TNF-R1 stimulation mediates the majority of the injurious effects of TNF-α, whereas TNF-R2 appears to be responsible for protective effects on the heart. The adverse effects of TNF-α include cardiac myocyte hypertrophy, contractile dysfunction, myocyte apoptosis, and extracellular matrix remodeling. A high level of TNF-α has been detected in EMBs from adult patients with myocarditis. The level of TNF-α expression significantly correlated with impaired cardiac function and its persistence, as demonstrated in the follow-up EMB or in the heart from patients who died or underwent heart transplantation. Thus TNF-α has been implicated in the progression of myocarditis and the development of irreversible heart failure.[113]

IL-6 signaling through the IL-6 receptor promotes increased cardiac contractility, but excessive IL-6 signaling can lead to maladaptive hypertrophy. In contrast, by interacting with TNF-α and IL-1, IL-6 can also depress cardiac function.

IL-18, like IFN-γ, is a proinflammatory cytokine in humans generated during infectious and other inflammatory processes. In synergy with IL-12, it stimulates IFN-γ production by T-cells, natural killer cells, and macrophages. IL-18 can also act through IFN-γ-independent mechanisms. It is associated with cardiac myocyte hypertrophy and eventual functional decompensation as well as increased cardiac myocyte apoptosis.

IFNs play a central role in antiviral responses. Administration of IFN-β reduces myocardial lesions when given in the acute stage of CVB3-induced myocarditis. Conversely, inhibiting IFN-β ameliorates disease only when given in a later stage, suggesting that the favorable effects of type I IFN may be time-dependent. Survival is markedly reduced in mice lacking either the IFN-β gene or the type I IFN receptor (thus deficient in both IFN-α and IFN-β signaling). Mortality and viral titers were not affected in IFN-γ-deficient animals, although pancreatic overexpression of IFN-γ protects against cardiac damage, viral replication, and mortality following CVB3 infection.[114]

GM-CSF stimulates the production of neutrophils, eosinophils, and monocytes. Monocytes exit the circulation and migrate into tissue, where they mature into macrophages and dendritic cells. Thus, GM-CSF is central to the immune/inflammatory cascade.[115]

IL-4 is the prototypic Th2 cytokine. These responses are mostly associated with allergic reaction and their role in myocarditis is often overlooked. Afanasyeva et al.[116] found that blocking IL4 with anti-IL4 monoclonal antibody significantly reduced the severity of myocarditis. It was associated with a shift from a Th2 to Th1 phenotype as shown by a reduction in the cardiac myosin specific IgG. IL-5 stimulates B-cell growth and increases immunoglobulin secretion, and it is also a key mediator in eosinophil activation. IL-10 is a pleiotropic immunomodulatory cytokine that functions at different levels in the immune response. It down-regulates the expression of Th1 cytokines, MHC class II antigens, and co-stimulatory molecules on macrophages. It also enhances B-cell survival, proliferation, and antibody production. IL-10 can block NF-κB activity. IL-10 is capable of inhibiting synthesis of pro-inflammatory cytokines such as IFN-γ, TNF-α, IL-2, and IL-3.

IL-13 protects from myocarditis induced by cardiac myosin peptide or viral infection. In the absence of IL13 a severe disease occurs characterized by increased leukocytic infiltration, elevated cardiac myosin antibodies, and subsequently greater cardiac fibrosis. In contrast IL-23 is required for autoimmune myocarditis. IL-23 is a heterodimeric cytokine consisting of two subunits: p40, which is also a component of the IL-12 cytokine, and p19, which is considered the IL-23 alpha subunit. IL-23 promotes upregulation of MMP9, increases angiogenesis and reduces CD8+ T-cell infiltration. In conjunction with IL-6 and TGF-β1, IL-23 stimulates naive CD4+ T-cells to differentiate into a novel subset of cells called Th17 cells.[110]

Finally, IL-17 increases in the circulation and heart during both acute and chronic myocarditis suggesting a role in disease. Blocking IL-17 using a neutralizing antibody, reduced acute myocarditis and viral replication in the heart.[110]

Toll-Like Receptors

TLRs recognize pathogen-associated molecular patterns. To date, ten members of TLRs have been identified in humans. TLRs are activated by a variety of ligands that are generally associated with infectious pathogens. TLRs are ubiquitously expressed at low levels in non-immune cells, but relatively high-level expression can be observed in innate immune cells. In the human heart, the mRNA expression level of TLR7, TLR8, and TLR9 is negligible; however, relatively low but clear mRNA expression of TLR3 and TLR4 has been reported using quantitative polymerase chain reaction.

The activation of distinct TLRs signaling cascades following pathogen recognition leads to beneficial or adverse immune responses. The majority of TLRs signal initiates through a common downstream adaptor molecule called myeloid differentiation primary response gene 88 (MyD88), which typically leads to activation of nuclear factor kappa B (NF-κB). An important exception is TLR3, which recognizes double-stranded RNA (ds RNA) and signals through TIR domain-containing adapter-inducing interferon-β (TRIF), which activates interferon regulatory factor 3 (IRF3) to up-regulate both type I and type II IFN expression.

Although not all TLRs have been investigated in knockout models, both TLR4- and TLR9-deficient mice are protected against CVB-induced cardiac injury with uncompromised viral clearance. In agreement with these studies, deficiency of the common adaptor protein MyD88 leads to nearly absent inflammation in combination with suppressed viral levels in cardiac tissue. These data indicate that the effects of the MyD88-mediated TLR signaling pathways contribute to cardiac pathology in myocarditis while bearing little importance for viral elimination. In contrast, viral replication and cardiac damage are aggravated in TLR3 knockout animals, and transgenic overexpression of TLR3 decreases mortality even in type I IFN-deficient animals through stimulation of type II IFN signaling. In agreement, mice lacking the TLR3-adaptor TRIF have impaired type I IFN responses and develop severe myocarditis with high viral loads, resulting in 100% mortality (Figure 8.8).[100,108]

Suppressor of Cytokine Signaling

Suppressor of cytokine signaling (SOCS) plays a key role in the negative regulation of both TLR- and cytokine receptor-mediated signaling, which is involved in innate and subsequent adaptive immunity. The SOCS family is composed of eight members: cytokine-inducible Src homology 2 domain-containing proteins, and SOCS1 to SOCS7. Coxsackievirus infects cardiomyocytes and induces the expression of SOCS1 and SOCS3 in cardiomyocytes, which can result in evasion of immune responses and facilitation of virus replication by inhibition of JAK–STAT signaling. The expression of SOCS1 is induced by various cytokines, including IFN-γ, IL-4, and IL-6, and also by TLR ligands, such as LPS and CpG-DNA.[117]

Chemokines

Chemokines are a family of small chemotactic cytokines causing directed migration of leukocytes into inflamed tissue and have been shown to directly induce chemotaxis. These inflammatory mediators could contribute to myocardial injury indirectly (e.g. via recruitment and activation of infiltrating leukocytes) and directly (e.g. through modulation of apoptosis, fibrosis, and angiogenesis within the failing myocardium). Chemokines have contrasting effects during myocarditis: the myocardial damage is reduced in knockout mice deficient for macrophage inflammatory protein-1α (MIP-1α, also known as CCL3) as well as in animals treated with inhibitors for IFN-γ-induced protein-10 (IP-10, also known as CXCL10). On the contrary, the damage is increased in animals deficient for IP-10, and more research is required to draw conclusions about the roles of chemokines during viral myocarditis.[100,118]

Matrix Metalloproteinases

The MMPs play critical functions in the inflammatory response, including matrix degradation, facilitation of leukocyte migration, and cytokine processing and regulation.[119] Contrasting data have been obtained from mice deficient in single MMPs. In MMP-8 knockout mice, viral load and inflammation remain unchanged, whereas MMP-9 knockout mice show increased myocardial damage and dysfunction associated with higher viral loads and infiltration. Similarly MMP-3-deficient animals also exhibited increased myocardial damage related to expression of macrophage chemoattractant protein-3.[119] Further research is required for a better understanding of the role of MMPs during viral myocarditis.

MicroRNA

MicroRNAs (miRNAs) are endogenous, conserved, single stranded, small (22 nucleotides in length), noncoding RNAs that control the expression of entire networks of complementary transcripts.[120] In the heart, miRNAs have been shown to be necessary for normal embryonic development and homeostasis. MiRNAs have been linked to specific cardiac diseases and recently also to viral myocarditis. In particular, Corsten et al.[121] demonstrated that a set of miRNAs is consistently up-regulated during acute inflammation in both human and mouse viral myocarditis, including miR-155, miR-146b, miR-21 (which have central functions in immune activation and inflammation), and less well known miR-130b

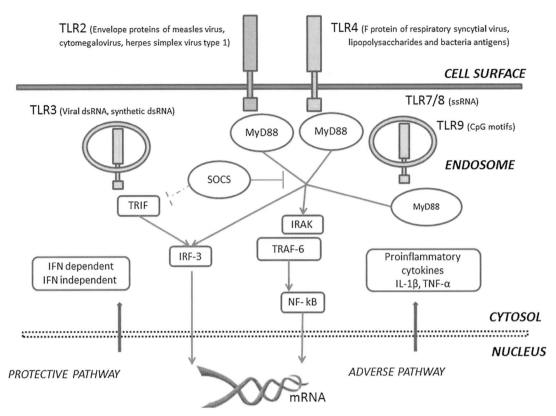

FIGURE 8.8   Effects of Toll-like receptor signaling pathway in viral myocarditis. IL, interleukin; INF, interferon; IRF, interferon regulatory factor; IRAK, Interleukin-1 associated kinase; MyD88, myeloid differentiation primary response gene 88; NFκB, nuclear factor kappa B; SOCS, suppressor of cytokine signaling; TLR, Toll-like receptor; TN-Fα, tumor necrosis factor-alpha; TRAF, TNF receptor associated factor; TRIF, TIR domain-containing adapter-inducing interferon β.

(which is a virus-induced inhibitor of target cell apoptosis in humans). MiR-155 is a known pro-inflammatory regulator of myeloid and lymphoid immune cell function, and is mainly expressed by recruited inflammatory cells. It is consistently up-regulated in human and experimental models of viral myocarditis during the acute phase. In lymphocytes, miR-155 promotes Th1 responses as well as germinal center formation, while favoring proliferation and pro-inflammatory cytokine secretion in myeloid cells. MiR-155 also mediates cross-talk between innate and adaptive immune cells. A further set of miRNAs was identified to be down-regulated for immune functions; miR-107, for example, is down-regulated in multiple cell types in response to TLR-4 signaling and is required to facilitate macrophage adhesion. In contrast, inhibition of miR-99b reduces dendritic cell differentiation.

In addition, a set of miRNAs was identified exclusively associated with viral infection in resistant C57Bl6/N mice, including miR-483, miR-411, miR-135, miR-300 and miR-140, miR-93, and miR-17. While these miRNAs are relatively unknown, their likely involvement in the primary response to viral infection and disease susceptibility needs further investigation. Targeted therapeutics for acute viral myocarditis using miR-155 inhibition or other miRNA therapies may provide novel approaches to this potentially devastating disease.

Gene Expression Profile in Myocarditis

Gene expression profiling by microarray technologies has been successfully applied to study the transcriptional changes that occur in tissues. Gene expression analyses engage the entire pathophysiological microenvironment associated with the myocarditic disease and may provide directions of greater relevance to the pathogenesis and ultimate treatment of viral myocarditis.

Taylor et al.[122] examined temporal changes in gene expression profiles in the mouse model of CVB3-induced myocarditis at days 3, 9, and 30 after infection. A total of 169 genes had a level of expression significantly different at one or more post-infection time points as compared with baseline. In the early myocarditis stage, there was diffuse down-regulation of mitochondrial and metabolic gene expression, which has important implications for cardiac function, considering the central role of mitochondria in energy production for contractile myocardial cells and in cell death. Mitochondria participate in CVB3-induced cell death, by releasing cytochrome c subsequent to caspase activation, and this cell death likely plays an important role in the release of progeny virus. Genes directly involved in viral replication, like poly(A)-binding protein (PABP), ubiquitin-specific protease UBP41, and inorganic pyrophosphatase genes were up-regulated in infected mice in the early phase. PABP transcript up-regulation is seemingly compensatory for viral degradation of PABP, which is cleaved by the CVB3 viral genome-encoded protease 2Apro.

Microarray technology has been applied to study myocarditis of various etiologies in human patients. For example, gene expression analysis provided novel insight into the pathogenesis of GCM by showing consistent up-regulation of genes involved in immune response, transcriptional regulation, and metabolism.[123] Cunha-Neto et al.[124] compared Chagas' cardiomyopathy gene expression profiles with those of patients with DCM in order to investigate selective disease pathways and potential therapeutic targets. The authors showed a selective up-regulation of genes involved in immune response, lipid metabolism, and mitochondrial oxidative phosphorylation in patients with Chagas' cardiomyopathy.[124]

Ruppert et al.[125] have investigated the expression profiles from EMB specimens obtained from patients with DCM, with or without inflammation, acute myocarditis, and pericarditis. Differences in gene expression between the groups were found for approximately 45% of genes tested, which had been selected based on likely involvement in inflammatory processes. A set of 42 genes was considered to be significantly different between inflammatory versus non-inflammatory cardiomyopathy. For the majority of the genes, the expression level was higher in patients with myocarditis and pericarditis as compared to DCM.

Recently, Heidecker et al.[126] performed gene expression analysis of EMB specimens in matched cohorts of patients with idiopathic cardiomyopathy and myocarditis. Genes encoding the TLR family were overexpressed in the patients with myocarditis. From these data a transcriptome-based biomarker set consisting of 62 genes was identified that distinguished myocarditis with 100% sensitivity and 100% specificity from cardiomyopathy. After multiple classification algorithms and quantitative RT-PCR, the subset of diagnostically relevant genes was further reduced to a highly robust signature of 13 genes, which still performed with 100% accuracy in the diagnostic distinction between different types of inflammatory heart disease from a single EMB. These differentially expressed genes may function as a highly accurate transcriptome-based biomarker with regard to the differential diagnosis of inflammatory heart diseases. The identification of transcriptome-based biomarkers may increase the diagnostic accuracy of EMB and may be of critical importance for developing specific therapies for inflammatory conditions in the heart.

TREATMENT

The current therapy guidelines for patients affected by active myocarditis include treatment for heart failure and arrhythmias as well as life-style modifications.[101] Patients with advanced refractory cardiac failure may be candidates for heart transplantation, but the possibility of disease recurrence, the poor prognosis, and the high rejection rate may reduce the benefits of this therapeutic approach.[127] When there is not significant improvement in ventricular function with a traditional 'symptomatic' therapy, an etiology-specific treatment may be considered. Treatment with non-steroidal anti-inflammatory drugs has not proven to be effective in several murine models, and in fact the use of these medications in murine models of acute viral myocarditis resulted in increased inflammation and higher mortality.

Viral Myocarditis

Since a common cause of myocarditis is viral infection, treatment with antiviral medications would seem reasonable. Different viruses, viral load, and the type of the infected cell may influence the clinical response to immunomodulatory treatment. Data regarding antiviral treatment in myocarditis are limited to murine models and a few case series in human patients, but results have been promising. In the single case series of antiviral therapy use in humans with fulminant myocarditis, ribavirin therapy did not prove effective.[128] However, most patients with acute myocarditis were diagnosed several weeks after viral infection, so it is unlikely that antiviral therapy administered once myocarditis has been confirmed would provide much benefit. In fact, some human case reports have indicated successful use of specific antiviral drugs (i.e. ganciclovir for CMV) in the early phases of disease.[129]

Recently an experimental gene therapy study used a soluble virus receptor against CVB3 in CVB3-infected mice. The authors showed improved cardiac contractility and diastolic relaxation in treated vs non-treated infected mice, so that the cardiac function indices in the treated mice were similar to those of healthy control animals.[130]

A broad range of pro- and anti-inflammatory cytokines have been studied as therapeutic targets in persistent viral myocarditis where the immune response plays a critical role. Etenarcepet, a recombinant human TNF-α antagonist, was considered a potential therapeutic agent for patients with myocarditis. However, two multicenter clinical trials were conducted using etenarcept in patients with chronic heart failure, but the results showed no clinical benefit.[131] Noteworthy, the patients enrolled in both studies were not evaluated for TNF-α serum load or for TNF-α gene or protein expression on EMB using immunohistochemistry and molecular techniques.

INF-β and α have been used to treat chronic viral heart disease.[132,133] INFs protect myocytes against injury and decrease inflammatory cell infiltrates. The European-wide multicenter BICC study used INF-β1β in 143 patients with inflammatory cardiomyopathy and myocardial viral infection. Preliminary data showed that treatment is associated with a significantly reduced viral load for enterovirus and adenovirus in the myocardium. While PVB 19 and HHV6 were not completely eliminated, clinical parameters including the New York Heart Association (NYHA) functional class and patient global assessment did improve.[134]

Autoimmune Myocarditis

In the treatment of myocarditis, controversial results were obtained with immunosuppressive treatment (azathioprine, prednisone, and cyclosporine). In the past, several clinical trials of immunosuppressive therapy failed to show improvement of left ventricle systolic function or survival.[135,136] In these trials patients were not evaluated for the presence of infections by molecular techniques, thus many of them may have harbored viral material, not detected by routine histology. Immunosuppressive therapy has been demonstrated to have a beneficial effect only in patients with EMB-proven virus-negative inflammatory cardiomyopathy. In fact, Frustaci et al. confirmed that immunosuppressive therapy is of no benefit in patients with molecular-proven viral myocarditis.[137] TIMIC was a clinical trial in which all EMB specimens were studied by histological, immunohistochemical, and molecular methods in order to exclude viral infection. A significant improvement in ejection fraction and decrease in left ventricular dimension were observed with immunosuppressive therapy with prednisone and azathioprine in virus-negative myocarditis.[138]

The use of intravenous immunoglobulin (IVIG) in acute myocarditis is not effective and appears to confer no advantage over steroid therapy alone. However, the immune-absorption of AHA and immune complex removal from patient's plasma seem to improve the cardiac function, hemodynamic parameters, clinical and humoral markers, and to reduce myocardial inflammation in DCM.[139]

Prevention

Specific strategies for the prevention of myocarditis are not yet available. Vaccination against measles, mumps, rubella, polio, and influenza has made myocarditis secondary to these diseases quite rare, thus raising the question whether vaccinations against other cardiotropic viruses could prevent myocarditis. Murine models have demonstrated that vaccination is protective against viral infection and prevents myocardial damage.[101] In particular, since CVB3, along with other enteroviruses, is involved in the majority of myocarditis and in many cases of DCM, prevention of CVB3 infection is highly

TABLE 8.7 Endomyocardial Biopsy Diagnostic Potential and Grading of Recommendation in Clinically Suspected Myocarditis/Inflammatory Cardiomyopathy (AECVP/SCVP consensus)

Histological Notes	Technical Aspects and Tissue Triage	Grading of Recommendation
Definite diagnosis: lymphocytic, granulocytic, polymorphous, eosinophil, necrotizing eosinophilic, giant cell, granulomatous myocarditis, with or without associated myocyte damage/necrosis.	In addition to ≥ three formalin fixed samples, two fragments (or one if greater than 3 mm²) could be snap-frozen or preserved in RNAlater for virus molecular investigation. One peripheral blood sample (5–10 ml) collected in EDTA or sodium citrate tubes.	Published peer-reviewed evidence supports the utility of the test
Histological findings alone or together with molecular techniques may be able to specify the etiology of the inflammatory disease.	Timing and number of EMB and serial histological sections are important for detection of myocarditis.	

(Modified from Reference 141)

desirable. To this regard, Zhang et al. have recently demonstrated that vaccination with CVB3-like particles elicits a humoral immune response and protects mice against myocarditis.[140] However, these studies are experimental, and there have been no vaccination trials in humans.

CONSENSUS STATEMENT ON EMB FROM THE ASSOCIATION FOR EUROPEAN CARDIOVASCULAR PATHOLOGY AND THE SOCIETY FOR CARDIOVASCULAR PATHOLOGY

An international task force of the AECVP and the Society for Cardiovascular Pathology (SCVP) recently published a position paper concerning the current role of EMB in the diagnosis of cardiac diseases and its contribution to patient management, mainly focusing on pathological issues. In this document, it has been emphasized that pathological analysis of EMB specimens is complex and requires standard protocols (Table 8.7).[141,142] Therefore, it is vital that the pathologist is provided with clinical information as an aid for diagnosis and to guide in the choosing of the most appropriate technical procedures to perform on the biopsy specimen. For myocarditis, optimal specimen procurement and triage indicates at least three, preferably four, EMB fragments, each 1–2 mm in size, be obtained and fixed in 10% buffered formalin at room temperature for light microscopic examination. In suspected focal myocardial lesions, such as myocarditis, additional sampling is recommended.

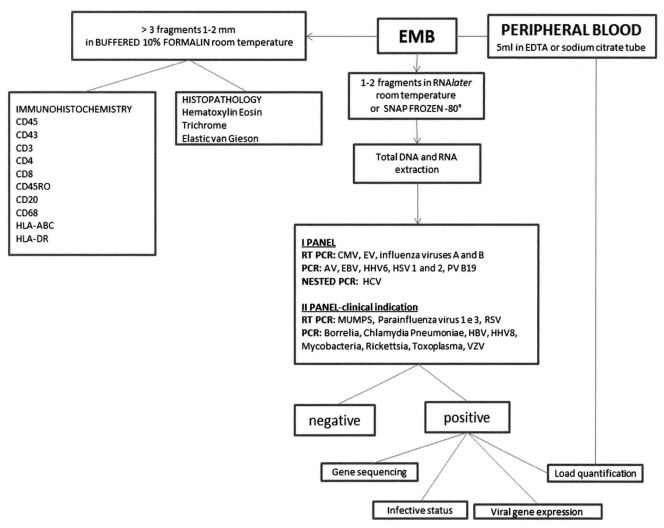

FIGURE 8.9 Endomyocardial biopsy protocol in patients with clinically suspected myocarditis. AV, adenovirus; CMV, cytomegalovirus; EBV, Epstein-Barr virus; EV, enterovirus; HCV, hepatitis C virus; HHV6, human herpes virus 6; HHV8, human herpes virus 8; HSV, herpes simplex virus; MUMPS, paramyxovirus; PV B19, Parvovirus B19; RSV, respiratory syncytial virus; VZV, varicella zoster virus; PV B19, parvovirus B19.

Moreover, in this clinical setting, one or two specimens should be snap-frozen in liquid nitrogen and stored at −80°C if required for molecular tests or specific stains. Otherwise, fragments may be stored in RNA-later (a solution preventing RNA degradation) at room temperature. A sample of peripheral blood (5–10 ml) in EDTA or citrate from patients with suspected myocarditis allows molecular testing for the same viral genomes sought in the myocardial tissue (Figure 8.9).

Concerning the diagnostic potential of EMB in myocarditis, the AECVP-SCVP panel of experts recognizes that EMB is the only way to reach a definite diagnosis of myocarditis based upon published peer-reviewed evidence. Furthermore, the authors emphasize the unique role of EMB in identifying lymphocytic, granulocytic, polymorphous, eosinophilic, necrotizing eosinophilic, giant cell, and granulomatous myocarditis, with or without associated myocyte damage/necrosis; these histological findings on EMB, together with molecular techniques, are the only way to address the etiology of the inflammatory disease.[142] Consideration of the timing and number of EMBs, as well as performing serial histological sections, are important factors that can help to maximize the diagnostic sensitivity of this technique. These indications have been included in the recent position statement of the European Society of Cardiology working group on myocardial and pericardial diseases.[143]

References

1. Richardson P, McKenna WJ, Bristow M, et al. Report of the 1995 WHO/ISFC task force on the definition of cardiomyopathies. *Circulation* 1996;**93**:841–2.
2. Maron BJ, Towbin JA, Thiene G, et al. Contemporary definitions and classification of cardiomyopathies. An American Heart Association Scientific statement from the Council on Clinical Cardiology, Heart Failure and Transplantation Committee; quality of Care and Outcome research and Functional Genomics and Translational Biology Interdisciplinary Working Groups: and Council on Epidemiology and Prevention. *Circulation* 2006;**113**:1807–16.

3. Basso C, Calabrese F, Corrado D, Thiene G. Postmortem diagnosis in sudden cardiac death victims: macroscopic, microscopic and molecular findings. *Cardiovasc Res* 2001;**50**:290–300.

4. Basso C, Calabrese F, Angelini A, Carturan E, Thiene G. Classification and histological, immunohistochemical, and molecular diagnosis of inflammatory myocardial disease. *Heart Fail Rev* 2012 Oct 25.

5. Mason JW, O'Connell JB, Herskowitz A, et al. Clinical trial of immunosuppressive therapy for myocarditis. The Myocarditis Treatment Trial Investigators. *N Engl J Med* 1995;**333**:269–75.

6. Felker GM, Hu W, Hare JW, Hruban RH, Baughman KL, Kasper EK. The spectrum of dilated cardiomyopathy. The Johns Hopkins experience in 1278 patients. *Medicine* 1999;**78**:270–83.

7. Towbin JA, Lowe AM, Colan SD, et al. Incidence, causes, and outcomes of dilated cardiomyopathy in children. *JAMA* 2006;**296**:1867–76.

8. Cooper Jr LT. *Myocarditis N Engl J Med* 2009;**360**:1526–38.

9. Friedrich MG, Sechtem U, Schulz-Menger J, et al. International Consensus Group on Cardiovascular Magnetic Resonance in Myocarditis. Cardiovascular magnetic resonance in myocarditis: A JACC White Paper. *J Am Coll Cardiol* 2009;**53**:1475–87.

10. Pankuweit S, Klingel K. Viral myocarditis: from experimental models to molecular diagnosis in patients. *Heart Fail Rev* 2012 Oct 16.

11. Calabrese F, Thiene G. Myocarditis and inflammatory cardiomyopathy: microbiological and molecular biological aspects. *Cardiovasc Res* 2003;**60**:11–25.

12. Mahrholdt H, Wagner A, Deluigi CC, et al. Presentation, patterns of myocardial damage, and clinical course of viral myocarditis. *Circulation* 2006;**114**:1581–90.

13. Pauschinger M, Dörner G, Meissner G, et al. Enteroviral RNA detection by polymerase chain reaction (PCR) in cardiac muscle biopsies of patients with myocarditis or dilated cardiomyopathy. *J Am Coll Cardiol* 1994;**2**:880–4.

14. Bowles NE, Ni J, Kearney DL, Pauschinger M, et al. Detection of viruses in myocardial tissues by polymerase chain reaction. Evidence of adenovirus as a common cause of myocarditis in children and adults. *J Am Coll Cardiol* 2003;**42**:466–72.

15. Kühl U, Pauschinger M, Seeberg B, et al. Viral persistence in the myocardium is associated with progressive cardiac dysfunction. *Circulation* 2005;**112**:1965–70.

16. Martin AB, Webber S, Fricker FJ, et al. Acute myocarditis. Rapid diagnosis by PCR in children. *Circulation* 1994;**90**:330–9.

17. Calabrese F, Rigo E, Milanesi O, et al. Molecular diagnosis of myocarditis and dilated cardiomyopathy in children: clinicopathologic features and prognostic implications. *Diagn Mol Pathol* 2002;**11**:212–21.

18. Grist NR, Reid D. Epidemiology of viral infections of the heart. In: Banatvala JE, editor. *Viral infections of the heart*. London: Hodder and Stoughton; 1993. p. 23–311.

19. Gdynia G, Schnitzler P, Brunner E, et al. Sudden death of an immunocompetent young adult caused by novel (swine origin) influenza A/H1N1-associated myocarditis. *Virchows Arch* 2011;**458**:371–6.

20. Caforio AL, Calabrese F, Angelini A, et al. A prospective study of biopsy-proven myocarditis: prognostic relevance of clinical and aetiopathogenetic features at diagnosis. *Eur Heart J* 2007;**28**:1326–33.

21. Schönian U, Crombach M, Maisch B. Assessment of cytomegalovirus DNA and protein expression in patients with myocarditis. *Clin Immun Immunpathol* 1993;**68**:229–33.

22. Chang YL, Parker ME, Nuovo G, Miller JB. Human herpesvirus 6–related fulminant myocarditis and hepatitis in an immunocompetent adult with fatal outcome. *Hum Pathol* 2009;**40**:740–5.

23. Schowengerdt K, Ni J, Denfield S, et al. Association of PV B19 genome in children with myocarditis and cardiac allograft rejection. Diagnosis using the polymerase chain reaction. *Circulation* 1997;**96**:3549–54.

24. Mahrholdt H, Wagner A, Deluigi CC, et al. Presentation, patterns of myocardial damage, and clinical course of viral myocarditis. *Circulation* 2006;**114**:1581–90.

25. Bültmann BD, Klingel K, Sotlar K, et al. Fatal parvovirus B19-associated myocarditis clinically mimicking ischemic heart disease: an endothelial cell-mediated disease. *Hum Pathol* 2003;**34**:92–5.

26. Frustaci A, Calabrese F, Chimenti C, Pieroni M, Thiene G, Maseri A. Lone hepatitis C virus myocarditis responsive to immunosuppressive therapy. *Chest* 2002;**122**:1348–56.

27. Amonkar G, RupaniA Shah V, Parmar H. Sudden death in tuberculous myocarditis. *Cardiovasc Pathol* 2009;**18**:247–8.

28. Jansen TL, Joosten P, Brouwer J. Cardiac failure following group A streptococcal infection with echocardiographically proven pericarditis, still insufficient arguments for acute rheumatic fever: a case report and literature update. *Neth J Med* 2003;**61**:57–61.

29. Costello JM, Alexander ME, Greco KM, Perez-Atayde AR, Laussen PC. Lyme carditis in children: presentation, predictive factors, and clinical course. *Pediatrics* 2009;**123**:e835–41.

30. Rassi Jr A, Rassi A, Little WC, et al. Development and validation of a risk score for predicting death in Chagas' heart disease. *N Engl J Med* 2006;**355**:799–808.

31. Wreghitt TG, Hakim M, Gray JJ, et al. Toxoplasmosis in heart and heart and lung transplant recipients. *J Clin Pathol* 1989;**42**:194–9.

32. Nosanchuk JD. Fungal myocarditis. *Front Biosci* 2002;**7**:1423–38.

33. Taliercio CP, Olney BA, Lie JT. Myocarditis related to drug hypersensitivity. *Mayo Clin Proc* 1985;**60**:463–8.

34. Morgan J, Roper MH, Sperling L, et al. Myocarditis, pericarditis, and dilated cardiomyopathy after smallpox vaccination among civilians in the United States, January–October 2003. *Clin Infect Dis* 2008;**46**:S242–50.

35. Spear GS. Eosinophilic explants carditis with eosinophilia: hypersensitivity to dobutamine infusion. *J Heart Lung Transplant* 1995;**14**:755–60.

36. Pereira NL, Soon JP, Daly RC, Kushwaha SS, Edwards WD. De Novo development of eosinophilic myocarditis with left ventricular assist device support as bridge to transplant. *Ann Thorac Surg* 2010;**90**:1345–7.

37. Okura Y, Dec GW, Hare JM, et al. A clinical and histopathologic comparison of cardiac sarcoidosis and idiopathic giant cell myocarditis. *J Am Coll Cardiol* 2003;**41**:322–9.

38. Cooper Jr LT, Berry GJ, Shabetai R. Idiopathic giant-cell myocarditis—natural history and treatment. Multicenter Giant Cell Myocarditis Study Group Investigators. *N Engl J Med* 1997;**336**:1860–6.

39. Larsen BT, Maleszewski JJ, Edwards WD, et al. Atrial giant cell myocarditis: a distinctive clinicopathologic entity. *Circulation* 2013;**127**:39–47.

40. Basso C, Thiene G. When giant cell myocarditis affects only the atria. *Circulation* 2013;**127**:8–9.

41. Youssef G, Beanlands RS, Birnie DH, Nery PB. Cardiac sarcoidosis: applications of imaging in diagnosis and directing treatment. *Heart* 2011;**97**:2078–87.

42. Vasaiwala SC, Finn C, Delpriore J, et al. Prospective study of cardiac sarcoid mimicking arrhythmogenic right ventricular dysplasia. *J Cardiovasc Electrophysiol* 2009;**20**:473–6.

43. Hervier B, Masseau A, Bossard C, Agard C, Hamidou M. Vasavasoritis of the aorta and fatal myocarditis in fulminant Churg-Strauss syndrome. *Rheumatology* 2008;**47**:1728–9.

44. Frustaci A, Cuoco L, Chimenti C, et al. Celiac disease associated with autoimmune myocarditis. *Circulation* 2002;**105**:2611–8.

45. Corssmit EP, Trip MD, Durrer JD. Loffler's endomyocarditis in the idiopathic hypereosinophilic syndrome. *Cardiology* 1999;**91**: 272–6.

46. Courand PY, Croisille P, Khouatra C, Cottin V, Kirkorian G, Bonnefoy E. Churg-Strauss syndrome presenting with acute myocarditis and cardiogenic shock. *Heart Lung Circ* 2012;**21**:178–81.

47. Setoguchi M, Okishigi K, Sugiyama K, et al. Sudden cardiac death associated with Churg-Strauss Syndrome. *Circ J* 2009;**73**:2355–9.

48. Leung WH, Wong KK, Lau CP, Wong CK, Cheng CH, So KF. Myocardial involvement in Churg–Strauss syndrome: the role of endomyocardial biopsy. *J Rheumatol* 1989;**16**:828–31.

49. Rezaizadeh H, Sanchez-Ross M, Kaluski E, Klapholz M, Haider B, Gerula C. Acute eosinophilic myocarditis: diagnosis and treatment. *Acute Cardiac Care* 2010;**12**:31–6.

50. Ginsberg F, Parrillo J. Eosinophilic myocarditis. *Heart Failure Clin* 2005;**1**:419–29.

51. Simmons A, Vacek JL, Meyers D. Anthracycline-induced cardiomyopathy. *Postgrad Med* 2008;**120**:67–72.

52. Phillips A, Luk GS, Soor GS, Abraham JR, Leong S, Butany J. Cocaine cardiotoxicity: a review of the pathophysiology, pathology, and treatment options. *Am J Cardiovasc Drugs* 2009;**9**:177–96.

53. Aretz HT, Billingham ME, Edwards WD, et al. Myocarditis: a histopathologic definition and classification. *Am J Cardiovasc Pathol* 1987;**1**:3–14.

54. Baughman KL. Diagnosis of myocarditis: death of Dallas criteria. *Circulation* 2006;**113**:593–5.

55. Billingham ME. Acute myocarditis: is sampling error a contraindication for diagnostic biopsies? *J Am Coll Cardiol* 1989;**14**:921–2.

56. Edwards WD, Holmes DR, Reeders GS. Diagnosis of active lymphocytic myocarditis by endomyocardial biopsy: quantitative criteria for light microscopy. *Mayo Clin Proc* 1982;**57**:419–25.

57. Edwards WD. Myocarditis and endomyocardial biopsy. *Cardiol Clin* 1984;**2**:647–56.

58. Edwards WD. Current problems in establishing quantitative histopathologic criteria for the diagnosis of lymphocytic myocarditis by endomyocardial biopsy. *Heart Vessels* 1985;**1**:138–42.

59. Chow LH, Radio SJ, Sears TD, McManus BM. Insensitivity of right ventricular endomyocardial biopsy in the diagnosis of myocarditis. *J Am Coll Cardiol* 1989;**14**:915–20.

60. Hauck AJ, Kearney DL, Edwards WD. Evaluation of postmortem endomyocardial biopsy specimens from 38 patients with lymphocytic myocarditis: implications for role of sampling error. *Mayo Clin Proc* 1989;**64**:1235–45.

61. Thiene G, Bartoloni G, Poletti A, Boffa GM. Tecniche, indicazioni ed utilità della biopsia endomiocardica. In: Baroldi G, editor. *Thiene G (eds) Biopsia endomiocardica—Testo atlante*. Padova: Edizioni Piccin; 1996. p. 5–45.

62. Burke AP, Farb A, Robinowitz M, Virmani R. Serial sectioning and multiple level examination of endomyocardial biopsies for the diagnosis of myocarditis. *Mod Pathol* 1991;**4**:690–3.

63. Frustaci A, Bellocci F, Osen EG. Results of biventricular endomyocardial biopsy in survivors of cardiac arrest with apparently normal hearts. *Am J Cardiol* 1994;**74**:890–5.

64. Calabrese F, Angelini A, Carturan E, Thiene G. Myocarditis and inflammatory cardiomyopathy: histomorphological diagnosis. In: Schultheiss HP, Kapp JF, Grotzbach G, editors. *Chronic viral and inflammatory cardiomyopathy*. Berlin: Springer; 2006. p. 305–21.

65. Linder J, Cassling RS, Rogler WC, et al. Immunohistochemical characterization of lymphocytes in uninflamed ventricular myocardium. Implications for myocarditis. *Arch Pathol Lab Med* 1985;**109**:917–20.

66. Schnitt SJ, Ciano PS, Schoen FJ. Quantitation of lymphocytes in endomyocardial biopsies: use and limitations of antibodies to leukocyte common antigen. *Hum Pathol* 1987;**18**:796–800.

67. Tazelaar HD, Billingham ME. Myocardial lymphocytes. Fact, fancy or myocarditis? *Am J Cardiovasc Pathol* 1987;**1**:47–50.

68. Maisch B, Bultmann B, Factor S. Dilated cardiomyopathy with inflammation or chronic myocarditis: variability and consensus in the diagnosis. *Eur Heart J* 1998;**19**:647 (Abs).

69. Kuhl U, Noutsias M, Seeber B, Shultheiss HP. Immunohistological evidence for a chronic intramyocardial inflammatory process in dilated cardiomyopathy. *Heart* 1996;**75**:295–300.

70. Angelini A, Crosato M, Boffa GM, et al. Active versus borderline myocarditis: clinicopathological correlates and prognostic implications. *Heart* 2002;**87**:210–5.

71. Maisch B, Richter A, Sandmoller A, Portig I, Pankuweit S; BMBF-Heart Failure Network. Inflammatory dilated cardiomyopathy (DCMI). *Herz* 2005;**30**:535–44.

72. Angelini A, Calzolari V, Calabrese F, et al. Myocarditis mimicking acute myocardial infarction: role of endomyocardial biopsy in the differential diagnosis. *Heart* 2000;**84**:245–50.

73. Herskowitz A, Ahmed-Ansari A, Neumann DA, et al. Induction of major histocompatibility complex antigens within the myocardium of patients with active myocarditis: a nonhistologic marker of myocarditis. *J Am Coll Cardiol* 1990;**15**:624–32.

74. Basso C, Carturan E, Corrado D, Thiene G. Myocarditis and dilated cardiomyopathy in athletes: diagnosis, management, and recommendations for sport activity. *Cardiol Clin* 2007;**25**:423–9.

75. Basso C, Burke M, Fornes P, et al. Guidelines for autopsy investigation of sudden cardiac death. *Virchows Arch* 2008;**452**:11–8.

76. Calabrese F, Carturan E, Thiene G. Cardiac infections: focus on molecular diagnosis. *Cardiovasc Pathol* 2010;**19**:171–82.

77. Rotbart H. Human enterovirus infections: molecular approaches to diagnosis and pathogenesis. In: Semler B, Ehrenfeld E, editors. *Molecular aspects of picornavirus infection and detection*. Washington (DC): American Society for Microbiology; 1989. p. 243–64.

78. Mahfoud F, Gärtner B, Kindermann M, et al. Virus serology in patients with suspected myocarditis: utility or futility? *Eur Heart J* 2011;**32**:897–903.

79. Gerzen P, Granath A, Holmgren B, Zetterquist S. Acute myocarditis. A follow-up study. *Br Heart J* 1972;**34**:575–83.

80. Woodruff JF. Viral myocarditis. *Am J Pathol* 1980;**101**:425–84.

81. Bowles N, Richardson P, Olsen E, Archard L. Detection of coxsackie-B-virus-specific RNA sequences in myocardial biopsy samples from patients with myocarditis and dilated cardiomyopathy. *Lancet* 1986;**1**:1120–3.

82. Kandolf R, Ameis D, Kirschner P, Canu A, Hofschneider PH. In situ detection of enteroviral genomes in myocardial cells by nucleic acid hybridization: an approach to the diagnosis of viral heart disease. *Proc Natl Acad Sci U S A* 1987;**84**:6272–6.

83. Akhtar N, Ni J, Stromberg D, Rosenthal GL, Bowles NE, Towbin JA. Tracheal aspirate as a substrate for polymerase chain reaction detection of viral genome in childhood pneumonia and myocarditis. *Circulation* 1999;**99**:2011–8.

84. Carturan E, Milanesi O, Kato Y, et al. Viral detection and tumor necrosis factor alpha profile in tracheal aspirates from children with suspicion of myocarditis. *Diagn Mol Pathol* 2008;**17**:21–7.

85. Dunn JJ, Chapman NM, Tracy S, Romero JR. Genomic determinants of cardiovirulence in coxsackievirus B3 clinical isolates: localization to the 5' non-translated region. *J Virol* 2000;**74**:4787–94.

86. Höfling K, Kim KS, Leser JS, et al. Progress toward vaccines against virus that cause heart disease. *Herz* 2000;**25**:286–90.

87. Lam KM, Oldenburg N, Khan MA, et al. Significance of reverse transcription polymerase chain reaction in the detection of human cytomegalovirus gene transcripts in thoracic organ transplant recipients. *J Heart Lung Transplant* 1998;**17**:555–65.

88. Pauschinger M, Phan MD, Doerner A, et al. Enteroviral RNA replication in the myocardium of patients with left ventricular dysfunction and clinically suspected myocarditis. *Circulation* 1999;**99**:889–95.

89. Okabe M, Fukuda K, Arakawa K, Kikuchi M. Chronic variant of myocarditis associated with hepatitis C virus infection. *Circulation* 1997;**96**:22–4.

90. Donoso Mantke O, Meyer R, Prösch S, et al. High prevalence of cardiotropic viruses in myocardial tissue from explanted hearts of heart transplant recipients and heart donors: a 3-year retrospective study from a German patients' pool. *J Heart Lung Transplant* 2005;**24**:1632–8.

91. Schenk T, Enders M, Pollak S, Hahn R, Huzly D. High prevalence of human parvovirus B19 DNA in myocardial autopsy samples from subjects without myocarditis or dilative cardiomyopathy. *J Clin Microbiol* 2009;**47**:106–10.

92. Wang X, Zhang G, Liu F, Han M, Xu D, Zang Y. Prevalence of human parvovirus B19 DNA in cardiac tissues of patients with congenital heart diseases indicated by nested PCR and in situ hybridization. *J Clin Virol* 2004;**31**:20–4.

93. Stewart GC, Lopez-Molina J, Gottumukkala RV, et al. Myocardial parvovirus B19 persistence: lack of association with clinicopathologic phenotype in adults with heart failure. *Circ Heart Fail* 2011;**4**:71–8.

94. De Salvia A, De Leo D, Carturan E, Basso C. Sudden cardiac death, borderline myocarditis and molecular diagnosis: evidence or assumption? *Med Sci Law* 2011;**51**:S27–9.

95. Bock CT, Klingel K, Kandolf R. Human parvovirus B19-associated myocarditis. *N Engl J Med* 2010;**362**:1248–9.

96. Chimenti C, Russo A, Pieroni M, et al. Intramyocyte detection of Epstein-Barr virus genome by laser capture microdissection in patients with inflammatory cardiomyopathy. *Circulation* 2004;**110**:3534–9.

97. Caforio AL, Iliceto S. Genetically determined myocarditis: clinical presentation and immunological characteristics. *Curr Opin Cardiol* 2008;**23**:219–26.

98. Caforio AL, Bonifacio E, Stewart JT, et al. Novel organ-specific circulating cardiac autoantibodies in dilated cardiomyopathy. *J Am Coll Cardiol* 1990;**15**:1527–34.

99. Caforio AL, Keeling PJ, Zachara E, et al. Evidence from family studies for autoimmunity in dilated cardiomyopathy. *Lancet* 1994;**344**:773–7.

100. Corsten MF, Schroen B, Heymans S. Inflammation in viral myocarditis: friend or foe? *Trends Mol Med* 2012;**18**:426–37.

101. Blauwet LA, Cooper LT. *Myocarditis Prog Cardiovasc Dis* 2010;**52**:274–88.

102. Jun EJ, Ye JS, Hwang IS, Kim YK, Lee H. Selenium deficiency contributes to the chronic myocarditis in coxsackievirus-infected mice. *Acta Virol* 2011;**55**:23–9.

103. Shen H, Thomas PR, Ensley SM, et al. Vitamin E and selenium levels are within normal range in pigs diagnosed with mulberry heart disease and evidence for viral involvement in the syndrome is lacking. *Transbound Emerg Dis* 2011;**58**:483–91.

104. Tu Z, Chapman NM, Hufnagel G, et al. The cardiovirulent phenotype of coxsackievirus B3 is determined at a single site in the genomic 5′ nontranslated region. *J Virol* 1995;**69**:4607–18.

105. Stadnick E, Dan M, Sadeghi A, Chantler JK. Attenuating mutations in coxsackievirus B3 map to a conformational epitope that comprises the puff region of VP2 and the knob of VP3. *J Virol* 2004;**78**:13987–4002.

106. He Y, Chipman PR, Howitt J, et al. Interaction of coxsackievirus B3 with the full length coxsackievirus-adenovirus receptor. *Nat Struct Biol* 2001;**8**:874–8.

107. Badorff C, Lee GH, Lamphear BJ, et al. Enteroviral protease 2A cleaves dystrophin: evidence of cytoskeletal disruption in an acquired cardiomyopathy. *Nat Med* 1999;**5**:320–6.

108. YajimaT Knowlton KU. Viral myocarditis: from the perspective of the virus. *Circulation* 2009;**119**:2615–24.

109. Broere F, Apasov SG, Sitkovsky MV, van Eden W. T-cell subsets and Tcell-mediated immunity. In: Nijkamp FP, Parnham MJ, editors. *Principles of Immunopharmacology: 3rd revised and extended edition*. Basel: BirkhauserVerlag; 2011. p. 15–27.

110. Xie Y, Chen R, Zhang X, et al. The role of Th17 cells and regulatory T-cells in Coxsackievirus B3-induced myocarditis. *Virology* 2011;**421**:78–84.

111. Ayach B, Fuse K, Martino T, Liu P. Dissecting mechanisms of innate and acquired immunity in myocarditis. *Curr Opin Cardiol* 2003;**18**:175–81.

112. Rose NR. Critical Cytokine Pathways to Cardiac Inflammation. *J Interferon Cytokine Res* 2011;**31**:705–10.

113. Calabrese F, Carturan E, Chimenti C, et al. Overexpression of tumor necrosis factor (TNF)alpha and TNFalpha receptor I in human viral myocarditis: clinicopathologic correlations. *Mod Pathol* 2004;**17**:1108–18.

114. Deonarain R, Cerullo D, Fuse K, Liu PP, Fish EN. Protective role for interferon-beta in coxsackievirus B3 infection. *Circulation* 2004;**110**:3540–3.

115. Blyszczuk P, Behnke S, Lüscher TF, Eriksson U, Kania G. GM-CSF promotes inflammatory dendritic cell formation but does not contribute to disease progression in experimental autoimmune myocarditis. *Biochim Biophys Acta* 2013;**1833**:934–44.

116. Afanasyeva M, Wang Y, Kaya Z, et al. Experimental autoimmune myocarditis in A/J mice is an interleukin-4-dependent disease with a Th2 phenotype. *Am J Pathol* 2001;**159**:193–203.

117. Yasukawa H, Yajima T, Duplain H, et al. The suppressor of cytokine signaling-1 (SOCS1) is a novel therapeutic target for enterovirus-induced cardiac injury. *J Clin Invest* 2003;**111**:469–78.

118. Weinzierl AO, Szalay G, Wolburg H, et al. Effective chemokine secretion by dendritic cells and expansion of cross-presenting CD4−/CD8+ dendritic cells define a protective phenotype in the mouse model of coxsackievirus myocarditis. *J Virol* 2008;**82**:8149–60.

119. Papageorgiou AP, Heymans S. Interactions between the extracellular matrix and inflammation during viral myocarditis. *Immunobiology* 2012;**217**:503–10.

120. Bartel DP. Micrornas: target recognition and regulatory functions. *Cell* 2009;**105**:2111–6.

121. Corsten MF, Papageorgiou A, Verhesen W, et al. MicroRNA profiling identifies microRNA-155 as an adverse mediator of cardiac injury and dysfunction during acute viral myocarditis. *Circ Res* 2012;**111**:415–25.

122. Taylor LA, Carthy CM, Yang D, et al. Host gene regulation during coxsackievirus B3 infection in mice: assessment by microarrays. *Circ Res* 2000;**87**:328–34.

123. Kittleson MM, Minhas KM, Irizarry RA, et al. Gene expression in giant cell myocarditis: Altered expression of immune response genes. *Int J Cardiol* 2005;**102**:333–40.

124. Cunha-Neto E, Dzau VJ, Allen PD, et al. Cardiac gene expression profiling provides evidence for cytokinopathy as a molecular mechanism in Chagas' disease cardiomyopathy. *Am J Pathol* 2005;**167**:305–13.

125. Ruppert V, Meyer T, Pankuweit S, et al. Gene expression profiling from endomyocardial biopsy tissue allows distinction between subentities of dilated cardiomyopathy. *J Thorac Cardiovasc Surg* 2008;**136**:360–9.

126. Heidecker B, Kittleson MM, Kasper EK, et al. Transcriptomic biomarkers for the accurate diagnosis of myocarditis. *Circulation* 2011;**123**:1174–84.

127. Calabrese F, Valente M, Thiene G, et al. Enteroviral genome in native hearts may influence outcome of patients who undergo cardiac transplantation. *Diagn Mol Pathol* 1999;**8**:39–46.

128. Ray CG, Icenogle TB, Minnich LL, Copeland JG, Grogan TM. The use of intravenous ribavirin to treat influenza virus-associated acute myocarditis. *J Infect Dis* 1989;**159**:829–36.

129. Dehtiar N, Eherlichman M, Picard E, et al. Cytomegalovirus myocarditis in a healthy infant: Complete recovery after ganciclovir treatment. *Pediatr Crit Care Med* 2001;**2**:271–3.

130. Pinkert S, Westermann D, Wang X, et al. Presentation of cardiac dysfunction in acute coxsackieviru B3 cardiomyopathy by inducible expression of a soluble coxsackiervirus-adenovirus receptor. *Circulation* 2009;**120**:2358–66.

131. Mann DL, McMurray JJ, Packer M, et al. Targeted anticytokine therapy in patients with chronic heart failure: results of the Randomized Etanercept Worldwide Evaluation (RENEWAL). *Circulation* 2004;**109**:1594–602.

132. Kühl U, Pauschinger M, Schwimmbeck PL, et al. Interferon-beta treatment eliminates cardiotropic viruses and improves left ventricular function in patients with myocardial persistence of viral genomes and left ventricular dysfunction. *Circulation* 2003;**107**:2793–8.

133. Daliento L, Calabrese F, Tona F, et al. Successful treatment of enterovirus-induced myocarditis with interferon-alpha. *J Heart Lung Transplant* 2003;**22**:214–7.

134. Schultheiss HP, Piper C, Sowade K, et al. The effect of subcutaneous treatment with interferon-beta-1b over 24 weeks on safety, virus elimination and clinical outcome in patients with chronic viral cardiomyopathy. *Circulation* 2008;**118**:3322 (abstr).

135. Mason JW, O'Connell JB, Herskowitz A. A clinical trial of immunosuppressive and immunomodulatory treatment for myocarditis. *N Engl J Med* 1995;**333**:269–75.

136. Maish B, Hufnagel G, Schonian U, Hengstenberg C. for the ESETCID investigators. The European study of epidemiology and treatment of cardiac inflammatory disease (ESETCID). *Eur Heart J* 1995;**S16**:172–5.

137. Frustaci A, Calabrese F, Chimenti C, Pieroni M, Thiene G, Maseri AA. Immunosuppressive therapy for active lymphocythic myocarditis: virologic and immunologic profile of responders versus non responders. *Circulation* 2003;**107**:857–63.

138. Frustaci A, Russo MA, Chimenti C. Randomized study on the efficacy of immunosuppressive therapy in patients with virus-negative inflammatory cardiomyopathy: the TIMIC study. *Eur Heart J* 2009;**30**:1995–2002.

139. Felix SB, Staudt A, Landsberger M, et al. Removal of cardiodepressant antibodies in dilated cardiomyopathy by immunoadsorption. *J Am Coll Cardiol* 2002;**39**:646–52.

140. Zhang L, Parham NJ, Zhang F, Aasa-Chapman M, Gould EA, Zhang H. Vaccination with coxsackievirus B3 virus-like particles elicits humoral immune response and protects mice against myocarditis. *Vaccine* 2012;**30**:2301–8.

141. Leone O, Veinot JP, Angelini A, et al. 2011 consensus statement on endomyocardial biopsy from the Association for European Cardiovascular Pathology and the Society for Cardiovascular Pathology. *Cardiovasc Pathol* 2012;**21**:245–74.

142. Thiene G, Bruneval P, Veinot J, Leone O. Diagnostic use of the endomyocardial biopsy: a consensus statement. *Virchows Arch* 2013;**463**:1–5.

143. Caforio AL, Pankuweit S, Arbustini E, et al. Current state of knowledge on aetiology, diagnosis, management, and therapy of myocarditis: a position statement of the European Society of Cardiology Working Group on Myocardial and Pericardial Diseases. *Eur Heart J* 2013;**34**:2636–48.

Calcific and Degenerative Heart Valve Disease

Elena Aikawa, MD, PhD, Frederick J. Schoen, MD, PhD

Brigham and Women's Hospital, Harvard Medical School, Boston, MA, USA

INTRODUCTION

The collective global impact of the spectrum of valvular heart diseases is a serious but under-recognized health problem. In the developed and rapidly developing regions of the world, calcific and degenerative valve diseases are the most prevalent valvular abnormalities necessitating surgical intervention. Since these conditions occur progressively with age and lifespan is increasing in these areas, the number of new cases is projected to rise sharply in the next few decades. Calcific aortic valve disease (CAVD), the most common type of surgically treated heart valve disease overall,[1] manifests clinically as aortic sclerosis (defined as an increased leaflet thickness without restriction of leaflet motion) or, more significantly, aortic stenosis (defined as thickened leaflets with significantly reduced systolic opening causing obstruction). Degenerative mitral valve disease (DMVD), manifesting as mitral valve regurgitation, is also common. Although rheumatic heart disease continues to be an important problem in developing countries, the prevalence of this disease has declined overall.[2] Therefore, this chapter will focus on evolving insights into the mechanisms underlying CAVD and DMVD. The mechanisms of the broader spectrum of etiologies of valvular heart disease have recently been reviewed.[3]

NORMAL VALVE FUNCTION, BIOMECHANICS, AND STRUCTURE

The coordinated opening and closing of the heart valves required for unidirectional blood flow occurs approximately 40 million times per year (3 billion cycles in a 75-year lifespan). The valves of the heart comprise the semilunar valves (aortic and pulmonary) and the atrioventricular valves (mitral and tricuspid). To withstand extensive repetitive mechanical stress during the cardiac cycle, healthy heart valves maintain dynamic function and durability by mechanisms that depend on a complex and highly differentiated tissue macro- and microstructure. Although the functional macroscopic morphology differs between the semilunar and atrioventricular valves, all four cardiac valves have a layered microscopic architecture, with each layer containing cells and characteristic extracellular matrix (ECM).[4,5] Macro- and microscopic features of the heart valves are illustrated in Figures 9.1 and 9.2.

Aortic Valve Cusps

Aortic valve (AV) cusps are attached to the aortic wall in a crescentic/semilunar manner, each ascending to the commissures and descending to its basal attachment to the aortic wall. Coaptation of the cusps during valve closure (diastole) occurs along a surface near the free edge of the cusps. The non-coapting portion of the cusp, which separates aortic from ventricular blood, is called the belly. During diastole, when the valve is closed, the cusps are stretched via backpressure of 80 mm Hg (the cuspal area increases by about 30%), filling the orifice; during the opened phase (systole), the cuspal tissue becomes relaxed and moves out of the central flow stream. Within the normal aortic root and behind the cusps are three bulging aortic sinuses of Valsalva. Pulmonary valve structure is analogous to that of the AV, but the sinuses are not prominent, and the cusps are thinner (than in the left ventricle and aorta),[6] consistent with the lower blood pressure in the right ventricle and pulmonary circulation.

The structural architecture, composition and integrity of valve ECM (composed of collagen, elastin, and glycosaminoglycans) are the major determinants of valve durability. For the AV, the mechanism of function is largely mediated by the intrinsic architecture. Immediately below the endothelium on the outflow (aortic) surface is the fibrosa layer, composed predominantly of densely packed and microscopically crimped collagen fibers arranged largely parallel to the free edge of the leaflet. Collagen, the

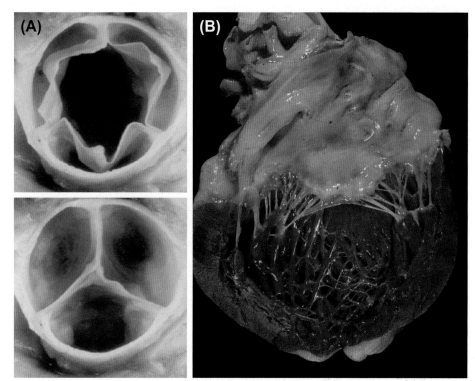

FIGURE 9.1   Normal aortic and mitral valves. **(A)** Outflow aspect of aortic valve in open (top) and closed (bottom) configurations, corresponding to systole and diastole, respectively. **(B)** Normal mitral valve and associated structures, after opening left ventricle. (A) Reproduced from Schoen FJ, Edwards WD. Valvular heart disease: General principles and stenosis. In: *Cardiovascular Pathology* 3rd Ed., Silver MD, Gotlieb AI, Schoen FJ (eds.), WB Saunders; 2001, pp. 402–442.

major stress-bearing component of the cusps, is abundant in the fibrosa, and is essential for the maintenance of valve durability. Individual collagen fibers can withstand high tensile forces but cannot be compressed. The ventricularis layer, near to the ventricular face, is rich in elastin, and provides tissue elasticity. When the valve is closed during cardiac diastole, the collagen is pulled taut, permitting maximum coaptation of the cusps without prolapse and shifting the load from the cusps to the aortic wall. During systole, the cusps, which were stretched during diastole, become relaxed due to elastin recoil. With its high compliance and bonds that link it to the adjacent fibrous layers, the proteoglycan-rich, central spongiosa layer facilitates the rearrangements of the fibrosa and thereby provides cushioning and shear absorbing function during the cardiac cycle. The predominant type of collagen in the valves, type I collagen, and elastin comprise ~80% of total valvular protein.[7] AV cusps have anisotropic mechanical properties (i.e., different in the radial and circumferential directions), with compliance and therefore stretch greater in the radial direction.[8] In addition, AV demonstrate (1) layer-specific directionality (i.e., the stiffer fibrosa dominates in the circumferential direction, and more compliant ventricularis dominates in the redial direction) and (2) regional heterogeneity (i.e., the cuspal belly is stiffer than the commissural region).[8]

Mitral Valve Leaflets

In cardiac systole, the back pressure on the closed mitral valve (MV) is approximately 120 mmHg, thereby engendering substantial forces tending to push the leaflets backward into the left atrium. Thus, normal MV function requires assistance from structures extrinsic to and largely below the valve leaflets, specifically a complex and coordinated interaction of leaflets, annulus, chordae tendineae, and papillary muscle (collectively, the MV apparatus). The MV has two leaflets, the anterior below the aortic valve and the posterior embedded into the posterolateral valve annulus, and composed of three scallops. At the leaflet edges of the MV, collagen-rich chordae tendineae extend from the fibrosa to link the leaflets to the papillary muscles.

Similar to the AV cusps, normal MV leaflets have well-defined tissue layers: fibrosa (closest to the ventricle), spongiosa and atrialis (near the atrium).[4] As with the AV discussed above, the anatomy and ECM composition of the MV are critical to understanding its material behavior. The anterior leaflet has heterogeneous anatomy in which the tissue layers vary in thickness from the mitral annulus to the free edge.[9] The thickest region, located close to the annulus, is called the 'clear zone' owing to the lack of chordal basal attachments. This region withstands high load when the valve is closed. Toward the middle of the leaflet the relative thickness of the fibrosa is diminished, contributing to the radial extensibility. The thickness of the spongiosa also increases, thereby promoting compressive load-bearing in the region of coaptation. The complex microstructure of the mitral valve leaflets results in marked non-linear anisotropic mechanical properties within leaflets and throughout individual layers, similar to AV cusps.[10]

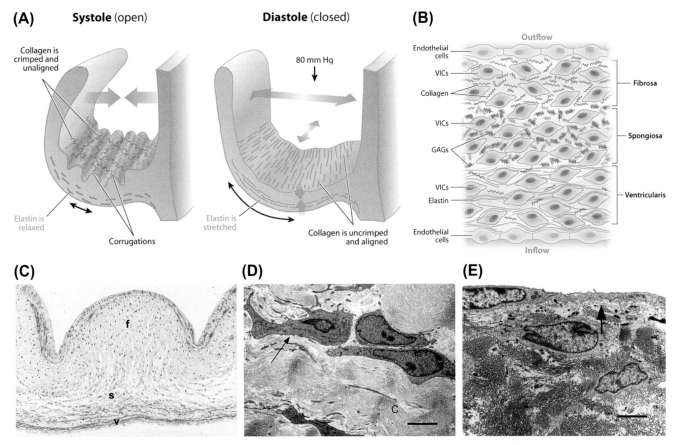

FIGURE 9.2 Aortic valve functional structure. **(A)** Schematic representation of architecture and configuration of aortic valve cusp in cross-section and of collagen and elastin in systole and diastole. **(B)** Schematic diagram of the detailed cellular and extracellular matrix architecture of a normal aortic valve. **(C)** Tissue architecture, shown as low-magnification photomicrograph of cross-section cuspal configuration in the non-distended state (corresponding to systole), emphasizing three major layers: ventricularis **(v)**, spongiosa **(s)**, and fibrosa **(f)**. The outflow surface is at the top. Original magnification: 100×. Movat pentachrome stain (collagen, yellow; elastin, black). **(D)** Transmission electron photomicrograph of relaxed fresh porcine aortic valve (characteristic of the systolic configuration), demonstrating the fibroblast morphology of valvular interstitial cells (VICs, indicated by arrow); the dense, surrounding closely apposed collagen with wavy crimp; and the potential for VIC–collagen and VIC–VIC interactions. Scale bar: 5 μm. **(E)** Transmission electron photomicrograph at the surface of the aortic valve demonstrating valvular endothelial cell (VEC, arrow) and proximity of deeper VICs and potential for VEC–VIC interactions. Scale bar: 5 μm. (A, B, D, and E) Reproduced with permission from Schoen FJ. Mechanisms of function and disease in natural and replacement heart valves. *Annual Review of Pathology: Mechanisms of Disease* 2012; 7:161–183. (C) Reproduced with permission from Schoen FJ, Edwards WD. Valvular heart disease: General principles and stenosis, In: *Cardiovascular Pathology* 3rd Ed., Silver MD, Gotlieb AI, Schoen FJ (eds.). WB Saunders; 2001, pp. 402–442.

Valvular Endothelial and Interstitial Cells

A continuous endothelial cell layer covers the cusps and leaflets on both inflow and outflow aspects. While valvular endothelial cells (VECs) resemble endothelial cells (ECs) elsewhere in the circulation, they have distinctive phenotypic features.[11] For example, in response to fluid shear stress in vitro, VECs align perpendicularly to flow in contrast to vascular ECs, which align parallel to flow.[12] In addition, VECs from the AV express different transcriptional profiles on the arterial vs. ventricular sides.[13] Indeed, VECs covering the fibrosa have elevated expression of proteins expected to promote calcification,[13] while VECs subjected to the ventricular hemodynamic waveform (i.e., experienced in vivo by the VEC lining the ventricularis) have increased expression of the 'atheroprotective' transcription factor Kruppel-like factor2 (KLF2),[14] expected to inhibit calcification. These valve-side-specific VEC phenotypic differences may contribute to the typical pattern of initiation and predominance of calcification near the AV outflow surface.

Moreover, a key property of VECs is the ability to undergo an epithelial to mesenchymal transition (EMT).[5,15] Emerging evidence suggests that VEC mesenchymal differentiation can be specifically directed toward osteogenic, chondrogenic, and adipogenic phenotypes.[16,17] This multi-lineage differentiation potential of VEC combined with a robust self-renewal capacity supports the hypothesis that VECs are capable of serving as progenitor cells.[11]

Deep to the VECs, heart valve cusps and leaflets contain valvular interstitial cells (VICs), a heterogeneous,

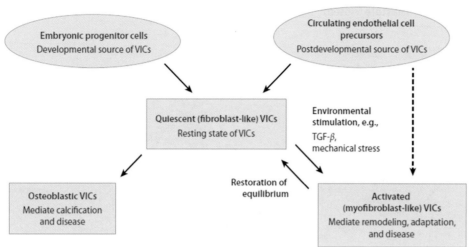

FIGURE 9.3 Spectrum of valvular interstitial cell (VIC) phenotypes. VIC functions can be conveniently organized into five phenotypes: embryonic progenitor endothelial/mesenchymal cells, quiescent VICs, activated VICs, adult, circulating stem-cell-derived progenitor VICs, and osteoblastic VICs. TGF, transforming growth factor. Modified by permission from Schoen FJ. Evolving concepts of heart valve dynamics. The continuum of development, functional structure, pathology and tissue engineering. *Circulation* 2008; 118:1864–1880, and reproduced with permission from Schoen FJ. Mechanisms of function and disease in natural and replacement heart valves. Ann Rev Path Mech Dis 2012; 7:161–183.

dynamic and highly plastic population of resident cells.[18,19] Adult VICs are predominantly quiescent non-contractile, alpha-smooth muscle actin-negative, vimentin-positive fibroblast-like cells; few are activated. In response to a need for either adaptation to changing conditions or repair of ECM functional damage (e.g., valve development, disease, abrupt changes in the mechanical stress, or surgical intervention), large numbers of VICs transition to an activated myofibroblast-like phenotype. In these situations, VIC express smooth muscle contractile proteins, matrix remodeling enzymes (e.g., matrix metalloproteinases [MMP-1, MMP-2, MMP-9, MMP-13] and their inhibitors [tissue inhibitors of metalloproteinases]), cathepsins (cathepsin K, cathepsin S)[4,6,20] and cytokines (IL-1, TGFβ)[4]. Five specialized VIC phenotypes have been described: embryonic progenitor (eVICs), quiescent (qVICs), activated (aVICs), progenitor (pVICs), and osteoblastic (oVICs) (Figure 9.3).[4,19,21] The transition from a quiescent to an activated VIC phenotype is thought to be induced by changes in the mechanical environment of the VICs and/or TGF-β signaling. In healthy valves, VIC plasticity contributes to adaptation to dynamic changes in the valve and the surrounding milieu. Following adaptation or repair; this transition may be reversible and VICs may revert to a quiescent phenotype if equilibrium is established. In contrast, when adaptation to a new environmental condition is not possible, e.g., with ongoing, progressive and excessive mechanical stress, valve disease can result.[4,5]

VALVE DEVELOPMENT, POST-DEVELOPMENTAL ADAPTATION, AND AGING

Development and Maturation

There is increasing evidence that the regulatory mechanisms involved in normal valve morphogenesis and

development are relevant to the pathobiology of valvular diseases. However, it is unknown to what extent and under what circumstances reactivation (or potentially failure) of these developmental mechanisms mediates repair or contributes to disease progression.

Valve formation is a complex and highly regulated process.[22] Valve development begins with the formation of the endocardial cushions, the precursors of the cardiac valves, which occurs by the separation of the endocardium and overlying myocardium in the outflow tract and in the atrioventricular canal by expansion of the acellular ECM called cardiac jelly. Subsets of endothelial cells change their phenotype into mesenchymal cells expressing alpha-smooth-muscle cell actin; they lose cell–cell contacts, and migrate into the cardiac jelly via the process of EMT discussed above. Migration of neural crest cells to the primordial aortic valve in the outflow tract occurs later. Swelling of the endocardial cushions and induction of EMT also contribute to the morphogenesis of the semilunar valve leaflets. EMT is induced by signaling molecules, including TGFβ1–3, BMP 2–4, and Notch 1–4, originating from the underlying myocardium.[22] Additionally, several transcription factors are important in endocardial cushion formation and EMT.[22] Stimulation of the Notch pathway depends on the transcription factor RBPJ, which activates transcription factor snail1 (Snai1). Loss of Snai1 in endothelial cells inhibits endocardial cushion formation.[23] Gain- and loss-of-function studies have demonstrated critical roles for Twist1, Tbx20, and Sox9 in proliferation of mesenchymal cells in the endocardial cushions. After endocardial cushion formation, Twist1 is down-regulated and cell proliferation is decreased. Moreover, in normal adult valves, cell proliferation is very low,[18] and expression of valve developmental transcription factors such as Twist1 and Sox9 is negligible. Also, emerging evidence suggests that VICs in adult valves are continuously replenished via EMT, with this process potentially playing an important role in

| Fetal 14-19 weeks | Fetal 20-39 weeks | 2-16 years | Adult valve |

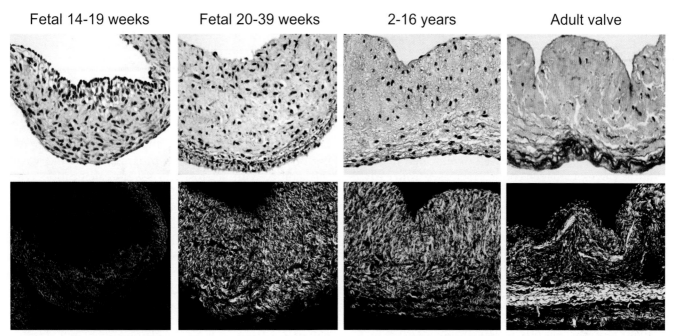

FIGURE 9.4 Evolving ECM structural composition of human cardiac valves. At 14–19 weeks of gestation, fetal valve ECM is composed mostly of glycosaminoglycans. At 20–39 weeks, fetal valves have a bilaminar structure with elastin in the ventricularis and increased unorganized collagen in the fibrosa. A trilaminar structure becomes apparent in children's valves but remains incomplete compared with normal adult valve layered architecture with collagen in the fibrosa, glycosaminoglycans in the spongiosa, and elastin in the ventricularis. Top, Movat pentachrome (collagen, yellow; glycosaminoglycans, blue-green; elastin, black); bottom, picrosirius red under circular polarized light. Original magnification ×200. Adapted from Aikawa E, Whittaker P, Farber M, Mendelson K, Padera RF, Aikawa M, Schoen FJ. Human semilunar cardiac valve remodeling by activated cells from fetus to adult: Implications for postnatal adaptation, pathology, and tissue engineering. *Circulation* 2006;113:1344–1352.

valve healing and remodeling under physiological and pathological conditions.[11]

The later stages of valve development and maturation are characterized by gradual thinning and elongation of pre-valvular tissue; this process is concurrent with the differentiation of the cushion mesenchymal cells into VEC and VIC, with concomitant ECM remodeling. Post-EMT valve development involves cell migration and proliferation associated with RANKL, cathepsin K, NFAT, periostin, cadherin, Notch, MMP-2, MMP-13, and VEGF expression. Cathepsin K, a matrix remodeling enzyme expressed during heart valve elongation, is regulated by the RANKL/NFATc1 pathway, and involved in differentiation and function of osteoclasts.[24]

Beyond calcification-promoting proteins discussed earlier, some signaling molecules and pathways important in valve pathophysiology are expressed differentially on the two sides of the valves. Notch signaling is localized on the surface adjacent to the AV ventricularis, while Wnt/β-catenin signaling is found throughout the valve interstitium in late gestation.[25] Periostin, required for collagen remodeling, is expressed in the fibrosa layer. Loss of periostin in mice leads to valve malformation and cardiac dysfunction.[26] Mutations in Collagen 1a2 or elastin insufficiency affect aortic valve morphogenesis leading to valve disease.[27] MMP-1, MMP-2, MMP-9, and MMP-13 are also expressed

during late valve morphogenesis and involved in ECM maturation and organization.[18] In certain disease conditions these transcriptional regulatory and signaling mechanisms are reactivated and have been implicated in aortic and mitral valve diseases.

Dynamic changes in ECM architecture and VEC and VIC phenotypes in response to environmental stimuli continue throughout fetal and postnatal development and throughout life. Studies of human valves obtained from second- and third-trimester fetuses, neonates, children, and adults have shown that valve structural architecture evolves over a lifetime (Figure 9.4). This likely effects a progressive adaptation of valvular cells to continuous hemodynamic changes and potential injury.[18] Embryonic VECs possess an activated phenotype throughout fetal development (i.e., expression of VCAM-1, ICAM-1), and VICs concurrently show a myofibroblast-like phenotype (alpha-smooth muscle actin, vimentin, desmin), abundant embryonic myosin (SMemb) and MMP-1, MMP-2, MMP-9, and MMP-13 expression, indicating an immature/activated phenotype engaged in rapid matrix remodeling. This compares with the quiescent fibroblast-like VICs usually present in adults. In addition, VIC density, proliferation and apoptosis are significantly higher in fetal than adult valves.[18] Indeed, cell density in adult valves is reduced to ~10% and cell proliferation is below 2% of that in utero.

At birth, after the single fetal circulation separates into pulmonary and systemic circulations, pulmonary pressure decreases and aortic pressure increases. This abrupt hemodynamic change is accompanied by increased aortic VIC activation and increased alpha-smooth-muscle actin expression consistent with increased mechanical stress. Moreover, a tri-laminar architecture evolves in human valves over time and becomes well-defined histologically only after 36 weeks of gestation, but the structure at that stage is less well demarcated than that of adult valves. Collagen content increases from early to late fetal stages, but is subsequently stable, while mature elastic fibers significantly increase only post-natally.

Post-Natal Adaptation

Under physiologic conditions of valve repair, adaptation, remodeling, and in a pathological environment, cardiac valves respond to changes in mechanical stress by cell activation leading to matrix remodeling. Accumulating evidence suggests that during adaptation (as evidenced by the pulmonary-to-aortic allograft surgical procedure),[6] disease (CAVD, DMVD),[4,5,28] mechanical stretching (mitral valve chordae displacement),[29] or remodeling (tissue-engineered valves),[19] many VICs acquire an activated myofibroblast-like phenotype. Nevertheless, as stated above, owing to high plasticity, these changes in VIC phenotype are potentially reversible, and VICs can return to a quiescent state after mechanical equilibrium is achieved by adaptive ECM remodeling. Thus, cardiac valves can adapt to pathological conditions by reversible phenotypic modulation of quiescent VICs to activated, myofibroblast-like VICs; moreover, analogous molecular mechanisms may direct both physiological and pathological cell activation.

Aging

Progressive age-associated decrease in cell density, cell turnover and reduction in the active capacity for remodeling, as described above, accompanies aging. In some AVs after the age of 30 years, histological analysis demonstrates atheromatous changes (lipid association with foamy macrophages) and increased extracellular matrix. Although these lipid-rich lesions are similar in some respects to fatty streaks and early atherosclerotic plaques found in the vasculature, they do not develop complications of thrombosis or ulceration/rupture.[30] Moreover, it is uncertain whether these lesions are precursors of CAVD. Also potentially important to the development of valve disease is the fact that the valve assumes a progressively more stretched configuration and collagen fibers become increasingly more aligned with increasing age, correlating with decreasing valve mechanical compliance.

CALCIFIC AORTIC VALVE DISEASE (CAVD)

Clinicopathologic Features

As illustrated in Figure 9.5, the morphologic hallmark of CAVD is heaped-up calcified masses within the aortic cusps that ultimately protrude through the outflow surfaces into the sinuses of Valsalva, preventing the opening of the cusps. CAVD can progress from mild valve thickening, with sclerosis, to severe calcification with impaired leaflet motion with stenosis. The calcific deposits distort the cuspal architecture, primarily at the cuspal bases; the free cuspal edges are usually not involved. The calcific process begins in the valvular fibrosa, at the points of maximal cusp flexion (the margins of attachment), and the microscopic layered architecture is often largely preserved. Cartilaginous and osseous metaplasia, with mature lamellar bone and maturing trilineage hematopoietic marrow, are frequently observed deep in valvular calcific nodules.[31] Calcium-associated rigidity, structural abnormalities and fibrotic thickening may induce not only functional valve abnormalities but via increased hemodynamic stresses, they may further contribute to calcification and fibrosis and internal microstructural rearrangements.[28]

The incidence and severity of CAVD increase with age, with only 5% of individuals younger than 45 years but more than 20% of subjects older than 75 years having AV calcification in an autopsy study.[32] Clinical data show that aortic sclerosis is present in ≈25% of people 65–74 years of age and in 48% of people older than 84 years, and aortic stenosis is present in 2–9% of elderly adults.[33,34] Most patients who require surgery for CAVD who had initially normal anatomy are >70 years of age. In the Euro Heart Survey, which included 4910 patients in 25 countries, aortic stenosis accounted for 43% of all patients who had valvular heart disease.[33] Aortic valve stenosis is the most common valvular heart disease in the aging population of the developed world with projected disease burden expected to increase from 2.5 million in 2000 to 4.5 million in 2030. Calcification of the mitral annulus (CMA) is thought to derive from a pathophysiology similar to that of CAVD. CMA increases with age, with calcification being present in 17% of men and 44% of women after 90 years. As in CAVD, histological examination reveals amorphous calcification, often with inflammation.

Although CAVD is more common with advanced age, it is not an inevitable consequence of aging. Accumulating evidence suggests that CAVD is an actively regulated disease process that cannot be characterized as purely 'senile' or 'degenerative' but some aspects of the process may be similar to degenerative calcification under certain circumstances.[35] Epidemiological studies suggest

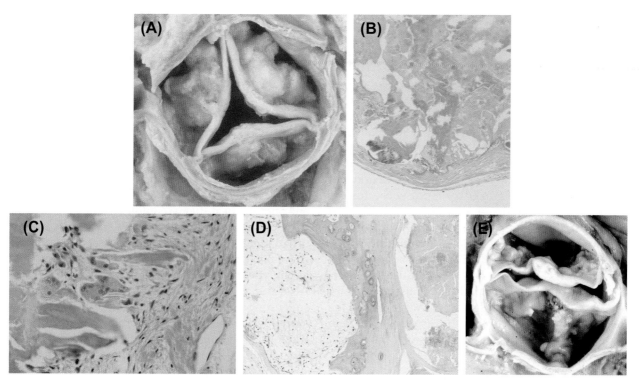

FIGURE 9.5   Calcific aortic valve disease (CAVD). **(A)** Calcific aortic stenosis of a previously normal valve having three cusps (viewed from aortic aspect). Nodular masses of calcium are heaped-up within the sinuses of Valsalva. **(B)** Low-magnification histologic appearance of CAVD/ aortic stenosis similar to lesion in **(A)** showing large amorphous calcium mass that obliterated fibrosa and grew toward but did not penetrate the ventricular surface at bottom. **(C, D)** Ossification, focal in **(C)** and extensive with bone marrow formation in **(D)**. **(E)** Congenitally bicuspid aortic valve with aortic stenosis. The conjoined cusp at the bottom of the photo has a line of failed separation at its center, called a raphe.

that CAVD and coronary atherosclerosis share similar risk factors, such as age, male gender, hypercholesterolemia, hypertension, metabolic syndrome, diabetes, and chronic kidney disease.[35] In addition, calcification of aortic valves may be affected by genetic factors, as revealed by population studies and case control comparisons for specific polymorphisms, including the vitamin D receptor, estrogen receptor, apolipoprotein E4, and interleukin 10 alleles.[36–39]

Diagnosis

Owing to physiological compensatory mechanisms, particularly left ventricular hypertrophy, symptoms of CAVD typically do not appear until aortic stenosis has developed, correlating with an orifice diameter less than 1.0 cm^2 (normal approximately 3–4 cm^2). Symptoms usually include shortness of breath, syncope, and chest pain. Aortic stenosis is typically confirmed using ultrasound echocardiography, which characterizes hemodynamics. The severity of aortic valve dysfunction is determined by measurements of aortic jet velocity, mean transvalvular pressure gradient, and continuity equation valve area.[40] Mild aortic valve stenosis is generally defined by restricted opening of the valve cusps, with a mean pressure gradient of less than 25 mmHg; moderate

aortic valve stenosis by a mean gradient between 25 and 40 mmHg; and severe aortic valve stenosis by a mean gradient above 40 mmHg.[41] However, echocardiography itself does not have the resolution for quantifying aortic valve calcium, thereby reducing usefulness for monitoring early-stage CAVD. In contrast, computed tomography (CT) is more sensitive for visualization and quantification of aortic calcium, and thus has emerged as a useful tool for studying aortic valve stenosis.[35] Moreover, emerging evidence indicates that positron emission tomography (PET) imaging combined with CT may simultaneously and non-invasively assess early calcification and inflammation in patients with CAVD.[42] Nevertheless, because direct imaging of the AV still involves technical challenges, echocardiography remains the clinical gold standard for assessing the severity of aortic valve dysfunction.[43]

Role of the Congenitally Bicuspid Aortic Valve (BAV)

BAV is a common congenital cardiac abnormality, which affects approximately 2% of the population with higher prevalence in men than women (2:1 male/female ratio). An autosomal dominant condition affecting up to 10% of first-degree relatives,[44] a BAV usually exhibits normal function at birth and during early life. However, by age 60 years, 85% of individuals who have a BAV develop stenosis and 15% have regurgitation. Indeed, owing to the high penetrance of superimposed CAVD, a BAV is present in approximately 50% of adults undergoing valve replacement for severe CAVD. Calcific stenosis of a BAV is generally accelerated, appearing 10–15 years earlier than in tricuspid aortic valves, and a calcified BAV often becomes clinically important in patients as young as 50 years old.

Anatomically, BAVs have either two cusps of unequal size, where the larger of two cusps represents a fusion, or incomplete separation of two cusps during fetal development. If the latter, the inadequately joined third commissure is generally located in the middle of the fused cusp (which is larger than a normal cusp, and the area of fusion is called a raphe). BAVs induce an abnormal, turbulent flow pattern and higher tissue stresses, which are concentrated in the larger cusps and at the raphe. Calcium deposition and fibrosis predominate in the raphe, when present, and at the bases of the cusps.

MECHANISMS OF CAVD

As discussed earlier, although some aortic valves show morphologic features that resemble those of atherosclerosis, and calcification is often an important feature of atherosclerosis, the morphology of established human CAVD is morphologically distant from that of atherosclerosis. Key controversies in the pathogenesis of CAVD relate to the extent to which its mechanisms are shared with those of atherosclerosis, the extent to which degenerative processes play a role, and the extent to which the initiation and progression of calcification are actively regulated.[35,45] In CAVD, the increased mechanical stress on resident VICs induced by aging-related valvular remodeling, inflammation and other mechanical and biochemical processes could play an important role in early cell injury (apoptosis or necrosis) or osteogenic differentiation of VICs.[28,46] Apoptosis/necrosis enabled dystrophic calcification mechanisms, in which cell injury is an important and early event, are exemplified by the failure of bioprosthetic substitute heart valves, in which calcification is initiated primarily within residual porcine aortic valve or bovine pericardial cells that have been devitalized by glutaraldehyde pretreatment[5] and thus are incapable of excluding excess calcium ions. These chemically fixed cells may share pathophysiologic features of cells damaged by 'wear and tear' that could initiate CAVD. Calcification of dead or damaged cells is called dystrophic calcification, and is considered a passive phenomenon. Conversely, numerous studies supporting an active mechanism in CAVD showed that VICs in calcified AVs express markers of osteoblastic activity,

such as runt-related transcription factor-2 (Runx2), a master regulatory osteogenic transcription factor, osteopontin, osteocalcin, bone sialoprotein, alkaline phosphatase, and bone morphogenetic protein (BMP)-2 and -4 (markers that characterize osteoblasts in bone).[28,47,48]

Evidence suggests that these two mechanistic pathways (dystrophic and active osteoblastic mineralization) comprise a range of responses of VICs to biomechanical and biochemical stimulation and these processes may not be mutually exclusive. Moreover, it is unclear whether active and dystrophic calcification processes can occur simultaneously although independently of one another, or if they are interdependent, and whether VICs can attain osteoblastic function directly or only by progressing through a myofibroblast phenotype. The core of calcific nodules generated in vitro contains different types of mineralization as well as apoptotic cells.[49] Also, not all ossifying nodules contain apoptotic cells, suggesting that different types of nodules may form concurrently.[50] In addition, a potential mechanism of calcification via the release of calcified extracellular vesicles was recently described,[51,51a,51b] consistent with the ultrastructural observation many years ago of matrix vesicles in calcified human aortic valves.[52]

Progress in understanding the mechanisms involved in human calcific valve disease has had three limitations: (1) the inability of human tissue studies to verify that early valve pathology is a precursor of clinically evident disease; (2) the inability to generate animal models that faithfully recapitulate the morphology and progression of human valve disease; and (3) the inability of imaging modalities to resolve and quantify early calcific changes. Nevertheless, several potentially interrelated factors and influences that have been shown to be associated with, contribute to and/or regulate calcification and other pathological features of valve disease are discussed below. The key role of mechanical factors has recently been reviewed and will not be discussed in detail in this communication.[53]

Endothelial Dysfunction

Valvular endothelium is subjected to constant hemodynamic forces with different mechanical stresses exerted on each side. As mentioned earlier in this chapter, the outflow surface of the aortic valve,[54] below which is the fibrosa, where early calcific lesions are localized, appears to be the valve surface more susceptible to inflammation. For example, high pulsatile shear stress on the fibrosa side induces up-regulation of adhesion molecules (VCAM-1, E-selectin) for circulating monocytes and leukocytes.[54,55] In addition, stretch loading induces pro-inflammatory bone morphogenic protein (BMP2/4) expression on the outflow surface, thus rationalizing a potential role for BMPs in early CAVD lesion formation.[56] Additional mechanisms may include endothelial damage by oxidized lipids leading to a loss of VEC alignment and up-regulation of cell adhesion molecules, which permits inflammatory infiltration.[57] Furthermore, inhibition of eNOS expression results in increased cusp stiffness, which, in parallel with reduced expression of eNOS on the outflow (fibrosa) side in CAVD,[58] suggests that eNOS may have a protective effect. Emerging evidence indicates that mechanical stress may also induce the osteogenic potential of VECs, providing support that the valvular endothelium harbors a reserve of progenitor cells that can repopulate the leaflet with osteogenic-like cells capable of differentiation that can contribute to calcification.[16]

Inflammation-Dependent Calcification

Clinical studies have found that aortic stenosis is associated with systemic signs of inflammation, including elevated serum C-reactive protein levels.[59] Histological assessment of human valves revealed that some early lesions contain inflammatory cells.[30] Moreover, experimental hypercholesterolemic rabbit[60] and atherosclerotic mouse models of CAVD[28,61–63] develop lesions of the aortic valve that contain inflammation. These investigations, in conjunction with innovative molecular imaging studies that applied multimodality imaging nanoprobes to the calcifying aortic valves of ApoE-deficient mice,[20,28] have established the concept of inflammation-dependent calcification.[64] A major limitation in the field has been the inability to spatially resolve and quantify the dynamic changes in the aortic valves from the earliest stages of disease. However, rapid advances in molecular imaging have permitted simultaneous visualization in experimental systems in vivo, and key cellular events in early CAVD, including endothelial cell activation, macrophage accumulation/activation, expression of matrix remodeling enzymes, and osteogenic activity.[20,28,64–66]

In atherosclerotic mouse and rabbit models, proinflammatory factors and mechanical forces activate VECs and promote recruitment of inflammatory monocytes/macrophages to the aortic valve (Figure 9.6). In molecular imaging studies of these animals, the VCAM-1-targeted molecular imaging agent was concentrated in valve commissures, the area that is known to encounter the greatest mechanical stress.[28] Activated macrophages produce a variety of cytokines, growth factors, and matrix-degrading enzymes. Proteolytic enzyme action may cause ECM remodeling and thickening of the leaflets, resulting in valvular dysfunction. Furthermore, elastolytic enzymes may induce elastin degradation, which provides a nidus for hydroxyapatite crystal formation.[20] Molecular imaging studies validated by histology have visualized proteolytic activity and activated myofibroblasts and thus provided a

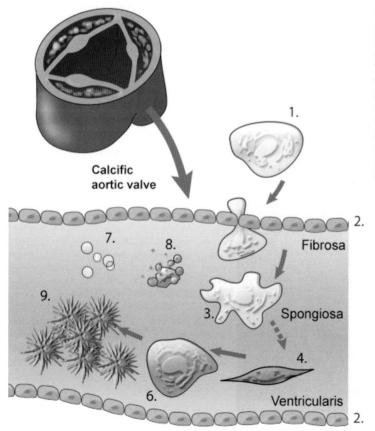

FIGURE 9.6 Schematic diagram describing hypothetical inflammatory mechanisms contributing to CAVD. Proinflammatory monocytes (1) are recruited to a site via activated endothelial cells (2) followed by subsequent macrophage (3) accumulation. Tissue macrophages then release pro-osteogenic cytokines, which stimulate the differentiation of VIC myofibroblasts (4) into osteoblast-like cells (6) resulting in generation of calcified matrix vesicles (7) or apoptotic bodies (8) followed by formation of micro- and macrocalcifications (9). Modified from New EP and Aikawa E. Molecular imaging insights into early inflammatory stages of arterial and aortic valve calcification. *Circulation Research* 2011;108:1381–1391. Illustration credit: Cosmocyte/Ikumi Kayama.

Calcific
aortic valve

biological readout of inflammation and matrix degradation. The altered mechanical stresses and disturbed flow patterns in early disease likely further induce inflammation and differentiation of VICs into activated myofibroblasts and, subsequently, as mentioned above, into osteoblast-like cells through up-regulation of osteogenic pathways, including osteogenic differentiation, apoptosis, and release of matrix vesicles.[51,64] Thus, inflammatory mechanisms could also promote deposition of calcium primarily in regions of high mechanical stress (Figure 9.7).

Oxidative Stress

Oxidative stress generated by oxylipids is associated with progression of CAVD. In vitro oxidative stress promotes vascular cell osteogenesis. In vivo, the levels of superoxide and hydrogen peroxide were markedly increased around calcium deposits in human[67] and mouse[63] aortic valves and valve leaflets of rabbits treated with a high-cholesterol and vitamin D diet.[68] Oxidative stress promotes osteogenic differentiation by induction of Runx2, a master regulatory osteogenic transcription factor.

Lipoproteins and Wnt/Lrp5 Signaling Pathway

Studies of human specimens have demonstrated the presence of lipids in the valvular tissue. Oxidative lipids can induce VIC calcification, stimulate oxidative stress in valve endothelium, and recruit inflammatory cells, which subsequently leads to the release of pro-inflammatory cytokines promoting VIC activation and osteogenic differentiation.[69] Lipoproteins promote atherosclerotic calcification and increase both the rate and severity of aortic valve lesions in animal models.[28] Moreover, a recent genome-wide association study demonstrated that a SNP in the Lipoprotein(a) (LPA) locus was significantly associated with clinical aortic-valve calcification; this SNP was prospectively associated with (and suggested to be causal to) aortic stenosis.[70]

The low-density lipoprotein (LDL) receptor-related protein 5 (Lrp5), a co-receptor of the LDL receptor family, has been discovered in skeletal bone. Lrp5 binds to the secreted glycoprotein wingless (Wnt), which activates β-catenin to induce bone formation. Canonical Wnt signaling acts through frizzled/Lrp5 receptors resulting in β-catenin nuclear localization. Wnt/β-catenin signaling is important for osteoblast maturation during embryonic development.[71] A number of studies have also shown that the Wnt/β-catenin pathway is reactivated in CAVD. Diseased human valves removed during surgical valve replacement or repair demonstrated increased expression of Lrp5, β-catenin, and Wnt3a ligand as compared to normal valves.[72] Increased Wnt signaling has also been observed in numerous atherosclerotic animal models. For example, rabbits maintained on an atherogenic diet develop aortic valve calcification and display increased expression of β-catenin and Lrp5.[60] In vitro studies also support the notion that Wnt/β-catenin signaling is involved in VIC activation, proliferation, and chondrogenesis.[22]

Bone Morphogenic Protein (BMP) Signaling

Examination of explanted human calcified aortic valves has shown increased BMP-2 and BMP-4 expression in areas associated with calcification.[31] BMP-2 and BMP-4 are potent morphogens that could regulate osteogenic differentiation of VICs and thus induce an active calcification process.[46] Active BMP signaling, as indicated by phospho-SMAD1/5/8 expression, a hallmark of canonical BMP signaling, has been identified in both diseased human AV and animal models with atherosclerotic AV lesions.[73] SMAD1/5/8 expression was consistently found on the fibrosa side of the disease human aortic valves, the primary location of CAVD. In vitro studies showed that BMP-2 stimulation promotes VIC calcification and expression of the osteogenic factors Runx2, osteopontin, and alkaline phosphatase.[74] In hypercholesterolemic mice, phospho-SMAD1/5/8

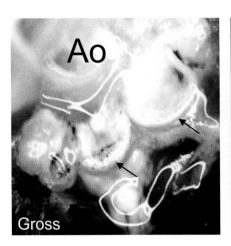

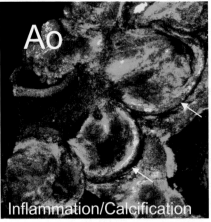

FIGURE 9.7 Gross morphology and molecular imaging (ex vivo near-infrared fluorescence microscopy; image stacks) of calcified aortic valves visualizes osteogenic activity (red fluorescence) in the areas of leaflet attachment to the aortic wall in inflamed valves (green fluorescence). Adapted from Aikawa E, Aikawa M, Libby P, Figueiredo JL, Rusanescu G, Iwamoto Y, Fukuda D, Kohler RH, Shi GP, Jaffer FA, Weissleder R. Arterial and aortic valve calcification abolished by elastolytic cathepsin S deficiency in chronic renal disease. *Circulation* 2009;119:1785-1794.

levels increase with progression of valvular calcification and impaired valve function.[75] Smad6-deficient mice develop early aortic ossification, suggesting that suppression of BMP signaling may be important in preventing cardiovascular calcification.[76]

The OPG/RANKL/RANK Signaling Pathway

Ligand of receptor activator of nuclear factor κB (RANKL), a member of the TNF-α superfamily, is expressed in osteoblasts, stromal cells, T cells, and endothelial cells and found in calcified valves. RANKL interacts with RANK, a transmembrane protein located on osteoclast precursors or mature osteoclasts, and induces osteoclastogenesis through NFκB. Mouse studies have shown that the RANKL/RANK pathway is the gatekeeper of osteoclast differentiation and activation and is critical in cardiovascular calcification. The interaction of RANKL with RANK also increases the binding of osteoblast transcription factor Runx2, which is essential for osteoblast differentiation.[77] Osteoprotegerin (OPG) is a soluble decoy receptor that binds to RANKL, thereby inhibiting the interaction of RANKL and RANK and thus preventing osteoclast differentiation. OPG is expressed by several cell types, including smooth muscle cells and endothelial cells, whereas RANKL and RANK are not expressed in cardiovascular tissue under physiological conditions.[78] RANKL-treated VICs have shown increased nodule formation and alkaline phosphatase activity, and enhanced DNA binding of Runx2.[79] TNF-α-treated VICs have shown similar effects.[80] Moreover, OPG-deficient mice display vascular calcification,[81] a finding which supports a protective role of OPG against calcification. The diversity of the OPG expression on the aortic and ventricular sides of the AV leaflets was also explored; induction of OPG occurred along the ventricular surface, consistent with the lack of calcification on the ventricular side of the aortic valve.[82]

Transforming Growth Factor-β (TGF-β)

TGF-β regulates VIC differentiation and activation via expression of alpha-smooth muscle actin in the AV.[83] TGF-β initiates signaling through Smad proteins, which in turn interact with the transcription factors FoxH1, c-jun, and c-fos. TGF-β can also regulate mitogen-activated protein kinase (MAPK) pathways. Both signaling pathways regulate cell cycle, proliferation, migration, cytokine secretion, and ECM synthesis and degradation – all of which are important in valve biology.[83] The role of TGF-β in aortic valve calcification is not clear, however. TGF-β1 induces rapid dystrophic calcification in vitro through an apoptosis mediated process.[84] Some studies showed that TGF-β induces calcified nodule formation on stiff matrix via cell apoptosis,[85] while another study suggests that TGF-β-mediated stimulation of VICs

plated on a less stiff collagen does not promote osteogenic differentiation.[46] Although some data suggest that TGF-β may actively suppress osteogenesis in skeletal osteoblasts in vivo, the role of TGF-β in osteogenic signaling remains controversial.

Renin-Angiotensin System (RAS)

Several studies suggest that components of the renin-angiotensin system (RAS), particularly Angiotensin II (ANG II), are involved in CAVD via induction of oxidative stress and inflammation. Studies have demonstrated the presence of AngII receptors on VICs and showed that the density of these receptors is higher in valves with CAVD compared to healthy AVs. Angiotensin-converting enzyme (ACE) is produced by monocytes/macrophages and co-localizes with LDL in calcified aortic valves.[86] Blockage of angiotensin type 1 receptor (AT1r) prevents inflammatory cell infiltration, endothelial disruption and VIC activation in aortic valves of hypercholesterolemic rabbits.[87] However, while some clinical studies showed slowing of progression of CAVD in patients taking ACE inhibitors,[88] other studies have yielded conflicting results.[89]

Notch Signaling

The Notch pathway has emerged as a central mechanism involved in cardiac development, which is reactivated during cardiovascular calcification. The Notch receptor family comprises four members, Notch 1, 2, 3, and 4, expressed as transmembrane molecules on the cell surface that allows signaling in a contact-dependent manner. Notch activation suppresses the Wnt/β-catenin pathway and Runx2 transcription factor activity thereby inhibiting osteogenesis.[90] It was suggested that Notch inhibits osteoblast differentiation through direct binding of Notch 1 intracellular domain (N1ICD) to β-catenin, which counteracts Wnt-mediated induction of osteogenesis. In addition, the Notch target genes Hey1 and Hey2 encode transcriptional repressors that inhibit the activity of Runx2.[91] Moreover, while Notch is required for chondrocyte proliferation, increased Notch activity may inhibit terminal chondrocyte differentiation and endochondral ossification.[92] Inactivation of Notch catalyzes the progression of Runx2-mediated calcification, and Notch mutations have been implicated in CAVD. However, whether mechanisms also involve inhibition of Wnt/β-catenin signaling, similar to bone mineralization, is unknown and requires further investigation.

Interestingly, Notch gene mutations were associated with a spectrum of developmental AV abnormalities and severe calcification in two families with non-syndromic familial AV disease including BAV.[93] Nevertheless, although Notch and TGF-β1 type II receptor gene mutations are associated with BAV, some individuals with these mutations do not have BAV, and these gene defects are rare relative to the incidence of BAV.[94] Moreover, Notch signaling represses osteoblast-like calcification pathways.[95]

ANIMAL MODELS OF CAVD

Experimental animal models are used to understand disease progression and to monitor the effects of new therapies. Even though well-established human CAVD is pathologically distinct from atherosclerosis, as emphasized earlier, elevated cholesterol levels promote CAVD. Thus, investigation of the AV lesions that occur in hypercholesterolemic animals (particularly Apolipoprotein E [ApoE]-deficient and low-density lipoprotein receptor [LDLr] knockout mice) has been useful to understand many aspects of the mechanisms of CAVD.

In ApoE-deficient mice with spontaneous hypercholesterolemia, AV regurgitation, increased transvalvular velocity, and valvular calcification occur progressively with increasing age.[62] In mice fed a high-cholesterol diet, disease develops earlier. The initial changes are characterized by activated valve endothelium, macrophage accumulation, proteolytic enzyme activity, and osteogenesis, which can be readily detected by novel molecular imaging approaches.[28] Moreover, these studies have shown that ApoE-deficient mice exhibit early AV lesions characterized by thickened leaflets with macrophage-rich lesions, followed by formation of calcium deposits on the aortic side of the valve in late stages. Recent studies have demonstrated that chronic kidney disease induced by 5/6 nephrectomy significantly accelerates cardiovascular calcification, CAVD, and osteoporosis.[20,66]

LDLr knockout mice also develop hypercholesterolemia through inhibition of the removal of circulatory LDL. Fed a high-cholesterol diet, LDLr knockout mice develop extreme hyperlipidemia and hyperglycemia, characteristics of the metabolic syndrome. This mouse model develops increased AV cuspal thickness, macrophage accumulation, myofibroblast activation, and osteogenesis after 16 weeks on a high-fat diet,[96] providing a useful tool to study CAVD in diabetes and other metabolic disorders.

Since some early human AV lesions have pathological features similar to those of inflamed atherosclerotic plaques, it is reasonable to hypothesize that cholesterol lowering may improve various features associated with CAVD. Reversa mice were used to examine the effects of lipid lowering on histological and biochemical properties of the AV.[75] Reduction of blood lipids by lipid-lowering therapy reduced valvular inflammation and lipid content, BMP and Wnt/β-catenin signaling, and reduced valvular calcium.[73,75] Cholesterol lowering by statins also showed reduction of inflammation-dependent cardiovascular calcification in both mouse and rabbit models.[60,65] Although these findings have stimulated interest in the possibility that the statin drugs, which lower systemic cholesterol and decrease inflammation in atherosclerosis, may decrease the rate of aortic stenosis progression in humans, this has not been demonstrated in clinical studies.[97–99]

Rabbits and pigs have also been used in CAVD research because their valves resemble human valves, including tri-layer valve morphology, they form valve lesions in response to a high-cholesterol diet, and they have similar lipid profiles and lipoprotein metabolism. The rabbit model is disadvantageous owing to the very high cholesterol levels required to form lesions, and vitamin D2 administration to cause accelerated calcification. In contrast, porcine models require a relatively high expenditure in animal care and maintenance.

DEGENERATIVE MITRAL VALVE DISEASE (DMVD)

The predominant form of DMVD is the displacement of enlarged, thickened, redundant mitral leaflet(s) into left atrium during systole.[100] Potential serious complications of DMVD include heart failure, mitral regurgitation, bacterial endocarditis, thromboembolism, and atrial fibrillation. DMVD is the most common indication for surgical repair or replacement of the mitral valve.

Barlow Disease and Fibroelastic Deficiency (FED)

Degenerative mitral valve disease (DMVD) is characterized by an excess of valvular tissue resulting in billowing, floppy leaflets (usually more prominent in the posterior leaflet), or rupture of the chordae tendineae resulting in a flail leaflet, either of which can lead to MV incompetence/regurgitation, often with dilatation of the mitral annulus.[101] In DMVD, the biomechanical properties of both leaflets and chordae tendineae are compromised; the leaflets are approximately one-third less stiff and twice as extensible as normal leaflets,[102] and the chordae are about two-thirds less stiff and 75% less strong than normal chordae.[103] In the 1980s, Carpentier characterized myxomatous DMVD, now called Barlow's disease, and fibroelastic deficiency (FED) based on clinical patterns, echocardiographic findings, and gross pathology.[104] Two decades later the histopathologic features of these two entities were described,[101] followed by studies addressing the mechanisms of DMVD.[4]

Patients with Barlow's disease, also known as 'floppy' mitral valve, often have a long clinical history of mitral valve regurgitation and are usually younger compared to patients with FED. Grossly, the leaflets in myxomatous DMVD are thickened with a soft, gelatinous consistency.

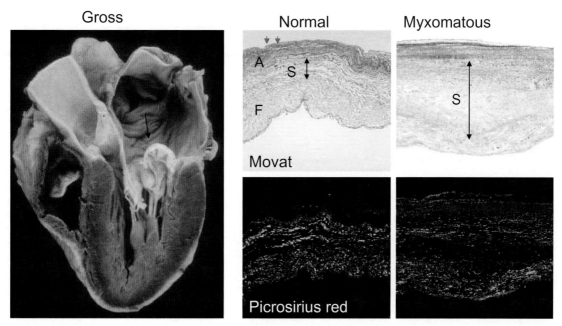

FIGURE 9.8   Myxomatous degeneration of the mitral valve. Left panel: Long axis of left ventricle demonstrating hooding with prolapse of the posterior mitral leaflet into the left atrium (arrow). (Courtesy of William D. Edwards, MD, Mayo Clinic, Rochester, MN). Morphological features of normal and myxomatous mitral valves. Right panel: Normal mitral valve (left) and valve with myxomatous degeneration (right). Myxomatous valves have an abnormal layered architecture: loose collagen in fibrosa, expanded spongiosa strongly positive for proteoglycans, and disrupted elastin in atrialis (top). Top, Movat pentachrome stain (collagen stains, yellow; proteoglycans, blue-green; and elastin, black). Bottom, Picrosirius red staining viewed under polarized light detected disruption and lower birefringence of collagen fibers in myxomatous leaflets. Magnification ×100. Left panel reproduced by permission from Schoen FJ, Mitchell RN: The heart. In: Robbins/Cotran *Pathologic Basis of Disease*, 8th Ed., Kumar V, Fausto N, Aster JC, Abbas A (eds.), Philadelphia: W.B. Saunders, 2010, pp. 529–587. Right panel adapted from Rabkin E, Aikawa M, Stone JR, Fukumoto Y, Libby P, Schoen FJ. Activated interstitial myofibroblasts express catabolic enzymes and mediate matrix remodeling in myxomatous heart valves. *Circulation* 2001;104:2525–2532.

Microscopically, myxomatous DMVD is characterized by expansion of the spongiosa by amorphous glycosaminoglycans and proteoglycans (Figure 9.8). Collagen in the fibrosa is fragmented and appears less dense than normal. Picrosirius red staining under polarized light shows that individual collagen fibers are disoriented, disrupted, and coiled. Elastic fibers are fragmented. Myxomatous degeneration is also frequently found in chordae tendineae, which may represent the background for chordae rupture. Myxomatous mitral valves often have a layer of superficial plaque characterized by the accumulation of stellate and spindle-shaped cells, predominantly on the ventricular aspect of the leaflet; this plaque probably is a secondary feature owing to flow abnormalities. Owing to both expansion of the spongiosa and superficial plaque formation/fibrosis, myxomatous leaflets are thicker than normal.

In contrast, FED is characterized morphologically by connective tissue deficiency, with thinning of leaflet tissue.[105] The three-layer architecture of the leaflet tissue is preserved. Secondary pathological changes in the prolapsing segments may result in deposition of myxomatous material with resultant thickening and expansion. Histologically, elastic fiber alterations are more prevalent in FED compared with Barlow's disease. The etiology of

the connective tissue deficiency in FED is unknown but may be age-related. Acute presentation with rupture of thin, deficient chords is the usual mechanism of regurgitation in affected patients.

Immunohistochemistry demonstrates three different cell types of VICs in normal and diseased MVs: quiescent fibroblasts were characterized as vimentin-positive and alpha-smooth-muscle actin-negative cells; activated myofibroblasts were identified by expression of alpha-smooth-muscle actin and vimentin, but not SM1 or SM2, markers of differentiated smooth muscle cells; and smooth muscle cells located below the endothelium were characterized as alpha-smooth muscle actin and SM1/SM2 positive (Figure 9.9).[4]Activated VIC in MVD express high levels of proteolytic enzymes (metalloproteinases: MMP-1, MMP-2, MMP-9, MMP-13) and elastolytic cathepsins participating in matrix degradation while collagen synthesis (procollagen-I mRNA expression) remains unchanged.[4] These observations demonstrate an important role for excessive levels of collagenolytic and elastolytic enzymes expressed by VIC, as opposed to decreased collagen synthesis, in the distorted layered leaflet architecture and structural abnormalities of collagen and other connective tissue components, and suggest that these structural alterations could cause

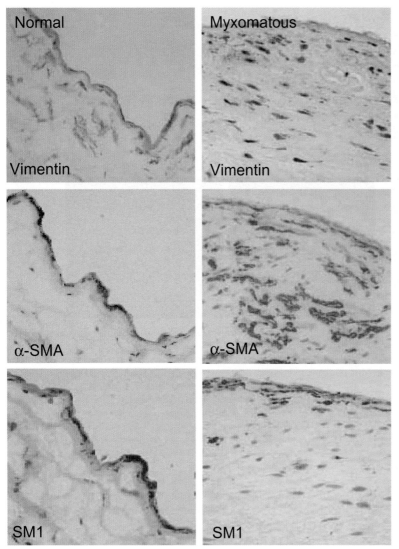

FIGURE 9.9 Characterization of quiescent VICs in normal mitral valves and activated myofibroblast-like VICs in myxomatous mitral valves. Interstitial cells in both normal and myxomatous leaflets stained positively for vimentin (top). Cells in normal valves showed low levels of α-SMA (middle left). However, interstitial cells in the spongiosa of myxomatous valves expressed vimentin and α-SMA, suggesting an activated phenotype (top and middle). In both groups, smooth muscle cells (detected by both α-SMA and SM1) accumulated in the subendothelial layer (middle and bottom). Note that interstitial cells in myxomatous valves showed undetectable levels of SM1 myosin (bottom right). Magnification ×400. Adapted from Rabkin E, Aikawa M, Stone JR, Fukumoto Y, Libby P, Schoen FJ. Activated interstitial myofibroblasts express catabolic enzymes and mediate matrix remodeling in myxomatous heart valves. *Circulation* 2001;104:2525–2532.

functional abnormalities in patients with DMVD. More recently, an investigation of chondrogenic remodeling in diseased MVs demonstrated hypertrophic chondrocytes in regions of myxomatous degeneration that were stained strongly for bone sialoprotein. Protein- and gene-level analyses also showed elevated expression of Lrp5, Runx2/Cbfa 1, SOX 9, cyclin, osteocalcin, and osteopontin compared with control valves, suggesting a role for chondrogenic differentiation, a process resembling early cartilage development.[72]

Genetics of DMVD

DMVD is associated with some heritable disorders of connective tissue, including Marfan syndrome, in which it is usually associated with mutations in Fibrillin-1 (FBN-1). However, most cases of DMVD are not associated with FBN-1 abnormalities; indeed, it is unlikely that more than 1–2% of patients with DMVD have associated clearly identifiable connective tissue disorder.[100] Several recent experimental and clinical findings implicate a key pathogenetic role for dysregulation of TGF-β, a key cytokine regulator of ECM assembly and remodeling in connective tissue as the common mechanism of Marfan-syndrome-related and possibly other syndromic and non-syndromic forms of DMVD.

The genetics of myxomatous degeneration is complex and uncertain and heterogeneous with multi-gene, multi-chromosomal autosomal dominance with incomplete penetrance. Myxomatous degeneration is likely a connective tissue disorder with altered ECM remodeling due to abnormal over-expression of matrix metalloproteinases, cysteine endoproteases, and tenomodulin.[106] Studies utilizing genetic linkage analysis have mapped families with autosomal dominant mitral valve prolapse to the X-chromosome, and chromosomes 11p15.4, 16p11.2-p12.1, and 13q31.3-q32.1. The latter locus is particularly interesting in that, of the at least 16 known genes in the region, several could be involved in valvular ECM remodeling and therefore are potentially good candidates for causation in some cases of DMVD.

MECHANISMS OF DMVD

The underlying pathological process in DMVD is myxomatous degeneration and ECM disruption (excess or deficiency). Nevertheless, whether and to what extent Barlow's disease and FED are distinct, both pathologically and mechanistically, is uncertain. These different clinical presentations may represent a spectrum of remodeling outcomes in response to a common substrate deficiency and/or abnormal mechanical environment.

Histologically, the essential change in DMVD is attenuation of the collagen-rich fibrosa layer of the valve, on which the structural integrity of the leaflet depends, accompanied by deposition of myxomatous material rich in proteoglycans. DMVD is associated with weakening of valvular connective tissue, characterized biomechanically by a decreased stiffness and an increase in extensibility, associated with increased glycosaminoglycans and abnormal fibrillar ECM organization in both leaflets and chordae.[100,102,107] The prevailing concept is that DMVD is caused by a defect in the mechanical integrity of the leaflet, which results from altered ECM synthesis and/or remodeling by VICs of the essential structural proteins and proteoglycans, which together with normal wear and tear leads to stretching, elongation, and other features of the clinical phenotype of DMVD. The connective tissue weakening may have as yet poorly understood implications for the durability of surgical procedures to repair DMVD. Furthermore, VICs in this disorder are activated, suggesting that a state of chronic mechanical disequilibrium exists because the adaptive ECM remodeling potential is exceeded.

Mitral VECs readily synthesize the basement membrane components laminin and type IV collagen, but not type I collagen and chondroitin sulfate.[108] VECs were also found to synthesize endothelial nitric oxide synthase (eNOS) both in vivo and in vitro, in contrast with VICs, for which only a minority of cells expressed neuronal NOS in vitro.[109] In vitro studies have also shown that growing mitral VECs on collagen-chondroitin sulfate scaffolds promoted a more in vivo like phenotype than did culture on collagen-only scaffolds.[108] In vitro studies have shown that mitral VICs treated with TGF-β have increased proliferation,[110] which was the opposite of the response by aortic VICs.[83] In the in vitro scratch wound model, the actively migrating mitral VICs showed increased fibroblast growth factor-2 (FGF-2) and fibroblast growth factor receptor 1 (FGF-1)[111] as well as greater expression of TGF-β, alpha-smooth muscle actin, and phospho-Smad2/3.[110] Mitral

VICs cultured within 3D collagen gels and subjected to mechanical loading showed that the stretching conditions modulated the deposition of glycosaminoglycans and proteoglycans by the VICs in a manner that reflected the magnitude and duration of applied stretch.[112]

ANIMAL MODELS OF MITRAL VALVE DISEASE

Sheep are used most commonly for investigations of mitral valve regurgitation. Mitral regurgitation can be induced by tachycardia-induced dilation, volume overload, coronary artery occlusion, transection of one of the valve leaflets, or chordae displacement.[29,113–116] Tachycardia-induced cardiomyopathy induces collagen and elastin turnover in the mitral valve leaflets.[117] Coronary artery ligation resulting in abnormal ventricular wall motion causes accumulation of procollagen within the mitral leaflets.[118] Active ECM remodeling within mitral valve leaflets occurs in a model of isolated mitral regurgitation without heart dysfunction,[119] suggesting that mitral regurgitation alone can promote leaflet remodeling. A cut-induced model of wound healing of the mitral anterior leaflet showed a robust healing response involving abundant production of glycosaminoglycans.[115] A stretch model using a displaced papillary muscle technique to impose additional stretch on the mitral valve leaflets resulted in leaflet thickening due to an expansion of the spongiosa, increased expression of alpha-smooth-muscle actin by VECs due to EMT, VICs activation, and reduced collagen fiber organization (Figure 9.10).[29]

Small animal models may also be useful to study MV remodeling. Rabbits in particular have been employed to investigate mechanistic aspects of numerous valve diseases, such as the suppression of the anti-angiogenic glycoprotein chondromodulin in infective endocarditis,[120] the transport of large macromolecules into the valve leaflets of hyperlipidemic rabbits,[121] and the ability of statins to blunt calcific remodeling in hypercholesterolemic rabbits.[122] Mouse models of DMVD are limited mainly due to a lack of knowledge of the genetics of DMVD. Mitral valve prolapse can occur in the context of genetic syndromes, including Marfan syndrome, an autosomal-dominant connective tissue disorder caused by mutations in fibrillin-1. Fibrillin-1 contributes to the regulated activation of the cytokine TGF-β. Mitral valves from fibrillin-1-deficient mice showed alterations in architecture, increased cell proliferation, decreased apoptosis, and excess TGF-β activation.[123] Expression analyses identified increased expression of numerous TGF-β-related genes that regulate cell proliferation and survival and possibly contribute to myxomatous valve disease. Development of new genetically modified mouse models will be valuable in studying of mechanistic insights on pathogenesis of DMVD.

Control MV   Stretched MV

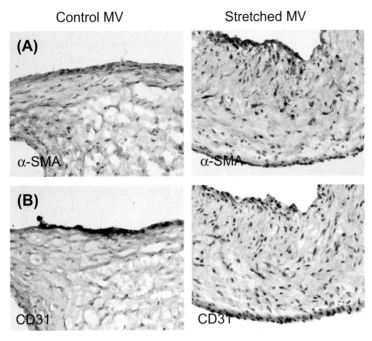

FIGURE 9.10 In the panel on the far left, a normal unstretched sheep mitral valve shows negative α-SMA staining along the CD31-positive endothelium. In the panel on the left, α-SMA-positive staining in the atrial endothelium (also CD31-positive) of a stretched mitral valve, with nests of α-SMA-positive cells appearing to penetrate the interstitium. Adapted from Dal-Bianco JP, Aikawa E, Bischoff J, Guerrero JL, Handschumacher MD, Sullivan S, Johnson B, Titus JS, Iwamoto Y, Wylie-Sears J, Levine RA, Carpentier A. Active adaptation of the tethered mitral valve: Insights into a compensatory mechanism for functional mitral regurgitation. *Circulation* 2009;120:334–342.

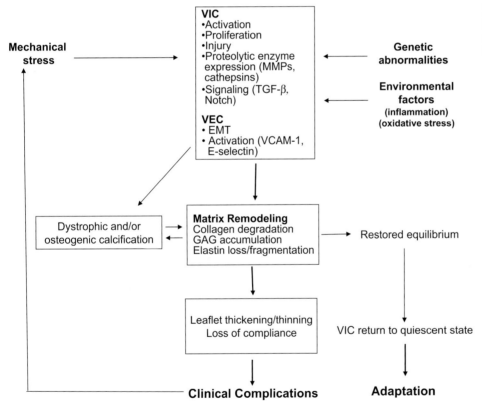

FIGURE 9.11 Schematic representation of general mechanisms of valve disease and adaptation in CAVD and DMVD.

FUTURE RESEARCH OPPORTUNITIES IN HEART VALVE DISEASE: KEY QUESTIONS

Figure 9.11 summarizes general mechanisms of valve disease and adaptation with emphasis on the following comprise key areas of contemporary research and future directions into the mechanisms of calcific and degenerative heart valve disease:

1. What are the relative roles of genetics, mechanical forces, and regulatory factors in heart valve development and post-developmental adaptation?
2. What are the mechanisms of VIC loss, proliferation, differentiation, and recruitment, and VIC–matrix interactions in health and disease?

3. How are structural changes transduced or regulated by VICs in the adaptive response to mechanical stimulation of immature and mature heart valves?

4. What determines the asymmetry between architectural features, gene expression, and pathophysiology across the thickness of the valve cusps and leaflets?

5. What is the role of the valvular endothelium in heart valve function, remodeling, pathology, and tissue engineering (i.e., inflow vs outflow surfaces)?

6. What is the role of EMT in the disease mechanisms associated with CAVD and DMVD?

7. What is the relationship between calcification of the aortic valve and bone mineralization, and how are these reciprocal processes regulated?

8. What are the most relevant animal models in which to study heart valve remodeling and disease?

9. Could high-resolution and high-sensitivity imaging modalities be developed to detect early and subclinical CAVD?

10. Could new therapeutic strategies for treatment of CAVD be identified to replace invasive and costly valve surgery?

References

1. Roger VL, Go AS, Lloyd-Jones DM, Benjamin EJ, Berry JD, Borden WB, et al. Heart disease and stroke statistics–2012 update: A report from the American Heart Association. *Circulation* 2012;**125** e2–e220.

2. Iung B, Vahanian A. Epidemiology of valvular heart disease in the adult. *Nat Rev Cardiol* 2011;**8**:162–72.

3. Schoen FJ. Mechanisms of function and disease of natural and replacement heart valves. *Annu Rev Pathol* 2012;**7**:161–83.

4. Rabkin E, Aikawa M, Stone JR, Fukumoto Y, Libby P, Schoen FJ. Activated interstitial myofibroblasts express catabolic enzymes and mediate matrix remodeling in myxomatous heart valves. *Circulation* 2001;**104**:2525–32.

5. Schoen FJ. Evolving concepts of cardiac valve dynamics: The continuum of development, functional structure, pathobiology, and tissue engineering. *Circulation* 2008;**118**:1864–80.

6. Rabkin-Aikawa E, Aikawa M, Farber M, Kratz JR, Garcia-Cardena G, Kouchoukos NT, et al. Clinical pulmonary autograft valves: Pathologic evidence of adaptive remodeling in the aortic site. *J Thorac Cardiovasc Surg* 2004;**128**:552–61.

7. Bashey RI, Torii S, Angrist A. Age-related collagen and elastin content of human heart valves. *J Gerontol* 1967;**22**:203–8.

8. Sacks MS, David Merryman W, Schmidt DE. On the biomechanics of heart valve function. *J Biomech* 2009;**42**:1804–24.

9. Kunzelman KS, Cochran RP, Murphree SS, Ring WS, Verrier ED, Eberhart RC. Differential collagen distribution in the mitral valve and its influence on biomechanical behaviour. *J Heart Valve Dis* 1993;**2**:236–44.

10. Grashow JS, Yoganathan AP, Sacks MS. Biaixal stress-stretch behavior of the mitral valve anterior leaflet at physiologic strain rates. *Ann Biomed Eng* 2006;**34**:315–25.

11. Bischoff J, Aikawa E. Progenitor cells confer plasticity to cardiac valve endothelium. *J Cardiovasc Transl Res* 2011;**4**:710–9.

12. Butcher JT, Penrod AM, Garcia AJ, Nerem RM. Unique morphology and focal adhesion development of valvular endothelial cells in static and fluid flow environments. *Arterioscler Thromb Vasc Biol* 2004;**24**:1429–34.

13. Simmons CA, Grant GR, Manduchi E, Davies PF. Spatial heterogeneity of endothelial phenotypes correlates with side-specific vulnerability to calcification in normal porcine aortic valves. *Circ Res* 2005;**96**:792–9.

14. Weinberg EJ, Mack PJ, Schoen FJ, Garcia-Cardena G, Kaazempur Mofrad MR. Hemodynamic environments from opposing sides of human aortic valve leaflets evoke distinct endothelial phenotypes in vitro. *Cardiovasc Eng* 2010;**10**:5–11.

15. Markwald RR, Norris RA, Moreno-Rodriguez R, Levine RA. Developmental basis of adult cardiovascular diseases: Valvular heart diseases. *Ann N Y Acad Sci* 2010;**1188**:177–83.

16. Wylie-Sears J, Aikawa E, Levine RA, Yang JH, Bischoff J. Mitral valve endothelial cells with osteogenic differentiation potential. *Arterioscler Thromb Vasc Biol* 2011;**31**:598–607.

17. Chakraborty S, Cheek J, Sakthivel B, Aronow BJ, Yutzey KE. Shared gene expression profiles in developing heart valves and osteoblast progenitor cells. *Physiol Genomics* 2008;**35**:75–85.

18. Aikawa E, Whittaker P, Farber M, Mendelson K, Padera RF, Aikawa M, et al. Human semilunar cardiac valve remodeling by activated cells from fetus to adult: Implications for postnatal adaptation, pathology, and tissue engineering. *Circulation* 2006;**113**:1344–52.

19. Rabkin-Aikawa E, Farber M, Aikawa M, Schoen FJ. Dynamic and reversible changes of interstitial cell phenotype during remodeling of cardiac valves. *J Heart Valve Dis* 2004;**13**:841–7.

20. Aikawa E, Aikawa M, Libby P, Figueiredo JL, Rusanescu G, Iwamoto Y, et al. Arterial and aortic valve calcification abolished by elastolytic cathepsin S Deficiency in Chronic Renal Disease. *Circulation* 2009;**119**:1785–94.

21. Liu AC, Joag VR, Gotlieb AI. The emerging role of valve interstitial cell phenotypes in regulating heart valve pathobiology. *Am J Pathol* 2007;**171**:1407–18.

22. Combs MD, Yutzey KE. Heart valve development: Regulatory networks in development and disease. *Circ Res* 2009;**105**:408–21.

23. Tao G, Levay AK, Gridley T, Lincoln J. MMP14 is a direct target of snail during endothelial to mesenchymal transformation and endocardial cushion development. *Dev Biol* 2011;**359**:209–21.

24. Lincoln J, Lange AW, Yutzey KE. Hearts and bones: Shared regulatory mechanisms in heart valve, cartilage, tendon, and bone development. *Dev Biol* 2006;**294**:292–302.

25. Alfieri CM, Cheek J, Chakraborty S, Yutzey KE. Wnt signaling in heart valve development and osteogenic gene induction. *Dev Biol* 2010;**338**:127–35.

26. Norris RA, Moreno-Rodriguez RA, Sugi Y, Hoffman S, Amos J, Hart MM, et al. Periostin regulates atrioventricular valve maturation. *Dev Biol* 2008;**316**:200–13.

27. Hinton RB, Adelman-Brown J, Witt S, Krishnamurthy VK, Osinska H, Sakthivel B, et al. Elastin haploinsufficiency results in progressive aortic valve malformation and latent valve disease in a mouse model. *Circ Res* 2010;**107**:549–57.

28. Aikawa E, Nahrendorf M, Sosnovik D, Lok VM, Jaffer FA, Aikawa M, et al. Multimodality molecular imaging identifies proteolytic and osteogenic activities in early aortic valve disease. *Circulation* 2007;**115**:377–86.

29. Dal-Bianco JP, Aikawa E, Bischoff J, Guerrero JL, Handschumacher MD, Sullivan S, et al. Active adaptation of the tethered mitral valve: Insights into a compensatory mechanism for functional mitral regurgitation. *Circulation* 2009;**120**:334–42.

30. Otto CM, Kuusisto J, Reichenbach DD, Gown AM, O'Brien KD. Characterization of the early lesion of 'degenerative' valvular aortic stenosis. Histological and immunohistochemical studies. *Circulation* 1994;**90**:844–53.

31. Mohler 3rd ER, Gannon F, Reynolds C, Zimmerman R, Keane MG, Kaplan FS. Bone formation and inflammation in cardiac valves. *Circulation* 2001;**103**:1522–8.

32. Pomerance A. Ageing changes in human heart valves. *Br Heart J* 1967;**29**:222–31.

33. Stewart BF, Siscovick D, Lind BK, Gardin JM, Gottdiener JS, Smith VE, et al. Clinical factors associated with calcific aortic valve disease. Cardiovascular Health Study. *J Am Coll Cardiol* 1997;**29**:630–4.

34. Otto CM, Lind BK, Kitzman DW, Gersh BJ, Siscovick DS. Association of aortic-valve sclerosis with cardiovascular mortality and morbidity in the elderly. *N Engl J Med* 1999;**341**:142–7.

35. Rajamannan NM, Evans FJ, Aikawa E, Grande-Allen KJ, Demer LL, Heistad DD, et al. Calcific aortic valve disease: Not simply a degenerative process: A review and agenda for research from the National Heart, Lung and Blood Institute Aortic Stenosis Working Group. Executive summary: Calcific aortic valve disease-2011 update. *Circulation* 2011;**124**:1783–91.

36. Ortlepp JR, Hoffmann R, Ohme F, Lauscher J, Bleckmann F, Hanrath P. The vitamin D receptor genotype predisposes to the development of calcific aortic valve stenosis. *Heart* 2001;**85**:635–8.

37. Nordstrom P, Glader CA, Dahlen G, Birgander LS, Lorentzon R, Waldenstrom A, et al. Oestrogen receptor alpha gene polymorphism is related to aortic valve sclerosis in postmenopausal women. *J Intern Med* 2003;**254**:140–6.

38. Novaro GM, Sachar R, Pearce GL, Sprecher DL, Griffin BP. Association between apolipoprotein E alleles and calcific valvular heart disease. *Circulation* 2003;**108**:1804–8.

39. Ortlepp JR, Schmitz F, Mevissen V, Weiss S, Huster J, Dronskowski R, et al. The amount of calcium-deficient hexagonal hydroxyapatite in aortic valves is influenced by gender and associated with genetic polymorphisms in patients with severe calcific aortic stenosis. *Eur Heart J* 2004;**25**:514–22.

40. Otto CM, Pearlman AS, Gardner CL. Hemodynamic progression of aortic stenosis in adults assessed by doppler echocardiography. *J Am Coll Cardiol* 1989;**13**:545–50.

41. Cosmi JE, Kort S, Tunick PA, Rosenzweig BP, Freedberg RS, Katz ES, et al. The risk of the development of aortic stenosis in patients with "benign" aortic valve thickening. *Arch Intern Med* 2002;**162**:2345–7.

42. Dweck MR, Jones C, Joshi NV, Fletcher AM, Richardson H, White A, et al. Assessment of valvular calcification and inflammation by positron emission tomography in patients with aortic stenosis. *Circulation* 2012;**125**:76–86.

43. Aikawa E, Otto CM. Look more closely at the valve: Imaging calcific aortic valve disease. *Circulation* 2012;**125**:9–11.

44. Huntington K, Hunter AG, Chan KL. A prospective study to assess the frequency of familial clustering of congenital bicuspid aortic valve. *J Am Coll Cardiol* 1997;**30**:1809–12.

45. Li C, Xu S, Gotlieb AI. The progression of calcific aortic valve disease through injury, cell dysfunction, and disruptive biologic and physical force feedback loops. *Cardiovasc Pathol* 2013;**22**:1–8.

46. Yip CY, Chen JH, Zhao R, Simmons CA. Calcification by valve interstitial cells is regulated by the stiffness of the extracellular matrix. *Arterioscler Thromb Vasc Biol* 2009;**29**:936–42.

47. Jian B, Jones PL, Li Q, Mohler 3rd ER, Schoen FJ, Levy RJ. Matrix metalloproteinase-2 is associated with tenascin-C in calcific aortic stenosis. *Am J Pathol* 2001;**159**:321–7.

48. Rajamannan NM, Subramaniam M, Rickard D, Stock SR, Donovan J, Springett M, et al. Human aortic valve calcification is associated with an osteoblast phenotype. *Circulation* 2003;**107**:2181–4.

49. Rodriguez KJ, Masters KS. Regulation of valvular interstitial cell calcification by components of the extracellular matrix. *J Biomed Mater Res A* 2009;**90**:1043–53.

50. Chen JH, Yip CY, Sone ED, Simmons CA. Identification and characterization of aortic valve mesenchymal progenitor cells with robust osteogenic calcification potential. *Am J Pathol* 2009;**174**:1109–19.

51. Kapustin AN, Davies JD, Reynolds JL, McNair R, Jones GT, Sidibe A, et al. Calcium regulates key components of vascular smooth muscle cell-derived matrix vesicles to enhance mineralization. *Circ Res* 2011;**109**:e1–12.

51a. New SEP, Goettsch C, Aikawa M, Marchini JF, Shibasaki M, Yabusaki K, et al. Macrophage-derived matrix vesicles: An alternative novel mechanism for microcalcification in atherosclerotic plaques. *Circulation Research* 2013;**113**:72–77.

51b. New SEP, Aikawa E. The role of extracellular vesicles in de novo mineralization: An additional novel mechanism of cardiovascular calcification. *ATVB* 2013;**33**:1753–1758.

52. Kim KM. Calcification of matrix vesicles in human aortic valve and aortic media. *Fed Proc* 1976;**35**:156–62.

53. Merryman WDSF, Schoen FJ. Mechanisms of calcification in aortic valve disease: Role of mechanokinetics and mechanodynamics. *Curr Cardiol Rep* 2013;**15**:355–361.

54. Sucosky P, Balachandran K, Elhammali A, Jo H, Yoganathan AP. Altered shear stress stimulates upregulation of endothelial VCAM-1 and ICAM-1 in a BMP-4 and TGF-beta1-dependent pathway. *Arterioscler Thromb Vasc Biol* 2009;**29**:254–60.

55. Mirzaie M, Schultz M, Schwartz P, Coulibaly M, Schondube F. Evidence of woven bone formation in heart valve disease. *Ann Thorac Cardiovasc Surg* 2003;**9**:163–9.

56. Balachandran K, Sucosky P, Jo H, Yoganathan AP. Elevated cyclic stretch induces aortic valve calcification in a bone morphogenic protein-dependent manner. *Am J Pathol* 2010;**177**:49–57.

57. Muller AM, Cronen C, Kupferwasser LI, Oelert H, Muller KM, Kirkpatrick CJ. Expression of endothelial cell adhesion molecules on heart valves: Up-regulation in degeneration as well as acute endocarditis. *J Pathol* 2000;**191**:54–60.

58. El-Hamamsy I, Balachandran K, Yacoub MH, Stevens LM, Sarathchandra P, Taylor PM, et al. Endothelium-dependent regulation of the mechanical properties of aortic valve cusps. *J Am Coll Cardiol* 2009;**53**:1448–55.

59. Galante A, Pietroiusti A, Vellini M, Piccolo P, Possati G, De Bonis M, et al. C-reactive protein is increased in patients with degenerative aortic valvular stenosis. *J Am Coll Cardiol* 2001;**38**:1078–82.

60. Rajamannan NM, Subramaniam M, Caira F, Stock SR, Spelsberg TC. Atorvastatin inhibits hypercholesterolemia-induced calcification in the aortic valves via the LRP5 receptor pathway. *Circulation* 2005;**112**:I229–34.

61. Drolet MC, Roussel E, Deshaies Y, Couet J, Arsenault M. A high fat/high carbohydrate diet induces aortic valve disease in C57B1/6J mice. *J Am Coll Cardiol* 2006;**47**:850–5.

62. Tanaka K, Sata M, Fukuda D, Suematsu Y, Motomura N, Takamoto S, et al. Age-associated aortic stenosis in apolipoprotein E-deficient mice. *J Am Coll Cardiol* 2005;**46**:134–41.

63. Weiss RM, Ohashi M, Miller JD, Young SG, Heistad DD. Calcific aortic valve stenosis in old hypercholesterolemic mice. *Circulation* 2006;**114**:2065–9.

64. New SE, Aikawa E. Molecular imaging insights into early inflammatory stages of arterial and aortic valve calcification. *Circ Res* 2011;**108**:1381–91.

65. Aikawa E, Nahrendorf M, Figueiredo JL, Swirski FK, Shtatland T, Kohler RH, et al. Osteogenesis associates with inflammation in early-stage atherosclerosis evaluated by molecular imaging in vivo. *Circulation* 2007;**116**:2841–50.

66. Hjortnaes J, Butcher J, Figueiredo JL, Riccio M, Kohler RH, Kozloff KM, et al. Arterial and aortic valve calcification inversely correlates with osteoporotic bone remodelling: A role for inflammation. *Eur Heart J* 2010;**31**:1975–84.

67. Miller JD, Chu Y, Brooks RM, Richenbacher WE, Pena-Silva R, Heistad DD. Dysregulation of antioxidant mechanisms contributes to increased oxidative stress in calcific aortic valvular stenosis in humans. *J Am Coll Cardiol* 2008;**52**:843–50.

68. Liberman M, Bassi E, Martinatti MK, Lario FC, Wosniak Jr J, Pomerantzeff PM, et al. Oxidant generation predominates around calcifying foci and enhances progression of aortic valve calcification. *Arterioscler Thromb Vasc Biol* 2008;**28**:463–70.

69. Cote C, Pibarot P, Despres JP, Mohty D, Cartier A, Arsenault BJ, et al. Association between circulating oxidised low-density lipoprotein and fibrocalcific remodelling of the aortic valve in aortic stenosis. *Heart* 2008;**94**:1175–80.

70. Thanassoulis G, Campbell CY, Owens DS, Smith JG, Smith AV, Peloso GM, et al. Genetic associations with valvular calcification and aortic stenosis. *N Engl J Med* 2013;**368**:503–12.

71. Long F. Building strong bones: Molecular regulation of the osteoblast lineage. *Nat Rev Mol Cell Biol* 2012;**13**:27–38.

72. Caira FC, Stock SR, Gleason TG, McGee EC, Huang J, Bonow RO, et al. Human degenerative valve disease is associated with up-regulation of low-density lipoprotein receptor-related protein 5 receptor-mediated bone formation. *J Am Coll Cardiol* 2006;**47**: 1707–12.

73. Miller JD, Weiss RM, Serrano KM, Castaneda LE, Brooks RM, Zimmerman K, et al. Evidence for active regulation of pro-osteogenic signaling in advanced aortic valve disease. *Arterioscler Thromb Vasc Biol* 2010;**30**:2482–6.

74. Osman L, Yacoub MH, Latif N, Amrani M, Chester AH. Role of human valve interstitial cells in valve calcification and their response to atorvastatin. *Circulation* 2006;**114**:I547–52.

75. Miller JD, Weiss RM, Serrano KM, Brooks 2nd RM, Berry CJ, Zimmerman K, et al. Lowering plasma cholesterol levels halts progression of aortic valve disease in mice. *Circulation* 2009;**119**: 2693–701.

76. Galvin KM, Donovan MJ, Lynch CA, Meyer RI, Paul RJ, Lorenz JN, et al. A role for smad6 in development and homeostasis of the cardiovascular system. *Nat Genet* 2000;**24**:171–4.

77. Kaden JJ, Bickelhaupt S, Grobholz R, Vahl CF, Hagl S, Brueckmann M, et al. Expression of bone sialoprotein and bone morphogenetic protein-2 in calcific aortic stenosis. *J Heart Valve Dis* 2004;**13**: 560–6.

78. Simonet WS, Lacey DL, Dunstan CR, Kelley M, Chang MS, Luthy R, et al. Osteoprotegerin: A novel secreted protein involved in the regulation of bone density. *Cellule* 1997;**89**:309–19.

79. Kaden JJ, Dempfle CE, Grobholz R, Fischer CS, Vocke DC, Kilic R, et al. Inflammatory regulation of extracellular matrix remodeling in calcific aortic valve stenosis. *Cardiovasc Pathol* 2005;**14**: 80–7.

80. Kaden JJ, Kilic R, Sarikoc A, Hagl S, Lang S, Hoffmann U, et al. Tumor necrosis factor alpha promotes an osteoblast-like phenotype in human aortic valve myofibroblasts: A potential regulatory mechanism of valvular calcification. *Int J Mol Med* 2005;**16**:869–72.

81. Nanes MS. Tumor necrosis factor-alpha: Molecular and cellular mechanisms in skeletal pathology. *Genewatch* 2003;**321**:1–15.

82. Miller JD, Weiss RM, Heistad DD. Calcific aortic valve stenosis: Methods, models, and mechanisms. *Circ Res* 2011;**108**:1392–412.

83. Walker GA, Masters KS, Shah DN, Anseth KS, Leinwand LA. Valvular myofibroblast activation by transforming growth factor-beta: Implications for pathological extracellular matrix remodeling in heart valve disease. *Circ Res* 2004;**95**:253–60.

84. Proudfoot D, Skepper JN, Hegyi L, Bennett MR, Shanahan CM, Weissberg PL. Apoptosis regulates human vascular calcification in vitro: Evidence for initiation of vascular calcification by apoptotic bodies. *Circ Res* 2000;**87**:1055–62.

85. Mohler 3rd ER, Chawla MK, Chang AW, Vyavahare N, Levy RJ, Graham L, et al. Identification and characterization of calcifying valve cells from human and canine aortic valves. *J Heart Valve Dis* 1999;**8**:254–60.

86. O'Brien KD, Shavelle DM, Caulfield MT, McDonald TO, Olin-Lewis K, Otto CM, et al. Association of angiotensin-converting enzyme with low-density lipoprotein in aortic valvular lesions and in human plasma. *Circulation* 2002;**106**:2224–30.

87. Arishiro K, Hoshiga M, Negoro N, Jin D, Takai S, Miyazaki M, et al. Angiotensin receptor-1 blocker inhibits atherosclerotic changes and endothelial disruption of the aortic valve in hypercholesterolemic rabbits. *J Am Coll Cardiol* 2007;**49**:1482–9.

88. Shavelle DM, Takasu J, Budoff MJ, Mao S, Zhao XQ, O'Brien KD. HMG CoA reductase inhibitor (statin) and aortic valve calcium. *Lancet* 2002;**359**:1125–6.

89. Rosenhek R, Rader F, Loho N, Gabriel H, Heger M, Klaar U, et al. Statins but not angiotensin-converting enzyme inhibitors delay progression of aortic stenosis. *Circulation* 2004;**110**:1291–5.

90. Deregowski V, Gazzerro E, Priest L, Rydziel S, Canalis E. Notch 1 overexpression inhibits osteoblastogenesis by suppressing WNT/beta-catenin but not bone morphogenetic protein signaling. *J Biol Chem* 2006;**281**:6203–10.

91. Hilton MJ, Tu X, Wu X, Bai S, Zhao H, Kobayashi T, et al. Notch signaling maintains bone marrow mesenchymal progenitors by suppressing osteoblast differentiation. *Nat Med* 2008;**14**:306–14.

92. Mead TJ, Yutzey KE. Notch pathway regulation of chondrocyte differentiation and proliferation during appendicular and axial skeleton development. *Proc Natl Acad Sci U S A* 2009;**106**:14420–5.

93. Garg V, Muth AN, Ransom JF, Schluterman MK, Barnes R, King IN, et al. Mutations in NOTCH1 cause aortic valve disease. *Nature* 2005;**437**:270–4.

94. Arrington CB, Sower CT, Chuckwuk N, Stevens J, Leppert MF, Yetman AT, et al. Absence of TGFBR1 and TGFBR2 mutations in patients with bicuspid aortic valve and aortic dilation. *Am J Cardiol* 2008;**102**:629–31.

95. Nigam V, Srivastava D. Notch1 represses osteogenic pathways in aortic valve cells. *J Mol Cell Cardiol* 2009;**47**:828–34.

96. Matsumoto Y, Adams V, Jacob S, Mangner N, Schuler G, Linke A. Regular exercise training prevents aortic valve disease in low-density lipoprotein-receptor-deficient mice. *Circulation* 2010;**121**:759–67.

97. Cowell SJ, Newby DE, Prescott RJ, Bloomfield P, Reid J, Northridge DB, et al. A randomized trial of intensive lipid-lowering therapy in calcific aortic stenosis. *N Engl J Med* 2005;**352**:2389–97.

98. Mohler 3rd ER, Wang H, Medenilla E, Scott C. Effect of statin treatment on aortic valve and coronary artery calcification. *J Heart Valve Dis* 2007;**16**:378–86.

99. Verma S, Szmitko PE, Fedak PW, Errett L, Latter DA, David TE. Can statin therapy alter the natural history of bicuspid aortic valves? *Am J Physiol Heart Circ Physiol* 2005;**288**:H2547–9.

100. Hayek E, Gring CN, Griffin BP. Mitral valve prolapse. *Lancet* 2005;**365**:507–18.

101. Fornes P, Heudes D, Fuzellier JF, Tixier D, Bruneval P, Carpentier A. Correlation between clinical and histologic patterns of degenerative mitral valve insufficiency: A histomorphometric study of 130 excised segments. *Cardiovasc Pathol* 1999;**8**:81–92.

102. Barber JE, Kasper FK, Ratliff NB, Cosgrove DM, Griffin BP, Vesely I. Mechanical properties of myxomatous mitral valves. *J Thorac Cardiovasc Surg* 2001;**122**:955–62.

103. Barber JE, Ratliff NB, Cosgrove 3rd DM, Griffin BP, Vesely I. Myxomatous mitral valve chordae. I: Mechanical properties. *J Heart Valve Dis* 2001;**10**:320–4.

104. Carpentier A, Chauvaud S, Fabiani JN, Deloche A, Relland J, Lessana A, et al. Reconstructive surgery of mitral valve incompetence: Ten-year appraisal. *J Thorac Cardiovasc Surg* 1980;**79**:338–48.

105. Anyanwu AC, Adams DH. Etiologic classification of degenerative mitral valve disease: Barlow's disease and fibroelastic deficiency. *Semin Thorac Cardiovasc Surg* 2007;**19**:90–6.

106. Guy TS, Hill AC. Mitral valve prolapse. *Annu Rev Med* 2012;**63**:277–92.

107. Grande-Allen KJ, Calabro A, Gupta V, Wight TN, Hascall VC, Vesely I. Glycosaminoglycans and proteoglycans in normal mitral valve leaflets and chordae: Association with regions of tensile and compressive loading. *Glycobiology* 2004;**14**:621–33.

108. Flanagan TC, Wilkins B, Black A, Jockenhoevel S, Smith TJ, Pandit AS. A collagen-glycosaminoglycan co-culture model for heart valve tissue engineering applications. *Biomaterials* 2006;**27**:2233–46.

109. Flanagan TC, Black A, O'Brien M, Smith TJ, Pandit AS. Reference models for mitral valve tissue engineering based on valve cell phenotype and extracellular matrix analysis. *Cells Tissues Organs* 2006;**183**:12–23.

110. Liu AC, Gotlieb AI. Transforming growth factor-beta regulates in vitro heart valve repair by activated valve interstitial cells. *Am J Pathol* 2008;**173**:1275–85.

111. Gotlieb AI, Rosenthal A, Kazemian P. Fibroblast growth factor 2 regulation of mitral valve interstitial cell repair in vitro. *J Thorac Cardiovasc Surg* 2002;**124**:591–7.

112. Gupta V, Werdenberg JA, Lawrence BD, Mendez JS, Stephens EH, Grande-Allen KJ. Reversible secretion of glycosaminoglycans and proteoglycans by cyclically stretched valvular cells in 3D culture. *Ann Biomed Eng* 2008;**36**:1092–103.

113. Nguyen TC, Itoh A, Carlhall CJ, Bothe W, Timek TA, Ennis DB, et al. The effect of pure mitral regurgitation on mitral annular geometry and three-dimensional saddle shape. *J Thorac Cardiovasc Surg* 2008;**136**:557–65.

114. Goetz WA, Lim HS, Pekar F, Saber HA, Weber PA, Lansac E, et al. Anterior mitral leaflet mobility is limited by the basal stay chords. *Circulation* 2003;**107**:2969–74.

115. Tamura K, Jones M, Yamada I, Ferrans VJ. Wound healing in the mitral valve. *J Heart Valve Dis* 2000;**9**:53–63.

116. Ryan LP, Jackson BM, Parish LM, Plappert TJ, St John-Sutton MG, Gorman 3rd JH, et al. Regional and global patterns of annular remodeling in ischemic mitral regurgitation. *Ann Thorac Surg* 2007;**84**:553–9.

117. Stephens EH, Timek TA, Daughters GT, Kuo JJ, Patton AM, Baggett LS, et al. Significant changes in mitral valve leaflet matrix composition and turnover with tachycardia-induced cardiomyopathy. *Circulation* 2009;**120**:S112–9.

118. Kunzelman KS, Quick DW, Cochran RP. Altered collagen concentration in mitral valve leaflets: Biochemical and finite element analysis. *Ann Thorac Surg* 1998;**66**:S198–205.

119. Stephens EH, Nguyen TC, Itoh A, Ingels Jr NB, Miller DC, Grande-Allen KJ. The effects of mitral regurgitation alone are sufficient for leaflet remodeling. *Circulation* 2008;**118**:S243–9.

120. Grammer JB, Eichinger WB, Bleiziffer S, Benz MR, Lange R, Bauernschmitt R. Valvular chondromodulin-1 expression is downregulated in a rabbit model of infective endocarditis. *J Heart Valve Dis* 2007;**16**:623–30 discussion 630.

121. Zeng Z, Nievelstein-Post P, Yin Y, Jan KM, Frank JS, Rumschitzki DS. Macromolecular transport in heart valves. Iii. Experiment and theory for the size distribution of extracellular liposomes in hyperlipidemic rabbits. *Am J Physiol Heart Circ Physiol* 2007;**292**:H2687–97.

122. Makkena B, Salti H, Subramaniam M, Thennapan S, Bonow RH, Caira F, et al. Atorvastatin decreases cellular proliferation and bone matrix expression in the hypercholesterolemic mitral valve. *J Am Coll Cardiol* 2005;**45**:631–3.

123. Ng CM, Cheng A, Myers LA, Martinez-Murillo F, Jie C, Bedja D, et al. TGF-beta-dependent pathogenesis of mitral valve prolapse in a mouse model of marfan syndrome. *J Clin Invest* 2004;**114**:1586–92.

Vasculogenesis and Angiogenesis

Joseph F. Arboleda-Velasquez, MD PhD, Patricia A. D'Amore, PhD

Harvard Medical School, Boston, MA, USA

INTRODUCTION

In light of the premise of angiogenesis as an organizing principle, it is not surprising that drugs that have been developed to modulate angiogenesis can be effective to treat apparently dissimilar conditions. For example, a blinding disease like wet age-related macular degeneration that affects primarily older individuals appears to have little in common with colorectal or lung cancer, yet these diseases may be treated using the same anti-angiogenesis therapies.[1]

Organizing principles also enable discoveries in one field to illuminate understanding in other fields. For instance, the study of large populations suffering from cerebral cavernous malformations, an inherited condition commonly associated with brain hemorrhage, uncovered a previously uncharacterized cell signaling mechanism mediating vascular development and stability in the brain.[2]

In the first section of this chapter, we briefly review our current understanding of the key principles governing vascular development. This discussion should familiarize the reader with the main cellular and molecular players involved in the process of vessel formation, maturation and remodeling, which are key for understanding the pathological conditions introduced thereafter. The coverage in this section is not intended to be comprehensive, as the main goal is to focus on molecular aspects of pathologic angiogenesis. We do, however, cite key review articles that the reader will find useful to read further on this topic.

The second part of this chapter will cover conditions that arise from abnormal vascular development with an emphasis on the genetics and molecular basis of these diseases. These include vascular malformations, hemangiomas, hereditary hemorrhagic telangiectasia, and cerebral cavernous malformations. Some of these conditions do not present in the newborn but instead become apparent or progress over time.

The third part of this chapter includes a discussion of the role of angiogenesis in the tissue response to disease for very prevalent conditions, including cancer, brain ischemia, myocardial infarction, critical limb ischemia, and neovascular eye disease. Here we emphasize the promise of angiogenesis as a therapeutic target for ischemic conditions.

VASCULAR DEVELOPMENT

The development of the vascular system involves the processes of vasculogenesis and angiogenesis.[3] Vasculogenesis is defined as the de novo formation of a blood vessel from angioblasts giving rise to the heart, the first primitive vascular plexus, and the yolk sac circulation. Angiogenesis involves the expansion of the vasculature from existing vessels through sprouting and intussusceptive microvascular growth (Figure 10.1).[4] The regulation of vascular development was first described in the chick in the 1930s where the phases of extraembryonic and embryonic vascularization were first documented.[5] These initial studies were largely descriptive and for decades the molecular bases of vascular development remained unknown.

The advent of mouse models of targeted gene deletion allowed the identification of specific gene products responsible for the process of vascular development. Accordingly, a large number of genes have been shown to be critical to proper vascular development and are the target of mutations associated with human disease.

Among the most dramatic vascular defects are seen with the knockout of vascular endothelial growth factor (VEGF; also known as vascular permeability factor, VPF) in mice.[6] Loss of a single allele of VEGF was shown to lead to embryonic lethality at embryonic day (E) 9.5 with nearly complete lack of aorta and the absence of yolk sac vasculature.[6] The fact that heterozygous deletion yields a phenotype nearly as complete as homozygous

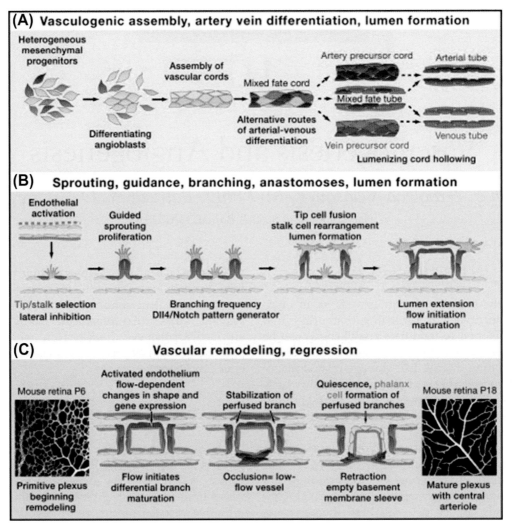

FIGURE 10.1   Hallmarks of vessel formation. (A) Angioblasts differentiate into endothelial cells, which form cords, acquire a lumen, and develop into arteries and veins. (B) Vessel sprouting: tip/stalk cell selection; tip cell navigation and stalk cell proliferation; branching coordination; stalk elongation, tip cell fusion, and lumen formation; and perfusion and vessel maturation. (C) Sequential steps of vascular remodeling from a primitive (left box) towards a stabilized and mature vascular plexus (right box) including adoption of a quiescent endothelial phalanx phenotype, basement membrane deposition, pericyte coverage, and branch regression. (Reproduced with permission from Potente M, Gerhardt, H and Carmeliet P. Basic and therapeutic aspects of angiogenesis. *Cell* 2011; 146: 873–87.)

knockout points to the importance of dosage in the effects of VEGF. Not surprisingly, targeted disruption of VEGFR2, the primary signaling receptor for VEGF, closely phenocopies the loss of VEGF.[7] Interestingly, loss of a second VEGF receptor VEGFR1 does not lead to a similar disruption of vascular development but rather results in an over-proliferation of vascular endothelial cells.[8,9] This observation has led to the suggestion that VEGFR1 may act as a 'decoy' for VEGF, sequestering soluble VEGF and thus preventing its mitogenic action on the endothelium. VEGF misregulation is central to vascular pathology during development and in adult tissues as discussed later.

Once a large vessel (such as the aorta) or a primitive capillary plexus is formed, the vessel undergoes a series of steps that lead to a mature, stable state. For large vessels, this consists of the recruitment of mesenchymal cells that are induced to differentiate into smooth muscle cells and then proliferate to generate vessels of appropriate wall thickness e.g. artery vs. vein.[10] At the level of the microvasculature, immature capillary tubes similarly recruit mesenchymal cells – with the primary difference being that pericytes, by definition, never fully cover the abluminal surface of the capillary nor do they overlap one another.[11] In spite of the significant structural difference between the largest arteries and the microvasculature, the same molecules and mechanisms are used in the process of their remodeling.[12] Tissue culture studies followed by targeted gene deletions have revealed the complexity of the remodeling and stabilization process. Using cocultures of vascular endothelial cells and undifferentiated mesenchymal cells, proliferating endothelial

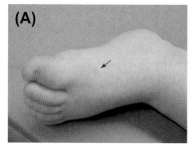 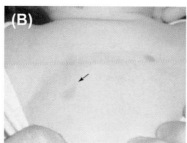 

FIGURE 10.2 Photographs of capillary malformation-arteriovenous malformations (CM-AVMs). **(A)** Small, oval, pink-red CM of foot. Note narrow white halo (arrow). **(B)** Small, reddish, oval CM on thorax (arrow). **(C)** Larger, red-purple CM of lateral leg above knee (arrow). (Reproduced with permission from Boon LM, Mulliken JB, Vikkula M. RASA1: variable phenotype with capillary and arteriovenous malformations. *Current Opinion in Genetics and Development* 2005; 15: 265–9.)

cells have been shown to release platelet-derived growth factor B (PDGF-B) that acts as both a chemoattractant and mitogen for the smooth muscle cell precursors.[13] Once the mesenchymal cells contact the endothelium, the mesenchymal cells are induced to differentiate to a smooth muscle cell/pericyte fate.

DEVELOPMENTAL ABNORMALITIES

Vascular Malformations

Developmental vascular malformations include lesions present in the individual at birth that generally grow with age and do not regress.[14] Vascular malformations primarily affect endothelial cells and are named after the type of vessel they affect, i.e. vein, artery, capillary, lymphatic, or arteriovenous malformation. In most cases, vascular malformations are sporadic, although rare familial forms have been associated with mutations in genes that have specific functions in vascular cells.[15]

The current information regarding the causative genes for these malformations has allowed the identification of links between misregulation in cell signaling and cellular pathology. This chapter includes a detailed discussion of capillary malformations, as they are the most common vascular malformations.[16]

It is worth noting that genetic analysis of tissue from vascular malformations in sporadic cases has revealed somatic mutations affecting the same genes that are associated with inherited forms of the disease.[15] Thus investigation of the genes involved in familial cases should provide insights for the design of rational therapies relevant to both inherited and sporadic forms.

Capillary malformations (CM) are the most common vascular malformations and are found in 0.3% of newborns.[16] These lesions, also known as port-wine stains, are characterized by abnormal capillary beds localized primarily in the skin (Figure 10.2). About 30% of the cases are associated with high-flow lesions containing arteriovenous malformations (AVM) or fistulas, often located in the neck or the head.[17,18] Genetic studies have identified mutations in the gene encoding for p120-RasGTPase-activating protein (p120-RasGAP or RASA1) as the cause of CM associated with AVM (CM-AVM).[19] This condition is inherited as an autosomal dominant trait with incomplete penetrance. Because of the localized nature of the lesions it has been proposed that a somatic mutation or 'second hit' is necessary for the development of the vascular malformations.[19]

Ras is a small GTP-binding protein that regulates growth and survival of cells by modulating multiple downstream effectors including the MAPK signaling pathway.[20] Extracellular signals sensed by receptor tyrosine kinases (RTKs) trigger the activation of Ras by promoting the binding of Ras to GTP through the recruitment of guanidine exchange factors (Ras-GEFs) to the plasma membrane.[21] Ras-GAPs proteins such as RASA1 are known to modulate the activity of Ras by promoting its GTPase activity, leading to conversion of active GTP-bound Ras into the inactive GDP-bound Ras.[22,23] Because most of the mutations associated with CM-AVMs are predicted to result in loss-of-function, it has been proposed that up-regulated Ras/MAPK-signaling activity in endothelial cells is crucial for the pathophysiology of CM-AVM.[19]

During development, VEGF secretion by keratinocytes and sensory nerves triggers the growth of vessels into the dermis.[24–26] A lack of proper levels of RASA1 in endothelial cells may result in unchecked proliferation, triggering the development of vascular malformations.[19] In light of the presumably pleiotropic effects of RASA1, it is unclear why mutations in this gene result in lesions restricted to the vascular system. Analyses of RASA1 mouse models indicate that this gene is essential for vascular development as well as for the maintenance of lymphatic vessels in the adult.[27,28] Mice lacking RASA1 during development show abnormal vessel growth and vessel leakage that leads to hemorrhage and embryonic lethality by day E10.5. However, conditional ablation of RASA1 in adult animals does not lead to capillary abnormalities but is instead associated with impaired development of lymphatic vessels including hyperplasia, leakage and lethality due to chylothorax (lymphatic fluid accumulation in the thorax cavity).[27] Consistent with this finding, some individuals with RASA1 mutations have been reported to develop chylothorax and chylous ascites.[18]

Hemangiomas

In contrast to vascular malformations, hemangiomas are small or absent at birth, grow with the individual in what is known as a 'proliferative phase' that could last for several years, and later start a 'phase of involution' that often occurs between 7–10 years of age.[29] For unknown reasons, hemangiomas are usually located in the head and neck, and are more frequent in premature girls (female to male ratio 2–4:1), and Caucasians.[30] Hemangiomas are reported as the most common soft tissue tumor of infancy affecting 5–10% of one-year-old children.[31] Hemangiomas are primarily composed of capillary-like layers of endothelial cells and pericytes that share a common basement membrane and have a narrow or compressed lumen.[32,33] The proliferative and involution phases are dynamic processes with different rates of cell division and apoptosis observed throughout (Figure 10.3).[32]

The origin of hemangiomas remains obscure although many theories have been tested over the years. The notion of placental embolization of endothelial cells as the origin of hemangiomas has recently been challenged by studies that ruled out maternal–fetal microchimerism within the lesions.[34–36] In other studies, genetic mapping linked locus 5q31–33 with hemangioma-segregating in families as an autosomal dominant trait.[37,38] Loss of heterozygocity at the same chromosomal region was later identified in sporadic hemangiomas compared to control tissue, although a specific gene remains to be mapped.[39] There

are also reports indicating that sporadic hemangiomas are associated with somatic mutations in genes known to modulate vascular development including VEGFR2 (FLK1/KDR), VEGFR3 (FLT4), and endothelial cell tyrosine kinase receptor Tie-2 (encoded by TEK).[40,41] Moreover, somatic mutations in isocitrate dehydrogenase 1 (IDH1) or IDH2 are associated with Maffucci syndrome, a skeletal condition characterized by enchondroma and spindle cell hemangiomas.[42] In Maffucci syndrome, hemangiomas are composed of thin-walled vessels mixed with areas composed of spindled cells.[43] IDH1 and IDH2 are metabolic enzymes involved in energy production through generation of α-ketoglutarate and NADPH. Gain of function mutations in these genes lead to increased production of the oncometabolite D-2-hydroxyglutarate (D2HG), which has been shown to cause hypermethylation and inhibition of genes responsible for terminal differentiation of cells.[44]

The isolation of hemangioma-derived cell lines (HemECs) and hemangioma-derived multipotential stem cells has provided significant insights about the origin of hemangiomas and provided researchers with new tools to test rational therapies.[45–47] Hemangioma-derived stem cells display unique multilineage differentiation capabilities in vitro and form human blood vessels with features of infantile hemangioma when injected subcutaneously into immunocompromised mice.[47] Characterization of non-random chromosome X inactivation patterns in HemECs supported the theory

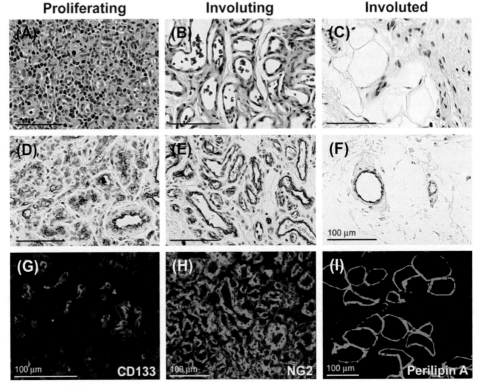

FIGURE 10.3 Histology of proliferating, involuting and involuted phases of hemangioma. (A–C) Hematoxylin and eosin staining. (D–F) Immunostaining for CD31, an endothelial cell marker. (G) Immunostaining for CD133, a stem and progenitor cell marker. (H) Immunostaining for NG-2, a pericyte marker. (I) Immunostaining for perilipin-A, an adipocyte marker. Scale bars are 100 μm. (Reproduced with permission from Boscolo E, Bischoff J. Vasculogenesis in infantile hemangioma. *Angiogenesis* 2009; 12: 197–207.)

of a clonal origin of hemangiomas.[40,48] Molecular studies of HemECs responsiveness to VEGF demonstrated constitutive activation of VEGFR2 signaling and down-regulation of VEGFR1, a receptor that is known to bind VEGF with high affinity, preventing it from activating VEGFR2.[49,50] Steroids and propranolol have been used as the first line of treatment for hemangiomas although their mechanisms of action are largely unknown. Analyses in hemangioma-derived stem cells showed that steroids lead to down-regulation of VEGF-A expression and suppression of their vasculogenic properties in mice.[51] For propranolol, a non-selective betablocker that is remarkably effective at treating complicated cases of hemangioma, the proposed mechanisms include vasoconstrictions due to decreased release of nitric oxide, blocking angiogenic signals, and induction of apoptosis.[52,53]

Hereditary Hemorrhagic Telangiectasia (HHT)

HHT is an autosomal dominant syndrome characterized by the presence of small dilated vessels in the skin or mucous membranes, and arteriovenous malformations in the pulmonary, hepatic, cerebral and spinal circulations.[54] Common symptoms include nosebleeds and gastrointestinal hemorrhages that can lead to anemia.[55] A systematic study of the natural history of the disease revealed that at early stages HHT lesions are characterized by dilation of postcapillary venules with increased investment by vascular smooth muscle cells (SMC). As lesions progress, abnormally dilated arterioles are observed, and these later connect directly to the abnormal venules – without any intervening capillary beds – thereby forming a blood shunt.[56]

HHT has been associated with mutations in endoglin (HHT type 1), activin receptor-like kinase (ALK1) (HHT type 2), and Smad4 (Mothers against decapentaplegic homolog 4-HHT in association with juvenile polyposis), and at least two other loci located in chromosomes 5 (HHT3) and 7 (HHT4).[54] Endoglin, ALK1, and Smad4 function as key elements of the TGF-β signaling pathway, which plays a key role during the angiogenesis process. The types of mutations in these genes associated with HHT strongly suggest loss-of-function-mechanisms.[54]

The TGF-β superfamily of ligands includes the bone morphogenetic proteins (BMPs), TGF-β, growth and differentiation factors (GDFs), anti-Müllerian hormone (AMH), activin, and nodal.[57] These ligands are secreted molecules that interact with transmembrane receptors located in nearby cells to regulate differentiation, proliferation and survival in multiple tissues including the vasculature. Binding of the ligand to a type II receptor triggers recruitment and phosphorylation of a type I receptor, which in turns activates a receptor-regulated SMAD (R-SMAD) also through phosphorylation. The active R-SMAD forms a complex with a coSMAD protein (SMAD4) to regulate gene expression. Another class of receptors, named type III, function as coreceptors that bind the ligand but lack a kinase domain in their intracellular domain thus requiring interactions with other receptors to activate downstream targets.[58]

Endothelial cells express ALK1, ALK5 (both type I receptors), and endoglin (type III coreceptor), which are all capable of binding to TGF-β1 and TGF-β3.[59–61] ALK1 and endoglin also appear to be sensitive to BMPs implicating these ligands in HHT pathophysiology.[62,63] ALK1/TGF-β receptor type II (TbR-II) activation leads to signaling through SMAD1 and SMAD5, which stimulates migration and proliferation of endothelial cells, whereas ALK 5/TbR-II activation signals through SMAD2 and SMAD3, which inhibit both processes.[61,64,65] Adding complexity to the system, endoglin can associate with ALK5/TbR-II and ALK1/TbR-II heteromeric complexes.[63,66] The stoichiometry of interactions between these receptors appears to dictate the direction of fates in vascular cells exposed to limiting amounts of ligands. The current view is that HHT lesions are caused by a disruption in the balance between endoglin, ALK5, and ALK1 in endothelial cells.[67]

In vitro studies in which endothelial cells were cocultured with vascular mural cells have shown that TGF-β signaling plays an essential role in vascular stability through inhibition of endothelial cell proliferation, the promotion of differentiation of vascular mural cells derived from mesenchymal precursors, and production of basement membrane.[13,68,69] Consistent with those observations, systemic inhibition of TGF-β signaling via overexpression of a soluble form of endoglin contributed to increased vascular permeability, hypertension, multiple end-organ ischemia, and proteinuria in humans with pre-eclampsia and in animal models.[70,71]

Mice lacking Alk1 or endoglin die between days 9.5–11.5 of embryonic development due to vascular abnormalities resembling HHT lesions, including dilated and leaky vessels with reduced pericyte investment. This phenotype is similar to that observed in mice lacking TbR-II and TbR-I receptors or the TGF-β1 ligand, clearly implicating this signaling pathway in the pathophysiology of HHT.[72–74]

Studies in animal models have also been helpful in identifying environmental and genetic factors that may influence clinical manifestations of HHT in patients. Analysis of Alk1-deficient animals revealed that both homozygous deletion of Alk1 and wound healing were necessary for the formation of HHT lesions in subdermal vessels.[75] Strain-specific penetrance of the phenotypes in TGF-β1 knockout mice led to the identification of a modifier locus suppressing prenatal vascular lethality.[67,76] Analysis of the analogous region in humans associated polymorphisms in PTPN14, which encodes the protein, non-receptor tyrosine phosphatase 14, with higher frequency of pulmonary arteriovenous malformations in HHT patients.[67]

Cerebral Cavernous Malformations

Cerebral cavernous malformations (CCM) are common vascular lesions affecting 1 in 200–250 and are usually diagnosed through magnetic resonance imaging (MRI).[77,78] Ultrastructurally, CCMs are characterized by dilated vascular cavities lined by a single layer of endothelium cells embedded in a collagenous matrix. Endothelial cells within CCMs lack tight junctions, and other blood–brain barrier components, such as astrocytic endfeet or pericytes, are rare.[79,80]

Although CCMs are primarily found in the brain, lesions are also observed with less frequency in the retina and the skin.[81] Depending upon their location, CCMs can manifest themselves with seizures, headaches, paralysis, and hemorrhage.[82] Sporadic cases are characterized by the presence of a single lesion in individuals without family history, accounting for about 80% of the cases. In familial cases, the lesions are multiple and progressive leading to more severe complications.[81]

Genetic analysis of families with CCM led to the identification of a previously unrecognized and highly complex molecular pathway regulating vascular development and homeostasis. Mutations in three genes, namely CCM1 (Krev1 interaction trapped, KRIT1), CCM2 (MGC4607, OSM, Malcavernin), or CCM3 (Programmed cell death 10, PDCD10) have been associated with CCM in more than 90% of familial cases. In every case, the pattern of inheritance is autosomal dominant and the mutations are predicted to result in gene loss-of-function.[83–87] Genetic analysis of human tissue and studies in animal models have clearly demonstrated a role for loss of heterozygosity in the pathobiology of familial CCM. Analysis of genetic material from endothelial cells derived from CCM lesions has shown biallelic mutations involving CCM1–3 genes.[88–90] Mice heterozygous for a Ccm1-null allele only develop CCM when they are null for a mismatch repair-deficient gene (Msh22), which makes them prone to somatic mutations.[91] In a more direct approach, Ccm2 or Ccm3 conditional knockout animals developed CCM when biallelic mutations were induced using tamoxifen-inducible cre-lox recombination postnatally.[92]

Molecular and cellular studies have demonstrated that all three CCM proteins operate in concert to regulate essential functions in endothelial cells including cell–cell junction, angiogenesis, lumen formation, and vascular permeability.[77,93,94] CCM proteins have been shown to execute many of these functions through regulation of the small GTPase RhoA and Rho kinases.[95–97] Consistent with this function, CCM protein deficiency in mouse models and human CCM tissue results in increased RhoA activity and is associated with increased myosin light-chain phosphorylation, a target of RhoA kinase activity.[93,95] CCM1 also operates as the effector of the small G protein Rap1 and as a scaffold for other proteins crucial to the stabilization of tight and adhering junctions in endothelial cells including VE-cadherin and beta-catenin.[98] There is evidence indicating that CCM3 operates within the circuitry of CCM1 and CCM2, although in vitro and in vivo evidence support the notion that it may also have additional functions.[94] Cell-autonomous and non-cell-autonomous functions have been defined for CCM3 in neural cells using a knockout mouse model.[99] Gfap- or Emx1-Cre-mediated neural deletion of CCM3 leads to increased proliferation, survival, and activation of astrocytes as well as vascular abnormalities resembling CCM.[99]

Studies in animal models have demonstrated a role for CCM proteins in angiogenesis and heart development. CCM1, CCM2, and CCM3 total and endothelial cell-specific knockout mouse models have revealed that these genes are essential for cardiovascular development, and vascular differentiation and maintenance (Figure 10.4). These effects appear to be mediated, at least partially, through interactions with other signaling pathways essential for vascular cell function including Notch, MAP kinase, and protein kinase B (AKT).[95,100] Lastly, studies in zebrafish and mice identified heart of glass (Heg1 in mammals) as a transmembrane receptor with strong molecular and functional interactions to the CCM proteins; albeit a ligand for Heg1 is yet to be identified.[101,102] Animals lacking the receptor show enlarged cardiac chambers resulting and other severe vascular abnormalities leading to lethality.[101,102] Although Heg1 function is unknown, structural similarities to mucin 13 suggest a role in signal transduction of extracellular signals linked to the regulation of actin cytoskeleton.[94]

ANGIOGENIC COMPONENT OF PATHOLOGIES

Cancer Angiogenesis

The ability to induce angiogenesis, sustain proliferative signaling, evade growth suppressors, resist cell death, enable replicative immortality, and activate invasion and metastasis are recognized hallmarks of cancer.[103,104] The current understanding of the principles governing tumor angiogenesis was greatly influenced by the work of Dr. Judah Folkman who defined cancer as an integrated ecosystem of tumor cells and vessels.[105] A large body of evidence now supports the concept that tumors induce the formation of vessels through long- and close-range signals; reciprocally, vessels promote the growth and expansion of tumors by providing

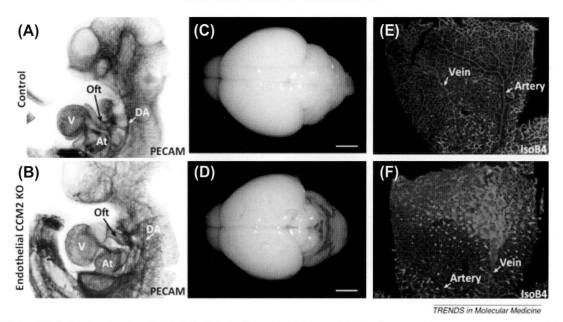

TRENDS in Molecular Medicine

FIGURE 10.4 CCM2 deletion in mice. **(A, B)** Endothelial cell-specific deletion of CCM2 during embryogenesis or at postnatal day 1 **(C–F)**. **(A, B)** CCM2 is needed for blood vessel formation and heart development. Endothelial CD31 staining on E10.5 embryos showing dilation of the heart and vessel abnormalities. At, atrium; DA, dorsal aorta; Oft, outflow track; V, ventricle. **(C–F)** Deletion of CCM2 after birth causes vascular malformations in the brain **(D)** and retina **(F)**. The venous vasculature in the retina is strongly affected by CCM2 loss, recapitulating human CCM lesions. (Reproduced with permission from Fischer A, Zalvide J, Faurobert E, Albiges-Rizo C, Tournier-Lasserve E. Cerebral cavernous malformations: from CCM genes to endothelial cell homeostasis. *Trends in Molecular Medicine* 2013; 19: 302–8.)

nourishment and cell signaling factors, and other mechanisms involving the immune system.[103,105]

Analysis of model organisms and humans indicate that tumor growth stalls at a stage of 2–3 mm in size unless new vessels are recruited.[103,105,106] Upon vascularization, neoplastic cells grow around vessels in cylindrical cords with approximately 150 μm radii, the threshold for oxygen and nutrient diffusion.[107] The classic experimental proof showed that a tumor implanted in the avascular cornea of a rabbit could grow only after recruiting vessels for the limbus.[108]

The ability to attract vessels does not appear to be a passive consequence of tumor growth but instead an acquired and regulated characteristic, known as the 'angiogenic switch,' that determines the fate of a tumor.[109] Seminal studies using the 'RIP-Tag' transgenic mice expressing the SV40 T antigen oncogene in the insulin-producing cells of the pancreas demonstrated that the angiogenic switch is temporally and mechanistically associated with the ability of a subset of hyperproliferating lesions to become solid tumors.[110] A substantial body of work supports the notion that this switch is turned on when pro-angiogenic signals counterbalance anti-angiogenic influences through complex interactions between tumors and their microenvironment.[103]

The development of a hypoxic microenvironment, the recruitment of immune cells, and the activity of oncogenes have been shown in animal models and humans

to be key drivers of tumor angiogenesis.[103] Tumor hypoxia results from the increase in metabolic demands and oxygen consumption by rapidly proliferating cells. The master regulator of hypoxia-induced angiogenesis is the transcription factor hypoxia-inducible factor 1 (HIF-1).[111] Consistent with this role, tumors lacking HIF-1 activity have failed to develop due to defective angiogenesis in a teratoma model.[112] Under normoxic conditions, oxygen-sensing enzymes hydroxylate HIF1α protein leading to its proteosomal degradation via binding to the E3 ubiquitin ligase von Hippel-Lindau protein (VHL). Under hypoxia, hydroxylation is inhibited, resulting in the accumulation of HIF1α, the formation of HIF1α/HIF-1β heterodimers and induction of the transcription of pro-angiogenic molecules including VEGF and nitric oxide synthase (NOS).[113] HIF-1 signaling also orchestrates the recruitment of macrophages to the hypoxic areas of the tumor through the expression of CCL-2 and endothelins.[114,115] Macrophages in tumors have been shown to produce VEGF and further increase its bioavailability through the production of MMP9, which triggers VEGF release from extracellular depots.[116,117] A population of macrophages that express Tie2, also a marker of mature endothelial cells, has been described in tumors and appears to play an essential role in angiogenesis.[114] Coinjection of these cells with tumor cells enhances angiogenesis whereas their ablation leads to defective angiogenesis in vivo.[118] HIF-independent mechanisms that drive VEGF expression in response to hypoxia include signaling through NF-κB and the KRAS oncogene.[119–121]

Before the first pro-angiogenic molecules were purified and cloned, it was believed that these molecules would be unique to tumors. However, it became clear that the molecules used by tumors are the same molecules that regulate angiogenesis during development; often their functions in tumors recapitulate their developmental role.[122] VEGF has been shown to play a key role in angiogenesis, permeability, immune cell recruitment (VEGF-VEGFR1/2), and metastasis through the lymphatic system (VEGFC–VEGFR3–NRP2). The misregulation of Notch signaling is associated with tip-cell hyperproliferation and dysfunctional tumor angiogenesis in animal models. PDGF-BB–PDGFRβ plays a role in recruitment of pericytes and other tumor-associated stromal cells. TGFβ1–TGFβrII induces pro-tumorigenic phenotypes and VEGF expression by tumor stromal cells, whereas ANGPT2–TIE2 plays a role in recruitment of tumor-associated TIE2-expressing monocytes and promotes neovascularization.[114,123]

The notion that endogenous anti-angiogenic molecules also regulate tumor growth arose from the observation that primary tumors had the ability to inhibit growth and vascularization of their metastasis.[124] In this model, primary tumor growth was thought to be driven by the imbalance of angiogenic factors over anti-angiogenic molecules, whereas vascularization and growth of micrometastasis were restrained by circulating anti-angiogenic factors with a longer half-life.[124] This hypothesis led to the identification of angiostatin, a 38-kDa fragment of plasminogen, as a factor capable of inhibiting angiogenesis and growth of metastases.[124] Subsequently, other endogenous proteins and small molecules with anti-angiogenic properties have been reported, some of which are being tested as therapeutic tools.[125,126] Many of these factors are fragments of extracellular matrix proteins released by metalloproteinases, elastases, and cathepsins, whereas others are metabolites of hormones, clotting factors, or immune system cytokines.[125,127] Endostatin, a prototypic anti-angiogenic molecule, is the C-terminal proteolytic fragment of collagen XVIII alpha 1.[128,129] Although a precise mechanism of action is yet to be defined, evidence indicates that some of endostatin's functions are mediated through interactions with integrin alpha beta 1 and E-selectin.[130–132] Expression-profiling studies have shown that endostatin regulates about 12% of the endothelial cell transcriptome, inducing the expression of anti-angiogenic genes while inhibiting the expression of molecules with angiogenic functions.[133]

Thalidomide, a drug when taken by women in their third trimester led to birth defects, was the first drug with anti-angiogenic properties approved for human use in the treatment of multiple myeloma. This was followed by bevacizumab (Genentech), a VEGF-specific humanized antibody, approved by the FDA for the treatment of colorectal cancer.[129] Anti-angiogenic drugs have been shown to have some modest benefit to patients in terms of slowing tumor growth and extending the lives of patients with some cancers. Notwithstanding, as is the case for most anti-cancer approaches, anti-angiogenic drug efficacy is often transient (increased survival of 2–5 months), requires combination therapy with cytotoxic agents (except in the case of renal cell carcinoma), and may lead to adverse effects including bowel perforations, thromboembolic events, and hemorrhage.[134] Moreover, it is speculated that long-term exposure to anti-VEGF therapy may lead to tumor vessel remodeling in ways that makes them ineffective for drug delivery. In addition, there is evidence indicating that tumors may escape anti-angiogenesis therapies by multiple mechanisms including the production of alternative angiogenic factors and through genomic instability.[134,135]

Angiogenesis and Arteriogenesis in Tissue Ischemia

Myocardial infarction (MI), stroke, and critical limb ischemia (CLI) are major causes of disability and death around the world. In most cases, tissue ischemia is the result of vessel occlusion leading to insufficient delivery of nutrients and oxygen to target tissues. Recent advancements in reperfusion therapy have greatly increased the chances of survival and fostered an interest in the development of therapies that prevent tissue degeneration and promote tissue repair.[136,137] Common pathophysiological mechanisms characteristic of the tissue response to ischemia include a switch from oxidative to glycolytic metabolic pathways, induction of angiogenesis and/or arteriogenesis, and the mobilization of progenitor cells that play roles in tissue repair.[138]

In animal models and in humans, HIF-1 expression is one of the earliest markers associated with deprivation of blood in the heart, brain, and skeletal muscle, observed less than 24 h after the insult.[139–142] Under hypoxia, HIF-1 drives the expression of pyruvate dehydrogenase kinase 1 and lactate dehydrogenase A leading to a suppression of the conversion of pyruvate to acetyl coenzyme A (the substrate of the tricarboxylic acid cycle) and facilitation of its conversion to lactate.[143,144] The transition from oxidative respiration to glycolysis, which allows the cells to produce ATP under hypoxic conditions, may also reduce the number of reactive oxygen species generated under conditions of inefficient electron transport.[145,146]

HIF-1 also operates as a key regulator of angiogenic responses to MI and CLI through the induction of angiogenic molecules and the recruitment of immune cells.[138]

In the myocardium, acute ischemic insult triggers an inflammatory response that facilitates the clearance of necrotic cardiomyocytes by neutrophils and macrophages. This is followed by a phase of rapid proliferation of endothelial cells and fibroblasts that form granulation tissue and angiogenesis of the peri-infarct region.[147] Evidence from animal models indicates that the coordination between angiogenesis and cardiac tissue remodeling is essential for the prevention of ventricular dilation and heart failure.[147,148] In limb ischemia, HIF-1 deficiency has been shown to result in diminished expression of angiogenic molecules including VEGF, angiopoietin 1, angiopoietin 2, and PDGF-B as well as impaired tissue recovery.[149]

Remodeling of pre-existing collateral arterioles to allow increased blood flow is also essential for positive outcomes to MI and CLI and involves proliferation of SMC and endothelial cells.[150] Most patients with critical stenosis of a coronary artery have one or more collaterals, and if an MI occurs these patients are likely to survive.[138,151] It has also been shown that individuals with critical coronary stenosis that lack collaterals are more likely to carry a proline to serine polymorphism in HIF-1α (P582S).[151] CLI occurs more commonly at the end stage of chronic peripheral arterial disease and is usually associated with atherosclerosis.[152] Some vessels downstream of the occluded areas have decreased wall thickness and decreased wall-to-lumen ratios while others that are upstream of the occlusion may be remodeled as collaterals.[153] Arteriolar vasodilation is a primary compensatory mechanism in ischemia of the limbs. In fact, there is evidence that vessels from these individuals are maximally dilated, a phenomenon called 'vasomotor paralysis,' which may contribute to edema.[153]

It has been shown that some of the vascular remodeling that occurs in MI and CLI is a direct consequence of changes in hemodynamics, which are sensed by endothelial cells as shear stress. Specifically, alterations in mechanical forces with resulting alterations in the endothelial cytoskeleton have been shown to lead to gene expression changes that govern the process of vascular remodeling including the induction of NOS isoforms, VEGF, MCP-1, and endothelial cell adhesion molecules ICAM-1 and ICAM-2.[154] Mononuclear cells recruited and activated by MCP-1 produce proteases that facilitate the digestion of extracellular matrix necessary for the proliferation and migration of vascular cells.[155,156] Altogether, the increase in cell numbers along with new matrix production deposited primarily by SMC that have changed from a contractile to a synthetic phenotype result in a 2–20-fold increase in vascular diameter. The functional recovery of blood flow is, however, generally insufficient, reaching less than 50% of maximal conductance.[150]

The role of circulating cells in mediating angiogenesis, arteriogenesis, and immune response to ischemia has been studied intensively. Endothelial cell progenitors were initially isolated from human blood by virtue of their expression of Flk-1 and CD34 and their ability to differentiate into cells expressing endothelial cell markers upon recruitment to the sites of injury.[157] It was later shown that specific growth factors including VEGF, placental growth factor (PlGF), stromal cell derived factor-1 (SDF-1), granulocyte colony stimulating factor (G-CSF) or granulocyte-monocyte colony stimulating factor (GM-CSF) can lead to the mobilization of bone-marrow-derived progenitors that may contribute to the formation of new vessels through differentiation into cells expressing vascular cell markers and release of angiogenic molecules.[158] Further analyses have shown that the number of progenitor cells incorporating into new vessels termed 'adult vasculogenesis' is relatively low and that the primary function of these cells, although depending upon differentiation, is mediated through secretion of paracrine signals.[159,160] The requirement for differentiation was demonstrated by introducing vectors expressing inducible 'suicide genes' under the control of endothelial-, vascular-smooth-muscle- or cardiomyocyte-specific promoters into bone marrow mononuclear cells prior to transplantation in a model of MI.[160]

The potential of therapeutic angiogenesis, arteriogenesis, and vasculogenesis, using growth factors, gene and cell therapies for the treatment of MI and CLI and other ischemic conditions has been thoroughly investigated.[136,158] However, to date none of these approaches has led to FDA-approved protocols for the treatment of ischemic disease. Notwithstanding early reports of angiomas that appeared in a patient injected with VEGF-expressing plasmids, evidence to date indicates that direct administration of recombinant growth factors, gene therapy approaches leading to endogenous overexpression of angiogenic molecules, and administration of purified endothelial progenitor cells or other progenitor cells is feasible and relatively safe in humans.[161,162] The major obstacle in the field is how to achieve clinical effects comparable to what has been observed in animal models of ischemia. Accumulating evidence indicates that administration of FGF or FGF/PDGF combinations may be more effective at inducing the formation of a stable and functional vessel compared with the leaky and dysfunctional vessels induced by VEGF administration.[163] As it has become clear that a single factor is not sufficient to trigger and maintain tissue recovery, gene therapy approaches are also moving in the direction of targets like HIF-1 that are capable of controlling the expression of many angiogenic factors.[138] Cell-based approaches tested in numerous clinical trials have also failed to achieve substantial clinical benefit, although meta-analysis of 13 trials has shown that bone marrow cell transplantation may lead to a small increase in left ventricular ejection fraction after MI.[164,165] Current efforts are focused on optimizing protocols for isolation and characterization of specific cell populations and the design of trials that may combine the administration of both cells and angiogenic factors.[164]

Neovascular Eye Diseases

Retinopathy of prematurity (ROP), proliferative diabetic retinopathy (PDR), and neovascular age-related macular degeneration (NVAMD) are prevalent causes of blindness in neonates, working adults, and aging individuals, respectively.[166] VEGF has been shown to play a key role in the pathophysiology of these conditions and is currently the target of effective therapeutic approaches for macular edema and NVAMD.[167] Analyses in non-human primates in which the retina was rendered hypoxic through photocoagulation of the veins demonstrated a clear association between angiogenesis in the iris and induction of VEGF.[168] In humans, high VEGF levels were also found in patients with diabetic retinopathy, neovascularization of the iris, and ischemic occlusion of the central retinal vein.[169]

ROP is prevalent among low-birth-weight infants who are placed in high oxygen to help with lung development. In ROP, the exposure of the developing vasculature to high oxygen leads to destabilization of the retinal blood vessels due to the down-regulation of VEGF. When the infants are removed from the high oxygen environment to normal air, the retina does not have enough blood vessels to be properly oxygenated, resulting in an increase in VEGF. With an increase in VEGF, the retinal blood vessels proliferate (revascularization phase), but generally to areas where they are not normally present, leading to vision loss.[170] The ability to save even lower-birth-weight infants has led to a resurgence of ROP. Close to 60% of the approximately 28 000 low-birth-weight babies born each year have some degree of ROP. Of these, about 1200 develop ROP that requires treatment and 400–600 become legally blind from ROP.[171] Treatments for ROP include laser therapy and cryotherapy in which the peripheral areas of the retina are damaged in order to diminish the metabolic demands of the tissue and prevent vessel growth.[172] These treatments are effective but also compromise side vision, thus anti-VEGF approaches are currently being tested for treatment of ROP with promising results.[172] Further studies are necessary to determine the long-term effects of anti-VEGF therapy in newborns and whether this approach is more effective than other established therapies for ROP.[173]

AMD is a degenerative condition characterized by progressive loss of central vision that is common in the aging population. The macula is the central region of the retina with the highest concentration of photoreceptors responsible for high-resolution vision. Some of the initial pathological changes in AMD include the accumulation of acellular debris called drusen underneath the retina between the retinal pigmented epithelium and the Bruch's membrane.[174] Degeneration of the retinal pigmented epithelium and photoreceptors as well as accumulation of drusen are characteristic of the dry form of AMD also known as geographic atrophy.[175] About 10–15% of people with AMD develop NVAMD also known as wet AMD, a very serious condition characterized by disruption of the Bruch's membrane and invasion of microvessels from the choriocapillaries into the retina (Figure 10.5).[174] NVAMD is the most common cause of blindness in people 50 years of age or older in the developed world.[176] Increased risk of AMD has been associated with age, smoking, Caucasian ethnicity, and genetic polymorphisms in the complement factor H gene and in a locus located between the genes encoding for Age-related Maculopathy Susceptibility 2 (ARMS2) and High-Temperature Requirement A Serine Peptidase 1 (HTRA1).[174] The ARMS2/HTRA1 locus is relevant because it is specifically associated with increased risk for NVAMD, although the pathophysiology is unclear.[177] The ARMS2 protein localizes to the mitochondria but its function is unknown, whereas HTRA1 is a protease known to cleave TGF-β.[174] There is evidence that both ARMS2 and HTRA1 may play a role in AMD pathophysiology, although the two loci are too closely linked to allow discerning of specific contributions.[177] In NVAMD, anti-VEGF was shown to slow progression of the disease and significantly improve visual acuity defining a paradigmatic example of successful anti-angiogenic therapy.

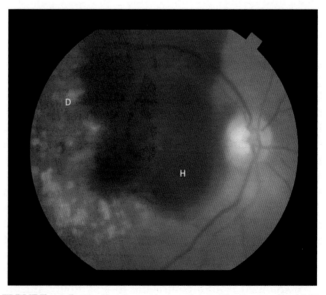

FIGURE 10.5 Fundus image in a patient with neovascular AMD. Note drusen temporally (D), and areas of atrophy and pigmentary changes within the macula. There is also a large subretinal hemorrhage due to choroidal neovascularization (H). Image courtesy of Dr. Leo Kim, M.D., Ph.D. (Mass Eye and Ear Infirmary).

However, there are concerns that long-term treatments may result in toxic effects to the eyes. Accumulating evidence from animal models indicates that basal levels of VEGF serve as a trophic factor for maintenance of the ciliary body, retinal neurons, and the choriocapillaries.[178–180]

Diabetic retinopathy (DR) is among the most common and devastating complications of diabetes and the most common cause of blindness in working adults in the US.[181] It was estimated that about one million individuals suffered from vision-threatening DR in 2007 in the US alone.[182] Basement membrane thickening, pericyte loss, microaneurysms, microhemorrhages, and discrete areas of ischemia called 'cotton wool spots' are early changes associated with the development of what is known as non-proliferative diabetic retinopathy.[183,184] As the disease progresses, increased ischemia as a result of vascular degeneration triggers angiogenesis in retinal vessels, particularly in those close to the optic nerve, a condition known as PDR (Figure 10.6). These new vessels often grow into the vitreous, are prone to hemorrhage and can lead to the formation of neovascular membranes and retinal detachments.[184] If left untreated, uncontrolled PDR can lead to neovascularization in the iris and the anterior chamber angle, which may result in increased intraocular pressure (glaucoma) due to impaired drainage of aqueous humor.[185] Macular edema leading to diminished visual acuity develops in about 25% of diabetic patients with or without other retinopathy changes and is associated with a break in the blood–retinal barrier.[184] Clinical and experimental data provide strong support for the concept that the microvascular complications are due in large part to hyperglycemia.[186] Specifically, tight glucose control has been shown to significantly reduce the development and progression of diabetic microangiopathy in diabetes.[186] There is also evidence that inflammation plays a key role in the pathophysiology of DR.[187] Diabetic mice lacking the endothelial cell adhesion molecules ICAM or its leukocyte-binding partner CD18 showed diminished leukostasis, less severe breakdown of the blood–retinal barrier and more preserved retinal microvasculature.[187,188] Increased levels of VEGF in the vitreous have been associated with the transition from non-proliferative DR to PDR in humans.[169] Accordingly, treatments for PDR include pan-retinal photocoagulation (PRP), therapy, surgical removal of neovascular membranes, and anti-VEGF therapy (off-label and only as a surgical adjuvant because of the risk of retinal detachment).[189] Anti-VEGF therapy is used successfully for treatment of macular edema in diabetic patients.[190]

References

1. Folkman J. Angiogenesis: an organizing principle for drug discovery? *Nat Rev Drug Discov* 2007 Apr;**6**(4):273–86.
2. Riant F, Bergametti F, Ayrignac X, Boulday G, Tournier-Lasserve E. Recent insights into cerebral cavernous malformations: the molecular genetics of CCM. *FEBS J.* 2010 Mar;**277**(5):1070–5.
3. Geudens I, Gerhardt H Coordinating cell behaviour during blood vessel formation. *Development.* 2011 Nov;**138**(21):4569–83.
4. Carmeliet P. Angiogenesis in life, disease and medicine. *Nature* 2005 Dec 15;**438**(7070):932–6.
5. Lange F, Ehrich W, Cohn AE. Studies on the blood vessels in the membranes of chick embryos: Part I. Absence of nerves in the vascular membrane. *J Exp Med* 1930 Jun 30;**52**(1):65–72.
6. Ferrara N, Carver-Moore K, Chen H, Dowd M, Lu L, O'Shea KS, et al. Heterozygous embryonic lethality induced by targeted inactivation of the VEGF gene. *Nature* 1996 Apr 4;**380**(6573):439–42.
7. Shalaby F, Rossant J, Yamaguchi TP, Gertsenstein M, Wu XF, Breitman ML, et al. Failure of blood-island formation and vasculogenesis in Flk-1-deficient mice. *Nature* 1995 Jul 6;**376**(6535):62–6.
8. Fong GH, Rossant J, Gertsenstein M, Breitman ML. Role of the Flt-1 receptor tyrosine kinase in regulating the assembly of vascular endothelium. *Nature* 1995 Jul 6;**376**(6535):66–70.
9. Fong GH, Zhang L, Bryce DM, Peng J. Increased hemangioblast commitment, not vascular disorganization, is the primary defect in flt-1 knock-out mice. *Development* 1999 Jul;**126**(13):3015–25.
10. Jain RK. Molecular regulation of vessel maturation. *Nat Med* 2003 Jun;**9**(6):685–93.
11. Rhodin JA. Ultrastructure of mammalian venous capillaries, venules, and small collecting veins. *J Ultrastruct Res* 1968 Dec;**25**(5):452–500.
12. Hirschi KK, D'Amore PA. Pericytes in the microvasculature. *Cardiovasc Res* 1996 Oct;**32**(4):687–98.
13. Hirschi KK, Rohovsky SA, D'Amore PA. PDGF, TGF-beta, and heterotypic cell-cell interactions mediate endothelial cell-induced recruitment of 10T1/2 cells and their differentiation to a smooth muscle fate. *J Cell Biol* 1998 May 4;**141**(3):805–14.
14. Donnelly LF, Adams DM, Bisset GS. 3rd. Vascular malformations and hemangiomas: a practical approach in a multidisciplinary clinic. *AJR Am J Roentgenol* 2000 Mar;**174**(3):597–608.

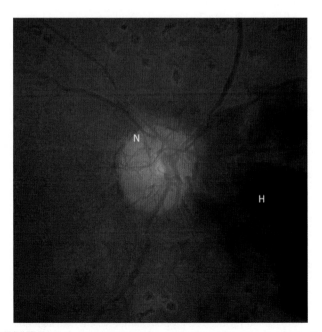

FIGURE 10.6 Fundus image in a patient with proliferative diabetic retinopathy. Note neovascularization of the disc with pre-retinal and vitreous hemorrhage nasal to the optic nerve head. Image courtesy of Dr. Leo Kim, M.D., Ph.D. (Mass Eye and Ear Infirmary).

15. Duffy K. Genetics and syndromes associated with vascular malformations. *Pediatr Clin North Am* 2010 Oct;57(5):1111–20.

16. Jacobs AH, Walton RG. The incidence of birthmarks in the neonate. *Pediatrics* 1976 Aug;58(2):218–22.

17. Brouillard P, Vikkula M. Genetic causes of vascular malformations. *Hum Mol Genet* 2007 Oct 15;16(Spec No. 2):R140–9.

18. Revencu N, Boon LM, Mulliken JB, Enjolras O, Cordisco MR, Burrows PE, et al. Parkes Weber syndrome, vein of Galen aneurysmal malformation, and other fast-flow vascular anomalies are caused by RASA1 mutations. *Hum Mutat* 2008 Jul;29(7):959–65.

19. Eerola I, Boon LM, Mulliken JB, Burrows PE, Dompmartin A, Watanabe S, et al. Capillary malformation-arteriovenous malformation, a new clinical and genetic disorder caused by RASA1 mutations. *Am J Hum Genet* 2003 Dec;73(6):1240–9.

20. Wennerberg K, Rossman KL, Der CJ. The Ras superfamily at a glance. *J Cell Sci* 2005 Mar 1;118(Pt 5):843–6.

21. Bos JL, Rehmann H, Wittinghofer A. GEFs and GAPs: critical elements in the control of small G proteins. *Cell* 2007 Jun 1;129(5):865–77.

22. Martin GA, Viskochil D, Bollag G, McCabe PC, Crosier WJ, Haubruck H, et al. The GAP-related domain of the neurofibromatosis type 1 gene product interacts with ras p21. *Cell* 1990 Nov 16;63(4):843–9.

23. Cullen PJ, Hsuan JJ, Truong O, Letcher AJ, Jackson TR, Dawson AP, et al. Identification of a specific Ins(1,3,4,5) P4-binding protein as a member of the GAP1 family. *Nature* 1995 Aug 10;376(6540):527–30.

24. Mukouyama YS, Shin D, Britsch S, Taniguchi M, Anderson DJ. Sensory nerves determine the pattern of arterial differentiation and blood vessel branching in the skin. *Cell* 2002 Jun 14;109(6):693–705.

25. Ballaun C, Weninger W, Uthman A, Weich H, Tschachler E. Human keratinocytes express the three major splice forms of vascular endothelial growth factor. *J Invest Dermatol* 1995 Jan;104(1):7–10.

26. Brown LF, Yeo KT, Berse B, Yeo TK, Senger DR, Dvorak HF, et al. Expression of vascular permeability factor (vascular endothelial growth factor) by epidermal keratinocytes during wound healing. *J Exp Med* 1992 Nov 1;176(5):1375–9.

27. Lapinski PE, Kwon S, Lubeck BA, Wilkinson JE, Srinivasan RS, Sevick-Muraca E, et al. RASA1 maintains the lymphatic vasculature in a quiescent functional state in mice. *J Clin Invest*. 2012 Feb 1;122(2):733–747.

28. Henkemeyer M, Rossi DJ, Holmyard DP, Puri MC, Mbamalu G, Harpal K, et al. Vascular system defects and neuronal apoptosis in mice lacking ras GTPase-activating protein. *Nature* 1995 Oct 26;377(6551):695–701.

29. Hoeger PH. Infantile haemangioma: new aspects on the pathogenesis of the most common skin tumour in children. *Br J Dermatol* 2011 Feb;164(2):234–5.

30. Holland KE, Drolet BA. Infantile hemangioma. *Pediatr Clin North Am* 2010 Oct;57(5):1069–83.

31. Drolet BA, Esterly NB, Frieden IJ. Hemangiomas in children. *N Engl J Med* 1999 Jul 15;341(3):173–81.

32. Lo K, Mihm M, Fay A. Current theories on the pathogenesis of infantile hemangioma. *Semin Ophthalmol* 2009 May-Jun;24(3):172–7.

33. Iwamoto T, Jakobiec FA. Ultrastructural comparison of capillary and cavernous hemangiomas of the orbit. *Arch Ophthalmol* 1979 Jun;97(6):1144–53.

34. Hoeger PH, Maerker JM, Kienast AK, Syed SB, Harper JI. Neonatal haemangiomatosis associated with placental chorioangiomas: report of three cases and review of the literature. *Clin Exp Dermatol* 2009 Jul;34(5):e78–80.

35. Pittman KM, Losken HW, Kleinman ME, Marcus JR, Blei F, Gurtner GC, et al. No evidence for maternal-fetal microchimerism in infantile hemangioma: a molecular genetic investigation. *J Invest Dermatol* 2006 Nov;126(11):2533–8.

36. North PE, Waner M, James CA, Mizeracki A, Frieden IJ, Mihm Jr MC. Congenital nonprogressive hemangioma: a distinct clinicopathologic entity unlike infantile hemangioma. *Arch Dermatol* 2001 Dec;137(12):1607–20.

37. Blei F, Walter J, Orlow SJ, Marchuk DA. Familial segregation of hemangiomas and vascular malformations as an autosomal dominant trait. *Arch Dermatol* 1998 Jun;134(6):718–22.

38. Walter JW, Blei F, Anderson JL, Orlow SJ, Speer MC, Marchuk DA. Genetic mapping of a novel familial form of infantile hemangioma. *Am J Med Genet* 1999 Jan 1;82(1):77–83.

39. Berg JN, Walter JW, Thisanagayam U, Evans M, Blei F, Waner M, et al. Evidence for loss of heterozygosity of 5q in sporadic haemangiomas: are somatic mutations involved in haemangioma formation? *J Clin Pathol* 2001 Mar;54(3):249–52.

40. Walter JW, North PE, Waner M, Mizeracki A, Blei F, Walker JW, et al. Somatic mutation of vascular endothelial growth factor receptors in juvenile hemangioma. *Genes Chromosomes Cancer* 2002 Mar;33(3):295–303.

41. Ye C, Pan L, Huang Y, Ye R, Han A, Li S, et al. Somatic mutations in exon 17 of the TEK gene in vascular tumors and vascular malformations. *J Vasc Surg* 2011 Dec;54(6):1760–8.

42. Pansuriya TC, van Eijk R, d'Adamo P, van Ruler MA, Kuijjer ML, Oosting J, et al. Somatic mosaic IDH1 and IDH2 mutations are associated with enchondroma and spindle cell hemangioma in Ollier disease and Maffucci syndrome. *Nat Genet* 2011 Dec;43(12):1256–61.

43. Kurek KC, Pansuriya TC, van Ruler MA, van den Akker B, Luks VL, Verbeke SL, et al. R132C IDH1 mutations are found in spindle cell hemangiomas and not in other vascular tumors or malformations. *Am J Pathol*. 2013 May;182(5):1494–1500.

44. Turcan S, Rohle D, Goenka A, Walsh LA, Fang F, Yilmaz E, et al. IDH1 mutation is sufficient to establish the glioma hypermethylator phenotype. *Nature* 2012 Mar 22;483(7390):479–83.

45. Mulliken JB, Zetter BR, Folkman J. In vitro characteristics of endothelium from hemangiomas and vascular malformations. *Surgery* 1982 Aug;92(2):348–53.

46. Boscolo E, Bischoff J. Vasculogenesis in infantile hemangioma. *Angiogenesis* 2009;12(2):197–207.

47. Khan ZA, Boscolo E, Picard A, Psutka S, Melero-Martin JM, Bartch TC, et al. Multipotential stem cells recapitulate human infantile hemangioma in immunodeficient mice. *J Clin Invest* 2008 Jul;118(7):2592–9.

48. Boye E, Yu Y, Paranya G, Mulliken JB, Olsen BR, Bischoff J. Clonality and altered behavior of endothelial cells from hemangiomas. *J Clin Invest* 2001 Mar;107(6):745–52.

49. Roberts DM, Kearney JB, Johnson JH, Rosenberg MP, Kumar R, Bautch VL. The vascular endothelial growth factor (VEGF) receptor Flt-1 (VEGFR-1) modulates Flk-1 (VEGFR-2) signaling during blood vessel formation. *Am J Pathol* 2004 May;164(5):1531–5.

50. Jinnin M, Medici D, Park L, Limaye N, Liu Y, Boscolo E, et al. Suppressed NFAT-dependent VEGFR1 expression and constitutive VEGFR2 signaling in infantile hemangioma. *Nat Med* 2008 Nov;14(11):1236–46.

51. Greenberger S, Boscolo E, Adini I, Mulliken JB, Bischoff J. Corticosteroid suppression of VEGF-A in infantile hemangioma-derived stem cells. *N Engl J Med*. 2010 Mar 18;362(11):1005–13.

52. Storch CH, Hoeger PH. Propranolol for infantile haemangiomas: insights into the molecular mechanisms of action. *Br J Dermatol*. 2010 Aug;163(2):269–74.

53. Leaute-Labreze C, Dumas de la Roque E, Hubiche T, Boralevi F, Thambo JB, Taieb A. Propranolol for severe hemangiomas of infancy. *N Engl J Med* 2008 Jun 12;358(24):2649–51.

54. Govani FS, Shovlin CL. Hereditary haemorrhagic telangiectasia: a clinical and scientific review. *Eur J Hum Genet* 2009 Jul;17(7):860–71.

55. Abdalla SA, Letarte M. Hereditary haemorrhagic telangiectasia: current views on genetics and mechanisms of disease. *J Med Genet* 2006 Feb;**43**(2):97–110.

56. Braverman IM, Keh A, Jacobson BS. Ultrastructure and three-dimensional organization of the telangiectases of hereditary hemorrhagic telangiectasia. *J Invest Dermatol* 1990 Oct;**95**(4):422–7.

57. Derynck R, Zhang YE. Smad-dependent and Smad-independent pathways in TGF-beta family signalling. *Nature* 2003 Oct 9;**425**(6958):577–84.

58. Pardali E, Goumans MJ, ten Dijke P. Signaling by members of the TGF-beta family in vascular morphogenesis and disease. *Trends Cell Biol*. 2010 Sep;**20**(9):556–67.

59. Attisano L, Carcamo J, Ventura F, Weis FM, Massague J, Wrana JL. Identification of human activin and TGF beta type I receptors that form heteromeric kinase complexes with type II receptors. *Cell* 1993 Nov 19;**75**(4):671–80.

60. ten Dijke P, Yamashita H, Ichijo H, Franzen P, Laiho M, Miyazono K, et al. Characterization of type I receptors for transforming growth factor-beta and activin. *Science* 1994 Apr 1;**264**(5155):101–4.

61. Goumans MJ, Valdimarsdottir G, Itoh S, Rosendahl A, Sideras P, ten Dijke P. Balancing the activation state of the endothelium via two distinct TGF-beta type I receptors. *EMBO J* 2002 Apr 2;**21**(7):1743–53.

62. David L, Mallet C, Keramidas M, Lamande N, Gasc JM, Dupuis-Girod S, et al. Bone morphogenetic protein-9 is a circulating vascular quiescence factor. *Circ Res* 2008 Apr 25;**102**(8):914–22.

63. Barbara NP, Wrana JL, Letarte M. Endoglin is an accessory protein that interacts with the signaling receptor complex of multiple members of the transforming growth factor-beta superfamily. *J Biol Chem* 1999 Jan 8;**274**(2):584–94.

64. Nakao A, Imamura T, Souchelnytskyi S, Kawabata M, Ishisaki A, Oeda E, et al. TGF-beta receptor-mediated signalling through Smad2, Smad3 and Smad4. *EMBO J* 1997 Sep 1;**16**(17):5353–62.

65. Oh SP, Seki T, Goss KA, Imamura T, Yi Y, Donahoe PK, et al. Activin receptor-like kinase 1 modulates transforming growth factor-beta 1 signaling in the regulation of angiogenesis. *Proc Natl Acad Sci U S A* 2000 Mar 14;**97**(6):2626–31.

66. Lux A, Attisano L, Marchuk DA. Assignment of transforming growth factor beta1 and beta3 and a third new ligand to the type I receptor ALK-1. *J Biol Chem* 1999 Apr 9;**274**(15):9984–92.

67. Benzinou M, Clermont FF, Letteboer TG, Kim JH, Espejel S, Harradine KA, et al. Mouse and human strategies identify PTPN14 as a modifier of angiogenesis and hereditary haemorrhagic telangiectasia. *Nat Commun* 2012 jan 3:616.

68. Antonelli-Orlidge A, Saunders KB, Smith SR, D'Amore PA. An activated form of transforming growth factor beta is produced by cocultures of endothelial cells and pericytes. *Proc Natl Acad Sci U S A* 1989 Jun;**86**(12):4544–8.

69. Neubauer K, Kruger M, Quondamatteo F, Knittel T, Saile B, Ramadori G. Transforming growth factor-beta1 stimulates the synthesis of basement membrane proteins laminin, collagen type IV and entactin in rat liver sinusoidal endothelial cells. *J Hepatol* 1999 Oct;**31**(4):692–702.

70. Venkatesha S, Toporsian M, Lam C, Hanai J, Mammoto T, Kim YM, et al. Soluble endoglin contributes to the pathogenesis of preeclampsia. *Nat Med* 2006 Jun;**12**(6):642–9.

71. Levine RJ, Lam C, Qian C, Yu KF, Maynard SE, Sachs BP, et al. Soluble endoglin and other circulating antiangiogenic factors in preeclampsia. *N Engl J Med* 2006 Sep 7;**355**(10):992–1005.

72. Dickson MC, Martin JS, Cousins FM, Kulkarni AB, Karlsson S, Akhurst RJ. Defective haematopoiesis and vasculogenesis in transforming growth factor-beta 1 knock out mice. *Development* 1995 Jun;**121**(6):1845–54.

73. Larsson J, Goumans MJ, Sjostrand LJ, van Rooijen MA, Ward D, Leveen P, et al. Abnormal angiogenesis but intact hematopoietic potential in TGF-beta type I receptor-deficient mice. *EMBO J* 2001 Apr 2;**20**(7):1663–73.

74. Arthur HM, Ure J, Smith AJ, Renforth G, Wilson DI, Torsney E, et al. Endoglin, an ancillary TGFbeta receptor, is required for extraembryonic angiogenesis and plays a key role in heart development. *Dev Biol* 2000 Jan 1;**217**(1):42–53.

75. Park SO, Wankhede M, Lee YJ, Choi EJ, Fliess N, Choe SW, et al. Real-time imaging of de novo arteriovenous malformation in a mouse model of hereditary hemorrhagic telangiectasia. *J Clin Invest* 2009 Nov;**119**(11):3487–96.

76. Bonyadi M, Rusholme SA, Cousins FM, Su HC, Biron CA, Farrall M, et al. Mapping of a major genetic modifier of embryonic lethality in TGF beta 1 knockout mice. *Nat Genet* 1997 Feb;**15**(2):207–11.

77. Fisher OS, Zhang R, Li X, Murphy JW, Demeler B, Boggon TJ. Structural studies of cerebral cavernous malformations 2 (CCM2) reveal a folded helical domain at its C-terminus. *FEBS Lett*. 2013 Jan 31;**587**(3):272–7.

78. Lehnhardt FG, von Smekal U, Ruckriem B, Stenzel W, Neveling M, Heiss WD, et al. Value of gradient-echo magnetic resonance imaging in the diagnosis of familial cerebral cavernous malformation. *Arch Neurol* 2005 Apr;**62**(4):653–8.

79. Tu J, Stoodley MA, Morgan MK, Storer KP. Ultrastructural characteristics of hemorrhagic, nonhemorrhagic, and recurrent cavernous malformations. *J Neurosurg* 2005 Nov;**103**(5):903–9.

80. Wong JH, Awad IA, Kim JH. Ultrastructural pathological features of cerebrovascular malformations: a preliminary report. *Neurosurgery* 2000 Jun;**46**(6):1454–9.

81. Labauge P, Denier C, Bergametti F, Tournier-Lasserve E. Genetics of cavernous angiomas. *Lancet Neurol* 2007 Mar;**6**(3):237–44.

82. Kondziolka D, Lunsford LD, Kestle JR. The natural history of cerebral cavernous malformations. *J Neurosurg* 1995 Nov;**83**(5):820–4.

83. Laberge-le Couteulx S, Jung HH, Labauge P, Houtteville JP, Lescoat C, Cecillon M, et al. Truncating mutations in CCM1, encoding KRIT1, cause hereditary cavernous angiomas. *Nat Genet* 1999 Oct;**23**(2):189–93.

84. Sahoo T, Johnson EW, Thomas JW, Kuehl PM, Jones TL, Dokken CG, et al. Mutations in the gene encoding KRIT1, a Krev-1/rap1a binding protein, cause cerebral cavernous malformations (CCM1). *Hum Mol Genet* 1999 Nov;**8**(12):2325–33.

85. Liquori CL, Berg MJ, Siegel AM, Huang E, Zawistowski JS, Stoffer T, et al. Mutations in a gene encoding a novel protein containing a phosphotyrosine-binding domain cause type 2 cerebral cavernous malformations. *Am J Hum Genet* 2003 Dec;**73**(6):1459–64.

86. Denier C, Goutagny S, Labauge P, Krivosic V, Arnoult M, Cousin A, et al. Mutations within the MGC4607 gene cause cerebral cavernous malformations. *Am J Hum Genet* 2004 Feb;**74**(2):326–37.

87. Bergametti F, Denier C, Labauge P, Arnoult M, Boetto S, Clanet M, et al. Mutations within the programmed cell death 10 gene cause cerebral cavernous malformations. *Am J Hum Genet* 2005 Jan;**76**(1):42–51.

88. Gault J, Shenkar R, Recksiek P, Awad IA. Biallelic somatic and germ line CCM1 truncating mutations in a cerebral cavernous malformation lesion. *Stroke* 2005 Apr;**36**(4):872–4.

89. Akers AL, Johnson E, Steinberg GK, Zabramski JM, Marchuk DA. Biallelic somatic and germline mutations in cerebral cavernous malformations (CCMs): evidence for a two-hit mechanism of CCM pathogenesis. *Hum Mol Genet* 2009 Mar 1;**18**(5):919–30.

90. Pagenstecher A, Stahl S, Sure U, Felbor U. A two-hit mechanism causes cerebral cavernous malformations: complete inactivation of CCM1, CCM2 or CCM3 in affected endothelial cells. *Hum Mol Genet* 2009 Mar 1;**18**(5):911–8.

91. McDonald DA, Shenkar R, Shi C, Stockton RA, Akers AL, Kucherlapati MH, et al. A novel mouse model of cerebral cavernous malformations based on the two-hit mutation hypothesis recapitulates the human disease. *Hum Mol Genet*. 2011 Jan 15;**20**(2):211–22.

92. Chan AC, Drakos SG, Ruiz OE, Smith AC, Gibson CC, Ling J, et al. Mutations in 2 distinct genetic pathways result in cerebral cavernous malformations in mice. J Clin Invest May 1871-81;**121**(5).

93. Stockton RA, Shenkar R, Awad IA, Ginsberg MH. Cerebral cavernous malformations proteins inhibit Rho kinase to stabilize vascular integrity. *J Exp Med.* 2010 Apr 12;207(4):881–96.

94. Faurobert E, Albiges-Rizo C. Recent insights into cerebral cavernous malformations: a complex jigsaw puzzle under construction. *FEBS J.* 2010 Mar;277(5):1084–1096.

95. Whitehead KJ, Chan AC, Navankasattusas S, Koh W, London NR, Ling J, et al. The cerebral cavernous malformation signaling pathway promotes vascular integrity via Rho GTPases. *Nat Med* 2009 Feb;**15**(2):177–84.

96. Crose LE, Hilder TL, Sciaky N, Johnson GL. Cerebral cavernous malformation 2 protein promotes smad ubiquitin regulatory factor 1-mediated RhoA degradation in endothelial cells. *J Biol Chem* 2009 May 15;**284**(20):13301–5.

97. Fidalgo M, Fraile M, Pires A, Force T, Pombo C, Zalvide J. CCM3/PDCD10 stabilizes GCKIII proteins to promote Golgi assembly and cell orientation. *J Cell Sci.* 2010 Apr 15;123(Pt 8):1274–84.

98. Glading A, Han J, Stockton RA, Ginsberg MH. KRIT-1/CCM1 is a Rap1 effector that regulates endothelial cell-cell junctions. *J Cell Biol* 2007 Oct 22;**179**(2):247–54.

99. Louvi A, Chen L, Two AM, Zhang H, Min W, Gunel M. Loss of cerebral cavernous malformation 3 (Ccm3) in neuroglia leads to CCM and vascular pathology. *Proc Natl Acad Sci U S A* 2011 Mar 1;108(9):3737–42.

100. Wustehube J, Bartol A, Liebler SS, Brutsch R, Zhu Y, Felbor U, et al. Cerebral cavernous malformation protein CCM1 inhibits sprouting angiogenesis by activating DELTA-NOTCH signaling. *Proc Natl Acad Sci U S A* 2010 Jul 13;107(28):12640–5.

101. Kleaveland B, Zheng X, Liu JJ, Blum Y, Tung JJ, Zou Z, et al. Regulation of cardiovascular development and integrity by the heart of glass-cerebral cavernous malformation protein pathway. *Nat Med* 2009 Feb;**15**(2):169–76.

102. Rosen JN, Sogah VM, Ye LY, Mably JD. CCM2-like is required for cardiovascular development as a novel component of the Heg-CCM pathway. *Dev Biol* 2013 Apr 1;376(1):74–85.

103. Hanahan D, Weinberg RA. Hallmarks of cancer: the next generation. *Cell* 2011 Mar 4;144(5):646–674.

104. Hanahan D, Weinberg RA. The hallmarks of cancer. *Cell* 2000 Jan 7;**100**(1):57–70.

105. Folkman J. Tumor angiogenesis: therapeutic implications. *N Engl J Med* 1971 Nov 18;285(21):1182–6.

106. Greene HS. Heterologous Transplantation of Mammalian Tumors: Ii. The Transfer of Human Tumors to Alien Species. *J Exp Med* 1941 Mar 31;**73**(4):475–86.

107. Thomlinson RH, Gray LH. The histological structure of some human lung cancers and the possible implications for radiotherapy. *Br J Cancer* 1955 Dec;**9**(4):539–49.

108. Gimbrone Jr MA, Cotran RS, Leapman SB, Folkman J. Tumor growth and neovascularization: an experimental model using the rabbit cornea. *J Natl Cancer Inst.* 1974 Feb;**52**(2):413–27.

109. Hanahan D, Folkman J. Patterns and emerging mechanisms of the angiogenic switch during tumorigenesis. *Cell* 1996 Aug 9;**86**(3):353–64.

110. Folkman J, Watson K, Ingber D, Hanahan D. Induction of angiogenesis during the transition from hyperplasia to neoplasia. *Nature* 1989 May 4;**339**(6219):58–61.

111. Liao D, Johnson RS. Hypoxia: a key regulator of angiogenesis in cancer. *Cancer Metastasis Rev* 2007 Jun;26(2):281–90.

112. Ryan HE, Lo J, Johnson RS. HIF-1 alpha is required for solid tumor formation and embryonic vascularization. *EMBO J* 1998 Jun 1;**17**(11):3005–15.

113. Semenza GL. Hydroxylation of HIF-1: oxygen sensing at the molecular level. *Physiol (Bethesda)* 2004 Aug;**19**:176–82.

114. Murdoch C, Muthana M, Coffelt SB, Lewis CE. The role of myeloid cells in the promotion of tumour angiogenesis. *Nat Rev Cancer* 2008 Aug;**8**(8):618–31.

115. Grimshaw MJ, Naylor S, Balkwill FR. Endothelin-2 is a hypoxia-induced autocrine survival factor for breast tumor cells. *Mol Cancer Ther* 2002 Dec;**1**(14):1273–81.

116. Lin MI, Sessa WC. Vascular endothelial growth factor signaling to endothelial nitric oxide synthase: more than a FLeeTing moment. *Circ Res* 2006 Sep 29;**99**(7):666–8.

117. Giraudo E, Inoue M, Hanahan D. An amino-bisphosphonate targets MMP-9-expressing macrophages and angiogenesis to impair cervical carcinogenesis. *J Clin Invest* 2004 Sep;**114**(5):623–33.

118. De Palma M, Venneri MA, Galli R, Sergi Sergi L, Politi LS, Sampaolesi M, et al. Tie2 identifies a hematopoietic lineage of proangiogenic monocytes required for tumor vessel formation and a mesenchymal population of pericyte progenitors. *Cancer Cell* 2005 Sep;**8**(3):211–26.

119. Pages G, Pouyssegur J. Transcriptional regulation of the Vascular Endothelial Growth Factor gene–a concert of activating factors. *Cardiovasc Res* 2005 Feb 15;**65**(3):564–73.

120. Mizukami Y, Li J, Zhang X, Zimmer MA, Iliopoulos O, Chung DC. Hypoxia-inducible factor-1-independent regulation of vascular endothelial growth factor by hypoxia in colon cancer. *Cancer Res* 2004 Mar 1;**64**(5):1765–72.

121. Mizukami Y, Jo WS, Duerr EM, Gala M, Li J, Zhang X, et al. Induction of interleukin-8 preserves the angiogenic response in HIF-1alpha-deficient colon cancer cells. *Nat Med* 2005 Sep;**11**(9):992–7.

122. Baeriswyl V, Christofori G. The angiogenic switch in carcinogenesis. *Semin Cancer Biol* 2009 Oct;**19**(5):329–37.

123. Chung AS, Lee J, Ferrara N. Targeting the tumour vasculature: insights from physiological angiogenesis. *Nat Rev Cancer* 2010 Jul;10(7):505–514.

124. O'Reilly MS, Holmgren L, Shing Y, Chen C, Rosenthal RA, Moses M, et al. Angiostatin: a novel angiogenesis inhibitor that mediates the suppression of metastases by a Lewis lung carcinoma. *Cell* 1994 Oct 21;**79**(2):315–28.

125. Ribatti D. Endogenous inhibitors of angiogenesis: a historical review. *Leuk Res* 2009 May;**33**(5):638–44.

126. Nyberg P, Xie L, Kalluri R. Endogenous inhibitors of angiogenesis. *Cancer Res* 2005 May 15;**65**(10):3967–79.

127. Folkman J. Endogenous angiogenesis inhibitors. *APMIS* 2004 Jul-Aug;**112**(7-8):496–507.

128. O'Reilly MS, Boehm T, Shing Y, Fukai N, Vasios G, Lane WS, et al. Endostatin: an endogenous inhibitor of angiogenesis and tumor growth. *Cellule* 1997 Jan 24;**88**(2):277–85.

129. Folkman J. Antiangiogenesis in cancer therapy–endostatin and its mechanisms of action. *Exp Cell Res* 2006 Mar 10;**312**(5):594–607.

130. Wickstrom SA, Alitalo K, Keski-Oja J. Endostatin associates with integrin alpha5beta1 and caveolin-1, and activates Src via a tyrosyl phosphatase-dependent pathway in human endothelial cells. *Cancer Res* 2002 Oct 1;**62**(19):5580–9.

131. Sudhakar A, Sugimoto H, Yang C, Lively J, Zeisberg M, Kalluri R. Human tumstatin and human endostatin exhibit distinct antiangiogenic activities mediated by alpha v beta 3 and alpha 5 beta 1 integrins. *Proc Natl Acad Sci U S A* 2003 Apr 15;**100**(8):4766–71.

132. Moulton KS, Olsen BR, Sonn S, Fukai N, Zurakowski D, Zeng X. Loss of collagen XVIII enhances neovascularization and vascular permeability in atherosclerosis. *Circulation* 2004 Sep 7;**110**(10):1330–6.

133. Abdollahi A, Hahnfeldt P, Maercker C, Grone HJ, Debus J, Ansorge W, et al. Endostatin's antiangiogenic signaling network. *Mol Cell* 2004 Mar 12;**13**(5):649–63.

134. Jain RK, Duda DG, Clark JW, Loeffler JS. Lessons from phase III clinical trials on anti-VEGF therapy for cancer. *Nat Clin Pract Oncol* 2006 Jan;**3**(1):24–40.

135. Azam F, Mehta S, Harris AL. Mechanisms of resistance to antiangiogenesis therapy. *Eur J Cancer*. 2010 May;46(8):1323–32.

136. Tongers J, Roncalli JG, Losordo DW. Therapeutic angiogenesis for critical limb ischemia: microvascular therapies coming of age. *Circulation* 2008 Jul 1;**118**(1):9–16.

137. Slevin M, Kumar P, Gaffney J, Kumar S, Krupinski J. Can angiogenesis be exploited to improve stroke outcome? Mechanisms and therapeutic potential. *Clin Sci (Lond)* 2006 Sep;**111**(3):171–83.

138. Semenza GL. Hypoxia-inducible factors in physiology and medicine. *Cell* 2012 Feb 3;148(3):399–408.

139. Wurzel J, Goldman BI. Angiogenesis factors in acute myocardial ischemia and infarction. *N Engl J Med* 2000 Jul 13;**343**(2):148–9.

140. Jurgensen JS, Rosenberger C, Wiesener MS, Warnecke C, Horstrup JH, Grafe M, et al. Persistent induction of HIF-1alpha and -2alpha in cardiomyocytes and stromal cells of ischemic myocardium. *FASEB J* 2004 Sep;**18**(12):1415–7.

141. Bergeron M, Yu AY, Solway KE, Semenza GL, Sharp FR. Induction of hypoxia-inducible factor-1 (HIF-1) and its target genes following focal ischaemia in rat brain. *Eur J Neurosci* 1999 Dec;**11**(12):4159–70.

142. Tuomisto TT, Rissanen TT, Vajanto I, Korkeela A, Rutanen J, Yla-Herttuala S. HIF-VEGF-VEGFR-2, TNF-alpha and IGF pathways are upregulated in critical human skeletal muscle ischemia as studied with DNA array. *Atherosclerosis* 2004 May;**174**(1):111–20.

143. Semenza GL, Jiang BH, Leung SW, Passantino R, Concordet JP, Maire P, et al. Hypoxia response elements in the aldolase A, enolase 1, and lactate dehydrogenase A gene promoters contain essential binding sites for hypoxia-inducible factor 1. *J Biol Chem* 1996 Dec 20;**271**(51):32529–37.

144. Kim JW, Tchernyshyov I, Semenza GL, Dang CV. HIF-1-mediated expression of pyruvate dehydrogenase kinase: a metabolic switch required for cellular adaptation to hypoxia. *Cell Metab* 2006 Mar;**3**(3):177–85.

145. Zhang H, Bosch-Marce M, Shimoda LA, Tan YS, Baek JH, Wesley JB, et al. Mitochondrial autophagy is an HIF-1-dependent adaptive metabolic response to hypoxia. *J Biol Chem* 2008 Apr 18;**283**(16):10892–903.

146. Chandel NS, Maltepe E, Goldwasser E, Mathieu CE, Simon MC, Schumacker PT. Mitochondrial reactive oxygen species trigger hypoxia-induced transcription. *Proc Natl Acad Sci U S A* 1998 Sep 29;**95**(20):11715–20.

147. Frangogiannis NG. The mechanistic basis of infarct healing. *Antioxid Redox Signal* 2006 Nov-Dec;**8**(11-12):1907–39.

148. Shiojima I, Sato K, Izumiya Y, Schiekofer S, Ito M, Liao R, et al. Disruption of coordinated cardiac hypertrophy and angiogenesis contributes to the transition to heart failure. *J Clin Invest* 2005 Aug;**115**(8):2108–18.

149. Bosch-Marce M, Okuyama H, Wesley JB, Sarkar K, Kimura H, Liu YV, et al. Effects of aging and hypoxia-inducible factor-1 activity on angiogenic cell mobilization and recovery of perfusion after limb ischemia. *Circ Res* 2007 Dec 7;**101**(12):1310–8.

150. Schaper W. Collateral circulation: past and present. *Basic Res Cardiol* 2009 Jan;**104**(1):5–21.

151. Resar JR, Roguin A, Voner J, Nasir K, Hennebry TA, Miller JM, et al. Hypoxia-inducible factor 1alpha polymorphism and coronary collaterals in patients with ischemic heart disease. *Chest* 2005 Aug;**128**(2):787–91.

152. Varu VN, Hogg ME, Kibbe MR. Critical limb ischemia. *J Vasc Surg* Jan;51(1):230–241.

153. Coats P, Wadsworth R. Marriage of resistance and conduit arteries breeds critical limb ischemia. *Am J Physiol Heart Circ Physiol* 2005 Mar;**288**(3):H1044–50.

154. Schirmer SH, van Nooijen FC, Piek JJ, van Royen N. Stimulation of collateral artery growth: travelling further down the road to clinical application. *Heart* 2009 Mar;**95**(3):191–7.

155. Ito WD, Arras M, Winkler B, Scholz D, Schaper J, Schaper W. Monocyte chemotactic protein-1 increases collateral and peripheral conductance after femoral artery occlusion. *Circ Res* 1997 Jun;**80**(6):829–37.

156. van Royen N, Hoefer I, Buschmann I, Kostin S, Voskuil M, Bode C, et al. Effects of local MCP-1 protein therapy on the development of the collateral circulation and atherosclerosis in Watanabe hyperlipidemic rabbits. *Cardiovasc Res* 2003 Jan;**57**(1):178–85.

157. Asahara T, Murohara T, Sullivan A, Silver M, van der Zee R, Li T, et al. Isolation of putative progenitor endothelial cells for angiogenesis. *Science* 1997 Feb 14;**275**(5302):964–7.

158. Renault MA, Losordo DW. Therapeutic myocardial angiogenesis. *Microvasc Res* 2007 Sep-Nov;**74**(2-3):159–71.

159. Fadini GP, Losordo D, Dimmeler S. Critical reevaluation of endothelial progenitor cell phenotypes for therapeutic and diagnostic use. *Circ Res* 2012 Feb 17;**110**(4):624–37.

160. Yoon CH, Koyanagi M, Iekushi K, Seeger F, Urbich C, Zeiher AM, et al. Mechanism of improved cardiac function after bone marrow mononuclear cell therapy: role of cardiovascular lineage commitment. *Circulation* 2010 May 11;121(18):2001-11.

161. Isner JM, Pieczek A, Schainfeld R, Blair R, Haley L, Asahara T, et al. Clinical evidence of angiogenesis after arterial gene transfer of phVEGF165 in patient with ischaemic limb. *Lancet* 1996 Aug 10;**348**(9024):370–4.

162. Gupta R, Tongers J, Losordo DW. Human studies of angiogenic gene therapy. *Circ Res* 2009 Oct 9;**105**(8):724–36.

163. Cochain C, Channon KM, Silvestre JS. Angiogenesis in the infarcted myocardium. *Antioxid Redox Signal*. 2013 Mar 20;18(9):1100–1113.

164. Silvestre JS. Pro-angiogenic cell-based therapy for the treatment of ischemic cardiovascular diseases. *Thromb Res*. 2012 Oct;130 (Suppl. 1):S90–4.

165. Martin-Rendon E, Brunskill SJ, Hyde CJ, Stanworth SJ, Mathur A, Watt SM. Autologous bone marrow stem cells to treat acute myocardial infarction: a systematic review. *Eur Heart J* 2008 Aug;**29**(15):1807–18.

166. Witmer AN, Vrensen GF, Van Noorden CJ, Schlingemann RO. Vascular endothelial growth factors and angiogenesis in eye disease. *Prog Retin Eye Res* 2003 Jan;**22**(1):1–29.

167. Kim LA, D'Amore PA. A brief history of anti-VEGF for the treatment of ocular angiogenesis. *Am J Pathol* 2012 Aug;**181**(2):376–9.

168. Miller JW, Adamis AP, Shima DT, D'Amore PA, Moulton RS, O'Reilly MS, et al. Vascular endothelial growth factor/vascular permeability factor is temporally and spatially correlated with ocular angiogenesis in a primate model. *Am J Pathol* 1994 Sep;**145**(3):574–84.

169. Aiello LP, Avery RL, Arrigg PG, Keyt BA, Jampel HD, Shah ST, et al. Vascular endothelial growth factor in ocular fluid of patients with diabetic retinopathy and other retinal disorders. *N Engl J Med* 1994 Dec 1;**331**(22):1480–7.

170. Sylvester CL. Retinopathy of prematurity. *Semin Ophthalmol* 2008 Sep-Oct;**23**(5):318–23.

171. Wheatley CM, Dickinson JL, Mackey DA, Craig JE, Sale MM. Retinopathy of prematurity: recent advances in our understanding. *Br J Ophthalmol* 2002 Jun;**86**(6):696–700.

172. Mutlu FM, Sarici SU. Treatment of retinopathy of prematurity: a review of conventional and promising new therapeutic options. *Int J Ophthalmol*.2013 6(2):228–36.

173. Wallace DK, Wu KY. Current and future trends in treatment of severe retinopathy of prematurity. *Clin Perinatol*. 2013 Jun;40(2):297–310.

174. Jager RD, Mieler WF, Miller JW. Age-related macular degeneration. *N Engl J Med* 2008 Jun 12;**358**(24):2606–17.

175. de Jong PT. Age-related macular degeneration. *N Engl J Med* 2006 Oct 5;**355**(14):1474–85.

176. Pascolini D, Mariotti SP, Pokharel GP, Pararajasegaram R, Etya'ale D, Negrel AD, et al. 2002 global update of available data

on visual impairment: a compilation of population-based prevalence studies. *Ophthalmic Epidemiol* 2004 Apr;**11**(2):67–115.

177. Francis PJ, Zhang H, Dewan A, Hoh J, Klein ML. Joint effects of polymorphisms in the HTRA1, LOC387715/ARMS2, and CFH genes on AMD in a Caucasian population. *Mol Vis* 2008;**14**:1395–400.

178. Saint-Geniez M, Maharaj AS, Walshe TE, Tucker BA, Sekiyama E, Kurihara T, et al. Endogenous VEGF is required for visual function: evidence for a survival role on muller cells and photoreceptors. *PLoS One* 2008;**3**(11):e3554.

179. Ford KM, Saint-Geniez M, Walshe TE, D'Amore PA. Expression and role of VEGF–a in the ciliary body. *Invest Ophthalmol Vis Sci* 2012 Nov;**53**(12):7520–27.

180. Saint-Geniez M, Kurihara T, Sekiyama E, Maldonado AE, D'Amore PA. An essential role for RPE-derived soluble VEGF in the maintenance of the choriocapillaris. *Proc Natl Acad Sci U S A* 2009 Nov 3;**106**(44):18751–6.

181. Cheung N, Mitchell P, Wong TY. Diabetic retinopathy. *Lancet* 2010 Jul 10;**376**(9735):124–36.

182. Zhang X, Saaddine JB, Chou CF, Cotch MF, Cheng YJ, Geiss LS, et al. Prevalence of diabetic retinopathy in the United States, 2005-2008. *JAMA*. 2010 Aug 11;**304**(6):649–56.

183. Cogan DG, Kuwabara T. Capillary Shunts in the Pathogenesis of Diabetic Retinopathy. *Diabetes* 1963 Jul-Aug;**12**:293–300.

184. Antonetti DA, Klein R, Gardner TW. Diabetic retinopathy. *N Engl J Med*. 2012 Mar 29;366(13):1227–39.

185. Iliev ME, Domig D, Wolf-Schnurrbursch U, Wolf S, Sarra GM. Intravitreal bevacizumab (Avastin) in the treatment of neovascular glaucoma. *Am J Ophthalmol* 2006 Dec;**142**(6):1054–6.

186. Retinopathy and nephropathy in patients with type 1 diabetes four years after a trial of intensive therapy. The Diabetes Control and Complications Trial/Epidemiology of Diabetes Interventions and Complications Research Group. *N Engl J Med* 2000 Feb 10;**342**(6):381–9.

187. Antonetti DA, Barber AJ, Bronson SK, Freeman WM, Gardner TW, Jefferson LS, et al. Diabetic retinopathy: seeing beyond glucose-induced microvascular disease. *Diabetes* 2006 Sep;**55**(9):2401–11.

188. Joussen AM, Poulaki V, Le ML, Koizumi K, Esser C, Janicki H, et al. A central role for inflammation in the pathogenesis of diabetic retinopathy. *FASEB J* 2004 Sep;**18**(12):1450–2.

189. Jardeleza MS, Miller JW. Review of anti-VEGF therapy in proliferative diabetic retinopathy. *Semin Ophthalmol* 2009 Mar-Apr;**24**(2):87–92.

190. Massin P, Bandello F, Garweg JG, Hansen LL, Harding SP, Larsen M, et al. Safety and efficacy of ranibizumab in diabetic macular edema (RESOLVE Study): a 12-month, randomized, controlled, double-masked, multicenter phase II study. *Diabetes Care*. 2010 Nov;33(11):2399–405.

Diseases of Medium-Sized and Small Vessels

J. Charles Jennette, MD[1], James R. Stone, MD, PhD[2]

[1]University of North Carolina at Chapel Hill, Chapel Hill, NC, USA; [2]Massachusetts General Hospital, Harvard Medical School, Boston, MA, USA

INTRODUCTION

Vascular diseases extract a large toll on the human population, both in terms of mortality and quality of life. Vascular disease is in fact the leading cause of death in developed countries. Numerous pathologic conditions alter the structure and/or function of blood vessels. This chapter will cover the mechanisms of common disorders affecting medium-sized and small blood vessels, which are not discussed elsewhere in this book. However, the basic mechanisms of vascular activation discussed in this chapter have relevance to many conditions discussed elsewhere in the book, including thrombosis (Chapter 15), angiogenesis (Chapter 10), atherosclerosis (Chapter 12), aneurysms (Chapter 13), and hypertension (Chapter 14). This chapter will only focus on several of the more common vascular diseases. Additional information on less common vascular pathologic conditions not covered in this chapter or elsewhere in this book can be found in the end-noted textbooks.[1,2]

NORMAL VESSEL WALL STRUCTURE

The primary components of normal blood vessels are endothelial cells, vascular smooth muscle cells, and extracellular matrix including collagen, elastin, and proteoglycans. Structurally, blood vessels are composed of three distinct concentric layers: intima, media, and adventitia. The intima refers to the inner portion of the vessel wall, and in normal vessels is composed primarily of a single layer of endothelium with its basement membrane and subendothelial connective tissue. In small and medium-sized muscular arteries, the intima is separated from the next layer, the media, by a discrete layer of elastic fibers called the internal elastic lamina. The media is composed primarily of vascular smooth muscle cells. In muscular arteries, the media is separated from the outer

adventitial layer by a second discrete layer of elastic fibers called the external elastic lamina. In elastic arteries, such as the aorta and its proximal branches, rather than discrete internal and external elastic lamina, there is a mesh of elastic fibers throughout the media, between the smooth muscle cells.

The adventitia is composed primarily of collagen and fibroblasts. It is becoming clear that the adventitia may serve as a reservoir of multipotent stromal cells, or stem cells, that can facilitate vascular responses to injury by differentiating into fibroblasts, endothelial cells, and vascular smooth muscle cells. Large vessels and some medium-sized vessels contain capillaries and other small vessels in the adventitia. In the aorta, this vasa vasorum infiltrates into the outer third of the media, contributing to the blood supply of the vessel wall. In normal medium-sized and small vessels, the media is typically not vascularized. The adventitia may also contain nerves, which innervate the vessel to regulate vasoconstrictor tone.

Proceeding from the aorta distally, the arteries branch to give rise to progressively smaller vessels. The medium-sized distributing arteries carry blood to the individual organs, and the small intraparenchymal arteries carry blood within the organs themselves. Both small arteries and medium-sized arteries are generally referred to as medium-sized vessels. The small arteries continue to branch until giving rise to arterioles, which contain only one to two layers of smooth muscle in the media. The arterioles branch to eventually become capillaries, which are essentially endothelial tubes lacking a defined media. Capillaries are surrounded and supported by interspersed pericytes. The structure of capillaries and their combined high luminal surface area facilitate diffusion of oxygen and nutrients to the surrounding tissues. The capillaries coalesce to form post-capillary venules, which themselves join to form veins. In contrast to arteries, veins have thinner walls relative to the

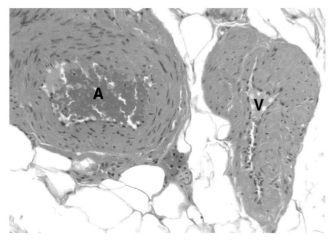

FIGURE 11.1 Comparison of normal artery and vein structure. Shown are histologic sections of a normal artery (A) and vein (V). The wall of the artery is composed primarily of dense concentrically arranged smooth muscle cells. The wall of the vein is in many places thinner than that of the artery, with the smooth muscle cells showing a less orderly arrangement. Arteries and veins are often present together as paired structures in tissue.

luminal diameter, and less compact smooth muscle in the media with more extracellular matrix (Figure 11.1). Medium-sized and large veins will also contain longitudinal bands of smooth muscle outside of the standard concentric layers of smooth muscle. Large veins will also contain valves on the luminal surface to facilitate unidirectional blood flow. Such venous valves are particularly important in the large veins of the legs.

VASCULAR CELL ACTIVATION

Vascular cells respond to injury and other stimuli by undergoing phenotypic changes referred to as activation. The endothelium plays important roles in vascular responses, as this cell sits at the border between the luminal blood and the vessel wall. Endothelial cells are critical for maintaining a non-thrombogenic surface in normal circumstances and for regulating vasodilation and vasoconstriction. Endothelial cells are responsive to the shear stress generated by the blood flowing across its surface. Under normal non-activating conditions of laminar shear stress, endothelial cells align with the direction of the blood flow, and primarily secrete agents that prevent coagulation and foster vasodilation, such as nitric oxide and prostacyclin. However, at branch points in the vasculature, the steady laminar flow of blood is disrupted resulting in an altered and abnormal pattern of shear stress on the endothelium. In these sites of disturbed flow, endothelial cells lose their orderly alignment with the flow of blood and secrete increased amounts of agents that promote coagulation and/or vasoconstriction such as endothelin, von

Willebrand factor, and superoxide. Endothelial cells are in fact the predominant source of reactive oxygen species in the normal vessel wall. Such endothelial activation is also termed endothelial dysfunction, since the endothelial cells become dysfunctional in facilitating vasodilation and preventing coagulation. The superoxide that is generated will inactivate nitric oxide. Nitric oxide would otherwise promote vasodilation and inhibit platelet aggregation.

The activated endothelium also secretes growth factors such as platelet-derived growth factor (PDGF), fibroblast growth factor (FGF), stromal cell-derived factor 1 (SDF-1), and another reactive oxygen species hydrogen peroxide. Hydrogen peroxide is rapidly formed from superoxide by the action of superoxide dismutase. These growth factors activate the underlying vascular smooth muscle cells. The activated endothelial cells also play important roles in regulating inflammatory responses in the vessel wall by expressing cell adhesion molecules, such as vascular cell adhesion molecule 1 (VCAM-1), intercellular adhesion molecule 1 (ICAM-1), P-selectin and E-selectin, which bind circulating leukocytes to facilitate their entry into the vessel wall. Endothelial cells also produce inflammatory mediators, such as interleukins (IL) including IL-1β, IL-6, and IL-8, and also tumor necrosis factor α (TNF-α), monocyte colony stimulating factor (M-CSF), granulocyte-monocyte colony stimulating factor (GM-CSF), and monocyte chemotactic protein 1 (MCP-1), which attract inflammatory cells and modulate their functions.

Activation of endothelial cells is orchestrated by multiple transcription factors. Endothelial responses to the shear stresses imparted by the flow of blood over the luminal surface are mediated primarily by two transcription factors, Kruppel-like factor 2 (KLF2) and nuclear factor erythroid-2-related factor-2 (Nrf2). Steady laminar shear stress, as is present in non-branching segments of the vasculature, stimulates expression of KLF2 and Nrf2, which help to maintain the endothelial cells in a relatively quiescent non-activated state.[3,4] KLF2 inhibits induction of adhesion molecules by cytokines such as IL-1β, thus limiting the binding of inflammatory leukocytes to the endothelial cell surface. KLF2 also promotes the expression of nitric oxide synthase, which generates nitric oxide. As mentioned above, nitric oxide promotes vasodilation and prevents coagulation. Nrf2 promotes the expression of several genes that combat oxidative stress. With disturbed non-laminar flow, the expression of KLF2 and Nrf2 is decreased in endothelial cells, allowing the cells to acquire an activated phenotype, characterized by enhanced surface adhesion molecule expression, decreased nitric oxide generation, and increased oxidative stress. These changes promote increased coagulation, vasoconstriction, and entry of inflammatory cells into the vessel wall.

In addition to disturbed laminar flow, endothelial cells are directly activated by other stimuli including infection, mechanical injury, radiation therapy, and cytokines released from inflammatory cells and other activated vascular cells. Cytokines such as TNF-α, IL-1, and IL-4 stimulate endothelial cells to express adhesion molecules for inflammatory cells, such as VCAM-1 and ICAM-1.[5] Many of the cytokine-stimulated effects on endothelial cells are mediated by the activation of the transcription factor NF-κB.[6] In endothelial cells, NF-κB promotes the expression of adhesion molecules such as ICAM-1, VCAM-1, and E-selectin. NF-κB also stimulates the expression of cytokines and immune modulators such as GM-CSF, M-CSF, MCP-1, IL-1, IL-6, and TNF-α. In addition to being activated by cytokines such as TNF-α, NF-κB is also activated by toll-like receptors. Toll-like receptors are pattern recognition receptors that bind universal danger signals, such as structural components present on bacterial and viral surfaces.

Like endothelial cells, vascular smooth muscle cells also undergo activation.[7,8] In the normal state, vascular smooth muscle cells are considered as displaying a contractile phenotype. The cells are rich in contractile apparatus proteins such as smooth muscle α-actin, smooth muscle myosin, SM-22α, smoothelin and h-caldesmon. The primary function of contractile vascular smooth muscle cells is to maintain blood pressure and normal vascular wall structure in the setting of pulsatile cyclic strain. Vascular smooth muscle cells are responsive to various stimuli including growth factors and cytokines released by the endothelium and infiltrating leukocytes, injured cells, and alterations in cyclic strain. Smooth muscle cells express receptors for endothelial-derived growth factors. These receptors include platelet-derived growth factor receptor and CXCR4, the receptor for SDF-1. Smooth cells respond to these stimuli by a phenotypic transition from the contractile state to a synthetic or activated state. The transition from the contractile to the synthetic phenotype is accompanied by a decrease in the expression of contractile proteins, and an increase in secretory vesicles. This phenotypic switch is accompanied by an increased production of extracellular matrix components, particularly proteoglycans such as versican and biglycan (Figure 11.2). Activated smooth muscle cells also secrete increased amounts of matrix metalloproteinases (MMPs), which facilitate extracellular matrix remodeling. Activated vascular smooth muscle cells show increased expression of heterogeneous nuclear ribonucleoprotein C (hnRNP-C), a nuclear mRNA binding protein that facilitates mRNA processing and nuclear export. Activated smooth muscle cells may also acquire a pro-inflammatory phenotype, expressing inflammatory cytokines, surface adhesion molecules such as VCAM-1, and the transcription factor NF-κB. Activated smooth muscle cells may demonstrate increased

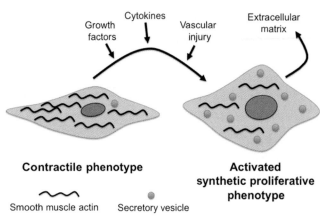

Vascular Smooth Muscle Cell Activation

FIGURE 11.2 Vascular smooth muscle cell activation. Vascular smooth muscle cells show distinct phenotypes. In the normal contractile phenotype, the cells have a spindle shape and contain large amounts of smooth muscle actin. These cells are activated by diverse stimuli including cytokines, growth factors, and vascular injury. The activated vascular smooth muscle cells may take a more polygonal shape and often contain less smooth muscle actin but an increased number of secretory granules. These activated cells have the ability to migrate and proliferate, and often secrete increased amounts of extracellular matrix material.

rates of proliferation and have increased expression of nuclear proteins that facilitate cell division such as proliferating cell nuclear antigen (PCNA), which promotes DNA synthesis by DNA polymerase.

Key activators of vascular smooth muscle cells include growth factors such as platelet-derived growth factor, which engage surface receptor tyrosine kinases. In addition, other mediators, such as angiotensin-II, activate G-protein-coupled receptors. Cell surface receptor activation stimulates intracellular protein kinase signaling cascades, including activation of mitogen-activated protein (MAP) kinases.[9,10] Hydrogen peroxide released from adjacent cells including endothelial cells also directly stimulates smooth muscle cells to proliferate, but the specifics of how this occurs are not completely understood.[11] In contrast, transforming growth factor β (TGF-β) drives smooth muscle cells towards the contractile phenotype by activating a group of transcription factors known as Smads, which up-regulate the expression of contractile proteins.

INTIMAL HYPERPLASIA

Intimal hyperplasia refers to a process in which the intima becomes thickened due to the presence of vascular smooth muscle cells and proteoglycan-rich extracellular matrix located between the endothelium and the internal elastic lamina (Figure 11.3). This pathologic change is also referred to as neointimal hyperplasia and

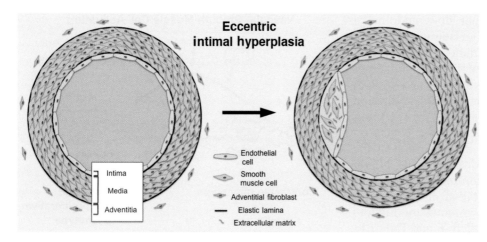

FIGURE 11.3 Formation of intimal hyperplasia. In the normal artery (left) the inner intimal layer is composed only of the endothelium and its underlying basement membrane. During the formation of intimal hyperplasia, the intima becomes expanded by the presence of smooth muscle cells and extracellular matrix between the endothelium and the internal elastic lamina. Intimal hyperplasia may form in an eccentric fashion, as shown here, or in some cases in a concentric fashion depending on the nature and duration of the stimulus.

intimal thickening, in different settings. Since this process is associated with an increase in the number of cells, it is by definition a form of hyperplasia. The formation of intimal hyperplasia is often linked with vascular cell activation. Numerous factors promote the formation of intimal hyperplasia such as vascular wall injury, aging and inflammation.[12–16] Non-laminar shear stress, particularly at branch points in the vasculature, results in a mild form of intimal hyperplasia often referred to as intimal thickening. Marked increases in blood pressure can stimulate intimal hyperplasia, such as occurs in the small arteries of the lungs in the setting of pulmonary hypertension.

Intimal hyperplasia may be either eccentric or concentric. Eccentric intimal hyperplasia often develops at branch sites due to the eccentric nature of altered shear stress at these locations. Such eccentric intimal hyperplasia induced by disturbed flow often progresses to a more concentric form with time. Vascular wall inflammation and injury may also be eccentric and stimulate the formation of eccentric intimal hyperplasia. In contrast, circumferential or concentric vascular wall injury or inflammation stimulates the formation of concentric intimal hyperplasia as occurs frequently in giant cell arteritis (see below).

The origin of the intimal smooth muscle cells in intimal hyperplasia is a topic of great interest. Originally, it was largely assumed that these cells derived from medial smooth muscle cells that migrated from the media into the intima. While medial smooth muscle cells do appear to be a source for intimal smooth muscle cells in many circumstances, it is becoming clear that in some settings, other sources may contribute to intimal smooth muscle cells, including either resident or circulating stem cells.[17,18] Models with which to study the development of intimal hyperplasia include animal models in which arterial flow is disturbed by ligating an artery, and animal models in which injury is induced to either the adventitia or to the endothelium,

using either a wire or balloon.[19] Intimal hyperplasia is also studied using ex vivo human arteries in organ culture.[20] This later model indicates that all of the cells necessary for the formation of intimal hyperplasia in this setting are present within the normal human artery wall. In these model systems, experimental perturbations that inhibit or prevent vascular smooth muscle cell activation generally prevent the development of intimal hyperplasia.

Intimal hyperplasia occurs in several distinct vascular diseases, and its severity and clinical significance vary greatly depending on the context. Nearly all human coronary artery atherosclerosis develops in the setting of pre-existing intimal hyperplasia. In humans, intimal hyperplasia forms in the coronary arteries within the first decade of life, and a particular vessel's propensity to develop intimal hyperplasia mirrors its propensity to develop atherosclerosis later in life. In this setting, the intimal hyperplasia usually develops first as eccentric lesions around branch points and then later progresses to concentric lesions. While this type of intimal hyperplasia is largely stimulated by disturbed flow, it is also enhanced by other risk factors for atherosclerosis including smoking and aging.[14] However, in this setting the intimal hyperplasia is usually mild and does not progress to significantly occlude the arterial lumen. However, this intimal hyperplasia likely serves as a precursor for the development of atherosclerosis by facilitating entrapment of LDL in the vessel wall.[21] In intimal hyperplasia the thickened intima contains large amounts of proteoglycans, particularly versican, biglycan, and perlecan.[22] Proteoglycans are highly negatively charged and avidly bind and retain positively charged LDL particles in the intima.

In other settings, intimal hyperplasia is not self-limited, and does progress to significantly narrow the lumen of the artery. In these settings the stimulus is usually more intense, and the development of the intimal hyperplasia is more rapid, occurring in weeks rather

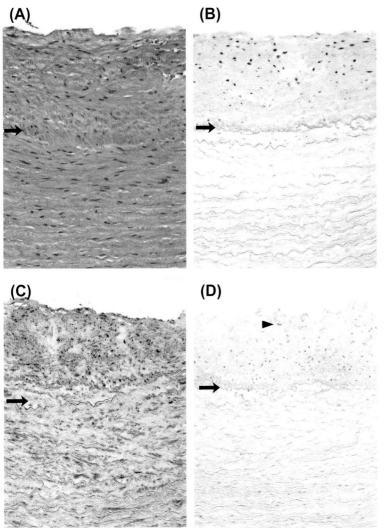

FIGURE 11.4 Activated intimal smooth muscle cells in intimal hyperplasia. Chronic intimal hyperplasia forms in the internal carotid artery due to turbulent flow caused by the carotid bifurcation. In the above images of the internal carotid artery, the arrows indicate the boundary between the intima (top) and media (bottom). The thickened intima appears to contain relatively mature smooth muscle cells on hematoxylin and eosin stain (A) and on an immunohistochemical stain for smooth muscle actin (C). However, unlike the smooth muscle cells in the media, many of the intimal smooth muscle cells show strong expression of the nuclear pre-mRNA binding protein hnRNP-C (B). Only rare CD68 expressing macrophages are present in the intima in this condition (arrowhead, D). (Reproduced with permission from Panchenko MP, Silva N, Stone JR. Upregulation of a hydrogen peroxide responsive pre-mRNA binding protein in atherosclerosis and intimal hyperplasia. *Cardiovascular Pathology* 2009;18:167–72.)

than years. One common example of this occlusive type of intimal hyperplasia is that seen after vascular injury such as balloon angioplasty or after surgical removal of an atherosclerotic plaque, a procedure referred to as endarterectomy. These procedures directly injure the intimal surface of the artery, provoking an inflammatory response and a strong smooth muscle cell proliferative response that may cause restenosis of the vessel. For this reason, coronary artery angioplasty is often now followed by placement of drug-eluting stents that secrete antiproliferative agents, such as rapamycin and paclitaxel, to prevent restenosis from intimal hyperplasia. Intimal hyperplasia is also the primary mechanism of vascular occlusion in some forms of vasculitis, such as giant cell arteritis.

The specific setting of the intimal hyperplasia will also influence the composition of the hyperplastic intima. In the rapid severe forms of intimal hyperplasia there is often an exuberant amount of extracellular matrix produced, often markedly exceeding the amount of extracellular matrix deposited in the more mild chronic forms of intimal hyperplasia. In the rapid severe forms, the intimal smooth muscle cells are actively dividing and express proliferation markers such as PCNA and Ki-67. In the mild chronic forms of intimal hyperplasia, the smooth muscle cells mostly do not express proliferation markers and are more densely packed together, often resembling the media itself. However, in the mild chronic forms of intimal hyperplasia, the intimal smooth muscle cells still possess an activated phenotype compared with the medial smooth muscle cells, as evidenced by nuclear hnRNP-C expression and enhanced extracellular matrix production (Figure 11.4).

The specific signaling pathways driving intimal hyperplasia depend on both the particular setting or model being studied and the species. For example, murine endothelial injury models not uncommonly show different results than are obtained with murine ligation models. In addition, many markers shown to be important in rodent models do not directly translate to human intimal hyperplasia. However, one model system that has been studied in some detail is the murine carotid artery ligation model.[23] In this model, intimal hyperplasia is enhanced by infusion of the vascular smooth muscle cell activator angiotensin-II. In the setting of angiotensin-II infusion, the intimal hyperplasia is largely independent of the leukocyte attractant MCP-1, suggesting that in this model, infiltrating leukocytes may not be important for the development of intimal hyperplasia. However in this model, NF-κB, TNF-α, and IL-1 have all been shown to promote the formation of intimal hyperplasia, likely in part by mediating vascular cell activation. In contrast to the murine carotid ligation model, the infiltration of inflammatory cells appears to play a more definitive role in promoting intimal hyperplasia in murine endothelial injury models.

DIABETIC VASCULOPATHY

Diabetes mellitus refers to a group of disorders characterized by elevated serum glucose levels or hyperglycemia. Serum glucose levels are regulated by the hormone insulin, which stimulates the uptake of glucose by target tissues such as skeletal muscle, adipose tissue, and the liver. Hyperglycemia results from either impaired insulin secretion from the pancreas or defective insulin action in target tissues such as skeletal muscle. In classic type 1 diabetes mellitus, there is an immune-mediated destruction of the β-cells in the pancreas. These cells are responsible for secreting insulin into the blood. In the more common adult-onset type 2 diabetes mellitus, genetic predisposition and/or obesity act to render target tissue relatively resistant to the effects of insulin. In this disorder, the pancreatic β-cells initially respond with compensatory hyperplasia and overall increased insulin production, but eventually the β-cells undergo failure, resulting in a relative insulin deficiency in the setting of insulin resistance.

The vascular complications of diabetes include both macrovascular disease and microvascular disease. The macrovascular disease is essentially accentuated atherosclerosis, which most severely impacts the heart and the peripheral arteries in the legs. Diabetics are at markedly increased risk for both myocardial infarction and peripheral vascular occlusive disease (PVOD). Involvement of cerebral vessels put diabetics at increased risk for strokes. In addition to promoting routine atherosclerosis, diabetes also results in an arteriosclerosis of smaller intraparenchymal arteries.[24] The microvascular disease of diabetes primarily affects capillaries and other small vessels in the kidneys, retina, and nerves resulting in the clinical disorders of diabetic nephropathy, diabetic retinopathy, and diabetic neuropathy respectively.[25–27]

While our understanding of the mechanisms underlying diabetic vascular disease continues to develop, most of the current evidence indicates that the complications are a result of hyperglycemia. Elevated blood glucose results in the formation of advanced glycation end products (AGEs). AGEs are created by the non-enzymatic reaction of protein amino groups with glucose-derived metabolites including glyoxal, methylglyoxal, and 3-deoxyglucosone. The vascular extracellular matrix is particularly prone to development of AGEs, which can result in crosslinking of matrix proteins with several deleterious consequences. Crosslinking of collagen renders the arteries stiffer and less elastic, altering shear stress and promoting endothelial activation and injury. AGE modification of the endothelial basement membrane impairs endothelial adhesion and promotes increased permeability. The increased permeability allows for increased serum proteins to enter into the intima. AGE-modified proteins are also relatively resistant to digestion by proteases. Since the amount of a protein at any given time is based on its rates of synthesis and degradation, this impaired degradation promotes an increase in the overall amount of vascular extracellular matrix in diabetes. The abundant AGE-modified extracellular matrix traps and retains serum proteins in the vessel wall. In large and medium-sized arteries, this relatively abundant cross-linked extracellular matrix promotes retention of LDL in the intima, spurring development of atherosclerosis. In smaller arteries, otherwise resistant to atherosclerosis, and in capillaries and other small vessels, the AGE-modified extracellular matrix binds other plasma proteins including albumin, which have entered the intima in relatively large amounts due to the increased endothelial permeability. The binding of plasma proteins to the extracellular matrix causes the vessel wall to thicken and to acquire a glassy or hyalinized appearance (Figure 11.5). Such diabetic vasculopathy can cause severe stenosis of the vessel and end-organ ischemia. The diseased vessels are also more fragile and are prone to rupture. In some vascular beds such as the retina, diabetic vascuopathy results in the death of pericytes around the capillaries, and the formation of microaneurysms. In the retina, the activated endothelial cells show increased proliferation resulting in proliferative diabetic retinopathy.

In addition to the ECM, other extracellular proteins are modified by AGEs. Some of these modified proteins will bind to cellular surface receptors for AGEs, such as RAGE, on endothelial cells.[28] Activation of RAGE results in endothelial activation, with increased secretion of reactive oxygen species, growth factors, and cytokines. These agents released from the endothelial cells stimulate vascular smooth muscle cell activation, promoting an increased production of extracellular matrix. Small vessels afflicted with diabetic vasculopathy contain

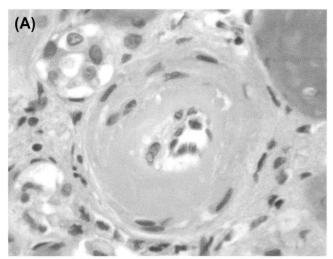

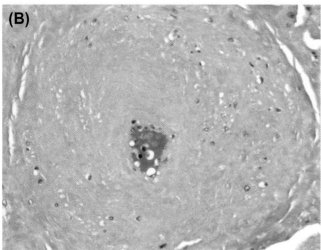

FIGURE 11.5 Diabetic vasculopathy. Shown are images of an arteriole in the kidney **(A)** and a small penetrating artery in the myocardium of the heart **(B)** in a patient with severe and long-standing diabetes mellitus. In both cases the vessel wall is thickened, and has a hyalinized amorphous appearance due to the accumulation of serum proteins and extracellular matrix.

increased amounts of collagen, fibronectin, and laminin. Activation of RAGE also leads to down-regulation of glyoxalase-1 (Glo1), an enzyme that metabolizes methylglyoxal. Thus, the interaction of AGE-modified proteins with RAGE fosters the generation of even more methylglyoxal and more AGEs, creating a vicious cycle leading to vascular compromise.

AMYLOID VASCULOPATHY

The amyloidoses are a group of disorders characterized by the formation of a specific type of protein deposit in tissues.[29] In amyloid deposits, there is a specific culprit protein that adopts an abnormal protein fold with an extended β-sheet conformation (Figure 11.6). These long β-sheets will wrap around each other in groups of

3–5 to form large insoluble fibrils. Amyloid fibrils can be deposited in almost any tissue, but the vasculature is particularly susceptible.

Currently there are at least 27 distinct proteins that are known to form amyloid.[30] The extended β-sheet structure of amyloid-fibrils allows Congo red dye to bind to the deposits in an orderly fashion, and in so doing enables the amyloid deposits to display green birefringence with plane-polarized light. In addition to the misfolded culprit protein, amyloid deposits also contain non-specific proteins including serum amyloid P, apolipoprotein E, and heparan sulfate proteoglycans.[31] The proteoglycans in amyloid deposits result in these deposits staining with sulfated alcian blue. Serum amyloid P is often utilized as a general tissue immunohistochemical marker for all amyloid deposits and in nuclear medicine as a radiolabeled probe for assessing amyloid deposits throughout the whole body.[32]

One mechanism leading to amyloidosis is the overproduction of a protein with a tendency to misfold and form these long β-sheet structures. Such proteins are referred to as being amyloidogenic. A common example of this process is the overproduction of immunoglobulin light chain by plasma cell neoplasms or myelomas. Plasma cell neoplasms are monoclonal and secrete a specific immunoglobulin. In some cases the light chain in the immunoglobulin is amyloidogenic, and the elevated levels of the amyloidogenic light chain released by the neoplasm result in amyloid deposition.[33–35] A similar mechanism is responsible for amyloidosis due to serum amyloid A. Serum amyloid A is an amyloidogenic protein that is up-regulated during inflammation. Patients with long-standing inflammatory diseases, such as rheumatoid arthritis, have chronically elevated levels of serum amyloid A, which can lead to amyloid formation.[36]

Another mechanism leading to the formation of amyloid deposits is the presence of a genetic alteration or polymorphism that enhances the amyloidogenic potential of a protein. Such genetic forms of amyloid typically display an autosomal dominant pattern of inheritance. A relatively common form of inherited amyloidosis results from mutations in the transthyretin gene.[37,38] Over 80 such admyloidogenic transthyretin mutations are known to exist.[39] In addition to transthyretin, mutations in other genes can also result in hereditary amyloidosis, including mutations in apolipoprotein A-I and apolipoprotein A-II.[40,41]

In very rare cases, amyloid can result from being 'infected' by a small protein already in the extended β-sheet structure. Such infectious protein particles are referred to as prions. The deposition of the environmentally derived amyloid fibril in the tissue causes the exposed patient's endogenous corresponding protein to misfold and deposit as amyloid along with

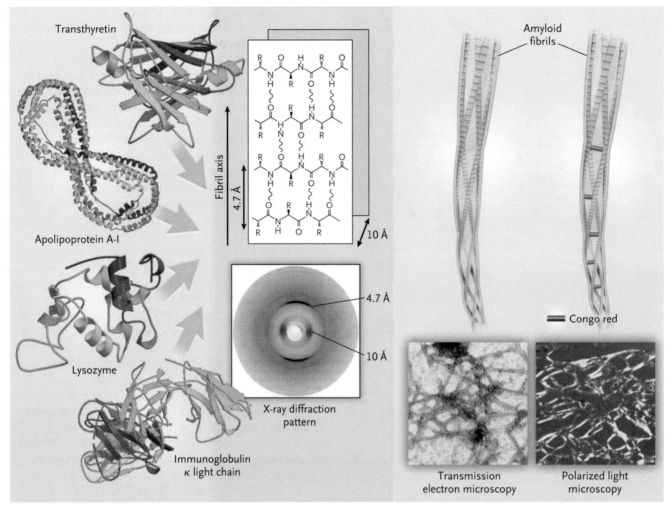

FIGURE 11.6 Mechanism of amyloid formation. Shown on the left are the ribbon diagrams of the three-dimensional structures of the amyloidogenic proteins transthyretin, apolipoprotein A-I, lysozyme, and immunoglobulin κ light chain. Each of these proteins normally contains α-helicies (spirals) and/or β-sheets (flat arrows). However, portions of each of these proteins may refold to form a complex extended β-sheet structure (middle), which by X-ray diffraction can show an interstrand distance of 4.7 Å and an intersheet distance of ~10 Å. Contiguous extended β-sheet peptides compose a protofilament. Four to six protofilaments wrap around each other to form an amyloid fibril (right). Amyloid fibrils are non-branching structures on electron microscopy with a diameter of 7.5–10 nm. The repeating structure of the β-sheet-containing fibrils enables the ordered binding of Congo red dye, which causes the amyloid fibrils to display apple-green birefringence when viewed by light microscopy with plane-polarized light. (Reproduced with permission from Merlini G, Bellotti V. Molecular mechanisms of amyloidosis. *New England Journal of Medicine* 2003;349:583–96.)

the externally derived amyloid fibril. Such prion diseases are predominantly neurodegenerative disorders such as mad cow disease and Creutzfeldt-Jakob disease. However, it should be noted that the vast majority of amyloid-related diseases are not believed to be infectious.

In many circumstances it remains unknown as to why the amyloid deposits occur. One common example of this situation is the formation of amyloid due to wild-type transthyretin.[42] This results in the relatively common disorder known as senile systemic amyloidosis, which as the name implies, is most often seen in older patients. This last type of amyloid is difficult to treat. While many forms of amyloid are essentially treated by reducing the level of the amyloidogenic protein, for senile systemic amyloidosis, efforts are focused at trying to prevent or reverse the misfolding of the protein.

Amyloid deposition in tissues such as the heart can lead to organ dysfunction, such as heart failure. Blood vessels are particularly susceptible to amyloid deposition (Figure 11.7). Once deposited in blood vessels, amyloid can result in vascular obstruction and impaired vasoreactivity leading to end-organ ischemia. In addition, amyloid in small blood vessels can increase vascular fragility leading to hemorrhage. For example, in the brain cerebral amyloid angiopathy can result in hemorrhagic stroke, particularly in older patients.

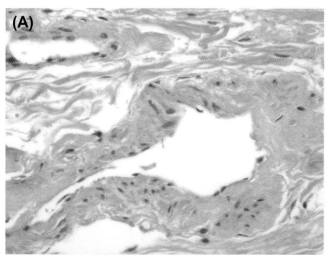

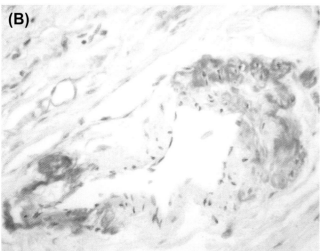

FIGURE 11.7 Amyloid vasculopathy. Depicted are portions of a small vein in the adventitia of the aorta in a patient with immunoglobulin light chain amyloidosis. On the hematoxylin-and-eosin-stained section, the wall of the vessel shows areas with an amorphous appearance **(A)**. These amorphous areas stain salmon-pink on Congo red stain indicating the presence of amyloid **(B)**.

SMALL VESSEL VASCULITIS

Vasculitis is inflammation of vessel walls. Different types of vasculitis preferentially affect different sizes of blood vessels (Figure 11.8). Small-vessel vasculitis predominantly affects small intraparenchymal arteries, arterioles, capillaries, and venules, although medium-sized arteries and veins may be affected.[43] Small-vessel vasculitis can be caused by the direct invasion of small vessel walls by infectious pathogens, such as rickettsia (e.g. Rocky Mountain spotted fever) and bacteria (e.g. Neisserial sepsis) that initiates an inflammatory response within the vessel wall. Most small-vessel vasculitis that is not caused by direct vessel wall infection falls into two immunopathologic categories: immune complex vasculitis and antineutrophil cytoplasmic antibody (ANCA)-associated vasculitis (Table 11.1).[43]

Immune Complex Small-Vessel Vasculitis

Immune complex small-vessel vasculitis is characterized by moderate to marked vessel wall deposits of immunoglobulin and/or complement components. This is in contradistinction to ANCA-associated vasculitis, which is characterized by few or no immune deposits in vessel walls.[43] Immune complex small-vessel vasculitis predominantly affects capillaries, venules, and arterioles. Most variants of immune complex small-vessel vasculitis only rarely affect arteries. An exception is hepatitis-B-associated vasculitis, which often affects arteries as well as venules and arterioles. Glomerulonephritis is a frequent component of immune complex small-vessel vasculitis.

There are many variants of immune complex small-vessel vasculitis (Table 11.1) with different clinical and pathologic phenotypes, different etiologies and different pathogenic mechanisms. All of them share the accumulation of substantial amounts of immunoglobulin and complement components in the walls of affected vessels, and share some of the same final common inflammatory pathways.[44] Variants include, but are not limited to, cryoglobulinemic vasculitis, IgA vasculitis (Henoch-Schönlein purpura), hypocomplementemic urticarial vasculitis (anti-C1q vasculitis), vasculitis secondary to antigen exposure (e.g. serum sickness, drug hypersensitivity), vasculitis secondary to infection (e.g. hepatitis B, hepatitis C), and vasculitis secondary to autoimmune disease (e.g. lupus, rheumatoid arthritis). Although it has a somewhat distinctive pathogenic mechanism that involves direct binding of autoantibodies to constituent antigens in vessel walls resulting in in situ immune complex formation, antiglomerular basement membrane (anti-GBM) disease can also be considered an immune-complex-mediated small-vessel vasculitis.[43]

By definition, all immune-complex vasculitis variants have substantial vessel wall localization of immunoglobulin and complement as constituents of immune complexes, but the origin, composition and mechanism for initiating inflammation vary. A common theme is that these immune deposits activate multiple inflammatory pathways, and recruit and activate leukocytes. In most if not all non-infectious small-vessel vasculitis, neutrophils are the predominant initial inflammatory cells, although these are quickly replaced by monocytes, macrophages, and eventually T-cells as the inflammation evolves. Immune complexes are able to activate complement through not only the classic pathway but also the alternative and lectin pathways, and the Fc regions of antibodies that are bound to antigens can recruit and activate leukocytes via Fc receptor engagement.[44,45] The presence of immune deposits in vessel walls activates endothelial cells and induces

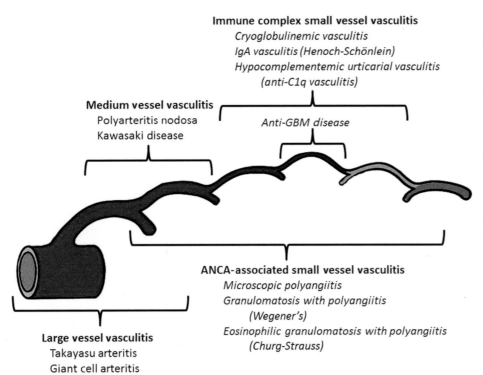

Immune complex small vessel vasculitis
Cryoglobulinemic vasculitis
IgA vasculitis (Henoch-Schönlein)
Hypocomplementemic urticarial vasculitis
(anti-C1q vasculitis)

Medium vessel vasculitis
Polyarteritis nodosa
Kawasaki disease

Anti-GBM disease

ANCA-associated small vessel vasculitis
Microscopic polyangiitis
Granulomatosis with polyangiitis
(Wegener's)
Eosinophilic granulomatosis with polyangiitis
(Churg-Strauss)

Large vessel vasculitis
Takayasu arteritis
Giant cell arteritis

FIGURE 11.8 Diagram showing the predominant distribution of vessel involvement by large-vessel vasculitis, medium-vessel vasculitis, and small-vessel vasculitis. There is substantial overlap with respect to arterial involvement, and an important concept is that all three major categories of vasculitis can affect arteries. Large-vessel vasculitis and medium-vessel vasculitis affect only arteries, whereas small-vessel vasculitis can affect any type of vessel, with a predilection for venules and capillaries. Most forms of immune-complex small-vessel vasculitis affect almost exclusively capillaries and venules, whereas ANCA-associated vasculitis more often affects small arteries as well as capillaries and venules. (Reproduced with permission from Jennette JC, Falk RJ, Bacon PA, et al. 2012 Revised international Chapel Hill Consensus Conference nomenclature of vasculitides. *Arthritis Rheum* 2013;65:1–11).

them to increase the surface display of adhesion molecules for leukocytes, which, along with the increased display of adhesion molecule ligands on leukocytes results in the margination, adhesion, and transmigration of leukocytes into vessel walls.[44,46–49] Table 11.2 lists some of the leukocyte receptors and endothelial

TABLE 11.1   Non-Infectious Small-Vessel Vasculitis (SVV)

IMMUNE COMPLEX SVV

Anti-GBM (anti-glomerular basement membrane) disease

Cryoglobulinemic vasculitis (CV)

IgA vasculitis (Henoch-Schönlein) (IgAV)

Hypocomplementemic urticarial vasculitis (Anti-C1q vasculitis)

Secondary to antigen exposure (e.g. serum sickness, drug hypersensitivity)

Secondary to infection (e.g. hepatitis B, hepatitis C)

Secondary to autoimmune disease (e.g. lupus, rheumatoid arthritis)

Paraneoplastic vasculitis

ANCA-ASSOCIATED VASCULITIS (AAV)

Microscopic polyangiitis (MPA)

Granulomatosis with polyangiitis (Wegener's) (GPA)

Eosinophilic granulomatosis with polyangiitis (Churg-Strauss) (EGPA)

Organ-limited ANCA-associated vasculitis (e.g. renal-limited, lung-limited)

ligands that are involved in recruiting inflammatory cells into the walls of vessels that have immune complex deposits.[44]

In addition to complement activation with C5a generation, many other chemoattractants recruit leukocytes to sites of immune-complex vasculitis including leukotriene B4, platelet-activating factor (PAF), and CXC chemokines (e.g. IL-8, CXCL8), platelet factor 4 (PF4, CXCL4), GRO1 oncogene (GROα, CXCL1), and stromal cell-derived factor-1 (SDF-1, CXCL12). The transition of neutrophil-rich vasculitis to vasculitis with a predominance of lymphocytes and monocytes is mediated by an adjustment in the chemokine microenvironment to a predominance of chemoattractants that preferentially recruit monocytes and T lymphocytes, such as macrophage inflammatory protein-1α (MIP-1α, CCL3), monocyte chemotactic protein-1 (MCP-1, CCL2), and regulated upon activation, normal T-cell expressed, and secreted (RANTES, CCL5).[44,47]

Once at the site of vasculitis, leukocytes release destructive factors such as reactive oxygen species and lytic and toxic enzymes that injure vessel walls and engender more inflammation and eventually scarring. Leukocytes and vessel wall cells undergo apoptosis and necrosis with karyorrhexis and generation of nuclear fragments (Figure 11.9). This vessel wall necrosis as well as matrix lysis allows plasma to spill into the vessel wall and perivascular tissue where the coagulation system is activated by thrombogenic surfaces and tissue factor. This results in the characteristic fibrinoid necrosis that is seen in the acute phase of many forms of vasculitis (Figure 11.9).[50]

TABLE 11.2   Adhesion Molecules that Mediate Leukocyte Endothelial Transmigration and Tissue Inflammation at Sites of Vasculitis

Leukocyte Event	Leukocyte Receptor	Endothelial Ligand
Rolling	L-selectin (CD62L)	PSGL-1 (CD162; on leukocytes not endothelium)
	PSGL-1 (CD162)	P-selectin (CD62P)
	PSGL-1 (CD162), ESL-1, CD44	E-selectin (CD62E)
	LFA-1 (CD11a/CD18) (β_2 integrin)	ICAM-1 (CD54)
	VLA-4 (CD49d/CD29) (β_1 integrin)	VCAM-1 (CD106)
Arrest	LFA-1 (CD11a/CD18) (β_2 integrin)	ICAM-1 (CD54)
	VLA-4 (CD49d/CD29) (β_1 integrin)	VCAM-1 (CD106)
	Mac-1 (CD11b/CD18) (β_2 integrin)	ICAM-1 (CD54)
Crawling	Mac-1 (CD11b/CD18) (β_2 integrin)	ICAM-1 (CD54)
	LFA-1 (CD11a/CD18) (β_2 integrin)	ICAM-1 (CD54)
Transmigration	Mac-1 (CD11b/CD18) (β_2 integrin)	ICAM-1 (CD54)
	LFA-1 (CD11a/CD18) (β_2 integrin)	JAM-A (CD321)
	VLA-4 (CD49d/CD29) (β_1 integrin)	JAM-B (CD322)
	Mac-1 (CD11b/CD18) (β_2 integrin)	JAM-C
	PECAM-1 (CD31)	PECAM-1 (CD31)
	MIC2 (CD99)	MIC2 (CD99)
TISSUE LIGAND		
Tissue infiltration	Mac-1 (CD11b/CD18) (β_2 integrin)	Fibrin, fibronectin
	PECAM-1 (CD31)	Basement membrane
	$\alpha_6\beta_1$ integrin (β_1 integrin; CD151)	Laminin
	VLA-4 (CD49d/CD29) (β_1 integrin)	Fibronectin
	VLA-5 (CD49e/CD29) (β_1 integrin)	Fibronectin
	CEACAM (CD66)	Fibronectin

PSGL-1, P-selectin glycoprotein ligand 1; LFA-1, lymphocyte function-associated antigen 1; ICAM-1, intercellular adhesion molecule 1; VLA, very late antigen; VCAM-1, vascular cell adhesion molecule 1; Mac-1, macrophage-1 antigen or complement receptor 3 (CR3); JAM, junctional adhesion molecule; PECAM-1, platelet endothelial cell adhesion molecule 1; MIC2, transmembrane glycoprotein p30/32[mic2]; CEACAM, carcinoembryonic antigen-related cell adhesion molecule. (Revised from Homeister J, Jennette JC, Falk RJ. Immunologic Mechanisms of Vasculitis in the Kidney: *Physiology and Pathophysiology,* 5th Ed., Robert J. Alpern RJ and Steven C. Hebert SC, Elsevier, 2013, Chapter 83, 2817–2846.)

This lytic component of vasculitis is mediated by enzymes released by leukocyte degranulation and synthesis by tissue cells.[44,51] Lytic enzymes derived from activated leukocytes include serine proteinases, such as elastase, proteinase-3, and cathepsin G.[51] Matrix metalloproteinases (MMPs), such as collagenase and gelatinase, are also produced by activated leukocytes and tissue cells.[52] In homeostatic conditions, serine proteinases are regulated by serine proteinase inhibitors (serpins) and MMPs are regulated by tissue inhibitors of MMPs (TIMPs).[53] However, in an inflammatory microenvironment, the release and activation of lytic enzymes overwhelms the regulatory mechanisms. Activated neutrophils are particularly effective at causing necrotizing inflammatory injury, but even after the neutrophil-rich acute phase of vasculitis has subsided, the subsequent macrophage and T-lymphocyte-rich inflammation can continue to cause vessel wall destruction through similar mechanisms. For example, activated T-lymphocytes produce MMPs and release granzymes that injure cells and degrade matrix.[54]

All of the generic inflammatory events reviewed thus far occur in all forms of immune-complex small-vessel vasculitis, and many also occur in ANCA-mediated vasculitis as well. Some of the pathogenic events that are unique to specific variants of immune-complex vasculitis will now be reviewed.

Cryoglobulinemic Vasculitis

Cryoglobulinemic vasculitis has cryoglobulin immune deposits affecting small vessels (predominantly capillaries, venules, or arterioles).[43] Skin, glomeruli, and

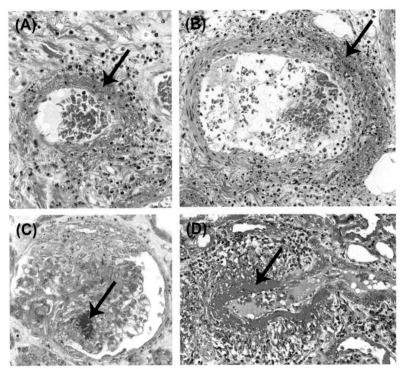

FIGURE 11.9 Photomicrographs of small vessel vasculitis (microscopic polyangiitis) affecting a venule in the small intestine **(A)** (H&E stain), small artery in the small intestine **(B)** (H&E stain), glomerulus **(C)** (Masson trichrome stain) and interlobular renal artery **(D)** (Masson trichrome stain). Panels A and B show segmental inflammation with transmural infiltration by leukocytes including numerous neutrophils. There is extensive leukocytoclasia with nuclear karyorrhectic fragments admixed with the intact leukocytes. There is localized fibrinoid necrosis (arrows) that is deeply acidophilic with the H&E staining, and has a finely fibrillary texture. Masson trichrome staining (panels C and D) shows segmental fibrinoid necrosis as irregular red (fuchsinophilic) material (arrows). The glomerulus in panel C has a cellular reaction in Bowman's space (crescent formation) above the inflamed glomerulus.

peripheral nerves are often involved. Cryoglobulinemic glomerulonephritis typically has glomerular hypercellularity caused by influx of macrophages, immune deposits in capillary walls, and large aggregates of immune complexes in capillary lumens ('hyaline thrombi') (Figure 11.10).

Cryoglobulins are aggregates of immunoglobulin that precipitate from serum at reduced temperatures.[55–57] There are three types of cryoglobulins. Type I

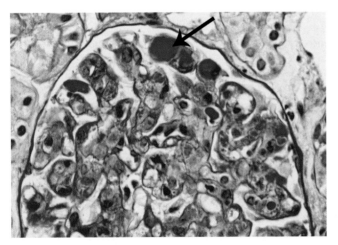

FIGURE 11.10 PAS-stained glomerulus from a renal biopsy of a patient with hepatitis-C-virus-induced cryoglobulinemic vasculitis demonstrating glomerular hypercellularity predominantly caused by influx of monocytes/macrophages, as well as thick capillary walls and PAS-positive intraluminal deposits (hyaline thrombi) that represent extensive accumulation of cryoglobulin immune complex deposits in the glomerulus.

cryoglobulins are composed of monoclonal immunoglobulin, usually IgM. Type II cryoglobulins, which are the most common type in patients with cryoglobulinemic vasculitis, usually are composed of complexes between polyclonal IgG and monoclonal IgM anti-IgG. The IgM anti-IgG binds to intact IgG, Fab fragments and Fc fragments, thus forming immune complexes.[56] Type III cryoglobulins have polyclonal IgG and polyclonal IgM in the cryoglobulin immune complexes. Types II and III are strongly associated with hepatitis C virus infection and type III is associated with autoimmune disease (e.g. systemic lupus erythematosus and rheumatoid arthritis).

The immunopathogenetic mechanism initiated by HCV may initially result in the production of polyclonal IgM lacking rheumatoid factor activity that evolves through somatic mutations to develop monoclonal IgM with anti-IgG reactivity.[56] This may result from HCV-activated B-cell proliferation with hypermutation.[56] Immune complexes in cryoglobulinemic vasculitis can contain non-enveloped nucleocapsid proteins and whole HCV virions.[56] The efficient engagement of complement protein by cryoglobulins is an important pathogenetic mechanism in the induction of vascular inflammation. Cryoglobulins in the circulation and vessels induce classical pathway complement activation resulting in the typical serologic finding of low or undetectable serum C4 but normal or near-normal C3.[57] Cryoglobulin-induced vascular inflammation is characterized by conspicuous accumulation of activated macrophages with up-regulated cytokine and chemokine synthesis.[58]

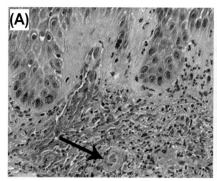
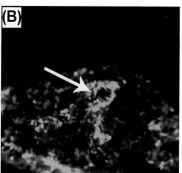

FIGURE 11.11 Photomicrographs of a skin biopsy from a patient with IgA vasculitis (Henoch-Schönlein purpura). **(A)** The panel on the left shows the light microscopic features in the upper dermis with a mixture of hemorrhage and leukocyte infiltration. There is extensive leukocytoclasia with nuclear karyorrhectic fragments admixed with the intact leukocytes, including numerous neutrophils. The small vessels (predominantly venules) involved by this leukocytoclastic angiitis are difficult to identify, but the arrow points to one that has mural fibrinoid necrosis. **(B)** The panel on the right shows the immunofluorescence microscopy finding of granular vessel wall staining with an antibody specific for IgA (arrow). An antibody for C3 also showed intense staining, but anti-IgG and anti-IgM produced only very weak staining.

IgA Vasculitis

IgA vasculitis (Henoch-Schönlein purpura) has IgA1-dominant immune deposits in vessel walls and glomeruli.[43] IgA vasculitis affects predominantly capillaries, venules, or arterioles (Figure 11.11). It often involves the skin and gastrointestinal tract, and frequently causes arthritis. Glomerulonephritis is a frequent component, and is pathologically indistinguishable from IgA nephropathy. IgA vasculitis is the most common form of systemic vasculitis in childhood, but can occur at any age.

IgA vasculitis and IgA nephropathy probably share a similar immunopathogenic mechanism that involves abnormally glycosylated IgA1.[59–62] Patients with IgA vasculitis and IgA nephropathy have increased serum levels of IgA1 with galactose-deficient hinge regions, and the vascular immune complexes contain this abnormal IgA1 but little or no IgA2. Abnormally glycosylated IgA1 is prone to self-aggregation and has an affinity to deposit in the microvasculature where the IgA1 complexes can activate complement via the alternative pathway and mediate inflammation. Vessel wall immune deposits in IgA vasculitis and IgA nephropathy contain not only IgA1 but also C3, C3c, C3d, and variable amounts of IgG and IgM, but little or no C1q or C4. Although not as well documented, there is evidence that the abnormally glycosylated IgA1 induces an autoimmune response with the development of anti-IgA antibodies, that can participate in immune complex formation in a fashion somewhat analogous to the anti-IgG antibodies in cryoglobulinemic vasculitis.[62]

Hypocomplementemic Urticarial Vasculitis

Hypocomplementemic urticarial vasculitis (anti-C1q vasculitis) is characterized by urticaria and hypocomplementemia, and affects predominantly capillaries, venules, and arterioles.[43,63–65] It is associated with anti-C1q antibodies, which may be involved in the pathogenesis. Glomerulonephritis, arthritis, obstructive pulmonary disease, and ocular inflammation are common. Some but not all patients with hypocomplementemic urticarial vasculitis also have systemic lupus erythematosus and thus have additional immunopathologic features.

Patients with hypocomplementemic urticarial vasculitis have reduced plasma levels of C1q, C2, C3, and C4 consistent with classic pathway complement activation.[63–65] They also have circulating anti-C1q antibodies, which may be involved in immune complex formation and the pathogenesis of vascular inflammation, but this has not been confirmed. The immune deposits at sites of vasculitis and glomerulonephritis contain C3, C1q, and IgG, which is consistent with complement-activating immune complexes containing C1q and anti-C1q; however, this has not been confirmed.

Secondary Immune-Complex Small-Vessel Vasculitis

Secondary immune-complex small-vessel vasculitis can result from an appropriate, hypersensitivity or autoimmune response to a circulating exogenous or endogenous antigen. For example, pathogenic immune complexes can result from immune complexes containing antigens from infectious pathogens (e.g. streptococci, staphylococci, hepatitis B virus, hepatitis C virus), therapeutic agents (e.g. drugs, monoclonal antibodies, genetically engineered biologically active molecules), other ingested agents (contaminated cocaine, herbal products), tumor neoantigens, and autoantigens (e.g. in systemic lupus erythematosus).

Probably the first description of immune-complex-mediated vasculitis was by Arnold Rich in the context of serum sickness and hypersensitivity reactions to drugs.[66] The ability of immune complexes containing antigens derived from infectious pathogens is best documented with hepatitis C and hepatitis B infections, but has been reported with many other infections.[55,56,67,68] Vasculitis as a component of a paraneoplastic syndrome is well documented and hypothesized to be caused by immune complexes containing tumor neoantigens, but this pathogenic mechanism is not well documented.[69,70] Vasculitis with vascular immune complex deposits occurs in a variety of systemic autoimmune diseases including systemic lupus erythematosus.[71] However, overt immune-complex vasculitis is surprisingly uncommon in lupus given the extent of immune complex deposition that occurs in vessels throughout the body in this disease. This suggests that vessel wall accumulation of immune complexes does not necessarily induce overt inflammatory injury without other synergistic events.

Anti-GBM disease is a form of vasculitis that affects glomerular capillaries and pulmonary capillaries.[43] It is mediated by autoantibodies directed against an epitope in the type 4 collagen vascular basement membranes of glomerular capillaries and alveolar capillaries.[72] Lung involvement causes pulmonary hemorrhage, and renal involvement causes glomerulonephritis with necrosis and crescents. Anti-GBM antibodies form immune complexes in situ, with resultant induction of inflammation that disrupts the capillaries. There is evidence that the autoimmune response requires a conformational change in the quaternary structure of the collagen molecule.[72] The in situ immune complexes induce inflammation by Fc receptor recognition by leukocytes and complement activation, although the former is more important in animal models.[73–75]

ANCA-Associated Vasculitis

ANCA-associated vasculitis (AAV) is necrotizing vasculitis, with few or no immune deposits, predominantly affecting small vessels, i.e., capillaries, venules, arterioles, and small arteries.[43] This paucity of immunoglobulin in vessel walls distinguishes AAV from immune-complex vasculitis, and suggests that these two immunopathologically distinct classes of vasculitis are also pathogenetically distinct. The major autoantigen specificities for ANCA are proteinase 3 (PR3) and myeloperoxidase (MPO). AAV is usually associated with MPO-ANCA or PR3-ANCA although a minority of patients are negative by current clinical laboratory assays. However, evidence is emerging that some if not all of these ANCA-negative AAV patients have ANCA with very restricted epitope specificity that can be detected using highly sensitive assay methods.[76]

AAV manifests clinically and pathologically as four clinicopathologic syndromes (Table 11.1): microscopic polyangiitis (MPA), granulomatosis with polyangiitis (Wegener's) (GPA), eosinophilic granulomatosis with polyangiitis (Churg-Strauss) (EGPA), and organ-limited ANCA-associated vasculitis (e.g. renal-limited, lung-limited AAV).[43] Table 11.3 shows the approximate frequencies of PR3-ANCA and MPO-ANCA in the different clinicopathologic phenotypes.

TABLE 11.3 Frequency of MPO-ANCA and PR3-ANCA in the Clinicopathologic Phenotypes of ANCA-Associated vasculitis. These are approximate frequencies for ANCA in North America and Europe. The frequencies vary among cohorts, and the relative proportion of PR3-ANCA-positive patients to MPO-ANCA in all phenotypes is higher in more northern locations and lower in more southern locations. In Asia, MPO-ANCA predominates in all phenotypes.

AAV Phenotype	ANCA Serology		
	PR3-ANCA	MPO-ANCA	ANCA-negative
MPA	70%	25%	5%
GPA	40%	50%	10%
EGPA	5%	40%	55%*
Renal-limited	20%	70%	10%

*The frequency of ANCA in EGPA patients with glomerulonephritis is >75%.

MPA is necrotizing vasculitis with few or no immune deposits predominantly affecting small vessels (i.e., capillaries, venules, or arterioles) (Figure 11.9).[43] Necrotizing arteritis involving small and medium-sized arteries may be present. Necrotizing glomerulonephritis is very common (Figure 11.9). Pulmonary capillaritis often occurs (Figure 11.12). Granulomatous inflammation is absent, which distinguishes MPA from GPA and EGPA.

GPA is characterized by necrotizing granulomatous inflammation that usually involves the upper and lower respiratory tract (Figure 11.12), and necrotizing vasculitis affecting predominantly small to medium-sized vessels (e.g., capillaries, venules, arterioles, arteries, and veins). Necrotizing glomerulonephritis is common, although less common than in MPA and more common than in EGPA.

EGPA is characterized by eosinophil-rich and necrotizing granulomatous inflammation often involving the respiratory tract, and necrotizing vasculitis predominantly affecting small to medium-sized vessels, and associated with asthma and eosinophilia. Many patients have little or no evidence for systemic vasculitis, especially during the initial phase of this disease that may be dominated by eosinophilic organ inflammation, such as eosinophilic pneumonia and eosinophilic gastroenteritis. ANCA are more frequent when glomerulonephritis or other overt evidence for vasculitis is present.

The pauci-immune necrotizing and crescentic glomerulonephritis (Figure 11.9) that occurs in MPA, GPA, and EGPA is indistinguishable among these systemic phenotypes, and also occurs as a renal-limited pauci-immune necrotizing and crescentic glomerulonephritis.

Pathogenesis of ANCA-Associated Vasculitis

The paucity of immunoglobulin in vessel walls in ANCA-associated vasculitis suggests that the pathogenesis

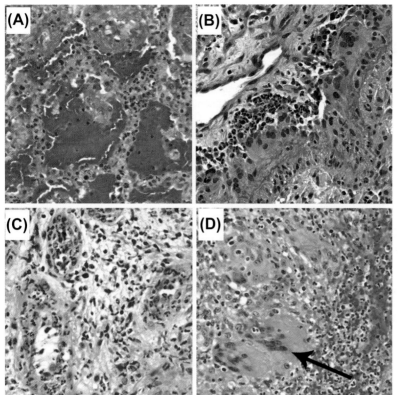

FIGURE 11.12 Photomicrographs of respiratory tract involvement by ANCA-associated vasculitis. Panel A shows pulmonary alveolar capillaritis in a patient with MPA with numerous neutrophils in the alveolar septa and massive hemorrhage into the air space. Panel B shows a small pulmonary artery in a patient with EGPA with transmural inflammation including numerous eosinophils in an expanded intima in the middle of the photomicrograph. Panel C shows nasal mucosa in a patient with GPA with numerous neutrophils and a few eosinophils infiltrating the walls of venules and the interstitial tissue. Panel D shows necrotizing granulomatous pulmonary inflammation in a patient with GPA with a hemorrhagic neutrophil-rich necrotic zone on the right and adjacent granulomatous inflammation on the left including multinucleated giant cells (arrow). Hematoxylin and eosin stains.

is different from that of immune-complex vasculitis. There is compelling evidence that ANCA-associated vasculitis is mediated primarily by direct activation of neutrophils and monocytes by ANCA with resultant inflammation in vessel walls, as well as activation of neutrophils and monocytes in extravascular tissue with resultant extravascular inflammation.[77–79] However, much of the evidence for pathogenic ANCA-induced events has come from the study of animal models of MPO-ANCA disease.[79] As yet, there are no widely accepted models of PR3-ANCA disease.

There is substantial evidence that ANCA IgG is able to activate neutrophils (and monocytes) both by Fab'2 binding to PR3 or MPO at the surface of the cells,[80] and by FcR engagement by ANCA complexed with antigen on the surface or in the microenvironment of neutrophils.[81] This is facilitated by priming of neutrophils, for example with proinflammatory cytokines (e.g. TNF)[82] or C5a,[83] that releases small amounts of PR3 and MPO to the cell surface where it can interact with ANCA. ANCA-activated neutrophils release factors that activate the alternative complement pathway, which amplifies the acute inflammatory process by attracting and activating more neutrophils (Figure 11.13).[83,84] C5a generation is particularly important because C5a is not only a strong chemoattractant for neutrophils but also primes neutrophils for activation by ANCA (Figure 11.13). Because alternative

pathway complement activation generates C3b, which is a component of the C3 convertase (i.e. C3bBbP) of the alternative pathway, an amplification loop is established that augments the inflammation (Figure 11.13). Once activated by ANCA, and juxtaposed to endothelial cells by adhesion molecules, neutrophils release toxic factors such as reactive oxygen species and lytic enzymes that destroy endothelial cells[85,86] and other vessel wall components resulting in necrotizing inflammation (Figure 11.13). Hypothetically, these same events occur at sites of extravascular necrotizing inflammation where primed neutrophils and monocytes are in contact with interstitial fluid ANCA causing them to undergo full activation.[77] This extravascular disease would require a synergistic inflammatory process that would position increased numbers of neutrophils in the extravascular tissue, such as a concurrent infection or allergic process.

The pathogenic events in ANCA-mediated vasculitis involve many of the same cytokines, chemokines, adhesion molecules, oxygen radicals, and proteases that were described earlier in the discussion of the pathogenesis of immune-complex-mediated vasculitis. However, ANCA-induced vascular inflammation is in general more aggressive than immune-complex-mediated vascular inflammation, causing more extensive vascular and perivascular inflammation and necrosis.

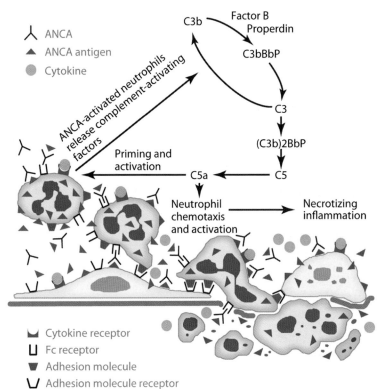

ANCA-activated neutrophils release complement-activating factors

- Ⅎ ANCA
- ▲ ANCA antigen
- ● Cytokine

- Cytokine receptor
- Fc receptor
- Adhesion molecule
- Adhesion molecule receptor

FIGURE 11.13 Diagram depicting putative pathogenic events in ANCA-associated vasculitis. Beginning on the left, circulating neutrophils are primed by cytokines to release small amounts of ANCA antigens (e.g. MPO and PR3) at their surface and into the microenvironment where it can interact with ANCA to activate the neutrophils. ANCA-activated neutrophils release factors that activate the alternative complement pathway, which generates C5a that amplifies the inflammation by attracting more neutrophils and priming the neutrophils to be activated by ANCA. The activated neutrophils adhere to, penetrate and destroy vessel walls. (Reproduced with permission from Jennette JC, Falk RJ, Hu P, Xiao H. Pathogenesis of anti-neutrophil cytoplasmic autoantibody associated small vessel vasculitis. *Annu Rev Pathol Mech Dis* 2013;8:139–60.)

KAWASAKI DISEASE

Kawasaki disease is an acute disease associated with fever, lymphadenopathy, and in some patients a necrotizing vasculitis of medium-sized arteries. It primarily afflicts infants and young children, preferentially those of Eastern descent. The incidence varies depending on geographic region, ranging from ~10 per 100 000 children under five years of age in the United States, up to 200 per 100 000 children under five years of age in Japan.[87,88] The patients often present with multiple signs that can include fever, cervical lymphadenopathy, conjunctivitis, erythema of the mouth and oropharynx with a strawberry tongue, red swollen hands and feet, and a macular or morbilliform rash on the trunk and extremities.

Up to one-quarter of untreated patients develop a necrotizing vasculitis of medium-sized arteries (Figure 11.8). Kawasaki arteritis preferentially involves the coronary arteries, and may lead to thrombotic coronary artery occlusion with infarction of the downstream myocardium. Injury to the walls of the coronary arteries may lead to formation of coronary artery aneurysms, which can rupture and lead to sudden death. Although the disorder is relatively uncommon, Kawasaki disease is considered to be the most common cause of acquired heart disease in children in the United States.

The vasculitis occurring in Kawasaki disease is characterized by a transmural inflammatory infiltrate (Figure 11.14), which is composed predominantly of lymphocytes and macrophages, with lesser and varying numbers of plasma cells, neutrophils, and eosinophils. The majority of the lymphocytes are CD8+ cytotoxic T-cells.[89] The cytotoxic T-cells and activated

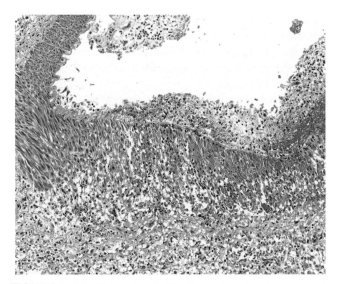

FIGURE 11.14 Kawasaki arteritis. Shown is an interlobar artery in the kidney of a young child who died of Kawasaki disease. There is segmental transmural necrotizing inflammation with a focally thickened intima (with fibrinoid necrosis at the far right), focal necrosis and lysis of the medial smooth muscle cells on the right (with intact media on the left), and adventitial inflammation.

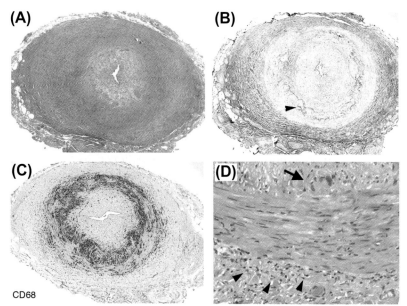

FIGURE 11.15 Giant cell arteritis. Depicted are portions of temporal artery biopsies from patients with giant cell arteritis. **(A)** At low magnification, the lumen of the artery is markedly narrowed due to intimal hyperplasia. **(B)** On elastic stain there is severe fragmentation, duplication, and loss of the elastic lamina. Only short segments of the original internal elastic lamina remain (arrowhead). **(C)** Immunohistochemical stain for the macrophage marker CD68 shows diffuse infiltration of the media by macrophages. The inner portion of the hyperplastic intima is relatively devoid of macrophages. **(D)** At higher magnification, in a case with a lesser degree of involvement, there is a macrophage giant cell at the internal elastic lamina (arrow), as well as multiple lymphocytes located in the adventitia (arrowheads).

macrophages are believed to cause smooth muscle cell injury and death in the media. In at least some cases, the majority of the plasma cells secrete IgA.[90] This is in contrast to most vascular inflammatory reactions in which the majority of plasma cells secrete IgG. Elevated expression of matrix metalloproteinases, such as MMP9 and MMP2, are seen in the inflamed vessel wall and likely contribute to vascular wall injury and aneurysm formation. The inflammatory process leads to scar formation in the vascular wall and fragmentation of the elastic lamina. There is also vascular cell activation with proliferation of myofibroblasts and smooth muscle cells and the formation of intimal hyperplasia.

It is generally believed that Kawasaki disease results from a reaction to an infectious agent in a genetically predisposed individual. Recent studies have suggested that the antibodies generated by the IgA-secreting plasma cells in Kawasaki arteritis recognize viral-like cytoplasmic inclusions within the respiratory epithelium in the lungs of patients with acute Kawasaki disease.[91] These particles or other infectious agents are thought to trigger an inflammatory reaction associated with elevation of inflammatory cytokines and chemokines including TNF-α, IL-1, IL-6, and IL-8. For reasons not completely understood, this inflammatory reaction results in a lymphocyte- and macrophage-predominant inflammatory filtrate preferentially targeting the coronary arteries. In a mouse model of coronary artery vasculitis, which involves peritoneal injections of Lactobacillus extract, selective expression of toll-like receptor 2 (TLR2) by coronary artery endothelium has been implicated in targeting the inflammatory reaction to the coronary arteries by enhancing the vascular expression of T-cell co-stimulatory molecules at this location.[88]

Fortunately, patients with Kawasaki disease are often treated effectively in the acute phase with intravenous immunoglobulin (IVIG) and aspirin. Not surprisingly, a primary aim of this treatment is to minimize coronary artery injury, which is responsible for many of the fatal outcomes of the disease. Unfortunately, patients who survive the acute phase of Kawasaki disease may sustain subclinical coronary artery injury, which can result in an increased risk for development of premature atherosclerotic coronary artery disease later in life.[92]

GIANT CELL ARTERITIS

Giant cell arteritis (GCA) is one of the most common forms of vasculitis. It primarily afflicts older individuals above the age of 50. The prevalence of clinically apparent GCA may be as high as 0.2% for people of northern European descent over the age of 50.[93] In a large autopsy study of European patients, the prevalence of GCA was reported to be 1.7% in older patients.[94] This would suggest that up to 90% of cases of pathologic GCA may be subclinical, i.e. not manifesting sufficient signs and symptoms to bring the condition to clinical attention. GCA is characterized pathologically by granulomatous inflammation in the vessel wall. (Figure 11.15), an inflammatory infiltrate containing numerous activated epithelioid macrophages. In GCA these activated macrophages often coalesce to form giant cells.[95,96] The granulomatous inflammation in GCA typically directly targets elastic laminae within the vessel walls of medium-sized and large arteries. In GCA there is usually also infiltration of the vessel wall by lymphocytes and lesser numbers of plasma cells, but these cells types may be largely restricted to the adventitia.[97]

Almost any artery can be affected by GCA but it typically involves large and medium-sized vessels (Figure 11.8). GCA most commonly involves the arteries of the head and neck, particularly the superficial temporal arteries. GCA also commonly involves the aorta resulting in its classification as a large-vessel vasculitis. In GCA the inflammatory infiltrate injures the arterial wall and causes vascular cell activation, leading to the formation of intense intimal hyperplasia. In fact, luminal obstruction in GCA is usually secondary to intimal hyperplasia rather than thrombosis. This contrasts with small-vessel vasculitides, in which luminal obstruction often results from thrombosis. In medium-sized arteries, such as the temporal arteries, overt necrosis is typically absent or minimal compared with that seen in ANCA-associated vasculitis. Thus, compared with many other forms of vasculitis, GCA can often be considered a relatively chronic insidious vasculitis, although episodic periods of enhanced activity do appear to occur clinically.

The specific etiology of GCA remains elusive. The three principal inflammatory cell types involved in GCA appear to be CD4+ T-cells, dendritic cells, and activated macrophages.[98,99] The activation of resident vascular adventitial dendritic cells may be an early event in the initiation of GCA. These dendritic cells are thought to sense and respond to danger signals, although the actual danger signals initiating GCA are unclear. Upon activation, adventitial dendritic cells produce cytokines and chemokines to attract other inflammatory cells to the vessel wall, including lymphocytes and macrophages. In GCA the lymphocytes are often primarily located in the adventitia, suggesting that this disorder begins in the adventitia. Later in the disease lymphocytes can also be located in the thickened intima. The lymphocytes in GCA are predominantly CD4+ helper T-cells, and these cells appear to also undergo some degree of clonal expansion within the wall of the inflamed artery. Early in the disease, there appear to be both type 1 helper T-cells (Th1 cells) secreting γ-interferon (γ-IFN) as well as type 17 helper T-cells (Th17 cells) secreting IL-17. However, later in the disease only Th1 cells are present. The γ-IFN is thought to contribute to macrophage activation and giant cell formation.

The activated macrophages play a central effector role in GCA. They secrete proinflammatory cytokines such as IL-1β, IL-6, and TNF-α, which contribute to the maintenance of the inflammatory response. Patients with GCA often have elevated levels of IL-6 both in the inflamed vessel wall and in the blood. Activated macrophages secrete MMPs, particularly MMP9, which degrade the vascular elastic lamina. In fact, the elastic laminae are a primary target in GCA. There is often marked elastic fiber degradation, and the giant cells are often seen ingesting fragments of elastic lamina. The activated macrophages also secrete large amounts of ROS, which injure medial smooth muscle cells. Activated macrophages secrete growth factors such as PDGF and FGF, which promote smooth muscle cell and myofibroblast proliferation, resulting in marked intimal hyperplasia.

Patients often present clinically with headache, fever, visual changes, and elevated serum inflammatory markers, such as erythrocyte sedimentation rate (ESR) and C-reactive protein (CRP).[100] Jaw pain with chewing or jaw claudication is a relatively specific but less common symptom of GCA, and is believed to result from stenosis of vessel perfusing the skeletal muscles. GCA is associated with another disorder called polymyalgia rheumatica (PMR), which manifests as musculoskeletal pain usually focused on the shoulder and hip girdles. A definitive pathologic diagnosis of GCA can be obtained by biopsy of the superficial temporal arteries. GCA can result in permanent vision loss from obstruction of the posterior ciliary, ophthalmic, or retinal arteries. GCA can also cause ischemic necrosis of the scalp and tongue, formation of aortic aneurysms, and life-threatening aortic dissection. Glucocorticoids are the primary treatment for GCA. Recently, patients refractory to glucocorticoids have also been treated with an antibody that blocks the receptor for IL-6.[101]

VASCULAR TRAUMA AND THE HYPOTHENAR HAMMER SYNDROME

Blood vessels are also susceptible to mechanical injury. While almost any blood vessel can be affected by penetrating injuries, small and medium-sized blood vessels located close to the skin are also susceptible to blunt force trauma. One classic example of the latter involves the superficial ulnar artery in the hand.[102,103] Once the superficial branch of the ulnar artery passes through the wrist and enters the hand, it is covered only by a thin palmaris brevis muscle, subcutaneous tissue, and skin, before it later passes below the dense palmar aponeuroses. In this relatively short 2 cm portion, the ulnar artery travels over the hook of the hammate bone. In the setting of blunt force trauma to the hypothenar portion of the hand, the hamate bone essentially serves as an anvil, against which the ulnar artery is pounded. Trauma-related injury to the distal ulnar artery can result in occlusion of the vessel or aneurysm formation (Figure 11.16). The trauma to this artery is often self-inflicted when one uses the hand as a hammer. Thus trauma-related injury to the distal ulnar artery is often referred to as the hypothenar hammer syndrome (HHS).[102] Typically the hand trauma in HHS is chronic in nature. Chronic blunt-force trauma resulting in the HHS has been linked to multiple occupational and recreational activities, particularly in carpenters, mechanics, industrial workers, and foresters.

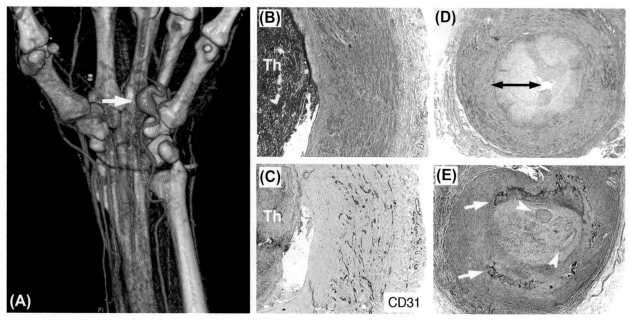

FIGURE 11.16   The hypothenar hammer syndrome. **(A)** Computed tomography imaging with three-dimensional reconstruction shows an ulnar artery aneurysm (arrow) in the hand of a truck driver who routinely used his hand as a hammer when securing loads on his truck. (Reproduced with permission from Modarai B, McIrvine A. The hypothenar hammer syndrome. *J Vasc Surg* 2008;47:1350.) Histologically in the early subacute phase of the hypothenar hammer syndrome **(B,C)**, thrombus (Th) is often found in the lumen of the vessel. On trichrome stain **(B)** a blue scar may be seen, indicating prior episodes of injury. There is also ingrowth of CD31 expressing capillaries into the media of the artery in response to the injury **(C)**. In the late chronic phase of the disease **(D)**, the lumen may be substantially narrowed from intimal hyperplasia (double headed arrow). In the chronic phase, elastic stain **(E)** often shows marked fragmentation of the internal elastic lamina (arrows). There may also be evidence of well-organized thrombus, with multiple small arteries in the thickened intima of the larger vessel, indicating recanalization (arrowheads).

The patients present clinically with evidence of arterial insufficiency in the hand, which may include cyanosis, pallor, cold and painful digits, paresthesias, Raynaud's phenomenon (see below), and/or ischemic ulcers. A tender mass may be present in the hypothenar region of the hand. The obstructed or aneurysmal segment is often resected and repaired surgically. The resected arteries show luminal occlusion with thrombosis in the more acute or subacute phases. There is also marked ingrowth of capillaries into the media of the artery, in response to the injured endothelial cells and vascular smooth muscle cells and the organizing luminal thrombus. With time the media will show scarring from the replacement of injured smooth muscle cells with collagen. There is often extensive destruction of the elastic lamina. The luminal thrombus may eventually be organized by the infiltration of fibroblasts, which lay down collagen, and by the ingrowth of small blood vessels, which partially recanalize the lumen. Intimal hyperplasia is usually present and may be extensive.

The HHS is generally thought to be a relatively uncommon disorder. Not surprisingly there is a striking male predominance, with a male to female ratio of more than 9:1. Behavioral differences are thought to underlie this impressive gender distribution. Also not surprising, the right hand is more commonly involved than the left hand. Although the HHS itself is uncommon, the activities that often cause it are not. Some population-based clinical and pathologic studies have suggested that in fact traumatic injury to this artery may be much more common than the actual HHS, with a significant portion of people having subclinical vascular injury.[104–106]

VASOSPASM AND RAYNAUD'S PHENOMENON

Blood flow through arteries is regulated by several processes including intrinsic vascular tone, nervous innervation, blood pressure, blood viscosity, and both locally secreted and circulating vasodilating and vasoconstricting agents. Vasconstriction is a normal physiologic process whereby a blood vessel will decrease its luminal diameter, often in order to regulate blood flow. However, vasospasm is a pathologic process characterized by excessive vasoconstriction, such that the lumen is obliterated, creating a situation where the target tissue may suffer ischemic injury. Vasospasm of a coronary artery can lead to myocardial infarction, and vasospasm of a cerebral artery can lead to stroke. One of the more common types of vasospasm is known as Raynaud's phenomenon.[107]

Raynaud's phenomenon entails episodic ischemia of the fingers and toes in response to cold or emotional stress.

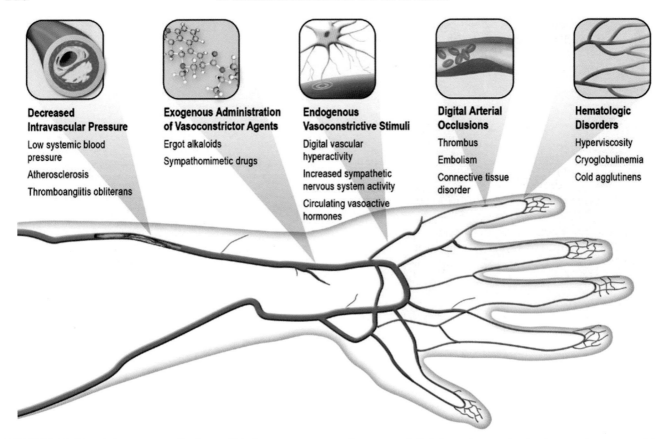

Decreased Intravascular Pressure

Low systemic blood pressure

Atherosclerosis

Thromboangiitis obliterans

Exogenous Administration of Vasoconstrictor Agents

Ergot alkaloids

Sympathomimetic drugs

Endogenous Vasoconstrictive Stimuli

Digital vascular hyperactivity

Increased sympathetic nervous system activity

Circulating vasoactive hormones

Digital Arterial Occlusions

Thrombus

Embolism

Connective tissue disorder

Hematologic Disorders

Hyperviscosity

Cryoglobulinemia

Cold agglutinens

FIGURE 11.17 Mechanisms contributing to digital artery vasospasm. Vasospasm of digital arteries may be due to excessive vasoconstriction and/or decreased intraluminal pressure. Excessive vasoconstriction may be due to vascular hyperreactivity, vasoactive drugs, increased sympathetic efferent activity and circulating or locally acting vasoactive hormones including endothelin 1, antiotensin II and thromboxane A2. Decreased blood pressure may be caused by low systemic blood pressure and by proximal arterial obstruction due to diseases such as atherosclerosis and thromboangiitis obliterans, a smoking-related inflammatory thrombotic condition. The smaller more peripheral arteries may also become obstructed in fibrosing collagen vascular diseases such as scleroderma or in disorders causing increased viscosity of the blood. (Reproduced with permission from Creager MA, Perlstein TS, Halperin JL. Raynaud's Phenomenon. *In Vascular Medicine: A Companion to Braunwald's Heart Disease* (Second Edition), Creager MA, Beckman JA, Loscalzo J, eds. Saunders 2012: 589–601).

Initially there is pallor or blanching of the skin due to vasospasm of digital arteries. Within the ischemic tissue, the small vessels dilate, and the digits acquire a bluish discoloration or cyanosis, due to accumulation of deoxygenated hemoglobin within the dilated small vessels. Upon resolution of the vasospasm, there is a rush of oxygenated blood into the dilated small vessels giving the digits a bright red color. This phase is referred to as reactive hyperemia, and is often associated with a throbbing sensation. The digits will then slowly regain their normal appearance.

Raynaud's phenomenon may occur as a primary idiopathic condition, or may occur as a secondary condition associated with diseases or conditions that induce vasospasm including trauma, use of vasoconstricting drugs, proximal arterial occlusive disease, blood hyperviscosity, and collagen vascular diseases such as scleroderma and systemic lupus erythematosus (Figure 11.17). Primary Raynaud's phenomenon affects women about five times more often than it affects men. It most commonly presents before the age of 40. Not surprisingly, it is more common in colder climates where up to 17% of

people may be affected, although overall about 3% to 5% of people are affected. The condition is usually relatively benign, with less than 1% of patients requiring amputation of part of a digit.

In many sites of the body, arterial tone is regulated by both vasoconstrictor and vasodilator sympathetic nerve fibers. However in the skin of the hands and feet, the vessels are innervated only by vasoconstrictor α-adrenergic nerve fibers. At these sites, vasodilation requires cessation or withdrawal of the sympathetic vasoconstrictor stimulus. Exposure to cold activates cutaneous receptors that initiate a reflex vasoconstriction. While such vasoconstriction is a normal response, in Raynaud's phenomenon, there is excessive vasoconstriction or vasospasm leading not just to reduction of the lumen, but transient obliteration of the lumen. Since arterial patency is dependent on both luminal blood pressure and arterial wall vasomotor tone, an imbalance in one of these two factors is thought to underlie Raynaud's phenomenon. However, primary Raynaud's phenomenon more likely results from increased vascular tone.

Possible etiologies for increased vasomotor tone in primary Raynaud's phenomenon include localized vascular hyperreactivity, increased sympathetic nervous system activity, and increased levels of hormones and exogenous agents that promote vasoconstriction. Of these possibilities, vascular hyperreactivity is the most likely underlying etiology. Sympathetic stimulation in the digital arteries is primarily through α_2-adrenergic receptors.[108] Cooling actually increases the activity of α_2-adrenergic receptors, and inhibitors of these receptors reverse the digital vasospasm in patients with Raynaud's phenomenon.[109] In addition, patients with Raynaud's phenomenon are more sensitive to the vasoconstrictor effects of α_2-adrenergic receptor agonists than are patients without Raynaud's phenomenon.[110,111] Estrogen actually increases the expression of α_2-adrenergic receptors in vascular smooth muscle, possibly explaining the higher prevalence of Raynaud's phenomenon in young women. However, the primary underlying reason for the increased α_2-adrenergic response in patients with Raynaud's phenomenon is not well understood.

References

1. Silver MD, Gotlieb AI, Schoen FJ. *Cardiovascular Pathology*. 3rd ed. Churchhill Livingston; 2001.
2. Stehbens WE, Lie JT. *Vascular Pathology*. Chapman & Hall Medical 1995.
3. Gimbrone MA, Garcia-Cardena G. Vascular endothelium, hemodynamics, and the pathobiology of atherosclerosis. *Cardiovasc Pathol* 2013;**22**:9–15.
4. Fledderus JO, Boon RA, Volger OL, Hurtilla H, Yla-Herttuala S, Pannekoek H, et al. KLF2 primes the antioxidant transcription factor Nrf2 for activation in endothelial cells. *Arterioscler Thromb Vasc Biol* 2008;**28**:1339–46.
5. Raines EW, Ferri N. Cytokines affecting endothelial and smooth muscle cells in vascular disease. *J Lipid Res* 2005;**46**:1081–92.
6. Vallabhapurapu S, Karin M. Regulation and function of NF-kappaB transcription factors in the immune system. *Ann Rev Immunol* 2009;**27**:693–733.
7. Beamish JA, He P, Kottke-Marchant K, Marchant RE. Molecular regulation of contractile smooth muscle cell phenotype: Implications for vascular tissue engineering. *Tissue Eng Part B Rev* 2010;**16**:467–91.
8. Rensen SS, Doevendans PA, van Eys GJ. Regulation and characteristics of vascular smooth muscle cell phenotypic diversity. *Neth Heart J* 2007;**15**:100–8.
9. Owens GK, Kumar MS, Wamhoff BR. Molecular regulation of vascular smooth muscle cell differentiation in development and disease. *Physiol Rev* 2004;**84**:767–801.
10. Berk BC. Vascular smooth muscle growth: autocrine growth mechanisms. *Physiol Rev* 2001;**81**:999–1030.
11. Stone JR, Yang S. Hydrogen peroxide: a signaling messenger. *Antioxidants Redox Signal* 2006;**8**:243–70.
12. Stary HC, Blankenhorn DH, Chandler AB, Glagov S, Insull W, Richardson M, et al. A definition of the intima of human arteries and of its atherosclerosis-prone regions. A report from the Committee on Vascular Lesions of the Council on Arteriosclerosis, American Heart Association. *Circulation* 1992;**85**:391–405.
13. Newby AC, Zaltsman AB. Molecular mechanisms in intimal hyperplasia. *J Pathol* 2000;**190**:300–9.
14. Cizek SM, Bedri S, Talusan P, Silva N, Lee H, Stone JR. Risk factors for atherosclerosis and the development of pre-atherosclerotic intimal hyperplasia. *Cardiovasc Pathol* 2007;**16**:344–50.
15. Ikari Y, McManus BM, Kenyon J, Schwartz SM. Neonatal intima formation in the human coronary artery. *Arterioscler Thromb Vasc Biol* 1999;**19**:2036–40.
16. Velican D, Velican C. Study of fibrous plaques occurring in the coronary arteries of children. *Atherosclerosis* 1979;**33**:201–15.
17. Tsai S, Butler J, Rafii S, Liu B, Kent KC. The role of progenitor cells in the development of intimal hyperplasia. *J Vasc Surg* 2009;**49**:502–10.
18. Torsney E, Xu Q. Resident vascular progenitor cells. *J Mol Cell Cardiol* 2011;**50**:304–11.
19. Abarbanell AM, Herrmann JL, Weil BR, Wang Y, Tan J, Moberly SP, Fiege JW, Meldrum DR. Animal models of myocardial and vascular injury. *J Surg Res* 2010;**162**:239–49.
20. Panchenko MP, Siddiquee Z, Dombkowski DM, Alekseyev YO, Lenburg ME, Walker JD, et al. Protein kinase CK1αLS promotes vascular cell proliferation and intimal hyperplasia. *Am J Pathol* 2010;**177**:1562–72.
21. Schwartz SM, deBlois D, O'Brien ER. The intima. Soil for atherosclerosis and restenosis. *Circ Res* 1995;**77**:445–65.
22. Talusan P, Bedri S, Yang S, Kattapuram T, Silva N, Roughley PJ, et al. Analysis of intimal proteoglycans in atherosclerosis-prone and atherosclerosis-resistant human arteries by mass spectrometry. *Mol Cell Proteomics* 2005;**4**:1350–7.
23. Zhang LN, Parkinson JF, Haskell C, Wang YX. Mechanisms of intima hyperplasia learned from a murine carotid artery ligation model. *Curr Vasc Pharmacol* 2008;**6**:37–43.
24. Hardin NJ. The myocardial and vascular pathology of diabetic cardiomyopathy. *Coronary Artery Dis* 1996;**7**:99–108.
25. He Z, King GL. Microvascular complications of diabetes. *Endocrinol Metab Clin N Am* 2004;**33**:215–38.
26. Cai J, Boulton M. The pathogenesis of diabetic retinopathy: old concepts and new questions. *Eye* 2002;**16**:242–60.
27. Cooper ME. Pathogenesis, prevention, and treatment of diabetic nephropathy. *Lancet* 1998;**352**:213–9.
28. Ramasamy R, Yan SF, Schmidt AM. The diverse ligand repertoire of the receptor for advanced glycation endproducts and pathways to the complications of diabetes. *Vasc Pharmacol* 2012;**57**:160–7.
29. Merlini G, Bellotti V. Molecular mechanisms of amyloidosis. *N Eng J Med* 2003;**349**:583–96.
30. Westermark P, Benson MD, Buxbaum JN, Cohen AS, Frangione B, Ikeda S, et al. A primer of amyloid nomenclature. *Amyloid* 2007;**14**:179–83.
31. Murphy CL, Wang S, Williams T, Weiss DT, Solomon A. Characterization of systemic amyloid deposits by mass spectrometry. *Meth Enzymol* 2006;**412**:48–62.
32. Hawkins PN. Serum amyloid P component scintigraphy for diagnosis and monitoring amyloidosis. *Cur Opin Nephrol Hyperten* 2002;**11**:649–55.
33. Walley VM, Kisilevsky R, Young ID. Amyloid and the cardiovascular system: a review of pathogenesis and pathology with clinical correlations. *Cardiovasc Pathol* 1995;**4**:79–102.
34. Shah KB, Inoue Y, Mehra MR. Amyloidosis and the heart: a comprehensive review. *Arch Intern Med* 2006;**166**:1805–13.
35. Wittich CM, Neben-Wittich MA, Mueller PS, Gertz MA, Edwards WD. Deposition of amyloid proteins in the epicardial coronary arteries of 58 patients with primary systemic amyloidosis. *Cardiovasc Pathol* 2007;**16**:75–8.
36. Rocken C, Shakespeare A. Pathology, diagnosis and pathogenesis of AA amyloidosis. *Virchows Arch* 2002;**440**:111–22.
37. Rapezzi C, Perugini E, Salvi F, Grigioni F, Riva L, Cooke RM, et al. Phenotypic and genotypic heterogeneity in transthyretin-related cardiac amyloidosis: towards tailoring of therapeutic strategies? *Amyloid* 2006;**13**:143–53.

38. Suhr OB, Svendsen IH, Andersson R, Danielsson A, Holmgren G, Ranlov PJ. Hereditary transthyretin amyloidosis from a Scandinavian perspective. *J Intern Med* 2003;**254**:225–35.

39. Jacobson DR, Pastore RD, Yaghoubian R, Kane I, Gallo G, Buck FS, et al. Variant-sequence transthyretin (isoleucine 122) in late-onset cardiac amyloidosis in black Americans. *N Eng J Med* 1997;**336**:466–73.

40. Asl LH, Liepnieks JJ, Asl KH, Uemichi T, Moulin G, Desjoyaux E, et al. Hereditary amyloid cardiomyopathy caused by a variant apolipoprotein A1. *Am J Pathol* 1999;**154**:221–7.

41. Yazaki M, Liepnieks JJ, Barats MS, Cohen AH, Benson MD. Hereditary systemic amyloidosis associated with a new apolipoprotein AII stop codon mutation Stop78Arg. *Kidney Internat* 2003;**64**:11–6.

42. Westermark P, Sletten K, Johansson B, Cornwell GG. Fibril in senile systemic amyloidosis is derived from normal transthyretin. *Proc Natl Acad Sci USA* 1990;**87**:2843–5.

43. Jennette JC, Falk RJ, Bacon PA, Basu N, Cid MC, Ferrario F, et al. 2012 Revised International Chapel Hill Consensus Conference Nomenclature of Vasculitides. *Arthritis Rheum* 2013;**65**:1–11.

44. Homeister J, Jennette JC, Falk RJ. Immunologic Mechanisms of Vasculitis in The Kidney: Physiology and Pathophysiology, 5th ed., Robert J, Alpern RJ, and Steven C, Hebert SC, Elsevier, 2013, Chapter 83, 2817–46.

45. Karsten CM, Köhl J. The immunoglobulin, IgG Fc receptor and complement triangle in autoimmune diseases. *Immunobiology* 2012;**217**:1067–79.

46. Ley K, Laudanna C, Cybulsky MI, Nourshargh S. Getting to the site of inflammation: The leukocyte adhesion cascade updated. *Nat Rev* 2007;**7**:678–89.

47. Weber C. Novel mechanistic concepts for the control of leukocyte transmigration: specialization of integrins, chemokines, and junctional molecules. *J Mol Med* 2003;**81**:4–19.

48. Middleton J, Patterson AM, Gardner L, Schmutz C, Ashton BA. Leukocyte extravasation: chemokine transport and presentation by the endothelium. *Blood* 2002;**100**:3853–60.

49. Cook-Mills JM, Deem TL. Active participation of endothelial cells in inflammation. *J Leukoc Biol* 2005;**77**:487–95.

50. Jennette JC. Implications for pathogenesis of patterns of injury in small and medium-sized vessel vasculitis. *Cleve Clin J Med* 2002;**69**:SII–33–8.

51. Owen CA, Campbell EJ. The cell biology of leukocyte-mediated proteolysis. *J Leukoc Biol* 1999;**65**:137–50.

52. Leppert D, Lindberg RL, Kappos L, Leib SL. Matrix metalloproteinases: multifunctional effectors of inflammation in multiple sclerosis and bacterial meningitis. *Brain Res Rev* 2001;**36**:249–57.

53. Hiemstra PS. Novel roles of protease inhibitors in infection and inflammation. *Biochem Soc Trans* 2002;**30**:116–20.

54. Esparza J, Vilardell C, Calvo J, Juan M, Vives J, Urbano-Marquez A, et al. Fibronectin upregulates gelatinase B (MMP-9) and induces coordinated expression of gelatinase A (MMP-2) and its activator MT1-MMP (MMP-14) by human T lymphocyte cell lines. A process repressed through RAS/MAP kinase signaling pathways. *Blood* 1999;**94**:2754–66.

55. Ferri C, Mascia MT. Cryoglobulinemic vasculitis. *Curr Opin Rheumatol* 2006;**18**:54–63.

56. Sansonno D, Dammacco F. Hepatitis C virus, cryoglobulinaemia, and vasculitis: immune complex relations. *Lancet Infect Dis* 2005;**5**:227–36.

57. Ferri C, Sebastiani M, Giuggioli D, Cazzato M, Longombardo G, Antonelli A, Puccini R, Michelassi C, Zignego AL. Mixed cryoglobulinemia: demographic, clinical, and serologic features and survival in 231 patients. *Semin Arthritis Rheum* 2004;**33**:355–74.

58. Rastaldi MP, Ferrario F, Crippa A, Dell'Antonio G, Casartelli D, Grillo C, D'Amico G. Glomerular monocyte-macrophage features in ANCA-positive renal vasculitis and cryoglobulinemic nephritis. *J Am Soc Nephrol* 2000;**11**:2036–43.

59. Lau KK, Wyatt RJ, Moldoveanu Z, Tomana M, Julian BA, Hogg RJ, et al. Serum levels of galactose-deficient IgA in children with IgA nephropathy and Henoch-Schönlein purpura. *Pediatr Nephrol* 2007;**22**:2067–72.

60. Kiryluk K, Moldoveanu Z, Sanders JT, Eison T, Suzuki H, Julian BA, et al. Aberrant glycosylation of IgA1 is inherited in both pediatric IgA nephropathy and Henoch-Schönlein purpura nephritis. *Kidney Int* 2011;**80**:79–87.

61. Boyd JK, Barratt J. Inherited IgA glycosylation pattern in IgA nephropathy and HSP nephritis: where do we go next? *Kidney Int* 2011;**80**:8–10.

62. Suzuki H, Fan R, Zhang Z, Brown R, Hall S, Julian BA, et al. Aberrantly glycosylated IgA1 in IgA nephropathy patients is recognized by IgG antibodies with restricted heterogeneity. *J Clin Invest* 2009;**119**:1668–77.

63. Grotz W, Baba HA, Becker JU, Baumgärtel MW. Hypocomplementemic urticarial vasculitis syndrome: an interdisciplinary challenge. *Dtsch Arztebl Int* 2009;**106**:756–63.

64. Balsam L, Karim M, Miller F, Rubinstein S. Crescentic glomerulonephritis associated with hypocomplementemic urticarial vasculitis syndrome. *Am J Kidney Dis* 2008;**52**:1168–73.

65. Jara LJ, Navarro C, Medina G, Vera-Lastra O, Saavedra MA. Hypocomplementemic urticarial vasculitis syndrome. *Curr Rheumatol Rep* 2009;**11**:410–5.

66. Rich AR. The role of hypersensitivity in periarteritis nodosa. As indicated by seven cases developing during serum sickness and sulfonamide therapy. *Bull Johns Hopkins Hosp* 1942;**71**:123–40.

67. Michalak T. Immune complexes of hepatitis B surface antigen in the pathogenesis of periarteritis nodosa. A study of seven necropsy cases. *Am J Pathol* 1978;**90**:619–32.

68. Weiss TD, Tsai CC, Baldassare AR, Zuckner J. Skin lesions in viral hepatitis: histologic and immunofluorescent findings. *Am J Med* 1978;**64**:269–73.

69. Hutson TE, Hoffman GS. Temporal concurrence of vasculitis and cancer: a report of 12 cases. *Arthritis Care Res* 2000;**13**:417–23.

70. Solans-Laqué R, Bosch-Gil JA, Pérez-Bocanegra C, Selva-O'Callaghan A, Simeón-Aznar CP, Vilardell-Tarres M. Paraneoplastic vasculitis in patients with solid tumors: report of 15 cases. *J Rheumatol* 2008;**35**:294–304.

71. Abdellatif AA, Waris S, Lakhani A, Kadikoy H, Haque W, Truong LD. True vasculitis in lupus nephritis. *Clin Nephrol* 2010;**74**:106–12.

72. Pedchenko V, Bondar O, Fogo AB, Vanacore R, Voziyan P, Kitching AR, et al. Molecular architecture of the Goodpasture autoantigen in anti-GBM nephritis. *N Engl J Med* 2010;**363**:343–54.

73. Mulligan MS, Johnson KJ, Todd 3rd RF, Issekutz TB, Miyasaka M, Tamatani T, et al. Requirements for leukocyte adhesion molecules in nephrotoxic nephritis. *J Clin Invest* 1993;**91**:577–87.

74. Suzuki Y, Shirato I, Okumura K, Ravetch JV, Takai T, Tomino Y, Ra C. Distinct contribution of Fc receptors and angiotensin II-dependent pathways in anti-GBM glomerulonephritis. *Kidney Int* 1998;**54**:1166–74.

75. Takai T, Ono M, Hikida M, Ohmori H, Ravetch JV. Augmented humoral and anaphylactic responses in Fc gamma RII-deficient mice. *Nature* 1996;**379**:346–9.

76. Roth AJ, Ooi JD, Hess JJ, van Timmeren MM, Berg EA, Poulton CE, et al. Epitope specificity determines pathogenicity and detectability in ANCA-associated vasculitis. *J Clin Invest* 2013;**123**:1773–83.

77. Jennette JC, Falk RJ, Hu P, Xiao H. Pathogenesis of Anti-neutrophil Cytoplasmic Autoantibody Associated Small Vessel Vasculitis. *Annu Rev Pathol Mech Dis* 2013;**8**:139–60.

78. Jennette JC, Falk RJ, Gasim AH. Pathogenesis of antineutrophil cytoplasmic autoantibody vasculitis. *Curr Opin Nephrol Hypertens* 2011;**20**:263–70.

79. Jennette JC, Xiao H, Falk R, Gasim AM. Experimental models of vasculitis and glomerulonephritis induced by antineutrophil cytoplasmic autoantibodies. *Contrib Nephrol* 2011;**169**:211–20.

80. Kettritz R, Jennette JC, Falk RJ. Cross-linking of ANCA-antigens stimulates superoxide release by human neutrophils. *J Am Soc Nephrol* 1997;**8**:386–94.

81. Porges AJ, Redecha PB, Kimberly WT, Csernok E, Gross WL, Kimberly RP. Anti-neutrophil cytoplasmic antibodies engage and activate human neutrophils via Fc gamma RIIa. *J Immunol* 1994;**153**:1271–80.

82. Falk RJ, Terrell RS, Charles LA, Jennette JC. Anti-neutrophil cytoplasmic autoantibodies induce neutrophils to degranulate and produce oxygen radicals in vitro. *Proc Natl Acad Sci U S A* 1990;**87**: 4115–9.

83. Schreiber A, Xiao H, Jennette JC, Schneider W, Luft FC, Kettritz R. C5a receptor mediates neutrophil activation and ANCA-induced glomerulonephritis. *J Am Soc Nephrol* 2009;**20**:289–98.

84. Xiao H, Schreiber A, Heeringa P, Falk RJ, Jennette JC. Alternative complement pathway in the pathogenesis of disease mediated by antineutrophil cytoplasmic autoantibodies. *Am J Pathol* 2007;**170**:52–64.

85. Ewert BH, Jennette JC, Falk RJ. Anti-myeloperoxidase antibodies stimulate neutrophils to damage human endothelial cells. *Kidney Int* 1992;**41**:375–83.

86. Savage CO, Gaskin G, Pusey CD, Pearson JD. Myeloperoxidase binds to vascular endothelial cells, is recognized by ANCA and can enhance complement dependent cytotoxicity. *Adv Exp Med Biol* 1993;**336**:121–3.

87. Burns JC, Glode MP. Kawasaki syndrome. *Lancet* 2004;**364**:533–44.

88. Yeung RSM. Kawasaki disease: Update on pathogenesis. *Curr Opin Rheumatol* 2010;**22**:551–60.

89. Brown TJ, Crawford SE, Cornwall ML, Garcia F, Shulman ST, Rowley AH. CD8 T lymphocytes and macrophages infiltrate coronary artery aneurysms in acute Kawasaki disease. *J Infect Dis* 2001;**184**:940–3.

90. Rowley AH, Eckerley CA, Jack HM, Shulman ST, Baker SC. IgA plasma cells in vascular tissue of patients with Kawasaki syndrome. *J Immunol* 1997;**159**:5946–55.

91. Rowley AH, Baker SC, Orenstein JM, Shulman ST. Searching for the cause of Kawasaki disease - cytoplasmic inclusion bodies provide new insight. *Nat Rev Microbiol* 2008;**6**:394–401.

92. Senzaki H. Long-term outcome of Kawasaki disease. *Circulation* 2008;**118**:2763–72.

93. Lee JL, Naguwa SM, Cheema GS, Gershwin ME. The geo-epidemiology of temporal (giant cell) arteritis. *Clin Rev Allergy Immunol* 2008;**35**:88–95.

94. Ostberg G. Arteritis with special reference to polymyalgia arteritica. *Acta Pathol Microbiol Scand A* 1973;**237**(Suppl.):1–59.

95. Allsop CJ, Gallagher PJ. Temporal artery biopsy in giant-cell arteritis. A reappraisal. *Am J Surg Pathol* 1981;**5**:317–23.

96. Lie JT. Illustrated histopathologic classification criteria for selected vasculitis syndromes. *Arthritis Rheum* 1990;**33**:1074–87.

97. Stone JR, Pless M, Salvarani C, Pipitone N, Lessell S, Stone JH. Giant cell arteritis and polymyalgia rheumatica. In: Stone JH, editor. *A Clinician's Pearls and Myths in Rheumatology*. New York, NY: Springer; 2009. p. 285–304.

98. Borches AT, Gershwin ME. Giant cell arteritis: A review of classification, pathophysiology, geoepidemiology and treatment. *Autoimmunity Rev* 2012;**11**:A544–54.

99. Weyand CM, Liao YJ, Gornzy JJ. The immunopathology of giant cell arteritis: Diagnostic and therapeutic implications. *J Neuro-ophthalmol* 2012;**32**:259–65.

100. Seo P, Stone JH. Large-vessel vasculitis. *Arthritis Rheum Arthritis Care Res* 2004;**51**:128–39.

101. Unizony S, Stone JH, Stone JR. New treatment strategies in large-vessel vasculitis. *Curr Opin Rheumatol* 2013;**25**:3–9.

102. Conn J, Bergan JJ, Bell JL. Hypothenar hammer syndrome: Post-traumatic digital ischemia. *Surgery* 1970;**68**:1122–8.

103. Spittell PC, Spittell JA. Occlusive arterial disease of the hand due to repetitive blunt trauma: A review with illustrative cases. *Int J Cardiol* 1993;**38**:281–92.

104. Little JM, Ferguson DA. The incidence of the hypothenar hammer syndrome. *Arch Surg* 1972;**105**:684–5.

105. Kaji H, Honma H, Usui M, Yasuno Y, Saito K. Hypothenar hammer syndrome in workers occupationally exposed to vibrating tools. *J Hand Surg* 1993;**18B**:761–6.

106. Stone JR. Intimal hyperplasia in the distal ulnar artery: Influence of gender and implications for the hypothenar hammer syndrome. *Cardiovasc Pathol* 2004;**13**:20–5.

107. Wigley FM. Raynaud's Phenomenon. *N Engl J Med* 2002;**347**:1001–8.

108. Flavahan NA, Cooke JP, Shepherd JT, Vanhoutte PM. Human postjunctional alpha-1 and alpha-2 adrenoceptors: differential distribution in arteries of the limbs. *J Pharmacol Exp Ther* 1987;**241**:361–5.

109. Vanhoutte PM, Cooke JP, Lindblad LE, Shepherd JT, Flavahan NA. Modulation of postjunctional alpha-adrenergic responsiveness by local changes in temperature. *Clin Sci* 1985;**68**:121s–3s.

110. Coffman JD, Cohen RA. Role of alpha-adrenoceptor subtypes mediating sympathetic vasoconstriction in human digits. *Eur J Clin Invest* 1988;**18** 309–3.

111. Cooke JP, Creager SJ, Scales KM, Ren C, Tsapatsaris NP, Beetham WP, et al. Role of digital artery adrenoceptors in Raynaud's disease. *Vasc Med* 1997;**2**:1–7.

Pathophysiology of Atherosclerosis

Michael A. Seidman, MD, PhD[1], Richard N. Mitchell, MD, PhD[1],
James R. Stone, MD, PhD[2]

[1]Brigham & Women's Hospital/Harvard Medical School, Boston, MA, USA; [2]Massachusetts General
Hospital/Harvard Medical School, Boston, MA, USA

INTRODUCTION

Credited to Felix Marchand in a 1904 publication, the name atherosclerosis roughly translates as 'hardened gruel,'[1] This colorful term, reminiscent of other culinary metaphors in pathology, underlies many of the most feared and lethal conditions in medicine. Atherosclerosis is primarily an arterial disorder, classically characterized by lipid deposition in the vessel intima, and associated with inflammation, scarring, and calcification. Eventually these lesions cause luminal stenosis and potentially culminate in thrombotic occlusion and/or embolism. Atherosclerosis is highly prevalent in the industrialized countries, with a growing frequency in all geographic regions of the world. Risk factors include elevated serum cholesterol, diabetes mellitus, smoking, male gender, advancing age, obesity, systemic chronic inflammation, and as yet incompletely defined genetic factors. Since the manifestations of atherosclerosis – including coronary artery disease (CAD), peripheral vascular disease (PVD), abdominal aortic aneurysm (AAA), renal artery stenosis (RAS), and carotid artery stenosis (CAS) – account for a significant fraction of worldwide morbidity and mortality, substantial effort has been expended to understand its pathogenesis. We have done well in identifying – and modifying – risk factors for atherosclerosis; modern medical intervention also does a credible job in treating the sequelae of this 'hardened gruel'. However, there is much that needs to be learned and a better understanding of the basic cellular mechanisms will allow us the opportunity to potentially prevent atherosclerosis before it even begins.

Atherosclerosis often begins relatively early in life. Autopsy and radiologic studies have revealed that even teenagers can have atherosclerotic plaque in their aortas and coronary arteries.[2,3] At the other end of the spectrum, older individuals may have arteries so calcified by this process that they appear similar to bone on routine radiologic imaging.

Pathologically, atherosclerotic lesions can appear very similar at different sites throughout the body. However, there is marked heterogeneity in the manner in which patients with atherosclerosis present with clinical symptoms. The inflammatory component of atherosclerosis can damage vessel walls to the extent that they can weaken and dilate, leading to aneurysm formation. In renal arteries, atherosclerotic occlusion can restrict kidney perfusion, leading to compensatory activation of the renin–angiotensin axis and poorly controlled systemic hypertension. Arteries in the legs may become so stenotic as to no longer serve the limb sufficiently, causing distal limb ischemia, and necessitating surgical vascular bypass operation or even amputation. Lesions in the carotid arteries may lead to ischemic or embolic stroke. In the heart, atherosclerosis can either slowly occlude arteries–leading to angina, or can suddenly rupture–causing downstream infarction or even sudden death.

Individuals can present with preferential involvement of one vascular site over another. Thus, some patients present primarily with carotid artery stenosis and stroke, while others present primarily with coronary artery disease and myocardial infarction. Similarly, one patient may have severe coronary artery disease and only mild aortic atherosclerosis, while another patient may have severe aortic atherosclerosis and only mild coronary artery involvement. It is likewise unclear why some branching arteries such as the internal mammary artery are relatively protected from atherosclerotic involvement, while other branching arteries, such as the carotid and coronary arteries, are considerably more susceptible. These observations highlight the fact that there is much we still do not understand.

EARLY LESIONS

The earliest visualizable lesion of atherosclerosis is the fatty streak, which is an accumulation of lipid-laden macrophages in the vascular intima (Figures 12.1, 12.2).[4] Fatty streaks can be appreciated grossly as focal yellow areas of discoloration of intimal surface. These lipid-laden macrophages are often referred to as foam cells because of their foamy appearance. These foam cells are believed to derive from the ingestion of lipids by macrophages within the intima. Research into the mechanisms leading to the formation of early atherosclerotic lesions is ongoing, but clearly a number of factors play important roles. In the current understanding of atherogenesis, the three most important factors, all interacting with each other, appear to be altered lipid metabolism, vascular cell activation, and inflammation.[5–11]

LIPIDS IN ATHEROSCLEROSIS

Accumulation of lipid in an interstitial space obviously requires a source of lipid. Thus circulating lipid, ostensibly to fulfill essential biologic functions, becomes trapped in the tissue of the blood vessel without ever reaching the intended target cells. This accumulation appears to be promoted in the context of increased levels of lipids in the bloodstream, and in particular, increased levels of cholesterol.[5,6,12] Cholesterol circulates in the blood as multiple distinct forms of lipoprotein particles. In particular, low-density lipoprotein (LDL)-associated cholesterol (so-called bad cholesterol) promotes atherosclerosis development. In contrast, high-density lipoprotein (HDL) particles (so-called good cholesterol) promote reverse cholesterol transport, removing cholesterol from the vessel wall. LDL cholesterol can thus be viewed as the fuel for the fire of atherosclerosis. There is no one critical threshold level of LDL that will trigger development of atherosclerosis, and since all humans have circulating LDL, all who live until adulthood will develop atherosclerosis to some extent. In most people, the atherosclerosis remains mild and subclinical, until late in life. However, the degree to which the atherosclerotic lesions in any one person will progress is impacted by the concentration of LDL in the blood of that person.

An important observation in the field of atherosclerosis research is that LDL particles that have entered the vessel wall are often chemically modified prior to being

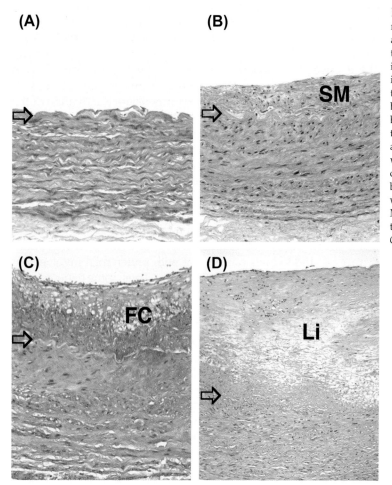

(A)

(B)

(C)

(D)

FIGURE 12.1 Early stages of atherosclerosis development. Shown are histologic images of human internal thoracic artery. The internal thoracic artery is normally highly resistant to atherosclerosis, but will develop atherosclerosis later in life. **(A)** The normal artery has a very thin intima. **(B)** In intimal thickening/intimal hyperplasia the intima becomes thickened due to the presence of smooth muscle cells (SM) beneath the endothelium. **(C)** The fatty streak is characterized by the appearance of macrophage foam cells (FC) within the thickened intima. **(D)** More advanced atherosclerotic lesions are characterized by the appearance of extracellular lipid (Li) which first appears interdigitating between the smooth muscle cells. The arrows indicate the internal elastic lamina, the boundary between the intima and the media. (Reproduced with permission from Cizek SM, Bedri S, Talusan P, Silva N, Lee H, Stone JR. Risk factors for atherosclerosis and the development of pre-atherosclerotic intimal hyperplasia. *Cardiovasc Pathol* 2007;16:344–50.)

engulfed by macrophages.[9,10] Thus LDL can undergo oxidative modifications (oxidized LDL) by the reactive oxygen species generated in the vessel wall at sites of vascular cell activation and inflammation. LDL may also undergo non-enzymatic glycation, particularly in patients with diabetes. Such modifications of LDL can facilitate its entrapment in the intima by altering its interaction with the extracellular matrix. More importantly, these modifications greatly enhance the degree to which macrophages will ingest the LDL particles. Macrophages can ingest modified LDL using specialized receptors called scavenger receptors. These receptors are distinct from the traditional LDL receptors used by the liver and other tissues to internalize normal LDL present in the serum.

Genetic variations in genes encoding proteins that control the metabolism of cholesterol and lipids can markedly impact a person's propensity to develop atherosclerosis.[13,14] Such genes include those coding for the LDL receptor (LDLR), apolipoprotein(a) (apo(a))

and apolipoprotein E (ApoE) (see Genetics of Atherosclerosis, below). The primary deleterious impact of many of these pro-atherogenic genetic variations is an increase in the levels of circulating lipoproteins. These findings in people have led to the development of genetically engineered mouse strains that develop marked hyperlipidemia and atherosclerosis, and are widely used to study the pathogenesis of atherosclerosis (see below).

ENDOTHELIAL ACTIVATION

Atherosclerosis results in large part from the interaction of lipoproteins with the vessel wall. Thus, changes in the cells of the vessel wall itself play an important role in the formation of atherosclerotic lesions. Activation of endothelial cells and vascular smooth muscle cells is a relatively general response of these cells to diverse stimuli (see Chapter 11). Activation of vascular cells is

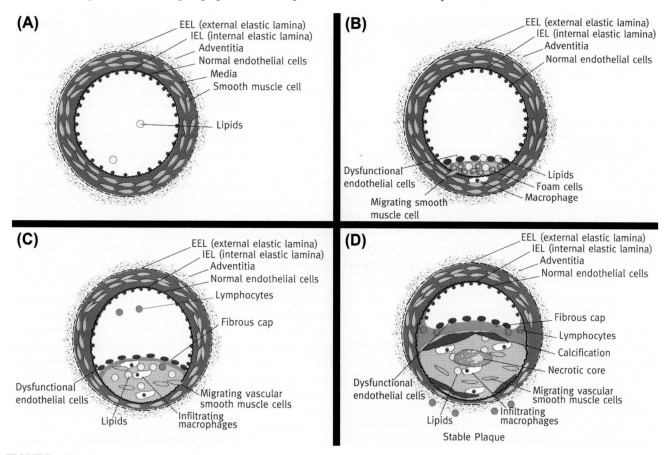

FIGURE 12.2 Progression of atherosclerotic plaques. (A) The normal vessel is characterized by a lack of inflammation, non-activated endothelium, and balanced levels of circulating lipids. (B) The fatty streak develops in the context of local endothelial dysfunction, lipid accumulation in the intima, and recruitment of monocytes that differentiate into macrophages and subsequently engulf lipid to become foam cells. (C) Plaque progression shows continued accumulation of lipid, macrophages and smooth muscle cells, with recruitment of other inflammatory cells such as lymphocytes. The recruited smooth muscle cells also synthesize collagen and other matrix proteins to form the nascent fibrous cap. (D) The stable fibroatheromatous plaque is characterized by a fibrous cap composed of smooth muscle cells and relatively dense extracellular matrix, separating the necrotic/lipid core from the lumen and a relative paucity of inflammation. Varying degrees of calcification are often present. (Reproduced with permission from Wang T, Palucci D, Law K, Yanagawa B, Yam J, Butany J. Atherosclerosis: pathogenesis and pathology. *Diag Histopathol* 2012;18:461–7.)

also an important component of atherogenesis. Since the endothelium sits at the border between the circulating blood and the vessel wall, it plays particularly important roles in this disease. In fact, phenotypic changes in endothelial cells impact the development of atherosclerosis in several distinct ways.

It is the endothelium that largely determines the specific section of an artery that will first develop atherosclerosis. Endothelial cells are responsive to biomechanical stimuli imparted by blood flow. Steady laminar shear stress promotes an 'atheroprotective state' in endothelial cells. However, disturbed blood flow, i.e. near branch sites in the vasculature, causes endothelial changes that promote atherosclerosis.[15,16]

Endothelium that experiences laminar shear stress within the blood stream promotes vasodilation and is anti-thrombotic. However, activated endothelial cells become less efficient in promoting vasodilation, in preventing platelet aggregation, and in suppressing coagulation. The shift toward such endothelial dysfunction is associated with increased reactive oxygen species production (e.g., superoxide and hydrogen peroxide), and a decreased generation of nitric oxide. This flow-related, site-selective activation of endothelial cells is believed to underlie the observation that atherosclerosis develops first and most severely at sites near vascular bifurcations, including the proximal left anterior descending artery of the heart, just after its bifurcation from the left main coronary artery.

The activated pro-atherogenic endothelium facilitates increased lipid uptake into the vessel wall through increased metabolism and altered transendothelial transport. Additionally, the activated endothelium expresses surface adhesion molecules, such as vascular cell adhesion molecule 1 (VCAM-1), intercellular adhesion molecule 1 (ICAM-1), P-selectin, and E-selectin. These adhesion molecules bind circulating inflammatory cells in the blood in order to recruit them to the site of vascular activation. The activated endothelial cells also secrete cytokines and chemokines that attract and activate these inflammatory cells. The activated endothelium secretes large amounts of growth factors, such as platelet-derived growth factor (PDGF), which promote vascular smooth muscle cell proliferation and extracellular matrix deposition. The accumulation of smooth muscle cells and a proteoglycan-rich extracellular matrix in the intima beneath the endothelium is a process referred to as intimal thickening or intimal hyperplasia (Figure 12.1). In humans, fatty streaks often form at sites where the intima is thus thickened. Within these lesions, the negatively charged proteoglycans are thought to bind the positively charged LDL particles, enhancing the retention of LDL within the intima and facilitating the phagocytosis of the LDL by macrophages.

INFLAMMATION IN ATHEROSCLEROSIS

Inflammatory responses play important roles in the development of atherosclerosis.[8–10,17] In atherosclerosis there is abnormal activation of inflammatory cells, which are largely directed towards the lipid deposited in the vascular wall. In general, the greater the systemic inflammatory response, the more likely there will be vascular-associated inflammatory reactions. Thus, atherosclerosis is amplified by a host of chronic inflammatory conditions including periodontal disease, rheumatoid arthritis, and systemic lupus erythematosus. In fact, current measures to stratify atherosclerotic risk include quantitation of non-specific inflammatory markers in the blood, such as C-reactive protein (CRP). People with a higher baseline CRP level have an increased risk for developing clinically significant atherosclerosis.[18]

In atherosclerosis, the leukocytes in the circulation are characteristically recruited into the vascular wall at the sites of lipid accumulation. Initially the leukocytes are primarily monocytes, which become activated to form tissue macrophages. This activation is stimulated by cytokines secreted from the resident vascular cells and is also driven by the accumulated lipids. Modified LDL, for example, binds to macrophage scavenger receptors, becomes internalized, and then accumulates within these cells due to an inability to be appropriately digested by lysosomal enzymes. This accumulation results in the foam cell, a macrophage swollen with lipid vacuoles. Foam cells have abnormal activity, and cannot readily egress the vascular intima or migrate to lymph nodes. Indeed, lipid loading of these macrophages can trigger cell death that contributes to the accumulation of thrombogenic material within the intima, including phospholipid and tissue factor.

As an abortive attempt to remove or sequester abnormal lipid (perceived as a 'danger signal'), foam cells recruit additional inflammatory cells, including pro-inflammatory cells such as T-helper type 1 (Th1) lymphocytes. Th1 lymphocytes secrete cytokines such as interferon-γ (IFNγ) and interleukin 2 (IL-2) that stimulate monocytes to form M1 polarized inflammatory macrophages.[9,17] In turn, these inflammatory macrophages secrete large amounts of reactive oxygen species and catabolic extracellular enzymes including matrix metalloproteinases (MMPs) and myeloperoxidase (MPO). Although the inflammatory process can be counter-balanced by regulatory T lymphocytes (T-regs), and in some cases antibody-producing B lymphocytes, leukocytes continue to accumulate in the generally pro-inflammatory milieu. Consequently, more foam cells form, and more cytokines, proteases, and reactive oxygen species are expressed. Such cytokine expression drives smooth muscle cell and fibroblast proliferation, and also promotes endothelial activation, creating a positive feedback loop. In addition, MMPs can degrade the extracellular matrix in the vessel wall, promoting further vessel dysfunction, with exacerbation of the smooth muscle cell and fibroblast proliferation. Subsequently, the reactive oxygen and nitrogen species generated by inflammatory cells oxidize the entrapped LDL, further driving macrophage uptake via scavenger receptors.

The continued lipid accumulation, endothelial activation, inflammation, and vascular smooth muscle cell proliferation leads to a positive feedback loop that drives the growth of the atherosclerotic plaque. In some cases, this process eventually may stabilize, while in other cases it can accelerate. The particular balance of these processes at different anatomic sites will determine the exact clinical sequelae.[5]

THE ATHEROSCLEROTIC PLAQUE

As lipid accumulates in foam cells, macrophage-derived cytokines, such as tumor necrosis factor α (TNF-α), further promote the recruitment and proliferation of smooth muscle cells and fibroblasts.[5,11,17] In turn, these cells secrete large amounts of extracellular matrix including collagen and proteoglycans. A key step in the transition from a fatty streak to a more advanced atherosclerotic plaque is the accumulation of extracellular lipid (Figure 12.1). Initially, this extracellular lipid interdigitates between the intimal smooth muscle cells. As the lesion progresses, the extracellular lipid coalesces to form large pools becoming the core of the atheroma. The core also contains necrotic material from dead foam cells and macrophages and is, thus, often referred to as the necrotic lipid core. Some of the lipid in the core originates from dead macrophage foam cells, while some entrapped LDL is likely deposited directly into the core. Within the lipid core cholesterol will crystallize to form sharp needle-like structures referred to as cholesterol clefts (Figure 12.3). Atherosclerotic plaques frequently show evidence of hemorrhage. The blood often comes from small vessels that have grown into the plaque from the adventitia. The cholesterol in red blood cell membranes can add substantially to the cholesterol pool in the necrotic lipid core.

Fibroblasts are recruited into the plaque, possibly from the adventitia or from circulating precursors. These cells secrete large amounts of collagen, causing fibrosis or scarring. As the plaque matures, the combined necrotic lipid core and the surrounding scar matrix form the characteristic fibroatheroma. The lipid and collagenous content in atherosclerotic plaques is quite variable, even within the same patient. Some plaques are predominantly deformable necrotic cores covered by a thin fibrous cap (see discussion below), while others are essentially all fibrous tissue, referred to as fibrous plaques. As we will see, the

actual architecture of plaques materially contribute to the subsequent pathologic outcomes. Advanced atherosclerotic plaques also undergo varying degrees of calcification.

Atherosclerotic plaques usually form with an eccentric geometry, that is, they are not uniformly distributed around the vascular circumference, but rather preferentially involve one portion of the vessel's luminal circumference. This is explicable based on the eccentric nature of the disturbed shear stresses being imparted on the endothelium at branch sites. Initially there is often outward or positive remodeling of the arterial wall due to changes in smooth muscle cell numbers, vessel tone, and

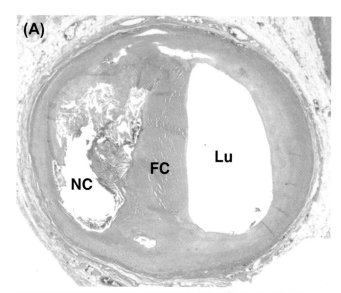

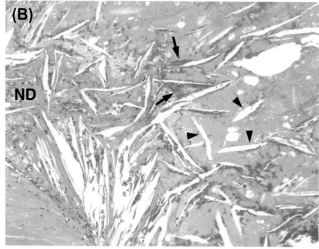

FIGURE 12.3 Stable atherosclerotic plaque. Shown are histologic sections of a stable fibroatheroma in a human coronary artery at autopsy. **(A)** At low magnification, the lumen (Lu) of the artery is seen to be well demarcated from the necrotic lipid core (NC) by a thick collagenous fibrous cap (FC). **(B)** On higher magnification, the necrotic lipid core can be seen to contain necrotic debris (ND), red blood cells (arrows), and white spike-like structures due to cholesterol esters (arrowheads). These latter structures are often referred to as cholesterol clefts. Cholesterol itself is extracted from tissue during routine processing leaving the open/white spike-like areas behind.

extracellular matrix deposition (Glagov phenomenon);[19] this maintains the size of the lumen despite the encroachment of the atherosclerotic plaque. Correspondingly, atherosclerotic vessels exhibit cellular activation markers in the medial smooth muscle cells.[20] However, there are limits to the adaptability of the vascular remodeling. When lesions occlude approximately 40% of the overall vessel diameter, the growth of the plaque begins to exceed the capacity to dilate, and there will be a diminished luminal cross-sectional area, so-called negative remodeling. At this point, such stenoses can be readily visualized by imaging techniques such as angiography and intravascular ultrasound.

The decreased luminal size restricts blood flow to the downstream tissue, although in most cases this will not become functionally limiting until 75% of the lumen is compromised. At this point, distal perfusion may be inadequate to supply the demands of the perfused tissue, leading to clinical symptoms, particularly when there is co-existent hypoxia, anemia, or hypotension.

Compromise of vascular flow in different sites becomes clinically apparent as distinct signs and symptoms. These include:

- Angina pectoris: insufficient blood flow to cardiac tissue classically results in chest pain; because of the vagaries of referred visceral pain (the pain sensation originates predominantly from pericardial innervation), angina may also be reflected by discomfort in the left arm or jaw. The pathogenesis of the pain is attributed to adenosine released from ischemic myocardium.
- Transient ischemic attacks (TIAs): insufficient blood flow to the brain can result in a host of neurologic sequelae including weakness, abnormal sensations, dysarthria (inability to speak), or syncope (loss of consciousness). If the insult is sufficiently mild or brief, there is full recovery. More significant compromise of perfusion leads to cerebrovascular accidents (stroke).
- Peripheral claudication: when blood flow to an extremity is compromised, there may be adequate perfusion at rest. However, any exertion (even simple ambulation) may lead to a supply–demand mismatch; with accumulating lactic acid (due to anaerobic metabolism), the muscles become painful.
- Increased blood pressure: diminished renal perfusion of the kidneys is perceived as a signal that there must be systemic hypotension. In such an instance, normal neuroendocrine feedback loops involving, for example, kidney renin release will lead to systemic hypertension by increasing vascular smooth muscle tone.

In some cases, however, the first symptoms relating to an atherosclerotic plaque may be catastrophic. General concepts are starting to emerge regarding the biological features of atherosclerosis that distinguish the gradual insidious forms of the disease from the sudden catastrophic manifestations. The current models distinguish between the stable atherosclerotic plaque as the source of slow progressive disease, and the unstable or vulnerable atherosclerotic plaque as the cause of sudden and catastrophic outcomes.[5,11]

Stable Plaques

A stable plaque is an atherosclerotic plaque that typically slowly accrues additional fibrous matrix and does not undergo a sudden change in its size or structure; it is therefore unlikely to cause sudden marked changes in luminal diameter. As discussed above, in some atherosclerotic plaques the amount of fibrosis or scarring may become extensive.[5] The necrotic lipid core may be completely covered by a thick fibrous cap, securely separating it from the overlying endothelium. These lesions may also calcify and in some cases undergo osseous metaplasia (bone formation) over time. Stable plaques also tend to have less inflammation than is present in unstable plaques. As the lesion grows in size, the vascular lumen narrows, leading to reduced downstream tissue perfusion. The symptoms in the heart, for example, would be reflected classically as stable angina, in which the same amount of exertion reliably triggers chest pain.

Unstable Plaques and Plaque Disruption

An unstable plaque is a plaque that undergoes a disruption of the endothelial surface causing a thrombus to form within the lumen of the vessel.[21–26] This thrombus can rapidly narrow the lumen of the artery causing sudden catastrophic events, such as stroke or sudden cardiac death (the latter most commonly due to ischemia-induced arrhythmias). Such unstable plaques are also referred to as vulnerable plaques, because they are prone to disruption and thrombosis.[5,8,9,17] Plaque disruptions can occur where there is previous severe stenosis of the lumen, with loss of around 75% or more of the cross-sectional area.[11] Distressingly, such plaque disruptions can also occur where there is less than 70% chronic occlusion,[27] and therefore when there is unlikely to be previous symptomatic ischemia. The resulting thrombus may entirely occlude the artery or only partially occlude the artery. In either case, fragments of the thrombus can embolize into the target tissue, obstructing smaller vessels and causing tissue ischemia.

Some of these plaque disruptions are due to full-thickness breaks, or ruptures, of the fibrous cap that overlies the necrotic and thrombogenic lipid core. In some fibroatheromas, the fibrous cap overlying the necrotic lipid core becomes markedly thinner due to the actions of matrix metalloproteinases. Such a plaque is often referred to as a thin cap fibroatheroma (TCFA) (Figure 12.4). Full-thickness ruptures of the fibrous cap usually occur at the site of a TCFA (Figures 12.5, 12.6). Upon rupture of the fibrous cap, the prothrombotic contents of the necrotic lipid core come into contact with the blood, triggering the clotting cascade and platelet aggregation. The ensuing thrombus markedly narrows the size of the previous lumen. Plaque ruptures are the most common form of plaque disruption, and since plaque ruptures most commonly occur in TCFAs, the TCFA is often considered to be the quintessential vulnerable plaque.

However, in addition to TCFAs, other forms of atherosclerotic plaque can also undergo acute plaque

disruption. This typically occurs without a full-thickness break of the fibrous cap, and in these cases the process is termed plaque erosion (Figure 12.7). In plaque erosions, the endothelial surface becomes disrupted and the underlying connective tissue triggers the formation of a thrombus. Unlike what happens in plaque ruptures, the entire necrotic lipid core does not come into contact with the blood. Thus, there is a tendency for the thrombus to be smaller in plaque erosions than it is in plaque ruptures. However, a plaque erosion can still cause significant acute narrowing of the vascular lumen as well as sudden catastrophic clinical events.

There is much interest in trying to understand the mechanisms triggering plaque rupture and erosion. Certainly in some patients, plaque disruptions appear to be related to an unusually exuberant inflammatory response in the atherosclerotic plaque (Figures 12.8, 12.9). In essence, what were smoldering embers may in some patients flame up into a firestorm. Exactly

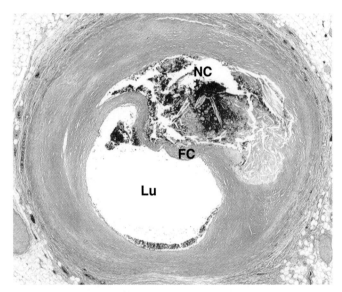

FIGURE 12.4 The vulnerable thin cap fibroatheroma. Shown is a histologic section of a thin cap fibroatheroma from a human coronary artery at autopsy. The necrotic lipid core (NC) is separated from the arterial lumen (Lu) by a relatively thin fibrous cap (FC). Compare with the stable plaque in Figure 12.3. Such thin cap fibroatheromas are considered to be vulnerable to rupture of the fibrous cap and sudden coronary artery occlusion. (Reproduced with permission from Stone JR. *Diag Histopathol* 2012;18:478–83.)

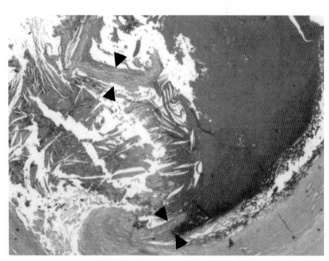

FIGURE 12.6 Acute plaque rupture. A thin cap fibroatheroma has ruptured allowing the necrotic lipid core on the left to contact the luminal blood on the right and trigger thrombus formation. Note the cholesterol clefts in the necrotic lipid core. The arrowheads indicate the ruptured thin fibrous cap. This plaque is from a coronary artery examined during the autopsy of a patient who died suddenly.

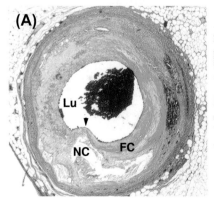

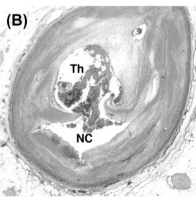

FIGURE 12.5 Rupture of a thin cap fibroatheroma. Shown are histologic sections from a coronary artery of a patient who died suddenly and unexpectedly. These sections were stained with Masson trichrome stain; collagen is stained blue. (A) The coronary artery contained areas of thin cap fibroatheroma (TCFA). The necrotic lipid core (NC) was separated from the arterial lumen (Lu) by a fibrous cap (FC), which in some areas was thin (arrow head). (B) In an adjacent section of the same coronary artery, there was acute rupture of the thin fibrous cap, such that the necrotic lipid core (NC) came into contact with the luminal blood, causing the formation of a thrombus (Th).

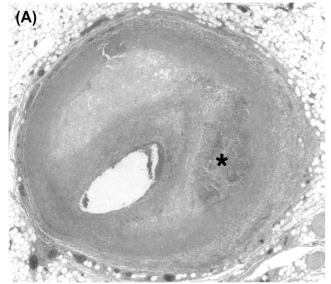

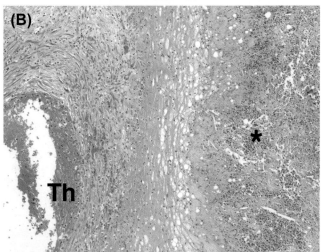

FIGURE 12.7 Erosion of an inflamed atherosclerotic plaque. Shown are histologic sections of a coronary artery from a man who died suddenly and unexpectedly. **(A)** At low magnification, there is severe narrowing of the artery due to atherosclerosis. There is focal bluish discoloration of the necrotic lipid core due to the intense infiltration by inflammatory cells (*). **(B)** At higher magnification the inflamed plaque is associated with surface erosion and luminal thrombus (Th). (Reproduced with permission from Stone JR. *Diag Histopathol* 2012;18:478–83.)

why and how this occurs is not known. The inflammatory response in these situations involves increased infiltration of the atherosclerotic plaque by macrophages, dendritic cells, and even neutrophils.[11,26,28–30] The inflammatory cells produce large amounts of MMPs that weaken the subendothelial connective tissue, in some cases causing rupture of the fibrous cap. In addition, the cytokines released by the inflammatory cells enhance endothelial dysfunction, promoting thrombosis. Many of the infiltrating inflammatory cells, both neutrophils and some macrophages, secrete

myeloperoxidase in an attempt to destroy a perceived foreign invader. These observations are spurring efforts to identify circulating inflammatory markers in the blood, such as myeloperoxidase, which may be useful in identifying which patients are undergoing an exuberant inflammatory reaction in their atherosclerotic plaques.[31,32]

While inflammation is clearly important in some cases of plaque disruption, sudden coronary artery thrombosis does occur in other patients by disruption of atherosclerotic plaques that contain few inflammatory cells. Thus, in these cases, alternative mechanisms of plaque disruption may be at work. Possible causes of plaque disruption in the absence of significant inflammation include vasospasm, endothelial degeneration, intraplaque hemorrhage, and physical protrusion of cholesterol crystals through the vessel wall elements. All of these mechanisms could disrupt the endothelial surface and promote the formation of luminal thrombus.

AORTIC ATHEROSCLEROSIS AND ATHEROSCLEROTIC ANEURYSMS

The aorta is a common site for the development of atherosclerosis.[33–38] However, the propensity to develop atherosclerosis is not uniformly equal throughout the length of the aorta. In humans, there is an increasing propensity to develop atherosclerosis as one proceeds distally from the ascending aorta towards the infrarenal abdominal aorta. In severe aortic atherosclerosis, the plaques are often chronically disrupted, giving a grossly ulcerated appearance to the aortic intimal surface (Figure 12.10). Complete luminal thrombosis can occur in the aorta, but it is uncommon due in part to the large luminal diameter. A much more common complication arising from these aortic atherosclerotic plaques is embolism. A fragment of the ulcerated surface material can break off, and migrate downstream to occlude a smaller vessel. This may involve either a fragment of thrombus (thromboembolism), or a fragment of the atherosclerotic plaque itself (atheroembolism), often involving a portion of the necrotic lipid core. Ulcerated plaques in the ascending aorta can produce atheroemboli that travel to the brain to cause strokes, or in rare cases can travel into the heart to cause myocardial ischemia and sudden death (Figure 12.10). However, since aortic atherosclerosis is most severe in the abdominal aorta, atheroemboli most commonly impact the kidneys and lower extremities.

Another common complication of atherosclerosis in the aorta and other large arteries is the formation of aneurysms.[39,40] As discussed above, early atherosclerosis is associated with positive remodeling, in which

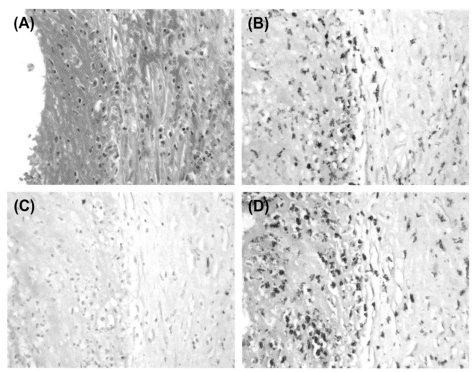

FIGURE 12.8 Macrophages and neutrophils in acute plaque erosion. Shown are high power images of the site of plaque erosion in Figure 12.7. **(A)** A routine hematoxylin and eosin stain shows infiltrating inflammatory cells with the morphology of macrophages and neutrophils. **(B)** Immunohistochemical stain for the macrophage marker CD68 (brown cells) highlights numerous macrophages. **(C)** Immunohistochemical stain for the T lymphocyte marker CD3 (brown cells) reveals only rare lymphocytes. **(D)** Immunohistochemical stain for myeloperoxidase (brown cells) shows numerous myeloperoxidase secreting neutrophils and macrophages. (Reproduced with permission from Stone JR. *Diag Histopathol* 2012;18:478–83.)

the outer circumference of the vessel wall enlarges in order to maintain the luminal diameter. However, this enlargement is markedly accentuated in the formation of an aneurysm. In this instance, there is often extensive destruction and scarring of the medial layer of the aorta, often due to the production of elastases and other MMPs. Consequently, much of the normal elasticity provided by the medial elastic fibers is lost. In addition there is often inflammation in the outer adventitia in the setting of severe aortic atherosclerosis. The adventitial inflammation consists primarily of macrophages, lymphocytes, plasma cells, and mast cells. The inflammatory cells secrete large amounts of digestive enzymes such as MMPs, which continually weaken the extracellular matrix. The end result is a localized marked enlargement of the vessel wall circumference. Abdominal aortic aneurysms are relatively common, affecting about 4% of men over the age of 65. The iliac arteries at the terminal bifurcation of the aorta are also prone to development of atherosclerotic aneurysms.

Aneurysms can lead to complications through multiple distinct mechanisms. The localized enlargement of the vascular wall creates regions of disturbed blood flow that promote thrombus formation. Abdominal aortic aneurysms are in fact often partially filled with poorly organized blood clots. The blood clot in an atherosclerotic aneurysm is prone to embolize, causing ischemia in downstream tissues. Another major complication of aneurysms is sudden rupture. The risk of rupture for an aortic aneurysm is related to its size; aortic aneurysms

greater than 5 cm diameter carry a 10% annual risk of rupture. Rupture of an aortic aneurysm can precipitate exsanguination, but even a circumscribed leak can lead to retroperitoneal hemorrhage, occasionally with extension into the abdominal cavity (hemoperitoneum). To prevent such catastrophes, aortic aneurysms greater than 5 cm in diameter are repaired, either by an open surgical procedure or by the transcatheter implantation of a stent graft.

THE GENETICS OF ATHEROSCLEROSIS

Not all people are equally predisposed to the development of atherosclerosis. In fact a positive family history is an important risk factor for predicting future atherosclerosis-related events for an individual. Population studies and animal models have confirmed several genetic variations that predispose to atherosclerotic plaque formation and progression.[14,41,42] While there are some notable exceptions, in most patients there is not a single genetic locus responsible for their atherosclerotic disease. Instead, a positive family history often reflects the complex interaction of several genes, each adding incrementally to the risk for disease development. Until recently, the roles of specific genetic variations in atherosclerosis were investigated essentially one at a time, largely based on knowledge of an atherosclerosis-related function of the protein product, such as the LDL receptor. More recently, genome-wide association studies (GWAS)

(A)

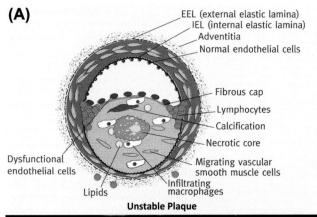

Unstable Plaque

(B)

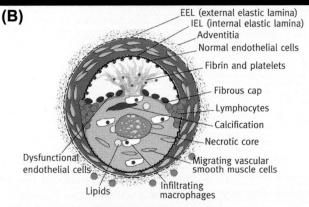

Unstable Plaque - with Erosion

(C)

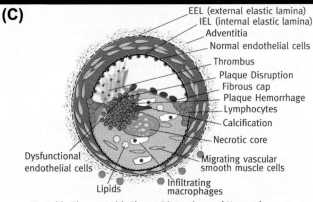

Unstable Plaque - with Plaque Disruption and Hemorrhage

FIGURE 12.9 Erosion of unstable atherosclerotic plaques.
(A) In some patients, unstable atherosclerotic plaques are characterized by increased amounts of inflammation, typically associated with increased production of matrix metalloproteinases and degradation of the extracellular matrix. **(B)** Erosion or disruption of the surface endothelium leads to luminal thrombus formation. **(C)** Disrupted plaques are often associated with hemorrhage into the plaque. The hemorrhage may originate from the luminal surface or from small blood vessels that have migrated into the plaque from the adventitia. (Reproduced with permission from Wang T, Palucci D, Law K, Yanagawa B, Yam J, Butany J. Atherosclerosis: pathogenesis and pathology. *Diag Histopathol* 2012;18:461–7.)

have greatly enhanced our understanding of the spectrum of genetic variations that may influence development of atherosclerosis.

ABO Blood Groups

One of the earliest sets of genetic variants to be associated with an elevated risk of atherosclerosis was the ABO blood group locus, and this finding has been confirmed in numerous studies.[14] Having a non-O blood group imparts increased risk of both myocardial infarction and peripheral vascular disease. Although the nature of the association is not fully elucidated, the ABO system correlates with thrombosis propensity, potentially through variation in the level of von Willebrand factor. The ABO blood group locus may also influence leukocyte–endothelial interactions early in inflammation. Thus, there is a connection between ABO blood group processing, and the expression of an endothelial surface adhesion molecule, E-selectin, which plays a role in leukocyte recruitment.

LDL Receptor (LDLR)

Low-density lipoprotein (LDL) is a plasma lipoprotein particle containing the protein apolipoprotein B (ApoB) and cholesterol esters. Unlike very-low-density lipoprotein (VLDL) and intermediate-density lipoprotein (IDL) particles, LDL does not contain significant triglycerides. In addition to ApoB, VLDL and IDL also contain an additional protein component, apolipoprotein E (ApoE). The liver secretes VLDL, which is converted to IDL and LDL as triglycerides are removed and taken up by target tissues. The LDL receptor (LDLR) is a cell surface protein that binds ApoB and mediates the uptake of LDL and IDL from the blood. Many cell types express the LDLR, but the majority of LDL is taken up by the liver.

Loss of the gene encoding the LDLR (or functional defects in the LDLR protein) results in a disease called familial hypercholesterolemia. Individuals who are heterozygous for this condition represent about one in every 500 people and display a two-fold to three-fold increase in plasma cholesterol levels compared with the general population. The heterozygotes are prone to the development of premature atherosclerosis in adulthood. People who are homozygous for the loss of the LDLR are more severely affected and may have five-fold to six-fold elevations in plasma cholesterol. These individuals have an even more accelerated rate of atherosclerosis development and may die from myocardial infarction before the age of 20. Familial hypercholesterolemia represents one of the few occasions in which atherosclerosis can be largely attributed to a single gene. The elucidation of the mechanisms underlying this disorder is a major cornerstone in our current understanding of the role of LDL in atherosclerosis.

In addition to complete loss of the LDLR gene, multiple polymorphisms have been identified in this gene, some of which lead to a pro-atherogenic state.[14,43] These

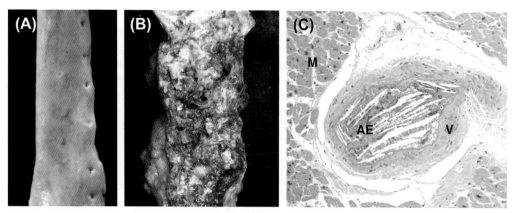

FIGURE 12.10   Aortic atherosclerosis and atheroemboli. The essentially normal aortic wall has a smooth tan inner luminal surface **(A)**. Atherosclerotic plaques in the aorta are often chronically disrupted on the surface, sometimes resulting in a grossly apparent thick carpet of ulcerated material **(B)**. This eroded surface material is prone to embolism, most commonly to the kidneys or lower extremities. Less commonly, eroded atherosclerotic plaques in the ascending aorta can send atheroemboli to the brain or even the heart. Shown at right **(C)** is a histologic image of the heart from a patient with severe ulcerated ascending aortic atherosclerosis who died suddenly and unexpectedly. Atheroemboli (AE) with cholesterol clefts were identified in multiple small vessels (V) within the myocardium (M).

pro-atherogenic polymorphisms are believed to either cause increased levels of LDL in the circulation or result in the abnormal processing of the LDL in the tissue. These LDL polymorphisms are often associated with a subtler phenotype than familial hypercholesterolemia, and in these cases the genetic variant is likely acting as one component of a polygenic disease.

Apolipoprotein E (ApoE)

ApoE is necessary for the uptake of chylomicron and VLDL remnants from the blood via the LDLR and the LDL receptor-related protein (LDR). Studies into another rare genetic disorder, primary dysbetalipoproteinemia (hyperlipoproteinemia type III), identified ApoE allelic variants as being involved in serum lipid metabolism and regulation.[14,42] Three common allelic variants of ApoE are now known to occur in the population and are referred to as E2, E3, and E4. The E3 allele is the most common, with a population frequency of about 60%. The E4 allele is the next most common allele, with a population frequency of 30%, and is associated with increased serum cholesterol levels compared with the E3 allele. In contrast the E2 allele has a population frequency of around 10% and is associated with decreased cholesterol levels. Together the ApoE alleles are thought to account for 5% of the genetic variation in cholesterol levels in the population.

Apolipoprotein a (LPA)

LPA is a gene encoding apolipoprotein a (apo(a)). Apo(a) is an important component of the lipoprotein particle Lp(a) and shows extensive homology to plasminogen. Lp(a) particles form from the covalent linking of ApoB on LDL particles with apo(a). Although the function of apo(a) is still poorly understood, Lp(a) particles may play a role in thrombosis. Lp(a) levels show tremendous variation throughout the population, due in large part to genetic polymorphisms in the LPA gene. Interestingly, Lp(a) arose rather recently in mammalian evolution and appears to only be found in humans and Old World monkeys. Increased levels of Lp(a) are associated with increased risk of atherosclerosis and associated diseases.[44,45]

Other Lipoprotein Genes

Variants in other apolipoprotein-related genes have been associated with altered atherosclerotic risk.[14,42] Some of the variants have been found in genes directly linked to LDL, such as the gene encoding ApoB. Other variants play important roles in triglyceride transport and metabolism. One example is the APOA5 gene encoding apolipoprotein A5 that is a component of chylomicrons, VLDL, and HDL, and which plays a major role in regulating plasma triglyceride levels. Another example is SORT1, which is involved in the synthesis of VLDL. These atherosclerosis-related variants indicate that dysregulation of multiple aspects of lipid metabolism can increase atherosclerosis risk.

Proprotein Convertase Subtilisin/Kexin type 9 (PCSK9)

PCSK9 is a protease that degrades the LDLR, thereby regulating its expression. Variants in the PCSK9 gene have been shown to alter LDLR levels and LDLR–LDL complex processing, leading to altered levels of serum and tissue cholesterol.[46,47] These PCSK9 variants have been found to affect the risk for atherosclerosis and coronary artery disease. An uncommon variant of PCSK9 is responsible for the rare autosomal dominant form of familial hypercholesterolemia. Other PCSK9 variants reduce cholesterol levels and the risk of atherosclerosis. Drugs that block PCSK9 activity may be beneficial in reducing cholesterol levels by increasing the expression of LDLR.

Cholesteryl Ester Transfer Protein (CETP)

CETP is an enzyme responsible for moving cholesterol esters and triglycerides between VLDL, LDL, and HDL. Lower CETP levels promote HDL formation. Since higher HDL levels are associated with decreased risk of atherosclerosis, the activity of CETP is thought to promote development of the disease by reducing HDL levels.[48] Likewise, rare variants in the CETP gene leading to increased CETP activity are associated with increased risk of atherosclerosis. Paradoxically, a D442G variant of CETP that decreases its activity and increases HDL levels is also associated with more severe atherosclerosis. Although there is much hope that CETP inhibition will raise HDL levels and reduce atherosclerosis, initial CETP inhibitor trials led to augmented HDL levels without improvement in atherosclerosis-related endpoints.

Inflammation-Related Genes

The synthesis of leukotrienes from arachidonic acid via lipoxygenase is a major pathway leading to small molecule regulators of inflammatory processes. ALOX5 is the gene encoding for arachidonate 5-lipoxygenase. Variants in ALOX5 can confer increased susceptibility to atherosclerosis. The proatherogenic 5-lipoxygenase variants appear to increase inflammation globally, and individuals with these variants have elevated serum levels of CRP.[14,49] Variants in other leukotriene-related genes, including ALOX5AP (encoding for the 5-lipoxygenase activating protein), LTA4H (leukotriene A4 hydrolase), and LTC4S (leukotriene C4 synthase) have also been linked with risk for atherosclerosis.[50,51] OX40, also known as CD134, is a T lymphocyte surface protein involved in T-cell activation. The ligand for OX40 (TNFSF4) is expressed by activated antigen-presenting cells. A variant in the gene encoding for the OX40 ligand has also been found to be associated with increased risk of myocardial infarction.[14]

Homocysteine-Related Genes

Elevated levels of homocysteine in the blood are associated with increased risk for vascular damage, atherosclerosis, thrombosis, coronary artery disease, and stroke, likely attributable to endothelial dysfunction.[52,53] Variants in a number of genes have been associated with elevated serum homocysteine levels and increased cardiovascular disease risk, including methylenetetrahydrofolate reductase (MTHFR), methionine synthase (MS), and cystathionine-beta-synthase (CBS). However, therapies to decrease circulating homocysteine levels in patients with elevated levels have not shown significant cardiovascular benefit to date.

Other Novel Atherosclerosis-Related Genetic Variations

Recent GWAS have identified novel genetic variants that influence susceptibility to atherosclerosis. The most consistent among these is a genomic locus located at 9p21. Surprisingly, this locus is not related to plasma lipid levels. The 9p21 locus is the site of a large non-coding RNA termed ANRIL (antisense non-coding RNA in the INK4 locus). The ANRIL transcript overlaps with several genes that regulate the cell cycle including cyclin-dependent kinase inhibitor 2A (CDKN2A), raising the possibility that the 9p21 locus may primarily affect cellular proliferation. However, enhancers important for interferon signaling are also located in this region.

A number of other genetic variants associated with atherosclerosis have been described in recent GWAS. One such variant is related to endothelin receptor type A (EDNRA), a member of the endothelin receptor family that is involved in maintaining vascular tone. Other novel genetic variants associating with atherosclerosis include genes encoding the metalloproteinase ADAMTS7, the cytochrome P450 member CYPA1, the cell cycle protein CNNM2, the 5'-nucleotidase NT5C2, and the zinc finger transcription factor ZHX2. Interestingly in these studies, variants that alter HDL levels (akin to CETP), tend not to associate with atherosclerosis risk.

CLINICAL IMPLICATIONS

Diagnosis

Clinical diagnosis of atherosclerosis-related illnesses most commonly occurs relatively late in the course of the disease process, well after the development of pathologic lesions. Before the onset of symptoms, risk for atherosclerosis is largely determined by assessing traditional risk factors including older age, male gender, family history of atherosclerosis-related events, elevated serum cholesterol, smoking, hypertension, diabetes mellitus, obesity, and sedentary lifestyle. More recently, non-specific serum inflammatory markers, such as CRP, are also being employed to predict risk of atherosclerosis-related events. However, while some serum biomarkers such as LDL and CRP may correlate with the degree of atherosclerosis in a population, there are no serum biomarkers currently in use that are specific for quantifying the atherosclerotic plaque burden in an individual. Thus, the definitive diagnosis of atherosclerosis still currently relies on a morphologic assessment.

Since arteries are necessary for the survival of the tissue they perfuse, arteries are not routinely biopsied to diagnose atherosclerosis. Consequently, the pre-mortem diagnosis of atherosclerosis is largely achieved through imaging studies. One commonly employed non-invasive imaging modality is ultrasound. Ultrasound is routinely used to assess for stenosis of carotid arteries in patients with symptoms of brain ischemia. In addition to identifying areas of stenosis, ultrasound does allow for some assessment of the total arterial wall thickness. Since ultrasound is relatively inexpensive and is non-invasive, it is being used to assess the carotid artery intimal medial thickness (CIMT) in patients without overt symptoms of carotid atherosclerosis. In this situation, an increased CIMT often correlates with traditional risk factors for atherosclerosis and inflammatory serum markers. It is likely that an increase in the CIMT by ultrasound often corresponds to the development of early atherosclerosis in that artery. Thus, measuring the CIMT may become a routine method to assess for atherosclerotic burden in asymptomatic patients.

However, routine non-invasive ultrasound is in general not useful for directly assessing atherosclerosis in the coronary arteries due to technical limitations. As indicated above, though, advanced atherosclerotic plaques can develop calcification, and this calcification can be detected by computed tomography (CT) radiologic evaluation. The amount of calcium in the coronary arteries also correlates with traditional risk factors for atherosclerosis and is predictive of future cardiac events. However, as might be anticipated based on individual plaque vulnerability, individuals can have cardiac events with little to no calcium in their atherosclerotic plaques, and some plaques can be loaded with calcium but are nevertheless quite stable and have low risk of a clinical event.

Angiography involves injecting a contrast agent into the arteries and imaging the lumen of the arteries with X-rays. Angiography may reveal stenoses or aneurysms consistent with atherosclerosis. Coronary artery angiography is the primary manner in which coronary artery atherosclerosis is currently diagnosed. Similar but more advanced techniques include computed tomography angiography (CTA) and magnetic resonance angiography (MRA).[54] A major limitation of traditional angiography is that it primarily images the lumen of the artery and not the vessel wall. This limits the sensitivity of angiography for detecting atherosclerosis, and also prevents an assessment of whether a plaque is stable or potentially vulnerable to disruption. MRA does offer some degree of assessment of the arterial wall itself, but often at a relatively low level of spatial resolution.

Newer imaging techniques are being employed which can reveal information concerning the structure of atherosclerotic plaques in living patients. The aim is to utilize these techniques to identify which patients have plaques likely to be vulnerable to disruption. Currently the most promising techniques are catheter-based modalities, where an imaging probe is inserted into the artery of interest. In these techniques, the close proximity of the imaging probe to the atherosclerotic plaque allows for enhanced spatial resolution. The most common intravascular catheter-based imaging modalities currently being utilized are radiofrequency intravascular ultrasound (VH-IVUS), optical coherence tomography (OCT), and near-infrared spectroscopy. Of these, VH-IVUS is currently the most widely used. The spatial resolution of IVUS is around 100–150 μm. Thus this technique cannot be used to measure the thickness of the fibrous caps in TCFAs, which are on the order of 10–100 μm. However, on IVUS, TCFAs are identified when the necrotic lipid core appears to be in contact with the lumen. OCT offers a better spatial resolution, around 10–15 μm. Thus OCT can be used to measure the cap thickness in a TCFA. Recently VH-IVUS has been shown to be able to predict, in some cases, which coronary atherosclerotic plaques may cause an acute coronary event.[55,56] These studies have revealed that a high degree of stenosis (loss of more than 70% cross-sectional area), small residual lumen, and a thin cap overlying the necrotic lipid core are all atherosclerotic plaque features associated with major adverse cardiac events, such as myocardial infarction. However, our ability to predict precisely which plaque will rupture or erode and in which patient is still far from ideal.

Treatment

Healthy diet and exercise are the classic recommendations for preventing atherosclerotic disease. In fact, much of the treatment for atherosclerosis is preventive, and focuses on the management of modifiable risk factors. Such preventive measures include a low-fat diet to reduce serum cholesterol levels, and a diet with a reasonable amount of calories along with exercise to prevent obesity. Smoking should be avoided. If necessary, hypertension is treated with antihypertensive medications including inhibitors of beta-adrenergic receptors (beta-blockers), calcium channel blockers, and angiotensin-converting enzyme (ACE) inhibitors. Diabetics are managed with oral agents that decrease serum glucose levels, and in some patients by insulin injection.

In addition to eating a low-fat diet, serum cholesterol levels can be reduced by a number of pharmacologic agents. Hydroxymethyl-glutaryl coenzyme A (HMG-CoA) reductase inhibitors, often referred to as statins, are currently a primary approach to pharmacologically prevent atherosclerosis-related events. Statins inhibit the endogenous synthesis of cholesterol in the liver, and have been shown to reduce atherosclerosis-related clinical events, such as myocardial infarction.[6] In addition to lowering cholesterol, there is growing evidence that statins can inhibit progression of atherosclerosis by a direct effect on the arterial wall. Statins inhibit the synthesis of isoprenoid lipid anchors, and thereby inhibit signaling through small G-proteins such as Ras and RhoA.[57] Statins also induce the expression of the atheroprotective endothelial transcription factors KLF2 and KLF4 (see Chapter 11).[58] Serum cholesterol levels can also be reduced by inhibiting the reabsorption of bile acids from the gastrointestinal tract with orally ingested resins, such as cholestyramine. Cholesterol absorption by intestinal epithelial cells can be inhibited with ezetimide. Niacin can be used to decrease hepatic secretion of VLDL.

Therapies aimed at inhibiting inflammation and coagulation have become more widely utilized in the war on atherosclerosis, particularly in secondary prevention where the patient is already identified as being at high risk for complications of atherosclerosis.[59] However, the benefits of inhibiting inflammation and coagulation must be balanced against the risks of increased infection and bleeding. Anti-inflammatory therapies are aimed at preventing the unstable plaque from progressing to plaque erosion or rupture, while anti-platelet and anti-thrombotic therapies are aimed at preventing lethal clots in response to plaque erosion or rupture.[8,9,11,59] The simplest and most common of these therapies is aspirin, which inhibits cyclo-oxygenase. Aspirin inhibits platelet aggregation by inhibiting production of thromboxane. Although aspirin also inhibits endothelial cell cyclo-oxygenase and thus anti-thrombotic prostacyclin production, endothelial cells have the capacity to regenerate the 'poisoned' enzyme (platelets do not), and the net effect is anticoagulant. Aspirin also alters the production of lipid mediators to favor production of anti-inflammatory lipoxins and resolvins, which are anti-inflammatory mediators that help clear or resolve chronic inflammation. It has been estimated that 40000 tons of aspirin are consumed worldwide each year.

When prevention fails, interventional therapies are required. A thrombus that forms over a ruptured plaque can be removed by direct enzymatic lysis, using 'clot busters' such as tissue type plasminogen activator (t-PA). Thrombi may also be removed mechanically using a catheter. The atherosclerotic plaque itself can be removed in some cases using an atherectomy catheter or by an open surgical procedure known as endarterectomy. An acutely occluded vessel can be opened using a balloon introduced through a catheter, a procedure known as balloon angioplasty. This procedure is often accompanied by the placement of a stent in the artery. Some of these stents are designed to deliver drugs locally (e.g., Taxol or rapamycin), which prevent smooth muscle cell proliferation and intimal hyperplasia. If catheter-based methods are not successful, patients with severe atherosclerotic coronary artery disease can also undergo coronary artery bypass grafting (CABG), in which vascular segments are sewn from the aorta to the more distal segments of the coronary arteries to bypass the proximal obstructions.

Despite all of these treatment approaches, atherosclerosis remains the number one cause of death in industrialized countries. The development of novel treatments will require a greater understanding of the underlying molecular mechanisms of the disease. Research into the nature of these molecular mechanisms relies not only on clinical observations and pathologic studies, but also on having model systems, which allow dissection of mechanistic aspects of the disease process.

LABORATORY AND ANIMAL MODELS

In vitro models using cultured endothelial cells, leukocytes, and vascular smooth muscle cells allow for dissection of cellular signaling pathways and processes relevant to atherosclerosis. Such models have elucidated many of the signaling pathways underlying vascular smooth muscle cell proliferation, extracellular matrix synthesis, cellular lipid accumulation, and foam cell formation.[60] For example, such cell culture models revealed that it was necessary for LDL to be modified by acetylation or oxidation in order to be efficiently phagocytized by macrophages via scavenger receptors. Endothelial culture models have elucidated cellular mechanisms of flow-dependent endothelial cell activation and leukocyte–endothelial interactions. Co-culture systems employing multiple cell types as well as models involving the culturing of intact segments of arteries allow for the characterization of even more complex processes, such as the formation of intimal hyperplasia.[61] One key advantage of these in vitro models is that they can be performed using human cells and tissues, thus overcoming some of the numerous species differences encountered when trying to translate results from animal models to human patients.

Currently, in vitro models do not accurately mirror all of the variables in an intact living person. Thus, studies with live animal models are important for assessing the effects of specific perturbations in the context of a living animal. The most widely used animal models in atherosclerosis research today are genetically engineered mice. The lab mouse was a challenge to the field of atherosclerosis research in the early years because wild-type mice simply do not develop atherosclerosis to any significant extent. A number of genetically modified variants now exist that develop atherosclerosis-like lesions in the aorta after receiving high-fat diets.[13] The most commonly used mice for these studies are deficient in either the LDLR or in ApoE. These models are widely available to researchers and are relatively inexpensive. These mice strains are often crossed with strains of mice that are genetically deficient in other genes to create double knockouts. The rapidly growing list of mice with specific gene deletions has resulted in an explosion of studies evaluating the role of specific genes in murine atherosclerosis. In fact well over 300 specific genes have been assessed in this fashion for their potential role in atherosclerosis, and we are beginning to understand murine atherosclerosis at a very detailed level.[62–69]

In general, in these mouse models of atherosclerosis, the genes that promote monocyte proliferation and macrophage activation, migration and phagocytosis enhance lesion development. Examples include monocyte chemoattractant protein 1 (MCP1), which regulates macrophage entry into the vessel wall, CCR2, which is the receptor on macrophages for MCP1, and the macrophage scavenger receptors CD36, scavenger receptor A (SR-A), and lectin-like oxidized LDL receptor 1 (LOX-1). In the mouse models, smooth muscle cells do not enter the intima until after lipid has accumulated, and they appear to do so as part of a reparative response. Thus in these gene deletion studies, genes that promote smooth muscle cell migration and proliferation tend to retard fatty lesion development. Examples include the chemokine/growth factor receptor CXCR4 and the matrix metalloproteinases MMP3 and MMP9. This latter observation does contrast somewhat with human pathologic studies in which MMPs are thought to promote plaque disruption and fatal coronary artery thrombosis. Along these lines, overexpression of MMP9 in a murine model was found to promote features of plaque instability.

A key issue concerning the mouse models is that it remains controversial as to whether the observations made with these models are directly relevant to the human disease. In these models the lesions develop quickly in the setting of very high serum cholesterol levels, and have a morphology that is distinct from that of human atherosclerosis. These may be very good models for understanding lipid storage disorders such as homozygous familial hypercholesterolemia, but it is unclear whether the results from these models can be extrapolated to the more general patient population having much more modest elevations in serum cholesterol.

Other animal species, including rabbits, pigs, and non-human primates, have also been utilized to study atherosclerosis.[13,70] When fed a high-fat diet, pigs and non-human primates will form coronary artery lesions that morphologically resemble human atherosclerosis. Moreover, in contrast to the mouse where the aorta is the site of atherosclerotic lesions, larger animal models can develop lesions in the coronary arteries or other smaller vascular beds.

Studies with non-human primates have shown that withdrawal of a high-fat diet after atherosclerotic lesion development leads to the formation of fibrous plaques with little to no lipid.[71] Thus, it is possible to 'dry out' or stabilize an atherosclerotic plaque by reducing serum cholesterol levels, at least in these animal systems. However, it is unclear whether all fibrous plaques found in human patients develop in such a manner, or if some patients simply develop fibrous plaques directly without going through a lipid-laden atheroma stage. Large animal models are also utilized in some instances to assess the effects of new pharmaceutical agents in preventing atherosclerosis.[72] An important caveat to such approaches is that the expense and ethical issues of utilizing larger animals can make such studies prohibitive.

CONCLUSIONS

Atherosclerosis is a complex pathophysiologic disease process resulting in the accumulation of lipids in the vascular wall with scarring and calcification. The disease may narrow arteries, slowly leading to insidious tissue ischemia or may do so rapidly causing catastrophic events such as stroke, myocardial infarction, and sudden cardiac death. While there are modifiable risk factors that can be targeted to delay the onset of the disease, a great number of people suffer regardless of risk factor optimization, particularly at advanced ages. Research has uncovered several molecular mechanisms involved in the disease process including mechanisms relating to vascular cell activation, foam cell formation, and fibrous cap disruption. These ongoing lines of investigation continue to spur on the development of novel treatments for the disease. However, despite these advances, cardiovascular diseases, particularly atherosclerosis, remain the number one cause of death in industrialized countries. Thus there is a great deal more work remaining to effectively combat this disease.

References

1. Schwartz CJ, Mitchell JR. The morphology, terminology and pathogenesis of arterial plaques. *Postgrad Med J* 1962;**38**:25–34.
2. Velican D, Velican C. Study of fibrous plaques occurring in the coronary arteries of children. *Atherosclerosis* 1979;**33**:201–15.
3. Tuzcu EM, Kapadia SR, Tutar E, Ziada KM, Hobbs RE, McCarthy PM, et al. High prevalence of coronary atherosclerosis in asymptomatic teenagers and young adults: Evidence from intravascular ultrasound. *Circulation* 2001;**103**:2705–10.
4. Stary HC, Chandler AB, Glagov S, Guyton JR, Insull W, Rosenfeld ME, et al. A definition of initial, fatty streak, and intermediate lesions of atherosclerosis. A report from the Committee on Vascular Lesions of the Council on Arteriosclerosis, American Heart Association. *Circulation* 1994;**89**:2462–78.

5. Wang T, Palucci D, Law K, Yanagawa B, Yam J, Butany J. Atherosclerosis: pathogenesis and pathology. *Diag Histopathol* 2012;**18**:461–7.

6. Aikawa M, Libby P. Lipid lowering therapy in atherosclerosis. *Semin Vasc Med* 2004;**4**:357–66.

7. Sitia S, Tomasoni L, Atzeni F, Ambrosio G, Cordiano C, Catapano A, et al. From endothelial dysfunction to atherosclerosis. *Autoimmun Rev* 2010;**9**:830–4.

8. Libby P. Inflammation in atherosclerosis. *Arterioscler Thromb Vasc Biol* 2012;**32**:2045–51.

9. Hansson GK. Inflammation, atherosclerosis, and coronary artery disease. *N Engl J Med* 2005;**352**:1685–95.

10. Ross R. Atherosclerosis - an inflammatory disease. *N Engl J Med* 1999;**340**:115–26.

11. Stone JR. Pathology of myocardial infarction, coronary artery disease, plaque disruption, and the vulnerable atherosclerotic plaque. *Diag Histopathol* 2012;**18**:478–83.

12. Rader DJ, Pure E. Lipoproteins, macrophage function, and atherosclerosis: beyond the foam cell? *Cell Metab* 2005;**1**:223–30.

13. Getz GS, Reardon CA. Animal models of atherosclerosis. *Arterioscler Thromb Vasc Biol* 2012;**32**:1104–15.

14. Lusis AJ. Genetics of atherosclerosis. *Trends Genet* 2012;**28**:267–75.

15. Dhawan SS, Avati Nanjundappa RP, Branch JR, Taylor WR, Quyyumi AA, Jo H, et al. Shear stress and plaque development. *Expert Rev Cardiovasc Ther* 2010;**8**:545–56.

16. Gimbrone MA, Garcia-Cardena G. Vascular endothelium, hemodynamics, and the pathobiology of atherosclerosis. *Cardiovasc Pathol* 2013;**22**:9–15.

17. Weber C, Noels H. Atherosclerosis: current pathogenesis and therapeutic options. *Nat Med* 2011;**17**:1410–22.

18. Ridker PM, Cushman M, Stampfer MJ, Tracy RP, Hennekens CH. Inflammation, aspirin, and the risk of cardiovascular disease in apparently healthy men. *N Eng J Med* 1997;**336**:973–9.

19. Glagov S, Weisenberg E, Zarins CK, Stankunavicius R, Kolettis GJ. Compensatory enlargement of human atherosclerotic coronary arteries. *N Eng J Med* 1987;**316**:1371–5.

20. Panchenko MP, Silva N, Stone JR. Upregulation of a hydrogen peroxide responsive pre-mRNA binding protein in atherosclerosis and intimal hyperplasia. *Cardiovasc Pathol* 2009;**18**:167–72.

21. Chapman I. Morphogenesis of occluding coronary artery thrombosis. *Archiv Pathol* 1965;**80**:256–61.

22. Constantinides P. Plaque fissures in human coronary thrombosis. *J Atherosclerosis Res* 1966;**6**:1–17.

23. Friedman M, Van den Bovenkamp GJ. The pathogenesis of a coronary thrombus. *Am J Pathol* 1966;**48**:19–44.

24. Fishbein MC. The vulnerable and unstable atherosclerotic plaque. *Cardiovasc Pathol* 2010;**19**:6–11.

25. Virmani R, Burke AP, Farb A, Kolodgie FD. Pathology of the vulnerable plaque. *J Am Coll Cardiol* 2006;**47**:C13–8.

26. Naruko T, Ueda M, Haze K, van der Wal AC, van der Loos CM, Itho A, et al. Neutrophil infiltration of culprit lesions in acute coronary syndromes. *Circulation* 2002;**106**:2894–900.

27. Davies MJ. A macro and micro view of coronary vascular insult in ischemic heart disease. *Circulation* 1990;**82**(Suppl. II):II-38–46.

28. Tavora F, Ripple M, Li L, Burke AP. Monocytes and neutrophils expressing myeloperoxidase occur in fibrous caps and thrombi in unstable coronary plaques. *BMC Cardiovasc Disord* 2009;**9**:27.

29. Sugiyama S, Okada Y, Sukhova GK, Virmani R, Helnecke JW, Libby P. Macrophage myeloperoxidase regulation by granulocyte macrophage colony-stimulating factor in human atherosclerosis and implications in acute coronary syndromes. *Am J Pathol* 2001;**158**:879–91.

30. Ionita MG, van den Borne P, Catanzariti LM, Moll FL, de Vries JP, Pasterkamp G, et al. High neutrophil numbers in human carotid atherosclerotic plaques are associated with characteristics of rupture-prone lesions. *Arterioscler Thromb Vasc Biol* 2010;**30**:1842–8.

31. Zhang R, Brennan ML, Fu X, Aviles RJ, Pearce GL, Penn MS, et al. Association between myeloperoxidase levels and risk of coronary artery disease. *JAMA* 2001;**286**:2136–42.

32. Baldus S, Heeschen C, Meinertz T, Zeiher AM, Eiserich JP, Munzel T, et al. Myeloperoxidase serum levels predict risk in patients with acute coronary syndromes. *Circulation* 2003;**108**:1440–5.

33. Holman RL, McGill Jr HC, Strong JP, Geer JC. The natural history of atherosclerosis: the early aortic lesions as seen in New Orleans in the middle of the 20th century. *Am J Pathol* 1958;**34**:209–35.

34. McGill Jr HC. Fatty streaks in the coronary arteries and aorta. *Lab Invest* 1968;**18**:560–4.

35. Tejada C, Strong JP, Montenegro MR, Restrepo C, Solberg LA. Distribution of coronary and aortic atherosclerosis by geographic location, race, and sex. *Lab Invest* 1968;**18**:509–26.

36. Virmani R, Avolio AP, Mergner WJ, Robinowitz M, Herderick EE, Cornhill JF, et al. Effect of aging on aortic morphology in populations with high and low prevalence of hypertension and atherosclerosis. Comparison between occidental and Chinese communities. *Am J Pathol* 1991;**139**:1119–29.

37. McGill Jr HC, McMahan CA, Herderick EE, Tracy RE, Malcom GT, Zieske AW. Strong JP, for the PDAY Research Group. Effects of coronary heart disease risk factors on atherosclerosis of selected regions of the aorta and right coronary artery. *Arterioscler Thromb Vasc Biol.* 2000;**20**:836–45.

38. van Dijk RA, Virmani R, von der Thusen JH, Schaapherder AF, Lindeman JHN. The natural history of aortic atherosclerosis: A systematic histopathological evaluation of the peri-renal region. *Atherosclerosis* 2010;**210**:100–6.

39. Imakita M, Yutani C, Ishibashi-Ueda H, Nakajima N. Atherosclerotic abdominal aortic aneurysms: Comparative data of different types based on the degree of inflammatory reaction. *Cardiovasc Pathol* 1992;**1**:65–73.

40. Nordon IM, Hinchliffe RJ, Holt PJ, Loftus IM, Thompson MM. Review of current theories for abdominal aortic aneurysm pathogenesis. *Vascular* 2009;**17**:253–63.

41. Milewicz DM, Seidman CE. Genetics of cardiovascular disease. *Circulation* 2000;**102**:IV103–11.

42. Lusis AJ, Fogelman AM, Fonarow GC. Genetic basis of atherosclerosis: part I: new genes and pathways. *Circulation* 2004;**110**:1868–73.

43. Raal FJ, Santos RD. Homozygous familial hypercholesterolemia: current perspectives on diagnosis and treatment. *Atherosclerosis* 2012;**223**:262–8.

44. Kamstrup PR. Lipoprotein(a) and ischemic heart disease–a causal association? A review. *Atherosclerosis* 2010;**211**:15–23.

45. Pati U, Pati N. Lipoprotein(a), atherosclerosis, and apolipoprotein(a) gene polymorphism. *Mol Genet Metab* 2000;**71**:87–92.

46. Cariou B, Le May C, Costet P. Clinical aspects of PCSK9. *Atherosclerosis* 2011;**216**:258–65.

47. Davignon J, Dubuc G, Seidah NG. The influence of PCSK9 polymorphisms on serum low-density lipoprotein cholesterol and risk of atherosclerosis. *Curr Atheroscler Rep* 2010;**12**:308–15.

48. Schwartz GG. New horizons for cholesterol ester transfer protein inhibitors. *Curr Atheroscler Rep* 2012;**14**:41–8.

49. Helgadottir A, Manolescu A, Thorleifsson G, Gretarsdottir S, Jonsdottir H, Thorsteinsdottir U, et al. The gene encoding 5-lipoxygenase activating protein confers risk of myocardial infarction and stroke. *Nat Genet* 2004;**36**:233–9.

50. Camacho M, Martinez-Perez A, Buil A, Siguero L, Alcolea S, Lopez S, et al. Genetic determinants of 5-lipoxygenase pathway in a Spanish population and their relationship with cardiovascular risk. *Atherosclerosis* 2012;**224**:129–35.

51. Sun H, Zhang J, Wang J, Sun T, Xiao H, Zhang JS. Association between genetic variants of the leukotriene biosynthesis pathway and the risk of stroke: a case-control study in the Chinese Han population. *Chin Med J (Engl)* 2013;**126**:254–9.

52. Smulders YM, Blom HJ. The homocysteine controversy. *J Inherit Metab Dis* 2011;**34**:93–9.

53. Trabetti E. Homocysteine, MTHFR gene polymorphisms, and cardio-cerebrovascular risk. *J Appl Genet* 2008;**49**:267–82.

54. Gallino A, Stuber M, Crea F, Falk E, Corti R, Lekakis J, et al. "In vivo" imaging of atherosclerosis. *Atherosclerosis* 2012;**224**:25–36.

55. Stone GW, Maehara A, Lansky AJ, de Bruyne B, Cristea E, Mintz GS, et al. A prospective natural-history study of coronary atherosclerosis. *N Engl J Med* 2011;**364**:226–35.

56. Calvert PA, Obaid DR, O'Sullivan M, Shapiro LM, McNab D, Densem CG, et al. Association between IVUS findings and adverse outcomes in patients with coronary artery disease: the VIVA (VH-IVUS in Vulnerable Atherosclerosis) study. *J Am Coll Cardiol Img* 2011;**4**:894–901.

57. Sadowitz B, Maier KG, Gahtan V. Basic science review: Statin therapy Part I: The pleiotropic effects of statins in cardiovascular disease. *Vasc Endovascular Surg* 2010;**44**:241–51.

58. Parmar KM, Nambudiri V, Dai G, Larman HB, Gimbrone Jr MA, Garcia-Cardena G. Statins exert endothelial atheroprotective effects via the KLF2 transcription factor. *J Biol Chem* 2005;**280**:26714–9.

59. Davi G, Patrono C. Platelet activation and atherothrombosis. *N Engl J Med* 2007;**357**:2482–94.

60. Rezvan A, Ni CW, Alberts-Grill N, Jo H. Animal, in vitro, and ex vivo models of flow-dependent atherosclerosis: role of oxidative stress. *Antioxid Redox Signal* 2011;**15**:1433–48.

61. Panchenko MP, Siddiquee Z, Dombkowski DM, Alekseyev YO, Lenburg ME, Walker JD, et al. Protein kinase CK1αLS promotes vascular cell proliferation and intimal hyperplasia. *Am J Pathol* 2010;**177**:1562–72.

62. Daugherty A, Rateri DL. Development of experimental designs for atherosclerosis studies in mice. *Methods* 2005;**36**:129–38.

63. Ni M, Chen WQ, Zhang Y. Animal models and potential mechanisms of plaque destabilisation and disruption. *Heart* 2009;**95**:1393–8.

64. Tedgui A, Mallat Z. Cytokines in atherosclerosis: pathogenic and regulatory pathways. *Physiol Rev* 2006;**86**:515–81.

65. Greaves DR, Gordon S. The macrophage scavenger receptor at 30 years of age: current knowledge and future challenges. *J Lipid Res* 2009;**50**(Suppl.):S282–6.

66. Moore KJ, Freeman MW. Scavenger receptors in atherosclerosis: beyond lipid uptake. *Arterioscler Thromb Vasc Biol* 2006;**26**:1702–11.

67. Newby AC, George SJ, Ismail Y, Johnson JL, Sala-Newby GB, Thomas AC. Vulnerable atherosclerotic plaque metalloproteinases and foam cell phenotypes. *Thromb Haemost* 2009;**101**:1006–11.

68. Kleemann R, Zadelaar S, Kooistra T. Cytokines and atherosclerosis: a comprehensive review of studies in mice. *Cardiovasc Res* 2008;**79**:360–76.

69. Zernecke A, Weber C. Chemokines in the vascular inflammatory response of atherosclerosis. *Cardiovasc Res* 2010;**86**:192–201.

70. Vilahur G, Padro T, Badimon L. Atherosclerosis and thrombosis: insights from large animal models. *J Biomed Biotechnol* 2011;**2011**:907575.

71. Stary HC. *Atlas of atherosclerosis progression and regression.* 2nd ed. New York, NY: Parthenon Publishing; 2003.

72. Wilensky RL, Shi Y, Mohler ER, Hamamdzic D, Burgert ME, Li J, et al. Inhibition of lipoprotein-associated phospholipase A2 reduces complex coronary atherosclerotic plaque development. *Nat Med* 2008;**14**:1059–66.

13

Genetic Diseases of the Aorta (Including Aneurysms)

Marc K. Halushka, MD, PhD

The Johns Hopkins Hospital, Baltimore, MD, USA

THE NORMAL AORTA: HISTOLOGY AND FUNCTION

The aorta is the largest artery, by caliber, in the human body. It is the conduit from the heart that extends from the thoracic cavity to the pelvic area where it bifurcates to create the iliac arteries. In adults, this vessel can measure around 490 cm in length.[1] This elastic artery is important in absorbing the systolic bolus of blood emanating from the contracting left vessel and supplying a continuous flow of blood into all of the body's organs and extremities. There is a slight vascular expansion that occurs in conjunction with the passing systolic wave followed by elastic recoil that occurs after the wave has passed. With recoil, the vessel returns to its normal caliber. The aortic wall is divided into three anatomic divisions relative to the vessel lumen: the tunica intima, tunica media, and tunica adventitia.

The tunica intima (intima) extends from the endothelium to the internal elastic lamina (IEL). The endothelium is a single layer of endothelial cells forming a barrier between the structural aorta and the passing blood. In healthy aortae, beneath these endothelial cells, is a small amount of connective tissue, primarily type III and type IV collagen.[2] Occasional mesenchymal cells, variably described as fibroblasts, myofibroblasts, or smooth muscle cells, are also present in this space. Endothelial cells are important mediators of vascular activity, providing paracrine control over the vascular smooth muscle cells (VSMCs) of the media (See Chapters 11 and 12). One key pathway of this regulation is the nitric oxide system. Nitric oxide, produced by endothelial cells, either constitutively by endothelial nitric oxide synthase (eNOS) or by the stimulated inducible NOS (iNOS), induces vasodilation by signaling underlying smooth muscle cells to relax.[3,4] Other endothelial function, or more accurately, endothelial dysfunction,

is thought to initiate pro-inflammatory signaling resulting in atherosclerosis. An important local factor causing endothelial dysfunction is non-laminar flow, found at branch points and bifurcations. This non-laminar flow results in pro-atherogenic shear stress. A force of 16 dynes/cm² and a pulsatile pattern wave form have been shown to be the most atherosclerosis-resistant flow across endothelium, while slower or more turbulent forms are pro-atherogenic.[5] Endothelial dysfunction can also be driven by systemic factors. Toxic metabolites of smoking, advanced glycation end products, cholesterol, hyperglycemic effects and dietary deficiencies all can result in increased endothelial signaling to increase inflammation and drive atherosclerosis.[4] Healthy and intact endothelial cells also prevent vascular thrombosis by acting as a barrier between extracellular matrix and circulating clotting factors.

The intima is sharply demarcated from the tunica media by the internal elastic lamina (IEL). The IEL is a continuous fenestrated elastic fiber. It is reasonably thick and is easily visible and distinguishable when elastic staining is performed on a histologic section. The IEL functions as a barrier preventing the movement of large molecules such as albumin and cholesterol from passing from the intima into the underlying media.[6] The IEL is a feature of all large- and medium-caliber arteries. Beneath the IEL is the tunica media.

The aortic media is a circumferential compaction of lamellar units (Figure 13.1). Each lamellar unit consists of a layer of elastic fibers overlying a layer of smooth muscle cells. Within the lamellar unit is a small amount of other extracellular matrix material including collagen and a fine elastic network of ground substance that is composed of mucopolysaccharides and glycosaminoglycans.[7] The dominant collagen type is type III, followed by type I and type IV.[2] Depending on the proximity to the annulus, there are ~60 lamellar units within

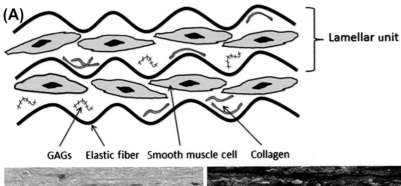

GAGs   Elastic fiber   Smooth muscle cell   Collagen

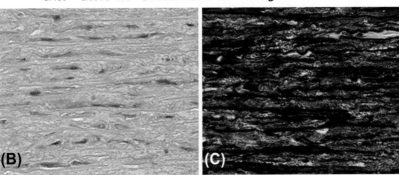

FIGURE 13.1   Schematic and histology of normal aortic media. **(A)** The media is composed of lamellar units, stacked layers of thick elastic fibers, and smooth muscle cells. Collagen and glycosaminoglycans (GAGs) are also present in the aortic wall. **(B)** A high-power view of the wall shows thin smooth muscle cells between difficult to observe elastic fibers (hematoxylin and eosin, 100 × original magnification). **(C)** A Movat Pentachrome highlights the dense elastic fibers (black) present in a normal aorta (100 × original magnification).

the human aorta stacked from the IEL to the outermost media. Interestingly, lamellar units correlate with animal body size with small animals such as a mouse having an average of five lamellar units and large animals such as pigs having an average of 72 lamellar units.[8] Healthy elastic fibers are roughly 1.5 µm in thickness.[7] The smooth muscle cells of the media are uniformly tapered cells oriented parallel to the arterial wall. They have a contractile apparatus, not visible on histologic slides, that controls vasorelaxation and vasoconstriction.

Embryologically, the VSMCs of the ascending and abdominal aortic media are derived from different founding cell populations. The VSMCs of the ascending aorta originate from ectodermal cardiac neural crest tissues. Conversely, the VSMCs of the abdominal aorta are mesodermally derived from somites.[9–11] This distinction is believed to be important in the role of TGF-β on the localization of syndromic aortic disease described below.[12]

Due to the thickness of the arterial wall, oxygen and other metabolites are unable to diffuse to the outermost lamellar units. Thus the aorta has a specialized vascular supply system to the outer third of the media. This system is known as the vasa vasorum which runs parallel to the aorta within the adventitia. The ascending aortic vasa vasorum arises from branching vessels of the coronary arteries, and the aortic arch has vasa vasorum derived from neck vessels. The vasa vasorum has small penetrating arteries and veins that enter the outer part of the media.[13]

The external elastic lamina separates the media from the adventitia. This elastic lamina is often less distinct from the IEL and is often recognizable on an elastic stain as the location where the normal lamellar unit structure

of the media segues into a more indistinct arrangement of elastic fibers. The adventitia functions to anchor the aorta to other adjacent structures and restrain the vessel from excessive extension and recoil.[14] The adventitia is comprised of a variety of extracellular matrix proteins providing loose and dense irregular connective tissues loosely arranged with intermixed adipocytes, lymphatic channels, nerves, thin-walled vessels and rare chronic inflammatory cells.[13] The adventitia can be variably sized from only a small amount of extracellular matrix material containing the vasa vasorum to a thick rind of fibrous tissue appreciated in some aortic diseases.

GROSS PATHOLOGIC CHANGES TO THE AORTA

The three main disease processes impacting on the aorta are atherosclerosis, inflammation, and aneurysm. Atherosclerosis, common to many large-caliber arteries, is described in detail in Chapter 12 and only briefly here. At the gross level, aortic atherosclerosis forms intimal plaques at areas of bifurcation, branch points, or areas of turbulent flow. The infrarenal aorta and aortic arch are the two sites that most commonly contain plaque. These intimal plaques appear as calcified, thickened, yellow-to-white areas in the aorta measuring up to several centimeters. Some severe plaques can have ulcerations with adherent thrombi. When atherosclerosis results in vessel aneurysm (typically located in the infrarenal aorta), mural thrombi can form in the vessel. This thrombus can appear gray and gelatinous-like.

Inflammatory processes, i.e. arteritis, are also a pathologic entity impacting on the aorta. Arteritis is described in Chapter 11. Grossly, arteritis of the aorta – aortitis – can cause adventitial thickening and/or intimal thickening as inflammatory cells drive fibrotic changes to the wall. Grossly, aortitis can have a 'tree bark' appearance in which the wall has wrinkles and ridges defining separate plaques.[15] Aortitis, from any cause, can also damage the media leading to its thinning and potential aneurysm.

Aneurysms have been defined as dilatations of the arterial wall, over 50% larger than the normal size, in which all layers – the intima, media, and adventitia – are involved. False aneurysms do not involve all three layers. The two major types of aneurysms – are saccular – outpouchings of the wall – and fusiform – concentric expansions of the wall. The so-called 'berry aneurysm' of the circle of Willis is a typical saccular aneurysm associated with autosomal dominant polycystic kidney disease (ADPKD), hypertension or smoking. Fusiform aneurysms are more frequent in both the ascending and abdominal aorta. In the ascending aorta, they are strongly associated with genetic influences and are frequently seen in syndromes involving mutation in a single gene.[15] Aortic aneurysms can also be the result of atherosclerosis or aortitis mentioned above. Aneurysms can occur at any location within the arterial tree but certain anatomic locations are more frequent. Common event sites include the ascending thoracic aorta and aortic arch, infrarenal abdominal aorta, carotid artery, renal artery, and the circle of Willis.

While aneurysms are worrisome clinical entities, they may cause no direct morbidity. Aneurysms in the ascending aorta can lead to insufficiency or regurgitation of the aortic valve. Large aneurysms may also rupture, leading to catastrophic events, such as hemoperitoneum and stroke. In contrast to ruptured aneurysms, dissections may occur in an artery either with or without a pre-existing aneurysm (Figure 13.2). Dissections cause mortality primarily through reduced arterial flow to critical organs such as the heart, brain, and abdominal viscera. Similar to aneurysm rupture, dissections can progress through the entire arterial wall allowing blood to exit the blood vessel and fill anatomic spaces resulting in hemothorax, hemoperitoneum, or cardiac compromise through hemopericardium. Even in the current era with excellent medical imaging and surgical techniques, dissections are still associated with a 13–30% rate of in-hospital mortality depending on their location.[16,17]

Two classification schemes exist to describe the directionality of dissection progression in the aorta. These two schemes, the DeBakey and the Stanford conventions, take into account the initial intimal location of the tear and the direction of their dissection. A dissection that begins in the ascending aorta and extends into the descending/abdominal aorta is a DeBakey I.

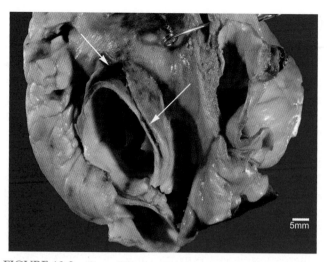

FIGURE 13.2 Gross dissection. In this postmortem heart, a dissection (arrows) was noted in the ascending aorta. The dissection is fixed blood (black) separating outer media and the adventitia.

If the dissection is confined to the ascending aorta it is a DeBakey II, and if it involves only the descending aorta it is a DeBakey III. In the Stanford criteria, any involvement of the ascending aorta is type A and any dissection not in the ascending aorta is a type B.[18]

DEMOGRAPHICS OF ANEURYSMS AND DISSECTIONS

As a result of dissections, thoracic aortic aneurysms precede major morbidity and mortality in industrialized nations, including the United States. Because preclinical aortic aneurysms are difficult to detect, the incidence and prevalence of these aortic lesions are incompletely described. However, aortic dissection, which is the life-threatening sequela of aortic aneurysm, is thought to occur at approximately ~3 cases per 100 000 patient years.[19] Dissection is slightly more likely to occur in males than females and among individuals in the sixth decade of life. Despite the overall advanced age of aneurysm patients, ascending aortic dissection has a bimodal pattern with a second population, harboring genetic mutations in a variety of aortopathic syndromes occurring in adolescents and young adults. This early appearance of aneurysm is associated with different etiologic causes and histopathologic appearances.

Classically, abrupt tearing chest pain is the main symptom of acute aortic dissection. Additional signs and symptoms of dissection include pulse differences between upper and lower extremities, acute congestive heart failure, altered consciousness, a widened mediastinum on chest X-ray and a periaortic hematoma by computed tomography (CT) or magnetic resonance imaging (MRI) scan.

HISTOPATHOLOGIC CHANGES TO THE AORTA

There are three important overlapping histopathologic observations that can be made in the aorta. These are atherosclerotic plaques, inflammation, and medial degeneration. As mentioned above, atherosclerosis is described in greater detail in Chapter 12. Briefly, atherosclerosis can be an important cause of aortic aneurysm. It can cause medial thinning in part likely due to anoxic injury to the smooth muscle cells below a large area of intimal atherosclerotic plaque. Once smooth muscle cell loss has occurred, leading to medial thinning, repeated systolic pulse pressure waves can dilate the wall locally in the absence of sufficient vascular strength and recoil. Atherosclerosis is a primary cause of abdominal aortic aneurysms. In the ascending aorta, extensive plaque burden is present in only a small subset of individuals and rarely in individuals harboring mutations in genes that predispose toward aneurysm. Atherosclerosis is generally not as frequent as the constellation of histopathologies known as medial degeneration in this second population. Atherosclerosis occurs throughout the full length of the aorta, but the plaque burden is usually more extensive in the perirenal abdominal aorta than ascending aorta. Extensive plaque burden in the abdominal area leads to a marked thinning of the media. Any degree of plaque severity, from pre-atheromas to complex and fissured plaques, may be present along with marked inflammation.[20] When associated with aneurysm, a large segment of the underlying VSMCs is lost with resultant laminar medial collapse. These histopathologies are described below. Atherosclerotic plaques are also associated with inflammation; generally chronic inflammatory cells present in the adventitia.

Aortic injury can also be the result of an inflammatory process. Several varieties of aortitis can occur including granulomatous (giant cell), lymphoplasmacytic, mixed inflammatory pattern and suppurative.[21] Vasculitis is described in greater detail in Chapter 11. When the inflammatory infiltrate causes medial injury, wall weakness can occur resulting in aneurysm. Sometimes heavy inflammation is seen in the presence of atherosclerotic plaque. A distinction between atherosclerosis and inflammatory disease can be challenging and likely represents a continuum.

Often, aneurysms can occur in the absence of atherosclerosis or inflammation. Generally this is in the setting of a genetic syndrome. The overarching term 'medial degeneration' is used to describe the collection of histopathologic changes seen in this situation. It is important to note that historically a variety of terms have been used to describe these histopathologies that have had overlapping meanings. Many of these terms, such as cystic medial necrosis, are understood by pathologists for the histopathologic change they represent, but are inaccurate to the event. For example, the histology described by cystic medial necrosis contains neither cysts nor necrosis. A report on specific parts of medial degeneration that represent histologic features of aneurysmal disease follows below. In general, these changes are non-specific, i.e. not indicative of any single genetic syndrome, but in general their presence in younger individuals does suggest an underlying genetic causality.

The most commonly recognized histopathologic finding is mucoid extracellular matrix deposition. This mucoid material, alternatively called glycosaminoglycans, glycoproteins, proteoglycans, or mucopolysaccharides increases in the arterial wall and causes either translamellar or intralamellar expansion. This mucoid material may be so abundant as to form 'pools' in the tissue, causing large translamellar expansions (Figure 13.3). Historically, this was referred to as cystic medial necrosis or cystic medial degeneration.[22] These terms are falling out of favor for reasons described above.[7,15] The distribution of mucoid material is a feature that may suggest different underlying genetic etiologies. Translamellar pools of mucoid material are classically described in Marfan

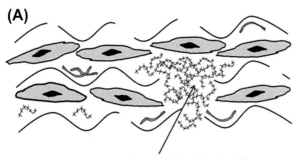
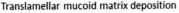

Translamellar mucoid matrix deposition

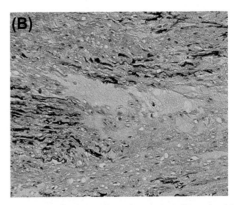

FIGURE 13.3 Translamellar mucoid matrix deposition. **(A)** A schematic shows a large collection of GAGs (mucopolysaccharides) extending across three lamellar units. There is also elastic fiber thinning, fragmentation and loss in this region. **(B)** A Movat Pentachrome identifies this 'pool' of mucoid material extending across lamellar units with concurrent elastic fiber loss (100 × original magnification).

syndrome (MFS) and other genetic syndromes.[23] Intralamellar collections of mucoid material are more subtle and appear to be more frequent in Loeys-Dietz syndrome (LDS) (Figure 13.4).[24] Mucoid extracellular matrix can be appreciated on hematoxylin and eosin slides and more easily on Movat Pentachrome stains. To date, it is not clear if there is a functional role of this material in aneurysms, or if it is just a marker of aortic disease.

A second histopathologic change, frequently encountered, is elastic fiber fragmentation and/or loss (Figure 13.5).[25] In syndromic aortic disease, elastic fibers are frequently fragmented, reduced in caliber or completely lost. To see these changes, an elastic fiber stain (Movat Pentachrome, Verhoeff-Van Gieson, or similar) must be performed on the tissue. This elastic fiber loss has a negative effect on aortic recoil after systolic ejection of blood from the heart. It is believed that over time, as the vessel (primarily the aortic root) is unable to achieve effective elastic recoil, the aneurysm expands.[26]

The loss of smooth muscle cells from the media is important in aneurysm formation (Figure 13.6). This loss is a common finding in atherosclerosis, aortitis, and some syndromic forms of aneurysm. VSMC loss is noted by the absence of VSMC nuclei in a segment of aortic media. This regional VSMC loss can be seen in a variety of pathologic states and important to note, occurs in normal aging. When the VSMC loss is extensive allowing the elastic fibers to collapse together in that region, the term laminar medial collapse is used.[15] Laminar medial collapse is best appreciated on an elastic strain (Figure 13.7). Injury to the vasa vasorum can specifically cause VSMC loss in the outer third of the aorta.

It is important to recognize that all of these histopathologic findings have been described as processes of aging[27] in addition to being precursors to aneurysm. Thus, it is the early onset of these histopathologies, rather than the pure uniqueness of these microscopic findings, that is important to appreciate in syndromic and nonsyndromic aortopathies. In older adults with ascending aneurysms but without genetic mutations, these findings are often present, generally concurrent to the presence of atherosclerotic plaques.

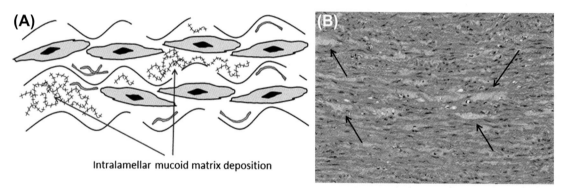

FIGURE 13.4   Intralamellar mucoid matrix deposition. **(A)** A schematic shows multiple areas with GAG expansion, sometimes with smooth muscle cell loss. The expansion occurs within a lamellar unit. **(B)** There are multiple intralamellar expansions seen in this image (arrows) from a Loeys-Dietz patient sample (H&E, 64 × original magnification).

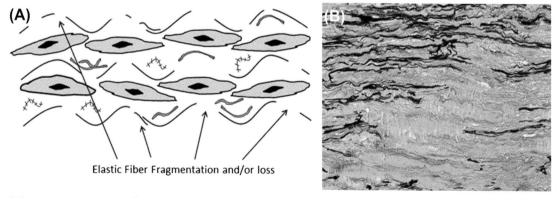

FIGURE 13.5   Elastic fiber fragmentation and/or loss. **(A)** A schematic shows thinning, fragmentation and reduplication of elastic fibers between smooth muscle cells. **(B)** This Movat Pentachrome stain shows areas of intact and lost elastic fibers. Where elastic fibers are present, many have been thinned or fragmented (100 × original magnification).

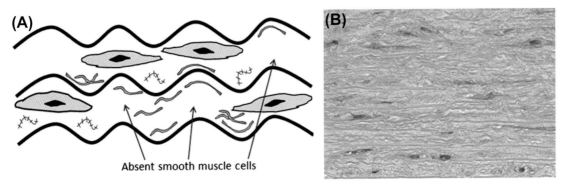

FIGURE 13.6 Smooth muscle cell loss. **(A)** A schematic shows a loss of smooth muscle cells. Sometimes, there is an increase in collagen in these areas. **(B)** H&E stain showing a lack of smooth muscle cells as noted by a decrease in nuclei.

SPECIFIC GENETIC SYNDROMES AND CAUSES OF ANEURYSM

Marfan Syndrome

Marfan syndrome (MFS) (OMIM# 154700), is the prototypical syndrome causing aortic aneurysms and dissections. MFS was first described in 1896 and was recognized as an autosomal dominant genetic syndrome in 1931. In 1991, the causative gene, Fibrillin-1 (FBN1), was identified (Table 13.1).[28] MFS is one of the more common causes of genetic aortic disease with a prevalence of 4–6 individuals per 100000 in the population.[29] MFS was classically diagnosed based on the physical features of pectus excavatum, arachnodactyly, tall stature, and lens ectopia. The Ghent criteria/nosology were created as a means of diagnosing MFS.[30] These criteria now incorporate aortic root dilatation, ectopia lentis, family history, genetic mutations, and a systemic scoring system of other findings. The aortic root size is based on a Z-score which standardizes aneurysm size to age and body size. Z-scores ≥2 are features of MFS.

MFS causes numerous cardiovascular problems affecting the heart and aorta. The most widely recognized and dangerous of these is aortic root dilatation leading to dissection. Without preemptive surgical repair, aortic dissection and rupture is the major cause of mortality in this population. Although still controversial and variable from institution to institution, aortas are generally surgically repaired when the diameter becomes ≥ 5.0 cm.[31] Mitral valve prolapse is also common in the population (~40%), but generally mild.[29] When mitral valve prolapse is severe and occurs early in life, it can lead to congestive heart failure, pulmonary hypertension, and failure to thrive. Over half of MFS patients have an atypical pattern of prolapse such as bileaflet involvement, making their pathology somewhat unique. Also, a generalized 'MFS cardiomyopathy' of reduced ejection fraction has been described,

affecting up to 25% of otherwise healthy MFS individuals.[32] Arrhythmias, particularly atrial fibrillation, are reported at higher rates in this population than matched controls.[33]

As stated above, mutations in fibrillin-1 (FBN1) cause MFS.[28] Mutations in other genes, including transforming growth factor beta receptor 2 (TGFBR2) have been suggested in the literature. However, in the current classification scheme, such other genes have been ascribed to different syndromes. FBN1 is a 350-kDa glycoprotein located in the extracellular matrix. Over 500 mutations have been reported in FBN1 with a loss of protein function believed to be the dominant mechanism. There is one variant form of MFS which causes a severe phenotype of accelerated aneurysm in neonates. It appears that FBN1 mutations causing this subtype of MFS may result in a protein with increased susceptibility to proteolytic cleavage and a resultant dominant negative effect on protein function, not typically seen in classical MFS variants.[34]

Of all syndromic forms of aortic aneurysm, MFS is the best characterized at the gross and microscopic levels. Despite that, the surgical pathology of MFS is not distinct. On gross examination the ascending aortic specimens (when intact) are generally dilated but otherwise unremarkable. If MFS is identified at autopsy, these aortae often have an annuloaortic ectasia in which both the aortic annulus and ascending aorta are enlarged and have a flask-like shape. The commonly described histopathology of the aorta is mucoid medial degeneration of the translamellar variety with marked elastic fiber fragmentation/loss and mucoid extracellular matrix material deposition.[23,35] VSMC loss, although reported, is not a common feature in younger individuals with MFS. The extent to which these histopathologic findings appear is highly variable based upon patient age, degree of aneurysm, and location of evaluated aortic tissue. Some individuals (particularly those undergoing prophylactic resection) can exhibit no histologic abnormalities in their aortae.

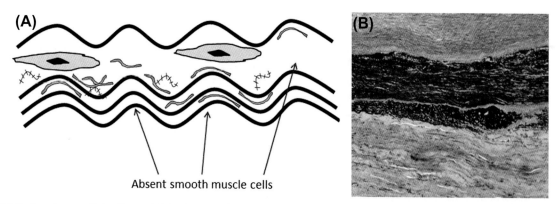

FIGURE 13.7 Laminar medial collapse. **(A)** A schematic demonstrating collapse of lamellar units due to the loss of intervening smooth muscle cells. **(B)** Marked laminar medial collapse of a segment of aorta. The elastic fibers are appropriately thick, but they are aligned atop of each other. The yellow area below is a thickened adventitia and the brown area above is intimal thickening (Movat Pentachrome, 40 × original magnification).

Loeys-Dietz Syndrome

Loeys-Dietz syndrome (LDS) (OMIM #609192, #61068, #610380, #613795, #614816) is a recently described syndrome of aortic aneurysm and dissection.[36] Phenotypically LDS has in common many extra-aortic features akin to MFS including skin striae, scoliosis, and flat feet. However, LDS also encompasses features including arterial tortuosity, bifid/broad uvula, club foot deformity, and craniosynostosis with variable levels of expressivity that are not part of MFS pathogenesis.[37] Ectopia lentis, dislocation of the eye lens, is not observed in LDS, which is why it is a useful Ghent criterion. LDS has been clinically distinguished into types 1 and 2 based on differences in the craniofacial findings. LDS type 1 is characterized mainly by craniofacial features including hypertelorism, craniosynostosis, and cleft palate. These individuals tend to also have skeletal features that overlap with MFS. LDS type 2 may not present with these features, but has cutaneous features that overlap more strongly with vascular Ehlers-Danlos syndrome.[37] Aortic aneurysms are common to both LDS types; however a more severe vascular phenotype is seen in subjects with LDS type 1 compared to those subjects with type 2. Individuals with LDS typically have an aggressive course with dissections at younger ages and at smaller aortic diameters than patients with MFS and a reduced median survival.

The original description of LDS described causal mutations in two genes: transforming growth factor beta receptor 1 (TGFBR1) and transforming growth factor beta receptor 2 (TGFBR2).[36] Two additional genes have now been implicated as additionally causing LDS. These receptors form a heterotetrameric complex that binds the TGF-β ligand and propagates pathway signaling. One gene, SMAD3, was initially described as causing aneurysm-osteoarthritis syndrome (AOS).[38] However, this AOS syndrome has come under the umbrella of LDS and is now known as LDS3 (OMIM#613795). Mutations

TABLE 13.1 Genetic Aneurysm Syndromes

Syndrome or Disease	Mutated Gene	Relative Frequency
Marfan	FBN1*	Common
Loeys-Dietz	TGFBR1*, TGFBR2*, SMAD3*, TGFB2*	Common
Shprintzen-Goldberg	SKI*	Rare
Arterial tortuosity	SLC2A10*	Rare
Bicuspid aortic valve with aneurysm	NOTCH1 and others	Common
Turner	Absent X chromosome	Common
Vascular Ehlers-Danlos	COL3A1	Rare
Autosomal dominant polycystic kidney	PKD1*	Rare
Noonan	PTPN11 and others	Rare
Tetralogy of Fallot	?	Rare
Familial thoracic aortic aneurysm and dissection	ACTA2, MYH11 and others	Common
Coarctation of the aorta	?	Rare
Autosomal recessive cutis-laxa	FBLN5*	Rare
X-linked Alport	COL4A5	Rare
Alagille	JAG1	Rare
Quadricuspid aortic valve with aneurysm	?	Rare

*= Thought to cause aortic aneurysm disease through altered TGF-β pathway activity.

in SMAD3 result in a shared phenotype of LDS including aortic root aneurysm, hypertelorism and bifid uvula. In AOS, osteoarthritis was a more prominent finding. Another recently described gene, TGFB2 – the TGFBR1/2 ligand, causes LDS (OMIM#614816) by loss of function mutations and has extended LDS to a fourth cause (LDS4).[39]

Gross pathology of aortae in LDS is generally similar to that seen in MFS. In contrast to translamellar mucoid matrix deposition, aortae from LDS patients have a more diffuse, intralamellar mucoid matrix deposition.[40] A significant correlation between histopathologic severity and echocardiographic Z-scores of the aortic root in LDS has also been reported.[24]

Shprintzen-Goldberg Syndrome

Shprintzen-Goldberg syndrome (SGS) (OMIM #182212) is a rare disorder with phenotypic overlap with both LDS and MFS. SGS is also known as 'craniosynostosis with arachnodactyly and abdominal hernias,' highlighting the similar phenotypic findings to the aforementioned diseases. Patients are severely affected and multigenerational pedigrees have yet been described likely due to decreased reproductive fitness in these individuals. Subjects may rarely demonstrate thoracic aortic aneurysms as a manifestation of the disease. This weak connection with MFS and LDS did suggest a shared mechanism which helped to recently identify a culprit gene. Mutations in the TGF-β repressor, SKI, were recently identified as a cause of classic SGS.[41]

Arterial Tortuosity Syndrome

Arterial tortuosity syndrome (ATS) (OMIM #208050) is a rare autosomal recessive disorder caused by mutations in the solute carrier family 2 member 10 (SLC2A10) gene that encodes for the facilitative glucose transporter GLUT10.[42] As shown in morpholino zebrafish studies, GLUT10 is an important link between mitochondrial function and TGF-β activity.[43] Subjects usually present early in their lives with characteristic dysmorphic features including hyperextensible skin, hyperextensible joints, and hernias.[42] Clinically, the disease has some overlap with vascular Ehlers-Danlos syndrome (vEDS), but ATS subjects do not have the structural collagen defect of vEDS.[44] Marked tortuosity of the aorta and other large arteries is invariably present in ATS and often has a striking appearance on radiographic imaging. This contrasts with tortuous vessels noted as a phenotype of LDS, where tortuosity is less frequent in presentation and often confined to neck or head vasculature. In ATS, 19–31% of patients develop aortic aneurysms.[45] Although the exact mechanism is unknown, the marked tortuosity may result in an increased shear stress that might predispose to arterial dissection. The mortality due to stroke and arterial dissection is quite high, even in young patients, although a recent review indicates a less severe cardiovascular prognosis than once believed.[45,46]

Grossly, major vessels including the aorta and carotid arteries appear thickened, elongated and tortuous. The elongation and stretching of these great arteries in older subjects is due primarily to loss of elasticity. Rare reports of histopathology in this disease describe fragmentation of the inner elastic membrane and fragmentation and loss of elastic fibers of the tunica media and external elastic membrane. The intima is often markedly thickened due to fibrosis, likely related to the altered shear stress caused by the tortuosity.

Bicuspid Aortic Valve Disease with Aneurysm

Bicuspid aortic valve disease with aneurysm (BAV/AA) (OMIM # 109730) represents a subset of individuals with the most common congenital heart abnormality – bicuspid aortic valve. BAV affects up to 2% of the population. Although mutations in NOTCH1 are associated with a subset of cases of BAV with characteristic aortic calcifications and variable aneurysm formation, the genetic basis of BAV/AA is still largely unknown.[47] Approximately 50% of young men with BAV have enlarged aortic dimensions consistent with aneurysm.[48] Approximately 5% of patients with BAV/AA will ultimately develop an aortic dissection.[49] In contrast to MFS, patients with BAV/AA do not have dilatation of the aortic sinuses. Subjects with BAV/AA tend to not have additional phenotypic features and often are identified on the basis of a heart murmur alone.[50]

Grossly, BAV/AA is characterized by a bicuspid aortic valve, which is often thickened, and a dilated aortic root. Histologically, mucoid matrix deposition has been reported in BAV/AA, but entirely histologically unremarkable aortae are also known.[50] It has also been reported that subjects with BAV have thinner elastic lamellae of the aortic media and greater distances between elastic lamellae than patients with tricuspid aortic valves.[51]

Turner Syndrome

Turner syndrome (TS) is a sex aneuploidy syndrome, exclusively among females, in which a single X chromosome is present (45,XO). The classic phenotypic findings of TS are short stature, webbed-neck and infertility. Cardiac anomalies are common in this population, with cardiovascular disease being a significant cause of mortality. In fact, only 1 in 100 Turner conceptions become live births primarily due to severe congenital cardiovascular defects at the fetal stage.[52] Cardiovascular diseases include congenital heart defects such as bicuspid aortic valves and a distinctive form of coarctation of the aorta, sometimes referred to as pseudocoarctation. TS females have an elongation of the transverse aortic arch with unusual kinking in the juxta-ductus region of the inferior curvature of the aortic arch.[53] Aortic dilation (or dissection) has been reported in conjunction with other cardiac anomalies (coarctation or bicuspid aortic valves) in ~1.5% of TS subjects.[54]

Aortic dilation in TS typically involves the root of the ascending aorta, occasionally extending through the aortic arch to the descending aorta. Aortic dilations and dissections occur in young females with TS. In one study, over 65% of subjects with aortic dilation were under 21 years of age.[54] In a separate review of aortic dissection cases, >50% of patients were less than 30 years of age.[55] In TS histopathology, mucoid matrix deposition has been reported similar to that described in MFS.[50]

Vascular Ehlers-Danlos Syndrome/Ehlers-Danlos Syndrome Type IV

Ehlers-Danlos syndrome (EDS) (OMIM#130050) is another long-appreciated genetic syndrome, first described in 1901. It was determined to be an autosomal dominant disease in 1949. EDS had subsequently been subdivided into six main types with significant vascular manifestations of the syndrome being encountered in a subset variably reported as EDS, type IV (EDS-IV), or vascular EDS. The most recent classification system of EDS has dropped EDS-IV in favor of vEDS.[56] vEDS has a prevalence of one individual in 250000 making it ten times less common than MFS. vEDS results from mutations in the collagen, type III, alpha-1 (COL3A1) gene which encodes for type III collagen.[57] This collagen is present in skin, vessel walls, and hollow organs (Table 13.1). Like other collagens, COL3A1 has a core triple-helical domain (Gly-X-Y) and disease-causing mutations invariably affect glycines or intron/exon splicing throughout the 139 kDa protein. COL3A1 mutations result in kinked collagen that is structurally compromised leading to weak blood vessels and organ walls. Unfortunately, the uterus is comprised of a significant amount of COL3A1 and the mortality rate among pregnant women with vEDS is ~15%.[58] Clinically, vascular EDS presents with characteristic facial features, thin skin, and rupture of vessels or viscera. The facial features, in particular, overlap with LDS.

Vascular EDS affects the entire vasculature, hollow organs, and the heart. Medium to large arteries of the thorax and abdomen have the greatest likelihood of rupture. Vessel dilation is not a necessary event prior to spontaneous dissection, showing the difficulty in preventing catastrophic events. Rarely, the left ventricle of the heart can spontaneously rupture causing death.

Spontaneous arterial tears of the aorta and its branches are considered hallmarks of this disease. The true incidence of aortic dissection in these patients is not known; however, a review of 112 cases of vEDS noted 10% of patients had an aortic dissection.[59] In general, rupture and dissection outweigh aneurysm and, interestingly, thoracic aortic root aneurysms in vEDS are quite rare relative to other locations of vascular compromise. vEDS subjects have a 25% risk of experiencing a major vascular complication by the age of 20 years and life expectancy is around 48 years.[58] Another difficulty for vEDS patients is that they have had historically low tolerance to emergency surgery owing to the extreme fragility of their vascular wall. A recent study has suggested that patients with vEDS can have good surgical outcomes if the surgery is planned.[60]

Histologic findings may be relatively subtle, non-specific, or even absent despite significant transmural tears in the aorta. When present, there have been reports of minimal medial degeneration of the aortic wall with partial disruption of elastic laminae and intervening organized fibrous tissue.[50]

Autosomal Dominant Polycystic Kidney Disease

Autosomal dominant polycystic kidney disease (ADPKD) (OMIM #173900) is a common hereditary disorder caused by mutations in the polycystic kidney disease 1 (PKD1) gene. PKD1 binds to PKD2 and serves as a membrane protein involved in cell-to-cell or cell-to-matrix interactions.[61] Vascular aneurysm is a well-appreciated complication of ADPKD. Additionally, left ventricular hypertrophy and valvular incompetence are common features. Saccular (berry) intracranial aneurysms are found in up to half of all patients with ADPKD and are prone to rupture.[62] Ruptured abdominal and/or thoracic aneurysms are also described, and their propensity for dissection is likely enhanced by the significant hypertension in this patient population. The histopathology of the excised aorta, when present, has been described as showing mucoid matrix deposition.[50]

Noonan Syndrome

Noonan syndrome (NS) (OMIM #163950) is a relatively common autosomal dominant cardiofacial syndrome within the RASopathy family of disorders. The phenotypic hallmarks of the disease are congenital heart defects, short stature, cognitive disabilities, pectus excavatum, webbed neck, and flat nasal bridge. The protein-tyrosine phosphatase, nonreceptor-type 11 (PTPN11) gene causes the majority of NS, although mutations in other genes in the RAS-MAPK pathway have also been described.[63] In blood vessels PTPN11 is activated by angiotensin receptor 1 (AT1R) to promote p190A phosphorylation which regulates the RhoA/Rock pathway.[64] PTPN11 mutations are clustered in the interacting portion of the amino N-SH2 domain and in the phosphotyrosine phosphatase domain.[63] Cardiac defects occur in over 50% of patients and include dysplastic pulmonic valves, hypertrophic cardiomyopathy, and, less commonly, pulmonary artery stenosis, atrial septal defects, or patent ductus arteriosus.[65] The syndrome is also associated with aneurysms of the sinuses of Valsalva and rarely with an actual aortic dissection. Medial degeneration has been described in the reported cases as a histologic finding suggesting a similar histopathologic phenotype of other described syndromes.[50]

Tetralogy of Fallot

Tetralogy of Fallot (ToF) (OMIM#187500) is a complex congenital heart disease that is so-named due to the four abnormalities of pulmonary stenosis, ventricular septal defects, rightward deviation of the aortic origin, and right ventricular hypertrophy. Although it is the most commonly encountered cyanotic congenital heart defect in infancy, insidious development of progressive aortic root dilatation has been seen in adult survivors who have undergone successful surgical repair.[66] The underlying pathophysiology of ascending aortic dilatation is currently unknown but is believed to be primary to the genetic abnormalities that also result in congenital heart disease. Intrinsic histological abnormalities in the aortic root and ascending aortic wall have been observed in various studies which are thought to be independent of previous long-standing volume overload. Histopathologic findings in ToF include abnormalities of smooth muscles, elastic fibers, collagen, and ground substance in the tunica media.[50]

Familial Thoracic Aortic Aneurysm and Dissection

As many as 20% of patients referred for repair of thoracic aneurysm or dissection have familial aneurysm clustering upon genetic counseling. Many of these families do not meet the clinical criteria for any of the above-mentioned hereditary disorders, but often have other constellations of phenotypes (OMIM %607086). In many of these families, the inheritance is autosomal dominant with incomplete penetrance and variable age-related onset of symptoms. Two genes with mutations that have been described within this larger cohort are ACTA2 and MYH11.[67, 68] Both of these mutant genes impact the function of smooth muscle cells, specifically the function of the smooth muscle contractile apparatus. AOS, or LDS3, was initially found using families that were within this FTAAD cohort. Further characterization and subclassification of this group will continue to decrease the percentage of individuals with genetically caused aortic dissections who fail to fall into any specific entity.

Grossly, aneurysms and dissections in FTAAD can occur anywhere along the ascending aorta and are not confined to the annulus, as appreciated in MFS.

Histologically, within aortae from individuals with MYH11 and ACTA2 mutations, there is focal mucoid matrix deposition with disorganization of smooth muscle cells, elastic fiber loss/fragmentation, and increased penetrance of vasa vasorum into the medial layer.[50]

Coarctation of the Aorta

Coarctation of the aorta occurs in conjunction with 5–10% of congenital heart disease. It is defined by a narrowing of the aorta just distal to the left subclavian artery or more rarely by a narrowing proximal to the left subclavian artery. Of congenital heart disease phenotypes, bicuspid aortic valve commonly occurs in conjunction with coarctation.[69] Depending on the severity of the coarctation, classic clinical findings will include a systolic murmur and high blood pressures in the upper extremities and low blood pressures with near-absent pulses in the lower extremities. In the current era of medical imaging, coarctation is generally detected and treated with surgical repair in infancy. Rarely in Western societies, but frequently in developing countries, coarctation is undiagnosed at birth. If not surgically corrected, the backpressure on the heart and aortic root frequently lead to congestive heart failure and aortic dissection.

Autosomal Recessive Cutis-Laxa

Autosomal recessive cutis-laxa (type 1A) (OMIM#219100) is caused by mutations in the fibulin 5 (FBLN5) gene.[54] FBLN5 regulates the adhesion of endothelial cells by interacting with integrins and the RGD motif.[70] The disorder is characterized predominantly by loose, sagging skin beginning at an early age. Less frequently, these patients have vascular diseases including fibromuscular dysplasia and pulmonary artery stenosis. Aortic aneurysm is a known, but rarely reported, complication in this entity. Histologically, one case report described degeneration of elastin, while a second case series identified a coarse outline of the elastic lamina in three subjects.[71,72]

X-linked Alport Syndrome

X-linked Alport syndrome (OMIM #301050) is predominantly a disease of early-onset renal failure, sensorineural hearing loss, and ocular anomalies. The renal failure is typically glomerulonephropathy. All of these pathologies are the result of mutations in the collagen 4A5 (COL4A5) gene.[73] COL4A5, like COL3A1, has a core triple-helical domain (Gly-X-Y) .COL4A5 is a major protein of basement membranes and is found in the renal basement membranes, vascular media, around visceral organs, cochlea, and eye. A subset of patients with Alport syndrome also develops ascending aortic aneurysm, which occurs in adolescents and young adults.[74] The presence of a defect in collagen suggests the etiology of X-linked Alport syndrome may be similar to vEDS. The histopathology in this disease has been reported as medial necrosis.[75]

Alagille Syndrome

Alagille syndrome (OMIM #118450) is an autosomal dominant disease caused by mutations in jagged1 (JAG1).[76] JAG1 is a ligand in the Notch signaling pathway. Alagille syndrome's key phenotypic findings are cholestasis, cardiovascular disease, skeletal abnormalities, ocular abnormalities, renal dysplasia, and a characteristic facial phenotype. The cholestasis is the result of a sharply reduced number of intrahepatic bile ducts. A variety of cardiovascular lesions have been described in Alagille syndrome including aortic coarctation, large artery aneurysms, moyamoya disease, and pulmonary artery stenosis.[77] Additionally, bleeding in general and intracranial bleeding specifically are increased in this population. In one study of 25 affected individuals, two had aortic aneurysms.[77]

Quadricuspid Aortic Valve with Aneurysm

While bicuspid aortic valve disease is a common congenital heart malformation, a quadricuspid aortic valve is a more rare entity. Its reported incidence ranges up to 0.033% of the population and represents between 0.55% and 1.46% of patients undergoing aortic repair for aneurysm.[78] Quadricuspid aortic valve disease is rarely associated with other congenital heart diseases but is highly associated with aortic regurgitation.[79,80] Currently, no specific gene has been implicated in this congenital heart disease and further delineation of quadricuspid aortic disease alone or in combination with aneurysm remains to be understood. Histologic specimens have shown no significant changes or mild mucoid extracellular matrix deposition.[78,81]

NON-GENETIC CAUSES OF AORTIC ANEURYSM

General Non-Specific Risk Factors for Aneurysm

The primary cause of aortic aneurysms among older individuals is still a heterogeneous collection of risk factors without a particular initiating factor. The incidence of thoracic aortic aneurysm is only 6–10 per 100000 patient years, while abdominal aortic aneurysms are found in up to 8% of older men.[15,60] The risk factors for aneurysm are patient age, hypertension, male sex,

atherosclerosis, smoking, and hypercholesterolemia. It is worth noting that these are similar risk factors for coronary artery disease, so there is nothing specific in these common risk factors that would suggest an individual is predisposed for an aneurysm. For other non-genetic causes of aneurysm, a causative or initiating factor can be described.

Aortitis

Arteritis is discussed in Chapter 11. However, it is worth a short mention here as well, that aortitis is a non-genetic cause of aortic aneurysm. The inflammatory infiltrate causes medial destruction with loss of VSMCs that leads directly to aneurysm formation. Often these aneurysms evolve and expand rapidly secondary to the destruction.

Cocaine

Cocaine abuse has been associated with numerous cardiovascular complications. These include myocardial infarction, coronary artery vasoconstriction, cardiomyopathy, arrhythmia, and aortic dissection. Cocaine abuse causes catecholamine release and stimulation of α and β adrenergic receptors. This can lead to vasoconstriction and spasm of any sized artery. Chronic cocaine abuse has been show to result in increased diastolic aortic diameter, loss of aortic elasticity, and increased aortic stiffness.[82] These chronic changes, coupled with other risk factors such as smoking and untreated hypertension, likely increase susceptibility to aortic dissections in a minority of cocaine abusers. This entity does have important regional variability as noted by a single inner city population in which 37% of aortic dissections were the result of cocaine abuse, predominantly in young male hypertensive smokers.[83] In other patient populations, this is a rare cause of dissection. Cocaine abuse has only been associated with 0.5% of cases in the International Registry of Aortic Dissection. Histologically, cystic medial degeneration has been described in a subject with dissection, suggesting the potential for an underlying aortic pathology.[50]

Weightlifting and Severe Physical Exertion

Weightlifting and severe physical exertion are additional causes of aortic dissections. One study reported 31 patients who developed acute aortic dissection in the context of severe physical exertion, mostly weightlifting.[84] Affected individuals ranged in age from 19 to 76 years of age with a strong male predominance (30:1). Most of these subjects had moderately enlarged aortas prior to the dissection (4–5 cm diameter) and >10% of subjects had a family history of aortic disease. Within this population, the etiology of dissection appears to be a rapid and dramatic elevation of blood pressure during extreme exertion (i.e. weightlifting) against a mildly dilated aorta. Histologically, mucoid medial deposition was described in a few cases further suggesting the possibility of an underlying condition predisposing to the dissection.[50]

EVIDENCE FOR THE TGF-β PATHWAY TO BE A UNIFYING MECHANISM OF AORTIC ANEURYSM

It has recently been determined that many, but not all, of the aortic aneurysm syndromes described above share a common aberrant TGF-β signaling pathway. Historically, this was not true. The initial histopathologic evidence of aneurysm – mucoid matrix deposition concurrent with elastic fiber fragmentation and loss – indicated that aneurysms were the failure of elastin homeostasis. Indeed, the discovery of mutations in FBN1, the gene encoding fibrillin-1 and genetic cause of MFS, was consistent with this notion. Fibrillin-1 is an extracellular matrix protein associated with elastin fibers and fulfilled these expectations of aortic structural weakness as causal in aortic aneurysm.

This 'construction failure' dogma lasted over a decade and was thought to offer a dismal prognosis for individuals with MFS and similar diseases, as it suggested that a structural defect during development was the root cause of disease. It is similar to suggesting that a tall building, built with an insufficient foundation, is susceptible to collapse at any time. Ultimately the dogma was challenged by the discovery that Fbn1 mutations in a mouse model resulted in a failure of distal alveolar septation during embryogenesis.[85] This finding was unrelated to elastin homeostasis and was incompatible with this 'construction failure' concept. Thus, another mechanism was sought. Further insights into the structure of Fibrillin-1 determined that it had secondary structural homology to latent TGF-β binding proteins. TGF-β signaling had previously been implicated in alveolar septation and a potential link was identified. Indeed, elevated TGF-β signaling was demonstrated in the developing lung and the disease phenotype of septation failure could be overcome with TGF-β neutralizing antibodies.[85] These lung observations led to the investigation of TGF-β signaling first within the mitral valve[86] and then ultimately the aorta of MFS mice.[87] When it was shown that antagonism of TGF-β signaling ameliorated these disease manifestations, a new paradigm was found.

The original observations for the role of TGF-β were made in MFS mouse models. Since then, multiple genes involved in TGF-β homeostasis have been implicated as causal in human aortic aneurysm by investigating other animal models,[36,38,39,41] indicating this is a common pathway that leads to aneurysmal disease. Within human subjects, dysregulation of TGF-β signaling has been identified in the aortic wall of patients with both syndromic and non-syndromic aneurysm, again expanding the likely number of individuals for whom altered TGF-β pathogenesis is important. This and additional discoveries have implicated aberrant TGF-β signaling as central to the pathogenesis of aneurysm.

TGF-β CANONICAL AND NON-CANONICAL SIGNALING IN THE ASCENDING AORTA

TGF-β is made and secreted by smooth muscle cells as one of three related proteins: TGF-β1, -β2 or -β3 (Figure 13.8). These are attached to an inactivating latency complex. The homodimeric TGF-β is sequestered in the extracellular matrix (microfibril fibers) and prevented from being activated by fibrillin (FBN) molecules. TGF-β can be cleaved from the latency complex by a number of proteases. When that occurs, TGF-β becomes activated and binds to the TGF-β receptor II (TGFBR2). TGFBR2 then becomes part of a heterotetrameric receptor complex with TGF-β receptor 1 (TGFBR1). Within the heterotetrameric complex the kinase domain of TGFBR2 phosphorylates TGFBR1. This activates the TGFBR1 kinase, allowing for activating phosphorylation of TGF-β effectors such as the Smad proteins. TGF-β binding and receptor activation initiates a cascade of signaling along both canonical and non-canonical pathways.[7,12,88]

Canonical (SMAD) pathway activation occurs upon the phosphorylation and activation of SMAD2 (to make phosphorylated SMAD2 or pSMAD2), or SMAD3 (pSMAD3). These pSMADs are translocated to the nucleus in a complex with SMAD4. This multi-protein complex acts as a transcription factor to increase expression of TGF-β regulated genes including connective tissue growth factor (CTGF), PAI-1, and matrix metalloproteinases (MMPs). High levels of pSMAD2 have generally been used to establish up-regulation of TGF-β pathway activity. However, it remains unclear if high pSMAD2 is causal or simply a marker of high TGF-β pathway activity in aneurysm pathogenesis. The protein SKI, described as the cause of SGS, acts as a repressor of this canonical signaling pathway.[12,41,88]

The non-canonical pathways involve the activation of a number of other well-known signaling cascades. TGF-β-activated kinase 1 (TAK1) is activated in a TGFBR complex-kinase independent fashion, which in turn activates MKK6 and/or MKK3. These further activate p38. TAK1 can alternatively activate MKK4, which results in JNK activation. TGF-β signaling can also activate the phosphatidylinositol-3-kinase and the

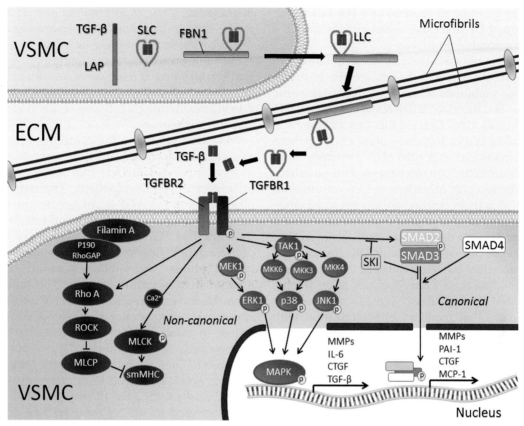

FIGURE 13.8 The TGF-β signaling pathway in the ascending aorta. TGF-β is expressed in vascular smooth muscle cells (VSMCs). Before leaving the VSMC, TGF-β is attached to a small latency peptide (SLC) and coupled to fibrillin (FBN1). It is then exported to the extracellular matrix (ECM) as a large latency peptide (LLC). The LLC then attaches to neighboring microfibrils sequestering TGF-β in the inactive state. TGF-β is released from the LLC by a number of proteases (elastase, MMPs, etc.). This unbound and active TGF-β binds to the TGF-β receptors 1 and 2 (TGFBR1 and TGFBR2). This heterotetrameric receptor complex then performs signaling along both canonical and non-canonical pathways. In one non-canonical pathway, TGFBR1/2 activates the Rho A/ROCK pathway and increases Ca2+ to ultimately impact on levels of smooth muscle myosin heavy chain (smMHC/MYH11). smMHC then regulates downstream contractility and transcriptional events. A second non-canonical pathway activates (via phosphorylation) MEK1 and TAK1 to ultimately activate the MAPK transcription factor. MAPK induces higher expression of matrix metalloproteases (MMPs), CTGF, interleukin-6 (IL-6), and TGF-β itself. In the canonical pathway, TGFBR1/2 activates (via phosphorylation) both SMAD2 and SMAD3 which form a complex and couple to SMAD4. This complex enters the nucleus, acts as a transcription factor and increases the expression of several genes. These genes includes the MMPs, CTGF, PAI-1, and MCP-1. SKI acts as a repressor of this canonical pathway.

RhoA-ROCK signaling cascades which have been impli-cated in numerous regulatory systems.[7,12] Each of these non-canonical pathways modulate a myriad of cellular functions related to cell adhesion, cell signaling, and cell turnover. TGF-β-mediated activation of ERK1 was recently identified as a potential mediator of aneurys-mal progression in an animal model of MFS,[89,90] and fur-ther work will be needed to determine if this pathway is as relevant to aneurysm progression in humans as it appears to be in mice.

TGF-β SIGNALING IN ASCENDING AORTIC DISEASES

It is now thought that elevated TGF-β signaling is a contributing pathologic factor to the progression of ascending aortic aneurysms of multiple etiologies. The primary data underlying this hypothesis were devel-oped from evidence in mouse models of Marfan and other syndromes. In MFS, the current hypothesis is that fibrillin mutations result in a failure to sequester inac-tive TGF-β, leading to unregulated TGF-β activation, and increased signaling through the TGF-β receptors, TGFBR1 and TGFR2. Additional evidence has come from other genetic syndromes that share mutations along the TGF-β pathway. Also, dysregulation of TGF-β signaling has been directly observed in other aneu-rysmal conditions. For instance, mutations in SLC2A10 increase expression of decorin (a TGF-β antagonist).[42] More recently, mutations in the gene encoding the proto-typical TGF-β repressor, SKI, the cause of SGS, resulted in up-regulation of the TGF-β pathway signalling result-ing in thoracic aortic aneurysm. Morpholino studies of SKI paralogs in zebrafish recapitulated human abnor-malities. Thus, in some conditions, TGF-β overactivity may be due to a failure of repression.[41] For other syndromes, such as LDS, the mechanism by which TGF-β overexpression occurs is less clear.

Classical LDS is caused by heterozygous mutations in TGFBR1 or TGFBR2. These mutations are almost exclusively missense mutations which tend to cluster in the kinase domain and therefore fail to allow canonical TGF-β signaling propagation. Despite this loss of canonical signaling function as seen in tissue culture assays, patient tissues still express elevated pSMAD2/3 staining, suggesting a different activation mechanism. This paradoxical effect on the receptor mutations in LDS, resulting in increased TGF-β activity by mechanisms yet to be elucidated, is likely related to increased non-canonical TGF-β signaling caused by loss of canonical feedback inhibition.

LDS can also be caused by mutations in other components of the canonical TGF-β signaling pathway, namely SMAD3 and TGFB2 (LDS types 3 and 4, respectively). Identified SMAD3 mutations were located within the MH2 domain that mediates oligomerization with SMAD4, an interaction necessary for transcriptional activation. Intuitively, these effects should decrease TGF-β signaling. However, both pSMAD2 and CTGF (a downstream reporter of canonical TGF-β activity) were paradoxically strongly increased in human aortas from affected patients.[38] Similarly, mutations in TGFB2 (coding for TGF-β2 ligand) also demonstrated over-activation of the TGF-β pathway as reported by both pSMAD2 and CTGF overexpression. This condition also correlated with up-regulation with the TGF-β1 isoform.[39]

TGF-β SIGNALING IN THE DESCENDING AORTA

Contrary to the role of TGF-β in the ascending aorta, a large amount of experimental evidence has indicated a protective role of TGF-β in abdominal aortic aneurysms. Abdominal aneurysms are generally observed in older men with significant infrarenal atherosclerotic plaque burden and a secondarily thinned media as described above. Abdominal aortas differ from thoracic ascending aortas by both the increased presence of inflammatory cells (macrophages and lymphocytes) in the disease state and by the embryologic origin of their VSMCs. VSMCs from different embryologic origins respond differently to TGF-β stimulation.

In experimental animal models, overexpression of TGF-β stabilizes abdominal aneurysms and reduces aortic dissection complications.[91] The mechanisms are generally related to suppressing inflammation and proteolytic enzyme activity. In one experiment, TGF-β neutralizing antibodies decreased levels of TGF-β, which led to an increased periaortic leukocyte infiltration. Leukocytes express MMPs and other enzymes that degrade the aortic ECM material and destabilize aortic walls, thus inflammation is pro-aneurysmal.[92] TGF-β can also directly down-regulate MMP-9 in macrophages, preventing proteolytic remodeling of the aortic wall.[93] Further, elevated TGF-β1 can prevent the loss of aortic smooth muscle cells in the abdominal region. A reduction in smooth muscle cells can result in wall weakening and aortic expansion.[94]

Most data related to the protective activity of TGF-β in abdominal aortic aneurysm have been obtained from murine aneurysm models that incorporate an intense inflammatory cell component. The detrimental effect observed through TGF-β antagonism in these models most likely relates to the known anti-inflammatory activity exerted by TGF-β pathway activation. The relevance of these data to human aneurysm is not completely correlated. Human abdominal lesions are certainly more inflammatory in nature than are their thoracic counterparts. However, anti-inflammatory therapies to prevent abdominal aortic aneurysm have not proven effective, unlike in other inflammatory vascular diseases (e.g. temporal arteritis). The human data supporting a role for decreased TGF-β in abdominal aortic aneurysms is particularly weak, but worthy of further discovery efforts. In one study, with only 18 subjects, TGFB2 expression was decreased in 92% of these patients in conjunction with abdominal aneurysms, and similar reports can be found in the literature.[95]

BIOMARKERS OF ANEURYSM

As described above, most patients with genetic ascending aortic aneurysms are identified by careful phenotypic characterization upon physical exam. For abdominal aneurysms, in which a large elderly population is potentially affected, and screening by imaging is expensive, there has been a considerable effort to identify useful blood-based biomarkers. D-dimer and fibrinogen have both been identified in multiple cross-sectional studies as being increased in the plasma of individuals with abdominal aneurysms.[96] However, neither biomarker has been developed as a diagnostic test as both are unlikely to have the sensitivity or specificity to be clinically relevant. Circulating serum TGF-β was found to be elevated in a MFS mouse model (Fbn1C1039G/+) and in human MFS patients.[97] Unfortunately, this potential biomarker is similarly elevated in a variety of other human pathologies and is unlikely to be specific for MFS. Soluble elastin fragments of medial elastic fibers and components of dead vascular smooth muscle cells have been reported as elevated shortly after an aortic dissection has begun.[98] These biomarkers may help distinguish between an aortic dissection and an acute myocardial infarction, but are not yet in the clinical domain.

TREATMENT FOR ANEURYSM

The definitive treatment for aortic aneurysm is surgical and represents the most effective therapy for patients with advanced aneurysmal disease. Traditionally this has required an open surgical procedure. Recently, placement of an aortic stent graft via a vascular catheter has been found to be effective in some patients. As aneurysms increase in size, the risk of catastrophic rupture increases significantly. Surgical replacement of the affected segment of the aorta is the cornerstone of care for aortic aneurysm and has clear lifespan benefits for many individuals affected by aneurysm. Recommendations for prophylactic ascending aortic surgery are based on genetic diagnosis and the rate of progression of aortic growth. For instance, in patients with Marfan syndrome, surgery has been recommended when the aortic root achieves a size of 5.0 cm, while for non-syndromic patients this dimension is 5.5 cm.[31]

Medical therapy has also been used in patients with ascending aortic aneurysm, particularly as a means to slow aneurysm formation and delay the need for surgery. Historically, patients were treated with a beta-blocker to lower blood pressure and decrease aortic wall stress. With the discovery of altered TGF-β activity as the cause of aneurysm, attention has turned to the potential of losartan, an angiotensin receptor type-1 antagonist, which has TGF-β antagonist functions. In a MFS animal model, losartan was able to prevent aneurysm formation by impacting on the TGF-β pathway.[87] Human clinical trials are nearing completion to evaluate if losartan is superior to beta blockers in delaying aneurysm formation.[99]

FUTURE DIRECTIONS

The last few years have seen an enormous amount of discovery resulting in a new paradigm of altered TGF-β pathway activity as a central cause of aortic aneurysm. The linking of mutations in SKI, SMAD3, SMAD4, TGFBR1, TGFBR2, and TGFB2 to syndromic aortic aneurysm diseases has helped to unify this key concept of altered TGF-β expression or function. Still these discoveries yield almost as many questions as they answer. Specifically – what are the roles of canonical and non-canonical TGF-β pathway mechanisms in aneurysm? Are they complementary? What is the best target in the pathway for therapeutics? How many additional genes along the TGF-β pathway will be found to be mutated among other FTAAD individuals to yield new syndromes or expand current syndromes? How do Mendelian diseases inform us on the pathogenesis of non-syndromic ascending and abdominal aneurysms? How do we reconcile the differing roles of TGF-β activity in the ascending and abdominal aortas? Will the human experience replicate what has been learned in animal models for abdominal aneurysms? Will losartan, whose use is based on rational mechanistic and animal data, prove to be better than beta blockers at slowing aneurysm growth?

Experiments to answer many of these exciting questions are underway. The coming years will reveal answers to many of these questions and likely yield more unexpected results. This is an exciting time for aneurysm research.

References

1. Sugawara J, Hayashi K, Yokoi T, Tanaka H. Age-associated elongation of the ascending aorta in adults. *JACC Cardiovasc Imaging* 2008;**1**(6):739–48.
2. Mayne R. Collagenous proteins of blood vessels. *Arteriosclerosis* 1986;**6**(6):585–93.
3. Alderton WK, Cooper CE, Knowles RG. Nitric oxide synthases: structure, function and inhibition. *Biochem J* 2001;**357**(Pt 3):593–615.
4. Davignon J, Ganz P. Role of endothelial dysfunction in atherosclerosis. *Circulation* 2004;**109**(23 Suppl. 1):III27–32.
5. Dai G, Kaazempur-Mofrad MR, Natarajan S, Zhang Y, Vaughn S, Blackman BR, et al. Distinct endothelial phenotypes evoked by arterial waveforms derived from atherosclerosis-susceptible and -resistant regions of human vasculature. *Proc Natl Acad Sci U S A* 2004;**101**(41):14871–6.
6. Smith EB. Transport, interactions and retention of plasma proteins in the intima: the barrier function of the internal elastic lamina. *Eur Heart J* 1990;**11**(Suppl. E):72–81.
7. Humphrey JD. Possible mechanical roles of glycosaminoglycans in thoracic aortic dissection and associations with dysregulated transforming growth factor-beta. *J Vasc Res* 2013;**50**(1):1–10.
8. Wolinsky H, Glagov S. A lamellar unit of aortic medial structure and function in mammals. *Circ Res* 1967;**20**(1):99–111.
9. Waldo KL, Hutson MR, Ward CC, Zdanowicz M, Stadt HA, Kumiski D, et al. Secondary heart field contributes myocardium and smooth muscle to the arterial pole of the developing heart. *Dev Biol* 2005;**281**(1):78–90.
10. Topouzis S, Majesky MW. Smooth muscle lineage diversity in the chick embryo. Two types of aortic smooth muscle cell differ in growth and receptor-mediated transcriptional responses to transforming growth factor-beta. *Dev Biol* 1996;**178**(2):430–45.
11. Nitzan E, Kalcheim C. Neural crest and somitic mesoderm as paradigms to investigate cell fate decisions during development. *Dev Growth Differ* 2013;**55**(1):60–78.
12. Lindsay ME, Dietz HC. Lessons on the pathogenesis of aneurysm from heritable conditions. *Nature* 2011;**473**(7347):308–16.
13. Veinot JP, Ghadially FN, Walley VM. *Cardiovascular Pathology*. 3rd ed. Philadelphia, Pennsylvania: Churchill Livingstone; 2001.
14. Tennant M, McGeachie JK. Blood vessel structure and function: a brief update on recent advances. *Aust N Z J Surg* 1990;**60**(10):747–53.
15. Halushka MK. Pathology of the Aorta. In: Winters GL, editor. *Surgical Pathology Clinics: Current Concepts in Cardiovascular Pathology*. Philadelphia, PA: Saunders; 2012. p. 417–33.
16. Tsai TT, Evangelista A, Nienaber CA, Trimarchi S, Sechtem U, Fattori R, et al. Long-term survival in patients presenting with type A acute aortic dissection: insights from the International Registry of Acute Aortic Dissection (IRAD). *Circulation* 2006;**114**(1 Suppl.):I350–6.
17. Tsai TT, Fattori R, Trimarchi S, Isselbacher E, Myrmel T, Evangelista A, et al. Long-term survival in patients presenting with type B acute aortic dissection: insights from the International Registry of Acute Aortic Dissection. *Circulation* 2006;**114**(21):2226–31.

18. LeMaire SA, Pannu H, Tran-Fadulu V, Carter SA, Coselli JS, Milewicz DM. Severe aortic and arterial aneurysms associated with a TGFBR2 mutation. *Nat Clin Pract Cardiovasc Med* 2007;**4**(3):167–71.

19. Tsai TT, Nienaber CA, Eagle KA. Acute aortic syndromes. *Circulation* 2005;**112**(24):3802–13.

20. van Dijk RA, Virmani R, von der Thusen JH, Schaapherder AF, Lindeman JH. The natural history of aortic atherosclerosis: a systematic histopathological evaluation of the peri-renal region. *Atherosclerosis* 2010;**210**(1):100–6.

21. Miller DV, Maleszewski JJ. The pathology of large-vessel vasculitides. *Clin Exp Rheumatol* 2011;**29**(1 Suppl. 64):S92–8.

22. Erdheim J. Medionecrosis aortae idiopathica cystic. *Virchows Arch Pathol Anat* 1930;**276**:187–229.

23. Homme JL, Aubry MC, Edwards WD, Bagniewski SM, Shane Pankratz V, Kral CA, et al. Surgical pathology of the ascending aorta: a clinicopathologic study of 513 cases. *Am J Surg Pathol* 2006;**30**(9):1159–68.

24. Maleszewski JJ, Miller DV, Lu J, Dietz HC, Halushka MK. Histopathologic findings in ascending aortas from individuals with Loeys-Dietz syndrome (LDS). *Am J Surg Pathol* 2009;**33**(2):194–201.

25. Schlatmann TJ, Becker AE. Pathogenesis of dissecting aneurysm of aorta. Comparative histopathologic study of significance of medial changes. *Am J Cardiol* 1977;**39**(1):21–6.

26. Belz GG. Elastic properties and Windkessel function of the human aorta. *Cardiovasc Drugs Ther* 1995;**9**(1):73–83.

27. Schlatmann TJ, Becker AE. Histologic changes in the normal aging aorta: implications for dissecting aortic aneurysm. *Am J Cardiol* 1977;**39**(1):13–20.

28. Dietz HC, Cutting GR, Pyeritz RE, Maslen CL, Sakai LY, Corson GM, et al. Marfan syndrome caused by a recurrent de novo missense mutation in the fibrillin gene. *Nature* 1991;**352**(6333):337–9.

29. Rybczynski M, Mir TS, Sheikhzadeh S, Bernhardt AM, Schad C, Treede H, et al. Frequency and age-related course of mitral valve dysfunction in the Marfan syndrome. *Am J Cardiol* 2010;**106**(7):1048–53.

30. Loeys BL, Dietz HC, Braverman AC, Callewaert BL, De Backer J, Devereux RB, et al. The revised Ghent nosology for the Marfan syndrome. *J Med Genet* 2010;**47**(7):476–85.

31. Gott VL, Greene PS, Alejo DE, Cameron DE, Naftel DC, Miller DC, et al. Replacement of the aortic root in patients with Marfan's syndrome. *N Engl J Med* 1999;**340**(17):1307–13.

32. Alpendurada F, Wong J, Kiotsekoglou A, Banya W, Child A, Prasad SK, et al. Evidence for Marfan cardiomyopathy. *Eur J Heart Fail* 2010;**12**(10):1085–91.

33. Savolainen A, Kupari M, Toivonen L, Kaitila I, Viitasalo M. Abnormal ambulatory electrocardiographic findings in patients with the Marfan syndrome. *J Intern Med* 1997;**241**(3):221–6.

34. Kirschner R, Hubmacher D, Iyengar G, Kaur J, Fagotto-Kaufmann C, Brömme D, et al. Classical and neonatal Marfan syndrome mutations in fibrillin-1 cause differential protease susceptibilities and protein function. *J Biol Chem* 2011;**286**(37):32810–23.

35. Nesi G, Anichini C, Tozzini S, Boddi V, Calamai G, Gori F. Pathology of the thoracic aorta: a morphologic review of 338 surgical specimens over a 7-year period. *Cardiovasc Pathol* 2009;**18**(3):134–9.

36. Loeys BL, Chen J, Neptune ER, Judge DP, Podowski M, Holm T, et al. A syndrome of altered cardiovascular, craniofacial, neurocognitive and skeletal development caused by mutations in TGFBR1 or TGFBR2. *Nat Genet* 2005;**37**(3):275–81.

37. Loeys BL, Schwarze U, Holm T, Callewaert BL, Thomas GH, Pannu H, et al. Aneurysm syndromes caused by mutations in the TGF-beta receptor. *N Engl J Med* 2006;**355**(8):788–98.

38. van de Laar IM, Oldenburg RA, Pals G, Roos-Hesselink JW, de Graaf BM, Verhagen JM, et al. Mutations in SMAD3 cause a syndromic form of aortic aneurysms and dissections with early-onset osteoarthritis. *Nat Genet* 2011;**43**(2):121–6.

39. Lindsay ME, Schepers D, Bolar NA, Doyle JJ, Gallo E, Fert-Bober J, et al. Loss-of-function mutations in TGFB2 cause a syndromic presentation of thoracic aortic aneurysm. *Nat Genet* 2012;**44**(8):922–7.

40. Iba Y, Minatoya K, Matsuda H, Sasaki H, Tanaka H, Morisaki H, et al. Surgical experience with aggressive aortic pathologic process in Loeys-Dietz syndrome. *Ann Thorac Surg* 2012;**94**(5):1413–7.

41. Doyle AJ, Doyle JJ, Bessling SL, Maragh S, Lindsay ME, Schepers D, et al. Mutations in the TGF-beta repressor SKI cause Shprintzen-Goldberg syndrome with aortic aneurysm. *Nat Genet* 2012;**44**(11):1249–54.

42. Coucke PJ, Willaert A, Wessels MW, Callewaert B, Zoppi N, De Backer J, et al. Mutations in the facilitative glucose transporter GLUT10 alter angiogenesis and cause arterial tortuosity syndrome. *Nat Genet* 2006;**38**(4):452–7.

43. Willaert A, Khatri S, Callewaert BL, Coucke PJ, Crosby SD, Lee JG, et al. GLUT10 is required for the development of the cardiovascular system and the notochord and connects mitochondrial function to TGFbeta signaling. *Hum Mol Genet* 2012;**21**(6):1248–59.

44. Franceschini P, Guala A, Licata D, Di Cara G, Franceschini D. Arterial tortuosity syndrome. *Am J Med Genet* 2000;**91**(2):141–3.

45. Wessels MW, Catsman-Berrevoets CE, Mancini GM, Breuning MH, Hoogeboom JJ, Stroink H, et al. Three new families with arterial tortuosity syndrome. *Am J Med Genet Part A* 2004;**131**(2):134–43.

46. Cartwright MS, Hickling WH, Roach ES. Ischemic stroke in an adolescent with arterial tortuosity syndrome. *Neurology* 2006;**67**(2): 360–1.

47. McKellar SH, Tester DJ, Yagubyan M, Majumdar R, Ackerman MJ, Sundt 3rd TM. Novel NOTCH1 mutations in patients with bicuspid aortic valve disease and thoracic aortic aneurysms. *J Thorac Cardiovasc Surg* 2007;**134**(2):290–6.

48. Borger MA, David TE. Management of the valve and ascending aorta in adults with bicuspid aortic valve disease. *Semin Thorac Cardiovasc Surg* 2005;**17**(2):143–7.

49. Warren AE, Boyd ML, O'Connell C, Dodds L. Dilatation of the ascending aorta in paediatric patients with bicuspid aortic valve: frequency, rate of progression and risk factors. *Heart* 2006;**92**(10): 1496–500.

50. Jain D, Dietz HC, Oswald GL, Maleszewski JJ, Halushka MK. Causes and histopathology of ascending aortic disease in children and young adults. *Cardiovasc Pathol* 2011;**20**(1):15–25.

51. Bauer M, Siniawski H, Pasic M, Schaumann B, Hetzer R. Different hemodynamic stress of the ascending aorta wall in patients with bicuspid and tricuspid aortic valve. *J Card Surg* 2006;**21**(3):218–20.

52. Stochholm K, Juul S, Juel K, Naeraa RW, Gravholt CH. Prevalence, incidence, diagnostic delay, and mortality in Turner syndrome. *J Clin Endocrinol Metab* 2006;**91**(10):3897–902.

53. Gotzsche CO, Krag-Olsen B, Nielsen J, Sorensen KE, Kristensen BO. Prevalence of cardiovascular malformations and association with karyotypes in Turner's syndrome. *Arch Dis Child* 1994;**71**(5):433–6.

54. Lin AE, Lippe B, Rosenfeld RG. Further delineation of aortic dilation, dissection, and rupture in patients with Turner syndrome. *Pediatrics* 1998;**102**(1):e12.

55. Carlson M, Silberbach M. Dissection of the aorta in Turner syndrome: two cases and review of 85 cases in the literature. *J Med Genet* 2007;**44**(12):745–9.

56. Beighton P, De Paepe A, Steinmann B, Tsipouras P, Wenstrup RJ. Ehlers-Danlos syndromes: revised nosology, Villefranche, 1997. Ehlers-Danlos National Foundation (USA) and Ehlers-Danlos Support Group (UK). *Am J Med Genet* 1998;**77**(1):31–7.

57. Superti-Furga A, Gugler E, Gitzelmann R, Steinmann B. Ehlers-Danlos syndrome type IV: a multi-exon deletion in one of the two COL3A1 alleles affecting structure, stability, and processing of type III procollagen. *J Biol Chem* 1988;**263**(13):6226–32.

58. Pepin M, Schwarze U, Superti-Furga A, Byers PH. Clinical and genetic features of Ehlers-Danlos syndrome type IV, the vascular type. *N Engl J Med* 2000;**342**(10):673–80.

59. Bergqvist D. Ehlers-Danlos type IV syndrome. A review from a vascular surgical point of view. *Euro J Surg = Acta chirurgica* 1996;**162**(3):163–70.

60. Brooke BS, Arnaoutakis G, McDonnell NB, Black 3rd JH. Contemporary management of vascular complications associated with Ehlers-Danlos syndrome. *J Vasc Surg* 2010;**51**(1):131–8; discussion 138–9.

61. Hanaoka K, Qian F, Boletta A, Bhunia AK, Piontek K, Tsiokas L, et al. Co-assembly of polycystin-1 and -2 produces unique cation-permeable currents. *Nature* 2000;**408**(6815):990–4.

62. Schievink WI, Torres VE, Piepgras DG, Wiebers DO. Saccular intracranial aneurysms in autosomal dominant polycystic kidney disease. *J Am Soc Nephrol* 1992;**3**(1):88–95.

63. Tartaglia M, Mehler EL, Goldberg R, Zampino G, Brunner HG, Kremer H, et al. Mutations in PTPN11, encoding the protein tyrosine phosphatase SHP-2, cause Noonan syndrome. *Nat Genet* 2001;**29**(4):465–8.

64. Bregeon J, Loirand G, Pacaud P, Rolli-Derkinderen M. Angiotensin II induces RhoA activation through SHP2-dependent dephosphorylation of the RhoGAP p190A in vascular smooth muscle cells. *Am J Physiol Cell Physiol* 2009;**297**(5):C1062–70.

65. Burch M, Sharland M, Shinebourne E, Smith G, Patton M, McKenna W. Cardiologic abnormalities in Noonan syndrome: phenotypic diagnosis and echocardiographic assessment of 118 patients. *J Am Coll Cardiol* 1993;**22**(4):1189–92.

66. Niwa K, Siu SC, Webb GD, Gatzoulis MA. Progressive aortic root dilatation in adults late after repair of tetralogy of Fallot. *Circulation* 2002;**106**(11):1374–8.

67. Guo DC, Pannu H, Tran-Fadulu V, Papke CL, Yu RK, Avidan N, et al. Mutations in smooth muscle alpha-actin (ACTA2) lead to thoracic aortic aneurysms and dissections. *Nat Genet* 2007;**39**(12): 1488–93.

68. Pannu H, Tran-Fadulu V, Papke CL, Scherer S, Liu Y, Presley C, et al. MYH11 mutations result in a distinct vascular pathology driven by insulin-like growth factor 1 and angiotensin II. *Hum Mol Genet* 2007;**16**(20):2453–62.

69. Nihoyannopoulos P, Karas S, Sapsford RN, Hallidie-Smith K, Foale R. Accuracy of two-dimensional echocardiography in the diagnosis of aortic arch obstruction. *J Am Coll Cardiol* 1987;**10**(5): 1072–7.

70. Nakamura T, Ruiz-Lozano P, Lindner V, Yabe D, Taniwaki M, Furukawa Y, et al. DANCE, a novel secreted RGD protein expressed in developing, atherosclerotic, and balloon-injured arteries. *J Biol Chem* 1999;**274**(32):22476–83.

71. Tsuji A, Yanai J, Miura T, et al. Vascular abnormalities in congenital cutis laxa–report of two cases. *Acta Paediatr Jpn; Overseas Ed* 1990;**32**(2):155–61.

72. Mehregan AH, Lee SC, Nabai H. Cutis laxa (generalized elastolysis). A report of four cases with autopsy findings. *J Cutan Pathol* 1978;**5**(3):116–26.

73. Barker DF, Hostikka SL, Zhou J, Chow LT, Oliphant AR, Gerken SC, et al. Identification of mutations in the COL4A5 collagen gene in Alport syndrome. *Science* 1990;**248**(4960):1224–7.

74. Earl TJ, Khan L, Hagau D, Fernandez AB. The spectrum of aortic pathology in alport syndrome: a case report and review of the literature. *Am J Kidney Dis* 2012;**60**(5):821–2.

75. Kashtan CE, Segal Y, Flinter F, Makanjuola D, Gan JS, Watnick T. Aortic abnormalities in males with Alport syndrome. *Nephrol Dial Transplant* 2010;**25**(11):3554–60.

76. Li L, Krantz ID, Deng Y, Genin A, Banta AB, Collins CC, et al. Alagille syndrome is caused by mutations in human Jagged1, which encodes a ligand for Notch1. *Nat Genet* 1997;**16**(3):243–51.

77. Kamath BM, Spinner NB, Emerick KM, Chudley AE, Booth C, Piccoli DA, et al. Vascular anomalies in Alagille syndrome: a significant cause of morbidity and mortality. *Circulation* 2004;**109**(11): 1354–8.

78. Tsukioka K, Nobara H, Takano T, Wada Y, Amano J. Quadricuspid aortic valve with ascending aortic aneurysm: a case report and histopathological investigation. *Ann Thorac Cardiovasc Surg* 2011;**17**(4):418–21.

79. Nucifora G, Badano LP, Iacono MA, Fazio G, Cinello M, Marinigh R, et al. Congenital quadricuspid aortic valve associated with obstructive hypertrophic cardiomyopathy. *J Cardiovasc Med* 2008;**9**(3):317–8.

80. Tutarel O. The quadricuspid aortic valve: a comprehensive review. *J Heart Valve Dis* 2004;**13**(4):534–7.

81. Attaran RR, Habibzadeh MR, Baweja G, Slepian MJ. Quadricuspid aortic valve with ascending aortic aneurysm: report of a case and discussion of embryological mechanisms. *Cardiovasc Pathol* 2009;**18**(1):49–52.

82. Bigi MA, Aslani A, Mehrpour M. Effect of chronic cocaine abuse on the elastic properties of aorta. *Echocardiography* 2008;**25**(3):308–11.

83. Hsue PY, Salinas CL, Bolger AF, Benowitz NL, Waters DD. Acute aortic dissection related to crack cocaine. *Circulation* 2002;**105**(13):1592–5.

84. Hatzaras I, Tranquilli M, Coady M, Barrett PM, Bible J, Elefteriades JA. Weight lifting and aortic dissection: more evidence for a connection. *Cardiology* 2007;**107**(2):103–6.

85. Neptune ER, Frischmeyer PA, Arking DE, Myers L, Bunton TE, Gayraud B, et al. Dysregulation of TGF-beta activation contributes to pathogenesis in Marfan syndrome. *Nat Genet* 2003;**33**(3):407–11.

86. Ng CM, Cheng A, Myers LA, Martinez-Murillo F, Jie C, Bedja D, et al. TGF-beta-dependent pathogenesis of mitral valve prolapse in a mouse model of Marfan syndrome. *J Clin Invest* 2004;**114**(11):1586–92.

87. Habashi JP, Judge DP, Holm TM, Cohn RD, Loeys BL, Cooper TK, et al. Losartan, an AT1 antagonist, prevents aortic aneurysm in a mouse model of Marfan syndrome. *Science* 2006;**312**(5770):117–21.

88. Halushka MK. Single gene disorders of the aortic wall. *Cardiovasc Pathol* 2012;**21**(4):240–4.

89. Holm TM, Habashi JP, Doyle JJ, Bedja D, Chen Y, van Erp C, et al. Noncanonical TGFbeta signaling contributes to aortic aneurysm progression in Marfan syndrome mice. *Science* 2011;**332**(6027):358–61.

90. Habashi JP, Doyle JJ, Holm TM, Aziz H, Schoenhoff F, Bedja D, et al. Angiotensin II type 2 receptor signaling attenuates aortic aneurysm in mice through ERK antagonism. *Science* 2011;**332**(6027):361–5.

91. Wang Y, Krishna S, Walker PJ, Norman P, Golledge J. Transforming growth factor-beta and abdominal aortic aneurysms. *Cardiovasc Pathol* 2013;**22**(2):126–32.

92. Wang Y, Ait-Oufella H, Herbin O, Bonnin P, Ramkhelawon B, Taleb S, et al. TGF-beta activity protects against inflammatory aortic aneurysm progression and complications in angiotensin II-infused mice. *J Clin Invest* 2010;**120**(2):422–32.

93. Zhang F, Banker G, Liu X, Suwanabol PA, Lengfeld J, Yamanouchi D, et al. The novel function of advanced glycation end products in regulation of MMP-9 production. *J Surg Res* 2011;**171**(2):871–6.

94. Losy F, Dai J, Pages C, Ginat M, Muscatelli-Groux B, Guinault AM, et al. Paracrine secretion of transforming growth factor-beta1 in aneurysm healing and stabilization with endovascular smooth muscle cell therapy. *J Vasc Surg* 2003;**37**(6):1301–9.

95. Biros E, Walker PJ, Nataatmadja M, West M, Golledge J. Downregulation of transforming growth factor, beta receptor 2 and Notch signaling pathway in human abdominal aortic aneurysm. *Atherosclerosis* 2012;**221**(2):383–6.

96. Golledge J, Tsao PS, Dalman RL, Norman PE. Circulating markers of abdominal aortic aneurysm presence and progression. *Circulation* 2008;**118**(23):2382–92.

97. Matt P, Schoenhoff F, Habashi J, Holm T, Van Erp C, Loch D, et al. Circulating transforming growth factor-beta in Marfan syndrome. *Circulation* 2009;**120**(6):526–32.

98. Shinohara T, Suzuki K, Okada M, Shiigai M, Shimizu M, Maehara T, et al. Soluble elastin fragments in serum are elevated in acute aortic dissection. *Arterioscler Thromb Vasc Biol* 2003;**23**(10):1839–44.

99. Lacro RV, Dietz HC, Wruck LM, Bradley TJ, Colan SD, Devereux RB, et al. Rationale and design of a randomized clinical trial of beta-blocker therapy (atenolol) versus angiotensin II receptor blocker therapy (losartan) in individuals with Marfan syndrome. *Am Heart J* 2007;**154**(4):624–31.

Blood Pressure Regulation and Pathology

Rhian M. Touyz, MD, PhD

University of Glasgow, Glasgow, UK

INTRODUCTION

Elevated blood pressure, also known as hypertension, affects ~30% of the adult population in both developed and developing countries and is a major cause of morbidity and mortality.[1,2] Hypertension is a major risk factor for many common chronic diseases, such as heart failure, myocardial infarction, stroke, and chronic kidney disease, and is increasingly being considered in vascular dementia. Epidemiological and clinical studies demonstrate that the morbid complications of hypertension-associated target organ damage and the risk of cardiovascular diseases and death rise gradually as blood pressure increases.[3] Clinical trials have clearly shown that blood pressure reduction leads to reduced morbidity and mortality with decreased incidence of stroke, heart failure, myocardial infarction, and end-stage kidney disease.[4] Blood pressure lowering is also associated with improved cognitive function in patients with vascular dementia. In spite of the large availability of excellent antihypertensive drugs, blood pressure control in the hypertensive population remains inadequate. Reasons for this are complex and multifactorial and relate, in part, to healthcare issues including poor access to care, lack of diagnosis, and sub-optimal management. In addition and of major significance, is the lack of understanding of the exact cause(s) of hypertension. This has led to an increase in incidence in age-adjusted stroke and end-stage renal disease and a rise in the prevalence of heart failure.[5] Moreover, it is predicted that by 2026, the prevalence of hypertension and associated co-morbidities will increase globally by 60%.[6]

Physiological control of blood pressure is based on Ohm's law modified for fluid dynamics, where blood pressure is proportional to cardiac output and resistance to blood flow in peripheral vessels (Figure 14.1). Blood flow depends on cardiac output and blood volume, whereas resistance is determined primarily by the contractile state of small peripheral arteries and arterioles.[7]

In general, cardiac output remains fairly stable, with increase in peripheral resistance being the major contributor to hypertension, particularly essential hypertension.[8] Many physiological systems influence blood pressure including (1) baroreceptors that sense acute changes in pressure in vessels; (2) the renin–angiotensin system (RAS) that influences vascular tone, salt handling, and volume homeostasis; (3) the adrenergic system which regulates heart rate, cardiac contraction, and vascular tone; and (4) vasoactive factors produced by the blood vessels that cause vasorelaxation, such as nitric oxide, or vasoconstriction, such as endothelin-1 and reactive oxygen species (ROS).[9–12] Many organs contribute to blood pressure control, including the central and peripheral nervous system, heart, and kidneys.[13,14] In addition, the vascular system is increasingly implicated as an important organ involved in blood pressure control (Figure 14.2).[15] These systems act in an integrated manner to maintain adequate perfusion of tissues and organs in spite of varying metabolic demands. Various risk factors, both non-modifiable (age, gender, genes) and modifiable (body mass index, diet, alcohol, salt, sedentary lifestyle) have an impact on blood pressure, probably by influencing physiological systems that control blood pressure.

DEFINITION OF ESSENTIAL (PRIMARY) HYPERTENSION

Blood pressure is a quantitative trait. In population studies, blood pressure has a bell-shaped distribution, which is slightly skewed to the right.[16] There is a positive and continuous correlation between blood pressure and the risk of cardiovascular disease, kidney disease, and stroke, even in the normal range of blood pressures. This correlation is more robust for systolic blood pressure than with diastolic blood pressure.[17] Because there is no clear-cut level of blood pressure where

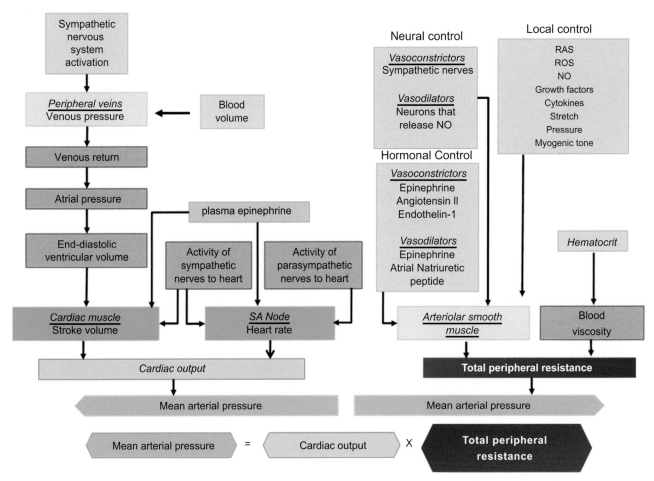

FIGURE 14.1   Physiological regulation of blood pressure. Blood pressure control is based on Ohm's law modified for fluid dynamics, where blood pressure is proportional to cardiac output and resistance to blood flow in peripheral vessels. Blood flow depends on cardiac output and blood volume, whereas resistance is determined mainly by the contractile state of small arteries. In general, cardiac output remains fairly stable, with increase in peripheral resistance being the major contributor to essential hypertension. Many physiological systems influence blood pressure including the sympathetic nervous system, hormones, vasoactive agents, and the renin–angiotensin system, amongst other complex interacting systems. RAS, renin–angiotensin system; ROS, reactive oxygen species; NO, nitric oxide.

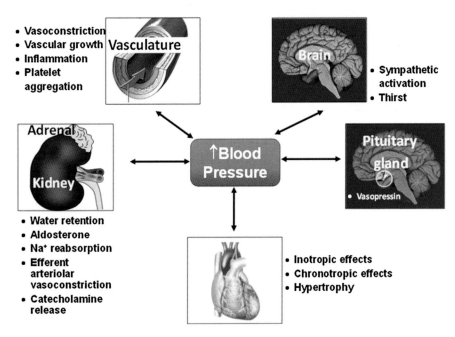

FIGURE 14.2   Mechanisms whereby different organ systems contribute to blood pressure elevation. While hypertension is a multiorgan disease, high blood pressure itself causes target organ damage, such as cardiac hypertrophy, vascular remodeling, renal dysfunction, and cerebral ischemia.

cardiovascular and renal complications start to occur, the definition of hypertension is arbitrary. As such, treatment of hypertension is largely empiric.

Essential, primary, or idiopathic hypertension is defined as high blood pressure in which secondary causes, such as primary hyperaldosteronism, phaechromocytoma, renovascular disease, kidney failure, and monogenetic causes of hypertension are excluded.[18] Essential hypertension is a heterogeneous condition, and accounts for ~ 95% of all patients with hypertension.[19]

The Seventh Report of the Joint National Committee on Prevention, Detection, Evaluation and Treatment of High Blood Pressure (JNC VII) defined and classified hypertension in adults (Table 14.1).[20] This report, to be updated in 2013–2014 as JNC VIII, provides current guidelines for hypertension prevention and management.

Some key points of JNC VII, which are not dissimilar to other major guidelines, include:

1. In persons older than 50 years, systolic blood pressure (BP) of more than 140 mmHg is a much more important cardiovascular disease (CVD) risk factor than diastolic BP.
2. The risk of CVD, beginning at 115/75 mmHg, doubles with each increment of 20/10 mmHg; individuals who are normotensive at 55 years of age have a 90% lifetime risk for developing hypertension.
3. Individuals with a systolic BP of 120–139 mmHg or a diastolic BP of 80–89 mmHg should be considered as prehypertensive and require health-promoting lifestyle modifications to prevent CVD.
4. Thiazide-type diuretics should be used in drug treatment for most patients with uncomplicated hypertension, either alone or combined with drugs from other classes. Certain high-risk conditions are compelling indications for the initial use of other antihypertensive drug classes (angiotensin-converting enzyme inhibitors, angiotensin-receptor blockers, beta-blockers, calcium channel blockers).
5. Most patients with hypertension will require two or more antihypertensive medications to achieve goal BP (<140/90 mmHg, or <130/80 mmHg for patients with diabetes or chronic kidney disease).
6. If BP is more than 20/10 mmHg above goal BP, consideration should be given to initiating therapy with two agents, one of which usually should be a thiazide-type diuretic.
7. The most effective therapy prescribed by the most careful clinician will control hypertension only if patients are motivated. Motivation improves when patients have positive experiences with and trust in the clinician. Empathy builds trust and is a potent motivator.

TABLE 14.1 Classification of Hypertension Based on JNC VII (20)

Blood pressure Classification	Systolic Blood pressure (mmHg)	Diastolic Blood pressure (mmHg)
Normal	<120	and <80
Pre-hypertension	120–139	or 80–89
Stage I hypertension	140–159	or 90–99
Stage 2 hypertension	>160	or >100

GENETICS OF HYPERTENSION

Primary hypertension clusters in families. Individuals with two or more first-degree relatives with hypertension younger than 55 years of age have an almost four times greater risk of developing hypertension before 50 years of age.[21] Epidemiological and family studies clearly demonstrate a significant heritability component for blood pressure. Blood pressure variability in populations attributed to genetic factors varies from 25–50%.[22] Twin studies show greater concordance of blood pressure of monozygotic than dizygotic twins and population studies document greater similarity of blood pressure within families than between families.[23] This aggregation within families is not only due to common environmental factors. Adoption studies demonstrate greater blood pressure concordance among biological siblings than adoptive siblings living in the same home.

Unravelling the genetics of hypertension is challenging because hypertension is a complex, polygenetic, quantitative disease, influenced by multiple environmental and physiological factors. Moreover, the two fundamental factors that determine blood pressure, namely cardiac output and total peripheral vascular resistance, are influenced by intermediary phenotypes, including the sympathetic nervous system, renin–angiotensin–aldosterone system, renal function, volume homeostasis, vascular structure, and many other systems.[24,25] To further complicate this, these intermediary phenotypes are also regulated by complex interacting systems, including blood pressure itself.[25]

Numerous approaches to study the genetics of hypertension have been used to identify putative blood-pressure-related genes. In the general population, studies of blood pressure variation are complicated by many factors that contribute to the trait in any single individual, including demographics, environment, and genetics. Moreover, blood pressure variation in the population is probably determined by numerous genes which each have a small, possibly additive effect. Accordingly very large sample sizes are required to identify variants using strategies such as genome-wide association studies (GWAS).[26–28] To date, a number of large GWAS, each comprising ~ 30 000 subjects have been completed. Together these GWAS identified at least 14 distinct loci having statistically significant associations with blood pressure traits.[28–30] However, each of the identified loci imparts only a small effect on blood pressure. Taken together, GWAS data have been disappointing, but certainly highlight the fact that many genetic variants influence blood pressure at the population level. Perhaps with the development of next-generation sequencing tools, which will provide opportunities to do whole-genome and exome sequencing, further insights into the genetics of hypertension will be possible.

While it has been very challenging to elucidate the genetics of hypertension in the general population, there has been enormous progress in the understanding of Mendelian forms of blood pressure variation in which mutations in single genes have large effects on blood pressure.[31–33] To date about 20 genes have been found to be responsible for blood pressure variation, with mutations in eight genes causing Mendelian forms of hypertension and nine genes that cause Mendelian forms of hypotension (Table 14.2). Typically, these single-gene mutations impart large effects on blood pressure. Of major significance, most of the described mutations influence renal salt reabsorption (Figure 14.3), highlighting the importance of the kidney in the pathophysiology of hypertension.

TABLE 14.2 Some Mendelian Forms of Blood Pressure Variation[25,30]

Genotype	Blood Pressure Phenotype
MUTATIONS AFFECTING MINERALOCORTICOID HORMONES	
Glucocorticoid-remediable aldosteronism	Hypertension
Defective aldosterone synthesis	Hypotension
Syndrome of apparent mineralocorticoid excess	Hypertension
MUTATIONS IN THE MINERALOCORTICOID RECEPTOR	
Hypertension exacerbated in pregnancy	Hypertension
Autosomal dominant pseudohypoaldosteronism type 1 (PHA1)	Hypotension
MUTATIONS ALTERING RENAL ION CHANNELS AND TRANSPORTERS	
Liddle syndrome	Hypertension
Recessive PHA1	Hypotension
Gitelman syndrome	Hypotension
Bartter syndrome	Hypotension
OTHER MENDELIAN FORMS OF HYPERTENSION	
Pseudohypoaldosteronism type II (PHAII)	Hypertension
Hypertension with brachydactyly	Hypertension
Dominant-negative missense mutations in PPARγ	Hypertension

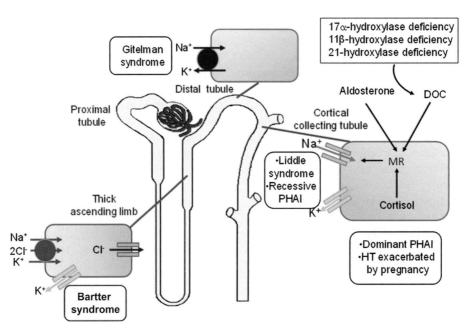

FIGURE 14.3 Mutations leading to changes in blood pressure. Numerous mutations in the nephron, the filtering unit of the kidney, have been described that result in hypertension or hypotension. Pathways regulating NaCl reabsorption in the thick ascending limb of the loop of Henle, the distal convoluted tubule, and the cortical collecting tubule are shown. Mutations in these locations result in altered Na⁺, K⁺ and volume balance. MR, mineralocorticoid receptors; DOC, deoxycorticosterone. (Adapted from References 25,30).

PHYSIOLOGICAL CONTROL OF BLOOD PRESSURE

The pressure required to maintain perfusion of all tissues and organs depends on cardiac output and peripheral resistance. These primary determinants of blood pressure are in turn influenced by various factors. Changes in cardiac output usually contribute to acute changes in blood pressure. Chronic or established hypertension is characterized hemodynamically by normal cardiac output and increased peripheral vascular resistance.[34] Major determinants of vascular resistance are structural thickening of the vessel wall and increased vascular tone due to increased vasoconstriction and/or decreased vasodilation (Figure 14.1).[35,36] Despite extensive research into the mechanisms that initiate these processes, the exact causes still remain elusive.[37] Nevertheless, it is clear that heredity plays a role, along with contributions of numerous environmental factors. At the pathogenic level, many interacting systems influence cardiac output and vascular resistance. Considering the major importance of vascular resistance in the pathogenesis of hypertension, this chapter will focus on processes underlying changes in vascular function and structure.

CARDIAC OUTPUT AND HYPERTENSION

Cardiac output is the product of stroke volume and heart rate. Accordingly, increased fluid volume (preload) or increased cardiac contractility or heart rate due to neural stimulation, are important. Although increased circulating fluid volume underpins increased preload, patients with established hypertension usually have a lower blood volume and total exchangeable sodium than normotensive individuals.[38] This may relate to increased translocation of fluid across the capillary bed into the interstitial space and possibly to increased intracellular fluid volume. Increased heart rate has been attributed to neurogenic mechanisms of increased sympathetic activity and/or decreased parasympathetic drive.

THE SYMPATHETIC NERVOUS SYSTEM AND HYPERTENSION

The sympathetic nervous system comprises the vasomotor center that activates efferent pathways, which innervate sympathetic ganglia. Activated sympathetic nerves secrete catecholamines (norepinephrine, epinephrine), which induce effects on the heart, kidneys, and blood vessels through presynaptic and post-synaptic receptors. Increased activity of the sympathetic

nervous system seems to play an important pathophysiological role in hypertension, particularly in the early stages.[39,40] This is evidenced by elevated plasma norepinephrine levels, increased norepinephrine spillover rate, increased heart rate and blood pressure variability, increased α-adrenergic vasoconstriction and increased vascular reactivity to norepinephrine.[41–43] Catecholamine-induced vasoconstriction of renal efferent arterioles influences renal sodium retention, which may further contribute to blood pressure elevation. Changes in other neurotransmitters, such as neuropeptide Y, a norepinephrine cotransmitter, adenosine, and dopamine in hypertension may also reflect sympathetic nervous system involvement.[44] Pathophysiological processes characterized by increased sympathetic activity and impaired baroreflex control include obesity, obstructive sleep apnea, and polycystic ovary syndrome, often associated with resistant hypertension.[45,46]

Other contributory mechanisms of the sympathetic nervous system involve resetting of the mechanoreceptors, particularly the sinoaortic baroreceptors (high pressure), that are activated by increased arterial pressure, and the cardiopulmonary baroreceptors (low pressure), that are activated by increased central venous pressure.[47] Baroreceptor activation leads to reduced heart rate and lower blood pressure by vagal stimulation and sympathetic inhibition.

Taken together, activation of the sympathetic nervous system and increased catecholamine secretion are major candidates underlying the cardiovascular pressor mechanisms that trigger blood pressure elevation and the trophic processes that maintain hypertension through effects on vascular structural changes, such as hypertrophy. Activation of the sympathetic nervous system also influences the RAS, a major player in established hypertension.[48] Moreover, sympathetic overactivity has been implicated in the increased cardiovascular morbidity and mortality associated with early morning blood pressure surges. This has been attributed, in part, to increased α-sympathetic activation that occurs during prewaking.[49] Targeting the sympathetic nervous system with anti-adrenergic drugs, anti-adrenergic devices (renal nerve denervation), and carotid baroreflex activation are increasingly being considered as effective antihypertensive therapies in patients with resistant hypertension.[50]

THE KIDNEY AND HYPERTENSION

The link between abnormal renal function and high blood pressure was made in the early 1830s, but it was some 130 years later in the 1960s when Guyton demonstrated the key role of the kidney in blood pressure regulation, when it was shown that the kidney increases urinary sodium excretion in response to elevated blood pressure (pressure-natriuresis), thereby reducing blood volume and normalizing blood pressure.[51,52] In patients with hypertension, a right-ward shift of the pressure–sodium excretion curve prevents the return of pressure to normal. Factors associated with this include increased renal vascular resistance, increased filtration fraction, increased peritubular oncotic pressure and consequent increased sodium reabsorption, which further contributes to sodium and fluid overload and increase in blood pressure. Although not conclusively demonstrated, it has also been suggested that a reduction in filtration surface, due to reduced number of nephrons or a reduction in the filtration surface area per glomerulus may limit renal sodium excretion.[53] Factors such as age and ischemia may contribute to reduced nephron number, which could amplify the impaired sodium excretion thereby further promoting hypertension.

Kidney cross-transplantation studies in experimental models support a role for the kidney in hypertension. Transplanting a donor kidney from a hypertensive rat caused hypertension in recipient control animals and conversely kidney transplant from a normal rat into a hypertensive recipient normalized the recipient's blood pressure.[54,55] Moreover, studies in mice showed that development of hypertension depends on the presence of angiotensin II receptor type I (AT_1R) in the kidney, particularly in the proximal tubule.[56] Other studies have shown that the kidney has a functional intrarenal RAS, which influences glomerular ultrafiltrate, nephron function, and hence blood pressure regulation.[57]

SODIUM AND HYPERTENSION

Extensive epidemiologic and experimental evidence implicates a role for excess dietary salt intake in elevated blood pressure.[58,59] High sodium intake induces hypertension by increasing fluid volume and preload, thereby increasing cardiac output.[60] Sodium excess may also increase blood pressure by influencing vascular reactivity and tone. Some individuals are particularly sensitive to sodium in that they have lower basal renin levels and blunted activation of the renin–angiotensin–aldosterone system with sodium restriction.[61] As such, sodium-sensitive hypertensive patients are usually volume expanded with suppressed renin–aldosterone activation. Mechanisms underlying this are unclear, but hereditary factors may be important.

Recent studies have challenged the dogma that retention of sodium is accompanied by commensurate retention of water, which expands the intravascular volume to promote blood pressure elevation. Titze and colleagues

suggest that sodium regulation involves the skin interstitium, which acts as a reservoir that buffers the impact of sodium accumulation on intravascular volume and blood pressure.[62,63] During high salt feeding, sodium accumulates in the subdermal interstitium at hypertonic concentrations, an effect mediated through interactions with proteoglycans. Local macrophages respond to this hypertonicity by secreting vascular endothelial growth factor-c (VEGF-C) via tonicity-responsive enhancer-binding protein (TonEBP), which promotes lymphatic hyperplasia, providing a buffer against increased blood pressure. Patients with refractory hypertension were found to have increased tissue sodium content compared with normotensive individuals.[64]

These researchers have further challenged the accepted concept that with an acute sodium load, steady-state sodium homeostasis is based on urinary sodium excretion. By fixing salt intake of men participating in space flight simulations for many months and testing for the predicted constancy in urinary excretion and total-body sodium content, it was found that at constant salt intake, daily sodium excretion exhibited aldosterone-dependent, weekly (circaseptan) rhythms, resulting in periodic sodium storage. Changes in total-body sodium exhibited longer infradian rhythm periods without parallel changes in body weight and extracellular water, and were directly related to urinary aldosterone excretion and inversely related to urinary cortisol, suggesting rhythmic hormonal control. These findings defined rhythmic sodium excretory and retention patterns independent of blood pressure or body water, which occur independent of salt intake.[65] These intriguing findings question the dogma of sodium homeostasis. However, these studies warrant further confirmation in physiological conditions and in the context of hypertension.

THE RENIN–ANGIOTENSIN SYSTEM (RAS)

The renin–angiotensin system was originally described as a hemodynamic regulator that increases blood pressure acutely by vasoconstriction and chronically through aldosterone-mediated volume expansion.[66] It is considered an endocrine system where Ang II is the product of sequential enzymatic cleavage of angiotensinogen. Renin is produced, stored and secreted from the renal juxtaglomerular cells within the wall of the afferent arteriole, which is contiguous with the macula densa. Circulating renin converts hepatic-derived angiotensinogen to Ang I, which in turn is cleaved by angiotensin-converting enzyme (ACE) to form Ang II.[67] Renal renin release is stimulated by low volume states, high salt in the distal tubules, renal sympathetic nerve activity and reduced renal perfusion. Ang II is the major effector peptide of the

RAS and induces its biological effects through two types of receptors, Ang II type I receptor (AT₁R) and Ang II type 2 receptor (AT₂R) (Figure 14.4).[68] Most of the known biological actions of Ang II are mediated via AT₁R. In the vascular system AT₁R activation by Ang II leads to vasoconstriction and a sustained increase in blood pressure, while in the adrenal gland, Ang II stimulates production of aldosterone, which in turn influences sodium reabsorption and volume homeostasis.[69,70]

The traditional view of the RAS has undergone major changes (Figure 14.5). In particular, (1) a functionally active tissue-based RAS has been described, (2) new components of the RAS have been discovered, including the renin/prorenin receptor and ACE2, and (3) Ang II-derived peptides are now known to be functionally active.[15]

Tissue-Based RAS

Tissue-based RAS is controlled independently from circulating RAS.[71] It is characterized by RAS components, including angiotensinogen, renin, ACE, Ang I, Ang II, and Ang II receptors, within tissues. Tissue RAS has been described in the heart, vessels, kidney, adrenal gland, pancreas, central nervous system, reproductive system, lymphatic and adipose tissue (Figure 14.2).[72] Components of the RAS have also been identified in the eye, which may be important physiologically in maintaining ocular pressure and pathologically in vasculopathies associated with retinopathy of hypertension and diabetes.[73] The local RAS acts in an autocrine/intracrine, paracrine and endocrine fashion. Elevated tissue levels of RAS components may occur in cardiovascular diseases independently of BP elevation.[74] An intracellular RAS has also been described.[75] In experimental models, overexpression of intracellular Ang II correlates with elevated BP and kidney pathology. However the significance in human hypertension is unclear.

New Components of the RAS: Renin/Prorenin Receptor and ACE2

Renin is the key enzyme of the RAS because of its rate-limiting hydrolytic activity. Renin also acts as a RAS agonist by binding to the renin/prorenin receptor ((P)RR).[76] The (P)RR is a new RAS member that binds renin

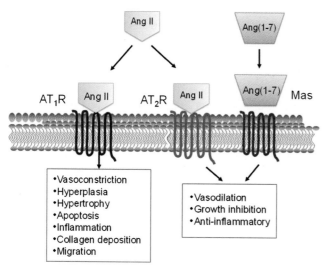

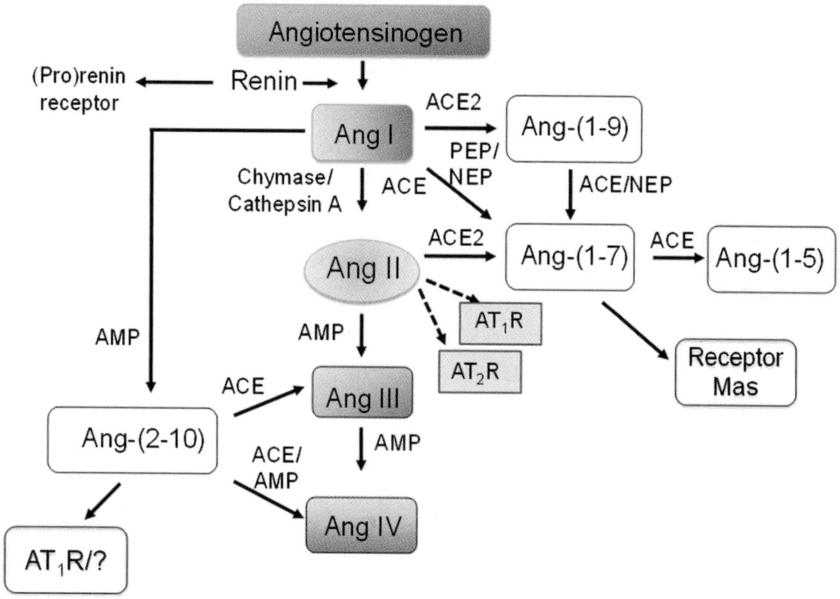

FIGURE 14.4 Vascular effects of Ang II and its receptors. Angiotensin II (Ang II) mediates effects through multiple G protein-coupled receptors (GPCR), including the AT₁ receptor (AT₁R) and the AT₂ receptor (AT₂R). The Ang II-derived peptide, Ang-(1-7), mediates effects via Mas receptor, also a GPCR. Most of the pathological actions of Ang II are induced through the AT₁R. Effects mediated via AT₂R and Mas receptor generally oppose those of AT₁R.

FIGURE 14.5 The renin–angiotensin system (RAS). The classical RAS involves renin production, which increases angiotensin I formation from angiotensinogen and ACE/chymase-induced angiotensin II formation from angiotensin I. New concepts related to the RAS include (1) pro-renin/renin and its receptor; (2) Ang-(1-12) formation; (3) ACE II-induced formation of Ang-(1-7) and Ang-(1-9); Ang III formation and its interaction with AT₁R; and (4) Ang IV formation. Angiotensin-converting enzyme, ACE; angiotensin-converting enzyme 2, ACE2; phosphoenolpyruvate, PEP; nNeutral endopeptidase, NEP; adenosine monophosphate, AMP; angiotensin II type 1 receptor, AT₁R; angiotensin II type 2 receptor, AT₂R.

and prorenin. Prorenin bound to (P)RR undergoes a conformational change and becomes catalytically active to induce activation of signaling pathways, e.g. ERK1/2 and p38MAP kinase, which promote cell growth and fibrosis independently of Ang II in cardiomyocytes, mesangial cells, podocytes, distal tubular cells, vascular endothelial cells, and vascular smooth muscle cells.[77,78] However, the exact (patho)physiological significance of this receptor remains unclear, but it may be important in early development, especially neuronal development. Transgenic animals overexpressing (P)RR develop high BP or glomerulosclerosis, and increased expression of (P)RR occurs in models of hypertension.[27] However, definitive proof of a role for this receptor in cardiovascular disease, especially in humans, is still lacking, with studies failing to demonstrate a significant function of (P)RR in hypertension and target organ damage.[78]

A truncated form of the receptor, M8-9, may be an accessory protein of vacuolar H^+-ATPase (V-ATPase) involved in non-RAS-related functions.[79] V-ATPase plays a role in physiological and biochemical processes by regulating intracellular pH, an effect that is enhanced by (pro)renin binding to (P)RR. The (P)RR also acts as an adaptor protein linking V-ATPase and Wnt receptors and signals through Wnt in a (pro)renin-independent manner. Wnt signaling is involved in cell proliferation, polarity, and fate determination during embryonic development and tissue homeostasis.

Another new RAS member is ACE2. ACE2 catalyzes Ang I and Ang II to generate the Ang peptides Ang-(1-9) and Ang-(1-7), which mediate, in general, opposite effects to those of Ang II.[80,81] Reduced ACE2 and consequent decreased Ang-(1-7) production is associated with vasoconstriction, vascular remodeling and oxidative injury in hypertension, diabetes, and kidney disease.[82] On the other hand, increased ACE2 activity promotes vasodilation and as such is currently being tested as a potential therapeutic target.

Ang II-Derived Peptides: Ang-(1-7), Ang III, Ang IV, and Other Ang Peptides

Ang-(1-7) is formed from Ang II by prolylendopeptidase, prolyl-carboxipeptidase, ACE2, or directly from Ang I through hydrolysis by prolylendopeptidase and endopeptidase 24.11 (Figure 14.5).[83] It is metabolized by ACE to Ang-(1-5). Ang-(1-7) is present in the circulation and in tissues (brain, heart, kidneys, vessels, liver, reproductive organs), and binds to the Mas receptor, a G-protein-coupled receptor.[84] Ang-(1-7)/Mas mediates vasodilation, growth-inhibition, anti-inflammatory responses, antiarrhythmogenic and antithrombotic effects.[85,86] At the molecular level, these responses involve nitric oxide synthase (NOS)-derived nitric oxide (NO) production, activation of protein tyrosine

phosphatases, inhibition of MAPK and inhibition of NADPH oxidase. The ACE2-Ang-(1-7)-Mas axis is now considered as a putative counter-regulatory system to the ACE-Ang II-AT_1R axis.[87]

Ang III, formed from Ang II by aminopeptidase A, is a biologically active peptide with actions similar to those of Ang II.[88] Ang III infusion in experimental models and in humans increases blood pressure, promotes vasoconstriction and stimulates aldosterone production. In vitro, it stimulates vascular growth, production of proinflammatory mediators and deposition of extracellular matrix proteins. Ang III seems to bind to AT_1R and to AT_2R.

Another small peptide derived from Ang II is Ang IV, generated by aminopeptidase N, which removes arginine from the N-terminus of Ang III.[89] Ang IV can also be formed directly from Ang II by the enzyme D-aminopeptidase. Ang IV increases renal blood flow, induces vasodilation and improves cardiac function.[89] Some of these actions are mediated via the AT_1R, but others are due to binding of Ang IV to a specific binding site, AT_4R, which has been identified as the transmembrane enzyme insulin-regulated aminopeptidase (IRAP).[90] The biological significance of Ang IV/IRAP remains to be elucidated. Other angiotensin-derived peptides, including Ang-(3-7), Ang-(1-9) and Ang-(1-12) have been identified. However the (patho)physiological significance of these peptides, especially in humans, is unknown.

ANGIOTENSIN RECEPTORS AND SIGNALING

Ang II binds with high affinity to the AT_1R and AT_2R and mediates effects via complex signaling pathways (Figure 14.6).[91,92] These receptors are members of the G-protein-coupled receptor (GPCR) superfamily. The AT_2R is the predominant Ang II receptor in the fetus. After birth, the AT_1R becomes the dominant subtype. AT_2R are re-expressed in pathological conditions including hypertension, myocardial infarction, cardiac failure, renal failure, cerebral ischemia, and diabetes.

Signaling Through AT_1R

Ang II promotes vasoconstriction by increasing intracellular free Ca^{2+} concentration ($[Ca^{2+}]_i$) and by activating RhoA and its downstream target Rho kinase.[93] RhoA is abundantly expressed in vascular smooth muscle cells and participates in vasoconstriction via phosphorylation of the myosin light chain and sensitization of contractile proteins to Ca^{2+}. Pharmacological inhibition of Rho kinase with fasudil or Y27632 suppresses acute pressor responses of Ang II, but does not seem to reduce blood pressure chronically.[93]

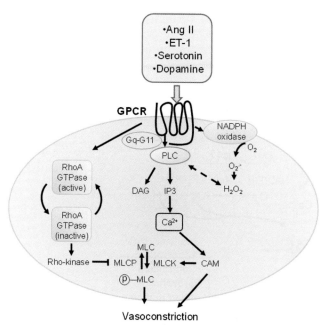

FIGURE 14.6 Activation of G-protein-coupled receptors (GPCR) by vasoactive agents regulates vasomotor tone through multiple mechanisms. Ligand binding to GPCR induces coupling to heterometric Gq proteins, to activate PLC, leading to generation of IP3 and DAG, resulting in increased $[Ca^{2+}]i$ that triggers phosphorylation of MLC20 and stimulation of contraction. GPCR activation also induces contraction through the RhoA/Rho kinase pathway that increases Ca^{2+} sensitivity by inhibiting MLCP. NADPH oxidase (Nox)-derived reactive oxygen species also influence Ca^{2+}-sensitive pathways that stimulate contraction. PLC, phospholipase C; DAG, diacylglycerol; CAM, calmodulin; MLCP, myosin light-chain phosphatase; MLC, myosin light chain; MLCK, myosin light-chain kinase; O_2^-, superoxide; H_2O_2, hydrogen peroxide; p, phosphorylation; IP3, inositol trisphosphate; GPCR, G-protein-coupled receptor; Ang II, angiotensin II; ET-1, endothelin-1.

Ang II stimulates cell growth through phosphorylation of MAPKs (ERK1/2, p38MAPK, JNK, ERK5) and tyrosine kinases (c-Src, JAK, FAK, Pyk2, p130Cas) and by increasing $[Ca^{2+}]_i$. In cardiac, renal and vascular tissue from hypertensive rats and vascular smooth muscle cells from hypertensive patients, Ang II-stimulated activation of tyrosine kinases and MAPKs is augmented.[94] These processes contribute to vascular wall thickening and tone, which impact vascular resistance and blood pressure (Figure 14.7).

Ang II also activates receptor tyrosine kinases, even though it may not bind directly to these receptors. This process of transactivation is demonstrated for the epidermal growth factor receptor (EGFR), platelet-derived growth factor receptor (PDGFR), and insulin-like growth factor 1 receptor (IGF-1R)[95] and occurs through tyrosine kinases, redox-sensitive processes, and through a metalloprotease (ADAM17)-dependent shedding of heparin-binding EGF-like growth factor (HB-EGF).[96] Moreover Ang II stimulates production of vasoactive hormones and growth factors in hypertension, such as ET-1, PDGF, transforming growth factor β (TGFβ), basic fibroblast growth factor (bFGF), and IGF-1, which promote hyperplasia and fibrosis, further contributing to cell growth and vascular wall thickening in hypertension.

Another important molecular mechanism whereby Ang II modulates cardiovascular function is through generation of reactive oxygen species (ROS). ROS are highly reactive, bioactive, short-lived molecules derived from reduction of O_2.[97] Major ROS produced within vascular cells include superoxide ($\bullet O_2^-$) and hydrogen peroxide (H_2O_2).[97] Among the many ROS-generating enzymes, including cyclo-oxygenase, lipoxygenase, heme oxygenase, cytochrome P450 monooxygenase, and xanthine oxidase, NAD(P)H (nicotinamide adenine dinucleotide phosphate, reduced form) oxidases are of major importance in vascular cells.[98] NADPH oxidase (also known as Nox) catalyzes the production of $\bullet O_2^-$ by the one electron reduction of oxygen using NADPH as the electron donor: $2O_2 + NAD(P)H \rightarrow 2O_2^- + NAD(P)H + H^+$. Seven Nox isoforms have been identified of which Nox1, 2, 4, and 5 are functionally active in the vascular system.[99,100] All vascular Nox isoforms are regulated by Ang II. The functional significance of different Noxs awaits clarification, but their differential tissue distribution, cellular localization, and subcellular compartmentalization probably play a major role in Nox-specific actions.

Superoxide and H_2O_2 activate multiple signaling molecules, including MAP kinases, tyrosine kinases, protein tyrosine phosphatases and transcription factors (NF-κB, AP-1, and HIF-1),[101] which regulate cell growth, migration, inflammation, fibrosis, and contraction/dilation, and are important in vascular remodeling and hypercontractility in hypertension. ROS modifies signaling molecules in large part through oxidative modification of redox-sensitive proteins, e.g. protein tyrosine phosphatases.[102]

Taken together, the RAS plays a major role in cardiovascular pathophysiology of hypertension, diabetes, chronic kidney disease, and heart failure. As such, pharmacological agents that block key components of the RAS such as direct renin inhibitors, ACE inhibitors, and Ang II receptor blockers (ARBs) have gained wide clinical use for these indications.

THE VASCULAR SYSTEM AND HYPERTENSION

Small arteries (lumen diameter <300 μm) are responsible for blood pressure control and regional distribution of blood flow through effects on vascular resistance. Based on Poiseuille's law, vessel resistance is inversely proportional to the radius (lumen diameter) to the fourth power (r^4). Hence, small changes in lumen size result in significant changes in resistance. The lumen diameter of resistance arteries is a function of vasomotor tone (vasoconstriction/vasodilation) and the structural

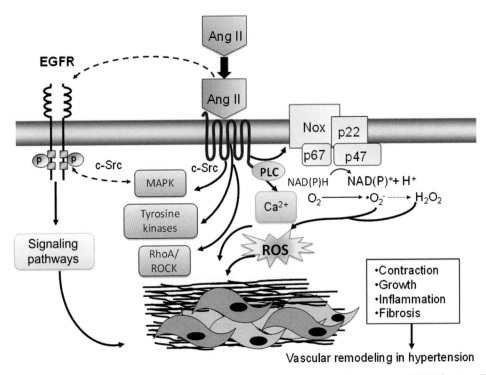

FIGURE 14.7 Molecular events regulating vascular remodeling in hypertension. Increased activation of AT₁R by Ang II leads to activation of multiple signaling pathways that stimulate cell growth, inflammation, fibrosis, and contraction. Signaling pathways include activation of mitogen-activated protein kinases (MAP), tyrosine kinases, RhoA/Rho kinase and phospholipase C (PLC)/Ca²⁺. Ang II is a potent stimulator of NADPH oxidase (Nox), a multisubunit enzyme, that when activated induces generation of reactive oxygen species (ROS), such as superoxide (O_2^-) and hydrogen peroxide (H_2O_2). ROS act as second messengers, which influence redox-sensitive signaling processes. Ang II also induces activation of receptor tyrosine kinases, such as epidermal growth factor receptor (EGFR), through transactivation pathways that may involve the tyrosine kinase c-Src. These processes lead to vascular smooth muscle cell contraction, growth, inflammation, and fibrosis that contribute to arterial remodeling in hypertension.

characteristics of the vessel (Figure 14.8).[104] Vasomotor control underlies acute rapid adaptation of vessel diameter, due mainly to vasoconstriction exerted by the active contraction of vascular smooth muscle cells, whereas alterations in structure (vascular wall thickening) represent a dynamic process in response to chronic hemodynamic variations. Initially, structural changes are adaptive, but subsequently become decompensated leading to alterations in media thickness and lumen diameter (called remodeling).[105]

Acute regulation of vascular diameter depends on the activation/deactivation of the contractile machinery involving actin–myosin interaction in vascular smooth muscle cells. Changes in [Ca²⁺]ᵢ, ion fluxes and membrane potential lead to calcium-calmodulin-mediated phosphorylation of the regulatory myosin light chains and actin-myosin cross-bridge cycling with consequent rapid vasoconstriction. Calcium-independent mechanisms associated with changes in calcium sensitization through RhoA/Rho kinase and actin filament remodeling contribute to intermediate and more chronic processes regulating vascular lumen diameter.

The lumen diameter of resistance arteries is governed not only by vasoconstriction, but also by the structural characteristics of the vessel wall.[105] At the molecular and cellular levels, remodeling involves changes in cytoskeletal organization, cell-to-cell connections, and altered growth/apoptosis, senescence, calcification, inflammation and rearrangement of VSMCs. At the extracellular level, remodeling is influenced by changes in matrix protein composition and reorganization of proteoglycans, collagens (types I and III) and fibronectin, which provide tensile strength, and elastin, responsible for vascular elasticity.[106]

With respect to resistance arteries in hypertension, the most widely used classification of vascular remodeling is that described by Mulvany et al.,[107] where changes in the passive luminal diameter may be increased (outward remodeling) or decreased (inward remodeling) and media mass (cross-sectional area) may be increased, unchanged or reduced (hypertrophic, eutrophic, hypotrophic remodeling, respectively). Remodeling allows arteries to withstand an increased pressure load, and under physiological conditions (e.g. aging and exercise) is adaptive. Pathological remodeling occurs when the adaptive process is overwhelmed, resulting in rigid, stiff and poorly compliant vessels typically observed in hypertension.

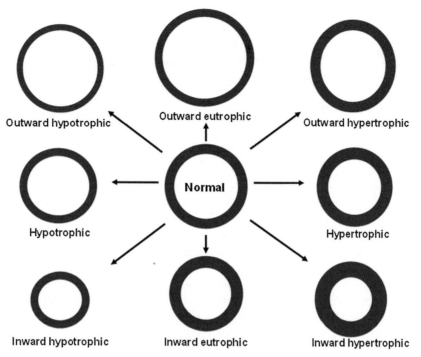

FIGURE 14.8 Remodeling of vessels. Changes in lumen diameter and media mass (cross-sectional area) define the different patterns of vascular remodeling.[5,8,9] Vessel narrowing with increased wall thickness occurs in chronic hypertension (hypertrophic remodeling), while mild hypertension is associated with smaller lumen and no increase in cross-sectional area (eutrophic remodeling). (Adapted from References 103,107).

Vascular Contraction

The major factor regulating vascular tone is vascular smooth muscle cell contraction, which is increased in experimental and clinical hypertension. Non-invasive approaches (such as vascular ultrasound, pulse wave analysis, peripheral arterial tone) to study endothelial function and vascular tone in humans have demonstrated that patients with hypertension exhibit impaired endothelium-dependent vasodilation, enhanced vascular reactivity, and increased contractility.[108] Vascular contraction is regulated primarily by an increase in $[Ca^{2+}]_i$.[108] Physiologically vascular smooth muscle cells maintain their contractile phenotype, although they have the potential to differentiate into a 'non-contractile' form as an adaptive response to changes in the local environment or in response to injury. They also undergo a phenotypic switch in pathological conditions such as in hypertension, aneurysms, angiogenesis, and atherosclerosis. However, factors triggering the shift from a contractile to a non-contractile phenotype still remain poorly understood but high blood pressure itself may play a role.

The major event in vascular smooth muscle contraction is a rise in $[Ca^{2+}]_i$ in response to agonist-induced activation of receptors, such as Ang II-AT_1R, coupled to phopholipase C (PLC), which leads to generation of the second messengers insitol trisphosphate (IP_3) and diacylglycerol (DAG) (Figure 14.6).[109] IP_3 stimulates intracellular Ca^{2+} release from the sarcoplasmic reticulum and DAG causes protein kinase C (PKC) activation. Ca^{2+} influx channels, such as voltage-operated (VOC), receptor-operated (ROC), store-operated (SOC) Ca^{2+} channels and Ca^{2+}-permeable nonselective cation channels (NSCC), are also involved in the elevation of $[Ca^{2+}]_i$. The primary target protein influenced by increased $[Ca^{2+}]_i$ is the calcium-binding protein calmodulin. $Ca^{2+}/$ calmodulin binding initiates a conformational change in the calmodulin molecule, which interacts with myosin light chain kinase (MLCK) leading to its activation. Activated MLCK triggers phosphorylation of the regulatory light chains of myosin (MLC20) at Ser19, promoting cycling of myosin cross-bridges with actin and consequent contraction.[110] Vascular smooth muscle cell relaxation is induced by dephosphorylation of MLC20 by myosin light chain phosphatase (MLCP). Hence the magnitude of MLC20 phosphorylation, and vascular smooth muscle tone, is determined by the relative activities of MLCK and MLCP.[111]

In addition to changes in $[Ca^{2+}]_i$, vascular smooth muscle is regulated in a Ca^{2+}-independent manner through a process of 'calcium sensitization.' This phenomenon involves kinases such as Rho kinase (Figure 14.6).[111] These Ca^{2+}-independent processes influence contraction by increasing Ca^{2+} sensitization, which maintains force generation by promoting MLC20 phosphorylation in a MLCK-independent manner and by reducing activity of

MLCP. Rho kinases (ROCK1 and ROCK2, also known as ROKβ and ROKα, respectively) are serine/threonine kinases and are downstream effectors of the small GTPase RhoA. RhoA, a member of the Rho family of small GTPase binding proteins, is abundantly expressed in vascular smooth muscle cells and is rapidly activated by vasoconstrictors that signal through GPCR. Increased Rho/Rho kinase activity leads to decreased MLCP activation and consequent sustained vasoconstriction.

Activated ROCK phosphorylates several targets of the contractile machinery, with the most important being MLC20, which when phosphorylated, inhibits MLCP activity.[112] The effect of RhoA/ROCK activity is enhancement of actomyosin interaction, prevention of actin depolymerization and consequent contraction. Vasoactive agonists that activate RhoA and also increase $[Ca^{2+}]_i$ include endothelin-1 (ET-1), Ang II, serotonin, dopamine, sphingosine-1-phosphate and adrenergic neurotransmitters, all of which are implicated in hypertension.[113] In addition, aldosterone, through its mineralocorticoid receptor, and receptor tyrosine kinases, such as PDGFR and EGFR, can activate RhoA/ROCK.[114]

Dysregulation of RhoA/ROCK signaling has been demonstrated in many disorders, including hypertension, atherosclerosis, stroke, diabetes, erectile dysfunction, heart failure, and pulmonary hypertension.[115] Most of these pathologies are associated with altered vascular smooth muscle contraction, although other RhoA-dependent systems, such as cell proliferation and migration, probably also play a role. The RhoA/ROCK pathway may be an important therapeutic target in cardiovascular medicine. Clinically, the beneficial effects of fasudil, a Rho kinase inhibitor, have been demonstrated for the treatment of several cardiovascular diseases in humans, particularly in cerebral vasospasm after subarachnoid haemorrhage.[116]

Endothelial Function and Hypertension-Associated Vascular Changes

Endothelial cells normally regulate vascular tone by releasing relaxing and constricting factors such as nitric oxide synthase (NOS)-derived nitric oxide (NO), arachidonic acid metabolites, ROS, and vasoactive agents. They also produce endothelial-derived hyperpolarizing factors (EDHF) that induce endothelium-dependent relaxation through hyperpolarization of underlying VSMCs independently of NO. EDHF-mediated responses are important in hypertension, where they provide a vasorelaxation reserve for endothelial dysfunction due to decreased NO bioavailability.[117] Endothelial dysfunction is a hallmark of hypertension and may reflect the premature aging of the intima exposed to chronic blood pressure increase. This dysfunction is characterized by impaired vasomotor responses, vascular smooth muscle

cell (VSMC) proliferation and migration, extracellular matrix (ECM) protein deposition, platelet activation, vascular permeability and a pro-inflammatory and pro-thrombotic phenotype.[118]

Vascular Smooth Muscle, Inflammation, and Vascular Remodeling in Hypertension

Low-grade inflammation in the vascular wall is increasingly recognized as an important contributor to the pathophysiology of hypertension,[119] to the initiation and progression of atherosclerosis, and the development of cardiovascular disease. Increased expression of adhesion molecules (VCAM-1, ICAM-1) on the endothelial cell membrane, accumulation of monocyte/macrophages, dendritic cells, natural killer (NK) cells, and B and T lymphocytes are some of the mechanisms that participate in the inflammatory response in the vascular wall.[120] Activators of nuclear receptors, such as PPARs, down-regulate the vascular inflammatory response in experimental animals and decrease serum markers of inflammation in humans. Innate immunity has been implicated to contribute to the low-grade inflammatory response in hypertension where different subsets of T lymphocytes may be involved in processes leading to inflammation. An imbalance between the pro-inflammatory (Th1, Th2, and Th17) and the anti-inflammatory T regulatory (Treg) subsets of T lymphocytes[120,121] has been demonstrated in experimental models of hypertension. Confirmation in humans is still awaited.

Vascular Aging, Remodeling, and Hypertension

The age-associated changes in blood vessels that occur in healthy individuals include increased arterial wall thickness, reduced compliance, increased stiffness and decreased lumen diameter, features typically found in hypertension.[122] These structural changes are associated with impaired endothelial function, caused by decreased production of vasodilators, such as NO and prostacyclins, and increased bioavailability of ROS.[122] Consistent with a dysfunctional endothelium is increased vasoconstriction and decreased fibrinolysis. Arterial aging is a predominant risk factor for the onset of cardiovascular diseases, such as hypertension, and is associated with activation of the RAS, increased vascular stiffness, intima-media thickening, calcification, and a proinflammatory phenotype. Moreover, vascular repair systems become progressively impaired with aging. These features are characteristic of the vascular phenotype in hypertension and in fact hypertension is considered an important factor in accelerated aging of the vasculature.

Cellular and molecular mechanisms underlying age-associated changes of the vascular system are unclear, but vascular cells, including stem and/or progenitor cells, undergo senescence. Senescent cells enter irreversible growth arrest, exhibit a flattened and enlarged phenotype and express genes, such as negative cell cycle regulators p53 and p16, that differ from those normally expressed.[123] Factors implicated in cellular senescence include decreased telomerase activity and telomere shortening, DNA damage, and genomic instability. These processes are modulated by Ang II and are redox-sensitive. There are now extensive data indicating that vascular ROS bioavailabilty and RAS activation are increased in aging, as well as in hypertension.[124]

Increased aortic stiffness, typical of vascular changes associated with aging and hypertension, has been correlated with the expression of genes involved in increased vascular tone and remodeling. These include protein phosphatase-1, the catalytic subunit of myosin light chain phosphatase, members of the family of A-kinase anchor protein and PKCβ-1, involved in long-term sustained contraction. There are relationships also between extracellular matrix molecules and aortic stiffness. Vascular calcification, which is associated with mineralization of the internal elastic lamina and elastic fibers within the media, also leads to age- and hypertension-related stiffening of vessels. Vascular calcification is a tightly regulated process similar to bone formation and VSMCs play a critical role in its processes.[125] VSMCs have a remarkable capacity to undergo phenotypic differentiation. Factors that trigger and promote VSMC osteogenic induction include abnormalities in mineral metabolism, particularly hyperphosphatemia and hypercalcemia, driven by up-regulation of transcription factors such as cbfa1 (core-binding factor alpha(1))/Runx2, MSX-2, and bone morphogenetic protein 2 (BMP-2), involved in normal bone development, and which control the expression of osteogenic proteins, including osteocalcin, osteonectin, alkaline phosphatase, collagen-1, and bone sialoprotein.[126] Another mechanism contributing to vascular mineralization is loss of calcification inhibitors, such as fetuin-A, matrix Gla protein, pyrophosphate, and osteopontin.[127] Molecular processes underlying this remain to be fully defined but Ang II, ET-1, and urotensin may be important modulators of vascular calcification. Ang II influences calcification by redox-sensitive pathways that stimulate expression of bone morphogenetic protein 2 (BMP2) and the osteoblast transcription factor Runx2/Cbfa1, and through modulation of Ca^{2+} and Mg^{2+} transport through cation channels, such as TRPM7.[128]

REACTIVE OXYGEN SPECIES, OXIDATIVE STRESS, AND HUMAN HYPERTENSION

Reactive oxygen species are signaling molecules important in the regulation of many physiological processes including host defense, aging, and cellular homeostasis.

Increased ROS bioavailability and altered redox signaling (oxidative stress), due in part to aberrant regulation of ROS production, have been implicated in the onset and/or progression of chronic diseases including hypertension.[129] Although oxidative stress may not be the only cause of hypertension, it amplifies blood pressure elevation in the presence of other pro-hypertensive factors, such as salt loading, activation of the RAS and sympathetic hyperactivity, at least in experimental models.

Hypertensive patients exhibit higher levels of plasma H_2O_2, increased plasma and urine markers of oxidative stress such as thiobarbituric acid reactive substances (TBARS), oxidized low density lipoprotein, and eight iso-prostane, and reduced antioxidant capacity compared with normotensive subjects.[130,131] Treatment with antihypertensive drugs reduces oxidative stress biomarkers, in some cases independently of blood pressure lowering. Normotensive subjects with a family history of hypertension have greater ROS production than blood-pressure-matched subjects without a family history of hypertension, suggesting a genetic component associated with elevated production of free radicals.[132]

Reactive oxygen species production is increased in vascular smooth muscle cells from resistance arteries of hypertensive patients and this is associated with up-regulation of vascular NADPH oxidase.[133,134] The importance of NADPH oxidase in human cardiovascular disease is supported by studies showing that polymorphisms in NADPH oxidase subunits are associated with increased atherosclerosis and hypertension. In particular, the −930(A/G) polymorphism in the p22phox promoter may be a novel genetic marker associated with hypertension. An association between the p22phox −930 G polymorphism has been associated with blood pressure in normotensive subjects.[135] The C242T CYBA polymorphism is associated with essential hypertension and hypertensive patients carrying the CC genotype of this polymorphism exhibit features of NADPH oxidase-mediated oxidative stress and endothelial damage and are prone to cerebrovascular disease. The T allele of the p22-phox C242T polymorphism is also associated with higher left ventricular mass/height and increased NADPH-oxidase activity in Brazilian hypertensive patients, suggesting that genetic variation within NADPH-oxidase components may modulate left ventricular remodeling in subjects with hypertension. In a Japanese population the G(−930)A polymorphism of CYBA was confirmed to be important in the pathogenesis of hypertension. Polymorphisms of the xanthine oxidase gene (−337GA and 565+64CT) have also been shown to be related to blood pressure and oxidative stress in hypertension.[136]

Oxidative stress is also found in secondary hypertension. Patients with primary hyperaldosteronism exhibit increased levels of plasma ROS and markers of subclinical inflammation compared with essential hypertensive patients.[137] Hypertension during pregnancy was also found to be associated with oxidative stress as evidenced by increased TBARS levels during labor. Elderly patients (~75 years) who exhibit endothelial dysfunction and decreased antioxidant capacity responded positively to oral antioxidant therapy, with improved flow-mediated vasodilation, decreased plasma TBARS levels, and increased antioxidants.

Decreased antioxidant defense mechanisms also contribute to oxidative stress in human hypertension.[138] Patients with essential hypertension exhibit reduced activity and decreased plasma levels of antioxidant enzymes, including superoxide dismutase (SOD), glutathione peroxidase, and catalase. Decreased levels of antioxidant vitamins A, C, and E have been shown in newly diagnosed, untreated hypertensive patients compared with normotensive controls.

Considering the possible pathophysiological role of oxidative stress in hypertension and the convincing experimental data, it is reasonable to imagine that reducing ROS bioavailability through antioxidants, ROS scavengers, and Nox inhibitors would have protective and blood-pressure-lowering effects. Antioxidants that have been tested in clinical trials include vitamins A, C, and E, co-enzyme Q, beta carotene, polyphenols, and flavonoids. However, findings are inconsistent and clinical trial data are inconclusive, with most large antioxidant clinical trials failing to demonstrate beneficial cardiovascular or antihypertensive effects.[139] As such, general consensus is that antioxidants are not used in the prevention or treatment of hypertension.

NEW DRUGS, PROCEDURES, AND DEVICES IN THE MANAGEMENT OF HYPERTENSION

According to Chobanian, the treatment of hypertension has been one of medicine's major successes, since advances in therapy have the potential for lowering blood pressure in almost every patient with hypertension.[140] However, hypertension continues to be a major public health burden, with an increasing prevalence globally. In addition, the number of individuals with sub-optimal blood pressure control is increasing, in spite of therapeutic advances. This has been termed the 'hypertension paradox.'[140] Factors contributing to this include poor lifestyle, lack of adherence/compliance, suboptimal choice of drugs and possibly inadequate available therapies. Moreover, 5–30% of the population have resistant hypertension requiring at least three different antihypertensive drugs for management.

A large number of classes of antihypertensive drugs are now available for the treatment of hypertension (Table 14.3).[141] Theoretically these drugs should successfully treat virtually all patients with hypertension. However, a significant proportion of treated patients still have suboptimal control. Accordingly, the need for improved and novel antihypertensive therapies continues to be important in cardiovascular medicine.

Innovative approaches to better treat hypertension show promise and include new classes of pharmacological agents, and new procedures and devices (Table 14.4). New pharmacological classes and molecules include dual vasopeptidase inhibitors (e.g. neprilysin (also called neutral endopeptidase) and ACE inhibitor; neprilysin and endothelin-converting enzyme (ECE) inhibitor), endothelin receptor antagonists, aldosterone synthase inhibitors, and nitric oxide donors.[142] New procedures and devices,

TABLE 14.3   Oral antihypertensive agents

Angiotensin-converting enzyme inhibitors (ACEi)
Angiotensin receptor blockers (ARB)
Adrenergic inhibitors
   Beta adrenergic blockers
   Beta blockers with sympathomimetic activity
   Alpha-beta blockers
   Alpha receptor blockers
Calcium channel blockers (CCB)
   Dihydropyridines
   Nondihydropyridines
Diuretics
   Thiazides
   Loop diuretics
   Potassium-sparing diuretics
Direct vasodilators
Centrally acting alpha$_2$ agonists
Peripherally acting adrenergic antagonists
Combination drugs
   Diuretic/diuretic
   Diuretic/beta blocker
   Diuretic/ACEi
   Diuretic/ARB
   ACEi/CCB

TABLE 14.4   New Drugs, Procedures, and Devices for Hypertension

NEW DRUGS
Dual vasopeptidase inhibitors
Aldosterone synthase inhibitor
Endothelin antagonists
Nitric oxide donor
VACCINES
Angiotensin 1 vaccine
Angiotensin 2 vaccine
DEVICES
Baroreceptor stimulator
Catheter-based renal denervation

such as baroreceptor stimulation and catheter-based renal denervation, target the sympathetic nervous system and show promise in the management of patients with resistant hypertension.[143,144] Vaccines for hypertension offer another possible strategy.[145] Two Ang-based vaccines are in development. An advantage of this strategy is long-lasting immunity, which would obviate the need for chronic daily dosing. Despite the potential of these new approaches, the clinical development of novel antihypertensive drugs has been slow and these still need to be tested in humans, and effectiveness and safety of new devices, procedures, and vaccines still warrant outcome trials.

CONCLUSIONS

Physiological regulation of blood pressure depends primarily on cardiac output and peripheral resistance. These fundamental processes are influenced by multiple complex interacting physiological systems, including the renal system, RAS, sympathetic nervous system, and ROS-generating systems. Dysregulation, or failure, of one or more of these systems impacts cardiac output and/or peripheral resistance, which predisposes to blood pressure elevation. In general, increased total peripheral vascular resistance, determined by increased vascular tone and structural remodeling (decreased lumen diameter, thickened media), is the major pathophysiologic process underlying essential hypertension. Accordingly, factors that influence vascular contractility and structural remodeling, such as the RAS, are major determinants of hypertension. Since the early 2000s there have been many advances in hypertension research that have described new paradigms in the pathogenesis of hypertension. Some of these include (1) new components of the RAS, (2) the role of inflammation and oxidative stress, (3) involvement of the immune system, (4) interstitial salt as a regulator of sodium balance and (5) identification of Mendelian forms of blood pressure variation where rare gene mutations impart large effects on blood pressure. Data from large GWAS studies have thus far been disappointing with no variants identified that account for a large fraction of the variation in blood pressure in the general population. This further highlights the fact that hypertension is a complex, polygenic quantitative disease. Although we still know little about the genetics of essential hypertension in the general population, and despite the complexity underlying the pathogenesis of hypertension, some common pathways and systems, e.g. the kidney, the RAS, and the sympathetic nervous system, are now well recognized to be important in the development of hypertension. Some of these systems have been used as therapeutic targets. For example, many commonly used antihypertensive drugs target renal sodium excretion, sympathetic nervous

system activation, endothelial function, and vascular contractility. A better understanding of molecular and genetic causes underlying the development of hypertension will lead to improved diagnostic and therapeutic approaches that should ameliorate the growing burden of hypertension and its complications. Future progress will derive from human molecular genetic studies, next-generation sequencing tools, animal models of blood pressure variation, and experimental medicine studies. Disease-targeted therapeutic strategies, such as vaccines, devices, siRNA, and novel molecules are promising new approaches that may improve management of hypertension. To date, these are still in the discovery and research phases and are not yet available for routine clinical use.

Acknowledgments

Quoted studies by the author were funded through grants from the Canadian Institutes of Health Research, the Heart and Stroke Foundation of Canada, the Kidney Foundation of Canada/Pfizer, and the Juvenile Diabetes Research Foundation.

References

1. Sliwa K, Stewart S, Gersh BJ. Hypertension: a global perspective. *Circulation* 2011;**123**(24):2892–6.
2. Kakar P, Lip GY. Towards understanding the aetiology and pathophysiology of human hypertension: where are we now? *J Hum Hypertens* 2006;**20**(11):833–6.
3. Lawes CM, Vander Hoorn S, Rodgers A. Global burden of blood-pressure related disease. *Lancet* 2008;**371**:1513–8.
4. The VA Cooperative Study Group. Effects of treatment on morbidity of hypertension. Results in patients with diastolic blood pressures averaging 115 through 129 mm Hg. *J Am Med Assoc* 1967;**202**:1028–34.
5. Franco OH, Peeters A, Bonneux L, de Laet C. Blood pressure in adulthood and life expectancy with cardiovascular disease in men and women. *Hypertension* 2005;**46**:280–6.
6. Kearney PM, Whelton M, Reynolds K, Muntner P, Whelton PK, He J. Global burden of hypertension: analysis of world-wide data. *Lancet* 2005;**365**:217–23.
7. Coffman T. Under pressure: the search for the essential mechanisms of hypertension. *Nat Med* 2011;**17**:1402–9.
8. Christensen KL, Mulvany MJ. Location of resistance arteries. *J Vasc Res* 2001;**38**:1–12.
9. Crowley SD. Distinct roles for the kidney and systemic tissues in blood pressure regulation by the renin-angiotensin system. *J Clin Invest* 2005;**115**:1092–9.
10. Touyz RM, Briones AM. Reactive oxygen species and vascular biology: implications in human hypertension. *Hypertens Res* 2011;**34**(1):5–14.
11. Versari D, Daghini E, Virdis A, Ghiadoni L, Taddei S. Endothelium-dependent contractions and endothelial dysfunction in human hypertension. *Br J Pharmacol* 2009;**157**(4):527–36.
12. Martinez-Lemus LA, Hill MA, Meininger GA. The plastic nature of the vascular wall: a continuum of remodeling events contributing to control of arteriolar diameter and structure. *Physiology* 2009;**24**:45–57.
13. Esler M. Pathophysiology of the human sympathetic nervous system in cardiovascular diseases: the transition from mechanism to medical management. *J Appl Physiol* 2010;**108**:227–37.

14. Carey RM. The intrarenal renin-angiotensin and dopaminergic systems: control of renal sodium excretion and blood pressure. *Hypertension* 2013;**61**(3):673–80.

15. Touyz RM. Advancement in hypertension pathogenesis: some new concepts. *Curr Opin Nephrol Hypert* 2011;**20**(2):105–6.

16. Carretero OA, Oparil S. Essential hypertension. *Circulation* 2000;**101**:329–35.

17. Cutler JA. High blood pressure and end organ damage. *J Hypertens* 1996;**14**:S3–6.

18. Rossi GP, Seccia TM, Pessina AC. Secondary hypertension: the ways of management. *Curr Vasc Pharmacol* 2010;**8**(6):753–68.

19. Safar ME, Balkau B, Lange C, Protogerou AD, Czernichow S, Blacher J, Levy BI, Smulyan H. Hypertension and vascular dynamics in men and women with metabolic syndrome. *J Am Coll Cardiol* 2013;**61**(1):12–9.

20. Chobanian AV, Bakris GL, Black HR, et al. National Heart, Lung, and Blood Institute Joint National Committee on Prevention, Detection, Evaluation, and Treatment of High Blood Pressure; National High Blood Pressure Education Program Coordinating Committee, The Seventh Report of the Joint National Committee on Prevention, Detection, Evaluation, and Treatment of High Blood Pressure: the JNC-7 Report. *JAMA* 2003;**289**(19):2560–72.

21. Delles C, McBride MW, Graham D, Padmanabhan S, Dominiczak AF. Genetics of hypertension: from experimental animals to humans. *Biochim Biophys Acta* 2010;**1802**(12) 1299–30.

22. Simino J, Rao DC, Freedman BI. Novel findings and future directions on the genetics of hypertension. *Curr Opin Nephrol Hypertens* 2012;**21**(5):500–7.

23. Feinleib M, Garrison RJ, Fabsitz R, Christian JC, Hrubec Z, Borhani NO, Kannel WB, Rosenman R, Schwartz JT, Wagner JO. The NHLBI twin study of cardiovascular disease and risk factors:methodology and summary of results. *Am J Epidemiol* 1977;**106**:284–5.

24. Longini IM, Higgins MW, Hinton PC, Moll PC, Keller JB. Environmental and genetic sources of familial aggregation of blood pressure in Tecumseh. *Michigan Am J Epidemiol* 1984;**120**:131–44.

25. Lifton RP. Genetic dissection of human blood pressure variation: common pathways from rare phenotypes. *Harvey Lect* 2004-2005;**100**:71–101.

26. Levy D. Genome-wide association study of blood pressure and hypertension. *Nat Genet* 2009;**41**:677–87.

27. Newton-Cheh C. Genome-wide association study identifies eight loci associated with blood pressure. *Nat Genet* 2009;**41**:666–76.

28. Kato N. Meta-analysis of genome-wide association studies identifies common variants associated with blood pressure variation in east Asians. *Nat Genetics* 2011;**43**:5310–538.

29. The International Consortium for Blood Pressure Genome-Wide Association Studies, et al. Genetic variants in novel pathways influence blood pressure and cardiovascular disease risk. *Nature* 2011;**478**:103–9.

30. Lifton RP, Gharavi A, Geller D. Molecular mechanisms of human hypertension. *Cell* 2001;**104**:545–56.

31. Dluhy RG, Powers M, Rich GM, Cook S, Ulick S, Lifton RP. A chimaeric 11 beta-hydroxylase/aldo-synthase gene causes glucocorticoid-remediable aldosteronism. *Nature* 1992;**355**:262–5.

32. Lifton RP, Dluhy RG, Powers M, Rich GM, Gutkin M, Fallo F, Gill J. Hereditary hypertension caused by chimaeric gene duplications and ectopic expression of aldosterone synthase. *Nat Genet* 1992;**2**:66–74.

33. Flatman PW. Cotransporters, WNKs and hypertension: important leads from the study of monogenetic disorders of blood pressure regulation. *Clin Sci (Lond)* 2007;**112**(4) 203–1.

34. Sonoyama K, Greenstein A, Price A, Khavandi K, Heagerty T. Vascular remodeling: implications for small artery function and target organ damage. *Ther Adv Cardiovasc Dis* 2007;**1**(2):129–37.

35. Secomb TW, Dewhirst MW, Pries AR. Structural adaptation of normal and tumour vascular networks. *Basic Clin Pharmacol Toxicol* 2012;**110**(1):63–9.

36. Tuna BG, Bakker EN, VanBavel E. Smooth muscle biomechanics and plasticity: relevance for vascular calibre and remodeling, *Basic Clin Pharmacol Toxicol* 2012;**110**(1):35–41.

37. Burger D, Nishigaki N, Touyz RM. New insights into molecular mechanisms of hypertension. *Curr Opin Nephrol Hypertens* 2010;**19**(2):160–2.

38. Julius S. Transition from high cardiac output to elevated vascular resistance in hypertension. *Aam Herat J* 1988;**116**:611–6.

39. DiBona GF. Sympathetic nervous system and hypertension. *Hypertension* 2013;**61**(3):556–60.

40. Grassi G, Bombelli M, Brambilla G, Trevano FQ, Dell'oro R, Mancia G. Total cardiovascular risk, blood pressure variability and adrenergic overdrive in hypertension: evidence, mechanisms and clinical implications. *Curr Hypertens Rep* 2012;**14**(4):333–8.

41. Parati G, Esler M. The human sympathetic nervous system: its relevance in hypertension and heart failure. *Eur Heart J* 2012;**33**(9):1058–66.

42. Friberg P, Meredith I, Jennings G, Lambert G, Fazio V, Essler M. Evidence of increased renal noradrenaline spillover rate during sodium restriction in man. *Hypertension* 1990;**16**:121–30.

43. Malpas SC. Sympathetic nervous system overactivity and its role in the development of cardiovascular disease. *Physiol Rev* 2010;**90**:513–57.

44. Lob HE, Schultz D, Marvar PJ, Davisson RL, Harrison DG. Role of the NADPH oxidases in the subfornical organ in angiotensin II-induced hypertension. *Hypertension* 2013;**61**(2):382–7.

45. Moser M, Setaro JF. Resistant or difficult-to-control hypertension. *N Engl J Med* 2006;**355**:385–92.

46. Granada JF, Buszman PP. Renal denervation therapies for refractory hypertension. *Curr Cardiol Rep* 2012;**14**(5):619–25.

47. Hering D, Lambert EA, Marusic P, Walton AS, Krum H, Lambert GW, Esler MD, Schlaich MP. Substantial reduction in single sympathetic nerve firing after renal denervation in patients with resistant hypertension. *Hypertension* 2013;**61**(2):457–64.

48. Khayat A, Gonda S, Sen S, Smeby RR. Responses of juxtaglomerular cell suspensions to various stimuli. *Hypertension* 1981;**3**(2):157–67.

49. Somers VK, Dyken ME, Mark AL, Abboud FM. Sympathetic nerve activity during sleep in normal subjects. *N Engl J Med* 1993;**328**:303–7.

50. Volpe M, Rosei EA, Ambrosioni E, Cottone S, Cuspidi C, Borghi C, et al., Renal artery denervation for treating resistant hypertension: definition of the disease, patient selection and description of the procedure. *High Blood Press Cardiovasc Prev* 2012;**19**(4):237–44.

51. Milliez P, Fritel D, Samarcq P, Blondeau J, Quichaud J, Tallone JC. Permanent arterial hypertension; unique kidney appearance; discovery of a renal outline on the opposite side by retropneumoperitoneum. *Bull Mem Soc Med Hop Paris* 1957;**73**(22-24):680–5.

52. Guyton AC. Blood pressure control—special role of the kidneys and body fluids. *Science* 1991;**252**:1813–6.

53. Walker KA, Cai X, Caruana G, Thomas MC, Bertram JF, Kett MM. High nephron endowment protects against salt-induced hypertension. *Am J Physiol Renal Physiol* 2012; **15 303**(2): F253–F8.

54. Rettig R, Grisk O. The kidney as a determinant of genetic hypertension: evidence from renal transplantation studies. *Hypertension* 2005;**46**: 463–46.

55. Crowley SD, et al. Angiotensin II causes hypertension and cardiac hypertrophy via its receptors in the kidney. *Proc Natl Acad Sci U S A* 2006;**103**:17985–90.

56. Siragy HM, Carey RM. Role of the intrarenal renin-angiotensin-aldosterone system in chronic kidney disease. *Am J Nephrol* 2010;**31**(6):541–50.

57. Gurley SB. AT1A Angiotensin receptors in the renal proximal tubule regulate blood pressure. *Cell Metab* 2011;**3**:469–75.

58. Meneton P, Jeunemaitre X, de Wardener HE, MacGregor GA. Links between dietary salt intake, renal salt handling, blood pressure, and cardiovascular diseases. *Physiol Rev* 2005;**85**(2): 679–15.

59. Frisoli TM, Schmieder RE, Grodzicki T, Messerli FH. Salt and hypertension: is salt dietary reduction worth the effort? *Am J Med* 2012;**125**(5):433–9.

60. Shimosawa T, Mu S, Shibata S, Fujita T. The kidney and hypertension: pathogenesis of salt-sensitive hypertension. *Curr Hypertens Rep* 2012;**14**(5):468–72.

61. Stolarz-Skrzypek K, Liu Y, Thijs L, Kuznetsova T, Czarnecka D, Kawecka-Jaszcz K, et al., Blood pressure, cardiovascular outcomes and sodium intake, a critical review of the evidence. *Acta Clin Belg* 2012;**67**(6):403–10.

62. Kopp C, Linz P, Hammon M, Schöfl C, Grauer M, Eckardt KU, et al., Seeing the sodium in a patient with hypernatremia. *Kidney Int* 2012;**82**(12):1343–4.

63. Kopp C, Linz P, Wachsmuth L, Dahlmann A, Horbach T, Schöfl C, et al., 23Na magnetic resonance imaging of tissue sodium. *Hypertension* 2012;**59**(1):167–72.

64. Kopp C, Linz P, Dahlmann A, Hammon M, Jantsch J, Müller DN, et al., 23Na magnetic resonance imaging-determined tissue sodium in healthy subjects and hypertensive patients. *Hypertension* 2013;**61**(3):635–40.

65. Rakova N, Jüttner K, Dahlmann A, Schröder A, Linz P, Kopp C, et al., Long-term space flight simulation reveals infradian rhythmicity in human Na(+) balance. *Cell Metab* 2013;**17**(1):125–31.

66. Steckelings UM. The evolving story of the RAAS in hypertension, diabetes and CV disease: moving from macrovascular to microvascular targets. *Fundam Clin Pharmacol* 2009;**23**(6):693–703.

67. Sequeira Lopez ML, Gomez RA. Novel mechanisms for the control of renin synthesis and release. *Curr Hypertens Rep* 2010;**12**(1):26–32.

68. Stegbauer J, Coffman TM. New insights into angiotensin receptor actions: from blood pressure to aging. *Curr Opin Nephrol Hypertens* 2011;**20**(1):84–8.

69. Nguyen Dinh Cat A, Touyz RM. Cell signaling of angiotensin II on vascular tone: novel mechanisms. *Curr Hypertens Rep* 2011;**13**(2):122–8.

70. Hattangady NG, Olala LO, Bollag WB, Rainey WE. Acute and chronic regulation of aldosterone production. *Mol Cell Endocrinol* 2012;**350**(2):151–62.

71. Nguyen Dinh Cat A, Touyz RM. A new look at the renin-angiotensin system–focusing on the vascular system. *Peptides* 2011;**32**(10):2141–50.

72. De Mello WC, Frohlich ED. On the local cardiac renin angiotensin system. Basic and clinical implications. *Peptides* 2011;**32**(8):1774–9.

73. Thatcher S, Yiannikouris F, Gupte M, Cassis L. The renin-angiotensin system in retinal health and disease: Its influence on neurons, glia and the vasculature. *Prog Retin Eye Res* 2010;**29**(4):284–311.

74. Bader M. Tissue renin-angiotensin-aldosterone systems: Targets for pharmacological therapy. *Annu Rev Pharmacol Toxicol* 2010;**50**:439–65.

75. Kumar R, Singh VP, Baker KM. The intracellular renin-angiotensin system in the heart. *Curr Hypertens Rep* 2009;**11**(2):104–10.

76. Nguyen G, Muller DN. The biology of the (pro)renin receptor. *J Am Soc Nephrol* 2010;**21**(1):18–23.

77. Hitom H, Liu G, Nishiyama A. Role of (pro)renin receptor in cardiovascular cells from the aspect of signaling. *Front in Biosc* 2010;**2**:1246–9.

78. Reudelhuber TL. Prorenin, Renin, and their receptor: moving targets. *Hypertension* 2010;**55**(5):1071–107.

79. Jansen EJ, Martens GJ. Novel insights into V-ATPase functioning: distinct roles for its accessory subunits ATP6AP1/Ac45 and ATP6AP2/(pro) renin receptor. *Curr Protein Pept Sci* 2012;**13**(2):124–33.

80. Santos RA, Ferreira AJ, Verano-Braga T, Bader M. Angiotensin-converting enzyme 2, Angiotensin-(1–7) and Mas: new players of the Renin Angiotensin System. *J Endocrinol* 2013;**216**(2): R1–7.

81. Heeneman S, Sluimer JC, Daemen MJ. Angiotensin-converting enzyme and vascular remodeling. *Circ Res* 2007;**101**(5):441–54.

82. Rentzsch B, Todiras M. Iliescu R, Popova E, Campos LA. Transgenic angiotensin-converting enzyme 2 overexpression in vessels of SHRSP rats reduces blood pressure and improves endothelial function. *Hypertension* 2008;**52**(5):967–73.

83. Silva DM, Vianna HR, Cortes SF, Campagnole-Santos MJ, Santos RA. Evidence for a new angiotensin-(1-7) receptor subtype in the aorta of Sprague-Dawley rats. *Peptides* 2007;**28**(3):702–7.

84. Ferrario CM. New physiological concepts of the renin-angiotensin system from the investigation of precursors and products of angiotensin I metabolism. *Hypertension* 2010;**55**(2):445–52.

85. Sampaio WO, Souza dos Santos RA, Faria-Silva R, da Mata Machado LT, Schiffrin EL. "b" Angiotensin-(1–7) through receptor Mas mediates endothelial nitric oxide synthase activation via Akt-dependent pathways. *Hypertension* 2007;**49**(1):185–92.

86. Sampaio WO, Henrique de Castro C, Santos RA, Schiffrin EL, Touyz RM. "c" Angiotensin-(1–7) counterregulates angiotensin II signaling in human endothelial cells. *Hypertension* 2007;**50**(6):1093–8.

87. Santos RA, Ferreira AJ, Simões E, Silva AC. Recent advances in the angiotensin-converting enzyme 2-angiotensin(1-7)-Mas axis. *Exp Physiol* 2008;**93**(5):519–27.

88. Rabelo LA, Alenina N, Bader M. ACE2-angiotensin-(1-7)-Mas axis and oxidative stress in cardiovascular disease. *Hypertens Res* 2011;**34**(2):154–60.

89. Yang R, Walther T, Gembardt F, Smolders I, Vanderheyden P, et al. Renal vasoconstrictor and pressor responses to angiotensin IV in mice are AT1a-receptor mediated. *J Hypertens* 2010;**28**(3):487–94.

90. Lew RA, Mustafa T, Ye S, McDowall SG, Chai SY, et al. Angiotensin AT4 ligands are potent, competitive inhibitors of insulin regulated aminopeptidase (IRAP). *J Neurochem* 2003;**86**(2):344–50.

91. Touyz RM, Schiffrin EL. Signal transduction mechanisms mediating the physiological and pathophysiological actions of angiotensin II in vascular smooth muscle cells. *Pharmacol Rev* 2000;**52**:639–72.

92. Herrera M, Coffman TM. The kidney and hypertension: novel insights from transgenic models. *Curr Opin Nephrol Hypertens* 2012;**21**:171–8.

93. Savoia C, Tabet F, Yao G, Schiffrin EL, Touyz RM. Negative regulation of RhoA/Rho kinase by angiotensin II type 2 receptor in vascular smooth muscle cells: role in angiotensin II-induced vasodilation in stroke-prone spontaneously hypertensive rats. *J Hypertens* 2005;**23**(5):1037–45.

94. Touyz RM, El Mabrouk M, He G, Wu XH, Schiffrin EL. Mitogen-activated protein/extracellular signal-regulated kinase inhibition attenuates angiotensin II-mediated signaling and contraction in spontaneously hypertensive rat vascular smooth muscle cells. *Circ Res* 1999;**84**(5):505–15.

95. Touyz RM, Wu X-H, He G, Salomon S, Schiffrin EL. Increased angiotensin II-mediated Src signaling via epidermal growth factor receptor transactivation is associated with decreased C-terminal Src kinase activity in vascular smooth muscle cells from spontaneously hypertensive rats. *Hypertension* 2002;**39**(2):479–85.

96. Touyz RM, He G, Wu XH, Park JB, Mabrouk ME, Schiffrin EL. Src is an important mediator of extracellular signal-regulated kinase 1/2-dependent growth signaling by angiotensin II in smooth muscle cells from resistance arteries of hypertensive patients. *Hypertension* 2001;**38**(1):56–64.

97. Montezano AC, Touyz RM. Molecular Mechanisms of Hypertension-Reactive Oxygen Species and Antioxidants: A Basic Science Update for the Clinician. *Can J Cardiol* 2012;28(3):288–95.

98. Touyz RM, Schiffrin EL. Increased generation of superoxide by angiotensin II in smooth muscle cells from resistance arteries of hypertensive patients: role of phospholipase D-dependent NAD(P)H oxidase-sensitive pathways. *J Hypertens* 2001;19(7):1245–54.

99. Montezano AC, Buger D, Ceravolo GS, Yusuf H, Montero M, Touyz RM. Novel Noxes Homologues in the Vasculature: Focusing on Nox4 and Nox5. *Clin Sc* 2011;120(4):131–41.

100. Touyz RM, Briones AB, Sedeek M, Burger B, Montezano AC. NOX Isoforms and Reactive Oxygen Species in Vascular Health. *Mol Interv* 2011;11:27–35.

101. Droge W. Free radicals in the physiological control of cell function. *Physiol Rev* 2002;82(1):47–95.

102. Tabet F, Schiffrin EL, Callera G, He Y, Yao G, Ostman A, et al., Redox-Sensitive Signaling by Angiotensin II Involves Oxidative Inactivation and Blunted Phosphorylation of Protein Tyrosine Phosphatase SHP-2 in Vascular Smooth Muscle Cells From SHR. *Circ Res* 2008;103(2):149–58.

103. Mulvany MJ, Baumbach GL, Aalkjaer C, Heagerty AM, Korsgaard N, Schiffrin EL, et al., Vascular remodeling. *Hypertension* 1996;28(3):505–6.

104. Schiffrin EL. Remodeling of resistance arteries in essential hypertension and effects of antihypertensive treatment. *Am J Hypertens* 2004;17(12 Pt 1):1192–200.

105. Fisher SA. Vascular smooth muscle phenotypic diversity and function. *Physiol Genomics* 2010;42A(3):169–87.

106. Touyz RM, Briones AM, Sedeek M, Burger D, Montezano AC. NOX isoforms and reactive oxygen species in vascular health. *Mol Interv* 2011;11(1):27–35.

107. Mulvany MJ. Small artery remodeling and significance in the development of hypertension. *News Physiol Sci* 2002;17:105–9.

108. Matchkov VV, Kudryavtseva O, Aalkjaer C. Intracellular Ca^{2+} signalling and phenotype of vascular smooth muscle cells. *Basic Clin Pharmacol Toxicol* 2012;10(1):42–8.

109. Fukami K, Inanobe S, Kanemaru K, Nakamura Y. Phospholipase C is a key enzyme regulating intracellular calcium and modulating the phosphoinositide balance. *Prog Lipid Res* 2010;49(4):429–37.

110. Erickson JR, He BJ, Grumbach IM, Anderson ME. CaMKII in the cardiovascular system: sensing redox states. *Physiol Rev* 2011;91(3):889–915.

111. Kaneko-Kawano T, Takasu F, Naoki H, Sakumura Y, Ishii S, et al. Dynamic Regulation of Myosin Light Chain Phosphorylation by Rho-kinase. *PLoS One* 2012;7(6):e39269.

112. Loirand G, Pacaud P. The role of Rho protein signaling in hypertension. *Nat Rev Cardiol* 2010;7(11):637–47.

113. Loirand G, Guerin P, Pacaud P. Rho kinases in cardiovascular physiology and pathophysiology. *Circ Res* 2006;98:322–34.

114. Montezano AC, Callera GE, Yogi A, He Y, Tostes RC, et al. Aldosterone and angiotensin II synergistically stimulate migration in vascular smooth muscle cells through c-Src-regulated redox-sensitive RhoA pathways. *Arterioscler Thromb Vasc Biol* 2008;28(8):1511–8.

115. Uehata M. Calcium sensitization of smooth muscle mediated by a Rho-associated protein kinase in hypertension. *Nature* 1997;389(6654):990–4.

116. Satoh K, Fukumoto Y, Shimokawa H. Rho-kinase: important new therapeutic target in cardiovascular diseases. *Am J Physiol Heart Circ Physiol* 2011;301(2):H287–96.

117. Virdis A, Ghiadoni L, Taddei S. Human endothelial dysfunction: EDCFs. *Pflugers Arch* 2010;459(6):1015–23.

118. Flammer AJ, Lüscher TF. Three decades of endothelium research: from the detection of nitric oxide to the everyday implementation of endothelial function measurements in cardiovascular diseases. *Swiss Med Wkly* 2010;140:w13122.

119. De Ciuceis C, Amiri F, Brassard P, Endemann DH, Touyz RM. Schiffrin ELCohn JS, et al., Reduced vascular remodeling, endothelial dysfunction, and oxidative stress in resistance arteries of angiotensin II-infused macrophage colony-stimulating factor-deficient mice: evidence for a role in inflammation in angiotensin-induced vascular injury. *Arterioscler Thromb Vasc Biol* 2005;25:2106–13.

120. Barhoumi T, Kasal DAB, Li MW, Shbat L, Laurant P, et al., T regulatory lymphocytes prevent angiotensin II-induced hypertension and vascular injury. *Hypertension* 2011;57:469–76.

121. Guzik TJ, Hoch NE, Brown KA. Role of T cell in the genesis of angiotensin II induced hypertension and vascular dysfunction. *J Exp Med* 2007;204:2449–60.

122. Savoia C, Burger D, Nishigaki N, Montezano A, Touyz RM. Angiotensin II and the vascular phenotype in hypertension. *Expert Rev Mol Med* 2011;13:e11–20.

123. Kortlever RM, Brummelkamp TR, van Meeteren LA, Moolenaar WH, Bernards R. Suppression of the p53-dependent replicative senescence response by lysophosphatidic acid signaling. *Mol Cancer Res* 2003;6(9):1452–60.

124. Nilsson PM, Lurbe E, Laurent S. The early life origins of vascular ageing and cardiovascular risk: the EVA syndrome. *J Hyperten* 2008;26(6):1049–57.

125. Serrano CV, Oranges M, Brunaldi V, de M, Soeiro A, Torres TA, et al., Skeletonized coronary arteries: pathophysiological and clinical aspects of vascular calcification. *Vasc Health Risk Man* 2011;7:143–51.

126. Alam MU, Kirton JP, Wilkinson FL, Towers E, Sinha S, Rouhi M, et al., Calcification is associated with loss of functional calcium-sensing receptor in vascular smooth muscle cells. *Cardiovasc Res* 2009;81:260–8.

127. Trebak M, Ginnan R, Singer HA, Jourd'heuil D. Interplay between calcium and reactive oxygen/nitrogen species: an essential paradigm for vascular smooth muscle signaling. *Antioxid Redox Signal* 2010;12:657–74.

128. Montezano AC, Zimmerman D, Yusuf H, Chignalia AZ, Wadhera V, Touyz RM. Vascular smooth muscle cell differentiation to an osteogenic phenotype involves TRPM7 modulation by magnesium. *Hypertension* 2010;56(3):453–62.

129. Marvar PJ, Thabet SR, Guzik TJ, Lob HE, McCann LA, Weyand C, et al., Central and peripheral mechanisms of T-lymphocyte activation and vascular inflammation produced by angiotensin II-induced hypertension. *Circ Res* 2010;107(2):263–70.

130. Chen K, Xie F, Liu S, Li G, Chen Y, Shi W, Hu H, Liu L, Yin D. Plasma reactive carbonyl species: Potential risk factor for hypertension. *Free Radic Res* 2011;45(5):568–74.

131. Lacy F, Kailasam MT, O'Connor DT, Schmid-Schonbein GW, Parmer RJ. Plasma hydrogen peroxide production in human essential hypertension: role of heredity, gender, and ethnicity. *Hypertension* 2000;36(5):878–84.

132. Serg M, Kampus P, Kals J, Zagura M, Zilmer M, Zilmer K, Kullisaar T, Eha J. Nebivolol and metoprolol: long-term effects on inflammation and oxidative stress in essential hypertension. *Scand J Clin Lab Invest* 2012;72(5):427–32.

133. Touyz RM, Yao G, Quinn MT, Pagano PJ, Schiffrin EL. p47phox associates with the cytoskeleton through cortactin in human vascular smooth muscle cells: role in NAD(P)H oxidase regulation by angiotensin II. *Arterioscler Thromb Vasc Biol* 2005;25(3):512–8.

134. Montezano AC, Burger D, Paravicini TM, Chignalia AZ, Yusuf H, Almasri M, He Y, Callera GE, et al., Nicotinamide adenine dinucleotide phosphate reduced oxidase 5 (Nox5) regulation by angiotensin II and endothelin-1 is mediated via calcium/calmodulin-dependent, rac-1-independent pathways in human endothelial cells. *Circ Res* 2010;106:1363–73.

135. Moreno MU, Jose GS, Fortuno A, Beloqui O, Diez J, Zalba G. The C242T CYBA polymorphism of NADPH oxidase is associated with essential hypertension. *J Hypertens* 2006;24(7):1299–306.

136. Zalba G, San Jose G, Moreno MU, Fortuno A, Diez J. NADPH oxidase-mediated oxidative stress: genetic studies of the p22(phox) gene in hypertension. *Antioxid Redox Signal* 2005;**7**(9-10):1327–36.

137. Stehr CB, Mellado R, Ocaranza MP, Carvajal CA, Mosso L, Becerra E, et al., Increased levels of oxidative stress, subclinical inflammation, and myocardial fibrosis markers in primary aldosteronism patients. *J Hypertens* 2010;**28**(10):2120–6.

138. Moran JP, Cohen L, Greene JM, Xu G, Feldman EB, Hames CG, et al. Plasma ascorbic acid concentrations relate inversely to blood pressure in human subjects. *Am J Clin Nutr* 1993;**57**(2):213–7.

139. Schiffrin EL. Antioxidants in hypertension and cardiovascular disease. *Mol Interv* 2010;**10**(6):354–62.

140. Chobanian AV. Shattuck Lecture. The hypertension paradox–more uncontrolled disease despite improved therapy. *N Engl J Med* 2009;**361**(9):878–87.

141. Tousoulis D, Androulakis E, Papageorgiou N, Stefanadis C. Novel therapeutic strategies in the management of arterial hypertension. *Pharmacol Ther* 2012;**135**(2):168–75.

142. Laurent S, Schlaich M, Esler M. New drugs, procedures, and devices for hypertension. *Lancet* 2012;**380**(9841):591–600.

143. Chinushi M, Izumi D, Iijima K, Suzuki K, Furushima H, Saitoh O, et al., Blood pressure and autonomic responses to electrical stimulation of the renal arterial nerves before and after ablation of the renal artery. *Hypertension* 2013;**61**(2):450–6.

Venous and Arterial Thrombosis

Evi X. Stavrou, MD[1, 2], *Alvin H. Schmaier, MD*[1, 3]

[1]Hematology and Oncology Division, Department of Medicine, Case Western Reserve University, [2]Louis Stokes
Veterans Administration Hospital, [3]University Hospital Case Medical Center, Cleveland, OH, USA

INTRODUCTION

The pathogenesis of venous and arterial thrombosis is broad and at times disparate. Virchow originally described venous thrombosis under low flow (shear) with red clots, occurring around and propagating through venous valves, and consisting of red cells and fibrin strands. Arterial thrombosis occurs under high shear stress in large (carotids, coronary, femoral arteries) and small vessels and initially consists of platelets and leukocytes with few red cells and little fibrin (white clot). However, the pathogenesis of these vessel occlusions (thrombosis) may be variations on basically similar mechanistic schema. A developing concept is that venous thromboembolism (VTE) is part of a pan-cardiovascular syndrome that includes coronary artery disease, peripheral artery disease, and cerebrovascular disease.[1] Classically, venous thrombosis has been associated with cancer, surgery, immobilization, fractures, thrombophilia, pregnancy, and estrogens. Arterial thrombosis (atherothrombosis) is associated with smoking, hypertension, diabetes, obesity, and hyperlipidemia.[2] Risk factors common to both venous and arterial thrombosis include aging, obesity, dyslipidemia, hypertension, diabetes, smoking, thrombophilia (antiphospholipid antibody syndrome, elevations of homocysteine, fibrinogen, factor VII, factor VIII, lipoprotein(a), factor V Leiden, and prothrombin 20210 polymorphism), and hormonal therapy (estrogens).[2] In the sections that follow, the mechanism(s) contributing to venous and arterial thrombosis are discussed in the context of clinical disease, with an emphasis on the structural differences between the vessels and how these anatomical differences impact on the complexity of processes that lead to the occlusive events. In both the venous and arterial sections, the anatomy and normal physiology of the vessel are discussed, so as to set the stage for the various elements involved in the pathophysiology of venous and arterial thrombosis, respectively. A brief discussion of the diagnosis and principles of management based upon the mechanistic basis of the disorder concludes each section.

VENOUS THROMBOSIS

Normal Venous Anatomy and Physiology

The venous wall is composed of three layers: the intima, media, and adventitia. In contrast to their arterial counterparts, veins have less smooth muscle and elastin. The venous intima consists of an endothelial cell layer resting upon a basement membrane. The adjacent media is composed of smooth muscle cells and elastin connective tissue, while the outer adventitia contains adrenergic neuronal fibers, particularly in the cutaneous veins. Central sympathetic discharge and brainstem thermoregulatory centers alter venous tone, as do other stimuli, such as temperature changes, pain, and volume changes. The histologic features of veins vary, depending on their caliber. Venules are the smallest veins, ranging from 0.1–1 mm and contain mostly smooth muscle cells, whereas larger extremity veins contain relatively few smooth muscle cells. Larger veins have limited contractile capacity. The venous valves prevent retrograde flow; it is their failure or valvular incompetence that leads to reflux and symptoms associated with venous stasis. Venous valves are most prevalent in the distal lower extremity, whereas as one proceeds proximally, the number of valves decreases to the point that no valves are present in the superior vena cava and inferior vena cava (IVC). Most of the capacitance of the vascular

Cellular and Molecular Pathobiology of Cardiovascular Disease
http://dx.doi.org/10.1016/B978-0-12-405206-2.00015-6

tree is in the venous system. Since veins do not have significant amounts of elastin, they can withstand large volume shifts with comparatively small changes in pressure. A vein has a normal elliptical configuration until the limit of its capacitance is reached, at which point the vein assumes a round configuration.

Pathophysiology of Venous Thrombosis

Venous thrombosis results from the sum of risk factors in an individual, including inflammation, age, immobility, and inheritance. The triad of venous stasis, endothelial injury, and hypercoagulable state, first posited by Virchow in 1856 as increasing the chances of thrombosis, has essentially held true more than a century and a half later. A venous clot usually arises in a low-flow portion of a vessel behind a venous valve and propagates proximally to occlude the vein. On gross section, a venous clot is a reddish gelatinous material consisting primarily of red cells and fibrin. While arterial and venous thrombi contain platelets and fibrin, the proportions differ. Venous thrombi, which form under low shear, contain relatively few platelets and consist mostly of fibrin and trapped red blood cells. Venous thrombi appear red because of their high red blood cell content. Unlike an arterial thrombus, a leading edge of leukocytes and platelets that most probably contributed to the initiation of the thrombus is usually absent.

Our current understanding of the cellular basis of venous thrombosis is distinctly different from that classically taught. However, the cellular and molecular events leading to impaired venous blood flow and thrombogenesis remain poorly characterized. In contrast to arterial thrombi, venous thrombi rarely form at sites of vascular disruption. Instead, they usually originate around the valves or in muscular sinuses, where stasis (slow blood flow) is more prominent,[3] resulting in diminished oxygen supply to the lining of affected vein segments. Experimentally, hypoxemia recapitulates the induction of thrombosis in the lung vasculature of mice. The underlying mechanism is hypothesized to be mediated by tissue factor induction in macrophages following activation of the transcription factor early-growth-response (Egr-1) gene.[4] In vitro, these responses lead to pro-inflammatory and pro-adhesive changes.[5] The finding that localized hypoxemia within valve sinuses triggers local activation of endothelial cells, represents a new concept in our understanding of venous thrombosis.[6]

Recent findings also suggest a novel role for innate immune cells in the pathogenesis of venous thrombosis. This contribution by the innate immune cells has been coined in a term to describe thrombosis as a thromboinflammatory state.[7] When fluorescently labeled microparticles derived from mouse monocytes are infused into a mouse before laser injury, the monocytes accumulate within the leading edge of a developing thrombus.[8] These monocyte-derived microparticles express tissue factor and fail to accumulate into thrombi when infused into P-selectin-null mice, demonstrating that the accumulation of monocyte microparticles bearing tissue factor in the thrombus is dependent on the interaction of platelet P-selectin and its counter-receptor, P-selectin glycoprotein ligand 1 (PSGL-1), expressed on the surface of the monocyte microparticles.

In addition to monocytes, neutrophils also contribute to venous thrombosis by their role in the formation of neutrophil extracellular traps (NETs). Originally described in 2004, NETs are comprised of DNA, histones, and antimicrobial proteases such as elastase. NETs are released as a defensive mechanism by activated neutrophils in response to pathogens in a process known as NETosis, and it complements other defensive mechanisms such as phagocytosis and oxidative burst.[9] NETs also have a second role in the pathogenesis of venous thrombosis. They provide a focal point for the activation of coagulation by inflammatory cells of innate immunity.[10,11] Upon venous constriction, but not complete occlusion, a carpet of neutrophils and monocytes adhere and grow over 6–24 hours as a venous thrombosis develops.[12] It is not yet clear whether neutrophil NETs or platelet thrombus initiates venous thrombus. Experimentally, the initiator of thrombosis may be model-dependent. It is believed that NETS are not an epiphenomenon because peptidylarginine deiminase 4 (PAD4) deficiency protects mice from venous thrombosis.[13] PAD4 is a nuclear enzyme that converts specific arginine residues to citrulline (citrullinization) on histones.[13] The level of plasma NETs is a risk factor for thrombosis in cancer and other conditions.[14,15] NETs released by adherent neutrophils promote further platelet as well as red cell binding. In experimental murine models of deep vein thrombosis (DVT) induced by partial occlusion/stenosis of the inferior vena cava, endothelial von Willebrand factor also is released and leads to platelet recruitment.[16] P-selectin expression on activated endothelium and platelets then mediates the recruitment and binding of neutrophils.[16,17]

NETs also promote thrombin generation via autoactivation of factor XII, thereby implicating the 'intrinsic' coagulation pathway in thrombus propagation.[12] Factor XII has been known to be a participant in Gram-negative sepsis for decades. Understanding its relation to NETs provides a mechanistic basis for this interaction. Depletion of platelets or neutrophils – or the genetic absence of P-selectin, von Willebrand factor, or PAD4 – is protective against the development of thrombosis. Similarly, systemic administration of DNase I, which degrades the chromatin backbone of NETs, prevents DVT formation in mice.[12,13] Activated platelets are well known to contribute to thrombin generation through the exposure of negatively charged phospholipids essential for coagulation enzymatic complex assembly.[18] More recently, activated cells and bacteria have been reported to release long-chain polyphosphates that provide a co-factor for activation of factor XII and factor XI by thrombin.[19,20] Finally, factor-XII-deficient mice are protected from venous thrombosis, which has been proposed to be due to the reduction in NETs' contribution to autoactivation.[12,21] Recent studies indicate that leukocyte factor XII deficiency is associated with both neutrophil and macrophage adhesion and chemotaxis defects as well.

These studies illustrate a novel role for innate immune cells in the development of venous thrombosis. These early reports implicating NETs in human thromboembolic disease provide a promising target by which thrombosis risk can be modified by adjuvant, non-anticoagulant-based treatments. Nucleosomes – consisting of a DNA segment coiled around four histone protein cores – are derived from released NETs, and may be detected in the circulation. A recent case–control study demonstrated that plasma levels of nucleosomes and complexes of elastase-α1-antitrypsin were associated with a three-fold increased risk of DVT, with an apparent dose–response relationship.[15] Finally, the dilemma of why some venous thrombi embolize might be a function of nucleosome-mediated stabilization of the thrombus through factor XII activation.

Specific Disorders Contributing to Venous Thrombosis

Venous thrombosis occurs in individuals with both acquired and inherited risk factors (Table 15.1). The majority of risk factors for venous thrombosis are acquired, such as advanced age, obesity, or cancer, which are accentuated when an individual becomes immobile. In all series of patients, over 50% of individuals with VTE are obese. This observation suggests that VTE will be an increasing problem in the developed world. The number of cases of thromboembolism for which an identifiable etiology can be recognized

is about 20% of individuals without a family history, and 40% if a family history is present. However, often there is no single etiology for venous thrombosis; the presence of thrombosis is the summation of multiple risk factors. Inherited and acquired factors combine to establish one's intrinsic risk of thrombosis.[1] Superimposed trigger factors, such as surgery, pregnancy, or hormonal therapy modify these risks and thrombosis occurs when the combination of genetic, acquired, and trigger factors exceeds a critical threshold.[2] It is a challenge for clinical laboratory medicine to develop assays to assess 'critical thresholds' for thrombosis risk to guide intensity of prophylaxis and, perhaps, treatment for thromboembolic risk. Figure 15.1 is an integrated presentation of some of the inherited genes and factors that influence the biochemistry of blood coagulation, anticoagulation, and fibrinolytic systems. Despite the interplay between acquired states and inherited traits, common familial (genetic) disorders that contribute to venous thrombosis have been described and are discussed next.

Inherited Prothrombotic States

FACTOR V LEIDEN

The factor V Leiden mutation (a polymorphism in factor V R506Q, FVL) accounts for most cases of activated protein C resistance (APCR) (due to resistance to proteolyze (i.e. inactivate) factor V by changing an arginine to a glutamine) and is inherited in an autosomal

TABLE 15.1   Etiologies of Venous Thrombosis

Inherited	Acquired
Factor V Leiden	Antiphospholipid antibody syndrome
Prothrombin 20210	Disseminated intravascular coagulation
Protein C deficiency	Heparin-induced thrombocytopenia
Protein S deficiency	Pregnancy
Antithrombin deficiency	Malignancy
Lipoprotein(a)	Myeloproliferative disorders
Dysfibrinogenemias	Paroxysmal nocturnal hemoglobinuria

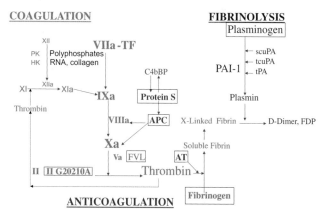

FIGURE 15.1   Biochemistry of venous thrombosis. This figure is a juxtaposition of the proteins involved in the coagulation (green), fibrinolytic (blue), and anticoagulation (red) systems. The boxes around proteins indicate that known defects in that protein are associated with venous thrombosis. PK, prekallikrein; HK, high-molecular-weight kininogen; TF, an abbreviation for tissue factor; FVL, an abbreviation for factor V Leiden; scuPA, single chain urokinase plasminogen activator; tcuPA, two-chain urokinase plasminogen activator; tPA, tissue plasminogen activator; AT, antithrombin; APC, activated protein C; C4bBP, C4b-binding protein; PAI-1, plasminogen activator inhibitor-1. Coagulation factors XII, XI, IX, X, VII and V are represented by their roman numeral alone. The presence of an 'a' after the roman numeral represents an 'activated' protein.

dominant fashion. A diagnosis of APCR is established using a functional assay that actually indicates the mechanistic basis of the disorder based on the ratio of the activated partial thromboplastin time (aPTT) of the patient versus control after activated protein C addition.[22,23] The prevalence of the mutation ranges from 2% to 5% in Caucasians, but is rare in Asians and Africans. In Caucasians, it may account for 20–60% of recognized etiologies for thrombosis. The prevalence of factor V Leiden homozygous states is about 1 in 2500.[22] Patients with factor V Leiden have thrombotic complications but are at a lower risk than those with deficiencies of antithrombin, protein C, or protein S (see below). Factor V Leiden heterozygotes have an annual risk of thrombosis of 0.5–0.7%, but this is a 5–7-fold increased risk over the normal risk. The presence of heterozygous FVL dramatically increases the risk of thrombosis with pregnancy or with the use of estrogen-containing oral contraceptives (35-fold). Homozygous FVL has an 80-fold increased risk for venous thromboembolism (VTE).

PROTHROMBIN GENE MUTATION

A G to A nucleotide transition at position 20210 in the 3′-untranslated region of the prothrombin gene (FIIG20210A) results in elevated levels of prothrombin due to slower clearance from plasma.[24] This defect is not structural, rather it is only associated with increased levels of plasma prothrombin. It occurs in 2–4% of the population, but in 6–8% of patients with primary venous thrombosis. Increased prothrombin, in turn, increases the risk of thrombosis by amplifying the generation of thrombin[24,25] or by inhibiting factor Va inactivation by activated protein C (APC).[26] FIIG20210A is found in 1–6% of Caucasians and may segregate with factor V Leiden, in which case the risk for venous thromboembolic phenomena summates.

HOMOCYSTEINEMIA

Homocysteinemia is a hypercoagulable state that occurs due to the combination of inherited and acquired factors.[27,28] Its elevation is associated with both venous and arterial thrombosis. Homocysteine is remethylated to methionine or catabolized to cystathionine (Figure 15.2). The major remethylation pathway requires folate and cobalamin (vitamin B12) and involves the action of methylenetetrahydrofolate reductase (MTHFR); a minor remethylation pathway is mediated by betaine-homocysteine methyltransferase. Vitamin B12 deficiency is associated with hyperhomocysteinemia and can present with venous thromboembolism before there are hematologic or neurologic defects. Paradoxically, polymorphisms in MTHFR are usually not associated with hyperhomocysteinemia or VTE. Alternatively, homocysteine is converted to

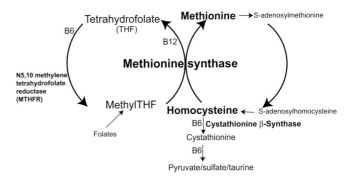

FIGURE 15.2 Homocysteine metabolism. Homocysteine is metabolized by two enzymes, methionine synthase and cystathionine β-synthase. N^5,N^{10} methylene tetrahydrofolate reductase makes an essential cofactor (MethylTHF) for methionine synthesis. B12 is a cofactor for methionine synthase and B6 is a cofactor for cystathionine β-synthase.

cystathionine in a trans-sulfuration pathway catalyzed by cystathionine β-synthase (CBS), with pyridoxine (B6) used as a cofactor.[29] As will be discussed below with respect to arterial thrombosis, deficiencies or defects in CBS are associated with severe venous and arterial thrombosis and mental retardation. Like vitamin B12, nutritional deficiency of folate or vitamin B6 is associated with thrombosis.[30] Replacement with folate, vitamin B12, and vitamin B6 reduces homocysteine levels, but most trials to date, however, have not shown that such therapy reduces the risk of recurrent cardiovascular events in patients with coronary artery disease, stroke or venous thromboembolic events.[31,32]

PROTEIN C DEFICIENCY

Activated protein C is a natural anticoagulant that becomes operative when thrombin is generated. Protein C, a vitamin-K-dependent zymogen, is activated by thrombin when both bind to endothelial cell thrombomodulin (TM). Initial thrombin formation results in an anticoagulant effect through protein C activation because the affinity for thrombin to activate protein C when bound to TM is tighter than to proteolyze fibrinogen and other coagulation proteins. The limiting factor in protein C activation and the anticoagulation effect of thrombin is the finite number of thrombomodulin-binding sites on endothelium. The endothelial protein C receptor (EPCR) also binds protein C and contributes to the localization of thrombomodulin-bound thrombin where it generates activated protein C (APC). With its cell membrane localizing cofactor, protein S (PS), APC binds to endothelium and activated platelet membranes functioning as an anticoagulant by proteolytically degrading thrombin-activated factor Va and VIIIa, thus attenuating further thrombin formation. Protein C deficiency can be inherited or acquired. Inherited protein C deficiency is usually divided into two subtypes.[33] The most common form of hereditary protein C deficiency is the classic or type

I deficiency state. This disorder is the result of reduced synthesis of a normal protein, and is characterized by a parallel reduction in protein C antigen and activity. Type II protein C deficiency results from production of a dysfunctional protein and is characterized by normal protein C antigen with reduced functional activity. Acquired protein C deficiency can be caused by decreased synthesis as in liver disease, increased consumption as in disseminated intravascular coagulation and loss as seen in nephrotic syndrome (see below). Functional deficiencies of protein C primarily cause venous thrombosis. The risk of thrombosis ranges from 0.5–2.5%/year. Protein C deficiencies are found in <5% of hypercoagulable patients. It is a serious prothrombotic risk disorder. Seventy-five percent of individuals with heterozygous protein C deficiency will suffer venous thromboembolism in their lifetime. When VTE occurs, 70% are spontaneous or non-provoked and 30% are provoked by pregnancy, oral contraceptives, inflammation or infection, surgery, trauma, etc. Homozygous protein C deficiency is associated with neonatal purpura fulminans, which leads to death if not recognized within three days of birth.

PROTEIN S DEFICIENCY

Protein S, a vitamin-K-dependent protein, serves as a cofactor for APC and enhances its capacity to inactivate factors Va and VIIIa. In plasma, it circulates in complex with C4b-binding protein. C4b-binding protein inversely regulates the availability of protein S to serve as a cofactor for APC function. Protein S deficiency can be inherited or acquired. Three types of inherited protein S deficiency have been identified. Type I or classic deficiency results from decreased synthesis of a normal protein, and is characterized by reduced levels of total and free protein S antigen together with reduced protein S functional activity. Type II protein S deficiency is characterized by normal levels of total and free protein S, associated with reduced protein S activity. This type of protein S defect is usually associated with a genetic polymorphism in the exons of the gene. Type III protein S deficiency is characterized by normal levels of total protein S, but low levels of free protein S associated with reduced protein S activity. This kind of protein S deficiency is due to the excessive binding of protein S to C4b-binding protein, which occurs in inflammatory states such as systemic lupus erythematosus or inflammatory bowel disease. Acquired protein S deficiency can be caused by decreased synthesis (liver disease), increased consumption (DIC), shift of free protein S to the C4b-bound form (inflammation), or true loss as seen in nephrotic syndrome. The risk of thrombosis from protein S deficiency may be up to 3.5%/year. It is primarily associated with venous thrombosis. Protein S deficiency is found in <5% of patients with hypercoagulable states and has an incidence of 1 in 2500 in the general population. Heterozygous protein S deficiency is also a serious disorder associated with a 74% lifetime risk of having VTE.

ANTITHROMBIN DEFICIENCY

Antithrombin, a member of the serine protease inhibitor (serpin) superfamily synthesized in the liver, plays a critical role in regulating coagulation by forming a covalent complex with thrombin, factor Xa, and all the other activated coagulation factors. Homozygous antithrombin deficiency is incompatible with mammalian gestation and delivery, so the inherited form is found in heterozygosity. The rate of antithrombin interaction with its target proteases is accelerated by heparin.[34] The frequency of symptomatic antithrombin deficiency in the general population has been estimated to be between one in 2000 and one in 5000. Among all patients seen with venous thromboembolism (VTE), antithrombin deficiency is detected in less than 1–2%.

Two forms of antithrombin deficiency exist: an inherited and an acquired form. Patients with type I antithrombin deficiency have proportionately reduced plasma levels of antigenic and functional antithrombin that result from a quantitative deficiency of the normal protein. Impaired synthesis (liver disease), defective secretion, or instability of antithrombin in type I antithrombin-deficient individuals is caused by major gene deletions, single nucleotide changes, or short insertions or deletions in the antithrombin gene. Patients with type II antithrombin deficiency have normal or nearly normal plasma antigen accompanied by low activity levels, characteristics indicative of a functionally defective molecule. Type II deficiency is usually caused by specific point mutations leading to single amino acid substitutions that produce a dysfunctional protein. Acquired antithrombin deficiency can reflect decreased antithrombin synthesis (i.e. liver disease), increased consumption (major surgery, acute thrombosis, sepsis, disseminated intravascular coagulation [DIC], malignancy) or enhanced clearance (heparin use, nephrotic range proteinuria). Antithrombin deficiency is a serious risk factor for VTE with 50% lifetime risk in heterozygous individuals. Onset of thrombosis is usually seen after puberty.

LIPOPROTEIN(A)

Lipoprotein(a) [LP(a)] consists of low-density lipoprotein (LDL) held together with a disulfide bond to apolipoprotein(a), a protein that has 85% sequence homology to kringle IV of plasminogen. LP(a) binds to fibrin and cells where plasminogen and α-2-antiplasmin bind. LP(a) inhibits plasminogen binding and activation, and brings LDL to endothelial cell surfaces to contribute to atherogenesis. It is commonly elevated in certain ethnic groups (e.g. South Asians). Epidemiologically, its elevation is associated with venous thromboembolism and ischemic cardiovascular disease. However, it is not considered a serious risk factor for thrombosis. At present, there is no known means to lower its levels or if lowering it will ameliorate cardiovascular disease risk.

DYSFIBRINOGENEMIAS

It has been recently shown that elevation of plasma fibrinogen alone alters the density of fibrin fibers and itself increases thrombosis risk.[35] Although most dysfibrinogenemias (abnormal fibrinogens) are associated with bleeding, some are associated with thrombosis. In dysfibrinogenemias, defective fibrinogen forms clots that are difficult for naturally occurring fibrinolytic (clot lysing) mechanisms to degrade. This fact results in both venous and arterial thrombosis. These entities are uncommon, but occur more commonly than antithrombin deficiency but less so than protein C or S deficiencies. They are not considered serious risk factors for thrombosis, but may contribute to an individual's overall thrombosis risk, e.g. in the cancer patient.

Acquired Hypercoagulable States
ANTIPHOSPHOLIPID ANTIBODY SYNDROME

The antiphospholipid antibody syndrome (APS) refers to a heterogeneous group of antibodies which are directed against proteins that can bind phospholipids. Diagnosis of APS requires that a patient has both a clinical event (thrombosis or pregnancy loss) and the persistent presence of antiphospholipid antibodies (aPL), documented by a solid phase serum assay [anticardiolipin or anti-β2-glycoprotein I (anti-β2GPI) immunoglobulin (IgG, IgM or IgA)], an abnormal coagulation assay (inhibitor of phospholipid-dependent clotting – the lupus anticoagulant test), or both, twice with a minimal time between testing of three months. β2GPI, normally present in serum at a concentration of 200 mg/mL, is a member of the complement control protein family, and has five repeating domains and several alleles. In vivo, β2GPI binds to phosphatidylserine on activated or apoptotic cell membranes, including those of platelets and endothelial cells. Under physiologic conditions, β2GPI functions in the elimination of apoptotic cells and as a natural anticoagulant.[36,37] In experimental animal models, passive or active immunization with viral peptides[38] and heterologous β2GPI,[39] induces polyclonal aPLs. These data suggest that pathologic autoimmune aPL induced in susceptible humans by infection via molecular mimicry reduces the anticoagulant effect of β2GPI. Low levels of aPLs may be present normally; one of the functions of normal aPLs may be to participate in the physiologic removal of oxidized lipids. Antiphospholipid antibodies influence thrombosis risk through multiple mechanisms. The process begins with activation or apoptosis of platelets and endothelial cells during which phosphatidylserine (a negatively charged phospholipid) migrates from the inner to the normally electrically neutral outer cell membrane. Circulating β2GPI binds to phosphatidylserine, and then aPL bind to a β2GPI dimer.[38] APL antibody–β2GPI dimer binding activates the complement cascade extracellularly, initiates an intracellular signaling cascade, probably through the C5a and β2GPI surface receptors, and recruits and activates inflammatory effector cells. These include monocytes, neutrophils, and platelets, whose activation leads to the release of proinflammatory products [e.g., tumor necrosis factor (TNF), oxidants, proteases] and the induction of a prothrombotic phenotype.[39] Other possible contributory mechanisms of aPL-mediated thrombosis include inhibition of coagulation cascade reactions (inhibition of protein C and S activation), induction of tissue factor expression on monocytes, reduced endothelial cell prostacyclin formation, and reduction of fibrinolysis.

Clinical manifestations range from asymptomatic aPL positivity (no history of vascular or pregnancy events) to catastrophic APS (multiple thromboses occurring over days). Thus patients should not be evaluated and managed as if they have a single disease. APS has to be considered in all patients with VTE regardless of other medical conditions. Overall, retrospective studies show that 31% of patients with APS have venous thrombosis. Patients with SLE have a thrombosis rate of 42%. The relative risk for venous thrombosis is 5.3 for patients with SLE who have persistent elevations of antiphospholipid antibodies. Hence, diagnosis of the condition in a thrombosis patient requires lifelong anticoagulation. Besides venous thrombosis, antiphospholipid antibodies have also been implicated in the etiology of arterial thrombosis (especially strokes) and in recurrent miscarriages.

DISSEMINATED INTRAVASCULAR COAGULOPATHY (DIC)

DIC is a consumption coagulopathy in which proinflammatory cytokines lead to excess thrombin formation and massive activation of the coagulation system.[40] A variety of disorders, including infections, inflammatory conditions, obstetrical emergencies, tissue destruction injuries, and malignant disease can lead to DIC.[41] An acute hemorrhagic state can result from the depletion of platelets and clotting factors and excessive plasmin formation. A subacute, chronic state associated with venous thromboembolism occurs when there is partial compensation for the consumption coagulopathy and excessive thrombin formation persists.[42] Both clinical evaluation and laboratory tests are necessary to diagnose DIC. Suggestive screening tests include elevated PT and aPTT, low platelet count, elevated D-dimer or fibrin degradation products, and low fibrinogen, protein C, and antithrombin levels. Serial tests may reveal abnormal trends and are more helpful than single results. D-dimer and fibrin degradation products are non-specific and often elevated in trauma or recent surgery. Because fibrinogen is an acute-phase reactant, serum levels may be normal despite an ongoing consumption coagulopathy. Schistocytes may be seen in the peripheral smear but are a non-specific finding and are not diagnostic for DIC. The presence of a disorder known to be associated with DIC, combined with abnormal test results, increases the likelihood of DIC. The diagnosis combines clinical information with clinical laboratory data. Scoring systems to aid the diagnosis have been developed but await validation.[43] DIC usually is associated with a high mortality rate caused by the underlying disorder. Sequelae of DIC including thrombosis, acral cyanosis, and limb ischemia, can lead to significant morbidity, which can be reduced by early recognition and treatment. Management includes treatment of the underlying disease and, as needed, blood products using platelets, fresh frozen plasma, or cryoprecipitate if the fibrinogen level is less than 100 mg/dL (with a maintained goal of 150 mg/dL or higher).[42]

HEPARIN-INDUCED THROMBOCYTOPENIA AND THROMBOSIS SYNDROME (HITTS)

HITTS is a complication of heparin therapy characterized by a decreased platelet count after treatment with heparin, platelet activation, and thrombosis. HITTS is an immune-mediated process that produces an acquired, transient hypercoagulable state. Immunoglobulin G antibodies form against a complex of heparin and platelet factor 4.[44] These antibodies then bind to the Fc receptors on platelets, resulting in platelet activation, increased thrombin generation, and an increased risk of thromboembolism. As a result of thrombin generation, there is acquired protein C deficiency that magnifies the prothrombotic risk of the condition. The incidence of true HITTS ranges from 0.3% to 3% depending on the clinical scenario in which it is studied. It is more common with prior exposure to unfractionated heparin, rather than low-molecular-weight heparin.[45] It is also more common among surgical patients (in particular after cardiac surgery) than in medical patients. In heparin-naive patients with HITTS, the platelet count usually does not fall before day 5 of heparin administration, although thrombosis can occur before the onset of true (platelet count < 150 000/μL) thrombocytopenia. A more rapid decline in platelet count may be seen in patients with heparin exposure in the previous 100 days.[46] Typically, platelet counts drop to less than 100 000/μL or decrease by more than 50% of the pretreatment level. The diagnosis of HITTS is supported by assays that rely on the platelet-activating properties of HITTS antibodies. The platelet serotonin release assay is the gold standard for the diagnosis of HITTS.[47] Enzyme immunoassays for

detection of antibodies against PF4 are more sensitive, but are less specific than the serotonin release assay.[48] The first step in management of HITTS is the avoidance of all heparin products (including intravenous line flushes). However, heparin cessation alone is not sufficient. Alternative anticoagulation should be initiated in patients with suspected HITTS even in the absence of clinically apparent thrombosis. Alternative anticoagulation with parenteral direct thrombin inhibitors (bivalirudin, argatroban) or factor Xa inhibitors (fondaparinux or danaparoid) is essential until the platelet count increases to normal. Oral anti-factor Xa inhibitors should also be effective. Since there is an acquired protein C deficiency in patients with HITTS, warfarin should be started only after the patient has achieved therapeutic anticoagulation on a parenteral anticoagulant and the thrombocytopenia has resolved with normalization of the D-dimer.

PREGNANCY

Venous thromboembolic disease is presently the leading cause of maternal morbidity and mortality in the developed world. About one in 1000 pregnancies are complicated by venous thromboembolism and about one in 1000 women develop venous thromboembolism in the postpartum period.[49] The individual risk of venous thromboembolism in pregnancy and the puerperium is influenced by patient-related factors; these include age older than 35 years, body mass index greater than 29, cesarean delivery, thrombophilia, or family history of venous thromboembolism.[50,51] A personal history of venous thromboembolism and multiparity are other risk factors. More than 90% of deep vein thrombosis in pregnancy occurs in the left leg because the enlarged uterus further compresses the left iliac vein by placing pressure on the overlying right iliac and ovarian arteries.[50,51] Systemic factors also contribute to hypercoagulability. The levels of circulating procoagulant proteins, such as factor VIII, fibrinogen, and von Willebrand protein, increase in the third trimester of pregnancy. In parallel, there is suppression of natural anticoagulant pathways. Thus, there is an acquired resistance to activated protein C related, at least in part, to reduced levels of free protein S.[52]

MALIGNANCY

Some patients who present with unprovoked venous thromboembolism have occult cancer. About 1–5% of patients who present with idiopathic VTE have occult malignancy. Cancer patients who develop venous thromboembolism have reduced survival compared with those who do not develop venous thromboembolism. Patients with pancreatic, gastric, lung, brain and advanced ovarian or prostate cancer have particularly high rates of venous thromboembolism.[53] The thrombotic tendency of patients with cancer is most often multi-factorial. Tumor cells often express tissue factor or other procoagulant factors that can initiate coagulation.[54,55] As mentioned above, circulating DNA levels also correlate with thrombosis in the cancer patient.[14,15] Patient-related factors that contribute to venous thromboembolism include mechanical factors, such as immobility, indwelling central venous catheters, or a bulky tumor mass compressing vessels, and comorbid conditions, such as sepsis, surgery, liver dysfunction secondary to metastases, and the prothrombotic effects of certain antineoplastic agents. In addition, tamoxifen and selective estrogen receptor modulators (SERM) induce an acquired hypercoagulable state by reducing the levels of natural anticoagulant proteins. No data yet show that anticoagulation treatment alters overall survival in the cancer patient.

MYELOPROLIFERATIVE DISORDERS

Polycythemia vera (PV), essential thrombocythemia (ET), and primary myelofibrosis (PMF) are categorized as myeloproliferative neoplasms (MPNs) in the 2008 World Health Organization (WHO) classification system for hematologic malignancies. In 2005, a Janus kinase 2 gain-of-function mutation (*JAK*2V617F) was reported in MPN,[56] and subsequent studies using sensitive assays have revealed the presence of this mutation in approximately 95% of patients with PV and 60% of those with ET or PMF.[57–59] The JAK-STAT pathway is normally activated by the occupancy of hematopoietic growth factor receptors by their respective ligands (erythropoietin (Epo), thrombopoietin). Thrombosis occurs in about 40% of patients with MPN, most commonly arterial thrombosis. Venous thrombosis can occur in unusual sites, such as mesenteric or hepatic vessels, and hepatic vein thrombosis (Budd-Chiari syndrome) or portal vein thrombosis may be the presenting manifestation before the full hematologic picture of PV has emerged. Potential culprits in the pathogenesis of thrombosis in MPNs include leukocytosis, leukocyte activation, rheologic abnormalities due to the raised red blood cell or platelet mass, abnormal platelet function, and a prothrombotic endothelial phenotype.[60]

PAROXYSMAL NOCTURNAL HEMOGLOBINURIA (PNH)

PNH is an acquired form of hemolytic anemia in which a somatic mutation in a hematopoietic stem cell renders erythrocytes susceptible to the lytic components of complement, thereby causing intravascular hemolysis in the absence of an anti-red blood cell antibody. The disease begins in a single hematopoietic stem cell when the PIGA gene on the short arm of the active X chromosome acquires a mutation. The PIGA gene encodes PIGA, an enzyme that is essential for the synthesis of glycosylphosphatidylinositol (GPI).[61] This lipid forms a peptide link with the C-terminal amino acid of numerous proteins, anchoring them to the red cell membrane. Molecular genetic studies indicate that the PNH clone up-regulates genes that make the PNH clone resistant to apoptosis and immune attack. The membrane inhibitor of reactive hemolysis (CD59) and CD55, an inhibitor of C3 convertase, are two of the many proteins that GPI anchors to the red cell. They prevent polymerization of C9, the final step in assembly of the membrane attack complex that begins with cleavage of C5 to C5b. The deficiencies of CD59 and CD55 allow the unopposed assembly of the membrane attack complex on the erythrocyte surface, thereby initiating intravascular hemolysis. Paroxysms of hemoglobinuria that occur on a background of chronic, low-grade intravascular hemolysis constitute the classic manifestations of the disease. In about a third of cases, venous thrombosis occurs and can cause Budd-Chiari syndrome by obstructing the hepatic veins. PNH and its thrombosis manifestations are effectively managed today with anti-C5 antibody therapy and anticoagulation.

SITUATIONAL HYPERCOAGULABLE STATES

Surgery is a large risk factor for thrombosis. Postoperative thrombosis is caused by a combination of local mechanical factors, including decreased venous blood flow in the lower extremities, and systemic changes in coagulation. The level of risk for postoperative thrombosis depends largely on the type of surgery performed. It is probably compounded by coexisting risk factors, such as an underlying inherited primary hypercoagulable state or malignancy, advanced age, and prolonged immobilization. Postoperative deep vein thrombosis and pulmonary embolism, the most common thrombotic complications, are often asymptomatic but detectable by non-invasive studies. The incidence of deep vein thrombosis after general surgical procedures is about 20–25%, with almost 2% of these patients having clinically significant pulmonary embolism. The risk for deep vein thrombosis after hip surgery and knee reconstruction ranges from 45–70% without prophylaxis, and clinically significant pulmonary embolism occurs in 20% of patients undergoing hip surgery. Prophylaxis in these situations reduces the risk to under 10%.

Acute deep vein thrombosis (DVT) poses several risks and has significant morbid consequences. The thrombotic process initiated in a venous segment may propagate, resulting in edema, pain, and immobility in the affected limb. The most dreaded sequel of acute DVT is that of pulmonary embolism (PE), a potentially lethal condition. Late consequences of lower extremity DVT can be chronic venous insufficiency (CVI) and ultimately post-thrombotic syndrome (PTS) as a result of venous valvular dysfunction in the presence of luminal obstruction.

Diagnostic Considerations

Venous thromboembolism occurs for the first time in approximately 100 persons/100 000 each year in the United States. This frequency increases with aging, with an incidence of 500/100 000 persons at 80 years of age. More than two-thirds of these patients have DVT alone, and the rest have evidence of pulmonary embolism. The recurrence rate with anticoagulation has been noted to be 10% in the first two years and 3% per year up to 10 years. Overall after 10 years, 30% of patients will have a recurrence. There are about 400 000 new cases annually in the United States and thromboembolism causes 75 000 deaths annually. A 28-day case-fatality rate of 9.4% after first-time DVT and of 15.1% after first-time pulmonary thromboembolism has been observed. Aside from pulmonary embolism, secondary effects resulting from DVT are significant in terms of cost, morbidity, and lifestyle limitations.

Clinical Diagnosis

The diagnosis of DVT requires a high index of suspicion. The extent of venous thrombosis in the lower extremity is an important factor in the manifestation of symptoms, although approximately 40% of patients with venous thrombosis do not have any clinical manifestations of the condition. Asymptomatic unilateral leg edema is the most common single sign, although there are many non-thrombotic conditions that can produce this. Major venous thrombosis involving the iliofemoral venous system results in a massively swollen leg, with pitting edema, pain and blanching, a condition known as phlegmasia alba dolens. With further progression of disease, there may be such massive edema that arterial inflow can be compromised. This condition results in a painful cyanotic leg, a condition called phlegmasia cerulea dolens. With this evolution of the condition, venous gangrene can develop unless flow is restored.

IMAGING

The current diagnostic test of choice for the diagnosis of limb DVT is duplex ultrasound, a modality that combines Doppler ultrasound and color flow imaging. The advantage of this test is that it is non-invasive, comprehensive, and without any risk of reaction to contrast angiography. This test is also highly operator-dependent, which is one of its potential drawbacks. Doppler ultrasound is based on the principle of the impairment of an accelerated flow signal caused by an intraluminal thrombus. If any segment of the venous system being examined fails to demonstrate augmentation on compression, venous thrombosis is suspected.

Computed tomography (CT) has come to the forefront of imaging for proximal venous disease and pulmonary emboli. It entails use of intravenous contrast material. It is particularly useful for imaging the iliac veins, IVC, and pulmonary artery vasculature, an area where the use of duplex ultrasound is limited.

LABORATORY EVALUATION

The D-dimer test measures the neo-epitope of plasmin-liberated, insoluble cross-linked fibrin. It is a surrogate for both thrombin by the formation of cross-linked insoluble fibrin (clot), and plasmin by the liberation of the D–D complex from fibrin. In combination with clinical evaluation and assessment, the sensitivity may exceed 90–95%.[62] In a prospective trial for the diagnosis of deep venous thrombosis in patients who presented with leg pain, the D-dimer along with soluble P-selectin, and leukocyte microparticles showed 73% specificity, 81% specificity, and 77% accuracy for diagnosis.[63] In the postoperative patient, D-dimer is causally elevated because of surgery and, as such, a positive D-dimer assay for evaluating for DVT is not useful. Additionally, genetic and functional and antigen assays can be performed for each of the genetic polymorphisms and protein alterations associated with VTE risk.

Principles of Treatment

Any venous thrombosis involving the femoral–popliteal system is treated with full anticoagulation. Antiplatelet agents are less effective for the treatment of venous thrombotic events, because of the limited platelet content of venous thrombi. Anticoagulants are the mainstay for prevention and treatment of venous thromboembolism because fibrin is the main component of venous thrombi. Traditionally the treatment of DVT has centered on heparin treatment, followed by warfarin therapy to obtain an international normalized ratio (INR) of 2.0–3.0. If unfractionated heparin is used, it is important to follow a weight-based dosing therapy. The incidence of recurrent venous thromboembolism increases if the time to reach therapeutic anticoagulation is prolonged. Therefore, it is important to reach therapeutic levels within 24 hours. Warfarin is started on the same day. If warfarin is initiated without heparin, the risk for a transient hypercoagulable state exists because protein C levels fall before the other vitamin-K-dependent factors are depleted.[64] With the advent of low-molecular-weight heparin (LMWH), it is no longer necessary to admit patients for i.v. heparin therapy. It is currently accepted practice to administer LMWH on an outpatient basis, as a bridge to warfarin therapy. Alternatively, the double-blind MATISSE trial concluded that use of fondaparinux, an injectable Xa-inhibitor, is equally effective as LMWH and has similar bleeding side-effects.[65]

Newer anticoagulants have recently been approved for both the prevention and treatment of deep vein thrombosis. These include the direct Xa-inhibitor

Rivaroxaban,[66] and the direct thrombin inhibitor Dabigatran.[67] Each agent is orally available but with distinct pharmacokinetic and pharmacodynamic properties. Systematic reviews of the available randomized trials (e.g., EINSTEIN-DVT, EINSTEIN-PE, RE-COVER) on this subject have concluded that all-cause mortality and VTE-related mortality did not differ significantly between these two agents versus adjusted-dose warfarin when used in the long-term management of VTE.[68] However, oral availability without the need for diet change and frequent monitoring makes them attractive agents for use.

Thrombolysis

There is increased interest in thrombolysis for DVT. The proposed benefit is preservation of valve function, with a subsequent lesser chance of developing chronic venous insufficiency (CVI). However, there have been few prospective, controlled studies to support the use of thrombolytic therapy for DVT. One exception is the patient with phlegmasia cerulea dolens, for whom thrombolysis is advocated for relief of significant venous obstruction. In this condition, thrombolytic therapy probably results in better relief of symptoms and less long-term sequelae than heparin anticoagulation alone.[69] The alternative for this condition is surgical venous thrombectomy. No matter which treatment is chosen, long-term anticoagulation is indicated. The incidence of major bleeding is higher with lytic therapy.

Vena Cava Filters

The most worrisome and potentially lethal complication of DVT is pulmonary embolism. The symptoms of pulmonary embolism, ranging from dyspnea, chest pain, and hypoxia to acute cor pulmonale, are non-specific and require a high index of suspicion. The diagnostic gold standard remains pulmonary angiography but, increasingly, this has been displaced by computed tomography angiography (CTA). Adequate anticoagulation is usually effective for stabilizing venous thrombosis but, if a patient develops a pulmonary embolism in the presence of adequate anticoagulation, a vena cava filter is indicated. Modern filters are placed percutaneously over a guide-wire. The Greenfield filter, most extensively used and studied, has a 95% patency rate and a 4% recurrent embolism rate. This high patency rate allows for safe suprarenal placement if there is involvement of the IVC up to the renal veins or if it is placed in a woman in her childbearing years. Device-related complications are wound hematoma, migration of the device into the pulmonary artery, and inferior vena caval occlusion caused by trapping of a large embolus. Although generally safe, IVC filters are not without risk and significant morbidity.[70] Therefore, permanent placement of a caval filter, particularly in a young patient who may only require short-term caval protection, is generally not justified.

Retrievable filters entered the field as a potential solution for the patient with temporary indications for pulmonary embolus prophylaxis. All can be deployed from the internal jugular vein or femoral vein and retrieved from the right jugular. Patient groups that may benefit from retrievable filters include multiple-trauma patients and high-risk surgical patients.

ARTERIAL THROMBOSIS

The following discussion of arterial thrombosis is divided into several sections. First, a brief overview of the contributors to arterial thrombosis is presented. Next, the majority of the remainder of this chapter will introduce the many different factors that contribute to vascular inflammation that lead to increased thrombosis risk. These include endothelial cell and vascular smooth muscle cell (VSMC) transcription factors, chemical substances, and vasculoprotective genes and proteins (Table 15.2). In particular, this section discusses how

TABLE 15.2 Vascular Cell Factors that Influence Arterial Inflammation Leading to Thrombosis

Cell	Component	Factors
Endothelial cell		
	Transcription factors	KLF2, KLF4, Nrf2, SIRT1
	Chemical mediators	NO, $O_{2'}$, PGI_2, homocysteine HS, CO
	Proteins	eNOS, TM, TF, PAI-1, CD39 CD73, FVII, fibrinogen, LP(a), cytokines, leukocyte adhesion molecules
Vascular smooth muscle cell		
	Transcription factors	KLF4, Nrf2, SIRT1
	Chemical mediators	NO, cAMP, cGMP, H_2O_2
	Proteins	TF, cyclophilin, PAR2
Platelet		
	Chemical mediators	NO, histamine, serotonin, TXA_2, 12-HETE, PAF
	Proteins	GPIb, GPVI, adhesion molecules (P-selectin, CD40L), alarmins, cytokines, IL-7, IL-1β, chemokines, growth factors

KLF2, KLF4, Kruppel-like factor 2, 4; Nrf2, nuclear factor E2; SIRT1, sirtuin 1; NO, nitric oxide; $O_2.$, reactive oxygen species; PGI_2, prostacyclin; HS, hydrogen sulfide; CO, carbon monoxide; LP(a), lipoprotein(a); cAMP, cyclic adenosine monophosphate; cGMP, cyclic guanosine monophosphate; H_2O_2, hydrogen peroxide; TF, tissue factor; PAR2, protease activated receptor 2; FVII, factor VII; TM, thrombomodulin; LP(a), lipoprotein(a); CD39, ectoADPase; CD73, 5'nucleotidase; TXA2, thromboxane A_2; 12-HETE, 12-hydroxyeicosatetraenoic acid; PAF, platelet-activating factor; GPIb, glycoprotein 1b; GPVI, glycoprotein VI.

these factors (genes, proteins, chemicals, cells) maintain vascular health and when disordered contribute to vascular inflammation and dysfunction. Although the major disorder of vascular inflammation is atherosclerosis, this topic will not be discussed here and the reader is referred to Chapter 12, Pathophysiology of Atherosclerosis, for a discussion of the pathogenesis of atherosclerosis. The focus here is on the changes in the vasculoprotective transcription factors, chemical mediators, and proteins that lead to inflammation that promotes atherosclerosis and arterial thrombosis. Finally, a brief reference is made to the clinical aspects of arterial thrombosis that manifest as myocardial infarction, stroke, peripheral arterial vessel occlusion and microvascular arterial disease as seen in diabetes mellitus.

Contributors to Arterial Thrombosis

In most individuals, arterial thrombosis is manifested as myocardial infarction (MI) or stroke, the number 1 and 2 killers of man the world over, respectively. MI and stroke occur subsequent to years of vascular damage and inflammation from common offenders such as hypertension and atherosclerosis. The pathogenetic contributions to arterial thrombosis by hypertension and atherosclerosis are not completely understood, but are likely to include vascular reactive oxygen species (ROS), shear changes, and many of the parameters described in the sections below. Oxidative stress alters the anticoagulant nature of the vasculature making it prothrombotic. Several antioxidant systems such as thioredoxin, xanthine oxidase, catalase, and superoxide dismutase contribute to vascular protection.[71] In atherosclerosis, the major event leading to thrombosis is vulnerable plaque rupture. Plaques that rupture often have thin fibrous caps (50–65 μm thick), large lipid cores, abundant inflammatory cells, and punctate calcifications. One hypothesis is that plaque macrophages may disrupt the plaque collagen leading to acute rupture, and exposure of the blood components to procoagulant molecules, triggering thrombosis. T-cells influence macrophage function in this setting by enhancing CD40 binding, MMP-1, 8, 13 expression, and tissue factor expression.[72]

Modulators of Vascular Health

In order to understand arterial disease pathogenesis, it is important to appreciate the entities that contribute to vascular well-being. These entities are the vascular cells and cell matrix whose properties are altered by vascular injury, resulting in loss of the anticoagulant, profibrinolytic nature of the vasculature, leading to atherosclerosis and thrombosis. Contributions to vascular health derive from the endothelium, vascular smooth muscle cells, and platelets. Alteration or disruption of these contributions

sets the stage for major arterial thrombotic events, resulting in myocardial infarction and stroke.

The Endothelium

From all points of view, the endothelium is critical for vascular health. Complex regulatory systems make endothelium constitutively anticoagulant. These consist of numerous transcription factors, chemical mediators, and protein products, each of which will be discussed in the following sections.

TRANSCRIPTION FACTORS

Several transcription factors have precise roles in the maintenance of vascular well-being. One major pair of transcription factors are the Kruppel-like factors, KLF2 and KLF4. These entities are zinc finger DNA-binding transcription factors. KLF2 expression is induced by laminar shear stress and is inhibited by interleukin-1β.[73] Its up-regulation induces transcription of endothelial nitric oxide synthase (eNOS) and thrombomodulin, and inhibits transcription of vascular cell adhesion molecule-1, E-selectin, tissue factor, plasminogen activator inhibitor-1, and protease activator receptor 1.[74–76] KLF2 is up-regulated through the MEK5/ERK5/MEF2 pathways and is down-regulated through p66shc and p53.[77–81] Like KLF2, KLF4 regulates endothelial cell inflammation by up-regulating eNOS and thrombomodulin and down-regulating TNF-α, VCAM-1, and tissue factor.[82] Endothelial cell KLF4 expression is modulated by MEK5/ERK5 activation.[81] In vivo, the presence of KLF4 reduces thrombosis risk and decreases risk for atherosclerosis in mice.[82]

Another vasculoprotective transcription factor is nuclear factor-erythroid 2-related factor (Nrf2), which functions to reduce oxidative stress in endothelium. Nrf2 is elevated by KLF2 and prolonged laminar shear stress, and it up-regulates transcription of NAD(P)H dehydrogenase quinone (NQO1) and heme oxygenase (HO-1).[83] Oscillatory and pulsatile blood flow that occurs as the result of the development of atherosclerotic plaque has an opposite influence on Nrf2 and KLF2 expression.[84] They lead to a decrease in KLF2 and Nrf2 with activation of proinflammatory and procoagulant transcription factors activator protein 1 (AP-1) and nuclear factor κB (NFκB).[85,86]

The class III histone deacetylase sirtuin 1 (SIRT1) catalyzes NAD$^+$-dependent deacetylation of ϵ-aminoacetylated lysine residues in protein substrates.[87] SIRT1 at high concentrations modulates cardiovascular oxidative stress by enhancing the expression of catalase to convert H_2O_2 into H_2O, and glutathione peroxidase (GPx1) to convert peroxynitrite (ONOO$^-$) to N_2O.[88] At lower levels, SIRT1 enhances expression of superoxide dismutase 1, 2, and 3 (SOD1-3).[89] SIRT1 activators also prevent reactive oxygen species generation via preventing eNOS uncoupling by enhancing GTP cyclohydrolase I (GCH1) and tetrahydrobiopterin (BH$_4$) expression.[90] Low SIRT1 levels are rescued by peroxisome proliferator-activated receptor (PPAR) δ.[91] High SIRT1 lowers vascular tissue factor and elevates thrombomodulin expression. SIRT1 also is an important modulator of vascular inflammation and risk for atherosclerosis. Its presence blocks ICAM-1 and VCAM-1 expression and subsequent monocyte adherence, it interferes with tissue factor release caused by plaque rupture, and it suppresses the expression of macrophage Lox-1, a scavenger receptor for oxLDL.[92]

CHEMICAL MODULATORS

The chemical modulators nitric oxide (NO), superoxide, prostacyclin, homocysteine, hydrogen sulfide (HS), and carbon monoxide are also important influences on vascular well-being. The NO-ROS system is a central signaling pathway regulating vascular function in vivo.[93] NO, the first recognized endothelium-derived relaxing factor, is produced by a family of NO synthase (NOS) enzymes that includes eNOS in endothelial cells, nNOS in neuronal cells, and iNOS induced in monocytes/macrophages. These NO synthases function as a dimer, and enzymatically produce an electron transfer for oxidation of L-arginine. This reaction requires several substrates and cofactors that include L-arginine, tetrahydrobiopterin (BH4), nicotinamide-adenine-dinucleotide phosphate (NADPH), flavin adenine dinucleotide, and flavin mononucleotide. Altered concentrations or function of any of these reactants modulates NO production and superoxide formation in a mutually reciprocal fashion. For example, the uncoupling of eNOS to a monomer reduces NO and increases oxygen free radical (O_2^-) and peroxynitriate ($ONOO^-$) generation. The physiologic target of NO is soluble quanylate cyclase which when bound elevates cyclic GMP. In vasculature, this results in vasodilation by acting on vascular smooth muscle. eNOS-deficient mice are hypertensive, but not prothrombotic.[94] As indicated above, eNOS expression is transcriptionally regulated by KLF2.[73] The balance between NO bioactivity and oxidative stress has an important role in regulating vascular contractility, thrombosis risk, development of atherosclerosis, and vascular inflammation. Oxidative stress causes endothelial dysfunction characterized by proinflammatory, prothrombotic, proliferative and vasoconstrictor processes. Several recent reviews examine these processes in detail.[95–97] Loss of NO production is associated with increased reactive oxygen species production in the vasculature. Mitochondrial dysfunction from hypertension, shear, angiotensin II, depleted prolylcarboxypeptidase, heme oxygenase I and II, and a myriad of other etiologies modulates endothelial NO and superoxide generation, which may lead to endothelial cell dysfunction.[98] Increased endothelial superoxide influences

expression, production, and function of several proteins that modulate to the constitutive anticoagulant nature of endothelium. These specific proteins will be discussed in the next section. Native and oxidized LDL (oxLDL) uncouples eNOS, induces monocyte adhesion to endothelium, induces smooth muscle cell migration and proliferation, decreases L-arginine, and increases the availability and activity of macrophage arginase II, an inflammatory marker. All of these activities that alter NO biology influence the ROS contribution to the pathogenesis of atherosclerosis. Heme oxygenase I, propylcarboxypeptidase or glutathione peroxidase-3 deficiencies are associated with generation of vascular ROS and increased risk for arterial thrombosis.[99–101]

Prostacyclin (PGI$_2$) and the prostaglandin system in general are important modulators of both vascular and platelet function, altering arterial thrombosis risk.[102,103] PGI$_2$ is produced in endothelial cells, and it inhibits platelet function and leads to vascular dilatation. Its influence is intimately tied to the effects of NO and cGMP.[104,105] However, recent data have begun to separate its precise influences on platelets and vascular endothelium. Two- to three-fold elevations of plasma prostacyclin induce a platelet defect in spreading on integrins and a glycoprotein VI activation defect without influence on thrombin- or ADP-induced platelet activation.[106] Lower elevation only induces a vascular effect without influence on platelets. Combined Ptgs1$^{+/-}$/Ptgs2$^{-/-}$ mice have enhanced arterial thrombosis in a murine Rose Bengal carotid artery thrombosis study, likely due to elevated plasminogen activator inhibitor-1 (PAI-1) activity.[107] Murine Ptgs2 disruption is also associated with reduced urinary PGI-M, the major metabolite of PGI$_2$, systemic hypertension and increased thrombosis risk.[108] Ptgs2$^{-/-}$ mice have reduced PGI$_2$, increased tissue factor (TF), and decreased SIRT1 expression.[91,108] Conversely, exogenous PGI$_2$ or a PPARδ agonist corrects thrombosis risk, normalizes SIRT1 levels, and reduces TF expression in Ptgs2$^{-/-}$ mice. Cyclooxygenase 2 depletion also reduces eNOS expression and NO formation.

Homocysteine production is a by-product of methionine metabolism from S-adenosyl-homocysteine (Figure 15.2). Homocysteine is metabolized by two enzymes, methionine synthase and cystathionine β-synthase (CBS). Homocysteine is degraded to cystathionine by CBS, and B6 (pyridoxine) is an essential cofactor for CBS. Hydrogen sulfide (H$_2$S) is a by-product of this reaction (see below). Deficiencies or defects in CBS cause homocystinemia, and are associated with mental retardation and severe recurrent arterial and venous thrombosis in man, but not in the mouse.[109,110] Plasma homocysteine levels higher than 11 μM are associated with cardiovascular disease. Elevated homocysteine is associated with increased vascular ROS that results in uncoupled eNOS and reduced NO production,

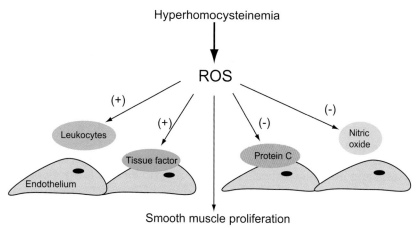

FIGURE 15.3   The relationship of homocysteine and endothelial dysfunction. In hyperhomocysteinemic states, there is increased production of reactive oxygen species (ROS). Regardless of the etiology, increased concentrations of ROS are detrimental to the anticoagulant function of endothelium. ROS transcriptionally regulates expression and inactivates thrombomodulin, reducing protein C activation, and causing a loss of this constitutive anticoagulant protein. ROS transcriptionally regulate eNOS (NOS3) expression and uncouple endothelial NO synthetase, allowing for reduced NO production and increased ROS production. ROS also increase tissue factor and PAI-1 (not shown) mRNA and protein expression in endothelium. In addition, ROS stimulate leukocyte migration and infiltration, and smooth muscle proliferation. Together, these activities lead to vascular dysfunction and increased risk for arterial thrombosis.

increased TF expression, reduced protein C activation, and increased leukocyte infiltration consistent with vascular endothelium dysfunction (Figure 15.3). Elevated homocysteine is also associated with smooth muscle proliferation, lipid peroxidation, and oxidation of LDL, factors that contribute to atherosclerosis. APOE-deficient mice fed a high-homocysteine diet, a high-fat diet, or both are prothrombotic.[111] The importance of homocysteine as a risk factor for venous and arterial thrombosis still needs clarification since there are no human data that show lowering of homocysteine ameliorates thrombosis risk.

Hydrogen sulfide is generated by both cystathionine β-synthase (CBS) and cystathionine γ-lyase (γ-cystathionase) (CSE). Animals deficient in CSE are hypertensive and have diminished endothelium-dependent vasorelaxation.[112] H_2S can initiate several physiological processes such as vasorelaxation, inflammation, and cell survival. H_2S signals by sulfhydration of reactive Cys residues in target proteins. This process is reversed by the thioredoxin system making sulfhydration another post-translational modification process. H_2S signaling may account for 25% of the production of vascular endothelium-derived relaxing factors (EDRF).[113] H_2S also induces vasorelaxation by changing the membrane potential (hyperpolarization) via alteration of ion channel function.[113]

The last EDRF is carbon monoxide (CO). CO is generated from heme cleavage by two distinct heme oxygenases. Inducible heme oxygenase-1 (HO-1) is newly synthesized in response to stress. Heme oxygenase-2 (HO2) is constitutive and highly expressed in the brain.[114,115] CO rescues HO-1-deficient mice from arterial

thrombosis, and up-regulation of HO-1 expression prevents platelet-dependent thrombosis.[116,117]

PROTEIN MODULATORS

Several proteins contribute to the constitutive anticoagulant nature of the arterial intravascular compartment, as well as the previously discussed venous compartment. Antithrombin is produced in the liver and binds to membrane glycosaminoglycans lining the intravascular compartment where it inactivates thrombin. Thrombomodulin (TM), which is transcriptionally regulated by KLF2 and KLF4, is expressed in endothelium and serves as a binding site for protein C, allowing it to be activated by thrombin. ROS inactivates a critical methionine in TM such that it cannot bind protein C for activation.[118] ROS in endothelium also lead to increased tissue factor and plasminogen activator inhibitor-1 expression resulting in increased thrombin formation and reduced fibrinolysis, respectively.

The endothelium also has unique proteins that contribute to the antithrombotic state. CD39 (ectoADPase) and CD73 (5 nucleotidase) are cell surface enzymes that degrade ATP and ADP to AMP, and AMP to adenosine, respectively. These enzymes reduce platelet activation on or about the vessel wall. Pancreatic islets from a transgenic mouse that over expressed CD39, when incubated in blood, delayed the time to clot formation.[119] Transgenic mice expressing CD39 are resistant to ferric-chloride-induced carotid artery thrombosis, and treatment with a non-hydrolyzable ADP analog negated the thrombosis resistance.[120] Use of a soluble CD39 molecule targeted to the epitope of the activation complex of platelet $\alpha_{2b}\beta_3$ integrin, reduced experimental thrombus formation in vivo.[121]

It is well-recognized that elevation of plasma factor VII and fibrinogen are associated with increased risk for myocardial infarction and stroke.[122] Factor VII elevation promotes increased thrombin formation with its cofactor tissue factor. As already mentioned, fibrinogen elevation alone is recognized to change the character and thrombogenicity of the physical clot.[35] Likewise, lipoprotein(a) elevation in plasma is associated with increased risk for myocardial infarction.[123] In addition, oxidized phospholipids on plasminogen, that correlate with elevated LP(a) levels, influence fibrinolysis and atherothrombosis risk in individuals.

Vascular Smooth Muscle Cells

Vascular smooth muscle cells (VSMC) are protected from injury by an intact endothelium. NO from the endothelium diffuses to VSMC where it elevates cGMP, leading to vasodilation. It, along with cAMP, also contributes to prevention of VSMC proliferation. However, VSMC retain the ability to dramatically modulate their phenotype in response to other extracellular cues and stimuli. For example, in mice deficient in microsomal prostaglandin E synthase-1, endothelial cell denudation from wire injury causes an exaggerated neointimal hyperplastic response.[124] Many of the factors that modulate endothelial cell biology also influence VSMC, but their effects may be quite different on these cell types.

The Kruppel-like transcription factors modulate VSMC function as well as endothelial cell function. KLF2 deficiency is embryonic lethal, and KLF2[-/-] embryos fail to develop beyond day 13.5 due to lack of blood vessel stabilization. In these embryos, there is arrest of vascular maturation due to a smooth muscle cell growth defect.[125] Murine embryonic fibroblasts prepared from KLF2[-/-] tissue also have a significant growth defect. In VSMC, platelet-derived growth factor-B (PDGF-B) induces KLF2 expression through a Src signaling pathway that activates sphingosine 1-phosphate.[126] Although KLF2 and KLF4 work in concert on endothelium, their actions diverge on VSMC. Normal VSMC do not constitutively express KLF4. However, with vascular injury such as altered shear, release of PDGF-B or transforming growth factor-β (TGF-β) increases expression of transcription factor Sp1, which binds to a regulatory unit on KLF4 to up-regulate KLF4 expression.[127] KLF4, a repressor of VSMC marker gene expression, inhibits myocardin expression, reduces binding of TGF-β control element (TCE) and transcription factor SRF to VSMC gene promoters, and recruits histone deacetylases to induce changes in chromatin structure that lead to transcriptional silencing.[127] KLF4 directly binds to the transforming growth factor-β receptor in the nucleus to phosphorylate Smad2/3. Smad 2/3, with Smad4 and p38 mitogen-activated protein kinase, activate several transcription programs, leading to VSMC proliferation.

Several other transcription factors also influence VSMC phenotype. Nrf2[-/-] mice have decreased thrombosis times. In CD36[-/-] mice, protection from arterial thrombosis is associated with reduced VSMC ROS and increased expression of the antioxidant enzymes peroxiredoxin-2 (Prdx-1) and heme oxygenase-1 (HO-1).[128] CD36 ligation results in Fyn phosphorylation and subsequent phosphorylation and degradation of Nrf2. However, with oxidative stress produced by microparticles or oxidized LDL, cytoplasmic Nrf2 migrates to the nucleus to bind to the promoters of prdx-1 and HO-1.[128] Additionally, SIRT1 modulates neointima formation after vascular injury.[129] SIRT1 overexpression inhibits VSMC proliferation and migration, and induces cell cycle arrest at the G1/S transition in vitro. Knockdown of SIRT1 has an opposite effect.

As in endothelial cells, oxidative stress modulates VSMC phenotype as indicated above. In VSMC, ROS stimulate cell growth and expression of the proto-oncogenes c-myc and c-fos.[130] Hydrogen peroxide (H_2O_2) specifically stimulates VSMC DNA synthesis and neointimal formation after arterial injury. In fact, H_2O_2 modulates VSMC function in a concentration-dependent manner.[131] At high concentrations (0.5–1 mM), H_2O_2 induces apoptosis; at moderate concentrations, H_2O_2 induces cell cycle arrest at the G1/S interface. Similarly, reduction of ROS by overexpression of catalase inhibits VSMC proliferation while increasing apoptosis.[132] Cyclophilin A up-regulation in VSMC, in response to ROS, contributes to proliferation and migration of the VSMC, and inflammatory cell recruitment.[133]

Modulation of arterial thrombosis risk, by the effects of ROS on VSMC tissue factor expression and mitogenic potential, has been shown in animal models. When arterial vessels are injured chemically by ferric chloride, increased tissue factor from VSMC is critical for the observed thrombotic response. Crossing TF[flox/flox] mice with SM22αCre transgenic mice produces mice with 94–96% reduction only in smooth muscle cell tissue factor and this change reduces induced aortic thrombus formation.[134] PAR1, which is under reciprocal control of endothelial cell KLF2, promotes a mitogenic response in VSMC. Its effect on VSMC proliferation is aided by PAR2 and abolished in PAR1[-/-] or PAR2[-/-] mice.[135] Finally, VSMC in normal vessels are able to withstand apoptosis and maintain vascular integrity. However, in atherosclerotic vessels, VSMC apoptosis makes the vessel vulnerable to plaque rupture.[136] Thus, many disparate factors contribute to the thrombosis risk profile of the vasculature.

Platelets

Platelets and their products influence both endothelial cell and VSMC phenotype. In flowing blood, platelets adhere and become activated in areas of high shear stress and at sites of vascular injury. Platelet adherence and activation to the vessel wall is initiated by endothelial cell von Willebrand factor (VWF) binding to platelet glycoprotein Ib (GPIb). Constitutively, VWF does not bind GPIb, but shear and oxidation of VWF allows its unfolding, which promotes its binding to platelet GPIb. The first step in laser-induced ROS-stimulated vessels is the exocytosis, unfolding, and binding of VWF to platelet GPIb. Additionally, the slowing of platelets via VWF binding promotes VWF binding to collagen.[137] These effects allow for additional vascular collagen/platelet interactions via platelet collagen receptors, glycoprotein VI (GPVI) and $\alpha_2\beta_1$ integrin. Platelet adherence to GPVI or collagen allows them to spread, forming the initial platelet plug. During this process, they are activated and release their granule contents that recruit more platelets to form a platelet plug, as well as liberate their growth factors and mitogens to promote an inflammatory response.

The release of platelet low-molecular-weight growth factors also contributes to the inflammatory response associated with arterial injury and thrombosis. The platelet alarmin, S100A8/S100A9 or MRP8/14 (myeloid-related protein8/myeloid-related protein 14), is a good example of how platelets contribute to vascular injury and the inflammatory response. MRP8/14 from myeloid cells or platelets initiates the inflammatory response by binding to their receptors (Toll-like receptors 2 and 4, RAGE) to stimulate NF-κB-dependent release of various cytokines (TNF-α, IL-6, IL-1β, IL-8, INF-γ) that promote inflammation.[138] In endothelial cells, MRP8/14 induces a thrombogenic, inflammatory response by increasing transcription of proinflammatory chemokines (IL-8, CXCL1, CXCL2, monocyte secretory protein) and adhesion molecules (VCAM-1, ICAM-1), whose expression levels closely correlate with inflammation.[139] The absence of MRP8/14 in vascular lesions inhibits leukocyte oxidative metabolism, limiting their ability to incite the inflammatory response. Plasma protein and platelet mRNA levels of MRP8/14 correlate with acute coronary artery disease and systemic inflammation.[140] In a model of femoral artery injury, MRP14$^{-/-}$ mice that lack the MRP8/14 complexes have reduced leukocyte accumulation, cell proliferation, and neointima formation compared to wild-type mice, indicating that MRP8/14 enhances vascular inflammation and the leukocyte response to injury.[141] In contrast, deficiency of prolyl-carboxypeptidase (PRCP$^{gt/gt}$), a protein important for angiotensin-(1-7) formation and endothelial cell growth, angiogenesis, and protection from ischemia/reperfusion injury, is associated with increased VSMC proliferation and leukocyte infiltration after endothelial cell denudation by wire injury.[142] Mating PRCP$^{gt/gt}$ mice with MRP-14$^{-/-}$ mice ameliorates the VSMC proliferation and inflammation.[142]

Platelets also contribute to the inflammatory process through intimate associations with innate immune cells. In transfusion-related lung injury, NET formation was reduced when platelet inhibitors such as aspirin or antagonists to the integrin receptor $\alpha_{2b}\beta_3$ were used, resulting in reduced lung injury similar to the protection conferred using DNAse I or a histone-blocking antibody.[143] Additionally, increased oxidative stress contributes to platelet activation and increased thrombosis risk. Glutathione peroxidase-3 deficiency results in shortened bleeding times, heightened platelet responses to ADP, higher plasma levels of P-selectin, reduced plasma cGMP, increased ADP-induced pulmonary emboli, and larger cerebral infarctions.[101]

Platelets and their progenitors also contribute to arterial thrombosis risk by their mass. Hypercholesterolemia and increased platelet production promote atherothrombosis. Recent data indicate that cholesterol efflux from megakaryocytes modulates thrombopoietin induction of platelet production.[144] The inability of megakaryocyte golgi to export cholesterol to HDL results in LYN kinase phosphorylation and disruption of a negative-feedback regulation of the thrombopoietin receptor, allowing for increased platelet production and relative thrombocytosis.[144] Higher platelet counts, along with increased platelet reactivity in high-cholesterol states, contribute to atherothrombosis.[145]

C-reactive protein (CRP) has also been shown to contribute to atherothrombosis via effects on platelets. CRP binds to the Fc$_\gamma$ receptor on platelets, triggering the production of ROS, which leads to decreased platelet aggregation but increased endothelial platelet adhesion. In addition, CRP-mediated production of ROS contributes to atherothrombosis by increasing ICAM-1, VCAM-1, and MCP-1 expression on endothelial cells, increasing oxLDL uptake by endothelial cells and macrophages, and increasing expression of tissue factor, tissue factor pathway inhibitor, and plasminogen activator inhibitor-1 by VSMC.[146] Additional contributions to atherothrombosis by platelets have been recently reviewed.[147]

Monocytes, Macrophages, Neutrophils, and T-Cells in Vascular Inflammation

The vascular changes described above result in vascular inflammation and dysfunction that drives the pathogenesis of atherosclerosis, which includes the recruitment of various types of immune cells.[148] The recruitment and accumulation of leukocytes is associated with the development of vulnerable plaques (see Chapter 12). Macrophages account for the majority of leukocytes. Numerous cytokines and chemokines, as well as many of the transcription factors discussed above, modulate the macrophage in its role as scavenger and producer of interleukins and tissue factor that contribute to atherothrombosis.[147] Experimental evidence shows that hypercholesterolemia itself leads to elevation of oxLDL, and that macrophage phagocytosis of oxLDL induces tissue factor expression via engagement of toll-like receptor 4/6 (TLR4/TLR6 complex).[149] Severe atherosclerosis also is associated with pronounced neutrophilia, neutrophil hyperactivity, markedly increased oxidative stress, neutrophil intraplaque infiltration, and apoptosis.[150] Additionally, regulatory T-cells whose presence is documented in plaques have been recognized to promote acute ischemic stroke in mice by inducing dysfunction in the cerebral vasculature. Forkhead box P3-positive regulatory T-cells (Tregs) induce microvascular dysfunction by increasing LFA-1/ICAM-1 expression and platelet adhesion, exacerbating ischemic stroke in mice.[151] Depletion of Tregs reduces microvascular thrombus formation and improves cerebral reperfusion in stroke.[151]

Clinical Considerations

The clinical approach to diagnosis and management of arterial thrombosis, in essence the prevention and management of myocardial infraction, stroke, and peripheral vascular disease, is beyond the scope of this chapter. However, the future focus of therapies for both prevention and treatment need to incorporate means to reduce vascular ROS, inflammation, and inflammatory behavior of leukocytes and platelets. We have begun to recognize the precise molecular, chemical, and protein targets involved in these processes. Future therapies will incorporate targeting of these pathways and specific molecules into management protocols.

References

1. Goldhaber SZ. Venous thromboembolism: epidemiology and magnitude of the problem. *Best Pract & Res Clin Hematol* 2012;**25**:235–42.
2. Franchini M, Mannucci PM. Association between venous and arterial thrombosis: clinical implications. *Eurp J Int Med* 2012;**23**:333–7.
3. Reitsma PH, Versteeg HH, Middledorp S. mechanistic view of risk factors of venous thromboembolism. *Arterioscl Thromb Vasc Biol* 2012;**32**:563.
4. Lawson CA, Yan SD, Yan SF, Liao H, Zhou YS, Sobel J, et al. Monocytes and tissue factor promote thrombosis in a murine model of oxygen deprivation. *J Clin Invest* 1997;**99**:1729–38.
5. Nizet V, Johnson RS. Interdependence of hypoxic and innate immune responses. *Nat Rev Immunol* 2009;**9**:609–17.
6. Lopez JA. Chen J Pathophysiology of venous thrombosis. *Thromb Res* 2009;**123**(Suppl. 4):S30–4.
7. Nieswandt B, Kleinschnitz C, Stoll G. Ischaemic stroke: a thrombo-inflammatory disease? *J Physiol* 2011;**589**:4115–23.
8. Falati S, Liu Q, Gross P, Merrill-Skoloff G, Chou J, Vandendries E, et al. Accumulation of tissue factor into developing thrombi in vivo is dependent upon microparticle P-selectin glycoprotein ligand 1 and platelet P-selectin. *J Exp Med* 2003;**197**:1585–98.
9. Brinkmann V, Reichard U, Goosmann C, Fauler B, Uhlemann Y, Weiss DS, et al. Neutrophil extracellular traps kill bacteria. *Science* 2004;**303**:1532–5.
10. Fuchs TA, Brill A, Duerschmied D, Schatzberg D, Monestier M, Myers Jr DD, et al. Wagner DD Extracellular DNA traps promote thrombosis. *Proc Natl Acad Sci USA* 2010;**107**:15880–5.
11. Brill A, Fuchs TA, Savchenko AS, Thomas GM, Martinod K, De Meyer SF, et al. Neutrophil extracellular traps promote deep vein thrombosis in mice. *J Thromb Haemost* 2012;**10**:136–44.
12. von Bruhl ML, Stark K, Steinhart A, Chandraratne S, Konrad I, Lorenz M, et al. Monocytes, neutrophils, and platelets cooperate to initiate and propagate venous thrombosis in mice in vivo. *J Exp Med* 2012;**209**:819–35.
13. Martinod K, Demers M, Fuchs TA, Wong SL, Brill A, Gallant M, et al. Neutrophil histone modification by peptidylarginine deiminase 4 is critical for deep vein thrombosis in mice. *Proc Natl Acad Sci USA* 2013;**110**:8674–9.
14. Demers M, Krause DS, Schatzberg D, Martinod K, Voorhees JR, Fuchs TA, et al. Cancers predispose neutrophils to release extracellular DNA traps that contribute to cancer-associated thrombosis. *Proc Natl Acad Sci USA* 2012;**109**:13076–81.
15. van Montfoort ML, Stephan F, Lauw MN, Hutten BA, Van Mierlo GJ, Solati S, et al. Circulating nucleosomes and neutrophil activation as risk factors for deep vein thrombosis. *Arterioscler Thromb Vasc Biol* 2013;**33**:147–51.
16. Brill A, Fuchs TA, Chauhan AK, Yang JJ, De Meyer SF, Kollnberger M, et al. Von Willebrand factor-mediated platelet adhesion is critical for deep vein thrombosis in mouse models. *Blood* 2011;**117**:1400–7.
17. Kaplan MJ, Radic M. Neutrophil extracellular traps: double-edged swords of innate immunity. *J Immunol* 2012;**189**:2689–95.
18. van der Meijden PE, Munnix IC, Auger JM, Govers-Riemslag JW, Cosemans JM, Kuijpers MJ, et al. Dual role of collagen in factor XII-dependent thrombus formation. *Blood* 2009;**114**:881–90.
19. Choi SH, Smith SA, Morrissey JH. Polyphosphate is a cofactor for the activation of factor XI by thrombin. *Blood* 2011;**118**:6963–70.
20. Morrissey JH, Choi SH, Smith SA. Polyphosphate: an ancient molecule that links platelets, coagulation, and inflammation. *Blood* 2012;**119**:5972–9.
21. Renne T, Pozgajova M, Gruner S, Schuh K, Pauer HU, Burfeind P, et al. Defective thrombus formation in mice lacking coagulation factor XII. *J Exp Med* 2005;**202**:271–81.
22. Aiach M. Emmerich J Thrombophilia genetics. In: Colman RW, Marder VJ, Clowes AW, editors. *Hemostasis and thrombosis basic principles and clinical practice* 5th ed. Philadelphia: Lippincott, Williams and Wilkins; 2006. p. 79–93.
23. Dahlback B, Hildebrand B. Inherited resistance to activated protein C is corrected by anticoagulant cofactor activity found to be a property of factor V. *Proc Natl Acad Sci USA* 1994;**91**:1396–400.
24. Poort SR, Rosendaal FR, Reitsma PH, Bertina RM. A common genetic variation in the 3'-untranslated region of the prothrombin gene is associated with elevated plasma prothrombin levels and an increase in venous thrombosis. *Blood* 1996;**88**:3698–703.
25. Wolberg AS, Monroe DM, Roberts HR, Hoffman M. Elevated prothrombin results in clots with an altered fiber structure: a possible mechanism of the increased thrombotic risk. *Blood* 2003;**101**:3008–13.
26. Kyrle PA, Mannhalter C, Beguin S, Stümpflen A, Hirschl M, Weltermann A, et al. Clinical studies and thrombin generation in patients homozygous or heterozygous for the G20210A mutation in the prothrombin gene. *Arterioscler Thromb Vasc Biol* 1998;**18**:1287–91.
27. Cattaneo M. Hyperhomocysteinemia, atherosclerosis and thrombosis. *Thromb Haemost* 1999;**81**:165–76.

28. Ray J. Meta-analysis of hyperhomocysteinemia as a risk factor for venous thromboembolic disease. *Arch Intern Med* 1998;**158**: 2101–6.

29. Finkelstein JD. Methionine metabolism in mammals. *J Nutr Biochem* 1990;**1**:228–37.

30. Kluijtmans LA, van den Huevel LP, Boers GH, Frosst P, Stevens EM, van Oost BA, et al. Molecular genetic analysis in mild hyperhomocysteinemia: a common mutation in the methylenetetrahydrofolate reductase gene is a genetic risk factor for cardiovascular disease. *Am J Hum Genet* 1996;**58**:35–41.

31. Homocysteine Lowering Trialists' Collaboration. Lowering blood homocysteine with folic acid based supplements: meta-analysis of randomized trials. *Brit Med J* 1998;**316**:894–8.

32. Lonn E, Yusuf S, Arnold MJ, Sheridan P, Pogue J, Micks M, et al. Homocysteine lowering with folic acid and B vitamins in vascular disease. *N Engl J Med* 2006;**354**:1567–77.

33. Miletich JP. Laboratory diagnosis of protein C deficiency. *Semin Throm Haemost* 1990;**16**:169–76.

34. Olson ST, Bjork I. Regulation of thrombin activity by antithrombin and heparin. *Semin Thromb Hemost* 1994;**20**:373–409.

35. Machlus KR, Cardenas JC, Church FC, Wolberg AS. Causal relationship between hyperfibrinogenemia, thrombosis and resistance to thrombolytics in mice. *Blood* 2011;**117**:4953–63.

36. Mori T, Takeya H, Nishioka J, Gabazza EC, Suzuki K. Beta 2-glycoprotein I modulates the anticoagulant activity of activated protein C on the phospholipid surface. *Thromb Haemost* 1996;**75**: 49–55.

37. Gharavi AE, Sammaritano LR, Wen J, Elkon KB. Induction of antiphospholipid antibodies by immunization with beta 2 glycoprotein I (apolipoprotein H). *J Clin Invest* 1992;**90**:1105–9.

38. Lutters BC, Derksen RH, Tekelenburg WL, Lenting PJ, Arnout J, de Groot PG. Dimers of beta 2-glycoprotein 1 increase platelet deposition to collagen via interaction with phospholipids and the apolipoprotein E receptor 2'. *J Biol Chem* 2003;**278**:33831–8.

39. Simantov R, LaSala J, Lo SK, Gharavi AE, Sammaritano LR, Salmon JE, et al. Activation of cultured vascular endothelial cells by antiphospholipid antibodies. *J Clin Invest* 1996;**95**:2211–9.

40. Levi M, van der Poll T. Inflammation and coagulation. *Crit Care Med* 2010;**38**(Suppl. 2):S26–34.

41. Seligsohn U. Disseminated intravascular coagulation. In: Handin RI, Lux SE, Stossel TP, editors. *Blood: Principles and practice of hematology*. Philadelphia: J.B. Lippincott; 2000.

42. Schmaier AH. Disseminated intravascular coagulation – pathogenesis and management. *J Intensive Care Med* 1991;**6**:209–28.

43. Levi M, Toh CH, Thachil J, Watson HG. Guidelines for the diagnosis and management of disseminated intravascular coagulation. British Committee for Standards in Haematology. *Br J Haematol* 2009;**145**:24–33.

44. Amiral J, Bridey F, Dreyfus M, Vissoc AM, Fressinaud E, Wolf M, et al. Platelet factor 4 complexed to heparin is the target for antibodies generated in heparin-induced thrombocytopenia. *Thromb Haemost* 1992;**68**:95–6.

45. Warkentin TE, Levine MN, Hirsh J, Horsewood P, Roberts RS, Gent M, et al. Heparin-induced thrombocytopenia in patients treated with low-molecular-weight heparin or unfractionated heparin. *N Engl J Med* 1995;**332**:1330–5.

46. Warkentin TE, Kelton JG. Temporal aspects of heparin-induced thrombocytopenia. *N Engl J Med* 2001;**344**:1286–92.

47. Sheridan D, Carter C, Kelton JG. A diagnostic test for heparin-induced thrombocytopenia. *Blood* 1986;**67**:27–30.

48. Greinacher A, Amiral J, Dummel V, Vissac A, Kiefel V, Mueller-Eckhardt C. Laboratory diagnosis of heparin-associated thrombocytopenia and comparison of platelet aggregation test, heparin-induced platelet activation test, and platelet factor 4/heparin enzyme-linked immunosorbent assay. *Transfusion* 1994;**34**:381–5.

49. Greer IA. Thrombosis in pregnancy: maternal and fetal issues. *Lancet* 1999;**353**:1258–65.

50. McColl MD, Ramsay JE, Tait RC, Walker ID, McCall F, Conkie JA, et al. Risk factors for pregnancy associated venous thromboembolism. *Thromb Haemost* 1997;**78**:1183–8.

51. Clark P. Changes of hemostasis variables during pregnancy. *Semin Vasc Med* 2003;**3**:13–24.

52. Comp P, Thurnau GR, Welsh J, Esmon CT. Functional and immunologic protein S levels are decreased during pregnancy. *Blood* 1986;**68**:881–5.

53. Lee AY. Management of thrombosis in cancer: primary prevention and secondary prophylaxis. *Br J Haematol* 2006;**128**:291–302.

54. Ruf W. Molecular regulation of blood clotting in tumor biology. *Haemostasis* 2001;**31**(Suppl. 1):5–7.

55. Gale AJ, Gordon SG. Update on tumor cell procoagulant factors. *Acta Haematol* 2001;**106**:25–32.

56. Zhao R, Xing S, Li Z, Fu X, Li Q, Krantz SB, et al. Identification of an acquired JAK2 mutation in polycythemia vera. *J Biol Chem* 2005;**280**:22788–92.

57. Levine RL, Wadleigh M, Cools J, Ebert BL, Wernig G, Huntly BJ, et al. Activating mutation in the tyrosine kinase JAK2 in polycythemia vera, essential thrombocythemia, and myeloid metaplasia with myelofibrosis. *Cancer Cell* 2005;**7**:387–97.

58. Kralovics R, Passamonti F, Buser AS, Teo SS, Tiedt R, Passweg JR, et al. A gain-of-function mutation of JAK2 in myeloproliferative disorders. *N Engl J Med* 2005;**352**:1779–90.

59. James C, Ugo V, Le Couédic JP, Staerk J, Delhommeau F, Lacout C, et al. A unique clonal JAK2 mutation leading to constitutive signalling causes polycythaemia vera. *Nature* 2005;**434**: 1144–8.

60. Barbui T, Carobbio A, Rambaldi A, Finazzi G. Perspectives on thrombosis in essential thrombocythemia and polycythemia vera: is leukocytosis a causative factor? *Blood* 2009;**114**:759–63.

61. Takeda J, Miyata T, Kawagoe K, Iida Y, Endo Y, Fujita T, et al. Deficiency of the GPI anchor caused by a somatic mutation of the PIG-A gene in paroxysmal nocturnal hemoglobinuria. *Cell* 1993;**73**:703–11.

62. Kovacs MJ, MacKinnon KM, Anderson D, O'Rourke K, Keeney M, Kearon C, et al. A comparison of three rapid D-dimer methods for the diagnosis of venous thromboembolism. *Br J Haematol* 2001;**115**:140–4.

63. Rectenwald JE, Myers DD, Hawley AE, Long C, Henke PK, Guire KE, et al. D-dimer, P-selectin, and microparticles: novel markers to predict deep venous thrombosis. *Thromb Haemost* 2005;**94**:1312–7.

64. Ansell J, Hirsh J, Hylek E, Jacobson A, Crowther M, Palareti G. American College of Chest Physicians Pharmacology and management of the vitamin K antagonists: American College of Chest Physicians Evidence-Based Clinical Practice Guidelines (8th Edition). *Chest* 2008;**133**(Suppl. 6):160S–98S.

65. Büller HR, Davidson BL, Decousus H, Gallus A, Gent M, Piovella, et al.; Matisse Investigators. Fondaparinux or enoxaparin for the initial treatment of symptomatic deep venous thrombosis: a randomized trial. *Ann Intern Med* 2004;**140**:867–73.

66. Investigators EINSTEIN, Bauersachs R, Berkowitz SD, Brenner B, Buller HR, Decousus H, Gallus AS, et al. Oral rivaroxaban for symptomatic venous thromboembolism. *N Engl J Med* 2010;**363**:2499–510.

67. Schulman S, Kearon C, Kakkar AK, Mismetti P, Schellong S, Eriksson H, et al. RE-COVER Study Group Dabigatran versus warfarin in the treatment of acute venous thromboembolism. *N Engl J Med* 2009;**361**:2342–52.

68. Adam SS, McDuffie JR, Ortel TL, Williams Jr JW. Comparative effectiveness of warfarin and new oral anticoagulants for the management of atrial fibrillation and venous thromboembolism: a systematic review. *Ann Intern Med* 2012;**157**:796–807.

69. Sillesen H, Just S, Jorgensen M, Baekgaard N. Catheter-directed thrombolysis for treatment of iliofemoral deep venous thrombosis is durable, preserves venous valve function and may prevent chronic venous insufficiency. *Eur J Vasc Endovasc Surg* 2005;**30**: 556–62.

70. Decousus H, Leizorovicz A, Patent F, Page Y, Tardy B, Girard P, et al. A clinical trial of vena caval filters in the prevention of pulmonary embolism in patients with proximal deep-vein thrombosis. Prevention du Risque d'Embolie Pulmonaire par Interruption Cave Study Group. *N Engl J Med* 1998;**338**:409–15.

71. Madrigal J, Martinez-Pinna R, Fernandez-Garcia CE, Ramos-Mozo P, Burillo E, Egido J, et al. Cell stress proteins in atherosclerosis. *Oxidative Med Cell Longevity* 2012. doi:10.115/2012/232464.

72. Libby P. Mechanism of acute coronary syndromes and their implications for therapy. *N Engl J Med* 2013;**368**:2004–13.

73. SenBanerjee S, Lin Z, Atkins GB, Greif DM, Rao RM, Kumar A, et al. KLF2 is a novel transcriptional regulator of endolthelial pro-inflammatory activation. *J Exp Med* 2004;**199**:1305–15.

74. Lin Z, Kumar A, SenBanerjee S, Staniszewski K, Parmar K, Vaughan DE, et al. Kruppel-like factor 2 (KLF2) regulates endothelial thrombotic function. *Circ Res* 2005;**96**:e48–57.

75. Lin Z, Hamik A, Jain R, Kumar A, Jain MK. Kruppel-like factor 2 inhibits protease activated receptor-1 expression and thrombin-mediated endothelial activation. *Arterioscler Thromb Vasc Biol* 2006;**26**:1185–9.

76. Parmar KM, Larman B, Dai G, Zhang Y, Wang ET, Moorthy SN, et al. Integration of flow-dependent endothelial phenotypes by Kruppel-like factor 2. *J Clin Invest* 2006;**116**:49–58.

77. Kumar A, Hoffman TA, DeRicco J, Naqvi A, Jain MK, Irani K. Transcriptional repression of Kruppel like factor-2 by the adaptor protein p66shc. *FASEB J* 2009;**23**:4344–52.

78. Kumar A, Kim C-S, Hoffman TA, Naqvi A, DeRicco J, Jung S-B, et al. p53 impairs endothelial function by transcriptionally repressing Kruppel-like factor 2. *Arterioscler Thromb Vasc Biol* 2011;**31**:133–41.

79. Sunadome K, Yamamoto T, Ebisuya M, Kondoh K, Sehara-Fujisawa A, Nishida E. ERK5 regulates muscle cell fusion through Klf transcription factors. *Developmental Cell* 2011;**20**:192–205.

80. Hamik A, Lin Z, Kumar A, Balcells M, Sinha S, Katz J, et al. Kruppel-like factor 4 regulates endothelial inflammatiuon. *J Biol Chem* 2007;**282**:13769–79.

81. Ohnesorge N, Viemann D, Schmidt N, Czymai T, Spiering D, Schmolke M, et al. Erk5 activation elicits a vasoprotective endothelial phenotype via induction of Kruppel-like factor 4 (KLF4). *J Biol Chem* 2010;**285**:26199–210.

82. Zhou G, Hamik A, Nayak L, Tian H, Shi H, Lu Y, et al. Endothelial Kruppel-like factor 4 protects against atherothrombosis in mice. *J Clin Invest* 2012;**122**:4727–31.

83. Fledderus JO, Boon RA, Volger OL, Hurttila H, Yla-Herttuala S, Pannekoek H, et al. KLF2 primes the antioxidant transcription factor Nrf2 for activation in endothelial cells. *Arterioscler Thromb Vasc Biol* 2008;**28**:1339–46.

84. Lee D-Y, Lee C-I, Lin T-E, Lim SH, Zhou J, Tseng Y-C, et al. Role of histone deacetylases in transcription factor regulation and cell cycle modulation in endothelial cell in response to disturbed flow. *Proc Natl Acad Sci* 2012;**109**:1967–72.

85. Boon RA, Horrevoets AJG. Key transcription regulators of the vasoprotective effects of shear stress. *Hamostraseologie* 2009;**1**:39–43.

86. Brand K, Page S, Rogler G, Bartsch A, Brandl R, Knuechel R, et al. Activated transcription factor nuclear factor-kappa-B is present in the atherosclerotic lesion. *J Clin Invest* 1996;**97**:1715–22.

87. Milne JC, Denu JM. The Sirtuin family:therapeutic targets to treat diseases of aging. *Curr Opin Chem Biol* 2008;**12**:11–7.

88. Chen AF, Chen D-D, Daiber A, Faraci FM, Huige LI, Rembold CM, et al. Free radical biology of the cardiovascular system. *Clinical Science* 2012;**123**:73–91.

89. Xia N, Daiber A, Habermeier A, Closs EI, Thum T, Spanier G, et al. Reveratrol reverses endothelial nitric oxide synthase uncoupling in apolipoprotein E knockout mice. *J Pharmacol Exp Ther* 2010;**335**:149–54.

90. Carnicer R, Hale AB, Suffredini S, Liu X, Reilly S, Zhang MH, et al. Cardiomyocyte GTP-cyclohydrolase 1 and tetrahydrobiopterin increase NOS1 activity and accelerate myocardial relaxation. *Circ Res* 2012;**111**:718–27.

91. Barbieri SS, Amadio P, Gianellini S, Tarantino E, Zacchi E, Veglia F, et al. Cyclooxygenase-2-derived prostacyclin regulates arterial thrombus formation by suppressing tissue factor in a SIRT1-dependent manner. *Circ* 2012;**126**:1373–84.

92. Winnik S, Stein S, Matter CM. SIRT1 – an anti-inflammatory pathway at the crossroads between metabolic disease and atherosclerosis. *Current Vas Pharm* 2012;**10**:693–6.

93. Napoli C, Ignarro LJ. Nitric oxide and pathogenic mechanisms involved in the development of vascular diseases. *Arch Pharm Res* 2009;**32**:1103–8.

94. Liu VWT, Huang PL. Cardiovascular roles of nitric oxide: A review of insights from nitric oxide synthase gene disrupted mice. *Cardiovasc Res* 2008;**77**:19–29.

95. Leopold JA, Loscalzo J. Oxidative risk for atherothrombotic cardiovascular diease. *Free Radical Biology & Medicine* 2009;**47**: 1673–706.

96. Forstermann U. Nitric oxide and oxidative stress in vascular disease. *Pflugers Arch – Eur J Physiol* 2010;**459**:923–39.

97. Kvieys PR, Granger DN. Role of reactive oxygen and nitrogen species in the vascular responses to inflammation. *Free Radical Biology & Medicine* 2012;**52**:552–92.

98. Doughan AK, Harrison DG, Dikalov SL. Molecular mechanisms of angiotensin II mediated mitochondrial dysfunction. Linking mitochondrial oxidative damage and vascular endothelial dysfunction. *Circ Res* 2008;**102**:488–96.

99. True AL, Olive M, Boehm M, San H, Westrick RJ, Raghavachari N, et al. Heme oxygenase-1 deficiency accelerates formation of arterial thrombosis through oxidative damage to the endothelium, which is rescued by inhaled carbon monoxide. *Circ Res* 2007;**101**:893–901.

100. Adams GN, LaRusch GA, Stavrou E, Zhou Y, Nieman M, Jacobs G, et al. Murine prolylcarboxypeptidase depletion induces vascular dysfunction with hypertension and faster arterial thrombosis. *Blood* 2011;**117**:3929–37.

101. Jin RC, Mahoney CE, Anderson L, Ottaviano F, Croce K, Leopold JA, et al. Glutathione peroxidase-3 deficiency promotes platelet-dependent thrombosis in vivo. *Circ* 2011;**123**:1963–73.

102. Weiss HJ, Turitto VT. Prostacyclin (Prostglandin I2 PGI2) inhibits platelet adhesion and thrombus formation on subendothelium. *Blood* 1979;**53**:244–50.

103. Tateson JE, Moncada S, Vane JR. Effects of prostacyclin (PGX) on cyclic AMP concentrations in human platelets. *Prostaglandins* 1977;**13**:389–97.

104. Mellion BT, Ignarro LJ, Ohlstein EH, Pontecorvo EG, Hyman AL, Kadowitz PJ. Evidence for the inhibitory role of guanosine 3′,5′-monophosphate in ADP-induced human platelet aggregation in the presence of nitric oxide and related vasodilators. *Blood* 1981;**57**:946–55.

105. Carrier E, Brochu I, de Brum-Fernandes AJ, D'Orleans-Juste P. The inducible nitric-oxide synthase modulates endothelin-1-dependent release of prostacyclin and inhibition of platelet aggregation ex vivo in the mouse. *J Pharm Exp Ther* 2007;**323**:972–8.

106. Fang C, Stavrou E, Schmaier AA, Grobe N, Morris M, Chen A, et al. Angiotensin 1-7 and Mas decrease thrombosis in *Bdkrb2-/-* mice by increasing NO and prostacyclin to reduce platelet spreading and glycoprotein VI activation. *Blood* 2012;**121**:3023–32.

107. Riehl TE, He L, Zheng L, Greco S, Tollefsen DM, Stenson WF. Cox-1+/-COX-2-/- genotype in mice is associated with shortened time to carotid artery occlusion through increased PAI-1. *J Thromb Haemost* 2011;**9**:350–60.

108. Yu Y, Ricciotti E, Scalia R, Tang SY, Grant G, Yu Z, et al. *Vascular COX-2 modulates blood pressure and thrombosis in mice*; 2012. 4: 132ra54, May 2 www.ScienceTranslationalMedicine.org.

109. Harker LA, Slichter SJ, Scott CR, Homocystinemia Ross R. Vascular injury and arterial thrombosis. *N Engl J Med* 1974;**291**: 537–43.

110. Dayal A, Chauhan AK, Jensen M, Leo L, Lynch CM, Faraci FM, et al. Paradoxical absence of a prothrombotic phenotype in a mouse model of severe hyperhomocysteinemia. *Blood* 2012;**119**:3176–83.

111. Wilson KM, McCaw RB, Leo L, Arning E, Lhotak S, Bottiglieri T, et al. Prothrombotic effects of hyperhomocysteinemia and hypercholesterolemia in ApoE-deficient mice. *Arterioscler Thromb Vasc Biol* 2007;**27**:233–40.

112. Yang G, Wu L, Jiang B, Yang W, Qi J, Cao K, et al. *Science* 2008;**322**:587–90.

113. Paul BD, Snyder SH. H2S signalling through protein sulfhydration and beyond. *Nat Mol Cell Biol Rev* 2012;**13**:499–507.

114. Verma A, Hirsch DJ, Glatt CE, Ronnett GV, Synder SH. Carbon monoxide: a putative neural messenger. *Science* 1993;**259**: 381–4.

115. Zakhary R, Gaine SP, Dinerman JL, Ruat M, Flavahan NA, Snyder SH. Heme oxygnase 2: endothelial and neuronal localization and tole inendothelium-dependent relaxation. *Proc Natl Acad Sci USA* 1996;**93**:795–8.

116. Chen B, Guo L, Fan C, Bolisetty S, Joseph R, Wright MM, Agarwal A, George JF. Carbon monoxide rescues heme oxygenase-1-deficient mice from arterial thrombosis in allogenic aortic transplantation. *Amer J Path* 2009;**175**:422–9.

117. Peng L, Mundada L, Stomel JM, Liu JJ, Sun J, Yet S-F, et al. Induction of heme oxygenase-1 expression inhibits platelet-dependent thrombosis. *Antioxidant & Redox Signaling* 2004;**6**:729–35.

118. Glaser CB, Moser J, Clarke JH, Blasko E, McLean K, Kuhn I, et al. Oxidation of a specific methionine in thrombomodulin by activated neutrophil products blocks cofactor activity. *J Clin Invest* 1992;**90**:2565–73.

119. Dwyer KM, Mysore TB, Crikis S, Robson SC, Nadurkar H, Cowan PJ, et al. The transgenic expression of human CD39 on murine islets inhibits clotting of human blood. *Transplantation* 2006;**82**:428–32.

120. Huttinger ZM, Milks MW, Nickoli MS, Aurand WL, Long LC, Wheeler DG, et al. Ectonucleotide triphosphate diphosphohydrolase-1 (CD39) mediates resistance to occlusive arterial thrombus resistance to occlusive arterial thrombus formation after vascular injury in mice. *Amer J Path* 2012;**181**:322–33.

121. Hohmann JD, Wang X, Krajewski S, Selan C, Haller CA, Straub A, et al. Delayed targeting of CD39 to activated platelet GPIIb/IIIa via a single-chain antibody: breaking the link between antithrombotic potency and bleeding. *Blood* 2013;**121**:3067–75.

122. Heinrich J, Balleisen L, Schulte H, Assman G, van de Loo J. Fibrinogen and factor VII in the prediction of coronary risk. Results from the PROCAM study in healthy men. *Arterioscler Thromb* 1994;**14**:54–9.

123. Rhoads GG, Dahlen G, Berg K, Morton NE, Dannenberg AL. LP(a) lipoprotein as a risk factor for myocardial infarction. *JAMA* 1986;**256**:2540–4.

124. Chen L, Yang G, Xu X, Grant G, Lawson JA, Bohlooly M, et al. Cell selective cardiovascular biology of microsomal prostaglandin E synthase-1. *Circ* 2013;**127**:233–43.

125. Wu J, Bohanan CS, Neumann JC, Lingrel JB. KLF2 transcription factor modulates blood vessel maturation through smooth muscle cell migration. *J Biol Chem* 2008;**283**:3942–50.

126. Deaton RA, Gan Q, Owens GK. Sp1-dependent activation of KLF4 is required for PDGF-BB-induced phenotypic modulation of smooth muscle. *Am J Heart Circ Physiol* 2009;**296**:H1027–37.

127. Li H-X, Han M, Bernier M, Zheng B, Sun S-G, Su M, et al. Kruppel-like factor 4 promotes differentiation by transforming growth factor-β receptor-mediated Smad and p38 MSPK signaling in vascular smooth muscle cells. *J Biol Chem* 2010;**285**: 17846–56.

128. Li W, Febbraio M, Reddy SP, Yu D-Y, Yamamoto M, Silverstein RL. CD36 participates in a signaling pathway that regulates ROS formation in murine VSMCs. *J Clin Invest* 2010;**120**: 3996–4006.

129. Li L, Zhang H-N, Chen H-Z, Gao P, Zhu L-H, Li H-L, et al. SIRT1 acts as a modulator of neointima formation following vascular injury in mice. *Circ Res* 2011;**108**:1180–90.

130. Rao GN, Berk BC. Active oxygen species stimulate vascular smooth muscle cell growth and proto-oncogene expression. *Circ Res* 1992;**70**:593–9.

131. Deshpande NN, Sorescu D, Seshiah P, Ushio-Fukai M, Akers M, Yin Q, et al. Mechanism of hydrogen peroxide-induced cell cycle arrest in vascular smooth muscle. *Antioxidants & Redox Signaling* 2002;**4**:845–54.

132. Brown MR, Miller Jr FJ, Li W-G, Ellingson AN, Mozena JD, Chatterjee P, et al. Overexpression of human catalase inhibits proliferation and promotes apoptosis in vascular smooth muscle cells. *Circ Res* 1999;**85**:524–33.

133. Satoh K, Nigro P, Berk BC. Oxidative stress and vascular smooth muscle cell growth: a mechanistic linkage by cyclophilin A. *Antioxidants & Redox Signaling* 2010;**12**:675–82.

134. Wang L, Miller C, Swarthout RF, Rao M, Mackman N, Taubman MB. Vascular smooth muscle-derived tissue factor is critical for arterial thrombosis after ferric chloride-induced injury. *Blood* 2009;**113**:705–13.

135. Sevigny LM, Austin KM, Zhang P, Kasuda S, Koukos G, Sharifi S, et al. Protease-activated receptor-2 modulates protease-activated receptor-1-driven neointimal hyperplasia. *Arterioscler Thromb Vasc Biol* 2011;**31**:e100–6.

136. Clarke MCH, Figg N, Maguire JJ, Davenport AP, Goddard M, Littlewood TD, et al. Apoptosis of vascular smooth muscle cells induces features of plaque vulnerability in atherosclerosis. *Nat Med* 2006;**12**:1075–80.

137. Bernardo A, Bergeron AL, Sun CW, Guchhait P, Cruz MA, Lopez JA, et al. Von Willebrand factor present in fibrillar collagen enhances platelet adhesion to collagen and collagen-induced platelet aggregation. *J Thromb Haemost* 2004;**2**:660–9.

138. Chan JK, Roth J, Oppenheim JJ, Tracey KJ, Vogl T, Feldman M, et al. Alarmins awaiting a clinical response. *J Clin Invest* 2012;**122**:2711–9.

139. Viemann D, Strey A, Janning A, Jurk K, Klimmek K, Vogl T, et al. Myeloid-related proteins 8 and 14 induce a specific inflammatory response in human microvascular endothelial cells. *Blood* 2005;**105**:2955–62.

140. Healy AM, Pickard MD, Pradhan AD, Wang Y, Chen Z, Croce K, et al. Platelet expression profiling and clinical validation of myeloid-related protein-14 as a novel determinant of cardiovascular events. *Circ* 2006;**113**:2278–84.

141. Croce K, Gao H, Wang Y, Mooroka T, Sakuma M, Shi C, et al. Myeloid-related protein-8/14 is critical for the biological response to vascular injury. *Circ* 2009;**120**:427–36.

142. Adams GN, Stavrou EX, Fang C, Merkulova A, Alait MA, Nakajima K, et al. Prolylcarboxy-peptidase promotes angiogenesis and vascular repair. *Blood* 2013;**122**:1522–31.

143. Caudrillier A, Kessenbrock K, Gilliss BM, Nguyen JX, Marques MB, Monestier M, et al. Platelets induce neutrophil extracellular traps in transfusion-related acute lung injury. *J Clin Invest* 2012;**122**:2661–7.

144. Murphy AJ, Bijl N, Yvan-Charvet L, Welch CB, Bhagwat N, Reheman A, et al. Cholesterol efflux in megakaryocyte progenitors suppresses platelet production and thrombocytosis. *Nat Med* 2013;**19**:586–94.

145. Shattil SJ, Anaya-Galindo R, Bennett J, Colman RW, Cooper RA. Platelet hypersensitivity induced by cholesterol incorporation. *J Clin Invest* 1975;**55**:636–43.

146. Zhang Z, Yang Y, Hill MA, Wu J. Does C-reactive protein contribute to atherothrombosis via oxidant-mediated release of pro-thrombotic factors and activation of platelets? *Frontiers in Physiology* 2012;**3**:433.

147. Kaplan ZS, Jackson SP. The role of platelets in atherothrombosis. *Amer Soc Hematology Educational Book* 2011:51–61.

148. Woollard KJ, Geissman F. Monocytes in atherosclerosis: subsets and functions. *Nat Rev Cardiology* 2010;**7**:77–86.

149. Owens AP, Passam FH, Antoniak S, Marshall SM, McDaniel AL, Rudel L, et al. Monocyte tissue factor-dependent activation of coagulation in hypercholesterolemic mice and monkeys is inhibited by simvastatin. *J Clin Invest* 2012;**122**:558–68.

150. Borissoff JI, Otten JJT, Heenman S, Leenders P, van Oerle R, Soehnlein O, et al. Genetic and pharmacologic modifications of thrombin formation in apolipoprotein E-deficient mice determine atherosclerosis severity and atherothrombosis onset in a neutrophil-dependent manner. *PLOS ONE* February 2013;**8**:e55784.

151. Kleinschnitz C, Kraft P, Dreykluff A, Hagedorn I, Gobel K, Schuhmann MK, et al. Regulatory T-cells are strong promoters of acute ischemic stroke in mice by inducing dysfunction of the cerebral microvasculature. *Blood* 2013;**121**:679–91.

The Pericardium and its Diseases

Pooja Gupta, MBBS, DNB, Amar Ibrahim, MB, ChB, MSc,
Jagdish Butany, MBBS, MS, FRCPC
University of Toronto, Toronto, Ontario, Canada

THE PERICARDIUM AND ITS DISEASES

The pericardium is a fibrous, relatively avascular tissue surrounding the heart which connects the heart to the sternum, diaphragm, and posteriorly to the anterior mediastinum. It consists of two layers: (1) the inner serous or visceral pericardium, also known as the epicardium; and (2) the outer fibrous parietal pericardium.[1] The parietal pericardium consists of a fibrous sac that is less than 2mm thick comprised mainly of collagen and some elastin fibers.[1] Its fibrous nature and the relatively inelastic physical properties assist it to achieve its function of anchoring the heart in its position and preventing friction between the heart and the surrounding intrathoracic structures. In addition, the pericardium limits distention during the cardiac cycle and acts as an immunological barrier to infection. The inner visceral pericardium is comprised of a single layer of mesothelial cells attached to the epicardial surface of the heart muscle.[1] To facilitate cardiac function and maintain effective function of the pericardium, the two layers of pericardium are separated by a potential space that contains 15–50mL of serous fluid distributed mostly over the atrioventricular and interventricular grooves.[1] The pericardium is a well-innervated structure such that pericardial inflammation can lead to severe pain and initiate vagally mediated reflexes. It also secretes prostaglandins that control cardiac reflexes and coronary tone.[1] The incidence of pericarditis in postmortem studies ranges from 1–6%, whereas it is diagnosed antemortem only in about 0.1% of hospitalized patients and 5% of patients seen in emergency departments with chest pain but without myocardial infarction (MI).[2] In other words, pericarditis constitutes up to 5% of the cases presenting to accident and emergency departments with non-ischemic chest pain[3] and is therefore considered the most common form of pericardial disease. A prospective study published in 2008 reported the incidence of acute pericarditis was 27.7 cases per 100000 of the population per year in an urban area of Italy.[4]

Pericarditis is caused by a variety of etiological factors (Table 16.1). The most common etiological factor is idiopathic pericarditis, which constitutes 80–90% of cases.[5,6] The remaining 10% of cases have an infectious or non-infectious etiology. The most common infectious agent associated with pericarditis in the remaining 10% of cases is Mycobacterium tuberculosis. Other infectious agents include viruses such as echovirus, coxsackie virus, Epstein-Barr virus, cytomegalovirus, adenovirus and bacteria such as pneumococci, staphylococci, Haemophilus influenza, meningococci, gonococci, mycoplasma, Legionella and Chlamydia.[7] Fungal and parasitic infections of the pericardium may occur, although they are rare.

Among non-infectious causes, pericarditis usually follows a transmural MI. It can result from malignant cell invasion (neoplastic pericarditis); commonly due to metastasis from adjacent sites such as lung, breast, and lymphoma and rarely caused by primary pericardial tumors – mainly pericardial mesothelioma. Pericarditis may also develop due to the accumulation of blood in the pericardial space following a dissecting aortic aneurysm. It may also occur following sharp or blunt trauma to the chest as in penetrating thoracic injury or due to esophageal perforation. Iatrogenic causes of pericarditis are among the contemporary examples of post-cardiac injury syndromes and these occur following percutaneous coronary interventions, pacemaker insertion and catheter ablation. Chest radiation is another rare cause of pericarditis.

Autoimmune diseases predispose to the pericarditis. These include systemic lupus erythematosus (SLE), Sjögren syndrome, rheumatoid arthritis, systemic sclerosis, systemic vasculitides, Behçet syndrome, sarcoidosis, and familial Mediterranean fever. The most common metabolic causes of pericarditis include uremia and myxedema.[7] Certain drugs are also among the rare causes of pericarditis.

TABLE 16.1   Classification of Pericarditis

Idiopathic
Infectious
    Viral (most common echovirus and coxsackievirus, adenovirus,
      parvovirus B19, herpes virus, human immunodeficiency virus)
Bacterial (most common: tuberculous; other rare causes are Gram-
positive infections, Gram-negative infections and other bacterial
infections such as Borrelia, Listeria, Legionella and Campylobacter)
    Fungal (histoplasma, blastomycosis, cryptococcus, and Candida)
    Parasitic (amoeba, toxoplasmosis, echinococcus, trypanosomiasis,
      filariasis, schistosomiasis)
Non-infectious
    Associated with acute myocardial infarction
    Associated with systemic inflammatory disease
        Rheumatologic diseases (rheumatic fever, rheumatoid arthritis,
          systemic lupus erythematosus, scleroderma, mixed connective
          tissue disorders , Sjögren syndrome)
        Systemic vasculitis (Kawasaki disease, Churg Strauss
          syndrome and Wegener disease)
        Granulomatous disease (sarcoidosis)
    Associated with metabolic and endocrine disorders (uremia,
    myxedema, diabetes mellitus, gout)
    Associated with malignant neoplasms (commonly metastatic
    tumors and rarely primary pericardial tumors)
    Iatrogenic and traumatic pericarditis
        Post-cardiac injury syndrome (cardiovascular surgery,
          complications of cardiac catheterization)
        Penetrating/non-penetrating injury to the chest wall
        Radiation
        Drugs

These can lead to the formation of pericarditis through differing mechanisms. Some of the drugs, such as procainamide, hydralazine, isoniazid (INAH), and phenytoin may cause SLE-like syndromes, which predispose to pericarditis, whereas doxorubicin and daunorubicin may predispose to cardiomyopathy and pericardiopathy. Penicillins cause hypersensitivity pericarditis with eosinophilia.[7]

The clinical diagnosis of pericarditis is relatively easy to make, compared with the task of establishing its actual cause.[6–9] The clinician needs to identify the specific causes of pericarditis that require targeted therapies. The most common presentation of pericarditis is chest pain, which is typically sudden in onset, retrosternal, pleuritic in nature and is exacerbated by inspiration. It is also exacerbated when the patient is in the supine position and relieved by sitting up or leaning forward. The pain often radiates to the neck, arms, or left shoulder. Typically, pericardial pain is referred to the scapular ridge, presumably due to irritation of the phrenic nerves, which passes adjacent to the pericardium.[7] Therefore, if pain radiates to one or both trapezius muscle ridges, it is probably due to pericarditis, since the phrenic nerve (which also innervates these muscles) is irritated and passes adjacent to the pericardium.[10] Chest pain can be severe and debilitating. However, pericarditis may at times be asymptomatic, as in pericarditis associated with rheumatoid arthritis.

The pericardial friction rub is another classic finding of acute pericarditis. It is high-pitched and scratchy and can have up to three components. These components occur when the cardiac volumes are changing most rapidly during the different parts of the cardiac cycle. It corresponds temporally to the movement of the heart within the pericardial sac.[10] Patients with atrial fibrillation have one or two components. A pleural friction rub can be differentiated from the pericardial friction rub by its absence when the patient holds his breath for a short time, whereas the pericardial friction rub continues even when the patient holds his/her breath.[11]

The clinical diagnosis of pericarditis should be reserved for patients with an audible pericardial friction rub or chest pain with typical electrocardiographic findings, bearing in mind that the two most important conditions that may cause chest pain similar to that of pericarditis include myocardial infarction (MI) and pulmonary embolism.[10] The clinical features that indicate a high-risk group of patients that need admission in order to search for the specific cause of pericarditis include fever (greater than 38°C), subacute course (symptoms develop over several days to weeks), large pericardial effusion (diastolic echo free space >20 mm in width), the development of cardiac tamponade and failure of aspirin therapy or NSAIDs.

Cardiac Tamponade

The presence of systemic arterial hypotension, tachycardia, elevated jugular venous pressure, and pulsus paradoxus (a decrease in systolic arterial pressure of more than 10 mmHg with inspiration) suggests cardiac tamponade. Cardiac tamponade is a potentially lethal complication of pericarditis. It is commonly reported in patients with tuberculous or purulent pericarditis (60%) and less commonly (15%) in patients with idiopathic pericarditis.[8] The 12-lead electrocardiogram (ECG) in acute pericarditis typically shows diffuse concave ST-segment elevation and PR-segment depression.[12] There are four stages[13] of ECG abnormalities in pericarditis:

- Stage I: diffuse ST-segment elevation and PR-segment depression;
- Stage II: normalization of the ST and PR segments;
- Stage III: widespread T-wave inversions;
- Stage IV: normalization of the T waves.

There are several ECG changes that allow differentiation between MI and pericarditis. In MI, ST-segment elevations are usually convex (dome-shaped) and not concave as in pericarditis, and ST segment elevations are found in specific leads depending on the anatomical site of the MI. In contrast, the ST-segment changes are diffuse and widespread, affecting different leads in pericarditis. The ECG changes in pericarditis are less likely to be seen in certain leads such as aVR, and they are more likely seen in all other leads. In addition, PR-segment depression is more likely encountered in pericarditis and uncommon in MI. In contrast, Q-wave formation and loss of R-wave voltage is often seen in MI and not in pericarditis. Atrioventricular block or ventricular arrhythmias are commonly found in MI, and less likely to be found in association with the changes of pericarditis. In pericarditis, T-wave changes usually start after ST-segment elevations return to the base line (stage III), whereas in MI, it is common to find ST segments with concomitant T-wave inversions. In other words, T-wave inversions appear before the ST segments return to baseline. The most important and distinguishing feature is the ratio of ST-segment elevation (in millimeters) to T-wave amplitude (height in millimeters) in lead V6. If the ratio exceeds 0.24, acute pericarditis is almost always present.[14] Chest imaging is usually performed to estimate the size of the heart and exclude abnormalities in the mediastinum, lungs or other organs surrounding the heart that might be responsible for the pericarditis. Cardiomegaly on chest imaging would suggest a significant pericardial effusion (more than 250 mL).

SEROLOGICAL TESTS

Viral cultures and antibody titers are not useful clinically in diagnosing pericarditis.[5,6,8] White cell count, erythrocyte sedimentation rate, and serum C-reactive protein concentration are generally elevated in patients with acute pericarditis, however, these tests are not useful for identifying the cause of the disease and do not provide clues regarding therapy. A marked increase in white-cell count indicates a purulent pericarditis; therefore, it is reasonable to obtain a complete blood count in all patients. Routine serologic testis, including testing for antinuclear antibody (ANA) and rheumatoid factor (RF), reveal a source for the pericarditis in only 10–15% of patients, and in these cases other evidence typically suggests the underlying disease. Plasma troponin concentrations are elevated in 35–50% of patients with pericarditis, a finding that is thought to be caused by epicardial inflammation rather than myocyte necrosis.[15] The intensity of the elevation of serum troponin is related to the degree of ST-segment elevation. Serum troponin concentration usually returns to normal within two weeks after the diagnosis. An elevated troponin concentration does not predict an adverse outcome.[15,16] A prolonged elevation (lasting longer than two weeks) suggests associated myocarditis, and this indicates a poor prognosis. Although the concentrations of serum creatine kinase and its MB fraction may also be elevated in pericarditis, they are abnormal less often than is the troponin concentration.

ECHOCARDIOGRAPHY

Transthoracic echocardiography (TEE) is indicated in patients with suspected pericarditis, to confirm the diagnosis of pericarditis and to rule out cardiac tamponade, which necessitates pericardiocentesis.[17]

PERICARDIOCENTESIS AND BIOPSY

Pericardiocentesis and pericardial biopsy are generally not useful in patients with small or moderate pericardial effusions, as a definitive diagnosis may not be provided. According to a study of 231 patients with acute pericarditis of unknown cause, pericardiocentesis provided the cause of pericarditis in 6% of cases and pericardial biopsy provided the diagnosis in 5% only. In contrast, in cases with pericardial tamponade, pericardiocentesis provided the diagnosis in 29% and pericardial biopsy provided a diagnosis in 54% of cases.[6] Therefore, pericardiocentesis is often indicated in patients with pericardial tamponade and in patients with suspected purulent or neoplastic pericarditis. In pericardiocentesis, the pericardial fluid should be sent for analysis of red and white cell counts, cytology for cancer detection, and triglycerides to exclude a chylous effusion. Microscopic examination, in addition to culture and sensitivity of the pericardial fluid, is indicated to diagnose the specific microorganism that causes the pericardial effusion. Polymerase chain reaction (PCR) assays and adenosine deaminase activity levels (greater than 30 U per liter) are helpful in identifying Mycobacterium tuberculosis.[18–20] Measuring the levels of glucose, lactic dehydrogenase, protein and pH is often a common practice in pericardiocentesis. However, testing for these levels does not provide the diagnosis for the specific cause of pericarditis.

SPECIFIC FORMS OF PERICARDITIS

Infectious Pericarditis

Idiopathic

Acute non-specific idiopathic pericarditis accounts for 80–85% of cases of acute pericarditis.[21] It is seen in the fourth decade of life and shows a male predominance.[22] Patients usually present with upper respiratory tract infection, congestive heart failure, pleural/pericardial effusion, leukocytosis, and raised erythrocyte sedimentation rate.[23]

It has a self-limiting course. Rarely, patients may develop recurrence and constrictive pericarditis.[24] Grossly, the pericardium initially looks shaggy and later develops fibrinous exudate and fibrous adhesions. Serous or serosanguineous pericardial effusion may be seen. Microscopically, the pericardium shows congestion and mixed inflammatory infiltrate.[23]

Viral Pericarditis

Viral infections are responsible for 1–2% of cases of acute pericarditis.[25] The most common causative viruses are coxsackievirus B, influenza, echovirus, and polio virus. Others include adenovirus, enterovirus, mumps virus, human immunodeficiency virus (HIV), Epstein-Barr virus (EBV), cytomegalovirus (CMV), hepatitis virus, herpes simplex virus (HSV), parainfluenza virus, varicella zoster virus (VZC), measles virus, and respiratory syncytial virus (RSV).[23] Infections with the coxsackie virus and echovirus occur following the seasonal epidemic, while CMV infection is seen in the immunocompromised.[26,27]

Viral pericarditis is usually associated with myocarditis and lasts for 1–3 weeks. It is self-limiting and has a good prognosis.[28] Patients may present with sharp, stabbing chest pain, difficulty in breathing, dry cough, fever, fatigue, and swelling of ankles and feet. Occasionally, fibrous adhesions, constrictive pericarditis and recurrence can occur. Coxsackie viral pericarditis occurs following viral invasion of lungs and mesothelial cells and can cause post-infection dilated cardiomyopathy.[23,29]

Viral pericarditis is diagnosed by examining the pericardial effusion using PCR or by in situ hybridization. A significant (four-fold) rise in serum antibody levels along with high interleukin-6 (IL-6) and tumor necrosis factor alpha (TNF-α) and low transforming growth factor beta (TGFβ) in the pericardial fluid suggest the diagnosis of viral pericarditis.[23,30]

AIDS-Associated Pericarditis

Pericardial involvement in AIDS occurs due to infectious, non-infectious and neoplastic diseases. Mild non-specific pericarditis is seen in approximately 50% of patients with AIDS while less than 1% of patients develop symptomatic pericarditis.[31] Infective pericarditis can occur due to local HIV infection or due to opportunistic infections such as CMV, HSV, Mycobacterium tuberculosis, Mycobacterium avium-intracellulare (MAI) and Cryptococcus.[32] Small, asymptomatic pericardial effusion without cardiac tamponade is seen in 40% of cases.[33,34]

Bacterial Pericarditis

Bacterial pericarditis, also called purulent pericarditis, is a less commonly encountered condition today compared to the pre-antibiotic era.[35] It affects children more commonly than adults and usually occurs as a complication of infection affecting structures adjacent or contiguous to the pericardium. It may also occur due to infection that arises in distant organs and affects the pericardium by hematogenous spread.[36] In adults, certain conditions such as pericardial effusion, immunosuppression, chronic illnesses, such as rheumatoid arthritis or alcohol abuse, cardiac surgery and chest trauma, can act as predisposing factors for purulent pericarditis. Purulent pericarditis can also be associated with other myocardial infections such as ring bacterial endocarditis and ring abscesses of native or prosthetic valves.

Gram-positive organisms including staphylococci, pneumococci, and streptococci are most commonly associated with purulent pericarditis. Staphylococcal pericarditis is now more likely to be found in patients infected with HIV, due to the high bacterial colonization of the nasal cavities and skin,[37] and the high incidence of skin diseases in HIV-positive patients.[38] In addition, it is related to the increased use of IV catheters in HIV-positive patients compared to those not HIV-positive. It is a life-threatening condition that can lead to cardiac tamponade. Pneumococcal pericarditis usually occurs as a result of direct extension of a lung or pleural infection, with only a few cases of primary pneumococcal pericarditis reported so far.[39]

Infections with Gram-negative bacilli including Neisseria, E. coli, Haemophilus, Pseudomonas aeruginosa and Klebsiella pneumoniae are also among the common infectious organisms that predispose to purulent pericarditis. Primary infection of the pericardium is rare, and meningococcal pericarditis is frequently seen in epidemics. A study showed that meningococcal pericarditis was found in around 20% of patients with meningococcal meningitis.[40] Meningococcal pericarditis is more common in children than adults and accounts for 33% and 2–4% of cases, respectively.[41,42] Haemophilus influenza pericarditis is more commonly encountered in children than in adults.[43]

Other infectious agents that lead to purulent pericarditis include Legionella pneumophila,[44] Campylobacter,[45] Listeria monocytogenes,[46] Neisseria mucosae,[47] Nocardia asteroids,[48] and Chlamydia.[49] Although anaerobic organisms are occasionally associated with purulent pericarditis, the most frequent anaerobic organisms associated with purulent pericarditis include Bacteroides fragilis and Clostridia species.

Grossly, both parietal and visceral pericardial layers appear to be thickened and may be covered by a fibrinopurulent exudate. The pericardial space may contain serous, serosanguineous, or turbid, yellow, green fluid. Microscopically, abundant infiltration of polymorphs is usually seen in the epicardium and the collected pericardial fluid.

The course of purulent pericarditis is usually acute, fulminant and requires immediate pericardiocentesis. The pericardial fluid should be sent for Gram staining, acid-fast and fungal stains, in addition to culture and sensitivity of the pericardial and other body fluids. Microorganisms can be demonstrated by Gram staining of the pericardial fluid before initiating antimicrobial treatment. Rinsing of the pericardial cavity, frequent drainage, in addition to systemic antibiotic therapy are all mandatory for effective treatment of purulent pericarditis. Surgery is indicated in patients with purulent pericarditis who develop complications such as dense fibrous adhesions, loculated and thick purulent effusion, tamponade, persistent infection, areas of calcification and progression to constrictive pericarditis.[50] Purulent pericarditis is always fatal if left untreated.[51–53] However, even in medically treated patients the mortality rate may reach up to 77%, with the main causes of death being cardiac tamponade, toxicity, and constriction.[54]

Lyme Pericarditis

Lyme disease is caused by a spirochete (Borrelia burgdorferi), which is transferred by tick bite (Ixodes dammini). As Lyme disease develops, it progresses into three stages: local disease, early dissemination, and late dissemination. Cardiac involvement is seen in the second stage of the disease (the early dissemination stage) and the patient may present with arrhythmia, heart block, myopericarditis, and pancarditis. Microscopically mild interstitial fibrosis, lymphocytic infiltrate, myocarditis, pericarditis and conduction system abnormalities can be seen.[23]

Tuberculous Pericarditis

Mycobacterium tuberculosis (TB) is a common cause of pericarditis worldwide.[55] The prevalence of tuberculous pericarditis (TB pericarditis) has declined significantly in developed countries during the past four decades. During the last decade, it has been seen mainly in immunocompromised patients.[56] Other forms of mycobacteria have also been reported in association with pericarditis such as Mycobacterium chelonae, and Mycobacterium avium intracellulare, which is mainly reported in HIV-positive patients.[57] TB pericarditis occurs in less than 10% of patients affected by systemic tuberculosis. It is usually caused by spread of infection from the mediastinal, paratracheal, peribronchial lymph nodes or by hematogenous spread from other distant sites. It can also occur due to direct spread from a lesion affecting the lungs, sternum, or spine.[58]

Tuberculous pericarditis is characterized by three stages: (1) acute; (2) subacute; and (3) chronic. The acute stage can be further subdivided into the fibrinous and effusive stages. In the fibrinous stage, patients may develop a serosanguineous effusion and lymphocytic infiltrate; while in the effusive stage, lymphocytes can be seen in the pericardial fluid. The subacute stage is characterized by the presence of granulomatous inflammation with or without caseous necrosis. The Mycobacterium can usually be detected in this stage using special stains. The chronic stage is characterized by the presence

of marked pericardial thickening and fibrosis, calcification, obliteration of pericardial cavity and constrictive pericarditis.[23]

The clinical presentation of TB pericarditis is variable. Patients may either present during the early stage of the disease (the effusive stage) or later after the development of constrictive pericarditis.[59] During the effusive stage, patients may develop persistent fever, night sweats, dyspnea, abdominal discomfort, and jugular venous distention.[60] Chest pain is less likely to be encountered in TB pericarditis and it usually indicates a viral etiology.

When pericarditis develops in patients with systemic tuberculosis, it is strongly suggestive that the causative microorganism is Mycobacterium tuberculosis.[61] Sputum cultures and tuberculin skin tests can be done. The tuberculin skin test has less sensitivity and specificity with 25–33% false-negative[62] and 30–40% false-positive cases.[56] The enzyme-linked immunosorbent spot (ELISPOT) test is more accurate, it detects T-cells specific for Mycobacterium tuberculosis antigen.[63] The diagnosis is usually established by demonstrating the presence of a serosanguineous effusion, lymphocytes and rarely the microorganism in the pericardial fluid or tissue in the early effusive stage of the disease, and/or by demonstrating the presence of caseating granulomas, with or without necrosis and microorganisms (with special stains) in the tissue biopsy in the subacute stage of the disease (Figure 16.1A,B).[56] The diagnostic yield of pericardiocentesis in TB pericarditis ranges from 30–76%.[62,64] The pericardial fluid in TB pericarditis usually demonstrates high specific gravity, high protein levels, and high white-cell count (from $0.7–54 \times 10^9/L$).[64] High adenosine deaminase activity and interferon gamma concentration in the pericardial fluid are diagnostic of TB pericarditis, with a high sensitivity and specificity. Pericardial biopsy allows diagnosis with a better sensitivity than pericardiocentesis (100% versus 33%).[59] Other important diagnostic techniques include polymerase chain reaction (PCR), which can detect DNA of the Mycobacterium from only 1 μL of pericardial fluid in a very short time.[64,65] In addition, both pericardioscopy and pericardial biopsy have improved the diagnostic accuracy for TB pericarditis.[66]

Various antituberculous drug combinations of varying durations, ranging from 6–12 months, have been indicated.[61] The use of steroids in treatment of TB pericarditis remains controversial.[61] It has been suggested that antituberculous drug treatment combined with steroids might be associated with an infrequent need for pericardiocentesis or pericardiectomy, more rapid clinical improvement and reduced mortality.[67–69] The prognosis of untreated effusive TB pericarditis is poor; constrictive pericarditis occurs in almost all patients and the mortality rate reaches up to 85%.[70]

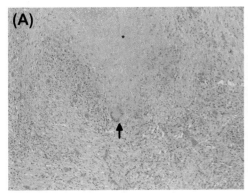
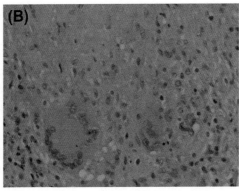

FIGURE 16.1 (A, B) Sections show a caseating (asterisks) granuloma, surrounded by chronic inflammation and a multinucleate giant cell (black arrow) in tuberculous pericarditis. (Stains: (A&B) hematoxylin and eosin; original magnification: A x10; B x20).

Fungal Pericarditis

Primary fungal infection of the pericardium is rare. The common fungal infectious agents that cause pericarditis include histoplasmosis, cryptococcosis, blastomycosis, and Candida species and rarely actinomycosis, cryptococcosis, and coccidioidomycoses. Fungal pericarditis usually occurs due to spread of infection from the adjacent structures such as lungs and pleural cavities. It can occur in patients with advanced malignancies or immunosuppressed state and caused by opportunistic infections including Candida, Aspergillus, and Blastomyces.[71,72] It can also be seen in people living in areas with endemic fungal infections such as histoplasmosis.[73,74]

HISTOPLASMOSIS

Fungal pericarditis due to Histoplasma capsulatum is rare, although it is commonly seen in endemic areas.[75] Grossly, it can lead to fibrinous exudate with or without serosanguineous effusion. Microscopically, mixed inflammatory cell infiltrate composed of polymorphonuclear leucocytes and lymphocytes is seen. Sometimes granulomata associated with cell necrosis may also be seen. Areas of organization, focal calcification, and fibrosis may develop, which can lead to constrictive pericarditis. The diagnosis of histoplasmosis is made by the demonstration of fungus in the pericardial fluid and/or the biopsy using special stains and culture techniques. Other tests that may assist in diagnosing histoplasmosis include the demonstration of antifungal antibodies in the serum, histoplasma skin tests, and histoplasma complement fixation tests.[59] Patients with pericarditis due to histoplasmosis do not need antifungal therapy, and they usually respond to non-steroidal anti-inflammatory drugs given during 2–12 weeks.[59]

CANDIDA

Candida pericarditis usually occurs in immunocompromised patients, especially those with deficient T-cell function. It can also occur in Candida endocarditis, especially following prosthetic valve surgery. The treatment involves the use of antifungal drugs such as fluconazole,

ketoconazole, itraconazole, amphotericin B, liposomal amphotericin B, or amphotericin B lipid complex.[59] Some studies have recommended that the combined use of medical therapy (Amphotericin B) along with surgery (pericardiectomy) is helpful for treatment.[76] In general, pericardiocentesis is usually indicated in hemodynamic impairment while pericardiectomy is indicated in constrictive pericarditis.[59] Purulent pericarditis due to *Candida* infection has a poor prognosis.[77]

PARASITIC PERICARDITIS

Pericardial involvement is rarely seen in parasitic infections.

AMOEBA

Amoebic pericarditis is extremely rare and a serious complication with an incidence of 1.3% and mortality of 29.6% in hepatic amoebiasis.[78] It is more commonly seen in children in association with intestinal amoebiasis.[79] Amoebic pericarditis can occur as a result of amoebic liver abscess rupture or from direct extension from the lung or pleura and present as purulent exudate, described characteristically as 'anchovy sauce' pus.[80] Patients may present with the sudden onset of cardiac tamponade or progressive effusion and have symptoms such as fever, chest pain, dyspnea, and shock.[79–81] The diagnosis is based on serology and histologic examination of acute inflammatory cells and trophozoites of Entamoeba histolytica in the pericardium. The treatment of choice is surgical drainage along with metronidazole therapy. Amoebic pericarditis is differentiated from tuberculous pericarditis by the predominance of neutrophils in the former.[79,80]

TOXOPLASMA

In immunocompetent patients, acute Toxoplasma infection is often asymptomatic, but latent infection due to cyst formation persists for life.[82] In immunocompromised patients as in AIDS, the prevalence of cardiac involvement in Toxoplasma gondii is less than 10%, which manifests as a myocarditis, pericardial effusion, constrictive pericarditis, arrhythmias, cardiac tamponade, and congestive heart failure.[83] Histologically, myocardium may show trophozoites and pericardium may show lymphocytes, plasma cells, fibrosis, and epithelioid granulomas. The diagnosis is generally made by serology.[23]

ECHINOCOCCUS

Cardiovascular involvement in hydatid cysts is seen in 0.5–3% cases. It commonly affects males in the second to fifth decades of life. Pericardial involvement is seen in 40% of cases of cardiac Echinococcus and occurs due to the spread from the initial location at the liver dome or due to the rupture of the left ventricular free wall cyst in the pericardium. Patients may present with arrhythmias, myocardial infarction, pulmonary hypertension, cardiac tamponade, pericardial effusion, purulent pericarditis, constrictive pericarditis, and sudden cardiac death. The effusion fluid may be clear or serous and may contain the hydatid cyst, germinal epithelium, daughter cysts, lymphocytes, and eosinophils. The pericardium may show secondary bacterial infection and may develop fibrinous pericarditis.[23,84,85]

CHAGAS DISEASE

Cardiac involvement in Chagas disease (caused by Trypanosoma cruzi) manifests as ventricular arrhythmias, congestive cardiac failure, and sudden death. Microscopically, myocardium may show multiple areas of scarring and chronic inflammatory infiltrate. Pericardium may show mild thickening of pericardial layers, patchy areas of fibrosis and occasional lymphocytic infiltrate.[86]

FILARIASIS

Filariasis cause by Wuchereria bancrofti rarely affects pericardium. It may present as pericardial thickening, fibrosis, mild lymphocytic infiltrate, pericardial effusion, and constrictive pericarditis. Rarely microfilaria can be detected in the pericardial fluid.[23]

SCHISTOSOMIASIS

Cardiac involvement is rarely seen in Schistosoma infection. Schistosoma primarily affects the liver causing hepatic fibrosis and portal hypertension and subsequently may lead to hepatopulmonary syndrome, pulmonary hypertension, right ventricular hypertrophy, cor pulmonale, cardiac arrhythmias, and sudden death. Microscopically, granulomatous myocarditis and pericarditis may be seen.[87,88]

Non-Infectious Pericarditis

Pericarditis Associated with Acute Myocardial Infarction

Pericarditis occurring post-acute myocardial infarction (AMI) occurs in two forms. The early form occurs due to the underlying transmural myocardial infarction and is seen in 5–20% of cases.[25] The delayed form (Dressler's syndrome) develops one week to several months after MI, does not require a transmural infarct and has an incidence of 0.5–5%.[89,90] Dressler's syndrome likely occurs due to viral activation and anti-myocardial antibodies and presents as fever, chest pain, leukocytosis, pleurisy, and pericardial friction rub.[91,92] Pericardial involvement correlates with larger infarct size, greater left ventricular dysfunction and greater mortality. For unexplained reasons, the incidence of Dressler's syndrome has decreased markedly.

Grossly, fibrinous pericarditis is seen overlying the area of transmural infarct in association with serous or serosanguineous effusion. Subsequently pericarditis organizes and fibrous adhesions are formed between the two pericardial layers. Finally, the parietal pericardium becomes thick and calcified in the region of the transmural infarct. Microscopically, the pericardium shows fibrin deposits, lymphocytes, macrophages, areas of fat necrosis, proliferation of capillaries, and fibroblasts.[93]

Occasionally, a pseudoaneurysm can form, from focal pericardial adhesions that localize the extravasated blood following post-infarct free wall rupture. Very rarely cardiac tamponade may occur due to the rupture of this pseudoaneurysm.[94]

Pericarditis Associated with Systemic Inflammatory Diseases

Systemic inflammatory diseases (SID) account for 2–7% cases of acute pericarditis, up to 10% cases of recurrent pericarditis and have a good prognosis.[95] Pericardial involvement in SID is associated with signs and symptoms of systemic involvement, active systemic disease and large pericardial effusions. Patients may develop mild asymptomatic pericardial effusion and rarely cardiac tamponade and constrictive pericarditis. Recurrence is seen in 20–30% of cases.[96] The usual screening tests include antinuclear antibody (ANA), rheumatoid factor, and autoantibody to Fc portion of IgG. The antibody titers of ≥1/640 are suspicious of autoimmune diseases.[95]

RHEUMATIC FEVER (RF)

Clinical pericarditis develops 7–10 days after the onset of fever and arthritis. It is more commonly seen in children and in people from the lower socioeconomic strata in association with endocarditis and myocarditis and has a poor prognosis. However, in adults, pericarditis is not associated with myocarditis or endocarditis and has a good prognosis. Pericardial effusion is commonly seen while cardiac tamponade is rare. Microscopically, fibrinous pericarditis (i.e. 'bread and butter pericarditis' due to its resemblance), foci of pericardial calcification and mild lymphocytic infiltrate can be seen.[23,97]

RHEUMATOID ARTHRITIS (RA)

Rheumatoid arthritis is a chronic systemic inflammatory disorder that may affect many tissues and organs but principally involves the joints. Pericarditis is seen in <10% of cases in patients with active rheumatoid disease and other extra-articular manifestations. Asymptomatic pericardial effusion is seen in one third of patients but myocarditis is uncommon. It usually presents as granulomatous or serofibrinous pericarditis. Patients have a positive serological test for rheumatoid factor.[98]

SYSTEMIC LUPUS ERYTHEMATOSUS (SLE)

SLE is an autoimmune multisystem disease characterized by the presence of autoantibodies, most commonly antinuclear antibodies (ANAs). Cardiovascular involvement is commonly seen in SLE and affects endocardium, myocardium, pericardium, and coronary arteries. Pericardial involvement is seen in >50% cases and presents as pericarditis, pericardial effusion, and rarely tamponade and constrictive pericarditis.[21] Pericarditis in SLE is fibrous or serofibrinous and causes partial or complete obliteration of the pericardial cavity. Microscopically pericardium shows edema, areas of vasculitis, inflammatory infiltrate, fibrinoid necrosis, hematoxylin bodies, pericardial fibrosis, and calcification. Pericarditis is seen in active disease in association with other serositis (pleural effusion and ascites). The pericardial fluid may be thick, hemorrhagic and may contain LE cells, ANA, and polymorphonuclear leukocytes. It may also have increased proteins, IL-6, TNF-α, and interferon gamma (IFN-γ) and low complement levels.[23,99–101]

SCLERODERMA

Scleroderma is a chronic inflammatory disease of unknown etiology characterized by vascular damage and collagen deposition in the skin and in multiple organs. Cardiac involvement is usually seen secondary to pulmonary hypertension and has a poor prognosis. Primary cardiac involvement manifests as arrhythmias, conduction system defects, myocarditis, myocardial fibrosis, heart failure, pericarditis, pericardial effusion, and cardiac tamponade. Pericarditis is seen in 7–20% of cases of scleroderma with an autopsy prevalence of 70–80% and is usually chronic.[102] The pericardial fluid has high proteins and specific gravity, normal glucose levels and contains few neutrophils.[23]

DERMATOMYOSITIS AND POLYMYOSITIS

These are idiopathic inflammatory myopathies that primarily affect skin and skeletal muscle. Pericardial involvement is seen in <10% of cases and presents as acute pericarditis, pericardial effusion, and cardiac tamponade.[21]

MIXED CONNECTIVE TISSUE DISORDER

These are the generalized connective tissue disorders with clinical features that are a mixture of SLE, systemic sclerosis, and polymyositis. The antibodies to U1-ribonucleoprotein are seen. Pericarditis is seen in 10–30% of cases.[103]

SJÖGREN SYNDROME

This is a chronic autoimmune connective tissue disorder that primarily affects salivary and lacrimal glands. Cardiac involvement usually manifests as pericarditis, which is seen in <30% of cases.[104]

VASCULITIS

Pericarditis is more commonly seen in medium-sized or small vessel vasculitis (Kawasaki disease, Churg, Strauss and Wegener syndromes) and rarely seen in large-vessel vasculitis.[21]

SARCOIDOSIS

This is a multisystem granulomatous disease of unknown etiology characterized by the formation of non-caseating granulomas (Figure 16.2A,B). Cardiac involvement is seen in about 25% of cases. Asymptomatic pericardial effusion is commonly seen while symptomatic pericarditis is rare.[105]

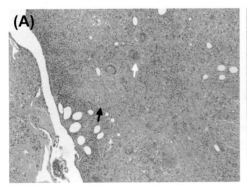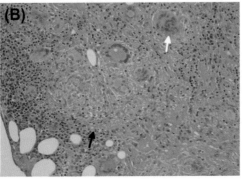

FIGURE 16.2 **(A, B)** Sections from pericardium show non-caseating granulomas (black arrow) with multinucleated giant cells from a case of pericardial sarcoidosis. (Stains: (A&B) hematoxylin and eosin; original magnification: A ×10; B ×20).

Pericarditis Associated with Endocrine and Metabolic Disorders

Uremic Pericarditis

This is seen in 6–10% of patients with acute or chronic renal failure.[106] It occurs due to the accumulation of metabolic toxins (urea, creatinine, methylguanidine, guanidinoacetate, parathyroid hormone, beta2-microglobulin, and uric acid), hemorrhagic diathesis, immune response and introduction of living agent during hemodialysis and also correlates with the degree of azotemia (blood urea nitrogen >60 mg/dL).[59,107] Grossly, fibrinous (i.e. resembling bread and butter) pericarditis is seen, along with thickened pericardium and pericardial adhesions. Microscopically, pericardium shows fibrinous exudate, inflammatory cells, and reactive mesothelial cells on the surface (Figure 16.3A–C). Large pericardial effusion is seen in up to 20% of cases of renal failure.[108] A small amount of asymptomatic pericardial effusion may be seen due to fluid overload. Hemorrhagic effusion may also develop due to platelet dysfunction and is associated with significant morbidity.

Myxedema

Pericardial effusion is seen in 5-30% of patients with hypothyroidism.[109] Patients may develop large pericardial effusion and cardiac tamponade, which resolves spontaneously once euthyroid state is restored. The effusion fluid is generally yellow, contains high proteins and cholesterol, and has high specific gravity. Chronic pericarditis is seen along with cholesterol deposits.[23]

DIABETES MELLITUS AND GOUT

A mild, non-specific type of pericarditis is seen in both of these conditions.

Pericarditis Associated with Malignant Neoplasms

Neoplastic pericarditis is 40 times more common in metastatic tumors than primary tumors of the pericardium.[59] The most common malignant tumors metastasizing to the pericardium are lung cancer, breast cancer, malignant melanoma, lymphomas, and leukemias, whereas mesothelioma is the most common primary malignant tumor of the pericardium associated with neoplastic pericarditis (Figures 16.4A–D, 16.5A,B). Metastatic tumors of the pericardium are found in 4% of autopsies and in up to 30% of autopsies performed on patients with malignant neoplasms.[110] The associated pericardial effusion may be either small or large, causing cardiac tamponade. Sometimes it can lead to recurrent episodes of pericardial effusion. Clinically, when there is a small pericardial effusion, patients are asymptomatic, however, cough, shortness of breath, and chest pain in addition to tachycardia and jugular venous distension, start to appear when pericardial fluid exceeds 500 mL. Chest radiography, CT, and MRI may reveal pleural effusion, widening of the mediastinum, or hilar masses.[59] Malignant neoplasms affecting the pericardium are likely to be associated with a serosanguineous effusion. The cytological analyses of pericardial fluid and pericardial or epicardial biopsies are important for confirming the diagnosis of malignant pericardial disease.

In neoplastic pericardial effusions, pericardiocentesis is performed to relieve symptoms and to establish a diagnosis. In addition, pericardial drainage is performed in patients with large effusions due to the possibility of a high recurrence rate (40–70%).[111,112] Local treatment by intrapericardial instillation of sclerosing or cytotoxic agents is performed to prevent recurrence. The type of agents used in the intrapericardial treatment usually depends on the type of the tumor affecting the pericardium. For example, the platinum-containing anti-cancer drug cisplatin is most effective in secondary lung cancer and an intrapericardial instillation of thioTEPA is more effective in pericardial metastases associated with breast cancer.[113,114] Systemic baseline antineoplastic treatment can prevent recurrences in up to 67% of cases.[115]

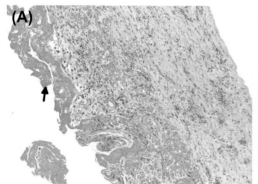 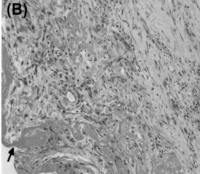 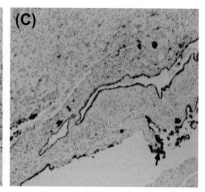

FIGURE 16.3 (A, B) Sections from a case of uremic pericarditis show thickened pericardium with fibrinous adhesions (black arrow) and chronic inflammatory cells. (C) Calretinin-positive mesothelial cells. (Stains: (A&B) hematoxylin and eosin; original magnification: A x10; B&C x20).

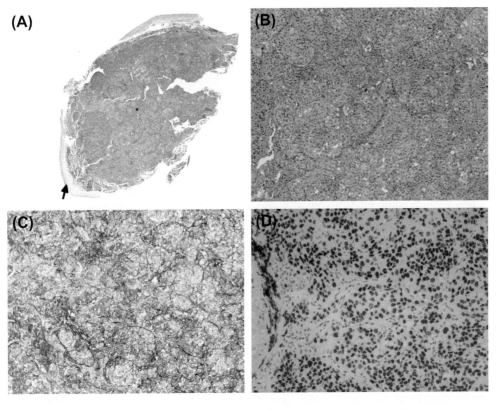

FIGURE 16.4 (A, B) Sections from pericardium (black arrow) show metastatic lung carcinoma (asterisks). (C) EGFR (epidermal growth factor receptor) positivity in the lung cancer. (D) TTF (thyroid transcription factor1) positivity in the tumor cells. (Stains: (A&B) hematoxylin and eosin; original magnification: A x2.5; B&C x10; D x20).

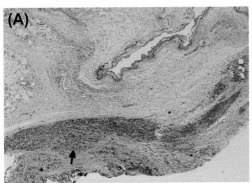
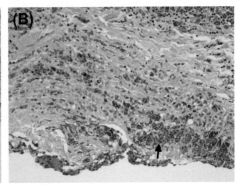

FIGURE 16.5 (A, B) Sections from pericardium (asterisks) show metastatic malignant melanoma (black arrow). (Stains: (A&B) hematoxylin and eosin; original magnification: A x5; B x10).

Iatrogenic and Traumatic Pericarditis

Post-Cardiac Injury Syndrome (Post-Pericardiotomy Syndrome)

The pathogenesis of post-cardiac injury/post-pericardiotomy syndrome is similar to Dressler's syndrome; both appear to be variants of a common immune pathological process. However, post-cardiac injury syndrome has a greater anti-heart antibody response (anti-sarcolemmal and anti-fibrillary). This could be related to the more extensive release of antigenic material.[116,117] It develops within days to months after an injury to the heart or the pericardium.[116] It is more commonly seen in patients who have undergone surgery that involves opening the pericardium such as implantation of epicardial pacemaker leads and transvenous

pacemaker leads, valve surgery, coronary artery bypass grafting (CABG), coronary stent implantation, and orthotopic heart transplantation. It is more commonly encountered after valve surgery than coronary artery bypass grafting (CABG) alone. This could be related to the preoperative use of anticoagulants.[118]

The syndrome is characterized by pericardial or pleuritic pain, effusion, friction rubs, pneumonitis, and cardiac tamponade. Pericardial effusion is seen in 21% of cases.[119] It can lead to constrictive pericarditis when warfarin is given to the patients with early postoperative pericardial effusion, following cardiac surgery. This may also increase the risk of developing pericarditis, especially in patients who did not have a pericardiocentesis and drainage of the effusion.[120]

The prevention of the post-pericardiotomy syndrome is still under investigation and it is achieved by using short-term perioperative steroid treatment or colchicine.[121] Treatment includes NSAIDs or colchicine for several weeks or months, even after disappearance of the effusion.[122] In refractory forms, oral corticosteroids for 3–6 months or pericardiocentesis and intra-pericardial instillation of triamcinolone (300 mg/m²) are good therapeutic options. Pericardiectomy is rarely indicated.

Traumatic Pericarditis

Pericardial injury can be induced by different mechanisms.[123,124] Procedure-related mortality was reported in only 0.05% in a worldwide study of more than 6000 cases and in none of the 2537 patients from the registry of an experienced reference center.[59] In addition, pacemaker leads may cause injury to the right ventricle and the epicardial electrodes may cause pericarditis with tamponade, adhesions, or constriction. Blunt chest trauma is another major risk factor for traumatic pericarditis. In this injury, the deceleration force may lead to contusion of the myocardium, with intrapericardial hemorrhage, cardiac rupture, or herniation. Pericardial laceration and partial extrusion of the heart into the mediastinum and pleural space may also occur after injury.[125] Pericardial injury may lead to pericarditis due to loss of mesothelial cells from the pericardial surfaces leading to pericardial effusion rich in fibrin.[126] As the healing process starts, the fibrin is gradually replaced by fibrous adhesions. However, the parietal pericardium is rarely thickened. Non-penetrating and penetrating chest trauma may lead to fibrinous reactions, whereas the presence of blood in the pericardial space may induce fibrous adhesions leading to pericardial thickening. Other early complications of chest trauma include rupture of the pericardium, hemopericardium, and acute tamponade.[91] Late complications include post-pericardiotomy syndrome, constrictive pericarditis, and cardiac herniation. Transesophageal echocardiography in the emergency room[127] or immediate computed tomography should be performed in traumatic injury to the pericardium.

Radiation-Induced Pericarditis

The development of radiation-induced pericarditis depends on the patient's age, dose of radiation, distance from the radiation source, duration of exposure to radiation and the body surface area exposed to radiation.[128] Radiation-induced pericarditis may occur within a short period after the radiation therapy of thoracic neoplasms or months to years later, following the radiation therapy (up to 15–20 years).

Patients may develop pericardial effusion and cardiac tamponade due to radiation-induced damage to endothelial cells of the lymphatics and capillaries. Radiation exposure may also cause pericarditis and damage to the epicardial coronary vessels, leading to atherosclerosis and myocardial infarction. Symptoms may be confused with the associated disease or the side effects of radiation and chemotherapy. Investigations should start with echocardiography, followed by cardiac CT or MRI if necessary. Grossly, serous or haemorrhagic pericardial effusion may be seen. Later, it may cause thickening of the two pericardial layers along with pericardial fibrinous adhesions, constriction, and fibrosis in the absence of tissue calcification. Histologically, two days post-radiation injury, the pericardium may show mononuclear cell infiltrate. Other common findings seen in radiation-exposed hearts are bizarre fibroblasts with abnormal nucleus and cytoplasm, sclerosis of small vessels, and pericardial fat necrosis.

Radiation-induced pericarditis (without tamponade) can be treated conservatively or by pericardiocentesis too, if other causes of pericarditis are ruled out. In 20% of patients, pericardial constriction may occur, which would indicate the need for a pericardiectomy. The five-year survival rate is low (1%) due to myocardial fibrosis and the operative mortality is high (up to 21%).

Drug-induced Pericarditis

Several drugs and toxic substances can lead to pericarditis, cardiac tamponade, adhesions, fibrosis, and constriction.[59] Procainamide and hydralazine can cause pericarditis through mechanisms that include drug-induced lupus reactions, whereas phenylbutazone, methysergide, minoxidil, cytarabine, and romocriptine lead to pericarditis through the mechanism of idiosyncratic reaction. Talc (Mg silicate) induces pericarditis through the mechanism of foreign substance reactions. Management mainly includes discontinuation of the causative agent and symptomatic treatment.

Constrictive Pericarditis

This is a relatively rare but life-threatening condition, resulting from chronic inflammation of the pericardium with thickening, fibrosis and sometimes calcification of both layers of the pericardium, which become adherent to the heart muscle, restricting diastolic filling of the ventricles and causing reduced ventricular function.[129]

Among the common causes of constrictive pericarditis are tuberculosis, chest radiation, and frequent surgical procedures.[130] In the past, tuberculosis was the most common cause. Since its incidence has declined, the common underlying cause of constriction is idiopathic.

Grossly, in constrictive pericarditis, the pericardial space is usually obliterated and the pericardial layers become thickened and develop focal or diffuse fibrous adhesions, with variable degrees of calcification (Figure 16.6A,B).

The clinical features are characterized by fatigue, shortness of breath, peripheral edema, and abdominal distension. More severe forms of the disease may also be associated with hepatomegaly, pleural effusions, ascites,

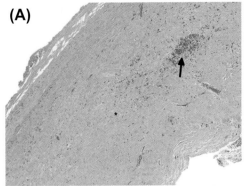
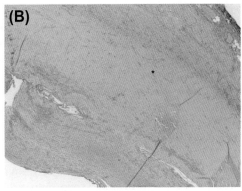

FIGURE 16.6 **(A, B)** Sections from constrictive pericarditis show thick and fibrotic pericardium (asterisks) with a focus of chronic inflammatory cells (black arrow). (Stains: (A) hematoxylin and eosin; (B) movat pentachrome; original magnification: A x10; B x12.5).

and hepatomegaly. These symptoms may be further aggravated by systolic dysfunction due to myocardial fibrosis and atrophy. The differential diagnosis includes acute dilatation of the heart, pulmonary embolism, right ventricular infarction, pleural effusion, chronic obstructive lung diseases, and restrictive cardiomyopathy.

Pericardiectomy is the only treatment for permanent constriction. If severe calcified adhesions are present between the pericardium and epicardium ('outer porcelain heart'), surgery may be high risk due to myocardial damage. In such situations, 'laser shaving' using an Excimer laser may be considered.[131] Some studies suggest that pericardiectomy has a mortality of 6–12%.[131,132]

Pericardial Tumors

Pericardial tumors can be benign or malignant. Of the benign pericardial tumors, pericardial cysts are the most common. Other benign pericardial tumors include angiomas, lymphangiomas, fibromas, teratomas, and lipomas. Mesothelial cardiac excrescences and mesothelial papillomas are among the benign tumors of the pericardium. They are small collections of mesothelial cells mixed with fat cells, macrophages with no intervening stroma. Histologically, mesothelial papillomas appear as a cuboidal epithelioid cell arising from the pericardial surface. It is also called an adenomatoid tumor. It is usually an incidental finding at autopsy.[133,134] Surgical excision of a benign pericardial tumor is usually curative. Metastatic pericardial tumors occur 20–40 times more commonly than the primary pericardial tumors.

Solitary Fibrous Tumor

Solitary fibrous tumors (SFTs) are rare neoplasms that most commonly involve the pleura, mediastinum, and lung. These tumors are believed to be submesothelial in origin.[135] They most often involve the pleura and mediastinum, but can be intrapulmonary, peritoneal, or involve other more uncommon sites such as the pericardium.[136] SFTs have been increasingly recognized in

several other locations outside the thoracic cavity, especially the soft tissues.[137] In 1996, the first SFT of the spinal cord was described by Carneiro et al.[137] In addition, SFTs have been reported in paracranial sites such as the orbit, meninges, and paranasal sinuses.[138] SFTs have been regarded as a form of mesothelioma; however, they are not associated with asbestosis[138] and are rarely associated with malignant pleural effusions.

SFTs of the pleura are usually asymptomatic, however, some patients may present with pain, cough, dyspnea, pulmonary osteoarthropathy, and hypoglycaemia.[139] SFTs of the pericardium are characterized by their mass effects on the pericardium. Chest radiographic findings of a SFT show a variable-size mass that is usually well defined and discovered accidentally during routine radiologic examinations.

Grossly, the tumor is usually well-circumscribed, firm, fleshy or white. However, diffuse involvement of the mesothelial surfaces has also been described. SFTs have a variety of histologic arrangements varying from a 'patternless pattern' to a hemangiopericytoma-like pattern or a diffuse sclerosing appearance.[135] SFTs in general show a predominant population of monotonous, bland-looking ovoid to spindle-shaped cells in a sclerotic stroma. The cells are characteristically arranged in short whorls or swirling fascicles. A prominent vascular pattern with a hemangiopericytoma-like appearance can sometimes be noted. A common pattern is that of alternating hypo- and hypercellular areas with a notable condensation of neoplastic cells around blood vessels. Occasionally, the tumors show an extremely hypocellular morphology traversed by dense collagen bands. Nuclear pleomorphism and relatively high mitotic figures can be noted. Hemorrhage and necrosis can occasionally be present in a SFT.[138] The differential diagnosis includes other monomorphic spindle cell tumors, including neurogenic tumors, spindle cell mesotheliomas, monophasic synovial sarcomas, and fibrosarcomas.[140] Recently, a desmoid tumor of the pleura has been added in the list of differential diagnostic considerations.[141] CD-34 may prove to be a useful positive marker to differentiate these neoplasms from mesotheliomas, synovial sarcomas, and fibrosarcomas, which are generally negative with this antibody[136] but these spindle cell tumors can usually be identified with an immunohistochemical panel using specific markers such as actin, desmin, and S-100 protein. Hemangiopericytomas do not express any of the markers that would allow distinction from SFTs and show positivity for CD-34 in some cases, which raise the question of a possible relationship between these two tumors.[136]

The single best prognostic factor appears to be encapsulation or pedunculation of the tumor and tumor resectability.[139] The prognosis is generally good. Recurrence and local spread have been reported. Criteria for malignancy of pleural tumors include necrosis and a mitotic count of greater than 4 per 10 high-powered fields, but the applicability of these criteria to tumors in the heart and pericardium is unknown.

Germ Cell Tumors

Germ cell tumors are the gonadal tumors derived from germ cells, which migrate from the yolk sac to the gonads during embryogenesis and deposit in the midline structures – mediastinum and nervous system. These are classified further as seminoma (dysgerminoma), embryonal carcinoma, yolk sac tumor, choriocarcinoma, and teratoma. Mediastinal germ cell tumors are very rare with approximately 100 cases reported in the literature so far. The most common of these are pericardial teratomas followed by yolk sac tumors.[142]

The normal age of presentation varies from intrauterine period (second and third trimester) to the late sixties.[143] Teratomas are commonly seen in infants and young children and show a slight female predominance, while malignant germ cell tumors are seen in adults.[144]

The diagnosis of teratoma in the prenatal period can be made as early as 20 weeks of gestation by transabdominal ultrasound, which shows pericardial effusion, cardiac tamponade, and non-immune hydrops.[145] In the post-natal period, echocardiography, computed tomography (CT) and magnetic resonance imaging (MRI) help in making the diagnosis and to identify tumor burden and the extent of organ involvement.[146] Patients with a teratoma have elevated serum levels of pro-inflammatory cytokines (interleukins-1, 6, 8) and alpha-fetoproteins.[147]

Grossly, the tumor can measure up to 15 cm. It can be located in the right or left side of the heart and displace the organs to the opposite side or be attached by a pedicle to the great vessels. Externally, the mass has a smooth and lobated surface. The cut surface shows solid and cystic areas and has a variegated appearance. The cysts are filled with gelatinous mucoid material and may sometimes contain hair and teeth. The diagnosis of teratoma is made histologically by the demonstration of structures derived from all three of the germ cell layers.[145]

The treatment of choice is surgical resection with careful dissection to prevent the risk of massive hemorrhage as teratomas get their blood supply from the aorta. In the prenatal period, transabdominal pericardiocentesis and open heart surgery are recommended in the second trimester and have a poor prognosis. In the third trimester, cesarean section is recommended followed by surgical excision soon after birth.[148] The tumor shows no recurrence so surgical excision is associated with excellent prognosis.

Malignant Mesothelioma

The definition of primary pericardial mesothelioma stipulates that no tumor is present outside the pericardium, with the exception of lymph node metastases. Pericardial mesothelioma is the third most common primary malignant tumor of the pericardium and heart. Nevertheless, it is considered a rare tumor.[149,150] It forms 0.7% of all malignant mesotheliomas. The majority of malignant mesotheliomas arise from the pleural and peritoneal surfaces. Similar to the cases of pleural mesothelioma, there is an increased association between the frequency of the tumor and previous exposure to asbestos. Pericardial mesotheliomas have been reported decades after exposure to pericardial dusting with asbestos and fiberglass as a treatment for angina pectoris. In addition, therapeutic radiation for breast cancer and mediastinal lymphoma has also been implicated in some situations. However, some patients with mesothelioma still have neither a previous history of exposure to asbestos nor radiation.

Grossly, the tumor usually forms grey-white nodules, variable in size and firm in consistency, usually on the surface of the pericardium and filling the pericardial cavity. The pericardial fluid associated with the tumor is usually serosanguineous. Mesothelioma rarely infiltrates the pericardium, however, involvement of the tricuspid valve has also been reported.

Histologically malignant mesotheliomas of the pericardium resemble pleural mesotheliomas. Microscopically, there are three types of malignant mesothelioma: (1) epithelioid; (2) sarcomatoid; and (3) biphasic (mixed). The epithelioid type is the most common type that accounts for about 50–60% of the malignant mesothelioma cases. In this, tumor cells are arranged in the form of tubules, papillary structures, and cords of infiltrating cells that can incite a desmoplastic response. It has a better prognosis than the sarcomatoid or biphasic subtypes.[151] Variants similar to those described in the pleura may be seen in the pericardium, e.g. microcystic, adenomatoid, deciduoid. The distinction between mesothelioma and pleural-based lung adenocarcinoma can be quite difficult, and is generally based on immunohistochemical findings. Malignant mesothelioma must be differentiated from metastatic adenocarcinoma. Intra-cytoplasmic periodic acid-Schiff (PAS)-positive vacuoles are rare in mesothelioma, whereas they are found in more than two-thirds of adenocarcinomas. Carcino-embryonic antigen (CEA) is positive in most adenocarcinomas; however, it is positive in a very small number of mesotheliomas. Leu M1 is positive in almost all cases of lung adenocarcinoma, whereas it is only positive in a small number of cases of mesothelioma. The prognosis of malignant mesothelioma is generally poor; patients usually survive for six months, however, in exceptional cases patients may survive for two years.

Pericardial Angiosarcoma

This involves the pericardium diffusely and produces a thick grey-black pericardium. Histologically the tumor shows endothelium-lined vascular channels, which are CD31-positive (Figure 16.7A–C).

Metastatic Tumors

Metastatic tumors to the heart are more common than primary and have an incidence of 2.3–18.3%.[152] The metastasis primarily involves the pericardium followed by myocardium, epicardium, and endocardium and can spread directly from the mediastinum, or through hematogenous, lymphatic route or intracavitary spread through IVC.[153] The metastatic pericardial tumors arise from primary tumors of lung, hematopoietic, gastrointestinal tract, breast, genitourinary system, and also from sarcomas (Figure 16.8A–E).[115,143] Patients with pericardial metastasis typically present with dyspnea, hypotension, tachycardia due to acute

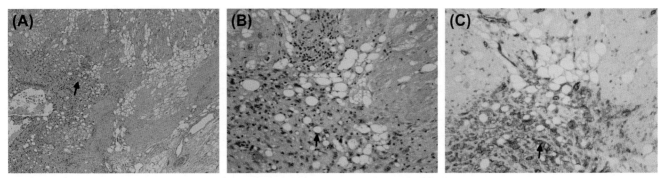

FIGURE 16.7 (A, B) Sections from primary angiosarcoma (black arrow) in the pericardium. (C) CD31 positivity in the tumor cells. (Stains: (A&B) hematoxylin and eosin; original magnification: A x5; B&C x20).

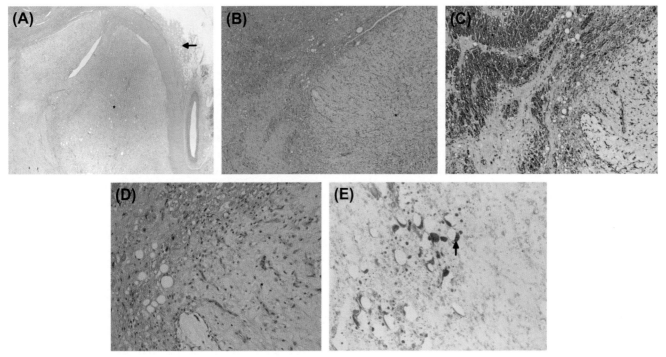

FIGURE 16.8 Sections show metastatic myxoid liposarcoma (asterisks) in the pericardium (black arrow). (Stains: (A, B, D) hematoxylin and eosin; (C) movat pentachrome; (E) oil red O; original magnification: A x1.2; B x2.5; C&E x10; D x20).

pericarditis, hemorrhagic or serosanguineous pericardial effusion, and cardiac tamponade.[115]

Repeated chest X-ray, ECG, CT scan, echocardiography, MRI, and cytological examination of pericardial fluid and pericardial biopsy helps in the diagnosis.[154] Breast carcinomas usually manifest as pericardial disease only after the primary tumor is known. However, in other cases when the primary neoplasm is not known, it becomes difficult to identify the primary site of the tumor. The examination of pericardial fluid, pericardial biopsy, and immunohistochemistry helps in the diagnosis of primary tumor.

Metastatic pericardial tumors or malignant pericardial effusion represent an advanced stage of the disease and have a poor prognosis.[155] The treatment is usually palliative and involves establishment of a pericardial window to drain the accumulating fluid, pericardiectomy, radiation therapy, and use of sclerosing agents such as tetracycline, cisplatin, bleomycin, and 5-fluorouracil to improve cardiac output and prevent cardiac tamponade.[156]

References

1. Little WC, Freeman GL. Pericardial disease. *Circulation* 2006; **113**(12):1622–32.
2. Shabetai R. Corticosteroids for recurrent pericarditis: On the road to evidence-based medicine. *Circulation* 2008;**118**(6):612–3.
3. Troughton RW, Asher CR, Klein AL. Pericarditis. *Lancet* 2004; **363**(9410):717–27.
4. Imazio M, Cecchi E, Demichelis B, et al. Myopericarditis versus viral or idiopathic acute pericarditis. *Heart* 2008;**94**(4):498–501.
5. Zayas R, Anguita M, Torres F, et al. Incidence of specific etiology and role of methods for specific etiologic diagnosis of primary acute pericarditis. *Am J Cardiol* 1995;**75**(5):378–82.
6. Permanyer-Miralda G, Sagrista-Sauleda J, Soler-Soler J. Primary acute pericardial disease: A prospective series of 231 consecutive patients. *Am J Cardiol* 1985;**56**(10):623–30.
7. Imazio M, Spodick DH, Brucato A, Trinchero R, Adler Y. Controversial issues in the management of pericardial diseases. *Circulation* 2010;**121**(7):916–28.
8. Permanyer-Miralda G. Acute pericardial disease: Approach to the aetiologic diagnosis. *Heart* 2004;**90**(3):252–4.
9. Imazio M, Trinchero R. Triage and management of acute pericarditis. *Int J Cardiol* 2007;**118**(3):286–94.
10. Lange RA, Hillis LD. Clinical practice. acute pericarditis. *N Engl J Med* 2004;**351**(21):2195–202.
11. Spodick DH. Acute pericarditis: Current concepts and practice. *JAMA* 2003;**289**(9):1150–3.
12. Wang K, Asinger RW, Marriott HJ. ST-segment elevation in conditions other than acute myocardial infarction. *N Engl J Med* 2003;**349**(22):2128–35.
13. Spodick DH. Diagnostic electrocardiographic sequences in acute pericarditis. Significance of PR segment and PR vector changes. *Circulation* 1973;**48**(3):575–80.
14. Ginzton LE, Laks MM. The differential diagnosis of acute pericarditis from the normal variant: New electrocardiographic criteria. *Circulation* 1982;**65**(5):1004–9.
15. Bonnefoy E, Godon P, Kirkorian G, Fatemi M, Chevalier P, Touboul P. Serum cardiac troponin I and ST-segment elevation in patients with acute pericarditis. *Eur Heart J* 2000;**21**(10):832–6.
16. Imazio M, Demichelis B, Cecchi E, et al. Cardiac troponin I in acute pericarditis. *J Am Coll Cardiol* 2003;**42**(12):2144–8.
17. Cheitlin MD, Armstrong WF, Aurigemma GP, et al. ACC/AHA/ASE 2003 guideline update for the clinical application of echocardiography: Summary article: A report of the American College of Cardiology/American Heart Association task force on practice guidelines (ACC/AHA/ASE committee to update the 1997 guidelines for the clinical application of echocardiography). *Circulation* 2003;**108**(9):1146–62.
18. Aggeli C, Pitsavos C, Brili S, et al. Relevance of adenosine deaminase and lysozyme measurements in the diagnosis of tuberculous pericarditis. *Cardiology* 2000;**94**(2):81–5.
19. Burgess LJ, Reuter H, Carstens ME, Taljaard JJ, Doubell AF. The use of adenosine deaminase and interferon-gamma as diagnostic tools for tuberculous pericarditis. *Chest* 2002;**122**(3):900–5.
20. Cegielski JP, Devlin BH, Morris AJ, et al. Comparison of PCR, culture, and histopathology for diagnosis of tuberculous pericarditis. *J Clin Microbiol* 1997;**35**(12):3254–7.
21. Imazio M. Pericardial involvement in systemic inflammatory diseases. *Heart* 2011;**97**(22):1882–92.
22. Scherl ND. Acute nonspecific pericarditis: A survey of the literature and a study of 30 additional cases. *J Mt Sinai Hosp N Y* 1956;**23**(3):293–305.
23. Butany J, Woo A. The pericardium and its disease. In: Silver M, Gotlieb A, Schoen F, editors. *Cardiovascular pathology*. 3rd ed ; 2001. p. 375.
24. Imazio M, Brucato A, Adler Y, et al. Prognosis of idiopathic recurrent pericarditis as determined from previously published reports. *Am J Cardiol* 2007;**100**(6):1026–8.
25. Khandaker MH, Espinosa RE, Nishimura RA, et al. Pericardial disease: Diagnosis and management. *Mayo Clin Proc* 2010;**85**(6):572–93.
26. Campbell PT, Li JS, Wall TC, et al. Cytomegalovirus pericarditis: A case series and review of the literature. *Am J Med Sci* 1995;**309**(4):229–34.
27. Saatci U, Ozen S, Ceyhan M, Secmeer G. Cytomegalovirus disease in a renal transplant recipient manifesting with pericarditis. *Int Urol Nephrol* 1993;**25**(6):617–9.
28. Friman G, Fohlman J. The epidemiology of viral heart disease. *Scand J Infect Dis Suppl* 1993;**88**:7–10.
29. Tsui CY, Burch GE. Coxsackie virus B4 pericarditis in mice. *Br J Exp Pathol* 1971;**52**(1):47–50.
30. Ristic AD, Pankuweit S, Maksimovic R, Moosdorf R, Maisch B. Pericardial cytokines in neoplastic, autoreactive, and viral pericarditis. *Heart Fail Rev* 2013;**18**(3):345–53.
31. Anderson DW, Virmani R. Emerging patterns of heart disease in human immunodeficiency virus infection. *Hum Pathol* 1990;**21**(3):253–9.
32. De Castro S, Migliau G, Silvestri A, et al. Heart involvement in AIDS: A prospective study during various stages of the disease. *Eur Heart J* 1992;**13**(11):1452–9.
33. Silva-Cardoso J, Moura B, Martins L, Mota-Miranda A, Rocha-Goncalves F, Lecour H. Pericardial involvement in human immunodeficiency virus infection. *Chest* 1999;**115**(2):418–22.
34. Chen Y, Brennessel D, Walters J, Johnson M, Rosner F, Raza M. Human immunodeficiency virus-associated pericardial effusion: Report of 40 cases and review of the literature. *Am Heart J* 1999;**137**(3):516–21.
35. Anderson JF, Johnson RC, Magnarelli LA, Hyde FW. Identification of endemic foci of lyme disease: Isolation of borrelia burgdorferi from feral rodents and ticks (dermacentor variabilis). *J Clin Microbiol* 1985;**22**(1):36–8.
36. Keersmaekers T, Elshot SR, Sergeant PT. Primary bacterial pericarditis. *Acta Cardiol* 2002;**57**(5):387–9.
37. Raviglione MC, Mariuz P, Pablos-Mendez A, Battan R, Ottuso P, Taranta A. High staphylococcus aureus nasal carriage rate in patients with acquired immunodeficiency syndrome or AIDS-related complex. *Am J Infect Control* 1990;**18**(2):64–9.

38. Fisher BK, Warner LC. Cutaneous manifestations of the acquired immunodeficiency syndrome. Update 1987. *Int J Dermatol* 1987;**26**(10):615–30.

39. Schlossberg D, Zacarias F, Shulman JA. Primary pneumococcal pericarditis. *JAMA* 1975;**234**(8):853.

40. Morse JR, Oretsky MI, Hudson JA. Pericarditis as a complication of meningococcal meningitis. *Ann Intern Med* 1971;**74**(2):212–7.

41. Beal LR, Ustach TJ, Forker AD. Meningococcemia without meningitis presenting as cardiac tamponade. survival with disseminated intravascular coagulation. *Am J Med* 1971;**51**(5):659–62.

42. Pierce HI, Cooper EB. Meningococcal pericarditis: clinical features and therapy in five patients. *Arch Intern Med* 1972;**129**(6):918–22.

43. Cheatham Jr JE, Grantham RN, Peyton MD, et al. Hemophilus influenzae purulent pericarditis in children: Diagnostic and therapeutic considerations. *J Thorac Cardiovasc Surg* 1980;**79**(6):933–6.

44. Feinstein V, Musher D, Young E. Purulent pericarditis in a patient with legionnaires disease. *Arch Intern Med* 1982;**142**:1234.

45. Lieber C, Rensimer E, Ericson C. Campylobacter pericarditis and myocardial abscess. *Am Heart J* 1982;**92**:18.

46. Tice AD, Nelson JS, Visconti EB. Listeria monocytogenes pericarditis and myocardial abscess. *R I Med J* 1979;**62**(4):135–8.

47. Feldman WE. Bacterial etiology and mortality of purulent pericarditis in pediatric patients. Review of 162 cases. *Am J Dis Child* 1979;**133**(6):641–4.

48. Hornick P, Harris P, Smith P. Nocardia asteroides purulent pericarditis. *Eur J Cardiothorac Surg* 1995;**9**(8):468–70.

49. Sutton GC, Morrissey RA, Tobin Jr JR, Anderson TO. Pericardial and myocardial disease associated with serological evidence of infection by agents of the psittacosis-lymphogranuloma venereum group (chamydiaceae). *Circulation* 1967;**36**(6):830–8.

50. Goodman LJ. Purulent pericarditis. *Curr Treat Options Cardiovasc Med* 2000;**2**(4):343–50.

51. Maisch B, Outzen H, Roth D, et al. Prognostic determinants in conventionally treated myocarditis and perimyocarditis–focus on antimyolemmal antibodies. *Eur Heart J* 1991;**12**(Suppl D):81–7.

52. Kinoshita Y, Shimada T, Murakami Y, et al. Ethanol sclerosis can be a safe and useful treatment for pericardial cyst. *Clin Cardiol* 1996;**19**(10):833–5.

53. Shabetai R. Acute pericarditis. *Cardiol Clin* 1990;**8**(4):639–44.

54. Rubin RH, Moellering Jr RC. Clinical, microbiologic and therapeutic aspects of purulent pericarditis. *Am J Med* 1975;**59**(1):68–78.

55. Suzman S. Tuberculous pericardial effusion. *Br Heart J* 1943;**5**(1):19–23.

56. Fowler NO. Tuberculous pericarditis. *JAMA* 1991;**266**(1):99–103.

57. Butany J, Silver M. Cardiovascular findings in AIDS: A report of 45 cases. *Lab Invest* 1989;**62**:13A.

58. Peel AA. Tuberculous pericarditis. *Br Heart J* 1948;**10**(3):195–207.

59. Maisch B, Seferovic PM, Ristic AD, et al. Guidelines on the diagnosis and management of pericardial diseases executive summary; the task force on the diagnosis and management of pericardial diseases of the European Society of Cardiology. *Eur Heart J* 2004;**25**(7):587–610.

60. Strang JI. Tuberculous pericarditis in Transkei. *Clin Cardiol* 1984;**7**(12):667–70.

61. Strang JI, Kakaza HH, Gibson DG, et al. Controlled clinical trial of complete open surgical drainage and of prednisolone in treatment of tuberculous pericardial effusion in Transkei. *Lancet* 1988;**2**(8614):759–64.

62. Sagrista-Sauleda J, Permanyer-Miralda G, Soler-Soler J. Tuberculous pericarditis: Ten year experience with a prospective protocol for diagnosis and treatment. *J Am Coll Cardiol* 1988;**11**(4):724–8.

63. Ewer K, Deeks J, Alvarez L, et al. Comparison of T-cell-based assay with tuberculin skin test for diagnosis of mycobacterium tuberculosis infection in a school tuberculosis outbreak. *Lancet* 2003;**361**(9364):1168–73.

64. Godfrey-Faussett P. Molecular diagnosis of tuberculosis: The need for new diagnostic tools. *Thorax* 1995;**50**(7):709–11.

65. Seino Y, Ikeda U, Kawaguchi K, et al. Tuberculous pericarditis presumably diagnosed by polymerase chain reaction analysis. *Am Heart J* 1993;**126**(1):249–51.

66. Nugue O, Millaire A, Porte H, et al. Pericardioscopy in the etiologic diagnosis of pericardial effusion in 141 consecutive patients. *Circulation* 1996;**94**(7):1635–41.

67. Mayosi BM, Ntsekhe M, Volmink JA, Commerford PJ. Interventions for treating tuberculous pericarditis. *Cochrane Database Syst Rev* 2002;(4):CD000526.

68. Strang JI. Rapid resolution of tuberculous pericardial effusions with high dose prednisone and anti-tuberculous drugs. *J Infect* 1994;**28**(3):251–4.

69. Strang JI, Kakaza HH, Gibson DG, Girling DJ, Nunn AJ, Fox W. Controlled trial of prednisolone as adjuvant in treatment of tuberculous constrictive pericarditis in Transkei. *Lancet* 1987;**2**(8573):1418–22.

70. Hageman JH, Esopo ND, Glenn WW. Tuberculosis of the pericardium. a long-term analysis of forty-four proved cases. *N Engl J Med* 1964;**270**:327–32.

71. Cishek MB, Yost B, Schaefer S. Cardiac aspergillosis presenting as myocardial infarction. *Clin Cardiol* 1996;**19**(10):824–7.

72. Rabinovici R, Szewczyk D, Ovadia P, Greenspan JR, Sivalingam JJ. Candida pericarditis: Clinical profile and treatment. *Ann Thorac Surg* 1997;**63**(4):1200–4.

73. Canver CC, Patel AK, Kosolcharoen P, Voytovich MC. Fungal purulent constrictive pericarditis in a heart transplant patient. *Ann Thorac Surg* 1998;**65**(6):1792–4.

74. Wheat J. Histoplasmosis. Experience during outbreaks in Indianapolis and review of the literature. *Medicine (Baltimore)* 1997;**76**(5):339–54.

75. Picardi JL, Kauffman CA, Schwarz J, Holmes JC, Phair JP, Fowler NO. Pericarditis caused by histoplasma capsulatum. *Am J Cardiol* 1976;**37**(1):82–8.

76. Kraus WE, Valenstein PN, Corey GR. Purulent pericarditis caused by candida: Report of three cases and identification of high-risk populations as an aid to early diagnosis. *Rev Infect Dis* 1988;**10**(1):34–41.

77. Kaufman LD, Seifert FC, Eilbott DJ, Zuna RE, Steigbigel RT, Kaplan AP. Candida pericarditis and tamponade in a patient with systemic lupus erythematosus. *Arch Intern Med* 1988;**148**(3):715–7.

78. Adams EB, MacLeod IN. Invasive amebiasis. II. amebic liver abscess and its complications. *Medicine (Baltimore)* 1977;**56**(4):325–34.

79. Li E, Stanley Jr SL. Protozoa. amebiasis. *Gastroenterol Clin North Am* 1996;**25**(3):471–92.

80. Shamsuzzaman SM, Hashiguchi Y. Thoracic amebiasis. *Clin Chest Med* 2002;**23**(2):479–92.

81. Mondragon-Sanchez R, Cortes-Espinoza T, Sanchez-Cisneros R, Parra-Silva H, Hurtado-Andrade H. Rupture of an amebic liver abscess into the pericardium. Presentation of a case and review of current management. *Hepatogastroenterology* 1994;**41**(6):585–8.

82. Sahasrabudhe NS, Jadhav MV, Deshmukh SD, Holla VV. Pathology of toxoplasma myocarditis in acquired immunodeficiency syndrome. *Indian J Pathol Microbiol* 2003;**46**(4):649–51.

83. Politi Okoshi M, Rubens Montenegro M. Pathology of the heart in AIDS. study of 73 consecutive necropsies. *Arq Bras Cardiol* 1996;**66**(3):129–33.

84. Rena O, Garavoglia M, Francini M, Bellora P, Oliaro A, Casadio C. Solitary pericardial hydatid cyst. *J Cardiovasc Surg (Torino)* 2004;**45**(1):77–80.

85. Bouraoui H, Trimeche B, Mahdhaoui A, et al. Echinococcosis of the heart: Clinical and echocardiographic features in 12 patients. *Acta Cardiol* 2005;**60**(1):39–41.

86. Rocha A, de Meneses AC, da Silva AM, et al. Pathology of patients with Chagas' disease and acquired immunodeficiency syndrome. *Am J Trop Med Hyg* 1994;**50**(3):261–8.

87. Denie C, Vachiery F, Elman A, et al. Systemic and splanchnic hemodynamic changes in patients with hepatic schistosomiasis. *Liver* 1996;**16**(5):309–12.

88. Mangoud AM, Morsy TA, Ramadan ME, Makled KM, Mostafa SM. The pathology of the heart and lung in Syrian golden hamsters experimentally infected with leishmania d. infantum on top of pre-existing schistosoma mansoni infection. *J Egypt Soc Parasitol* 1998;**28**(2):395–402.

89. Lichstein E. The changing spectrum of post-myocardial infarction pericarditis. *Int J Cardiol* 1983;**4**(2):234–7.

90. Spodick D. Postmyocardial infarction syndrome (Dressler's Syndrome). *ACC Curr J Rev* 1995;**4**:35–7.

91. Kossowsky WA, Lyon AF, Spain DM. Reappraisal of the postmyocardial infarction Dressler's Syndrome. *Am Heart J* 1981;**102**(5):954–6.

92. Dressler W. The post myocardial infarction syndrome. *JAMA* 1970;**160**:1379.

93. Charlap S, Greenberg S, Greengart A, et al. Pericardial effusion early in acute myocardial infarction. *Clin Cardiol* 1989;**12**(5):252–4.

94. Ersek RA, Chesler E, Korns ME, Edwards JE. Spontaneous rupture of a false left ventricular aneurysm following myocardial infarction. *Am Heart J* 1969;**77**(5):677–80.

95. Imazio M, Brucato A, Doria A, et al. Antinuclear antibodies in recurrent idiopathic pericarditis: Prevalence and clinical significance. *Int J Cardiol* 2009;**136**(3):289–93.

96. Knockaert DC. Cardiac involvement in systemic inflammatory diseases. *Eur Heart J* 2007;**28**(15):1797–804.

97. Spodick D. Pericardial disease in the vasculitis- connective tissue disease group. In: Spodick D, editor. *The pericardium. A comprehensive text book*. New York: Marcel Dekker; 1997. p. 314–33.

98. Voskuyl AE. The heart and cardiovascular manifestations in rheumatoid arthritis. *Rheumatology (Oxford)* 2006;**45**(Suppl 4):iv4–7.

99. Langley RL, Treadwell EL. Cardiac tamponade and pericardial disorders in connective tissue diseases: Case report and literature review. *J Natl Med Assoc* 1994;**86**(2):149–53.

100. Imazio M, Cecchi E, Demichelis B, et al. Indicators of poor prognosis of acute pericarditis. *Circulation* 2007;**115**(21):2739–44.

101. Vila LM, Rivera del Rio JR, Rios Z, Rivera E. Lymphocyte populations and cytokine concentrations in pericardial fluid from a systemic lupus erythematosus patient with cardiac tamponade. *Ann Rheum Dis* 1999;**58**(11):720–1.

102. Byers RJ, Marshall DA, Freemont AJ. Pericardial involvement in systemic sclerosis. *Ann Rheum Dis* 1997;**56**(6):393–4.

103. Lundberg IE. Cardiac involvement in autoimmune myositis and mixed connective tissue disease. *Lupus* 2005;**14**(9):708–12.

104. Gyongyosi M, Pokorny G, Jambrik Z, et al. Cardiac manifestations in primary Sjogren's syndrome. *Ann Rheum Dis* 1996;**55**(7):450–4.

105. Wyplosz B, Marijon E, Dougados J, Pouchot J. Sarcoidosis: An unusual cause of acute pericarditis. *Acta Cardiol* 2010;**65**(1):83–4.

106. Rostand SG, Rutsky EA. Pericarditis in end-stage renal disease. *Cardiol Clin* 1990;**8**(4):701–7.

107. Compty C, Cohen S, Shapiro F. Pericarditis in chronic uremia and its sequelae. *Ann Inter Med* 1973;**75**:173.

108. Colombo A, Olson HG, Egan J, Gardin JM. Etiology and prognostic implications of a large pericardial effusion in men. *Clin Cardiol* 1988;**11**(6):389–94.

109. Spodick D. Pericardial diseases. In: Braunwald E, Zippes D, Libby P, editors. *Heart disease*. 6th ed. Toronto: W.B. Saunders; 2001. p. 1823–76.

110. Thumber DL, Edwards JE, Achor RW. Secondary malignant tumors of the pericardium. *Circulation* 1962;**26**:228–41.

111. Fagan SM, Chan KL. Pericardiocentesis: Blind no more! *Chest* 1999;**116**(2):275–6.

112. Merce J, Sagrista-Sauleda J, Permanyer-Miralda G, Soler-Soler J. Should pericardial drainage be performed routinely in patients who have a large pericardial effusion without tamponade? *Am J Med* 1998;**105**(2):106–9.

113. Bishiniotis TS, Antoniadou S, Katseas G, Mouratidou D, Litos AG, Balamoutsos N. Malignant cardiac tamponade in women with breast cancer treated by pericardiocentesis and intrapericardial administration of triethylenethiophosphoramide (thiotepa). *Am J Cardiol* 2000;**86**(3):362–4.

114. Girardi LN, Ginsberg RJ, Burt ME. Pericardiocentesis and intrapericardial sclerosis: Effective therapy for malignant pericardial effusions. *Ann Thorac Surg* 1997;**64**(5):1422–7; discussion 1427–8.

115. Vaitkus PT, Herrmann HC, LeWinter MM. Treatment of malignant pericardial effusion. *JAMA* 1994;**272**(1):59–64.

116. Maisch B, Berg PA, Kochsiek K. Clinical significance of immunopathological findings in patients with post-pericardiotomy syndrome. I. relevance of antibody pattern. *Clin Exp Immunol* 1979;**38**(2):189–97.

117. Maisch B, Schuff-Werner P, Berg PA, Kochsiek K. Clinical significance of immunopathological findings in patients with post-pericardiotomy syndrome. II. The significance of serum inhibition and rosette inhibitory factors. *Clin Exp Immunol* 1979;**38**(2):198–203.

118. Kuvin JT, Harati NA, Pandian NG, Bojar RM, Khabbaz KR. Postoperative cardiac tamponade in the modern surgical era. *Ann Thorac Surg* 2002;**74**(4):1148–53.

119. Quin JA, Tauriainen MP, Huber LM, et al. Predictors of pericardial effusion after orthotopic heart transplantation. *J Thorac Cardiovasc Surg* 2002;**124**(5):979–83.

120. Matsuyama K, Matsumoto M, Sugita T, et al. Clinical characteristics of patients with constrictive pericarditis after coronary bypass surgery. *Jpn Circ J* 2001;**65**(6):480–2.

121. Finkelstein Y, Shemesh J, Mahlab K, et al. Colchicine for the prevention of postpericardiotomy syndrome. *Herz* 2002;**27**(8):791–4.

122. Horneffer PJ, Miller RH, Pearson TA, Rykiel MF, Reitz BA, Gardner TJ. The effective treatment of postpericardiotomy syndrome after cardiac operations. A randomized placebo-controlled trial. *J Thorac Cardiovasc Surg* 1990;**100**(2):292–6.

123. Nagy KK, Lohmann C, Kim DO, Barrett J. Role of echocardiography in the diagnosis of occult penetrating cardiac injury. *J Trauma* 1995;**38**(6):859–62.

124. Asensio JA, Murray J, Demetriades D, et al. Penetrating cardiac injuries: A prospective study of variables predicting outcomes. *J Am Coll Surg* 1998;**186**(1):24–34.

125. Buckman Jr RF, Buckman PD. Vertical deceleration trauma. principles of management. *Surg Clin North Am* 1991;**71**(2):331–44.

126. Cliff WJ, Grobety J, Ryan GB. Postoperative pericardial adhesions. the role of mild serosal injury and spilled blood. *J Thorac Cardiovasc Surg* 1973;**65**(5):744–50.

127. Chirillo F, Totis O, Cavarzerani A, et al. Usefulness of transthoracic and transoesophageal echocardiography in recognition and management of cardiovascular injuries after blunt chest trauma. *Heart* 1996;**75**(3):301–6.

128. Kumar PP. Pericardial injury from mediastinal irradiation. *J Natl Med Assoc* 1980;**72**(6):591–4.

129. McCaughan BC, Schaff HV, Piehler JM, et al. Early and late results of pericardiectomy for constrictive pericarditis. *J Thorac Cardiovasc Surg* 1985;**89**(3):340–50.

130. Reinmuller R, Gurgan M, Erdmann E, Kemkes BM, Kreutzer E, Weinhold C. CT and MR evaluation of pericardial constriction: A new diagnostic and therapeutic concept. *J Thorac Imaging* 1993;**8**(2):108–21.

131. Ling LH, Oh JK, Schaff HV, et al. Constrictive pericarditis in the modern era: Evolving clinical spectrum and impact on outcome after pericardiectomy. *Circulation* 1999;**100**(13):1380–6.

132. Yetkin U, Kestelli M, Yilik L, et al. Recent surgical experience in chronic constrictive pericarditis. *Tex Heart Inst J* 2003;**30**(1):27–30.

133. Luthringer DJ, Virmani R, Weiss SW, Rosai J. A distinctive cardiovascular lesion resembling histiocytoid (epithelioid) hemangioma. Evidence suggesting mesothelial participation. *Am J Surg Pathol* 1990;**14**(11):993–1000.

134. Hansen RM, Caya JG, Clowry Jr LJ, Anderson T. Benign mesothelial proliferation with effusion. Clinicopathologic entity that may mimic malignancy. *Am J Med* 1984;**77**(5):887–92.

135. Moran C, Suster S, Koss M. The spectrum of histologic growth patterns in benign and malignant fibrous tumours of the pleura. *Semin Diagn Pathol* 1992;**9**:169–80.

136. Van de Rijn, Lombard C, Rouse R. Expression of CD34 by solitary fibrous tumours of the pleura, mediastinum and lung. *Am J Surg Pathol* 1994;**18**:814–20.

137. Carneiro S, Scheithauer B, Nascimento A, et al. Solitary fibrous tumour of the meninges: A lesion distinct from fibrous meningioma; a clinicopathologic and immunohistochemical study. *Am J Clin Pathol* 1996;**106**:217–24.

138. Sayed Z, Vanada H, Sayed H, Robert H, Maureen F. Solitary fibrous tumour: A cytologic histologic study with clinical, radiologic and immunohistochemical correlations. *Cancer Cytopathol* 1997;**81**:116–21.

139. England D, Hochholzer L, McCarthy M. Localized benign and malignant fibrous tumours of the pleura. A review of 223 cases. *Am J Surg Pathol* 1989;**13**:640–58.

140. Miettinen MM, el-Rifai W, Sarlomo-Rikala M, Andersson LC, Knuutila S. Tumor size-related DNA copy number changes occur in solitary fibrous tumors but not in hemangiopericytomas. *Mod Pathol* 1997;**10**(12):1194–200.

141. Wilson RW, Gallateau-Salle F, Moran CA. Desmoid tumors of the pleura: A clinicopathologic mimic of localized fibrous tumor. *Mod Pathol* 1999;**12**(1):9–14.

142. Burke A, Virmani R. Tumors of the heart and great vessels, Fascicle 16. In: *Atlas of tumor pathology*, 3rd Series. Washington, D.C.: Armed Forces Institute of Pathology; 1996.

143. Coskun U, Gunel N, Yildirim Y, Memis L, Boyacioglu ZM. Primary mediastinal yolk sac tumor in a 66-year-old woman. *Med Princ Pract* 2002;**11**(4):218–20.

144. Burke A, Loire R, Virmani R. Pericardial tumors. In: Travis W, Brambilla E, Hermelink H, Harris C, editors. *Pathology and genetics, tumors of lung, pleura, thymus and heart*. Lyon: IARC; 2004. p. 285–8.

145. Roy N, Blurton DJ, Azakie A, Karl TR. Immature intrapericardial teratoma in a newborn with elevated alpha-fetoprotein. *Ann Thorac Surg* 2004;**78**(1):e6–8.

146. Luk A, Ahn E, Vaideeswar P, Butany JW. Pericardial tumors. *Semin Diagn Pathol* 2008;**25**(1):47–53.

147. Maeyama R, Uchiyama A, Tominaga R, Ichimiya H, Kuroiwa K, Tanaka M. Benign mediastinal teratoma complicated by cardiac tamponade: Report of a case. *Surg Today* 1999;**29**(11):1206–8.

148. Laquay N, Ghazouani S, Vaccaroni L, Vouhe P. Intrapericardial teratoma in newborn babies. *Eur J Cardiothorac Surg* 2003;**23**(4):642–4.

149. Suzuki Y. Pathology of human malignant mesothelioma. *Semin Oncol* 1981;**8**(3):268–82.

150. Suzuki Y, Kannerstein M. Ultrastructure of human malignant diffuse mesothelioma. *Am J Pathol* 1976;**85**(2):241–62.

151. Edwards JG, Abrams KR, Leverment JN, Spyt TJ, Waller DA, O'Byrne KJ. Prognostic factors for malignant mesothelioma in 142 patients: Validation of CALGB and EORTC prognostic scoring systems. *Thorax* 2000;**55**(9):731–5.

152. Bussani R, De-Giorgio F, Abbate A, Silvestri F. Cardiac metastases. *J Clin Pathol* 2007;**60**(1):27–34.

153. Butany J, Leong SW, Carmichael K, Komeda MA. 30-year analysis of cardiac neoplasms at autopsy. *Can J Cardiol* 2005;**21**(8):675–80.

154. Watanabe A, Sakata J, Kawamura H, Yamada O, Matsuyama T. Primary pericardial mesothelioma presenting as constrictive pericarditis: A case report. *Jpn Circ J* 2000;**64**(5):385–8.

155. Loire R, Hellal H. Neoplastic pericarditis. Study by thoracotomy and biopsy in 80 cases. *Presse Med* 1993;**22**(6):244–8.

156. Kralstein J, Frishman W. Malignant pericardial diseases: Diagnosis and treatment. *Am Heart J* 1987;**113**(3):785–90.

Index